Geostatistics

Geostatistics

Modeling Spatial Uncertainty

Second Edition

JEAN-PAUL CHILÈS
BRGM and MINES ParisTech

PIERRE DELFINER
PetroDecisions

A JOHN WILEY & SONS, INC., PUBLICATION

Published by John Wiley & Sons, Inc., Hoboken, New Jersey
Published simultaneously in Canada

For general information on our other products and services or for technical support, please contact our Customer Care Department within the United States at (800) 762-2974, outside the United States at (317) 572-3993 or fax (317) 572-4002.

Wiley also publishes its books in a variety of electronic formats. Some content that appears in print may not be available in electronic formats. For more information about Wiley products, visit our web site at www.wiley.com.

ISBN: 978-0-470-18315-1

Library of Congress Cataloging-in-Publication Data is available.

Printed in the United States of America

10 9 8 7 6 5 4 3 2

Contents

In memory of Georges MATHERON
(1930–2000)

Preface to the Second Edition

Twelve years after publication of the first edition in 1999, ideas have matured and new perspectives have emerged. It has become possible to sort out material that has lost relevance from core methods which are here to stay. Many new developments have been made to the field, a number of pending problems have been solved, and bridges with other approaches have been established. At the same time there has been an explosion in the applications of geostatistical methods, including in new territories unrelated to geosciences—who would have thought that one day engineers would krige aircraft wings? All these factors called for a thoroughly revised and updated second edition.

Our intent was to integrate the new material without increasing the size of the book. To this end we removed Chapter 8 (Scale effects and inverse problems) which covered stochastic hydrogeology but was too detailed for the casual reader and too incomplete for the specialist. We decided to keep only the specific contributions of geostatistics to hydrogeology and to distribute the material throughout the relevant chapters. The following is an overview of the main changes from the first edition and their justification.

Chapter 2 (Structural analysis) gives complements on practical questions such as spatial declustering and declustered statistics, variogram map calculation for data on a regular grid, variogram in a non-Euclidean coordinate system (transformation to a geochronological coordinate system). The Cauchy model is extended to the Cauchy class whose two shape parameters can account for a variety of behaviors at short as well as at large distances. The Matérn model and the logarithmic (de Wijsian) model are related to Gaussian Markov random fields (GMRF). New references are given on variogram fitting and sampling design. New sections propose covariance models on the sphere or on a river network. The chapter also includes new points on random function theory, such as a reference to the recent proof of a conjecture of Matheron on the characterization of an indicator function by its covariogram. The introductory example of variography in presence of a drift was removed to gain space.

The external drift model which was presented with multivariate methods is now introduced in Chapter 3 (Kriging) as a variant of the universal kriging model with polynomial drift. The special case of a constant unknown mean (ordinary kriging) is treated explicitly and in detail as it is the most common in applications. Dual kriging receives more attention because of its kinship with radial basis function interpolation (RBF), and its wide use in the design and analysis of computer experiments (DACE) to solve engineering problems. Three solutions are proposed to address the longstanding problem of the spurious discontinuities created by the use of moving neighborhoods in the case of a large dataset, namely covariance tapering, Gaussian Markov random field approximation, and continuous moving neighborhoods. Another important kriging issue, how to deal with outliers, is discussed and a new, relatively simple, truncation model developed for gold and uranium mines is presented. Finally a new form of kriging, Poisson kriging, in which observations derive from a Poisson time process, is introduced.

Few changes were made to Chapter 4 (Intrinsic model of order k). The main one is the addition of Micchelli's theorem providing a simple characterization of isotropic generalized covariances of order k. Another addition is an analysis of the structure of the inverse of the intrinsic kriging matrix. The Poisson differential equation $\Delta Z = Y$ previously in the deleted chapter 8 survives in this chapter.

Chapter 5 (Multivariate methods) was largely rewritten and augmented. The main changes concern collocated cokriging and space–time models. The chapter now includes a thorough review of different forms of collocated cokriging, with a clear picture of which underlying models support the approach without loss of information and which use it just as a convenient simplification. Collocated cokriging is also systematically compared with its common alternative, kriging with an external drift. As for space–time models, they were a real threat for the size of the book because of the surge of activity in the subject. To deal with situations where a physical model is available to describe the time evolution of the system, we chose to present sequential data assimilation and ensemble Kalman filtering (EnKF) in some detail, highlighting their links with geostatistics. For the alternative case where no dynamic model is available, the focus is on new classes of nonseparable space–time covariances that enable kriging in a space–time domain. The chapter contains numerous other additions such as potential field interpolation of orientation data, extraction of the common part of two surveys using automatic factorial cokriging, maximum autocorrelation factors, multivariate Matérn cross-covariance model, layer-cake estimation including seismic information, compositional data with geometry on the simplex.

Nonlinear methods and conditional simulations generally require a preliminary transformation of the variable of interest into a variable with a specified marginal distribution, usually a normal one. As this step is critical for the quality of the results, it has been expanded and updated and now forms a specific section of chapter 6 (Nonlinear methods). More elaborate methods than the simple normal score transform are proposed. The presentation of the change of support has been restructured. We now present each model at the global scale and then immediately continue with the local scale. Conditional expectation is more detailed and accounts for a locally variable mean. The most widely used change-of-support model, the discrete Gaussian model, is discussed in depth, including the variant that appeared in the 2000s. Practical implementation questions are examined: locally variable mean, selection on the basis of future information (indirect resources), uniform conditioning. Finally this chapter features a section on the change of support by the maximum, a topic whose development in a spatial context is still in infancy but is important for extreme-value studies.

Chapter 7 incorporates the numerous advances made in conditional simulations in the last decade. The simulation of the fractional Brownian motion and its extension to IRF–k's, which was possible only in specific cases (regular 1D grid, or at the cost of an approximation) is now possible exactly. A new insight into the Gibbs sampler enables the definition of a Gibbs propagation algorithm that does not require inversion of the covariance matrix. Pluri-Gaussian simulations are explained in detail and their use is illustrated in the Brent cliff case study, which has been completely reworked to reflect current practice (separable covariance models are no longer required). New simulation methods are presented: stochastic process-based simulation, multi-point simulation, gradual deformation. The use of simulated annealing for building conditional simulations has been completely revised. Stochastic seismic inversion and Bayesian approaches are up-to-date. Upscaling is also discussed in the chapter.

ACKNOWLEDGMENTS

Special acknowledgement is due to Christian Lantuéjoul for his meticulous reading of Chapters 6 and 7, numerous helpful comments and suggestions, and for writing the section on change of support by the maximum. We are also greatly indebted to Jacques Rivoirard for many contributions and insights. Thierry Coléou helped us with seismic applications and Henning Omre with Bayesian methods. Xavier Freulon provided the top-cut gold grades example and Hélène Beucher the revised simulation of the Brent cliff. Didier Renard carried out calculations for new figures and Philippe Le Caër redrew the cover figure. This second edition also benefits from fine remarks of some readers of the first edition, notably Tilmann Gneiting, and from many informal discussions with our colleagues of the Geostatistics group of MINES ParisTech.

We remain, of course, grateful to the individuals acknowledged in the Preface to the first edition, and especially to Georges Matheron, who left us in 2000, but continues to be a source of inspiration.

Fontainebleau   JEAN-PAUL CHILÈS
October 23, 2011   PIERRE DELFINER

Preface to the First Edition

This book covers a relatively specialized subject matter, geostatistics, as it was defined by Georges Matheron in 1962, when he coined this term to designate his own methodology of ore reserve evaluation. Yet it addresses a larger audience, because the applications of geostatistics now extend to many fields in the earth sciences, including not only the subsurface but also the land, the atmosphere, and the oceans.

The reader may wonder why such a narrow subject should occupy so many pages. Our intent was to write a short book. But this would have required us to sacrifice either the theory or the applications. We felt that neither of these options was satisfactory—there is no need for yet another introductory book, and geostatistics is definitely an applied subject. We have attempted to reconcile theory and practice by including application examples, which are discussed with due care, and about 160 figures. This results in a somewhat weighty volume, although hopefully more readable.

This book gathers in a single place a number of results that were either scattered, not easily accessible, or unpublished. Our ambition is to provide the reader with a unified view of geostatistics, with an emphasis on *methodology*. To this end we detail simple proofs when their understanding is deemed essential for geostatisticians, and we omit complex proofs that are too technical. Although some theoretical arguments may fall beyond the mathematical and statistical background of practitioners, they have been included for the sake of a complete and consistent development that the more theoretically inclined reader will appreciate. These sections, as well as ancillary or advanced topics, are set in smaller type.

Many references in this book point to the works of Matheron and the Center for Geostatistics in Fontainebleau, which he founded at the Paris School of Mines in 1967 and headed until his retirement in 1996. Without overlooking the contribution of Gandin, Matérn, Yaglom, Krige, de Wijs, and many others, it is from Matheron that geostatistics emerged as a discipline in its own right—a body of concepts and methods, a theory, and a practice—for the study of spatial phenomena. Of course this initial group spawned others, notably in

Europe and North America, under the impetus of Michel David and André Journel, followed by numerous researchers trained in Fontainebleau first, and then elsewhere. This books pays tribute to all those who participated in the development of geostatistics, and our large list of references attempts to give credit to the various contributions in a complete and fair manner.

This book is the outcome of a long maturing process nourished by experience. We hope that it will communicate to the reader our enthusiasm for this discipline at the intersection between probability theory, physics, and earth sciences.

ACKNOWLEDGMENTS

This book owes more than we can say to Georges Matheron. Much of the theory presented here is his work, and we had the privilege of seeing it in the making during the years that we spent at the Center for Geostatistics. In later years he always generously opened his door to us when we asked for advice on fine points. It was a great comfort to have access to him for insight and support. We are also indebted to the late Geoffrey S. Watson, who showed an early interest in geostatistics and introduced it to the statistical community. He was kind enough to invite one of the authors to Princeton University and, as an advisory editor of the Wiley Interscience Series, made this book possible. We wish he had been with us to see the finished product.

The manuscript of this book greatly benefited from the meticulous reading and quest for perfection of Christian Lantuéjoul, who suggested many valuable improvements. We also owe much to discussions with Paul Switzer, whose views are always enlightening and helped us relate our presentation to mainstream statistics. We have borrowed some original ideas from Jean-Pierre Delhomme, who shared the beginnings of this adventure with us. Bernard Bourgine contributed to the illustrations. This book could not have been completed without the research funds of Bureau de Recherches Géologiques et Minières, whose support is gratefully acknowledged.

We would like to express our thanks to John Wiley & Sons for their encouragement and exceptional patience during a project which has spanned many years, and especially to Bea Shube, the Wiley-Interscience Editor when we started, and her successors Kate Roach and Steve Quigley.

Finally, we owe our families, and especially our dear wives Chantal and Edith, apologies for all the time we stole from them, and we thank them for their understanding and forebearance.

La Villetertre   JEAN-PAUL CHILÈS
July 12, 1998   PIERRE DELFINER

Abbreviations

ALC−k	allowable linear combination of order k
c.d.f.	cumulative density function
CK	cokriging
DFT	discrete Fourier transform
DGM1, DGM2	discrete Gaussian model 1, 2
DK	disjunctive kriging
GC−k	generalized covariance of order k
GLS	generalized least squares
GV	generalized variogram
i.i.d.	independent and identically distributed
IRF	intrinsic random function
IRF−k	intrinsic random function of order k
KED	kriging with external drift
MM1, MM2	Markov model 1, 2
m.s.	mean square
m.s.e.	mean square error
OK	ordinary kriging
PCA	principal component analysis
p.d.f.	probability density function
RF	random function
SK	simple kriging
SRF	stationary random function
UK	universal kriging

Introduction

Geostatistics aims at providing quantitative descriptions of natural variables distributed in space or in time and space. Examples of such variables are

- Ore grades in a mineral deposit
- Depth and thickness of a geological layer
- Porosity and permeability in a porous medium
- Density of trees of a certain species in a forest
- Soil properties in a region
- Rainfall over a catchment area
- Pressure, temperature, and wind velocity in the atmosphere
- Concentrations of pollutants in a contaminated site

These variables exhibit an immense complexity of detail that precludes a description by simplistic models such as constant values within polygons, or even by standard well-behaved mathematical functions. Furthermore, for economic reasons, these variables are often sampled very sparsely. In the petroleum industry, for example, the volume of rock sampled typically represents a minute fraction of the total volume of a hydrocarbon reservoir. The following figures, from the Brent field in the North Sea, illustrate the orders of magnitude of the volume fractions investigated by each type of data ("cuttings" are drilling debris, and "logging" data are geophysical measurements in a wellbore):

Cores   0.000 000 001
Cuttings   0.000 000 007
Logging   0.000 001

By comparison, if we used the same proportions for an opinion poll of the 100 million US households (to take a round number), we would interview only

Geostatistics: Modeling Spatial Uncertainty, Second Edition. J.P. Chilès and P. Delfiner.
© 2012 John Wiley & Sons, Inc. Published 2012 by John Wiley & Sons, Inc.

between 0.1 and 100 households, while 1500 is standard. Yet the economic implications of sampling for natural resources development projects can be significant. The cost of a deep offshore development is of the order of 10 billion dollars. Similarly, in the mining industry "the decision to invest up to $1-2$ billion dollars to bring a major new mineral deposit on line is ultimately based on a very judicious assessment of a set of assays from a hopefully very carefully chosen and prepared group of samples which can weigh in aggregate less than 5 to 10 kilograms" (Parker, 1984).

Naturally, these examples are extreme. Such investment decisions are based on studies involving many disciplines besides geostatistics, but they illustrate the notion of *spatial uncertainty* and how it affects development decisions. The fact that our descriptions of spatial phenomena are subject to uncertainty is now generally accepted, but for a time it met with much resistance, especially from engineers who are trained to work deterministically. In the oil industry there are anecdotes of managers who did not want to see uncertainty attached to resources estimates because it did not look good—it meant incompetence. For job protection, it was better to systematically underestimate resources. (Ordered by his boss to get rid of uncertainty, an engineer once gave an estimate of proven oil resources equal to the volume of oil contained in the borehole!) Such conservative attitude led to the abandonment of valuable prospects. In oil exploration, profit comes with risk.

Geostatistics provides the practitioner with a *methodology* to quantify spatial uncertainty. Statistics come into play because probability distributions are the meaningful way to represent the range of possible values of a parameter of interest. In addition, a statistical model is well-suited to the apparent randomness of spatial variations. The prefix "geo" emphasizes the spatial aspect of the problem. Spatial variables are not completely random but usually exhibit some form of structure, in an average sense, reflecting the fact that points close in space tend to assume close values. G. Matheron (1965) coined the term *regionalized variable* to designate a numerical function $z(x)$ depending on a continuous space index x and combining high irregularity of detail with spatial correlation. Geostatistics can then be defined as "the application of probabilistic methods to regionalized variables." This is different from the vague usage of the word in the sense "statistics in the geosciences." In this book, geostatistics refers to a specific set of models and techniques, largely developed by G. Matheron, in the lineage of the works of L. S. Gandin in meteorology, B. Matérn in forestry, D. G. Krige and H. J. de Wijs in mining, and A. Y. Khinchin, A. N. Kolmogorov, P. Lévy, N. Wiener, A. M. Yaglom, among others, in the theory of stochastic processes and random fields. We will now give an overview of the various geostatistical methods and the types of problems they address and conclude by elaborating on the important difference between description and interpretation.

TYPES OF PROBLEMS CONSIDERED

The presentation follows the order of the chapters. For specificity, the problems presented refer to the authors' own background in earth sciences applications, but newcomers with different backgrounds and interests will surely find equivalent formulations of the problems in their own disciplines. Geostatistical terms will be introduced and highlighted by italics.

Epistemology

The quantification of spatial uncertainty requires a model specifying the mechanism by which spatial randomness is generated. The simplest approach is to treat the regionalized variable as deterministic and the positions of the samples as random, assuming for example that they are selected uniformly and independently over a reference area, in which case standard statistical rules for independent random variables apply, such as that for the variance of the mean. If the samples are collected on a systematic grid, they are not independent and things become more complicated, but a theory is possible by randomizing the grid origin.

Geostatistics takes the bold step of associating randomness with the regionalized variable itself, by using a *stochastic model* in which the regionalized variable is regarded as one among many possible *realizations* of a *random function*. Some practitioners dispute the validity of such probabilistic approach on the grounds that the objects we deal with—a mineral deposit or a petroleum reservoir—are uniquely deterministic. Probabilities and their experimental foundation in the famous "law of large numbers" require the possibility of repetitions, which are impossible with objects that exist unambiguously in space and time. The objective meaning and relevance of a stochastic model under such circumstances is a fundamental question of epistemology that needs to be resolved. The clue is to carefully distinguish the model from the reality it attempts to capture. Probabilities do not exist in Nature but only in our models. We do not choose to use a stochastic model because we believe Nature to be random (whatever that may mean), but simply because it is analytically useful. The probabilistic content of our models reflects our imperfect knowledge of a deterministic reality. We should also keep in mind that models have their limits and represent reality only up to a certain point. And finally, no matter what we do and how carefully we work, there is always a possibility that our predictions and our assessments of uncertainty turn out to be completely wrong, because for no foreseeable reason the phenomenon at unknown places is radically different than anything observed (what Matheron calls the risk of a "radical error").

Structural Analysis

Having observed that spatial variability is a source of spatial uncertainty, we have to quantify and model spatial variability. What does an observation at a point tell us about the values at neighboring points? Can we expect continuity in a mathematical sense, or in a statistical sense, or no continuity at all? What is the signal-to-noise ratio? Are variations similar in all directions or is there anisotropy? Do the data exhibit any spatial trend? Are there characteristic scales and what do they represent? Is the histogram symmetric or skewed?

Answering these questions, among others, is known in geostatistics as *structural analysis*. One key tool is a structure function, the *variogram*, which describes statistically how the values at two points become different as the separation between these points increases. The variogram is the simplest way to relate uncertainty with distance from an observation. Other two-point structure functions can be defined that, when considered together, provide further clues for modeling. If the phenomenon is spatially homogeneous and densely sampled, it is even possible to go beyond structure functions and determine the complete *bivariate distributions* of measurements at pairs of points. In applications there is rarely enough data to allow empirical determination of multiple-point statistics beyond two points, a notable exception being when the data are borrowed from training images.

Survey Optimization

In resources estimation problems the question arises as to which sampling pattern ensures the best precision. The variogram alone permits a comparison between random, random stratified, and systematic sampling patterns. Optimizing variogram estimation may actually be a goal in itself. In practice the design is often constrained by operational and economic considerations, and the real question is how to optimize the parameters of the survey. Which grid mesh should be used to achieve a required precision? What is the optimal spacing between survey lines? What is the best placement for an additional appraisal well? Does the information expected from acquiring or processing more data justify the extra cost and delay? What makes life interesting is that these questions must be answered, of course, prior to acquiring the data.

Interpolation

We often need to estimate the values of a regionalized variable at places where it has not been measured. Typically, these places are the nodes of a regular grid laid out on the studied domain, the interpolation process being then sometimes known as "gridding." Once grids are established, they are often used as *the* representation of reality, without reference to the original data. They are the basis for new grids obtained by algebraic or Boolean operations, contour maps, volumetric calculations, and the like. Thus the computation of grids deserves care and cannot rely on simplistic interpolation methods.

The estimated quantity is not necessarily the value at a point; in many cases a grid node is meant to represent the grid cell surrounding it. This is typical for inventory estimation or for numerical modeling. Then we estimate the mean value over a cell, or a block, and more generally some weighted average.

In all cases we wish our estimates to be "accurate." This means, first, that on the average our estimates are correct; they are not systematically too high or too low. This property is captured statistically by the notion of *unbiasedness*. It is especially critical for inventory estimation and was the original motivation for the invention of *kriging*. The other objective is precision, and it is quantified by the notion of *error variance*, or its square root the *standard error*, which is expressed in the same units as the data.

The geostatistical interpolation technique of kriging comes in different flavors qualified by an adjective: *simple kriging, ordinary kriging, universal kriging, intrinsic kriging,* and so on, depending on the underlying model. The general approach is to consider a class of unbiased estimators, usually linear in the observations, and to find the one with minimum uncertainty, as measured by the error variance. This optimization involves the statistical model established during the structural analysis phase, and there lies the fundamental difference with standard interpolation methods: These focus on modeling the interpolating surface, whereas geostatistics focuses on modeling the phenomenon itself.

Polynomial Drift

Unexpected difficulties arise when the data exhibit a spatial trend, which in geostatistical theory is modeled as a space-varying mean called *drift*. The determination of the variogram in the presence of a drift is often problematic due to the unclear separation between global and local scales. The problem disappears by considering a new structural tool, the *generalized covariance*, which is associated with increments of order k that filter out polynomial drifts, just like ordinary increments filter out a constant mean. When a polynomial drift is present, the generalized covariance is the minimum parametric information required for kriging. An insightful bridge with radial basis function interpolation, including thin plate splines, can be established.

Intrinsic random functions of order k (IRF$-k$), which are associated with generalized covariances, also provide a class of nonstationary models that are useful to represent the nonstationary solutions of stochastic partial differential equations such as found in hydrogeology.

Integration of Multiparameter Information

In applications the greatest challenge is often to "integrate" (i.e., combine) information from various sources. To take a specific example, a petroleum geologist must integrate into a coherent geological model information from cores, cuttings, open-hole well logs, dip and azimuth computations, electrical and acoustic images, surface and borehole seismic, and well tests. The rule of the game is: "Don't offend anything that is already known." Geostatistics and multivariate statistical techniques provide the framework and the tools to build a consistent model.

The technique of *cokriging* generalizes kriging to multivariate interpolation. It exploits the relationships between the different variables as well as the spatial structure of the data. An important particular case is the use of slope information in conjunction with the variable itself. When the implementation of cokriging requires a statistical inference beyond reach, shortcuts can be used. The most popular ones are the *external drift* method and *collocated cokriging*, which use a densely sampled auxiliary field to compensate for the scarcity of observations of the variable of interest.

Spatiotemporal Problems

Aside from geological processes which are so slow that time is not a factor, most phenomena have both a space and a time component. Typical examples are meteorological variables or pollutant concentrations, measured at different time points and space locations. We may wish to predict these variables at a new location at a future time.

One possibility is to perform kriging in a space–time domain using spatiotemporal covariance models. New classes of nonseparable stationary covariance functions have been developed in the recent years that allow space–time interaction.

Alternatively, if a physical model is available to describe the time evolution of the system, the techniques of *data assimilation* can be used—and in particular the *ensemble Kalman filter* (EnKF), which has received much attention.

Indicator Estimation

We are interested in the event: "at a given point x the value $z(x)$ exceeds the level z_0." We can think of z_0 as a pollution alert threshold, or a cutoff grade in mining. The event can be represented by a binary function, the *indicator function*, valued 1 if the event is true, and zero if it is false, whose expected value is the probability of the event "$z(x)$ exceeds z_0." Note that the indicator is a nonlinear function of the observation $z(x)$. The mean value of the indicator over a domain V represents the fraction of V where the threshold is exceeded. When we vary the threshold, it appears that indicator estimation amounts to the determination of the histogram or the cumulative distribution function of the values of $z(x)$ within V. The interesting application is to estimate this locally over a subdomain v to obtain a *local distribution function* reflecting the values observed in the vicinity of v. *Disjunctive kriging*, a nonlinear technique based on a careful modeling of bivariate distributions, provides a solution to this difficult problem.

Selection and Change-of-Support Problems

The *support* of a regionalized variable is the averaging volume over which the data are measured or defined. Typically, there are *point values* and *block values*, or high-resolution and low-resolution measurements. As the size of the support changes, the histogram of the variable is deformed, but there is no straightforward relationship between the distributions of values measured over two different supports, except under very stringent Gaussian assumptions. For example, ore sample grades and blocks grades cannot both be exactly log-normally distributed—although they might as approximations. Predicting the change of distribution when passing from one size of support to another, generally point to block, is the *change of support* problem. Specific *isofactorial models* are proposed to solve this problem.

Change of support is central in inventory estimation problems in which the resource is subject to *selection*. Historically, the most important application has been in mining, where the decision to process the ore or send it to waste, depending on its mineral content, is made at the level of a block, say a cube of 10-m side, rather than, say, a teaspoon. The recoverable resources then depend on the local distributions of block values. Modeling the effect of selection may be a useful concept in other applications, such as the delineation of producing beds in a petroleum reservoir, the remediation of contaminated areas, or the definition of pollution alert thresholds.

Simulation

Kriging, as any reasonable interpolation method, has a *smoothing effect*. It does not reproduce spatial heterogeneity. In the world of images we would say that it is not true to the "texture" of the image. This can cause significant biases when nonlinear effects are involved. To take a simple example, compare the length of an interpolated curve with the length of the true curve: It is much shorter—the true curve may not even have a finite length! Similarly, for the same average permeability, a porous medium has a very different flow behavior if it is homogeneous or heterogeneous.

This is where the stochastic nature of the model really comes into play. The formalism of random functions involves a family of alternative realizations similar in their spatial variability to the reality observed but different other-wise. By *simulation* techniques it is possible to generate some of these "virtual realities" and produce pictures that are true to the fluctuations of the phe-nomenon. A further step toward realism is to constrain the realizations to pass through the observed data, thus producing *conditional simulations*. By generating several of these digital models, we are able to materialize spatial uncertainty. Then if we are interested in some quantity that depends on the spatial field in a complex manner, such as modeling fluid flow in a porous medium, we can compute a result for each simulation and study the statis-tical distribution of the results. A typical application is the determination of scaling laws.

Iterative methods based on Markov chain Monte Carlo enable conditioning non-Gaussian random functions and constraining simulations on auxiliary information such as seismic data and production data in reservoir engineer-ing. These methods provide an essential contribution to stochastic inverse modeling.

Problems Omitted

A wide class of spatial problems concerns the processing and analysis of images. This is a world by itself, and we will not enter it, even though there will be occasional points of contact. An image analysis approach very much in line with geostatistics, and developed in fact by the same group of researchers, is Mathematical Morphology [see Serra (1982)]. Variables regionalized in time only will also be left out. Even though geostatistical methods apply, the types of problems considered are often of an electrical engineering nature and are better handled by digital signal processing techniques.

Finally, the study of point patterns (e.g., the distribution of trees in a forest) and the modeling of data on a lattice or on a graph are intentionally omitted from this book. The reader is referred to Cressie (1991) for a comprehensive overview of the first two approaches, to Guyon (1995) for a presentation of Markov fields on a lattice, and to Jordan (1998, 2004) for graphical models.

DESCRIPTION OR INTERPRETATION?

Geostatistical methods are goal-oriented. Their purpose is not to build an explanatory model of the world but to solve specific problems using the minimal prerequisites required, following the principle of parsimony. They are descriptive rather than interpretive models. We illustrate this important point with an example borrowed from contour mapping.

Mathematically inclined people—including the present authors—have long thought that computer mapping was the definitive, clean, and objective replacement of hand contouring. Hand-drawn maps are subjective; they can be biased consciously or unconsciously. Even when drafted honestly, they seem suspect: If two competent and experienced interpreters can produce different maps from the same data, why should one believe any of them? And of course there is always the possibility of a gross clerical error such as overlooking or misreading some data points. By contrast, computer maps have all the attributes of respectability: They don't make clerical mistakes, they are "objective," reproducible, and fast. Yet this comparison misses an important point: It neglects the semantic content of a map. For a geologist, or a meteorologist, a map is far more than a set of contours: It represents *the state of an interpretation*. It reflects the attempt of its author to build a coherent picture of the geological object, or the meteorological situation, of interest.

This is demonstrated in a striking manner by a synthetic sedimentological example constructed by O. Serra, a pioneer in the geological interpretation of well logs. He considered a regular array of wells (the favorable case) and assigned them sand thickness values, without any special design, in fact using only the round numbers 0, 10, 20, 30. From this data set he derived four very different isopach maps. Figure 0.1a pictures the sand body as a meandering channel; Figure 0.1b as an infill channel with an abrupt bank to the east; Figure 0.1c as a transgressive sand filling paleo-valleys; and Figure 0.1d as a

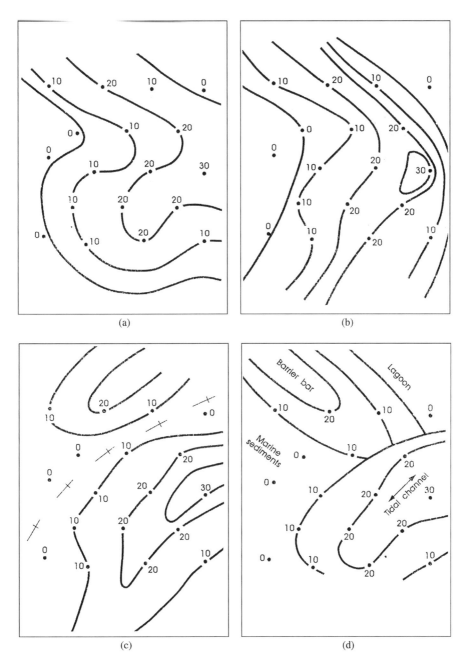

FIGURE 0.1 Four interpretations of the same synthetic data (hand-drawn isopach maps): (a) meandering channel; (b) infill channel; (c) transgressive sand filling paleo-valleys; (d) barrier bar eroded by a tidal channel. (From O. Serra, personal communication.)

barrier bar eroded by a tidal channel. Each of these maps reflects a different depositional environment model, which was on the interpreter's mind at the time and guided his hand.

Geostatistical models have no such explanatory goals. They model mathematical objects, a two-dimensional isopach surface, for example, not geological objects. The complex mental process by which a geologist draws one of the above maps can better be described as *pattern recognition* than as interpolation. Compared with this complexity, interpolation algorithms look pathetically crude, and this is why geological maps are still drawn by hand. To the geostatistician's comfort, the fact that widely different interpretations are consistent with the same data makes them questionable. For one brilliant interpretation (the correct one), how many "geofantasies" are produced?

Another way to qualify description versus interpretation is to oppose *data-driven* and *model-driven* techniques. Traditionally, geostatistics has been data-driven rather than model-driven: It captures the main structural features from the data, and knowledge of the subject matter does not have much impact beyond the selection of a variogram model. Therefore it cannot discriminate between several plausible interpretations. We can, however, be less demanding and simply require geostatistics to take external knowledge into account, and in particular an interpretation proposed by a physicist or a geologist. The current trend in geostatistics is precisely an attempt to include physical equations and model-specific constraints.

Hydrogeologists, who sought ways of introducing spatial randomness in aquifer models, have pioneered the research to incorporate physical equations in geostatistical models. Petroleum applications where data are scarce initially have motivated the development of *object-based* models. For example, channel sands are simulated directly as sinusoidal strips with an elliptic or rectangular cross section. This is still crude, but the goal is clear: Import geological concepts to the mapping or simulation processes. We can dream of a system that would find "the best" meandering channel consistent with a set of observations. Stochastic process-based models work in this direction.

To summarize, the essence of the geostatistical approach is to (a) recognize the inherent variability of natural spatial phenomena and the fragmentary character of our data and (b) incorporate these notions in a model of a stochastic nature. It identifies the structural relationships in the data and uses them to solve specific problems. It does not attempt any physical or genetic interpretations but uses them as much as possible when they are available.

CHAPTER 1

Preliminaries

1.1 RANDOM FUNCTIONS

G. de Marsily started the defense of his hydrogeology thesis by showing the audience a jar filled with fine sand and announced "here is a porous medium." Then he shook the jar and announced "and here is another," shook it again and said "and yet another." Indeed, at the microscopic scale the geometry is defined by the arrangement of thousands of individual grains with different shapes and dimensions, and it changes as the grains settle differently each time. Yet at the macroscopic scale we tend to regard it as the same porous medium because its physical properties do not change. This is an ingenious illustration of the notion of a random function in three-dimensional space.

Random functions are useful models for regionalized variables.

1.1.1 Definitions

Notations

Throughout this book the condensed notation x is used to denote a *point* in the n-dimensional space considered. For example, in 3D x stands for the coordinates (x_1, x_2, x_3) (usually called x, y, z). The notation $f(x)$ represents a function of x as well as its value at x. The notation f is used for short, and sometimes the notation $f(\cdot)$ is employed to emphasize that we consider the function taken as a whole and not its value at a single point. Since x is a point in $\mathbb{R}^n$, dx stands for an element of length ($n = 1$), of surface ($n = 2$), or volume ($n = 3$) and $\int_V f(x)dx$ represents the integral of $f(x)$ over a domain $V \subset \mathbb{R}^n$. For example, if $n = 2$ and V is the rectangle $[a_1, b_1] \times [a_2, b_2]$, we obtain

$$\int_V f(x)dx = \int_{a_1}^{b_1} dx_1 \int_{a_2}^{b_2} f(x_1, x_2)dx_2$$

Geostatistics: Modeling Spatial Uncertainty, Second Edition. J.P. Chilès and P. Delfiner.
© 2012 John Wiley & Sons, Inc. Published 2012 by John Wiley & Sons, Inc.

We will seldom need an explicit notation for the coordinates of a point; thus from now on, except when stated otherwise, x_1, x_2, ..., will represent distinct points in $\mathbb{R}^n$ rather than the coordinates of a single point.

Coming back to the sand jar, we can describe the porous medium by the indicator function of the grains, namely the function $I(x) = 1$ if the point x (in 3D space) is in a grain and $I(x) = 0$ if x is in a void (the pores). Each experiment (shaking the jar) determines at once a whole function $\{I(x) : x \in V\}$ as opposed to, say, throwing a die that only determines a single value (random variable). In probability theory it is customary to denote the outcome of an experiment by the letter ω and the set of all elementary outcomes, or events, by Ω. To make the dependence on the experiment explicit, a random variable is denoted by $X(\omega)$, and likewise our random indicator function is $I(x, \omega)$. For a fixed $\omega = \omega_0$, $I(x, \omega_0)$ is an ordinary function of x, called a *realization* (or *sample function*); any particular outcome of the jar-shaking experiment is a realization of the random function $I(x, \omega)$. On the other hand, for a fixed point $x = x_0$ the function $I(x_0, \omega)$ is an ordinary random variable. Thus mathematically a random function can be regarded as an infinite family of random variables indexed by x.

We can now give a formal definition of a random function [from Neveu (1970), some details omitted; see also Appendix, Section A.1]:

Random Function

Given a domain $D \subset \mathbb{R}^n$ (with a positive volume) and a probability space (Ω, $\mathcal{A}$, P), a random function (abbreviation: RF) is a function of two variables $Z(x, \omega)$ such that for each $x \in D$ the section $Z(x, \cdot)$ is a random variable on (Ω, $\mathcal{A}$, P). Each of the functions $Z(\cdot, \omega)$ defined on D as the section of the RF at $\omega \in \Omega$ is a realization of the RF. For short the RF is simply denoted by $Z(x)$, and a realization is represented by the lowercase $z(x)$.

In the literature a random function is also called a *stochastic process* when x varies in a 1D space, and can be interpreted as time, and it is called a *random field* when x varies in a space of more than one dimension.

In geostatistics we act *as though* the regionalized variable under study $z(x)$ is a realization of a parent random function $Z(x)$. Most of the time we will not be able to maintain the notational distinction between $Z(x)$ and $z(x)$, and we will get away with it by saying that the context should tell what is meant. The same is true for the distinction between an estimator (random) and an estimate (fixed).

Spatial Distribution

A random function is described by its *finite-dimensional distributions*, namely the set of all multidimensional distributions of k-tuples ($Z(x_1)$, $Z(x_2)$, ..., $Z(x_k)$) for all *finite* values of k and all configurations of the points $x_1, x_2, \ldots, x_k$. For short we will call this the *spatial distribution*.

In theory, the spatial distribution is not sufficient to calculate the probability of events involving an infinite noncountable number of points, such as the following important probabilities:

$\Pr\{\sup[Z(x) : x \in V] < z_0\}$ the maximum value in V is less than z_0
$\Pr\{\exists x \in V : Z(x) - 0\}$ a zero crossing occurs in domain V
$\Pr\{$every realization of $Z(\cdot)$ is continuous over $V\}$

This difficulty is overcome by adding the assumption of separability of the random function. A random function is separable if all probabilities involving a noncountable number of points can be uniquely determined from probabilities on countable sets of points (e.g., all points in $\mathbb{R}^n$ with rational coordinates), and hence from the spatial distribution. A fundamental result established by Doob (1953, Section 2.2) states that for any random function there always exists a separable random function with the same spatial distribution. In other words, among random functions that are indistinguishable from the point of view of their spatial distribution, we pick and work with the smoothest possible version (see footnote 3 in Section 2.3.1). For completeness let us also mention that tools more powerful than the spatial distribution are required to represent random sets [e.g., Matheron (1975a)] but will not be needed in this book.

Moments

The mean of the RF is the expected value $m(x) = \mathrm{E}[Z(x)]$ of the random variable $Z(x)$ at the point x. It is also called the *drift* of Z, especially when $m(x)$ varies with location. The (centered) covariance $\sigma(x, y)$ is the covariance of the random variables $Z(x)$ and $Z(y)$:

$$\sigma(x, y) = \mathrm{E}[Z(x) - m(x)][Z(y) - m(y)]$$

In general, this function depends on both x and y. When $x = y$, $\sigma(x, x) = \mathrm{Var}\, Z(x)$ is the variance of $Z(x)$. Higher-order moments can be defined similarly.

Naturally, in theory, these moments may not exist. As usual in probability theory, the mean is defined only if $\mathrm{E}\,|Z(x)| < \infty$. If $\mathrm{E}[Z(x)]^2$ is finite at every point, $Z(x)$ is said to be a second-order random function: It has a finite variance, and the covariance exists everywhere.

Convergence in the Mean Square

A sequence of random variables X_n is said to converge in the mean square (m.s.) sense to a random variable X if

$$\lim_{n \to \infty} \mathrm{E}|X_n - X|^2 = 0$$

Taking $X_n = Z(x_n)$ and $X = Z(x)$, we say that an RF $Z(x)$ on $\mathbb{R}^n$ is m.s. continuous if $x_n \to x$ in $\mathbb{R}^n$ implies that $Z(x_n) \to Z(x)$ in the mean square. This definition generalizes the continuity of ordinary functions.

1.1.2 Hilbert Space of Random Variables

It is interesting to cast the study of random functions in the geometric framework of Hilbert spaces. To this end, consider for maximum generality a family of complex-valued random variables X defined on a probability space $(\Omega, \mathcal{A}, P)$ and having finite second-order moments

$$\mathrm{E}|X|^2 = \int |X(\omega)|^2 P(d\omega) < \infty$$

These random variables constitute a vector space denoted $\mathrm{L}^2(\Omega, \mathcal{A}, P)$ which can be equipped with the scalar product $\langle X, Y \rangle = \mathrm{E}[X\overline{Y}]$ defining a norm[1] (or distance) $\|X\| = \sqrt{\mathrm{E}|X|^2}$ (the upper bar denotes complex conjugation). In this sense we can say that two random variables are *orthogonal* when they are *uncorrelated*. Then $\mathrm{L}^2(\Omega, \mathcal{A}, P)$ is a Hilbert space (every Cauchy sequence converges for the norm). An example is the infinite-dimensional Hilbert space of random variables $\{Z(x) : x \in D\}$ defined by the RF Z.

A fundamental property of a Hilbert space is the possibility of defining the orthogonal projection of X onto a closed linear subspace K as the unique point X_0 in the subspace nearest to X. This is expressed by the so-called *projection theorem* [e.g., Halmos (1951)]:

$$X_0 = \arg\min_{Y \in K} \|X - Y\| \quad \Leftrightarrow \quad \langle X - X_0, Y \rangle = 0 \quad \text{for all } Y \in K \quad (1.1)$$

Since $X_0 \in K$, it satisfies $\langle X - X_0, X_0 \rangle = 0$ so that

$$\|X - X_0\|^2 = \|X\|^2 - \|X_0\|^2 \quad (1.2)$$

This approximation property is the mathematical basis of kriging theory.

1.1.3 Conditional Expectation

Consider a pair of random variables (X, Y), and let $f(y \mid x)$ be the density of the conditional distribution of Y given that $X = x$. The *conditional expectation of Y given $X = x$* is the mean of that conditional distribution

$$\mathrm{E}(Y \mid X = x) = \int_{-\infty}^{+\infty} y f(y \mid x)\, dy$$

$\mathrm{E}(Y \mid X = x) = \phi(x)$ is a function of x only, even though Y appears in the expression. It is also known as the *regression function* of Y on X. When (X, Y) are jointly Gaussian,[2] this function is a straight line. If the argument of $\phi(\cdot)$ is the random variable X, $\phi(X)$ is itself a random variable denoted by $\mathrm{E}(Y \mid X)$. This definition carries over to the case where there are several conditioning variables $X_1, \ldots, X_N$.

---

[1] Strictly speaking, $\|X\| = 0$ implies that $X = 0$ only up to a set of probability zero, but as usual, equivalence classes of random variables are considered.

[2] "Gaussian" and "normal" will be used as synonyms.

It is possible to develop a theory of conditional expectations without reference to conditional distributions, and this is mathematically better and provides more insight. The idea is to find the best approximation of Y by a function of X. Specifically, we assume X and Y to have finite means and variances and pose the following problem: Find a function $\phi(X)$ such that $E[Y - \phi(X)]^2$ is a minimum. The solution is the conditional expectation $E(Y|X)$.

This solution is unique (up to an equivalence between random variables) and is *characterized* by the following property:

$$E\{[Y - E(Y \mid X)]H(X)\} = 0 \qquad \text{for all measurable } H(\cdot) \qquad (1.3)$$

In words, the error $Y - E(Y|X)$ is uncorrelated[3] with any finite-variance random variable of the form $H(X)$. Notice that this is a particular application of the projection formula (1.1).

In particular, when $H(X) \equiv 1$, we get

$$E[E(Y \mid X)] = E(Y) \qquad (1.4)$$

The conditional variance is defined by

$$\text{Var}(Y \mid X) = E(Y^2 \mid X) - [E(Y \mid X)]^2$$

from which we deduce the well-known *total variance* formula

$$\text{Var}(Y) = \text{Var}[E(Y \mid X)] + E[\text{Var}(Y \mid X)] \qquad (1.5)$$

The variance about the mean equals the variance due to regression plus the mean variance about regression.

For $H(X) = E(Y \mid X)$ we have

$$E\{[Y - E(Y \mid X)]E(Y \mid X)\} = 0$$

so that

$$\text{Cov}(Y, E(Y \mid X)) = \text{Var}(E(Y \mid X))$$

which shows that Y and $E(Y|X)$ are always positively correlated with

$$\rho^2 = \frac{\text{Var}(E(Y \mid X))}{\text{Var}(Y)} \qquad (1.6)$$

From (1.5) the residual variance takes the familiar form

---

[3] This does not imply independence between the error and X; if $X = Y^2$, Y symmetric about 0, $E(Y|X) = 0$, but Y is not independent of X.

$$E[\mathrm{Var}(Y \mid X)] = (1 - \rho^2)\mathrm{Var}(Y) \qquad (1.7)$$

(note that here ρ is not the correlation between Y and X but between Y and its regression on X).

In addition to the unbiasedness property, let us mention the property of *conditional unbiasedness*, which we will often invoke in this book in relation to kriging:

$$\phi(X) = E(Y \mid X) \qquad \Rightarrow \qquad E(Y \mid \phi(X)) = \phi(X)$$

The proof follows immediately from the characteristic property (1.3), since

$$E\{[Y - \phi(X)]H(X)\} = 0 \qquad \text{for all measurable } H(\cdot)$$

entails that $\phi(X)$ also satisfies

$$E\{[Y - \phi(X)]H(\phi(X))\} = 0 \qquad \text{for all measurable } H(\cdot)$$

Some Properties of Conditional Expectation

The following results can be derived directly from the characteristic formula and are valid almost surely (a.s.):

Linearity $E(aY_1 + bY_2 \mid X) = a E(Y_1 \mid X) + b E(Y_2 \mid X)$
Positivity $Y \geq 0 \text{ a.s.} \Rightarrow E(Y \mid X) \geq 0 \text{ a.s.}$
Independence X and Y are independent $\Rightarrow E(Y \mid X) = E(Y)$
Invariance $E(Y f(X) \mid X) = f(X) E(Y \mid X)$
Successive projections $E(Y \mid X_1) = E[E(Y \mid X_1, X_2) \mid X_1]$

1.1.4   Stationary Random Functions

Strict Stationarity

A particular case of great practical importance is when the finite-dimensional distributions are invariant under an arbitrary translation of the points by a vector h:

$$\Pr\{Z(x_1) < z_1, \ldots, Z(x_k) < z_k\} = \Pr\{Z(x_1 + h) < z_1, \ldots, Z(x_k + h) < z_k\}$$

Such RF is called *stationary*. Physically, this means that the phenomenon is homogeneous in space and, so to speak, repeats itself in the whole space. The sand in the jar is a good image of a stationary random function in three dimensions, at least if the sand is well sorted (otherwise, if the jar vibrates, the finer grains will eventually seep to the bottom, creating nonstationarity in the vertical dimension).

Second-Order Stationarity

When the random function is stationary, its moments, if they exist, are obviously invariant under translations. If we consider the first two moments only, we have for points x and $x + h$ of $\mathbb{R}^n$

$$E[Z(x)] = m,$$
$$E[Z(x) - m][Z(x + h) - m] = C(h)$$

The mean is constant and the covariance function only depends on the *separation h*. We will see in Section 2.3.2 that a covariance must be a *positive definite* function.

By definition, a random function satisfying the above conditions is *second-order stationary* (or weakly stationary, or wide-sense stationary). In this book, unless specified otherwise, stationarity will always be considered at order 2, and the abbreviation SRF will designate a second-order stationary random function.

An SRF is *isotropic* if its covariance function only depends on the length $|h|$ of the vector h and not on its orientation.

Intrinsic Hypothesis

A milder hypothesis is to assume that for every vector h the *increment* $Y_h(x) = Z(x + h) - Z(x)$ is an SRF in x. Then $Z(x)$ is called an *intrinsic random function* (abbreviation: IRF) and is characterized by the following relationships:

$$E[Z(x + h) - Z(x)] = \langle a, h \rangle,$$
$$\text{Var}[Z(x + h) - Z(x)] = 2\gamma(h)$$

$\langle a, h \rangle$ is the *linear drift* of the IRF (drift of the increment) and $\gamma(h)$ is its *variogram* function, studied at length in Chapter 2.

If the linear drift is zero—that is, if the mean is constant—we have the usual form of the intrinsic model:

$$E[Z(x + h) - Z(x)] = 0,$$
$$E[Z(x + h) - Z(x)]^2 = 2\gamma(h)$$

Gaussian Random Functions

A random function is Gaussian if all its finite-dimensional distributions are multivariate Gaussian. Since a Gaussian distribution is completely defined by its first two moments, knowledge of the mean and the covariance function suffices to determine the spatial distribution of a Gaussian RF. In particular, second-order stationarity is equivalent to full stationarity.

A Gaussian IRF is an IRF whose increments are multivariate Gaussian.

A weaker form of Gaussian behavior is when all *bivariate* distributions of the RF are Gaussian; the RF is then sometimes called *bi-Gaussian*. A yet weaker form is when only the marginal distribution of $Z(x)$ is Gaussian. This by no way implies that $Z(x)$ is a Gaussian RF, but this leap of faith is sometimes made.

1.1.5 Spectral Representation

The spectral representation of SRFs plays a key role in the analysis of time signals. It states that a stationary signal is a mixture of statistically independent sinusoidal components at different frequencies. These basic harmonic constituents can be identified physically by means of filters that pass oscillations in a given frequency interval and stop others. This can also be done digitally using the discrete Fourier transform.

In the case of spatial processes the physical meaning of frequency components is generally less clear, but the spectral representation remains a useful theoretical tool, especially for simulations. For generality and in view of future reference we will state the main results in $\mathbb{R}^n$, which entails some unavoidable mathematical complication.

Theorem. A real, continuous, zero-mean RF defined on $\mathbb{R}^n$ is stationary (of order 2) if and only if it has the spectral representation

$$Z(x) = \int e^{2\pi i \langle u, x \rangle} Y(du) \tag{1.8}$$

for some unique *orthogonal random spectral measure* $Y(du)$ (see Appendix, Section A.1). Here i is the unit pure imaginary number, $u = (u_1, \ldots, u_n)$ denotes an n-dimensional frequency vector, du is an element of volume in $\mathbb{R}^n$, $x = (x_1, \ldots, x_n)$ is a point of $\mathbb{R}^n$, and $\langle u, x \rangle = u_1 x_1 + \cdots + u_n x_n$ is the scalar product of x and u.

For any Borel sets B and B' of $\mathbb{R}^n$, the measure Y satisfies

$$E[Y(B)] = 0$$
$$E[Y(B)\overline{Y(B')}] = 0 \qquad \text{if} \quad B \cap B' = \varnothing$$
$$Y(B \cup B') = Y(B) + Y(B') \quad \text{if} \quad B \cap B' = \varnothing$$

$Z(x)$ being real, we have in addition the symmetry relation $Y(-B) = \overline{Y(B)}$, where $-B$ denotes the symmetric of B with respect to the origin. Note that the random variables associated with disjoint sets B and B' are uncorrelated, hence the name *orthogonal* measure.

Now define $F(B) = E|Y(B)|^2$. F is a positive bounded symmetric measure called the *spectral measure*. We have in particular

$$F(B \cup B') = F(B) + F(B') \quad \text{if} \quad B \cap B' = \varnothing$$
$$E[Y(B)\overline{Y(B')}] = F(B \cap B')$$

It follows readily from (1.8) and the symmetry of F that the covariance of $Z(x)$ has the spectral representation

$$C(h) = E[Z(x)\overline{Z(x+h)}] = \int e^{2\pi i \langle u, h \rangle} F(du)$$

For time signals, the power of the RF $Z(x)$, which is the energy dissipated per unit time, is generally proportional to $Z(x)^2$. If the SRF has zero mean, $C(0)$ is equal to $E[Z(x)^2]$ and plays the role of an average power, and the measure F represents the decomposition of this power into the different frequencies. Note that the integral $\int F(du)$ of the spectral measure is equal to the total power $C(0)$.

Real Spectral Representation

It is interesting to separate the real and imaginary parts of the random spectral measure Y in the form

$$Y(B) = U(B) - i\,V(B)$$

where U and V are two *real* random measures (notice the $-i$ in the definition of V). From the properties of Y, we can deduce the following properties that will be useful for simulations:

$$
\begin{array}{lll}
U(-B) = U(B) & V(-B) = -V(B) & \\
E[U(B)U(B')] = E[V(B)V(B')] = 0 & \text{if}\quad B\cap B' = B\cap(-B') = \varnothing & \\
E[U(B)V(B')] = 0 & \forall B, B' & (1.9) \\
E\big[U(B)^2\big] = E\big[V(B)^2\big] = F(B)/2 & \text{if}\quad \{0\} \notin B & \\
E\big[|U(\{0\})|^2\big] = F(\{0\}) & V(\{0\}) = 0 &
\end{array}
$$

Also, $Z(x)$ has the representation

$$Z(x) = \int \cos(2\pi\langle u, x\rangle) U(du) + \int \sin(2\pi\langle u, x\rangle) V(du)$$

1.1.6 Ergodicity

Ergodicity is an intimidating concept. The practitioner has heard that the RF should be ergodic, since "this is what makes statistical inference possible," but he or she is not sure how to check this fact and proceeds anyway, feeling vaguely guilty of having perhaps overlooked something very important. We will attempt here to clarify the issues. In practice, ergodicity is never a problem. When no replication is possible, as with purely spatial phenomena, we can safely choose an ergodic model. If the phenomenon is repeatable, typically time-dependent fields or simulations, averages are computed over the different realizations, and the only issue (more a physical than a mathematical one) is to make sure that we are not mixing essentially different functions.

A detailed discussion of ergodicity can be found in Yaglom (1987, Vol. I, Chapter 3), and an analysis of its meaning in the context of unique phenomena in Matheron (1978). We have summarized the most important results so that practitioners can pay their respects to ergodicity once for all and move on.

Ergodic Property

In order to carefully distinguish a random function from its realizations, we will revert, in this section only, to the full notation $Z(x, \omega)$, where ω is the random event indexing the realization. By definition, a stationary random function $Z(x, \omega)$ is ergodic (in the mean) if the spatial average of $Z(x, \omega)$ over a domain $V \subset \mathbb{R}^n$ converges to the expected value $m = E[Z(x, \omega)]$ when V tends to infinity:

$$\lim_{V \to \infty} \frac{1}{|V|} \int_V Z(x, \omega)\,dx = m \qquad (1.10)$$

In this expression the norming factor $|V|$ denotes the volume of the domain V, and the limit is understood, as we will always do, in the mean square sense. In $\mathbb{R}^n$ it is important to specify how V tends to infinity, since we may imagine that V becomes infinitely long in some directions only, but we exclude this and assume that V grows in *all* directions. For example, V may be the cube $[0, t]^n$, where $t \to \infty$. Of course the limit does not depend on the particular shape of V.

To gain insight into the meaning of this property, it is interesting to revisit the sand jar a last time and do a little thought experiment. We consider a point x at a fixed location relative to the jar and shake the jar repeatedly, recording each time a 1 if x falls in a grain and a 0 otherwise. From this we can evaluate the mean of $I(x, \omega)$, namely the probability that x is in a grain, which should not depend on x. It is intuitively obvious that we will get the same result if we keep the jar fixed and select the point x at random within the jar, the probability of landing in a grain being equal to the proportion of the space occupied by the grains.

The ergodic property can be extremely important for applications, since it allows the determination of the mean from a *single* realization of the stationary random function, and precisely most of the time we only have one realization to work with. Not all stationary random functions are ergodic. The classic counterexample is the RF $Z(x, \omega) \equiv A(\omega)$ whose realizations are constants drawn from the random variable A. Clearly for each realization the space integral (1.10) is equal to the constant level $A(\omega)$ but not to the mean of A. Another, more realistic example of nonergodic RF is to consider a family of different stationary and ergodic RFs and select one of them according to the outcome of some random variable A, thus defining the composite RF $Z(x, \omega; A)$. On each realization the space integral converges to the mean $m(a) = E(Z(x, \omega; A) \mid A = a)$ of the particular RF $Z(x, \omega; a)$, but this is different from the overall mean $E[m(A)]$. Here we have the most common source of stationary but nonergodic random functions arising in practice. As has been pointed out in the literature, nonergodicity usually means that the random function comprises an artificial union of a number of distinct ergodic stationary functions.

Ergodic Theorem

This theorem states that if $Z(x, \omega)$ is a stationary random function (of order 2), the space integral (1.10) *always* converges to some value $m(\omega)$, but this value in general depends on the realization ω: it is a random variable, not a constant:

$$\lim_{V \to \infty} \frac{1}{|V|} \int_V Z(x, \omega) dx = m(\omega) \qquad (1.11)$$

This result is a direct consequence of the stationarity of $Z(x, \omega)$ and again requires V to grow in all directions. The random variable $m(\omega)$ has mean m and a fluctuation equal to the atom at the origin of the random spectral measure associated with the RF $Z(x, \omega)$:

$$m(\omega) = m + Y(\{0\}, \omega)$$

Since $E|Y(\{0\}, \omega)|^2 = F(\{0\})$, it appears that $Z(x, \omega)$ possesses the ergodic property if and only if its spectral measure F has no atom at the origin.

An equivalent condition, known as Slutsky's ergodic theorem, is

$$\lim_{V \to \infty} \frac{1}{|V|} \int_V C(h)\,dh = 0 \qquad (1.12)$$

This is always satisfied if $C(h) \to 0$ as $h \to \infty$, as is usually the case, but the condition is not necessary. Moreover, if the integral of the covariance in $\mathbb{R}^n$ is finite, when V is very large the left-hand side of (1.12) is an approximation to the variance of the space integral (1.11), and we will revisit this in Section 2.3.5 with the notion of *integral range*.

Ergodicity in the Covariance

In the above we have only considered first-order ergodicity, or ergodicity in the mean. It is also important to be able to determine the covariance from a single realization. This implies second-order ergodicity, or ergodicity in the covariance. To establish the ergodicity of the covariance, the same theory can be applied to the product variable $Q_h(x) = Z(x)\,Z(x+h)$ considered as a second-order stationary random function of x with h fixed. This involves the stationarity of fourth-order moments. For Gaussian RFs the fourth-order moments depend on the second-order moments, and simple results can be obtained. The analogue of Slutsky's condition for the convergence of covariance estimates is then

$$\lim_{V \to \infty} \frac{1}{|V|} \int_V [C(h)]^2\,dh = 0 \qquad (1.13)$$

This condition is more restrictive than (1.12). Its equivalent spectral formulation is that the spectral measure F has no atom *anywhere*. In other words, the covariance has no sinusoidal component. Again the convergence $C(h) \to 0$ as $h \to \infty$ suffices to fulfill (1.13), but the proof is only valid for Gaussian RFs.

Now What?

In the case of a unique phenomenon, there is no way of knowing if the space integral would have converged to a different value on another realization, since there is, and can be, only one. As will be seen in a moment, ergodicity is not an objective property in the sense that it cannot be falsified. Therefore we *choose* to model $Z(x, \omega)$ as an ergodic random function whose mean is the limit of the space integral (1.10). Likewise, we take the limit of the regional covariance (a space integral) as the *definition* of the covariance of the parent RF. Any other choice would have no relevance to the situation considered.

Strictly speaking, there still is a problem. Recall that in practice, we work in a bounded domain and cannot let it tend to infinity. This is a matter of scale. If the domain is large enough for the integral (1.12) to be small, the mean can be estimated reliably. But if the variance is still large, due to a slow fall-off of the covariance, the estimation of the mean is difficult, and it is preferable to avoid using it at all and only consider increments. This is the justification for using the variogram instead of the covariance. The possibility of statistical inference of the variogram is discussed in Section 2.9.

When dealing with space–time phenomena observed at a fixed set of monitoring stations, we typically consider spatially nonstationary models and compute time-averaged estimates of spatial means and covariances. We are thus treating the data as a collection of (correlated) stationary and ergodic random functions of time (multiple time series). These assumptions have to be checked carefully.

Micro-Ergodicity

As we have noted earlier, it is impossible to extend the domain to infinity. Matheron (1978) introduced the notion of *micro-ergodicity*, also called *infill asymptotics* (Cressie, 1991), concerned with the convergence of space integrals when the domain D remains fixed but the sampling density becomes infinite. This concept is distinct from standard ergodicity. For example, neither the mean nor the variance of a stationary and ergodic RF is micro-ergodic, but the slope of the variogram at the origin is micro-ergodic if the variable is not too smooth (Section 2.9.2). Micro-ergodic parameters represent physically meaningful properties.

1.2 ON THE OBJECTIVITY OF PROBABILISTIC STATEMENTS

What sense does it make to speak of the probability of a unique event? When we are told that "there is a 60% chance of rain tomorrow," we know the next day if it rains or not, but how can we check that the probability of rain was indeed 60% on that day at a specific place? We can't. The only probabilistic statement that can be disproved is "there is a zero chance of rain tomorrow": If it does rain the next day, then clearly the forecast was wrong. The same problem essentially arises for spatial "prediction." What is the physical meaning of a statement such as "there is a 0.95 probability that the average porosity of this block is between 20% and 25%"? Potentially we could measure the porosity and check if it lies in the interval, but we will never know if the 0.95 was correct. Yet despite their unclear meaning, we tend to find probabilistic statements useful in giving us an appreciation of uncertainty.

In reality we establish the credibility of weather forecasts not from a single prediction but over time. Someone with enough motivation could check if out of 100 days associated with a forecast of a 60% chance of rain about 60 days were indeed rainy, and do this for all % chance classes. A successful track record, without proving the correctness of the forecast on any given day, proves at least that it is correct *on the average*. It validates the forecasting *methodology*.

One may object that since we introduced repetitions we are no longer really dealing with a unique phenomenon. But the distinction between unique and repeatable situations is not as clear-cut as it seems. Strictly speaking, it is impossible to repeat the "same" experiment: They always differ in some aspects; we simply judge those unimportant. On the other hand, even though every petroleum reservoir, every mine, and every forest is unique, they all belong to classes of situations, shaly sand reservoirs, copper deposits, or tropical woods that are similar enough to give rise to specific methodologies that over time can be validated objectively. This is "external" objectivity.

The practitioner who is interested in the evaluation of this specific deposit or that specific forest would rather have criteria for "internal" objectivity that are based on those unique situations. If we cannot pin down the meaning of a probabilistic statement on a singular event, then the question becomes, Which concepts, statements, and parameters have an objective, observable, measurable counterpart in reality? Matheron (1978) devotes a fascinating essay entitled "Estimating and Choosing" to this quest for objectivity. The central idea is this: *The only objective quantities are those that may be calculated from the values of a single realization over a bounded domain D.* Indeed, in the absence of repetitions, the maximum information we can ever get is the complete set of values $\{z(x) : x \in D\}$. Objective quantities are essentially space integrals of functions of $z(x)$, referred to as *regionals:* all the values of $z(x)$ itself, block averages, mean values above thresholds, and so on, along with the regional mean, variogram, or histogram. On the contrary, the expected value m, the (true) variogram γ, or the marginal distribution of $Z(x)$ are *conventional* parameters. To emphasize the difference, Matheron says that we *estimate* a regional whose exact value is unknown but nevertheless exists independently of us, namely is potentially observable, but we *choose* the value of a conventional parameter.[4]

These considerations lead to a striking reversal of point of view where regionals cease to be mere estimates of "true" parameters to become the physical reality itself, while their theoretical counterparts turn into conventional parameters. For example, the regional variogram γ_R should not be regarded as the regional version of γ but rather γ as being the theoretical version of γ_R. Likewise, the fluctuation variance (in the probabilistic model) of a regional is not indicative of the difficulty of the statistical inference of its expected value but rather of the lack of objective meaning of this parameter.

The objectivity of statements can be defined by two criteria. The stronger one is to regard a statement as objective if it is *decidable*, which means that it can be declared true or false once we know $z(x)$ for all $x \in D$. The weaker form of objectivity is K. Popper's demarcation criterion for scientific hypotheses: It must be possible to design experiments whose outcomes are liable to *falsify* predictions derived from these hypotheses, that is, events with probabilities (nearly) equal to 0 or 1 (in the model). If such attempts are successful, the

---

[4] In the statistical literature, *to predict* means *to estimate* in Matheron's sense and *to estimate* means *to choose*. In this book we will estimate observables in Matheron's sense but *fit* model parameters rather than choose them.

hypothesis is falsified. If it withstands testing, we may not conclude that it is true but only that it is corroborated (not refuted).

The statement "$z(x)$ is a realization of a random function $Z(x)$" or even "of a stationary random function" has no objective meaning. Indeed, since D is bounded, it is always possible by periodic repetitions and randomization of the origin to construct a stationary random function having a realization that coincides over D with the observed $z(x)$. Therefore no statistical test can disprove stationarity in general. We *choose* to consider $z(x)$ as a realization of $Z(x)$ over D. It does not mean that this decision is arbitrary—in practice, it is suggested by the spatial homogeneity of the data—but simply that it cannot be refuted. As stated earlier, ergodicity is also not an objective property.

If repetitions are the objective foundation of probabilities and if only regionals are physically meaningful, then in the case of a unique phenomenon the objectivity of our measures of uncertainty must be based on spatial repetitions. These are obtained by moving the configuration involved—for example, a block and its estimating data points—throughout a domain D_0. Denoting by Z_v the block value and by Z^* its estimator, the estimation variance of such block is interpreted as the spatial average of the squared error $(Z^* - Z_v)^2$ over D_0. By construction, this variance is *not localized* (i.e., is constant) within D_0, but neither is the kriging variance calculated in the stationary model (since it only depends on the geometry and on the variogram). Of course the domain D_0 can itself be local and correspond to a homogeneous subzone of the total domain D. However, D_0 should not be too small; otherwise, we will lack repetitions. We are tempted to say that there is a trade-off between objectivity and spatial resolution.[5]

1.3 TRANSITIVE THEORY

To avoid the epistemological problems associated with the uniqueness of phenomena, Matheron (1965) first developed an estimation theory in purely spatial terms which he named *transitive theory*. In this approach the regionalized variable $z(x)$ is deterministic and only assumed to be identically zero outside a bounded domain D; it represents a so-called transition phenomenon, a spatial equivalent of a transient phenomenon in time. We will focus here on the global estimation problem, namely the evaluation of the integral of $z(x)$ which typically represents the total amount of some resource. Initially applied to mineral resources, the transitive approach has received a renewed interest in the last two decades for the estimation of fish abundance when the areas of fish presence have diffuse limits (Petitgas, 1993; Bez, 2002).

---

[5] The image of "dithering" comes to mind. This is a binarization technique to transform a halftone image into a black-and-white image. A gray level is obtained by judiciously distributing black and white dots in the cells of a matrix: A 4×4 matrix allows 16 gray levels, and a 16×16 matrix is required to render 256 gray levels. Thus there is a trade-off between the representation of gray level amplitude and the spatial resolution.

In this model, randomness is introduced through sampling. The easiest would be to use the classic Monte Carlo method and select N samples randomly and independently, leading to an unbiased estimator with a variance equal to σ^2/N, where σ^2 is the spatial variance of $z(x)$. However, systematic sampling is usually more efficient. We will present the theory in this case, mainly for background, but also to justify a neat formula for surface estimation. The transitive theory can also be developed for local estimation, but it has no advantage over the more elegant random function approach. Transitive theory will not be used elsewhere in the book.

1.3.1 Global Estimation by Systematic Sampling

Consider the estimation of the integral

$$Q = \int z(x)dx$$

which is finite since $z(x)$ is zero outside the domain D. If $z(x)$ is a mineral grade (in g/ton) and if the ore density d is a constant, Qd is the quantity of metal in the deposit; if $z(x)$ is an indicator function, Q is the volume of D. We assume that the domain D is sampled on a rectangular grid that extends as far as needed beyond the boundaries of D.

As usual, we reason in $\mathbb{R}^n$ and denote by a the elementary grid spacing $(a_1, a_2, \ldots, a_n)$ and by $|a|$ its volume (i.e., the product $a_1 a_2 \ldots a_n$). The origin of the grid, which is one of its points, is denoted by x_0, and k denotes the set of positive or negative integers $k = (k_1, k_2, \ldots, k_n)$. The simplest estimate of Q is

$$Q^*(x_0) = |a| \sum_{k \in \mathbb{Z}^n} z(x_0 + ka)$$

where $\mathbb{Z}$ is the set of relative integers. If we select the origin x_0 at random and uniformly within the parallelepiped $\Pi = [0, a_1] \times [0, a_2] \times \cdots \times [0, a_n]$, then Q^* becomes a random variable whose expected value is

$$E(Q^*) = \frac{1}{|a|} \int_\Pi Q^*(x_0)dx_0 = \frac{1}{|a|} \int_\Pi dx_0 |a| \sum_k z(x_0 + ka) = \int z(x)dx = Q$$

It is unbiased. When we define the *transitive covariogram* $g(h)$ by

$$g(h) = \int z(x)z(x + h)\, dx$$

similar calculations show that

$$E(Q^*)^2 = |a| \sum_k g(ka) \qquad \text{and} \qquad Q^2 = \int g(h)\, dh$$

so that the variance of the error $Q^* - Q$, or *estimation variance*, is given by the formula

$$E(Q^* - Q)^2 = |a| \sum_k g(ka) - \int g(h)\, dh \qquad (1.14)$$

This estimation variance, denoted $\sigma^2(a)$, appears as the error incurred by approximating the integral $\int g(h)\, dh$ by a discrete sum over the grid. It decreases as the grid becomes finer and as the function $z(\cdot)$ becomes smoother.

This variance is always nonnegative provided that the covariogram $g(h)$ is modeled as a positive definite function ($g(h)$ is the convolution of $z(x)$ by $z(-x)$). The transitive covariogram plays the role of the covariance in an RF model, and in fact they are related, since the regional *noncentered* covariance over D is given by

$$C_R(h) = \frac{1}{K(h)} \int_{D \cap D_{-h}} z(x)z(x+h)dx, \qquad \text{where } K(h) = |D \cap D_{-h}|$$

so that $g(h) = K(h) \, C_R(h)$.

An expansion of formula (1.14) as a Euler–MacLaurin series leads, for small a, to a decomposition of the variance $\sigma^2(a)$ into two terms:

$$\sigma^2(a) = T_1(a) + T_2(a)$$

The first term, $T_1(a)$, is related to the behavior of $g(h)$ near the origin; the second one, $T_2(a)$, depends on its behavior near the range (the distance b beyond which $g(h)$ becomes identically zero). $T_2(a)$ is the *fluctuating term*, also called *Zitterbewegung* (the German for jittery motion). It is a periodic function of the remainder ε of the integer division b/a and cannot be evaluated from the grid data; since it has a zero mean, it is simply ignored.

The regular term $T_1(a)$ can be approximated from the expansion of $g(h)$. For example, in 2D and for an isotropic covariogram with a linear behavior near the origin, the explicit result is

$$\sigma^2(a) = \sigma^2(a_1, a_2) \approx -g'(+0)\left[\frac{1}{6}a_1^2 a_2 + 0.0609 a_2^3\right], \qquad a_1 \le a_2 \tag{1.15}$$

where $g'(+0)$ is the slope of $g(h)$ at the origin and is < 0.

1.3.2 Estimation of a Surface Area

If $z(x)$ is the indicator function of a geometric object, its covariogram is necessarily linear at the origin. More precisely, for a vector h in a direction α, we have $g(h) = g(0) - D_\alpha |h| + \cdots$, where D_α is the "total diameter" in the direction α. If the object is convex, D_α is simply the so-called tangent diameter (or caliper diameter); otherwise, $2D_\alpha$ is the total length of the contour of the object projected orthogonally along the direction α (see Section 2.3.4). Here we consider an object with surface area A and a total diameter D that is approximately the same in all directions. Replacing A by its estimate $A^* = N \, a_1 \, a_2$, where N is the number of positive samples, we can express the variance (1.15) in the dimensionless form σ_A^2 / A^2:

$$\frac{\sigma_A^2}{A^2} \approx \frac{D}{\sqrt{A}} \frac{1}{N^{3/2}}\left[\frac{1}{6}\sqrt{\lambda} + 0.0609\lambda^{-3/2}\right], \qquad \lambda = a_1/a_2 \le 1 \tag{1.16}$$

Note that the variance decreases like $1/N^{3/2}$ rather than $1/N$. We can evaluate D from the contour of the object by counting the number of boundary segments $2N_1$ and $2N_2$ respectively, parallel to a_1 and a_2, including possible holes in the contour (the total perimeter comprises $2(N_1+N_2)$ segments):

$$D = N_1 a_1 = N_2 a_2$$

and upon replacement in (1.16), we get

$$\frac{\sigma_A^2}{A^2} \approx \frac{1}{N^2}\left[\frac{1}{6}N_2 + 0.0609\frac{N_1^2}{N_2}\right], \qquad N_2 \le N_1 \tag{1.17}$$

This formula remains valid if the object is not isotropic but has a main direction of elongation parallel to one of the grid axes, which is the natural orientation for the grid. Indeed we can

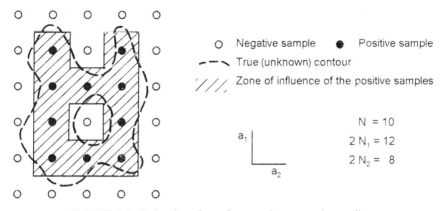

FIGURE 1.1 Estimation of a surface area by systematic sampling.

then restore isotropy, at least approximately, by an affine transformation parallel to one of the grid axes; this changes a_1 or a_2 as well as A and D_1 or D_2 but not N, N_1, N_2, or σ_A^2/A^2, which are dimensionless. Thus (1.17) is a simple, self-contained (no calculation and modeling of $g(h)$ needed), and yet theoretically founded formula for evaluating the error in the estimation of a surface area.

To illustrate its use, consider the example shown in Figure 1.1. We read from the figure:

$$
\begin{aligned}
N &= 10 \\
2N_1 &= 12 \quad \text{so that} \quad \frac{\sigma_A^2}{A^2} = \frac{1}{100}\left[\frac{4}{6} + 0.0609\frac{36}{4}\right] = \frac{1.21}{100} \\
2N_2 &= 8
\end{aligned}
$$

The relative error standard deviation on the surface area is therefore 11%.

An interesting indication can be derived concerning the optimal grid mesh. In case of an isotropic object the variance in (1.16) is minimized for $\lambda = 1$, that is, $a_1 = a_2$. If the object is not isotropic an affine transformation is applied to restore isotropy, for example multiplying lengths along D_2 by D_1/D_2. The new grid spacings become $a_1' = a_1$ and $a_2' = (D_1/D_2)a_2$, and in this isotropic case the optimal grid mesh satisfies $a_1' = a_2'$. Therefore optimum sampling is achieved when $D_1/a_1 = D_2/a_2$, or equivalently $N_1 = N_2$, that is, when the grid mesh is adapted to the anisotropy of the object. Formula (1.17) also shows that for the optimal grid mesh ($N_1 = N_2$) the estimation variance increases with the perimeter of the object.

CHAPTER 2

Structural Analysis

Au-delà de l'outil, et à travers lui, c'est la vieille nature que nous retrouvons, celle du jardinier, du navigateur, ou du poète.

Beyond our tools, and through them, it is old mother nature that we reach, an experience that we share with gardeners, sailors, or poets.

—Saint-Exupéry

2.1 GENERAL PRINCIPLES

2.1.1 Introduction

The theory of stochastic processes and random functions has been in use for a relatively long time to solve problems of interpolation or filtering. The methods proposed are based on the first two moments of the random functions. In the real world, however, these are never known and must be determined first. The present chapter is devoted to the analysis of the structural characteristics of spatial phenomena which are generally

- unique, nonreproducible,
- defined in a two- or three-dimensional domain,
- too complex for a precise deterministic description,
- known from samples taken at scattered locations.

These phenomena are regionalized variables $\{z(x): x \in D \subset \mathbb{R}^n\}$. We decide to regard them as realizations of random functions.

Geostatistics: Modeling Spatial Uncertainty, Second Edition. J.P. Chilès and P. Delfiner.
© 2012 John Wiley & Sons, Inc. Published 2012 by John Wiley & Sons, Inc.

A random function (RF) $\{Z(x): x \in \mathbb{R}^n\}$ is characterized by its finite-dimensional distributions (also called here *spatial distribution* for short), namely the set of all multidimensional distributions of k-tuples $(Z(x_1), Z(x_2), \ldots, Z(x_k))$ for all values of k and all configurations of the points $x_1, x_2, \ldots, x_k$. Even if a very large number of realizations of a random function were available, the combinatorial possibilities are such that, in practice, one could calculate sample multidimensional distributions only for the simplest k-tuples. When a single realization is available, which is the common case, these distributions cannot be determined, except under an assumption of stationarity which introduces repetition in space: Two configurations of points that are identical up to a translation are considered as statistically equivalent. Since the sample points are unevenly distributed, the only (nearly) identical configurations that can be found are pairs of sample points. A large part of the book is therefore dedicated to methods involving only the knowledge of two-point statistics. Multipoint distributions, implicitly or explicitly used when building conditional simulations, will be considered in Chapter 7. The complete knowledge of the bivariate distributions, required for the nonlinear techniques, will be presented in Chapter 6. For linear methods, which are the most widely used, it suffices to know the second-order moments. These are the focus of the present chapter.

The main tool is the variogram. We will distinguish three main definitions: (1) the variogram of the random function, or theoretical variogram, whose knowledge is required for kriging; (2) the variogram of the regionalized variable, or regional variogram, which could be calculated if we knew the value of the regionalized variable at every point of its domain of study; and (3) the sample variogram, which can be calculated from the data. Our task can therefore be split into two phases: (1) Compute a sample variogram that best approximates the regional variogram and (2) fit a theoretical model to this sample variogram.

In applications, a sample is usually not a point but a volume such as a core, that is, a piece of rock characterized by its shape, size, and location. Shape and size define the *support* of the sample. If it is very small and the same for all the data we can forget it and regard it as a point. We will adopt that point of view throughout this chapter and specify the support only when needed, for example when data relate to different supports.

2.1.2 Covariance Versus Variogram

We will consider two classes of random functions: stationary random functions and intrinsic random functions. Throughout the book, unless stated otherwise, stationarity means second-order stationarity.

Covariance of a Stationary Random Function

As we have seen in Section 1.1.4, a stationary random function (SRF) $Z(x)$ is characterized by its mean

$$m = \mathrm{E}[Z(x)]$$

and its covariance function (or covariance for short)

$$C(h) = E[Z(x) - m][Z(x+h) - m] \qquad (2.1)$$

A related function is the *correlogram* $\rho(h) = C(h)/C(0)$, which is the correlation coefficient between $Z(x)$ and $Z(x+h)$. The covariance and the correlogram show how this correlation evolves with the *separation*, or *lag*, h. Note that h is a vector. These functions therefore depend both on its length, which is the distance between x and $x+h$, and on its direction. When the covariance depends only on distance, it is said to be *isotropic*. A covariance is an even function, and by the Schwarz inequality it is bounded by its value at the origin (i.e., the variance of the SRF):

$$C(h) = C(-h), \qquad |C(h)| \le C(0)$$

More precisely, we will show that a covariance is a positive definite function. Since the random function has a finite variance, it fluctuates around the mean. Some phenomena do not show this behavior: If we compute the sample mean and variance over increasingly large domains, the sample mean does not stabilize, and the sample variance always increases. This motivates the next model.

Variogram of an Intrinsic Random Function

An intrinsic random function (IRF) is a random function whose *increments* are second-order stationary. It is characterized by its linear drift

$$m(h) = E[Z(x+h) - Z(x)] = \langle a, h \rangle$$

and its variogram

$$\gamma(h) = \tfrac{1}{2} \text{Var}[Z(x+h) - Z(x)] \qquad (2.2)$$

To prove that the drift is linear, start from the obvious relation

$$Z(x+h+h') - Z(x) = [Z(x+h) - Z(x)] + [Z(x+h+h') - Z(x+h)]$$

Passing on to the mathematical expectation gives $m(h+h') = m(h) + m(h')$, which implies that m is a linear function of the vector $h = (h_1, \ldots, h_n)'$, namely $m(h) = \langle a, h \rangle = a_1 h_1 + \cdots + a_n h_n$ for some gradient vector $a = (a_1, \ldots, a_n)'$. From now on, unless explicitly stated otherwise, we will consider that the IRF has no drift, or $m(h) \equiv 0$. The opposite case will be included in the model of universal kriging (see the introduction of Section 3.4).

The variogram shows how the dissimilarity between $Z(x)$ and $Z(x+h)$ evolves with the separation h. Like the covariance, it is in general anisotropic. Obviously the variogram is an even, nonnegative function valued 0 at $h = 0$:

$$\gamma(h) = \gamma(-h), \qquad \gamma(h) \ge 0, \qquad \gamma(0) = 0$$

It can be bounded or increase to infinity. Just like a covariance, a variogram cannot be an arbitrary function. We will show that $-\gamma(h)$ must be a conditionally positive definite function.

The variogram $\gamma(h)$ is sometimes called "theoretical" to remind us that it is a theoretical construct involving neither a particular realization nor a particular region. Let us mention that $\gamma(h)$ is also called "semivariogram." However, the term "variogram" tends to become established for its simplicity and can be supported by theoretical arguments (see the definition of the generalized variogram given in Section 4.7.1); it was already used by Jowett (1955a,c), where it represented the graph of the sample or experimental version of $\gamma(h)$.

Bounded Variograms and Stationarity

An SRF is obviously also an IRF and therefore has a variogram. In that case the varioram is linked to the covariance by the relation

$$\gamma(h) = C(0) - C(h) \tag{2.3}$$

Thus the variogram of an SRF is bounded by $2\,C(0)$. Equation (2.3) shows that if the covariance is known, the variogram is also known. Conversely, if the variogram of an IRF is bounded by a finite value, $\gamma(h)$ is of the form (2.3) for a stationary covariance $C(h)$ and the IRF only differs from an SRF by a random constant (Matheron, 1973a, pp. 454–457). If the variogram has a sill, the value of $C(0)$ must be chosen equal to or greater than the sill.[1] It is then equivalent to know $\gamma(h)$ or $C(h)$.

Variogram and Sample Variance

The theoretical variance of $Z(x)$ is equal to $C(0)$ if Z is an SRF, or does not exist (i.e., is infinite) if Z is an IRF but not an SRF. However, the sample variance of a finite number of values always has a finite expectation. Let us consider a particular realization $z(x)$ of the IRF $Z(x)$. If the N values $\{z(x_\alpha) : \alpha = 1, \ldots, N\}$ are known, we can define the sample variance, which characterizes the dispersion of the $z(x_\alpha)$ around their mean, as

$$s^2(0|N) = \frac{1}{N} \sum_{\alpha=1}^{N} [z(x_\alpha) - \bar{z}]^2 \quad \text{with} \quad \bar{z} = \frac{1}{N} \sum_{\alpha=1}^{N} z(x_\alpha)$$

The notation $s^2(0|N)$ will be generalized in Section 2.8.2 to other dispersion variances. The 0 means that the unit samples are considered as punctual (zero

[1] In the general case of a bounded variogram in $\mathbb{R}^n$, Gneiting et al. (2001) show that $A - \gamma(h)$ is a covariance function if and only if

$$A \geq \lim_{u \to \infty} (2u)^{-n} \int_{[-u, u]^n} \gamma(h)\, dh$$

volume), and N reminds us that the variance is related to a sampling pattern $\{x_\alpha : \alpha = 1, \ldots, N\}$. This variance can be expresssed in the form

$$s^2(0 \mid N) = \frac{1}{2N^2} \sum_{\alpha=1}^{N} \sum_{\beta=1}^{N} [z(x_\beta) - z(x_\alpha)]^2 \qquad (2.4)$$

Now, if the values $z(x_\alpha)$ are not available, $s^2(0 \mid N)$ cannot be calculated, but we can obtain its expected value $\sigma^2(0 \mid N)$ by randomizing (2.4) with respect to the realization. It is equal to

$$\sigma^2(0 \mid N) = \frac{1}{N^2} \sum_{\alpha=1}^{N} \sum_{\beta=1}^{N} \gamma(x_\beta - x_\alpha) \qquad (2.5)$$

In the case of an SRF, this expression is equivalent to

$$\sigma^2(0 \mid N) = C(0) - \frac{1}{N^2} \sum_{\alpha=1}^{N} \sum_{\beta=1}^{N} C(x_\beta - x_\alpha) = C(0) - \mathrm{Var}\left[\frac{1}{N} \sum_{\alpha=1}^{N} Z(x_\alpha)\right]$$

The expectation of the sample variance is always less than the theoretical variance. If the x_α are so far apart that the $Z(x_\alpha)$ are mutually uncorrelated, this amounts to

$$\sigma^2(0 \mid N) = \left(1 - \frac{1}{N}\right) C(0)$$

The discrepancy between the expectation of the sample variance and the theoretical variance vanishes as $N \to \infty$.

Variogram as a Structural Tool

Unless we are considering an SRF with a known mean, an exceptional situation, the above gives us two reasons to favor the variogram over the covariance. The first is theoretical: Since the class of IRFs includes the SRFs, the variogram is a more general tool than the covariance. This is why it was introduced in the 1940s for the study of turbulent flow [e.g., see Kolmogorov (1941a,b), Obukhov (1949a,b), and Gandin (1963), for applications to meteorology].

The second reason is practical: The variogram does not require the knowledge of the mean, whereas to compute the covariance the mean has to be estimated from the data, which introduces a bias. For example, the covariance at $h = 0$ will be approximated by the sample variance, which is biased downward as seen above. This bias cannot be corrected unless the covariance function, or at least the correlation function, is already known, which is not the case (unless the data can be considered uncorrelated, but this is a very special case). A similar bias can also corrupt the behavior of the covariance near the origin: Matheron (1970, Chapter 2, exercise 18) shows that in 1D, for a covariance that is linear over the data domain, the slope at the origin of the sample covariance is equal to 4/3 the true slope. Use of the sample covariance when the mean is not known can thus result in erroneous interpretations. The variogram is not affected by these problems since it automatically filters the mean. It was precisely for this reason that Jowett (1955a) used the sample variogram rather than the covariance. For completeness we should mention that as early as 1926 A. Langsæter had already used the variogram to characterize the variability of data derived from forest surveys [reported in Matérn (1960, p. 51)].

To summarize, we can say that even if the objective is the covariance, the structural tool is the variogram. This can justify why, following Obukhov (1949a,b), Yaglom (1987) calls the variogram the *structure function*, a phrase to which we give a more general meaning.

2.2 VARIOGRAM CLOUD AND SAMPLE VARIOGRAM

We now turn to the empirical aspects of variogram analysis.

2.2.1 Preliminary: Exploratory Data Analysis

Geostatistical tools do not replace, but complement, standard statistical tools. Before computing sample variograms and other spatial statistics, one should perform an exploratory data analysis (Tukey, 1977), namely compute usual univariate and bivariate statistics such as posted maps, histograms, scatterplots, and box plots. Multivariate techniques such as cluster analysis or principal component analysis can also be used to reduce multivariate problems to univariate ones. Since this book is devoted to geostatistical methods, we will not develop this point, but the relevance of a geostatistical study largely depends on the quality of this preliminary phase. At this stage, most inconsistencies in the data can be detected (gross errors on data coordinates, for example), as well as mixing of different populations that should be studied separately (bimodal histogram), or preferential sampling of particular areas such as the richest parts of an orebody or the most contaminated zones of a plume, which should require the use of declustering techniques. In a real study, it is not uncommon that the task of obtaining clean data and clear objectives is at least as long as the geostatistical study itself, if not longer. Examples of exploratory analysis abound in the literature; for example, see Isaaks and Srivastava (1989) with a synthetic data set, Webster and Oliver (1990) with applications to soil and land resources, Rossi et al. (1992) with applications to ecology, and Cressie (1991). The geostatistical tools we present now offer the opportunity to go further in the analysis by exploring the spatial relationships between data pairs [e.g., see the applications presented by Bradley and Haslett (1992)].

2.2.2 Variogram Cloud

Let us step for a moment outside the probabilistic context and consider a regionalized variable $\{z(x): x \in D \subset \mathbb{R}^n\}$, with known values $z_\alpha = z(x_\alpha)$ at N sample points $\{x_\alpha : \alpha = 1, \ldots, N\}$. Physical intuition suggests that two points that are close assume close values because these values were generated under similar physical conditions. A geologist would say that the two points have the same "geological environment." On the contrary, at long distances the genetic conditions are different, and greater variations are to be expected. This intuition of variability with distance can be quantified with *h-scattergrams* (also known as *lagged scatterplots*). Instead of describing the relationship between two variables, which an usual scatterplot does, an *h*-scattergram describes the relationship between the variable of interest at some location and the same variable at a location separated by a vector *h* representing some distance in a certain direction. If the *h*-scattergrams corresponding to a common distance and to various directions are similar, an *h*-scattergram based on distance only is considered. Figure 2.1 shows an example with three scattergrams: For $h = 1$m the points represent pairs $(z_\alpha, z_{\alpha+1})$ of neighbouring data, so that all points of the scattergram are close to the diagonal; for a small *h* value, the cloud of points

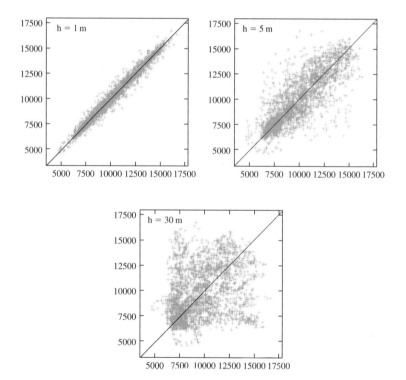

FIGURE 2.1 *h*-scattergrams of impedance in boreholes for $h = 1$m, 5m, and 30m. [From D. Aburto, personal communication.]

becomes diffuse but with limited scatter from the diagonal; for the largest h value the scatttergram is highly dispersed and expresses an absence of correlation of the data at such a distance. h-scattergrams are very useful for a detailed study of the spatial variability of the regionalized variable (see, e.g., Section 6.4.4). But modeling spatial variability directly from h-scattergrams is not a simple task because the various h-scattergrams must be modeled in a consistent manner.

A way to simplify the analysis is to only consider increments $z_\beta - z_\alpha$ rather than z_α and z_β separately, namely one variable instead of two. This makes it possible to represent the increment with respect to the separation between the two data points. This is the variogram cloud. This tool, first used by Gandin (1963, p. 47) for the study of meteorological fields, and reintroduced and systematically exploited by Chauvet (1982), is a plot of all sample point pairs (α, β) showing

- along the horizontal axis, distance $r_{\alpha\beta} = |x_\beta - x_\alpha|$;
- along the vertical axis, halved squared increment $\frac{1}{2}(z_\beta - z_\alpha)^2$.

Figure 2.2, from Bastin et al. (1984), shows a nice example of this. It concerns rainfall data (Figure 2.2a), a time-repeated phenomenon. Each point in Figure 2.2b is in fact the mean of the 1425 values obtained from 1425 six-hourly observations. When computed from a single realization, the cloud is of course much more dispersed.

The variogram cloud can show different behaviors along the different directions of the separation $h_{\alpha\beta} = x_\beta - x_\alpha$, namely display an anisotropy. This is frequent in 2D, and especially in 3D where vertical variability is rarely of the same nature as horizontal variability (layered media). The cloud is then calculated by classes of direction. The main anisotropy directions are often suspected from geological knowledge, and the variogram cloud computed along these directions. If this is not the case, it is necessary to compute the cloud in several directions to detect a possible anisotropy. In 2D, at least four equally spaced directions are usually considered (the coordinate axes and the diagonals). In subhorizontal layered media, the horizontal directions must be complemented by the vertical and oblique directions, and so on.

Note that the choice to represent halved squared increments is made by reference to the definition of the variogram of an IRF. It is of course possible, and even recommended, to similarily display other characteristics of the increment $z_\beta - z_\alpha$, for example the increment itself, which makes it possible to check the absence of drift, especially at the border of the domain of study (in that case one must be careful not to mix pairs with opposite directions).

2.2.3 Sample Variogram

From the variogram cloud it is possible to extract the following information:

- *The Sample Variogram.* This is the curve giving the mean of the halved squared increment as a function of distance; in practice, it is calculated by classes of distance, taking in each class the center of gravity of the sample points of the variogram cloud (Figure 2.2d).

- *Any Other Characteristic of the Cloud.* This is calculated by classes of distance (median, quartiles).
- *Box Plots.* These present several characteristics of the variogram cloud in a single figure, usually the mean, the median, and some other quantiles of the halved squared increments for each class of distance (Figure 2.2c).

Like the variogram cloud, the sample variogram can be anisotropic and is therefore calculated and displayed by classes of direction. The sample

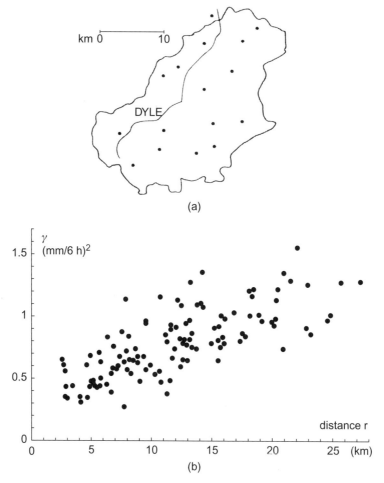

FIGURE 2.2   Variography of rainfall data (Dyle River Basin, Belgium): (a) locations of the rain gauges; (b) variogram cloud (each point represents the mean of 1425 values); (c) box plot showing mean (×), median (−), quartiles, extreme values, and number of pairs; and (d) sample variogram (circle area proportional to number of pairs). [From Bastin et al. (1984), © American Geophysical Union.]

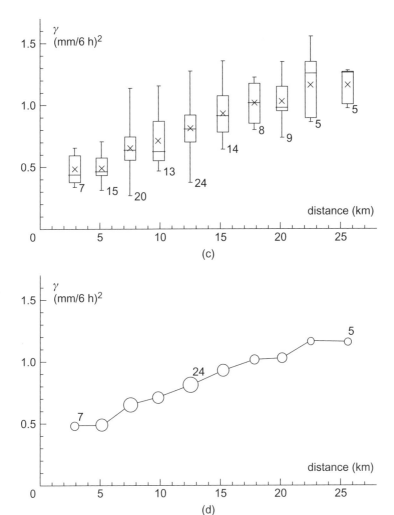

FIGURE 2.2 (Continued)

variogram can also be defined directly: denoting by N_h the count of pairs of points separated (approximately) by the lag h, it is defined by

$$\hat{\gamma}(h) = \frac{1}{2N_h} \sum_{x_\beta - x_\alpha \approx h} [z(x_\beta) - z(x_\alpha)]^2 \tag{2.6}$$

It is, of course, calculated for discrete values of h. Graphically, the sample variogram can be represented by several curves corresponding to several directions and, in the 2D case, by a *variogram map* or *variomap* figuring out the variogram value in each angular sector and distance class by a color coding. The latter representation is well adapted for the detection of anisotropies

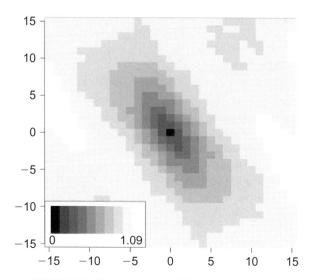

FIGURE 2.3 Variogram map of data on a regular grid.

whereas the former is better suited to variogram modeling. Note that the variogram map is symmetric with respect to the origin.

In the case of equally spaced data in 1D, the variogram can be computed for multiples of the spacing Δx. If N denotes the number of data, the sample variogram at lag $h = k\,\Delta x$ is

$$\hat{\gamma}(k\,\Delta x) = \frac{1}{2(N-k)} \sum_{\alpha=1}^{N-k} [z_{\alpha+k} - z_{\alpha}]^2$$

This definition can be generalized to data on a 2D or 3D grid, thus producing a sample variogram that is a 2D or 3D array, which can be represented as a variogram map with square or cubic cells (Figure 2.3). It can be calculated very efficiently with fast Fourier transforms, even if some data are missing, as shown by Marcotte (1996): to allow FFT calculations, N zero data are added to the sequence and missing data are replaced by zeros; the sum of the $z_{\alpha}\,z_{\alpha+k}$ can therefore be calculated regardless of the missing data; the number of pairs is obtained similarly by working with the indicator of the data.

There is a simple relationship between the sample variogram and the sample variance of the data: the average of all terms $\hat{\gamma}(h)$ for all possible lags including the lag $h = 0$, weighted by N_h, is the mean of $\frac{1}{2}(z_{\beta} - z_{\alpha})^2$ for all pairs (α, β), including those for which $\alpha = \beta$, and it coincides with the sample variance of the data.

2.2.4   Regional Variogram

When studying a domain D, the ideal sample variogram is that which can be calculated when the domain D is perfectly known. This variogram, named the *regional variogram*, is defined as an areal average of $\frac{1}{2}[z(x+h) - z(x)]^2$ by the formula

$$\gamma_R(h) = \frac{1}{2|D \cap D_{-h}|} \int_{D \cap D_{-h}} [z(x+h) - z(x)]^2 dx \qquad (2.7)$$

The actual domain of integration is not D but is, instead, the intersection of D with its translate by the vector $-h$, denoted $D \cap D_{-h}$. Indeed x and $x + h$ both belong to D if and only if x belongs to $D \cap D_{-h}$. This intersection usually shrinks as the modulus of h increases. $|D \cap D_{-h}|$ is the measure of $D \cap D_{-h}$, namely its length, area, or volume according to the dimensionality of the space. Since in practice only a limited number of sample points are available, one may wonder to what extent the sample variogram is representative of the regional variogram. It is usually considered that a sample variogram value is not reliable if it has been calculated from less than 50 pairs; this is, however, only a broad indication which, to be refined, needs to take into consideration the lag, the locations of the pairs, the shape of the variogram, the histogram of the data, the higher-order moments, and so on. Section 2.9.1 gives further insight into this question.

At this point, the usual question is: Which of the regional variogram and the theoretical variogram matters? From a statistical point of view, the theoretical variogram is more meaningful, since it is a property of the permanent process underlying the observations rather than a feature of incidental sample fluctuations. It is the essential rather than the anecdote. On the other hand, from an engineering point of view, the parent process is an abstraction, and what really matters is the behavior of a specific variable over a specific domain. We will abandon here this quasi philosophical discussion to simply note that geostatistical practice reconciles these points of view. The sample variogams that we calculate from the data are in fact discrete approximations of regional variograms and thus inherit their physical significance. But the interpretation of these empirical variograms is performed in reference to properties of theoretical variograms which alone can provide a common background to analyze diverse particular situations.

2.2.5   Robust Variograms

As is shown in Section 2.9.1, the sample variogram can give a poor estimate of the regional variogram if the histogram has a long tail. This motivates the search for other variogram estimators.

Robustness and Resistance

Let us consider a set of data all equal to 0, except one that has a value of 1 (Figure 2.4a). The corresponding sample variogram is of the form

$$\hat{\gamma}(h) = \frac{1}{2} \frac{n_h}{N_h}$$

where N_h is the total number of pairs of points separated by a lag h and n_h is the number of pairs involving the value 1.

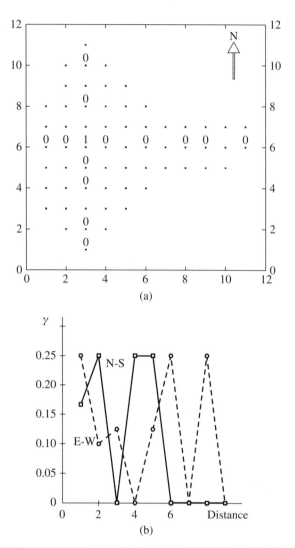

FIGURE 2.4 Synthetic example of an anomaly: (a) data; (b) variogram.

This is equal, up to the 1/2 factor, to the proportion of pairs of points that include the value 1. The variogram is therefore only representative of the geometric configuration of the sample points and of the position of the point valued 1, and it has a rather erratic look (Figure 2.4b). This may occur, in reality, in two types of situation:

1. When a data point is corrupted by a gross clerical error (relative to the dynamic range of the true data), such as an absent value that has been conventionally set to a large value (typically −9999).

2. When the data do include exceptional values, such as by the presence of a gold nugget in a sample while all the other samples are waste.

Even in the "ideal" case of a Gaussian RF, the variogram (2.6) defined as a sample mean of squared increments is not a stable estimator of the theoretical mean if we do not have a large number of increments. Indeed squared increments follow a chi-square distribution on one degree of freedom (a gamma distribution with shape parameter $\alpha = 1/2$), which has a fairly long tail. Hence the idea of seeking variogram estimators that are more stable than the classic estimator (2.6).

This type of problem is common in statistics and motivated the development of robust and resistant methods (Tukey, 1960; Huber, 1964, 1981; Mosteller and Tukey, 1977; Maronna et al., 2006). The concept of resistance appeared in the context of exploratory data analysis and does not involve distributional assumptions: A procedure is said to be *resistant* if its result is insensitive to changes, even large ones, in a few data (erroneous or anomalous data). For example, the sample median is resistant, whereas the sample mean is not.

Just like an optimal procedure, a *robust* procedure is defined in relation to a statistical model. The optimal procedure is the one that achieves maximum efficiency when the data conform to the model, but it can lead to erroneous results when it doesn't. A robust procedure is a prodedure that is slightly less efficient in the case where the data conform to the ideal model, but still gives good efficiency when the data depart slightly from this model, and does not give absurd results in case of larger deviations. A classic example is the contamination of a normally distributed batch of "good" observations by a small proportion of "bad" observations, also normally distributed, with the same mean but a much larger variance. The sample median is then a robust estimate of the theoretical mean whereas the sample mean is not. In practice, as pointed out by Huber (1981), the distinction between resistance and robustness is largely arbitrary, and from now on we will only refer to robustness.

Robust techniques are not a substitute for conscientiousness and will not permit a blind processing of any data set. It is always preferable to spot suspect data and deal with them appropriately. Tools as simple as posted maps, histograms, correlation plots, or variogram clouds are excellent for locating suspect data. Data proved to be grossly erroneous must then be corrected or eliminated. However, one must beware of automatically eliminating all extreme data: In many cases (geochemistry, pollution, bathymetry) it is precisely the anomalous data that are important, and if they are inconsistent with the model (stationary, Gaussian, etc.), then it is the model that must be reexamined and not the reality.

Robust Geostatistical Procedures

It is not possible here to give a complete account of all robust variogram proposals. We will only discuss the main principles and a few examples. But first we must point out an ambiguity in the expression "robust variogram" which can denote:

1. An alternative structure function to the variogram, also based on increments but lending itself to a more stable estimation.
2. An estimator of the ordinary variogram $\gamma(h)$, but more stable than the standard estimator $\hat{\gamma}(h)$ defined by (2.6).

Most robust variograms rest on one of, or variations of, the following principles:

- Replacing squared increment by lower-order powers, which are less dispersed and less sensitive to anomalous data.
- Replacing the average of the squared increments (for a given distance class) by their median, which is particularly insensitive to extreme values.
- Eliminating or clipping large increments.

From this point of view a typical robust variogram proposed by Dowd (1984) is the median of the magnitude of increments (for a given distance class). Variants proposed include the quantile variograms (Armstrong and Delfiner, 1980), the "Huberized" variogram (ibid.), and the variogram of order $1/2$ (Cressie and Hawkins, 1980). We can also consider the variogram of a transformed variable and the indicator variograms (see Section 2.5.3).

These variograms are better at revealing the possible structure of the phenomenon. But when we are dealing with a skewed distribution, such as that of squared increments, the sample mean and the sample median, say, do not estimate the same parameter. We nevertheless calculate a median variogram, or any other robust variogram, for its good properties, but the variogram thus obtained is no longer an estimate of the ordinary variogram $\gamma(h)$ required for kriging. Similarly the variogram of an indicator of the variable under study has usually a different behavior than the variogram of the variable itself. Two solutions can be considered:

1. To develop a robust geostatistics involving only the robust variogram: One then leaves the formalism of the L^2 norm (minimization of the quadratic error). This does not lead to tractable ways of constructing estimates and assessing their uncertainties, though solutions exist. For example, a large number of mathematical results are known in the scope of the L^1 norm (minimization of the absolute value of the error), in relation with methods of linear programming (Dantzig, 1963), and particularly in the context of an exponential distribution, which has a larger spread than a Gaussian distribution. But change-of-support models are no longer available, although necessary for many applications. Hawkins and Cressie (1984) and Dowd (1984) discuss attempts along these lines (cf. Section 3.7.3).

2. To find a means of passing from the robust variogram to the ordinary variogram $\gamma(h)$. This is possible provided that the distribution of the data pairs is specified. Most robust methods include this step under the assumption of a contaminated bi-Gaussian RF (one for which all bivariate distributions are Gaussian). These methods estimate the variogram of the noncontaminated variable rather than that of the whole (which in general should show an additional nugget term that needs to be evaluated separately, e.g., from the global variance of the data). Robustness against extreme data is achieved only within fairly restrictive hypotheses outside which bias problems persist.

Variogram of Order 1/2

Cressie and Hawkins (1980) have observed that if the RF $Z(x)$ is bi-Gaussian, the distribution of the increment of order 1/2,

$$\tfrac{1}{2}|Z(x+h) - Z(x)|^{1/2}$$

is close to a Gaussian distribution, and that approximately

$$\mathrm{E}\left[\tfrac{1}{2}|Z(x+h) - Z(x)|^{1/2}\right] = A\gamma(h)^{1/4}$$

with

$$A = \tfrac{1}{2}\Gamma\left(\tfrac{3}{4}\right)/\pi^{1/2} = 0.346$$

where Γ is the Euler gamma function. If G_h denotes the mean of the N_h terms $\tfrac{1}{2}|Z(x_\beta) - Z(x_\alpha)|^{1/2}$ such that $x_\beta - x_\alpha \approx h$, Cressie and Hawkins find that

$$\mathrm{E}\left[G_h^4\right] = \frac{1}{8}\left(0.457 + \frac{0.494}{N_h} + \frac{0.045}{N_h^2}\right)\gamma(h)$$

which allows the calculation of an unbiased estimator of $\gamma(h)$. This result is exact if the N_h terms are independent, which they are not. Cressie and Hawkins show that the interdependence among these terms seems to have a negligible impact for problems of practical interest.

   The application example presented by Genton (1998a) shows that this variogram estimator does not perform better than the classical estimator (2.6) in the Gaussian case and is not really more robust than the classical estimator in the case of the contaminated Gaussian model. These two variogram estimators belong to a common class of estimators of scale, the L^q M-estimators (Genton and Rousseeuw, 1995), which are not robust with respect to a contamination.

Median Variogram and Quantile Variograms

The classic variogram estimator (2.6) is

$$\hat{\gamma}(h) = \operatorname*{Mean}_{x_\beta - x_\alpha \approx h}\left\{\tfrac{1}{2}[z(x_\beta) - z(x_\alpha)]^2\right\}$$

where Mean represents the mean calculated from N_h pairs of points separated by the lag h. Since the median is more robust than the mean, it is worthwhile considering the *median variogram* taken, for each value of h, as the median of the sample halved squared increments. Armstrong and Delfiner (1980) define the *quantile variogram* $\hat{\gamma}_p(h)$ more generally as

$$\hat{\gamma}_p(h) = \underset{x_\beta - x_\alpha \approx h}{Q_p} \left\{ \tfrac{1}{2}[z(x_\beta) - z(x_\alpha)]^2 \right\}$$

where Q_p denotes the quantile associated with the proportion $p(0 < p < 1)$. Calculation of $\hat{\gamma}_p(h)$ simply requires us to sort for each value of h, the available halved squared increments by increasing values so as to construct the cumulative frequency curve.

Assuming that the bivariate distributions of the RF $Z(x)$ associated with the regionalized variable $z(x)$ are Gaussian, $Z(x + h) - Z(x)$ follows a Gaussian distribution with mean 0 and variance $2\gamma(h)$. Therefore $\tfrac{1}{2}[Z(x + h) - Z(x)]^2$ is distributed as $\gamma(h)\chi_1^2$, where χ_1^2 has a chi-square distribution on one degree of freedom:

$$\tfrac{1}{2}[Z(x + h) - Z(x)]^2 \overset{\mathcal{D}}{=} \gamma(h)\chi_1^2$$

where $\mathcal{D}$ represents equality in distribution. It follows that quantile variograms are all proportional to the variogram $\gamma(h)$, the proportionality factors being the quantiles of the distribution of χ_1^2; in particular, we have

$$Q_{0.25}(\chi_1^2) = 0.101, \qquad Q_{0.50}(\chi_1^2) = 0.455, \qquad Q_{0.75}(\chi_1^2) = 1.324$$

Similar properties are obtained with quantiles defined on halved absolute increments. Still assuming Gaussian bivariate distributions, this gives

$$\tfrac{1}{2}|Z(x + h) - Z(x)| \overset{\mathcal{D}}{=} \sqrt{\tfrac{1}{2}\gamma(h)}|U|$$

where U follows a standard normal distribution. The quantile variograms are thus proportional to $\sqrt{\tfrac{1}{2}\gamma(h)}$, the proportionality factors being this time the quantiles of the distribution of $|U|$. In particular,

$$Q_{0.25}(|U|) = 0.318, \qquad Q_{0.50}(|U|) = 0.674, \qquad Q_{0.75}(|U|) = 1.150$$

Improved variogram estimates may be obtained by linear combinations of quantile variograms.

Robust Scale Estimator

The sample variogram for a given h is the semi-variance of the increments $V_{\alpha\beta} = Z(x_\beta) - Z(x_\alpha)$ available for that distance. In the context of robust statistics, its estimation falls in the scope of scale estimation methods. The robust estimator of scale proposed by Rousseeuw and Croux (1993) leads to an estimator of $\gamma(h)$ based on the 25% quantile of the absolute value of differences between increments for that h value: Renumbering V_i, $i = 1, \ldots, N_h$, the increments $V_{\alpha\beta}$, the estimator is

$$\hat{\gamma}_{RC}(h) = 1.11 \underset{i<j}{Q_{0.25}} \left\{ |V_j - V_i| \right\}$$

Genton (1998a) shows, in the framework of the contaminated Gaussian model, that this estimator is much less sensitive to outliers than the classic variogram estimator (2.6), even if it remains influenced by these outliers. It usually clearly exhibits the structured part of the phenomenon but does not provide a fully unbiased estimate of the variogram of the noncontaminated variable.

2.2.6 Analysis of Heterogeneous Data

Whichever estimator is used, it must be calculated from a homogeneous population. Therefore it is important to treat cases separately:

1. Domains that exhibit different variabilities or are separated by discontinuities.
2. Data measured on different supports (which give neither the same histogram nor the same variogram).
3. Measurements of different qualities (corresponding to measurement errors with different magnitudes).
4. Pairs of points for which the distance is well known and those for which an uncertainty exists.

The variograms obtained in these different cases are of course different, but related, and must be analyzed in a consistent manner (see Section 2.4).

Spatial Declustering

Similarly, one must pay a particular attention to cases of preferential sampling because they are often in areas of high variability (e.g., one tends to oversample rich areas, which are often more variable). A variogram computed without care will be unusable. Two examples illustrate this point.

1. A deposit was investigated by E–W profiles. In the northern part the profiles were sampled on a 40-m grid, whereas in the southern part, which is narrower but richer, the grid was tightened to 20 m (Figure 2.5a). The E–W variograms calculated for each area have the same shape but differ in vertical scaling, which is indicative of a proportional effect (see the next section). If one had merely calculated a global E–W variogram of the deposit from all the profiles, the resultant curve would have been very poorly structured (Figure 2.5b).

2. A site contaminated by chemicals was investigated using a nearly regular sampling pattern. Additional samples were then taken in the anomalous areas, thus forming clusters of points (Figure 2.6a). If an ordinary variogram were calculated from all the data, these areas would be overrepresented, especially at short distances, and since they are areas of high variability, the variogram would be useless (Figure 2.6b).

When a clear-cut subdivision in homogeneous domains is not possible, as is the case in the second example, the data should be declustered. This consists in assigning a specific weight w_α to each datum z_α. Several procedures have been proposed for assigning these weights. The simplest ones are purely geometrical; for example, in 2D the weights can be chosen proportional to the area of influence of the data within the study area, limited to a maximum distance.

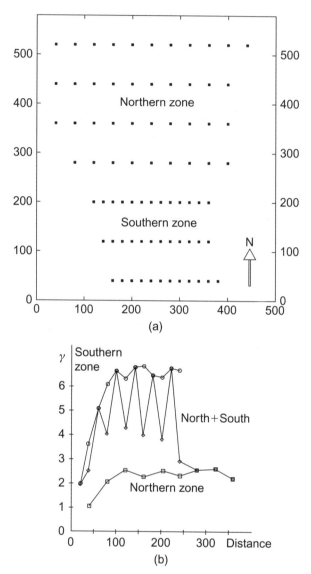

FIGURE 2.5 Effect of refining the sampling grid in the rich zone of a deposit (northern area): (a) data; (b) variograms of the northern and southern areas and variogram of the whole deposit calculated directly, all in the E–W direction.

More elaborate procedures are based on kriging. At our stage, where the "true" variogram is not known, these methods can be used heuristically with a standard variogram selected by the user. Declustering is also an issue when modeling the histogram for the nonlinear techniques presented in Chapter 6, and it will be discussed there in Section 6.2.

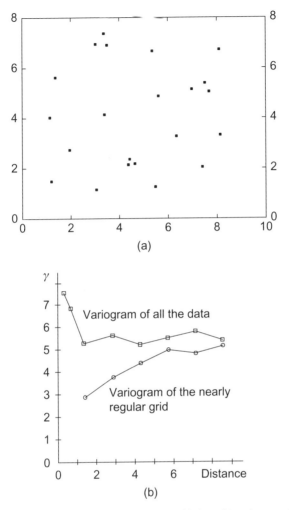

FIGURE 2.6   Effect of oversampling anomalous zones: (a) data; (b) variogram of the regular-grid data and variogram of all the data.

Once the weights are known, declustered statistics can be built. For example, formulas for the sample mean, variance, and the variogram become

$$\hat{z} = \frac{\sum\limits_{\alpha} w_{\alpha} z(x_a)}{\sum\limits_{\alpha} w_{\alpha}}, \qquad \hat{\sigma}^2 = \frac{\sum\limits_{\alpha} w_{\alpha}[z(x_{\alpha}) - \hat{z}]^2}{\sum\limits_{\alpha} w_{\alpha}}$$

$$\hat{\gamma}(h) = \frac{1}{2} \frac{\sum\limits_{x_{\beta} - x_{\alpha} \approx h} w_{\alpha} w_{\beta} \left[z(x_{\beta}) - z(x_{\alpha})\right]^2}{\sum\limits_{x_{\beta} - x_{\alpha} \approx h} w_{\alpha} w_{\beta}}$$

Using weights assigned to the data has the advantage of maintaining the consistency between the sample variogram and the sample variance. If we accept to separately evaluate the variance and the variogram, we can assign to each pair (α, β) a weight $w_{\alpha\beta}$ not necessarily of the form $w_{\alpha}w_{\beta}$. Presenting the problem in an estimation framework, Emery and Ortiz (2005b, 2007) obtain optimal weights for the pairs involved in the estimation of $\gamma(h)$ for a given h value by applying kriging to the random function $Q_h(x) = \frac{1}{2}[Z(x+h) - Z(x)]^2$. This approach requires the moments of order 4 of Z to be known and its use is limited to Gaussian random functions, an important special case however (see Section 2.9.1 for the expression of the variogram of Q_h as a function of the variogram of Z in that case). Variants of this approach are proposed by Reilly and Gelman (2007) and Emery (2007b).

Declustering, however, does not solve the problem for short distances since all pairs then come from oversampled areas.

Non-Euclidean Coordinate System

In certain cases the Euclidean distance is inappropriate and should be replaced, for example, by a "curvilinear" distance for measurements at sea along a very ragged coastline, or by geological distances "down dip," "along strike," and "across strike" for measurements in folded beds [e.g., see Figure 2.7, Dagbert et al. (1984), and Boisvert et al. (2009)]. However, it is not easy to develop a complete workflow consistent with such distances and practitioners first try to "flatten" the data. A nice situation is when the data can be replaced in a paleo-coordinate system before folding and/or faulting occurred. Mallet (2004, 2008) proposes a 3D geological modeling method (called "GeoChron") to put the data back in a geochronological coordinate system (u, v, t), where (u, v) are the geographical coordinates of the sediment particle when it was deposited and t is the geological time of deposition (up to an anamorphosis, due to the uncertainties in the sedimention and compaction rates required to reconstitute the geological time). The transformation of the present-day (x, y, z) coordinates to the (u, v, t) paleocoordinate system is called the *uvt* transform. The whole geostatistical study can be carried out in that (u, v, t) geochronological coordinate system, and the results transferred to the present (x, y, z) coordinate system at the end of the study (Figure 2.8). This assumes, of course, that the rock properties under study have not been affected by compaction and folding. Otherwise their transformation must be taken into account, for example in relation to the magnitude of the deformation (that deformation can be derived from the *uvt* transform).

2.2.7 Physical Interpretation of the Variogram

The graph of the sample variogram $\hat{\gamma}(h)$ against $|h|$, plotted for a given direction of h or for all directions taken together, generally shows the following behavior:

1. It starts at zero (for $h = 0$, $z(x+h) - z(x) = 0$).
2. It increases with $|h|$.
3. It continues to increase, or else stabilizes at a certain level.

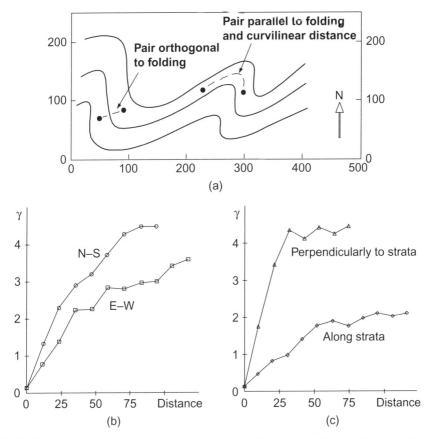

FIGURE 2.7 Calculation of a sample variogram in folded beds: (a) stratification pattern; (b) variogram calculated in the Euclidean system; (c) variogram calculated according to the stratification.

We now review its main features. In practice, the interpretation of a variogram must be made in relation to the contextual knowledge about the variable of interest, for example to geology [e.g., see Rendu (1984) and the enlightening examples presented by Rendu and Readdy (1980)].

Range and Sill

The rate of the variogram increase reflects the degree of dissimilarity of ever more distant samples. The variogram can increase indefinitely if the variability of the phenomenon has no limit at large distances [no "recall force"; see Figure 2.9 taken from Krige (1978), where the variogram is calculated from 3 to 1000 m]. If, conversely, the variogram stabilizes at a value, called the *sill*, it means that there is a distance beyond which $Z(x)$ and $Z(x+h)$ are uncorrelated. This distance is

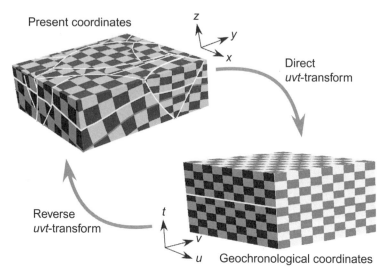

FIGURE 2.8   Graphical characterization of the direct and reverse *uvt*-transforms. [After Mallet (2008).]

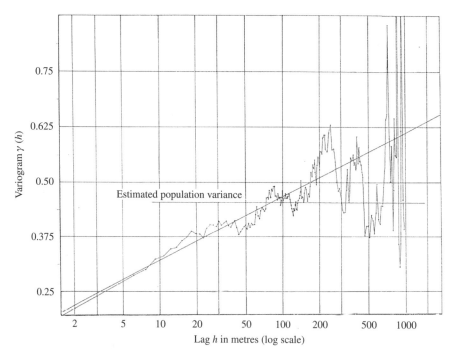

FIGURE 2.9   Variogram with no sill: gold accumulation at the President Steyn mine. Notice the logarithmic scale of the distance axis. [From Krige (1978).]

called the *range* (see Figure 2.10). It gives a precise meaning to the conventional notion of *area of influence* of a sample.

The variogram can reveal *nested structures*—that is, hierarchical structures, each characterized by its own range. Serra (1968), in his study of the Lorraine iron basin, exhibited up to seven nested structures ranging from the petrographic scale, due to oolites, up to the megastructure scale with a range of 10–20 km, passing by decimeter- to meter-size concretions and 100-m-size lenses. Figure 2.11 shows a simpler example, from Goovaerts et al. (1993), concerning environmental data (springwater solute contents): the variogram of

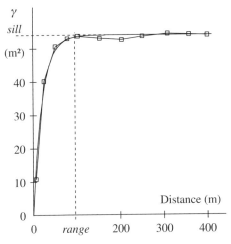

FIGURE 2.10   Range and sill on the variogram of the thickness of nickel-bearing garnieritic ore.

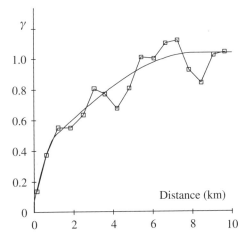

FIGURE 2.11   Nested structures on a variogram of environmental data (alkalinity of springwater, Dyle River Basin, Belgium). [From Goovaerts et al. (1993), © American Geophysical Union.]

alkalinity which is displayed here reveals two scales of variation: (1) a short-range component (1 km) corresponding to local sources of contaminants due to human activities and (2) a long-range component (9 km) interpreted as regional changes in the geologic characteristics of the aquifer. These interpretations make it possible to estimate each component through a factorial kriging analysis (see Section 5.6.7).

Behavior Near the Origin and Nugget Effect

Having just examined the variogram behavior at large distances, it is equally interesting to examine its behavior near the origin because it reflects the continuity and the spatial regularity of the regionalized variable. Figure 2.12 shows four typical behaviors:

a. *A parabolic behavior.* This characterizes a highly regular regionalized variable that is usually differentiable at least piecewise.

b. *A linear behavior.* The regionalized variable is continuous, at least piece-wise, but being no longer differentiable, it is less regular than in the previous case.

c. *A discontinuity at 0—nugget effect.* $\hat{\gamma}(h)$ does not seem to tend to zero when $h \to 0$. This means that the regionalized variable is generally not continuous and is thus very irregular. The origin of this denomination is as follows: In gold deposits, gold commonly occurs as nuggets of pure metal that are much smaller than the size of a sample. This results in strong grade variability in the samples, even when physically very close and therefore in a discontinuity of the variogram at the origin. By extension, the term "nugget effect" (in the wide sense) is applied to all discontinuities at the origin, even if their cause is different. In general, the nugget effect is due to:

- a microstructure or "geological noise," namely a component of the phenomenon with a range shorter than the sampling support (true nugget effect);
- a structure with a range shorter than the smallest interpoint distance;
- measurement or positioning errors.

The various sources of nugget effect are modeled in Section 2.4. In the absence of close sampling points, it is impossible to tell from the variogram itself which cause is applicable, especially since they can be mixed. Knowledge about the physics of the problem can help in discriminating the various causes.

d. *A flat curve—pure nugget effect or white noise.* There is no correlation between the two points, however close they may be. This is the extreme case of total absence of structure.

From a theoretical point of view, one must qualify the preceding conclusion, because absence of correlation does not necessarily imply independence and absence of structure. For example, if X and Y are two independent random variables valued -1 or $+1$ with equal probability, $X - Y$ and $X + Y$ have the

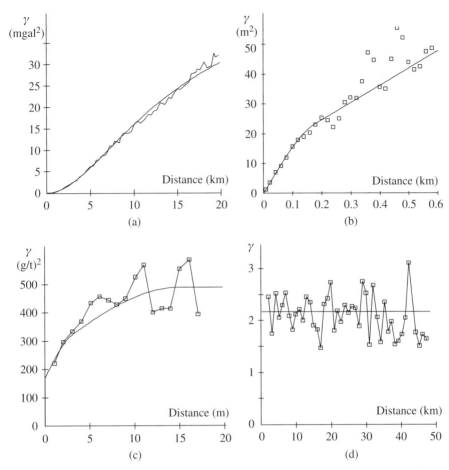

FIGURE 2.12 Examples of variogram behaviors near the origin: (a) parabolic (gravity); (b) linear (depth of a geological formation, Paris Basin); (c) with nugget effect (gold grade, Salsigne); (d) apparent pure nugget effect (logarithm of permeability, Paris Basin). [From Chilès and Guillen (1984) (a); B. Bourgine, personal communication (b); Chilès and Liao (1993), with kind permission from Kluwer Academic Publishers (c); J. C. Martin, personal communication (d).]

same distribution and can take the values -2, 0, $+2$. They are uncorrelated but not independent because the possible pairs $(X - Y, X + Y)$ necessarily include a zero value and a value equal to ±2. Figure 2.13 displays simulations with a flat variogram obtained by different methods (Matheron, personal communication). One is pure noise, but the other four exhibit patterns. The patchwork image, for example, is obtained by simulating 0–1 noise separately on two orthogonal discrete axes i and j, thus giving two one-dimensional simulations $Y_1(i)$ and $Y_2(j)$, and forming $Y(i, j) \equiv Y_1(i) + Y_2(j)$ mod 2. The patterning reflects the lack of independence of the $Y(i, j)$ (N^2 values are constructed from $2N$ values). These are, however, curiosities never found in the usual applications of geostatistics.

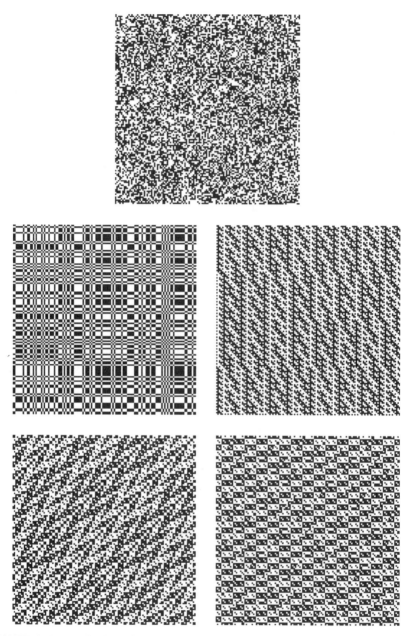

FIGURE 2.13 Realizations of binary random functions with a purely flat variogram. The one shown in the upper image is random noise, but the others exhibit patterning (grid of 128 × 128 pixels).

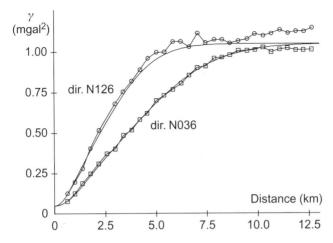

FIGURE 2.14 Geometric anisotropy (gravity residual). [From Simard (1980).]

Anisotropy

When the variogram does not vary with direction, it is said to be isotropic. It is then a function of the modulus of the vector h, namely of the distance between the points. In the opposite case, the variogram is anisotropic. Two typical cases of anisotropy are the following:

1. The sill is constant, but the range varies with the direction: The variogram of a lenticular formation can show a larger range in the direction of lens elongation than in the other directions. Figure 2.14 shows an example obtained with gravity data (Simard, 1980).
2. The variogram displays a lower sill in a specific direction. Figure 2.15 shows a 2D example where this is due to a fault: The variability is stronger in the direction normal to the fault (Champigny and Armstrong, 1989). In 3D the vertical generally plays a particular role: Variations are greater between strata than within a single stratum.

Hole Effect

The hole effect is characterized by the presence of one or more bumps on the variogram which correspond to an equivalent number of holes (negative values) on the covariance. It reflects a tendency for high values to be systematically surrounded by low values, and vice versa. One must be cautious, however, not to interpret mere fluctuations of the variogram as a hole effect. In general it is advisable to consider the presence of a hole effect only if there is a reasonable physical explanation. Figure 2.16 shows, as an example, variograms of the iron, CaO, and SiO_2 grades obtained by Serra (1967, pp. 85–91) in his study of the Lorraine oolitic iron ore; the data were reconstructed from 170

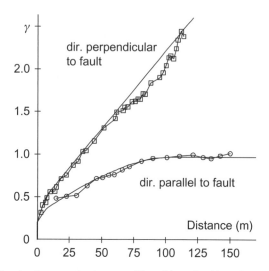

FIGURE 2.15 Zonal anisotropy (variogram of logarithm of gold grade calculated parallel and perpendicular to a fault). [From Champigny and Armstrong (1989), with kind permission from Kluwer Academic Publishers.]

cubic samples, with 4-cm sides, collected at 5-cm intervals in vertical channels. The vertical variograms show a very clear hole effect over a distance of 10 cm: It would seem that the migration of carbonates in this moderately reduced ore gave rise to calcite nodules about 10 cm thick (and 50 cm long). The hole effect can also arise from competition between plants, although, as noted by Matérn (1960, p. 62), the effect is commonly masked by strong correlations among soil properties or by a sampling area that is too large compared to the distances over which the competition is effective.

Periodicities

A particular case of the hole effect warrants a separate treatment: It is when the variogram shows a periodic, or at least a pseudoperiodic, behavior. Here again one must be sure that this behavior is really significant, especially when studying spatial variables. Time observations are frequently periodic due to the influence of the basic daily and yearly cycles on natural phenomena and human activities. Such clear periodicities scarcely exist in space, except in cases where time is involved indirectly. This is the case for some sedimentary rocks: Sedimentation is influenced by the climate, itself influenced by the distance of the earth to the sun, which displays Milankovitch cycles of about 100,000, 40,000, and 20,000 years. If the sedimentation and compaction rates are approximately constant, these cycles in the geological time are transferred to the sediments in periodicities of rock properties along the vertical. Figure 2.17 shows an example of that, where two periodicities can be observed, with a ratio of 5:1. They are

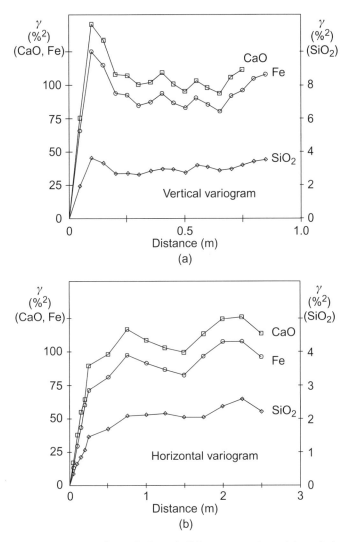

FIGURE 2.16 Variograms of Fe, CaO, and SiO_2 concentrations: (a) vertical variograms; (b) horizontal variograms. The vertical variograms show a hole effect resulting from a migration phenomenon. [From Serra (1967).]

due to the 100,000- and 20,000-year Milankovitch cycles. This example is part of a study where periodicities were used to estimate the duration of a geological series and identify possible hiatuses (Lefranc et al., 2008). Another example, cited by Matérn (1960), is the effect of waves on a beach. Note also that problems that are, by nature, clearly oscillatory seem to be better approached by Fourier methods (in the frequency domain), provided that the data are sampled on a regular grid.

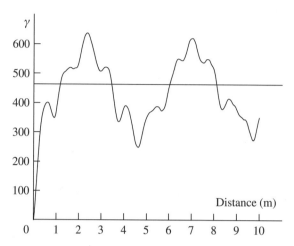

FIGURE 2.17 Pseudo-periodicity of the variogram of a geophysical log along a vertical borehole in argillites. Two periods can be observed. [From Lefranc et al. (2008).]

Proportional Effect

Variability is sometimes higher in areas with high average values than in areas with low average values. This may corrupt the sample variogram, which may reflect more the variations in the average value of the data used for each lag

$$\hat{m}(h) = \frac{1}{2N_h} \sum_{x_\beta - x_\alpha \approx h} \left[z(x_\beta) + z(x_\alpha) \right]$$

than the spatial variability itself [e.g., see the ecological data studied by Rossi et al. (1992)]. Then one should calculate local variograms. If all computed variograms look the same and differ only by a multiplicative factor, this is a case of proportional effect; the most common situation is when the sill is proportional to the square of the local mean (the local sample variance is then also proportional to the square of the local mean). This occurs when the regionalized variable has a lognormal histogram.

Presence of a Drift

Theoretically, the variogram at large distances should increase more slowly than a parabola (see Section 2.3.1). In practice, however, one can see sample variograms that increase as fast as $|h|^2$, if not faster. The sample variogram may also exhibit very strong and very complex variations with direction. In both cases this indicates the presence of a space-varying mean, or "drift," representing a trend in the data; for example, the depth of the seafloor increases with distance from the coastline. A variogram of the raw data is then of little use: What is needed is the underlying variogram, that of the phenomenon without its drift. Models other than IRFs (universal kriging, IRF–k) and other structural tools (variogram of residuals, generalized variogram) have to be used (see Sections 2.7 and 4.7).

2.3 MATHEMATICAL PROPERTIES OF THE VARIOGRAM

To proceed further in the applications, it is necessary to know the variogram $\gamma(h)$ for any value of h. We cannot use the regional variogram, and even less the sample variogram, instead of the theoretical variogram. Indeed we will show that $-\gamma(h)$ must be a conditionally positive definite function and that the regional variogram as well as the sample variogram do not possess this property in general, even in the case of regularly spaced data. Therefore, to pass from the sample variogram to the theoretical variogram, we turn to theoretical models that are known to be valid variogram functions. We will see in Section 2.6 how to fit a model to a sample variogram. But let us first review the main mathematical properties of variograms and characterize the class of functions that can represent variograms. Additional results can be found in Yaglom (1987).

2.3.1 Continuity and Differentiability

The degree of regularity of an SRF or IRF is directly related to the behavior of its variogram at the origin.

Continuity

The continuity most suited to second-order models is continuity in the mean square (m.s.): By definition, an RF $Z(\cdot)$ is m.s. continuous at point x if

$$\lim_{h \to 0} E[Z(x + h) - Z(x)]^2 = 0$$

It follows immediately that an IRF is m.s. continuous everywhere if and only if its variogram is continuous at $h = 0$. And it can easily be shown that if the variogram is continuous at 0, then it is continuous at all $h \in \mathbb{R}^n$.

On the other hand, if the variogram has a discontinuity at the origin, it may also have others elsewhere (Gneiting et al., 2001).[2] In practical applications, the possible discontinuity of the variogram is restricted to the origin: the nugget effect. Therefore, in this book we only consider that situation. In such a case, the IRF can be decomposed into the sum of two uncorrelated terms: (1) a random noise corresponding to the nugget effect and (2) a m.s. continuous IRF

---

[2] As an example, consider an SRF $Z(x)$ on $\mathbb{R}^n$ whose covariance is continuous except at the origin. The covariance of the SRF $Z_1(x) = Z(x) + Z(x + a)$, where a is a given vector of $\mathbb{R}^n$, is discontinuous at $h = 0$, a, and $-a$, and so is its variogram. Note that if $n > 1$, the covariance of Z_1 is not isotropic. According to a conjecture of Schoenberg (1938a), proved by Crum (1956), an isotropic covariance in $\mathbb{R}^n$, $n > 1$, cannot have a discontinuity except at the origin. The same is true for an isotropic variogram (Gneiting et al., 2001).

corresponding to the continuous variogram. In this section we focus on continuous variograms.

Since m.s. convergence does not imply almost-sure convergence, the m.s. continuity of an RF does not imply continuity of its realizations. These realizations can, for example, show discontinuities such as faults, and they can be continuous only in the compartments delimited by these faults. We will see numerous examples in the construction of simulations (e.g., random tessellations). For separable Gaussian IRFs, however, m.s. continuity implies continuity of almost every realization.[3] For more information on this subject, see, for example, Yaglom (1987, Vol. I, p. 65, and Vol. II, Chapter 1, Note 12), Sobczyk (1991, Section 15), and Adler and Taylor (2007, Chapter 1).

Differentiability

The notion of differentiability associated with second-order models is mean square differentiability. Let us first consider the case of an IRF defined on the line. By definition, the random variable $Z'(x)$ is the m.s. derivative of the random process $Z(\cdot)$ at point x if the finite difference $[Z(x+h) - Z(x)]/h$ converges to $Z'(x)$ in the m.s. sense when $h \to 0$:

$$\lim_{h \to 0} \mathrm{E}\left[\frac{Z(x+h) - Z(x)}{h} - Z'(x)\right]^2 = 0$$

It is shown in the theory of random processes that $Z'(x)$ exists if and only if the covariance of $[Z(x+h) - Z(x)]/h$ and $[Z(x+h') - Z(x)]/h'$ has a limit when h and h' tend to zero independently.[4]

An immediate consequence is that an IRF $Z(\cdot)$ in $\mathbb{R}$ is m.s. differentiable everywhere if and only if $\gamma(h)$ has a second derivative at 0. That condition concerns γ considered as a (symmetric) function of $h \in \mathbb{R}$ and not only of $h \in \mathbb{R}^+$. It implies a parabolic behavior at the origin.

Furthermore, $Z(\cdot)$ having stationary increments, its derivative $Z'(\cdot)$ is a stationary random process and has for covariance the second derivative of the variogram of $Z(\cdot)$:

$$\mathrm{Cov}(Z'(x), Z'(x+h)) = \gamma''(h)$$

---

[3] This is true at least for any separable Gaussian IRF whose variogram satisfies $\gamma(h) \leq c\,|h|^\delta$ when $|h| \leq r_0$ for some strictly positive constants c, δ, r_0, which is the case for all continuous variograms of practical use. In the general case, the following theorem due to Kolmogorov applies: If the separable random function Z is such that $\mathrm{E}|Z(x+h) - Z(x)|^\alpha \leq c\,|h|^{1+\beta}$ for small h for some strictly positive constants α, β, c, then almost every realization of Z is continuous. In all cases the assumption of separability is essential—and fully justified in practical applications, since a nonseparable version of a random function is really a curiosity. As an example, consider an Gaussian SRF $Z(x)$ with continuous realizations. Now select a random location X from any continuous distribution. The RF $Z_1(x)$ defined as equal to $Z(x)$ if $x \neq X$ and $Z(x) + 1$ if $x = X$ has the same finite-dimensional distributions as Z, but it is not separable and none of its realizations is continuous.

[4] A more rigorous presentation can be found, for example, in Yaglom (1987, Vol. I, pp. 66–67).

Thus, if $\gamma(h)$ has a second derivative at 0, it has a second derivative at all $h \in \mathbb{R}$.

These results are easily extended to the partial m.s. derivatives of an IRF $Z(\cdot)$ in $\mathbb{R}^n$. If all the partial derivatives exist, $Z(\cdot)$ is said to be m.s. differentiable. As in the continuity case, the m.s. differentiability of an RF does not imply differentiability of its realizations. For separable Gaussian IRFs, however, m.s. differentiability implies differentiability of almost every realization.[5]

Principal Irregular Term

The behavior of the variogram near the origin can be fully described by its series expansion about zero (Matheron, 1965). In the isotropic case, this expansion admits two types of terms: (1) even powers of $|h|$, called regular terms because they are infinitely differentiable, and (2) irregular terms of the form $|h|^\nu$, $\nu > 0$ and not an even integer, or in $|h|^\nu \log |h|$, ν an even integer. Approximation formulas for global estimation variances of a domain V from regularly sampled data only include the irregular terms (Matheron, 1970). When the data spacing is very small the irregular term with the lowest degree in $|h|$ plays the dominant role. This term is therefore called the principal irregular term (Stein, 1999).

Behavior of $\gamma(h)/|h|^2$

The variogram of an SRF is finite. That of an IRF that is not an SRF can increase to infinity, but not in an uncontrolled fashion, as is expressed by the following two properties (e.g., Yaglom, 1987, Vol. I, pp. 397–400):

1. If $\gamma(h)$ is the variogram of a m.s. continuous IRF, $\gamma(h)/|h|^2 \to 0$ as $|h| \to \infty$.
2. If $\gamma(h)$ is the variogram of a m.s. differentiable IRF, $\gamma(h)$ has a majorization of the form

$$\gamma(h) \le A |h|^2 \qquad \forall h \in \mathbb{R}^n$$

2.3.2 Conditional Positive Definiteness

The properties described above are not sufficient to characterize covariance or variogram functions. The only truly necessary condition for a function to be a covariance or a variogram is that all variance calculations lead to a nonnegative result.

---

[5] Like for continuity (see footnote 3), there are some conditions on the behavior of the variogram at the origin, which are met by usual m.s. differentiable separable Gaussian IRFs; see Adler and Taylor (2007, Chapter 1).

Covariance and Positive Definiteness

In the stationary case, let us consider any linear combination $\sum_{i=1}^{N} \lambda_i Z(x_i)$ of N terms. Its variance, necessarily positive or zero, can be expressed with the covariance $C(h)$ as

$$\text{Var}\left[\sum_{i=1}^{N} \lambda_i Z(x_i)\right] = \sum_{i=1}^{N}\sum_{j=1}^{N} \lambda_i \lambda_j C(x_j - x_i) \qquad (2.8)$$

By definition, a function $C(h)$ in $\mathbb{R}^n$ for which the right-hand side of relation (2.8) is always positive (or zero), for any choice of N, x_i, and λ_i, is a *positive definite*[6] function in $\mathbb{R}^n$. A covariance is thus necessarily a positive definite function. Conversely, if $C(h)$ is a positive definite function, one can construct a Gaussian SRF with $C(h)$ as its covariance (the knowledge of the covariance function is sufficient to define a consistent set of Gaussian finite-dimensional distributions; see, for example, Doob (1953, Sections II.3 and XI.3); see also the spectral representation of an SRF in Section 2.3.3). Thus the covariance functions in $\mathbb{R}^n$ and the positive definite functions in $\mathbb{R}^n$ are an identical class.

The family of covariance functions in $\mathbb{R}^n$ satisfies the following *stability properties*:

1. If $C_k(h)$, $k \in \mathbb{N}$, are covariances in $\mathbb{R}^n$, then $C(h) = \lim_{k \to \infty} C_k(h)$ is a covariance in $\mathbb{R}^n$, provided that this limit exists for all h.
2. If $C(h; t)$ is a covariance in $\mathbb{R}^n$ for all values $t \in A \subset \mathbb{R}$ of the parameter t, and if $\mu(dt)$ is a positive measure on A, then $\int C(h; t) \, \mu(dt)$ is a covariance in $\mathbb{R}^n$, provided that the integral exists for all h.
3. If $C_1(h)$ and $C_2(h)$ are two covariances in $\mathbb{R}^n$, then $C_1(h) C_2(h)$ is a covariance in $\mathbb{R}^n$.

Properties 1 and 2 result from the definition of positive definiteness. Property 3 is obtained by establishing the covariance of the product of two independent zero-mean SRFs Z_1 and Z_2 with covariances C_1 and C_2, respectively.

It is useful to know if the covariance $C(h)$ ensures that the variance (2.8) is always strictly positive (except of course when all the λ_i are zero), since the simple kriging system (3.2) then always has a unique solution. Such a covariance function will be said *strictly positive definite*. An example of a covariance that is not strictly positive definite is the cosine model $C(h) = \cos(\langle \omega, h \rangle)$, where ω is a given vector of $\mathbb{R}^n$.

Allowable Linear Combinations

In the stationary case any finite linear combination has a finite variance, given by (2.8). This is no longer true for random functions that have a variogram but no covariance. In the intrinsic case the only linear combinations for which one

---

[6] Positive definite is taken synonymously to *nonnegative definite*. To exclude the value zero, we refer to *strict* positive definiteness.

can calculate the variance are linear combinations of increments. These are called allowable linear combinations and are characterized by the condition

$$\sum_{i=1}^{N} \lambda_i = 0 \tag{2.9}$$

It is clear that all linear combinations of increments satisfy equation (2.9), since each increment satisfies it. And conversely, any linear combination $\sum_{i=1}^{N} \lambda_i Z(x_i)$ satisfying (2.9) is equal to $\sum_{i=1}^{N} \lambda_i [Z(x_i) - Z(x_0)]$, for any choice of the origin x_0, and is a linear combination of increments.

Conditional Positive Definiteness

The variance of the allowable linear combination $\sum_{i=1}^{N} \lambda_i Z(x_i)$ can be expressed in terms of the variogram by

$$\mathrm{Var}\left[\sum_{i=1}^{N} \lambda_i Z(x_i)\right] = -\sum_{i=1}^{N} \sum_{j=1}^{N} \lambda_i \lambda_j \gamma(x_j - x_i) \tag{2.10}$$

This result is a particular case of equation (2.11) which is proved below. A function $G(h)$ in $\mathbb{R}^n$, for which an expression of the form

$$\sum_{i=1}^{N} \sum_{j=1}^{N} \lambda_i \lambda_j G(x_j - x_i)$$

is always positive or zero, provided that $\sum_{i=1}^{N} \lambda_i = 0$, is said to be a *conditionally positive definite* function in $\mathbb{R}^n$. Thus, if $\gamma(h)$ is a variogram, then $-\gamma(h)$ is a conditionally positive definite function. This necessary condition is also sufficient, provided that $\gamma(0) = 0$ (see Section 2.3.3). Moreover, if $-\gamma(h)$ is a conditionally positive definite function, one can construct a Gaussian IRF with $\gamma(h)$ as its variogram (the reason is similar to that establishing the existence of Gaussian SRFs with a given covariance function).

Stability properties 1 and 2 of covariances carry over to variograms.

Positive Definiteness of Isotropic Functions

An isotropic covariance or variogram is expressed as a function of $r = |h|$. It is important, however, to keep in mind the dimension n of the space, since a positive definite isotropic function (conditionally or not) in $\mathbb{R}^n$ of course satisfies the same property in $\mathbb{R}^m$ for $m < n$ but not necessarily for $m > n$. Golubov (1981) proves that the truncated power function

$$C(r) = \begin{cases} \left(1 - \dfrac{r}{a}\right)^{\nu} & \text{if } r \le a \\ 0 & \text{if } r \ge a \end{cases} \qquad (a > 0)$$

is positive definite in $\mathbb{R}^n$ if and only if $\nu \geq (n+1)/2$. So the triangle function, which corresponds to $n = 1$, is a covariance in $\mathbb{R}$ but not in $\mathbb{R}^n$, $n \geq 2$. Armstrong and Jabin (1981) exhibit a linear combination in $\mathbb{R}^2$ for which the application of (2.8) leads to a negative variance.[7]

Covariance of Two Allowable Linear Combinations

The variogram also allows the calculation of the covariance of two allowable linear combinations. In the case of an SRF, the covariance of any two linear combinations $\sum_{i=1}^{N} \lambda_i Z(x_i)$ and $\sum_{j=1}^{N'} \mu_j Z(x_j)$, where the x_i do not necessarily represent the same points as the x_j, is expressed in terms of the covariance function by

$$\mathrm{Cov}\left(\sum_{i=1}^{N} \lambda_i \, Z(x_i), \sum_{j=1}^{N'} \mu_j \, Z(x_j)\right) = \sum_{i=1}^{N} \sum_{j=1}^{N'} \lambda_i \, \mu_j \, C(x_j - x_i)$$

Similarly in the case of an IRF, but considering this time only allowable linear combinations, this covariance is written in terms of the variogram as

$$\mathrm{Cov}\left(\sum_{i=1}^{N} \lambda_i \, Z(x_i), \sum_{j=1}^{N'} \mu_j \, Z(x_j)\right) = -\sum_{i=1}^{N} \sum_{j=1}^{N'} \lambda_i \, \mu_j \, \gamma(x_j - x_i) \qquad (2.11)$$

To see this, it suffices to introduce an arbitrary origin x_0. Since $\sum_{i=1}^{N} \lambda_i$ and $\sum_{j=1}^{N'} \mu_j$ are zero, the quantity sought is equal to

$$\sum_{i=1}^{N} \sum_{j=1}^{N'} \lambda_i \, \mu_j \, \mathrm{Cov}\left(Z(x_i) - Z(x_0), Z(x_j) - Z(x_0)\right)$$

The identity

$$(Z_j - Z_i)^2 = [(Z_j - Z_0) - (Z_i - Z_0)]^2$$
$$= (Z_i - Z_0)^2 - 2(Z_i - Z_0)(Z_j - Z_0) + (Z_j - Z_0)^2$$

---

[7] The restriction to $|h| \leq a$ of the function $C(h) = 1 - |h|/a$ is a permissible covariance model in the 2D or 3D ball of radius a, as a locally stationary covariance associated with a linear variogram (see Example 2 in Section 4.6.2). Gneiting (1999a) showed that this local covariance can be extended to a covariance in the whole 2D or 3D space, which is quite intriguing since the triangular covariance is not valid in these spaces. It works because the extended covariance is not equal to 0 when $|h| > a$ but oscillates around that value. This result derives from an extension theorem due to Rudin (1970) and from the turning bands operator: Any isotropic function which is positive definite in a ball of $\mathbb{R}^n$ can be extended to an isotropic positive-definite function in the whole space $\mathbb{R}^n$.

where Z_i, Z_j, and Z_0 stand for $Z(x_i)$, $Z(x_j)$, and $Z(x_0)$, makes it possible to calculate the product $(Z_i - Z_0)(Z_j - Z_0)$ from squared increments and, thus, to calculate the covariance from the variogram:

$$\text{Cov}\left(Z(x_i) - Z(x_0), Z(x_j) - Z(x_0)\right) = \gamma(x_i - x_0) + \gamma(x_j - x_0) - \gamma(x_j - x_i)$$

Inserting this into the double sum yields (2.11).

The restriction to allowable linear combinations is not a problem from the point of view of usual geostatistical applications. For example, if the value $Z(x_0)$ at an unmeasured point x_0 is estimated by the mean $\overline{Z}$ of the N sample points, $\overline{Z}$ is not an allowable linear combination but the estimation error $\overline{Z} - Z(x_0)$ is, and thus its variance can be calculated.

The computational gymnastics to express the variance of a linear combination or the covariance of two linear combinations is generally easier with the covariance function $C(h)$. If this does not exist, and if one is dealing with an IRF, it suffices to replace $C(h)$ by $-\gamma(h)$ in the result, provided, of course, that only allowable linear combinations are considered.

Stochastic Integrals

The preceding results can be extended to stochastic integrals. The stochastic integral of the SRF $Z(x)$ associated with a numerical weighting function $w(x)$ in $\mathbb{R}^n$ is defined by

$$Z_w = \int w(x)\, Z(x)\, dx$$

The conditions of its existence are given by the following theorem [e.g., Yaglom (1987, Vol. I, pp. 67–69)]: The stochastic integral Z_w exists, in the mean square sense, if and only if the integral $\iint w(x)C(x' - x)w(x')\, dxdx'$ is finite, and this integral is then the (finite) variance[8] of Z_w

$$\text{Var}(Z_w) = \iint w(x)C(x' - x)w(x')\, dx\, dx' \tag{2.12}$$

If the functions $w_1(x)$ and $w_2(x)$ define two stochastic integrals Z_{w_1} and Z_{w_2}, they have the covariance

$$\text{Cov}(Z_{w_1}, Z_{w_2}) = \iint w_1(x)C(x' - x)w_2(x')\, dx\, dx' \tag{2.13}$$

The definition of stochastic integrals can be extended to the case where $Z(x)$ is an IRF, provided that we only consider allowable weighting functions, namely

---

[8] Positive definiteness of the covariance ensures that this integral is positive or zero.

satisfying $\int w(x)\,dx = 0$. The stochastic integral Z_w then exists if and only if the integral $-\iint w(x)\gamma(x' - x)w(x')\,dxdx'$ is finite, and this integral is then the variance of Z_w. When the weighting functions are allowable and the corresponding stochastic integrals exist, it suffices to replace C by $-\gamma$ in (2.12) and (2.13).

From a practical point of view, the double integral (2.12) can be expressed as a simple integral using the change of variable $h = x' - x$ (*Cauchy algorithm*):

$$\iint w(x)C(x' - x)w(x')\,dx\,dx' = \int g(h)C(h)\,dh$$

where $g(h) = \int w(x)w(x + h)\,dx$ is the covariogram of w, whose properties are examined in Section 2.3.4.

2.3.3   Spectral Representation

Spectral or harmonic analysis is concerned with the decomposition of functions into Fourier series or integrals. It is widely used in the physical sciences. Fourier series apply to periodic functions, whereas Fourier integrals apply to functions that decay to zero rapidly enough at infinity. Many functions, however, belong to neither of these two categories. From this point of view, SRFs and IRFs have the advantage over ordinary functions that they always have a spectral representation, which in addition has a clear physical significance. The main results have been established by Kolmogorov (1940a) and Cramér (1942) in $\mathbb{R}$ and by Yaglom (1957) in $\mathbb{R}^n$. Here we will just state the results on the representation of continuous covariances and variograms, as well as that of the corresponding random functions, and draw a few conclusions. We refer the interested reader to the standard references in this field, notably Doob (1953) in $\mathbb{R}$, Monin and Yaglom (1965, Chapter 6) in $\mathbb{R}^3$, and the very complete book of Yaglom (1987).

Spectral Representation of a Covariance and an SRF

We have seen that there is an identity between the class of continuous covariance functions in $\mathbb{R}^n$ and the class of positive definite functions in $\mathbb{R}^n$. Bochner's theorem [e.g., Feller (1971, Chapter XIX)] identifies the characteristic functions $\Phi(u)$ of probability distributions in $\mathbb{R}^n$ with the positive definite continuous functions satisfying $\Phi(0) = 1$. Hence we have the following corollary, which has also been proved directly and independently by Khinchin (1934):

A continuous real function $C(h)$ defined in $\mathbb{R}^n$ is a covariance if and only if it is the (inverse) Fourier transform of a positive bounded symmetric measure $F(du)$:

$$C(h) = \int e^{2\pi i \langle u, h \rangle} F(du) = \int \cos(2\pi \langle u, h \rangle)\, F(du) \tag{2.14}$$

with

$$\int F(du) < \infty \tag{2.15}$$

where u represents a frequency[9] (i is here the unit pure imaginary number). The integral $\int F(du)$ of the spectral measure is equal to the total power $C(0)$. Note that the covariance is not necessarily a *strictly* positive definite function. A sufficient condition for this is that the support of the spectral measure F includes an open set of $\mathbb{R}^n$ (Dolloff et al., 2006). In the one-dimensional case, this implies that a valid covariance function is strictly positive definite if its spectral measure is not limited to a set of discrete points on the frequency axis.[10]

As we saw in Section 1.1.5, any SRF is also characterized by its spectral representation: A m.s. continuous RF is an SRF (of order 2) if and only if it is of the form

$$Z(x) = \int e^{2\pi i \langle u, x \rangle} Y(du) \tag{2.16}$$

where Y is an orthogonal complex random measure ($Y(du)$ and $\overline{Y(dv)}$ have zero correlation when du and dv are nonoverlapping) such that

$$E|Y(du)|^2 = F(du) \tag{2.17}$$

The above results are valid for complex random functions. In the case where $Z(x)$ is real-valued, the complex random measure Y satisfies $Y(-du) = \overline{Y(du)}$.

Example 1. If

$$Z(x) = \sum_{j=1}^{k} [a_j \cos(2\pi \langle u_j, x \rangle) + b_j \sin(2\pi \langle u_j, x \rangle)]$$

where $a_1, \ldots, a_k, b_1, \ldots, b_k$ are zero-mean uncorrelated random variables such that $E(a_j^2) = E(b_j^2) = \sigma_j^2 > 0$ and $u_1, \ldots, u_k$ are distinct vectors, $Z(x)$ is an SRF with covariance $C(h) = \sum_{j=1}^{k} \sigma_j^2 \cos(2\pi \langle u_j, h \rangle)$. $\square$

---

[9] Fourier expansions into harmonics $\exp(i \langle \omega, x \rangle)$ instead of $\exp(2\pi i \langle u, x \rangle)$ are also used, where ω is the angular frequency.

[10] If the support of the spectral measure is the whole space $\mathbb{R}^n$, any stochastic integral, and not only any finite linear combination, has a strictly positive variance. A weaker and sufficient criterion for this is the summability of $|C(h)|$ over $\mathbb{R}^n$, since then the spectral density is absolutely continuous with respect to the Lebesgue measure. We conjecture that a necessary and sufficient condition for stochastic integrals and linear combinations to have strictly positive variances is that the spectral measure is not concentrated on a set whose Lebesgue measure is zero.

Example 2. If $Z(x) = a\sqrt{2}\cos(2\pi\langle U, x\rangle + \Phi)$, where a, U, and Φ are three independent random variables, a with mean zero and variance $\sigma^2 = C(0)$, U with distribution $F(du)/\sigma^2$, and Φ with a uniform probability density on $[0, 2\pi[$, $Z(x)$ is an SRF whose covariance is given by (2.14). The realizations of this RF with random amplitude, random frequency, and random phase are sinusoids in $\mathbb{R}$, cylinders with a sinusoidal base in $\mathbb{R}^2$, and so on.   □

When $C(h)$ falls off sufficiently rapidly to ensure that $C(h)$ is absolutely integrable in $\mathbb{R}^n$, namely $\int |C(h)|\, dh < \infty$, which is the case for usual covariances, the measure F is the integral of a bounded continuous function $f(u)$ called the *spectral density* (or *power spectrum* abbreviated as *spectrum*):

$$F(du) = f(u)\, du$$

This spectral density is then the Fourier transform (in the sense of ordinary functions) of the covariance:

$$C(h) = \int e^{2\pi i\langle u, h\rangle} f(u)\, du = \int \cos(2\pi\langle u, h\rangle) f(u)\, du,$$
$$f(u) = \int e^{-2\pi i\langle u, h\rangle} C(h)\, dh = \int \cos(2\pi\langle u, h\rangle) C(h)\, dh \tag{2.18}$$

The condition of absolute integrability of $C(h)$ can be replaced by the less restrictive condition of square integrability, namely $\int |C(h)|^2 dh < \infty$, but in this case the function $f(u)$ will no longer be necessarily continuous and bounded (Yaglom, 1987, Vol. I, p. 104; in $\mathbb{R}$).

To check that a given absolutely integrable function $C(h)$ is a covariance, it suffices to calculate its Fourier transform and to verify that it is always non-negative. Thus in $\mathbb{R}$ the triangle function

$$C(h) = \begin{cases} 1 - \dfrac{|h|}{a} & \text{if } |h| \le a \\ 0 & \text{if } |h| \ge a \end{cases} \qquad (a > 0) \tag{2.19}$$

is a covariance because its Fourier transform is

$$f(u) = a\left(\frac{\sin(\pi au)}{\pi au}\right)^2$$

Conversely, as mentioned earlier this is not a positive definite function in $\mathbb{R}^n$, $n \ge 2$.

The function (2.19) is, up to a multiplicative factor, the autoconvolution $C(h) = (w * \breve{w})(h)$ of the function $w(x) = 1_{|x| < a/2}$ [$\breve{w}$ denotes the function $\breve{w}(x) = w(-x)$, which is here equal to $w(x)$]. The autoconvolution of a square integrable function is always a positive definite function, called a *covariogram* (see the next section).

As a consequence of the stability property 2 applied to the family (2.19), any function of the form

$$C(h) = \int_{|h|}^{\infty} (1 - |h|/t) \; \mu(dt) \tag{2.20}$$

where μ is a bounded positive measure on $\mathbb{R}_+$, is a covariance in $\mathbb{R}$. This formula generates the class of symmetric functions that are convex over $[0, \infty[$, namely satisfy

$$C\left(\lambda h_1 + (1 - \lambda)h_2\right) \le \lambda\, C(h_1) + (1 - \lambda)\, C(h_2), \quad h_1, h_2 > 0, \;\; 0 < \lambda < 1$$

and tend to zero as $h \to +\infty$. Therefore any function of this family is a covariance in $\mathbb{R}$. This result is often referred to as *Pólya's theorem* (Pólya, 1949). For example, $\exp(-|h|/a)$, $a > 0$, is a covariance in $\mathbb{R}$. We can also obtain this result directly by computing the Fourier transform of this function, which is $f(u) = 2a/(1 + 4\pi^2 a^2 u^2)$. We will see that $\exp(-|h|/a)$ is also a covariance in $\mathbb{R}^n$ for all n.

Spectral Representation of a Variogram and an IRF

The characterization of variograms was established by Schoenberg (1938b) and von Neumann and Schoenberg (1941):

If $\gamma(h)$ is a continuous function in $\mathbb{R}^n$, satisfying $\gamma(0) = 0$, the following three properties are equivalent:

1. $\gamma(h)$ is a variogram.
2. $e^{-t\,\gamma(h)}$ is a covariance for all $t > 0$.
3. $\gamma(h)$ is of the form

$$\gamma(h) = \int \frac{1 - \cos(2\pi\langle u, h\rangle)}{4\pi^2 |u|^2} \chi(du) + Q(h) \tag{2.21}$$

where $Q(h)$ is a positive quadratic form and χ a positive symmetric measure with no atom at the origin and satisfying

$$\int \frac{\chi(du)}{1 + 4\pi^2 |u|^2} < \infty \tag{2.22}$$

This expression is valid for the general case of an IRF with a random linear drift $m(h) = \langle a, h\rangle$ such that $E(a) = 0$ and for a noncentered definition of the variogram $\gamma(h) = \frac{1}{2} E[Z(x + h) - Z(x)]^2$. Consequently, it is possible to have a quadratic term $Q(h)$ that does not appear with the usual centered definition of $\gamma(h)$.

Proof. (1) ⇒ (2). Let $Z(x)$ be a Gaussian IRF without drift, or with a zero-mean random linear drift, and with the (noncentered) variogram $\gamma(h)$ (such IRFs do exist; see below). The complex RF $X(x) = \exp\{i(Z(x) - Z(0))\sqrt{t}\}$ has for noncentered covariance

$$E\left[X(x)\overline{X(x+h)}\right] = E\left[\exp\left(-i\left(Z(x+h) - Z(x)\right)\sqrt{t}\right)\right] = \exp\left(-t\,\gamma(h)\right)$$

$e^{-t\gamma(h)}$ is thus a covariance function for all $t > 0$.

(2) ⇒ (3). If one considers $e^{-t\gamma(u)}$ to be a characteristic function, it is the characteristic function of an infinitely divisible distribution. The general form of such characteristic functions is given by P. Lévy's theorem [e.g., Feller (1971, Chapter XVII)]. Considering only real-valued functions γ, this theorem states (3).

(3) ⇒ (1). When $\gamma(h)$ is of the form (2.21), $-\gamma(h)$ is a conditionally positive-definite function (straightforward proof).   □

For example, since $\exp(-t\,|h|)$ is a covariance in $\mathbb{R}^n$ for any n for all $t > 0$ (this will be shown later), the function $\gamma(h) = |h|$ is a variogram in $\mathbb{R}^n$.

Comparing equations (2.14) and (2.21), we can define the spectral measure of a variogram as $F(du) \equiv \chi(du)/(4\pi^2|u|^2)$. It is defined in $\mathbb{R}^n - \{0\}$. Condition (2.22) on χ is equivalent to the following two conditions on the spectral measure:

$$\int_{|u| < \varepsilon} |u|^2 F(du) < \infty, \qquad \int_{|u| > \varepsilon} F(du) < \infty$$

where ε is an arbitrary positive value. One also finds this type of condition for IRF–k, and the reader is referred to Section 4.5.3 for further discussion. Let us simply note that the low frequencies can contain an infinite energy since the condition $\int_{|u| < \varepsilon} F(du) < \infty$ is not required. We are thus dealing with the phenomenon known as an *infrared catastrophe*. When, on the contrary, the integral of the spectral measure is finite, $\gamma(h)$ is always of the form $C(0) - C(h)$, with $C(h) = \int \cos(2\pi\langle u, h\rangle)F(du)$, so that $F(du)$ is the spectral measure of a covariance. And when the integral of $|u|^2 F(du)$ over the whole space is finite, then $\gamma(h)$ is twice differentiable and $Z(x)$ is an m.s. differentiable IRF.

Pólya's theorem can be extended easily to variograms [see Matheron (1988)]: Any positive symmetric function $\{\gamma(h): h \in \mathbb{R}\}$ that satisfies $\gamma(0) = 0$ and is concave over $[0, \infty[$, namely is such that

$$\gamma\left(\lambda h_1 + (1 - \lambda)h_2\right) \geq \lambda\gamma(h_1) + (1 - \lambda)\gamma(h_2), \qquad h_1, h_2 > 0, \qquad 0 < \lambda < 1$$

is a variogram in $\mathbb{R}$.

The spectral characterization of SRFs extends to the IRFs in $\mathbb{R}$ and $\mathbb{R}^n$ (Kolmogorov, 1940a; Yaglom, 1957): An RF is an IRF if and only if it has the form

$$Z(x) = Z_0 + \int_{\mathbb{R}^n - \{0\}} e^{2\pi i\langle u, x\rangle} Y(du)$$

where Z_0 is a random variable (equal to $Z(0)$) and where Y is an orthogonal complex random measure such that

$$E|Y(du)|^2 = F(du) = \frac{\chi(du)}{4\pi^2|u|^2}$$

Spectral Representation of an Isotropic Covariance or Variogram

Let $C(h)$ be a covariance in $\mathbb{R}^n$, which for simplicity we presume to be absolutely integrable. The covariance and its spectral density $f(u)$ are then Fourier transforms of each other. Because the Fourier transform of an isotropic function is itself isotropic, if we assume that the covariance is isotropic, namely of the form $C(h) = C_n(r)$ with $r = |h|$, the spectral density is of the form $f(u) = f_n(\rho)$ with $\rho = |u|$. The index n recalls that these functions represent isotropic functions in $\mathbb{R}^n$, which is important since the Fourier transform $\mathcal{F}_n$ that relates C and f depends on n. Considered as relating two functions of a single variable C_n and f_n, $\mathcal{F}_n$ represents the Hankel transform of order n, and (2.18) takes the form

$$f_n(\rho) = 2\pi\rho^{1-n/2} \int_0^\infty r^{n/2} J_{n/2-1}(2\pi\rho r)\, C_n(r)\, dr,$$

$$C_n(r) = 2\pi r^{1-n/2} \int_0^\infty \rho^{n/2} J_{n/2-1}(2\pi\rho r) f_n(\rho)\, d\rho \tag{2.23}$$

where J_ν represents the Bessel function of the first kind of order ν (A.2). Despite this apparent symmetry, the two functions are usually not interchangeable: In particular, f_n is a nonnegative function, whereas C_n can take negative values.

If we concentrate $f(u)$ on the surface of the hypersphere with radius $1/(2\pi)$ and replace $f_n(\rho)\, d\rho$ in (2.23) by $A\, \delta_{1/(2\pi)}(d\rho)$ with $A = (4\pi)^{n/2-1}\Gamma(n/2)$, we obtain the *J-Bessel model*

$$\kappa_n(r) = 2^{n/2-1}\Gamma(n/2)r^{1-n/2}J_{n/2-1}(r)$$

The multiplicative factor A is chosen so that $\kappa_n(0) = 1$. The covariance $\kappa_n(r)$ exhibits a hole effect, this effect being less pronounced as n increases (for $n = \infty$, we obtain the Gaussian model).

The second equation in (2.23) expresses the following characterization: A necessary and sufficient condition for a function $C_n(r)$ to be an isotropic covariance in $\mathbb{R}^n$ is to be of the form

$$C_n(r) = \int_0^\infty \kappa_n\left(\frac{r}{t}\right)\mu(dt) \tag{2.24}$$

for a bounded positive measure μ on $\mathbb{R}_+$. Therefore

$$C_n(r)/C_n(0) \leq \inf_{s>0} \kappa_n(s)$$

No isotropic covariance in $\mathbb{R}^n$ can have a relative hole effect more pronounced than κ_n. More explicitly, for $n = 1, 2, 3$, the relations between C_n and f_n are

$$n = 1: \qquad C_1(r) = 2\int_0^\infty \cos(2\pi\rho r) f_1(\rho)\, d\rho,$$

$$n = 2: \qquad C_2(r) = 2\pi\int_0^\infty \rho J_0(2\pi\rho r) f_2(\rho)\, d\rho, \tag{2.25}$$

$$n = 3: \qquad C_3(r) = 2\pi \int_0^\infty \rho \sin(2\pi\rho r) f_3(\rho) \, d\rho, \tag{2.26}$$

Generally speaking, the formula involves trigonometric functions when n is odd and involves the somewhat intractable Bessel function of integer order when n is even.

As far as the variogram is concerned, if the measure χ is of the form $\chi(du) = \varphi_n(|u|) \, du$, relation (2.21) becomes in the isotropic case

$$\gamma_n(r) = \frac{\pi^{n/2-2}}{2\Gamma(n/2)} \int_0^\infty [1 - \Gamma(n/2)(\pi\rho r)^{1-n/2} J_{n/2-1}(2\pi\rho r)] \, \rho^{n-3} \varphi_n(\rho) \, d\rho$$

Recall that a positive definite function (conditionally or not) in $\mathbb{R}^n$ is not necessarily so in $\mathbb{R}^m$, $m > n$.

It is convenient to have isotropic models that are covariances in $\mathbb{R}^n$ for any n. An example is the *Gaussian model*

$$C(r) = \exp\left(-\frac{r^2}{a^2}\right) \qquad (a > 0)$$

Indeed the Hankel transform of order n exchanges the functions $\exp(-\pi r^2)$ and $\exp(-\pi\rho^2)$. This is a classic result. These functions are, respectively, the density and the characteristic function of the isotropic Gaussian multidimensional distribution with variance $\sigma^2 = 1/(2\pi)$. The Gaussian model is extremely regular, having derivatives of any order.

The following characterization holds in the general case: A necessary and sufficient condition for a function $C(r)$ to be an isotropic covariance in $\mathbb{R}^n$ for all n is that $C(r)$ is of the form

$$C(r) = \int_0^\infty \exp\left(-\frac{r^2}{t^2}\right) \mu(dt) \tag{2.27}$$

where μ is an arbitrary bounded positive measure on $\mathbb{R}_+$ (randomized scale parameter).

This condition is indeed sufficient, as results from the stability condition (2). It is also necessary, as shown by Schoenberg (1938a).[11] Note that (2.27) generates *all* types of behaviors near the origin because the base model is infinitely differentiable: randomization produces models that can be *less* regular than the base model but never more regular. For example, randomizing the scale parameter t in (2.27) by a one-sided Gaussian distribution leads to the *exponential model*[12]

$$C(r) = \exp\left(-\frac{r}{a}\right) \qquad (a > 0)$$

which is thus a covariance in $\mathbb{R}^n$ for any n. This model is a completely monotone function.[13] It constitutes the base model of isotropic covariances with this property, since any completely monotone function that is an isotropic covariance in $\mathbb{R}^n$ for all n is of the form

$$C(r) = \int_0^\infty \exp\left(-\frac{r}{t}\right) \mu(dt) \tag{2.28}$$

---

[11] The proof is based on the fact that $\kappa_n(\sqrt{2n}\, r)$ tends to $\exp(-r^2)$ as $n \to \infty$

[12] This results from an application of formula (7.4.3) in Gautschi (1972, p. 302):

$$\frac{1}{a\sqrt{\pi}} \int_0^\infty \exp\left(-\frac{t^2}{4a^2} - \frac{r^2}{t^2}\right) dt = \exp\left(-\frac{r}{a}\right), \qquad r \geq 0, \ a > 0$$

[13] A continuous function f on $[0, \infty[$ is completely monotone if it possesses derivatives $f^{(n)}(t)$ of all orders and if $(-1)^n f^{(n)}(t) \geq 0$ for $n = 0, 1, 2, \ldots$, and $t > 0$.

for a bounded positive measure μ on $\mathbb{R}_+$ [e.g., see Schoenberg (1938a) and Feller (1971, p. 439)]. The family of completely monotone isotropic covariances includes, for example, the *iterated exponential model* $\exp(\exp(-r/a)) - 1$ $(a > 0)$ and the *gamma model* $(1 + r/a)^{-\beta}$ $(a, \beta > 0$; for $\beta = 1$ this model is named the *hyperbolic model*).

Schoenberg also proves that $C(r)$ is a completely monotone isotropic covariance if and only if $C(r^2)$ is an isotropic covariance in $\mathbb{R}^n$ for all n. A consequence is for example that the *Cauchy model* $(1 + r^2/a^2)^{-\beta/2}$ $(a, \beta > 0)$ is a covariance whatever the space dimensionality.[14]

Let us end with a useful result whose proof can be found in Yaglom (1987, Vol. I, pp. 358–360): since an isotropic covariance in $\mathbb{R}^n$, $n > 1$, is also a covariance in $\mathbb{R}$, it has an n-dimensional spectral density f_n and a one-dimensional spectral density f_1 as well. These spectral densities are related by

$$f_1(\rho_1) = \frac{2\pi^{(n-1)/2}}{\Gamma\left((n-1)/2\right)} \int_{\rho_1}^{\infty} f_n(\rho)(\rho^2 - \rho_1^2)^{(n-3)/2} \rho \, d\rho \qquad (2.29)$$

This relation can be easily inverted when $n = 2$ or 3:

$$f_2(\rho) = -\frac{1}{\pi} \int_{\rho}^{\infty} \frac{df_1(\rho_1)}{d\rho_1} (\rho_1^2 - \rho^2)^{-1/2} d\rho_1 \qquad f_3(\rho) = -\frac{1}{2\pi\rho} \frac{df_1(\rho)}{d\rho}$$

To calculate the spectral density of an isotropic covariance in $\mathbb{R}^2$ or $\mathbb{R}^3$, it is often simpler to calculate the spectral density $f_1(\rho)$ in $\mathbb{R}$ and apply these relations rather than directly use relations (2.25) and (2.26). Note that the 1D spectral density associated with an isotropic 3D density is necessarily decreasing. Inverting (2.29) is difficult when $n > 3$, especially when n is even (see Yaglom, 1987, Vol. II, Chapter 4, Note 45).

The characterization of isotropic variogram models valid in $\mathbb{R}^n$ for all n is a special case of a theorem presented in Section 4.5.4. To conclude, it may also be interesting to exploit the *turning bands* operator of Section 7.4 which establishes a one-to-one mapping between isotropic covariances or variograms in $\mathbb{R}^n$ and covariances or variograms in $\mathbb{R}$.

2.3.4 Covariograms

Definition and Properties

Let $w(x)$ be a function in $\mathbb{R}^n$, both integrable and square integrable, and let $\varphi(u)$ be its Fourier transform. We can define the covariogram $g(h)$, $h \in \mathbb{R}^n$, as the convolution of $w(x)$ by the function $\breve{w}(x) = w(-x)$:

$$g(h) = (w * \breve{w})(h) = \int w(x) \, w(x + h) \, dx \qquad (2.30)$$

Since the Fourier transformation exchanges convolution and multiplication, the Fourier transform of $g(h)$ is $|\varphi(u)|^2$. It is thus always positive and $g(h)$ is a covariance in $\mathbb{R}^n$.

---

[14] The general result derives from the following relations where C denotes a completely monotone isotropic covariance and C_1 an isotropic covariance in $\mathbb{R}^n$ for all n:

$$C(r^2) = \int_0^{\infty} \exp\left(-\frac{r^2}{t}\right) \mu(dt) = \int_0^{\infty} \exp\left(-\left(\frac{r}{\sqrt{t}}\right)^2\right) \mu(d\sqrt{t}) = \int_0^{\infty} \exp\left(-\frac{r^2}{t^2}\right) \nu(dt) = C_1(r)$$

In addition to the properties of covariances, the covariogram satisfies the equation

$$\int g(h)\, dh = \left[\int w(x)\, dx\right]^2 \tag{2.31}$$

If the support of w is bounded, covariograms have a finite range, in the sense of the integral range defined in Section 2.3.5 and in the strict sense (zero value when $|h|$ exceeds a finite limit, which can vary according to the direction).

The covariogram has a particularly simple interpretation when $w(x)$ is the indicator function of a bounded domain B (Figure 2.18a):

$$w(x) = 1_{x \in B} = \begin{cases} 1 & \text{if } x \in B \\ 0 & \text{if } x \notin B \end{cases}$$

The product $w(x)\, w(x+h)$ is then equal to 1 if x and $x+h$ belong to B (i.e., if x belongs both to B and to the translate B_{-h} of B in the translation $-h$) and equal to 0 if not. Thus $g(h)$ is the measure of $B \cap B_{-h}$ (its volume in $\mathbb{R}^3$, its area

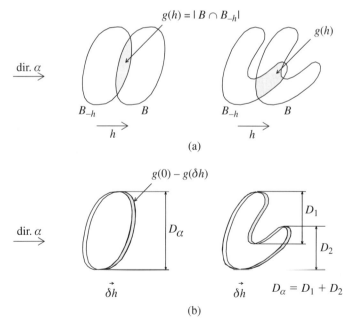

FIGURE 2.18   Geometric covariogram of a bounded domain B: (a) geometric interpretation of $g(h)$; (b) caliper diameter and total diameter in the direction α.

in $\mathbb{R}^2$, its length in $\mathbb{R}$). This measure will be denoted by $|B \cap B_{-h}|$. In this case, $g(h)$ is called the *geometric covariogram* of B.[15]

Again let us consider the case of an indicator function. As shown in Figure 2.18b, for a small displacement δh in the direction α, the difference $g(0) - g(\delta h)$ is the surface area swept by a vector δh included in B when its end point scans the boundary of B. For a convex domain B, this area is $D_\alpha |\delta h|$, where D_α is the tangent diameter (caliper diameter) in the direction α, and we can write in $\mathbb{R}^2$

$$g(0) - g(\delta h) = D_\alpha |\delta h|$$

This formula remains valid for a nonconvex (but nonfractal) B, but D_α now represents the *total diameter* in the direction α, namely the total length of the orthogonal projection along α of the portion of contour scanned by δh, or, equivalently, one-half the total projected length of the complete contour of B. In any case the partial derivative of $g(h)$ at 0 in the direction α is $g'_\alpha (0) = -D_\alpha$. The left and right partial derivatives at $h = 0$ are of opposite signs, and $g(h)$ is therefore not differentiable at $h = 0$. Differentiability of $g(h)$ at 0 is only possible for functions $w(x)$ that display less abrupt transitions than an indicator function. These properties are easily transposed to $\mathbb{R}^n$, where the measure of the total projection of B in $\mathbb{R}^{n-1}$ replaces the total diameter.

The covariogram is the basic tool of *transitive methods*, where $w(x)$ is the studied regionalized variable: In the case of a mineral deposit, it is the indicator function or the concentration of a constituent present only in the deposit. A brief presentation of this approach is given in Section 1.3. It will suffice here to consider the covariogram as a means of creating covariance and variogram models, and we will take ordinary analytical function for $w(x)$, the simplest example being the indicator of the sphere of $\mathbb{R}^n$ which gives the spherical model of $\mathbb{R}^n$ (see Section 2.5.1). By reference to the transitive theory, the obtained models are known as *transition models*.

Radon Transform

Let us explicitly write $w(x)$ as a function $w_n(x_1, x_2, \ldots, x_{n-1}, x_n)$ of the n coordinates of point $x \in \mathbb{R}^n$. By integration parallel to the x_n axis, we define a function $w_{n,1}$ in $\mathbb{R}^{n-1}$:

$$w_{n,1}(x_1, \ldots, x_{n-1}) = \int_\mathbb{R} w_n(x_1, \ldots, x_{n-1}, x_n) \, dx_n$$

The operation allowing one to pass from w_n to $w_{n,1}$ is a Radon transform [cf. Gel'fand et al. (1962, Section 1.1) and Santaló (1976, Section. IV.19.8)], also known in geostatistics as a transitive "*montée*" (upscaling) along the x_n axis.[16] It represents an accumulation along the

---

[15] The knowledge of B determines $g(h)$. Conversely, does the knowledge of $g(h)$ determine B, up to translations and reflections? Matheron (1986, p. 20) asked that question within the subset of convex bodies and conjectured that the answer would be positive in $\mathbb{R}^2$. The problem was addressed by numerous authors, and a complete proof in $\mathbb{R}^2$ has been provided by Averkov and Bianchi (2009). A positive answer was given in $\mathbb{R}^3$ for convex polyhedra. Counterexamples are known in $\mathbb{R}^n$, $n \geq 4$.

[16] In stereology, and particularly in tomographic imaging, the main concern is the inverse problem, namely the derivation of w_n knowing the value of its integral along any line or any plane.

direction of integration. By repeating this, we can define the Radon transform or transitive *montée* of order m ($m < n$) in the hyperplane defined by the axes of $x_{n-m+1}, \ldots, x_n$.

Also let $g_n(h_1, h_2, \ldots, h_{n-1}, h_n)$ denote the covariogram of w_n. It is readily seen from the definition of $w_{n,1}$ that its covariogram $g_{n,1}$ takes the form

$$g_{n,1}(h_1, \ldots, h_{n-1}) = \int_{\mathbb{R}} g_n(h_1, \ldots, h_{n-1}, h_n) \, dh_n$$

Thus the covariogram of the Radon transform of w_n is the Radon transform of g_n. Through iteration this property is generalized to the Radon transform of order m.

If the function w_n is isotropic, so are the Radon-transformed variables and all their covariograms. Let us denote by $g_n(r)$, $g_{n,1}(r)$, and $g_{n,2}(r)$ the initial covariogram and the Radon-transformed covariograms of orders 1 and 2 as functions of the modulus r of a vector of the $\mathbb{R}^n$, $\mathbb{R}^{n-1}$, and $\mathbb{R}^{n-2}$ space, respectively. Elementary calculations show that

$$g_{n,1}(r) = 2 \int_0^\infty g_n(\sqrt{r^2 + \rho^2}) \, d\rho$$

$$g_{n,2}(r) = 2\pi \int_r^\infty u \, g_n(u) \, du \tag{2.32}$$

Therefore a transform of order 2, and more generally all transforms of even order obtained by iteration of (2.32), reduce to a very simple integral, whereas a transform of odd order is generally cumbersome. The same relations also obviously apply between w_n and the transformed variables $w_{n,1}$ and $w_{n,2}$.

Still in the isotropic case, a Radon-transformed function is differentiable once more than the original function. Therefore, if we start from a covariogram g_n that is not twice differentiable (e.g., a geometric covariogram), $g_{n,2}$ is then twice differentiable and is a valid model for a differentiable SRF. More generally the covariogram $g_{n,2q}$ (provided that $2q < n$) is $2q$ times differentiable and constitutes a covariance model of q times differentiable SRFs. In applications, one must not forget that the Radon transform of order m of a covariogram in $\mathbb{R}^n$ produces a model that is only valid in $\mathbb{R}^{n-m}$. Matheron (1965, Chapter II) presents several families of Radon-transformed covariograms.

2.3.5 Integral Range

Let $Z(x)$ be an SRF with covariance $C(h)$, and let us consider the spatial average $Z_V = (1/|V|) \int_V Z(x) \, dx$. According to (2.12), its variance is

$$\text{Var}(Z_V) = \frac{1}{|V|^2} \int_V \int_V C(x' - x) \, dx \, dx' \tag{2.33}$$

If $C(h)$ has a finite range and if the support V is very large with respect to the range, the integral $\int_V C(x' - x) \, dx$ is equal to $\int C(h) \, dh$, except when x' is close to the boundary of V (at a distance less than the range). To a first approximation the variance of Z_V is then of the form

$$\text{Var}(Z_V) \approx \frac{A}{|V|} \sigma^2 \tag{2.34}$$

where $\sigma^2 = C(0)$ is the variance of $Z(x)$ and where

$$A = \frac{1}{\sigma^2} \int C(h) \, dh \tag{2.35}$$

Expression (2.34) remains valid for a covariance that reaches zero asymptotically when $|h| \to \infty$, provided that (2.35) is finite. We call A the *integral range*. In 1D a useful mnemonic is to remember that the integral range is the range of the triangular covariance having the same value at the origin and the same area under the curve as $C(h)$. Equivalently A is the integral of the correlogram $\rho(h) = C(h)/\sigma^2$. Alternative names given by Yaglom (1987) are "correlation time" (in a spatial context we would say "correlation length") or "integral time scale" in 1D, "correlation area" or "integral area scale" in 2D space.[17] Its dimension is that of a volume of the space $\mathbb{R}^n$. If, for example, $C(h)$ is the covariogram of a function $w(x)$ in $\mathbb{R}^n$, in view of (2.30) and (2.31) its integral range is

$$A = \left[\int w(x)dx \right]^2 \bigg/ \int [w(x)]^2 dx$$

Two special cases deserve mention: (1) If $w(x)$ is the indicator function of a bounded domain of $\mathbb{R}^n$, A is the measure of this domain; (2) if $w(x)$ is a function integrating to zero, we obtain $A = 0$, which means that the approximate formula (2.34) is no longer valid and that the variance of Z_V decreases faster than $1/|V|$.

If $C(h)$ is a continuous covariance of $\mathbb{R}^n$ with $C(h) = 0$ for $|h| \geq a$, then

$$A \leq 2^{-n} V_n a^n$$

where V_n is the volume of the unit-radius sphere of $\mathbb{R}^n$, given by (A.5) (Gorbachev, 2001; Kolountzakis and Révész, 2003). The equality is reached by the spherical covariance of $\mathbb{R}^n$.

If we write $N = |V|/A$, (2.34) takes the form

$$\mathrm{Var}(Z_V) \approx \frac{\sigma^2}{N}$$

This is the conventional formula for the variance of the arithmetic mean of N independent points (i.e., mutually located at distances greater than the range). In other words, from the point of view of the mean value, the domain V is equivalent to N independent samples.

The integral range is related to the ergodicity of the SRF: If it is finite, the SRF is ergodic, since the variance of Z_V tends to zero when V tends to infinity. However, the SRF may be ergodic in cases when the integral range does not exist. For example, if the integral $\int_V C(h)\,dh$ does not converge when V tends to infinity but remains bounded, Slutsky's condition (1.12) is satisfied and $Z(x)$ is ergodic. This case is exceptional in practice, but it occurs, for example, in $\mathbb{R}$ with the periodic covariance $C(h) = \cos(h/a)$. Also note that when $C(h)$ is a covariance in $\mathbb{R}^n$, $n > 1$, its integral range in a subspace of $\mathbb{R}^n$ does not coincide

---

[17]Yaglom's terminology relates to $2^{-n} A$, where n is space dimension. In the literature the term "correlation length" sometimes also means what we call range, and sometimes it refers to what we call scale parameter.

with its integral range in $\mathbb{R}^n$. In particular, $Z(x)$ may be ergodic in $\mathbb{R}^n$ but not in a subspace of $\mathbb{R}^n$.

From the above considerations, it appears that the integral range represents a yardstick by which we can judge how large a domain of study V is with respect to the scale of the phenomenon. It is an indicator of practical ergodicity. Lantuéjoul (1991) illustrates this with very demonstrative examples.

2.4 REGULARIZATION AND NUGGET EFFECT

The data usually relate to a sampling support that is not punctual. They are affected by microstructures and various error sources that are pooled together into the nugget effect. A delineation of the various components of the nugget effect is necessary for a sound application of geostatistics; an example, among others, is provided by Baghdadi et al. (2005), who create a digital elevation model of a tropical forest area by merging ground data, airborne laser and radar data, and remote sensing data, which are associated with very different supports, measurement errors and positioning accuracies. In mining, such delineation permits an improved data collection and sampling control [e.g., Sinclair and Vallée (1994)].

2.4.1 Change of Support—Regularization

Our data may be measured over different sampling supports (cores, channel samples, mine blocks, plots of different sizes, etc.) or represent a weighted average over a volume of investigation (e.g., geophysical tools, remote sensing data). This averaging process can be modeled as a convolution of a (perhaps hypothetical) point-support variable. From this it is easy to establish the relationship between covariances and variograms for different supports.

Let $Z(x)$ be a point-support RF, and define the regularized RF $Z_p(x)$ by the stochastic convolution

$$Z_p = Z * \breve{p}$$

where $p(u)$ is a sampling function and $\breve{p}(u) = p(-u)$, that is, $Z_p(x) = \int Z(x+u)p(u)\,du$. This function could be the indicator function of a sample v, normalized by its volume $|v|$, in which case $Z_p(x)$, generally denoted by $Z_v(x)$, is simply the mean value over the sample of support v centered at x. It could also be a more general weighting function that we assume integrable and square integrable so that $Z_p(x)$ exists.

If $Z(x)$ is an SRF with covariance $C(h)$, it follows from (2.13) and the Cauchy algorithm that $Z_p(x)$ is an SRF whose covariance $C_p(h)$ is given by

$$C_p = C * P \tag{2.36}$$

that is, $C_p(h) = \int C(h+u)P(u)\,du$, where $P = p * \breve{p}$ is the covariogram of p: C_p derives from C by regularization by the covariogram P. In particular, the variance of the SRF Z_p is $C_p(0) = \int C(h)P(h)\,dh$. When $Z_p(x)$ is a weighted moving average, namely when p sums to one, P also sums to one as shown by (2.31), and $C_p(0)$ is then a weighted average of $C(\cdot)$. The variance of Z_p is thus smaller than $C(0)$.

If $Z(x)$ is not an SRF but an IRF with variogram $\gamma(h)$, then similarly $Z_p(x)$ is an IRF whose variogram is given by

$$\gamma_p(h) = (\gamma * P)(h) - (\gamma * P)(0) \tag{2.37}$$

These transformations generally have little effect on the shape of the variogram at large distances but impart a more regular behavior at the scale of the support of the function $p(u)$. This behavior is not isotropic if the function $p(u)$ is not isotropic. Thus for a linear variogram $\gamma(h) = b\,|h|$ and a regularization along segments with the same orientation and same length l (regularization along cores of a drill hole, for example), we obtain a regularized variogram $\gamma_l(h)$ such that (with $r = |h|$):

- For a vector h with the same direction as the segments, we have

$$\gamma_l^{/\!/}(h) = \begin{cases} b\dfrac{|r|^2}{l^2}\left(l - \dfrac{|r|}{3}\right) & \text{if } |r| \leq l, \\[4mm] b\left(|r| - \dfrac{l}{3}\right) & \text{if } |r| \geq l \end{cases} \tag{2.38}$$

- For a vector h orthogonal to the cores, we have

$$\gamma_l^{\perp}(h) = b\left[\frac{1}{3}\sqrt{l^2 + r^2} + \frac{r^2}{l}\log\frac{l + \sqrt{l^2 + r^2}}{r} - \frac{2\,r^2}{3\,l^2}\left(\sqrt{l^2 + r^2} - r\right) - \frac{l}{3}\right] \tag{2.39}$$

- And more complex expressions for the other directions.

In all directions, $\gamma_l(h)$ asymptotically reaches the same line $b\,(|h| - l/3)$ when $|h| \to \infty$. We can see from Figure 2.19a that $\gamma_l(h)$ has a more regular behavior than the point variogram $\gamma(h)$. This behavior is parabolic along the direction of the cores; its great regularity is due to a partial overlap of the cores for $|h| < l$; in practice, the smallest distance for which we calculate a sample variogram is l, and for $|h| \geq l$ the regularized variogram differs from the point variogram only by a constant value; if we extrapolate it to the origin, we obtain an apparent negative nugget effect. Note that (2.39) can be used as an isotropic variogram model for $h \in \mathbb{R}^n$ for all n because it can be considered as the result of a regularization in $\mathbb{R}^{n+1}$.

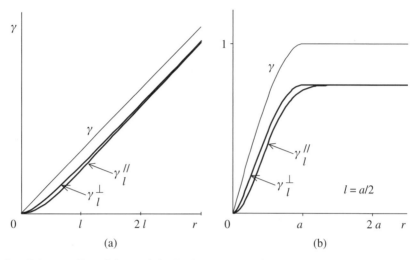

FIGURE 2.19 Effect of the regularization by a segment of length l on the variogram along the direction of segment ($\gamma_l^{/\!/}$) and perpendicular to segment ($\gamma_l^{\perp}$): (a) for a linear variogram; (b) for a spherical variogram with range $a = 2\,l$.

If the variogram has a finite range a, the regularization extends the range by l in the direction of the cores, whereas it reduces the sill by the amount

$$(\gamma * P)(0) = \frac{1}{l^2} \int_0^l \int_0^l \gamma(x' - x)\, dx\, dx' = \frac{2}{l^2} \int_0^l (l - u)\gamma(u)\, du$$

In the case of a spherical variogram, this quantity represents a proportion $l/(2\,a) - l^3/(20\,a^3)$ of the sill if $l \le a$ (Figure 2.19b). On the other hand for $l \gg a$ the sill of the regularized variogram is only about $\tfrac{3}{4}\,(a\,/\,l)$ of the original sill, thus this sill mainly reflects the contributions to variance of the long-range components. Regularizations over varying supports are a useful tool to analyze nested structures [an example with ranges from 12 m to 18 km is provided by Rivoirard et al. (2000, Section 4.3)]. Further results can be found in Journel and Huijbregts (1978, Section II.D).

Relation (2.37) is usually applied with a known sampling function. It is also possible to invert it to determine P and, to some extent, p (thus the volume investigated), knowing the point and regularized variograms (Royer, 1988).

2.4.2 Microstructure

The formalism of regularization provides a representation of the upscaling mechanism by which a microstructure turns up as a nugget effect at the macroscopic level. Consider a phenomenon made up of microscopic heterogeneities with some characteristic dimension that is very small in comparison with the size of the sampling unit. A good example is a piece of sedimentary rock, such as a core, typically made up of a large number of grains. At the grain level the phenomenon is characterized by a microstructure with a covariance $C_\mu(h)$

whose range is of the order of the diameter of the grains. Let us suppose that the observations are average values over a support v with indicator function $k(u)$ and geometric covariogram $K(h)$. Equivalently, they are weighted averages with the sampling function $p(u) = k(u)/|v|$, which has covariogram $P(h) = K(h)/|v|^2$. According to (2.36) the microstructure is regularized into a component with covariance

$$C_0(\cdot) = \frac{1}{|v|^2} C_\mu * K$$

Since the range of C_μ is small in relation to that of K, we have

$$C_0(h) = \frac{1}{|v|^2} \int C_\mu(u) K(h+u) \, du \approx \frac{1}{|v|^2} \left[\int C_\mu(u) \, du \right] K(h) = \frac{f_0 K(h)}{|v|^2}$$

where $f_0 = \int C_\mu(h) \, dh$ (the value of the spectral density at zero frequency). This covariance is equal to $f_0/|v|$ at the origin ($K(0) = |v|$) and vanishes as soon as h exceeds the dimensions of the sampling unit v. Since distances smaller than the sampling unit are not considered in practice (except of course at $h = 0$), the effect of the microstructure is to offset the sample variogram at the origin by the amount

$$C_0 = \frac{f_0}{|v|} \tag{2.40}$$

Thus the nugget effect is a reminiscence of a microstructure that is no longer perceptible at the scale of study, and it disappears as the sampling unit becomes larger. The macrostructure, made of medium- to long-range components, evolves very slowly at the scale of the sampling unit and remains practically unaffected by the regularization. Increasing the sample support enhances the contrast between micro- and macrostructures. Note that this theory assumes spatially homogeneous heterogeneities (else a location-dependent nugget effect should be considered).

From the point of view of geostatistical calculations where one considers supports and distances that are large compared with the range of the microstructure, this component appears to be purely random, namely without spatial correlation. In 1D this corresponds to the so-called white noise process, defined as a zero-mean process with a constant spectral density over all frequencies (the term "white" is by analogy with the flat spectrum of white light, as opposed to a preponderance of low frequencies for red light and of high frequencies for blue light).

Strictly speaking, such process cannot exist in the ordinary sense for its total power, the variance, is infinite. However, it can be modeled within the scope of *generalized random processes* [e.g., Yaglom (1987, Vol. I, pp. 117–120 and Section 24.1)], which are processes defined not by point values but by the values obtained by convolution with regular enough weighting functions. In this model the microstructure can be represented by the macroscopic covariance $f_0 \, \delta(h)$ where $\delta(h)$ is the Dirac delta function (Appendix, Section A.1). It is as though the integral $f_0 = \int C_\mu(h) \, dh$ were concentrated at the origin. The benefit of this formalism is to be able to standardize the treatment of nugget effect under change of support by adding a term $f_0 \, \delta$ in all formulas involving a covariance. In formulas involving the variogram the nugget effect is modeled as $f_0 \, [1 - \delta(h)]$.

The proportionality of the covariance $C_0(h)$ with the geometric covariogram of the support shows the possibility of modeling the true nugget effect (i.e., due to microstructures) as a number of points of a Poisson point process with intensity $\lambda = f_0/|v|^2$ (see Section 7.5.2). Such a model fits well with the idea or actual nuggets (stones) scattered randomly in the rock. It also highlights the isotropy of the true nugget effect. Indeed, the Poisson point process is isotropic, and an affine transformation of the space does not introduce any anisotropy but just changes the intensity of the process.[18]

2.4.3 Measurement Errors

Experimental data are often affected by measurement errors. Instead of giving the value $Z(x_\alpha)$ at location x_α, the measurement gives $Z_\varepsilon(x_\alpha) = Z(x_\alpha) + \varepsilon_\alpha$, where ε_α is a measurement error. The numerical value of this error being unknown, ε_α is considered as a random variable. If measurement errors at the sample points are independent of Z, mutually independent, with mean zero and the same variance σ_ε^2, the variogram $\gamma_\varepsilon(h)$ of Z_ε is

$$\gamma_\varepsilon(h) = C_\varepsilon + \gamma(h) \tag{2.41}$$

where $C_\varepsilon = \sigma_\varepsilon^2$ and $\gamma(h)$ is the variogram of Z. Formula (2.41) remains valid at $h = 0$ if one considers replicate measurements at the same location. Otherwise $\gamma_\varepsilon(0) = 0$ and the variogram of Z_ε shows an offset at the origin similar to that produced by a microstructure. Note that formula (2.41) remains valid if the mean error is not zero: The variogram cannot detect a systematic error.

The reproducibility (or precision) of measurements is often known. For example, in dividing up a pile of ore to obtain a sample for analysis, the variance of the sampling error (also known as *Gy's fundamental error*) can be calculated (Gy, 1975, 1979; François-Bongarçon, 1998). For the relationship between sampling theory and geostatistics, see Deverly (1984a,b) and François-Bongarçon (2004). Similarly, a chemical analysis technique or an instrument can be calibrated, and thus its precision can be estimated.

If the measurement errors are not with the same variance, the apparent nugget effect will be equal to the average variance. Marine or airborne data collected along profiles often include an error term that is almost constant along each profile and independent between profiles. It can be analyzed at the

---

[18] Incidentally, the persistence of the Poisson point process under a large variety of transformations rules out the possibility of detecting strain in a rock by analysis of the position of objects after deformation if the original object positions are "random" and mutually independent (Fry, 1979). But such detecion is possible if the objects themselves are deformed; for example, initially spherical markers become ellipsoids.

intersection of profiles, or by comparing the variogram along profiles with the variogram between profiles. More complex cases (e.g., correlation with Z) are handled with multivariate models.

2.4.4 Positioning Errors

Positioning uncertainty is typically associated with marine and aerial surveys. The advent of satellite positioning (GPS) has made this uncertainty very small[19] but perhaps still significant in applications where extreme precision is required, such as marine 3D seismic. The problem is also encountered in remote sensing applications when massive, irregularly spaced data have been relocated on regularly spaced grid nodes.

In case of a positioning error, the value believed to be at point x_α has in fact been measured at some other point $x_\alpha + U_\alpha$. Instead of studying the data $\{Z(x_\alpha) : \alpha = 1, \ldots, N\}$, we are in fact studying $Z_1(x_\alpha) = Z(x_\alpha + U_\alpha)$, where the U_α are random vectors. The problem often occurs with data originating from a dense sampling of profiles, where the errors are correlated. But let us first consider the case where the U_α are uncorrelated and have the same p.d.f. $p(u)$. If $Z(x)$ is an SRF or an IRF, the variogram of the observations is given by

$$\gamma_1(h) = \frac{1}{2} \iint E[Z(x + h + u') - Z(x + u)]^2 \, p(u) \, p(u') \, du \, du'$$
$$= \iint \gamma(h + u' - u) \, p(u) \, p(u') \, du \, du' \tag{2.42}$$

where γ is the variogram of Z, that is,

$$\gamma_1 = \gamma * P \qquad \text{with} \quad P = p * \check{p}$$

This formula is valid even for $h = 0$ if we consider pairs of distinct measurement points supposed to have the same location but subject to two different positioning errors. Of course, if we are considering twice the same value, $\gamma_1(0) = 0$. It is thus essential in the analysis of such data to compute the variogram at $h \approx 0$ from pairs of distinct points only.

A comparison with formula (2.37) shows that positioning errors act as a regularization plus the addition of a nugget effect

$$C_0 = \iint \gamma(u' - u) \, p(u) \, p(u') \, du \, du' = \int \gamma(h) \, P(h) \, dh$$

adding to the possible nugget effect of Z (Figure 2.20a). This discontinuity reflects the fact that two apparently close data points can in fact be significantly

---

[19] The GPS horizontal position is accurate to about 10 m and to 1 m or less using differential corrections.

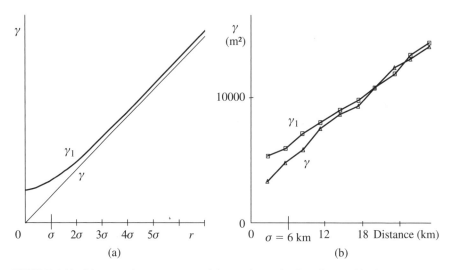

FIGURE 2.20 Linear variogram corrupted by an isotropic Gaussian positioning uncertainty U (2D) with mean square $E(|U|^2) = \sigma^2$: (a) model; (b) example in bathymetry (including measurement error and positioning uncertainty). [From Chilès (1976), with kind permission from Kluwer Academic Publishers.]

apart. Formula (2.42) can be easily generalized to correlated errors, provided that the joint p.d.f. of U and U' as a function of $h = x' - x$ is known.

These features are exploited in the analysis of data along marine survey lines (Chilès, 1976, 1977; Figure 2.20b): The sample variogram computed from pairs of points belonging to the same profile is very little affected by the positioning errors because they are highly correlated at short distances, whereas the sample variogram computed from pairs of points on different profiles conforms to the model of uncorrelated positioning errors and can be used to assess the distribution of this error.

2.5   VARIOGRAM MODELS

2.5.1   Isotropic Covariance or Variogram Models

We begin with isotropic models and present the covariances and variograms as functions of $r = |h|$ and the corresponding spectra, also isotropic, as functions of $\rho = |u|$. For variograms associated with a covariance, we give the analytical form of the covariance, the variogram being deduced from this by the equation $\gamma(h) = C(0) - C(h)$; conversely, all figures display the graph of the variogram function, which is the structural tool. Models are given in normalized form, namely $C(0) = 1$ for covariances (they are thus correlograms) and multiplicative coefficient equal to 1 for variograms.

Spherical Models and Derived Models

By autoconvolution of the indicator function of the sphere of $\mathbb{R}^n$ with diameter a, namely in terms of the modulus $\xi = |x|$, of the function

$$w_n(\xi) = \begin{cases} 1 & \text{if } \xi \leq a/2, \\ 0 & \text{if } \xi > a/2 \end{cases} \tag{2.43}$$

we obtain the spherical covariogram of $\mathbb{R}^n$, which we can consider as a function of $r = |h|$ (Matheron, 1965, pp. 56–57):

$$g_n(r) = \begin{cases} a^n v_{n-1} \int_{r/a}^1 (1 - u^2)^{(n-1)/2} du & \text{if } r \leq a, \\ 0 & \text{if } r \geq a \end{cases} \tag{2.44}$$

In these formulas, v_n is the volume of the unit-*diameter* ball of $\mathbb{R}^n$, which can be deduced from the volume V_n of the unit-*radius* ball of $\mathbb{R}^n$, whose expression is given by (A.5):

$$v_n = \frac{V_n}{2^n} = \frac{\pi^{n/2}}{2^{n-1} n \, \Gamma(n/2)}$$

For odd $n = 2p + 1$ the integrand of (2.44) is a polynomial, which gives after integration

$$g_{2p+1}(r) = a^{2p+1} \left[v_{2p+1} - v_{2p} \sum_{l=0}^p \frac{(-1)^l}{2l+1} \binom{p}{l} \left(\frac{r}{a}\right)^{2l+1} \right] \quad (r \leq a)$$

The models used in practice correspond to $n = 1, 2, 3$:

- *Triangular model*, also called *tent covariance*, valid in $\mathbb{R}$:

$$C_1(r) = \begin{cases} 1 - \dfrac{r}{a} & \text{if } r \leq a, \\ 0 & \text{if } r \geq a \end{cases} \tag{2.45}$$

- *Circular model*, valid in $\mathbb{R}^2$:

$$C_2(r) = \begin{cases} \dfrac{2}{\pi} \left[\arccos\left(\dfrac{r}{a}\right) - \dfrac{r}{a} \sqrt{1 - \dfrac{r^2}{a^2}} \right] & \text{if } r \leq a, \\ 0 & \text{if } r \geq a \end{cases} \tag{2.46}$$

• *Spherical model*, valid in $\mathbb{R}^3$:

$$C_3(r) = \begin{cases} 1 - \dfrac{3}{2}\dfrac{r}{a} + \dfrac{1}{2}\dfrac{r^3}{a^3} & \text{if } r \le a, \\ 0 & \text{if } r \ge a \end{cases} \tag{2.47}$$

The corresponding variograms exhibit a linear behavior near the origin and reach their sill at $r = a$, so that the scale parameter a of the spherical models coincides with the range. These variograms maintain a quasi-linear behavior up to the sill,[20] which can be related to the following property: Among all covariances of $\mathbb{R}^n$ with support included in the ball of radius a and with unit value at 0, the spherical covariance of $\mathbb{R}^n$ has the largest integral range $A = 2^{-n} V_n a^n$. When we speak of a spherical model without specifying n, we are referring to the model $C_3(r)$ (Figure 2.21a). Because of its validity in $\mathbb{R}^n$ for $n = 1$, 2, or 3, its well-marked range, and its ease of calculation, it is very widely used.

As was shown in Section 2.3.4, more regular models, corresponding to q-times m.s. differentiable SRFs, are obtained by Radon transform of order $2q$ ($2q < n$). The Radon transform of order $2q$ of the indicator function (2.43) of $\mathbb{R}^n$ is an isotropic function $w_{n,2q}$ in $\mathbb{R}^{n-2q}$. Considered as function of $\xi = |x|$ ($x \in \mathbb{R}^{n-2q}$), $w_{n,2q}(\xi)$ represents the volume of the sphere of $\mathbb{R}^{2q}$ with radius $\sqrt{\frac{1}{4}a^2 - \xi^2}$. Thus it depends only on q and is given by

$$w_{n,2q}(\xi) = \begin{cases} v_{2q}\left(a^2 - 4\xi^2\right)^q & \text{if } \xi \le a/2 \\ 0 & \text{if } \xi \ge a/2 \end{cases}$$

The corresponding covariograms can be obtained by recursive application of equation (2.32) to the expression (2.44) of g_n. The result is

$$g_{n,2q}(r) = \frac{\pi^q}{q!} \sum_{l=0}^{q} (-1)^l \binom{q}{l} \frac{v_{n-1}}{v_{n+2l-1}} \left(a^2 - r^2\right)^{q-l} g_{n+2l}(r) \qquad (r \le a) \tag{2.48}$$

and of course $g_{n,2q}(r) = 0$ for $r \ge a$. These models are valid in $\mathbb{R}^{n-2q}$. Considering only the lowest-degree models valid in $\mathbb{R}^3$, we obtain from formula (2.48) the following two models for $q = 1$, $n = 5$, and $q = 2$, $n = 7$, respectively (given here in their normalized form):

---

[20] We have seen in footnote 7 that a variogram with a perfectly linear behavior up to the sill exists in $\mathbb{R}^2$ and $\mathbb{R}^3$, but that model does not stay at the sill for $r > a$.

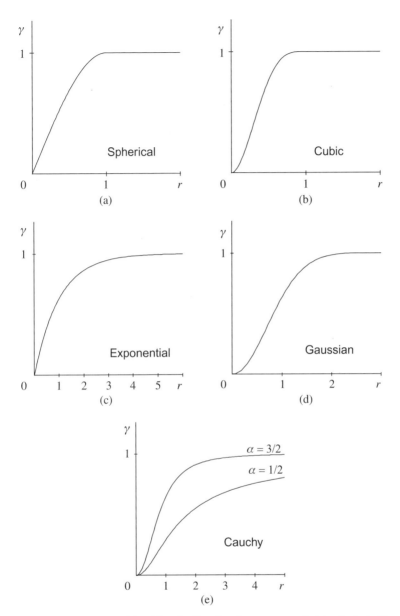

FIGURE 2.21 Variogram models with unit sill and scale parameter—1: (a) spherical; (b) cubic; (c) exponential; (d) Gaussian; (e) Cauchy.

1. *The model known as "cubic"* because its principal irregular term is in r^3. It corresponds to the Radon transform of order 2 of the spherical covariogram of $\mathbb{R}^5$ (Figure 2.21b):

$$
C(r) = \begin{cases} 1 - 7\dfrac{r^2}{a^2} + \dfrac{35}{4}\dfrac{r^3}{a^3} - \dfrac{7}{2}\dfrac{r^5}{a^5} + \dfrac{3}{4}\dfrac{r^7}{a^7} & \text{if } r \le a, \\[4mm] 0 & \text{if } r \ge a \end{cases} \tag{2.49}
$$

This model is used for differentiable variables such as pressure fields or geopotential in meteorology (Royer, 1975), as well as potential fields in geological modeling (Chilès et al., 2005).

2. *A model for a twice m.s. differentiable SRF*, which could be called the "pentamodel" because its principal irregular term is in r^5. It corresponds to the Radon transform of order 4 of the spherical covariogram of $\mathbb{R}^7$:

$$
C(r) = \begin{cases} 1 - \dfrac{22}{3}\dfrac{r^2}{a^2} + 33\dfrac{r^4}{a^4} - \dfrac{77}{2}\dfrac{r^5}{a^5} + \dfrac{33}{2}\dfrac{r^7}{a^7} - \dfrac{11}{2}\dfrac{r^9}{a^9} + \dfrac{5}{6}\dfrac{r^{11}}{a^{11}} & \text{if } r \le a, \\[4mm] 0 & \text{if } r \ge a \end{cases} \tag{2.50}
$$

Its graph differs very little from that of the cubic model, except in the immediate vicinity of the origin. This model is to be used only if the physical conditions ensure that the studied variable is twice differentiable, an exceptional case in the usual applications of geostatistics.

Exponential Model and Derived Models

The exponential model with scale parameter $a > 0$ is defined by

$$
C(r) = \exp\left(-\frac{r}{a}\right) \tag{2.51}
$$

This model is a covariance in $\mathbb{R}^n$ for any n, since it corresponds to the positive spectral density

$$
f_n(\rho) = \frac{2^n \pi^{(n-1)/2}\Gamma\big((n+1)/2\big)a^n}{(1 + 4\pi^2 a^2 \rho^2)^{(n+1)/2}}
$$

[e.g., see Yaglom (1987, Vol. I, pp. 362–363)]. As shown by Figure 2.21c, the variogram reaches its sill only asymptotically when $r \to \infty$, and its practical range (95% of the sill, or equivalently a correlation of only 5%) is about $3a$.

In $\mathbb{R}$ the exponential model is the covariance of continuous-time Markov processes that possess the property of conditional independence between the past and the future when the present is known. More generally, the distribution of $Z(x)$ conditional on $Z(x_1), \ldots, Z(x_i), Z(x_{i+1}), \ldots, Z(x_N)$, for $x_1 < \cdots < x_i < x < x_{i+1} < \cdots < x_N$, only depends on the two neighbors $Z(x_i)$ and $Z(x_{i+1})$.

Since the exponential model is valid in $\mathbb{R}^n$ for any n, Radon transforms of $C(r)$ provide differentiable covariances that are also valid in $\mathbb{R}^n$ for all n, and in particular:

- for the Radon transform of order 2:

$$C_2(r) = \left(1 + \frac{r}{a}\right) \exp\left(-\frac{r}{a}\right) \tag{2.52}$$

- for the Radon transform of order 4:

$$C_4(r) = \left(1 + \frac{r}{a} + \frac{1}{3}\frac{r^2}{a^2}\right) \exp\left(-\frac{r}{a}\right) \tag{2.53}$$

Gaussian Model

The Gaussian model with scale parameter $a > 0$ defined by

$$C(r) = \exp\left(-\frac{r^2}{a^2}\right) \tag{2.54}$$

is a covariance in $\mathbb{R}^n$ for any n. Its practical range is about $1.73\,a$, as shown in Figure 2.21d. This model is associated with an infinitely differentiable SRF and thus is extremely regular. After a Radon transform of order m, the covariance remains Gaussian, as can be seen by applying (2.32): It only differs from the initial model by the multiplicative factor $(\pi a^2)^{m/2}$. This regularity gives the SRF a deterministic character, in that knowing the value of the SRF at 0 and the values of its partial derivatives of all orders at 0 determines the value of the SRF at any location x. Such regularity is hardly ever encountered in the earth sciences, and its use, without any other component, can lead to unacceptable predictions [see the example provided by Stein (1999, Section 6.9)]. This model has been used in combination with a nugget effect representing a microstructure, in meteorology for geopotential fields (Delfiner, 1973; Schlatter, 1975; Chauvet et al., 1976), and in bathymetry in areas where the seafloor surface is smooth due to water flow, erosion, and sedimentation (Herzfeld, 1989).

Cauchy Model and Cauchy Class

The gravity and magnetic fields are governed by the laws of physics. If the geometry, density, and magnetism of the sources are known, the corresponding fields can be determined. However, short of knowing these characteristics exactly, a statistical model of the main parameters allows the determination, if not of the fields, at least of their spectra. This has been studied by several authors—in particular, Spector and Bhattacharyya (1966) and Spector and Grant (1970). The expression of the spectrum is complex, but simpler expressions can be derived by making specific assumptions about the geometry of the sources.

A particularly interesting case is that of sources formed of parallelepipeds of random height, length, width, and orientation located at a random depth but close to an average depth $d = a/2$. The 2D spectra f_G and f_M of the gravity and magnetic fields created at the surface are then approximately of the following form:

$$f_G(\rho) = \alpha \frac{e^{-2\pi a\rho}}{\rho}, \qquad f_M(\rho) = \beta\, e^{-2\pi a\rho}$$

Through application of (2.25) we can deduce the corresponding covariances:

$$C_G(r) = \alpha'\left(1 + \frac{r^2}{a^2}\right)^{-1/2}, \qquad C_M(r) = \beta'\left(1 + \frac{r^2}{a^2}\right)^{-3/2}$$

We note the remarkable physical significance of the scale parameter a: it is equal to twice the average depth of the sources. These variogram models enable us to take into account both the physics and the geology of the problem. They also enable optimal gravity data transformations (Marcotte and Chouteau, 1993) and an optimal decomposition of the total field into several components, either by spectral methods or by cokriging (Chilès and Guillen, 1984; see Section 5.6.7).

These two models are in fact particular cases of the *Cauchy model*

$$C(r) = \left(1 + \frac{r^2}{a^2}\right)^{-\beta/2} \qquad (a > 0,\ \beta \geq 0) \tag{2.55}$$

which is a model in $\mathbb{R}^n$ for all n because it can be expressed in the form (2.27) [e.g., see Yaglom (1987, Vol. I, p. 365)]. This type of model is very regular near the origin, since its Taylor expansion only contains even terms and it reaches its sill slowly (Figure 2.21e).

The *Cauchy class* generalizes the Cauchy model to behaviors in r^α at the origin, with a covariance of the form

$$C(r) = \left(1 + \frac{r^\alpha}{a^\alpha}\right)^{-\beta/\alpha} \qquad (a > 0,\ 0 < \alpha \leq 2,\ \beta \geq 0)$$

where α is a shape parameter whereas β parametrizes the dependence at large distances. It is a valid model in $\mathbb{R}^n$ for all n [see Gneiting (1997), in $\mathbb{R}$; and Gneiting and Schlather (2004), in $\mathbb{R}^n$].

Matérn Model

The isotropic function (2.55) is positive and integrable in $\mathbb{R}^n$ for all n. Its n-dimensional Fourier transform is therefore a covariance in $\mathbb{R}^n$. Considering the case $\beta = 2\nu + n$, $\nu \geq 0$, leads to the Matérn model (also known as the *K-Bessel model*):

$$C(r) = \frac{1}{2^{\nu-1}\Gamma(\nu)} \left(\frac{r}{a}\right)^{\nu} K_{\nu}\left(\frac{r}{a}\right) \qquad (a>0, \nu \geq 0) \qquad (2.56)$$

where K_{ν} is the modified Bessel function of the second kind of order ν defined by (A.4) [Blanc-Lapierre and Fortet (1953), in $\mathbb{R}$; Matérn (1960, p. 18), in $\mathbb{R}^n$]. This model can have any type of behavior near the origin, since its principal irregular term behaves like $r^{2\nu}$ if ν is not an integer and like $r^{2\nu} \log r$ if ν is an integer (Figure 2.22a).

A Gaussian SRF $Z(x)$ of $\mathbb{R}^n$ with a covariance (2.56) with $\nu>0$ is a solution to the linear fractional stochastic partial differential equation

$$(1/a^2 - \Delta)^{\alpha/2} Z(x) = \varepsilon(x), \qquad \alpha = \nu + n/2, \quad \nu > 0$$

where Δ is the Laplacian, $(1/a^2 - \Delta)^{\alpha/2}$ is a pseudodifferential operator defined through its spectral properties, and $\varepsilon(x)$ is Gaussian white noise with unit variance (Whittle, 1963). The case $\nu = 1/2$ corresponds to the exponential model, which is the covariance of a Markov process in $\mathbb{R}$ but not in $\mathbb{R}^2$. Seeking a 2D Markov SRF model, Whittle (1954) introduced the covariance $(r/a) K_1(r/a)$, namely the special case $\nu = 1$. Among its early applications, let us mention its use by Rodríguez-Iturbe and Mejía (1974) to describe rainfall variability in hydrology. More generally the Matérn covariance is that of a Gaussian Markov random field of $\mathbb{R}^n$ when $\nu + n/2$ is integer (Lindgren et al., 2011) and is therefore a reference model in the theory of these random fields. A thorough presentation of the emergence of the Matérn model in very different contexts is given by Guttorp and Gneiting (2006).

$|h|^{\alpha}$ Model

The *power-law model*

$$\gamma(r) = r^{\alpha} \qquad (2.57)$$

is a variogram, provided that $0 < \alpha < 2$. This has been shown by Schoenberg (1938b) and Kolmogorov (1940b) in $\mathbb{R}$, and, for example, by Yaglom (1987, Vol. I, pp. 406–407) in $\mathbb{R}^n$.

Indeed the function $|h|^{\alpha}$ is obtained through application of (2.21) with

$$\chi(du) = 4\pi^{2-\alpha-n/2} \frac{\Gamma\left(\frac{\alpha+n}{2}\right)}{-\Gamma\left(-\frac{\alpha}{2}\right)} |u|^{2-\alpha-n} \, du$$

where $\Gamma(\cdot)$ is the Euler gamma function (A.1). The condition $0 < \alpha < 2$ ensures that the measure χ is positive and that the condition (2.22) is satisfied.

This model does not have a sill (Figure 2.22b). The extreme cases of $\alpha = 0$ and $\alpha = 2$ correspond, respectively, to a pure nugget effect and to a linear RF

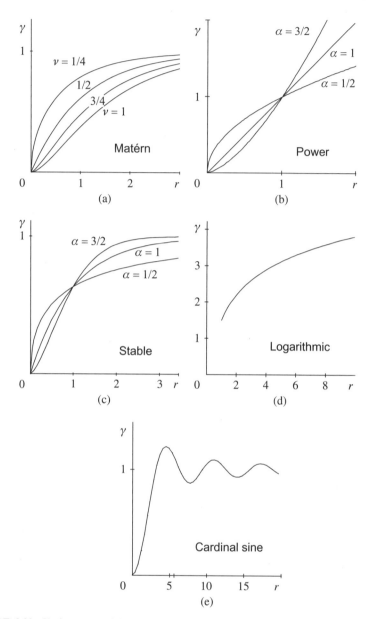

FIGURE 2.22 Variogram models with unit sill and scale parameter—2: (a) Matérn; (b) $|h|^{\alpha}$; (c) stable; (d) regularized logarithmic; (e) cardinal sine.

with a random slope. For $\alpha = 1$ we obtain the much used linear variogram $\gamma(h) = |h|$.

The $|h|^{\alpha}$ model satisfies a property of similarity: It is invariant under a change of the scale of observation. If we go from r to sr, where $s > 0$ is a scale factor, we have $\gamma(s\,r) = s^\alpha \gamma(r)$. Therefore we cannot associate a characteristic scale to the phenomenon. This model is the only one with such property.[21]

The $|h|^\alpha$ variogram has close ties with fractals. We have all seen spectacular computer-generated images of fractal landscapes (Mandelbrot, 1975b, 1977, 1982). They are extremely rugged surfaces with a noninteger Hausdorff dimension comprised between 2 and 2.5 while their topological dimension is 2. Fractal landscapes are simulations of Gaussian IRFs with variograms of type $|h|^\alpha$, which Mandelbrot named "fractional Brownian" random functions, because they generalize the Brownian motion, or Wiener process, obtained in 1D for $\alpha = 1$ [see Mandelbrot and Van Ness (1968)]. They are defined for $0 < \alpha < 2$, but $\alpha > 1$ is used for simulating relief (α is related to the standard Hurst exponent H of fractal theory by $\alpha = 2\,H$). A fractional Brownian random function of $\mathbb{R}^n$ in $\mathbb{R}^{n+1}$ (the space $\mathbb{R}^n$ plus the coordinate z) is a surface with fractal (or Hausdorff) dimension $D = n + 1 - \alpha/2$. The situation may be very different for non-Gaussian random functions. For example, the RF model based on Poisson hyperplanes presented at the end of Section 7.5.2 has a linear variogram, but its realizations are not fractal.[22]

While on the subject of fractals it may be useful to point out a common confusion between two different properties: (a) an extreme irregularity of detail and (b) the property to repeat itself at all scales. Because these properties are both present with the fractional Brownian random function and with many nonrandom fractal constructions, they are sometimes assumed to go together and lead to the erroneous conclusion, for example, that since a phenomenon is very irregular it ought to be self-similar or self-affine. In fact the fractal character is mathematically defined by a Hausdorff dimension exceeding the topological dimension—that is, by the irregular character of the surface. If we add any other component to a fractal $Z(x)$, be it a smooth SRF or another fractal with a different α, the result is still fractal but no longer self-affine; on the other hand, a plane is a self-similar surface that is obviously nonfractal. In the Gaussian case it is the $|h|^\alpha$ behavior *near the origin*, $0 < \alpha < 2$, which is related to the fractal character so that, for example, any realization of a Gaussian SRF with a spherical or exponential covariance is fractal without being self-affine. Again the situation is different for non-Gaussian random functions.[23]

---

[21] Note that this property is in fact *self-affinity* rather than *self-similarity* (Mandelbrot, 1985): If we consider the surface defined by $Z(x)$, $x \in \mathbb{R}^2$, in the Gaussian case (fractional Brownian random function) the invariance is ensured by performing two different scale changes: a factor s in the horizontal plane and a factor $s^{\alpha/2}$ in the vertical plane. Isoline curves, which are horizontal planar sections, are self-similar. Thus the compass method (also known as the yardstick method) can be applied to determine the fractal dimension of a fractional Brownian surface from its contours in horizontal cuts (cf. the well-known example of the length of coastlines) but not from vertical cross sections.

[22] It seems that the madogram, or variogram of order 1, is the right tool to diagnose the fractal character (then the realizations define fractal surfaces with fractal dimension $D = n + 1 - H$ if $\gamma_1(h) = b\,|h|^H$, $0 < H < 1$).

[23] The criterion should be based on the behavior of the madogram *at the origin*.

In 1D the Brownian motion or Wiener process ($\alpha = 1$) is a process with independent and stationary Gaussian increments. Its discrete-time equivalent is the random walk (Section 7.2.1). For $\alpha \neq 1$ the successive increments $Z(x) - Z(x-h)$ and $Z(x+h) - Z(x)$ have a correlation coefficient $\rho = 2^{\alpha-1} - 1$ regardless of the value of h, a remarkable property characterizing this model. Thus a fractional Brownian motion has long-term memory. For $\alpha > 1$ the successive increments are positively correlated, which corresponds to a phenomenon of persistence: As Mandelbrot phrases it, the curve tends "to persist in any direction upon which it has embarked." For $\alpha < 1$, on the contrary, we have antipersistence (limited to $\rho > -0.5$): The curve tends to turn back constantly toward the point it came from.

The $|h|^{\alpha}$ model exhibits a large variety of behaviors near the origin, but it has no range. The application of the theorem characterizing variograms (Section 2.3.3) provides us with a model that has the same behavior near the origin and reaches a sill (Figure 2.22c). Indeed the *powered exponential*

$$C(r) = \exp\left(-\left(\frac{r}{a}\right)^{\alpha}\right) \qquad (a > 0,\ 0 < \alpha \leq 2) \qquad (2.58)$$

is a covariance in $\mathbb{R}^n$ for all n. Since in $\mathbb{R}$ this function is the characteristic function of a stable random variable, this covariance is also named the *stable model* [see Schoenberg (1938b) and Yaglom (1987, Vol. II, Chapter 4, Note 50)]. The exponential and Gaussian models belong to this family.

The $|h|^{\alpha}$ model and its variant play an important role in the theory of turbulence and its application to meteorology: Kolmogorov (1941a,b) and Obukhov (1941, 1949b) have shown from theoretical considerations that in a fully developed turbulence the velocity components have variograms of type $|h|^{\alpha}$ with $\alpha = 2/3$ at short and medium distances, which has been confirmed experimentally [see also Monin and Yaglom (1965, Chapter 8)]. At larger distances, however, the variogram reaches a sill and the stable model or the Matérn model with $\nu = 1/3$ have been used (Gandin, 1963, p. 51).[24] The situation is similar for the geopotential height of isobaric surfaces, as reported by Gandin (1963, p. 44), this time with $\alpha = 5/3$ and $\nu = 5/6$, but twice-differentiable models are also used. Blanc-Lapierre and Fortet (1953, pp. 453–454) use the Matérn model for electrical noises.

Logarithmic Model

The logarithmic model

$$\gamma(r) = \log r \qquad (2.59)$$

is known under the name of *de Wijs model*. The function does not vanish at zero but has a value $-\infty$. In reality this model is used for describing not point

---

[24] At very short distances the variogram is no longer of type $|h|^{2/3}$ but is parabolic. This behavior can be modeled by regularization of an $|h|^{2/3}$, stable, or Matérn model.

variables but variables regularized by a sampling support. Since a measurement is always based on a support which, no matter how small, is not strictly a point, this restriction is not constraining. By application of (2.37), the variogram $\gamma_p(h)$ associated with such sampling function $p(x)$ is itself an ordinary variogram satisfying $\gamma_p(0) = 0$ and $\gamma_p(h) > 0$ when $|h| > 0$. Figure 2.22d shows the variogram regularized by a segment of length $l = 1$, calculated along a direction orthogonal to the segment. As soon as r is sufficiently large (e.g., greater than $2l$), this variogram is approximately

$$\gamma_p(r) \approx \log\left(\frac{r}{l}\right) + \frac{3}{2}$$

From a mathematical point of view, $\log r$ is a variogram model for a random distribution (in the sense of pseudofunctions), but as soon as one passes to regularized variables, this random distribution becomes a standard RF (Matheron, 1965, pp. 173, 242).

In $\mathbb{R}^2$ the logarithmic model has a Markov property similar to that of the linear model in $\mathbb{R}$: If the IRF is Gaussian and known over a closed contour $\mathcal{C}$, there is independence between the inside and outside of $\mathcal{C}$ (Matheron, 1970, Chapter 3, exercise 9). Besag and Mondal (2005) have shown that there is a close relationship between the de Wijsian Gaussian IRF in $\mathbb{R}^2$ and first-order intrinsic Gaussian Markov random fields on regular lattices in $\mathbb{Z}^2$, similar to the close correspondence between the Brownian motion and first-order random walks in the one-dimensional case. This property can have important applications since it enables a close approximation to the de Wijsian Gaussian IRF on a 2D lattice [also see Rue and Held (2005, p. 107)].

Because of its good analytical properties, the logarithmic model was widely used by the pioneers of geostatistics (Matheron, 1962; Carlier, 1964; Formery, 1964) in applications to deposits of gold and uranium and also of bauxite and base metals. This model has one remarkable property: If the variogram of the RF defined in $\mathbb{R}^n$ is

$$\gamma(r) = n\,\alpha\,\log r$$

and if v and V are two geometrically similar sets of $\mathbb{R}^n$, the dispersion variance within V of the grades of samples of size v (see Section 2.8.2) is given by

$$\sigma^2(v|V) = \alpha\,\log\left(\frac{|V|}{|v|}\right) \tag{2.60}$$

This expresses a property of very strong similarity: The variance of the small blocks v within the large block V depends only on the ratio $|V|/|v|$ of volumes, regardless of the scale (*de Wijs formula*). The parameter α is called the absolute dispersion, since it characterizes the dispersion of the phenomenon independently of the geometry of the domain V and the support v.

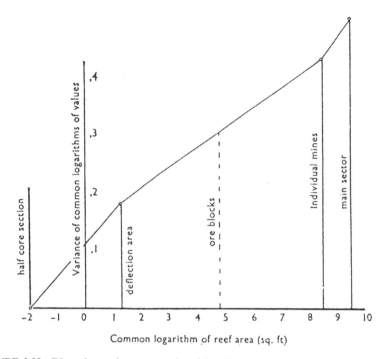

FIGURE 2.23 Dispersion variance versus size of domain: example of gold data from the Orange Free State. The horizontal axis represents the area of the domain in logarithmic scale, from 10^{-2} to 10^{10} ft^2. [From Krige (1952).]

This type of behavior was revealed by Krige (1952) when studying the large Orange Free State gold deposit which is subdivided into 10 mines: The variance of borehole grades (or more precisely the variance of the logarithms of the gold accumulations[25]) within the groups of boreholes, then within the mines, and finally within the whole deposit (300 km^2) increases as the logarithm of the surface area of the zone in which the variance is calculated (Figure 2.23). Most applications of the logarithmic model have been related to the gold deposits of South Africa (Krige, 1978).

The logarithmic model is called the de Wijsian model in honor of the work of H. J. de Wijs (1951, 1953) on the distribution of grades in mine deposits. De Wijs derived formula (2.60) by postulating a principle of similarity by which, when a block V with grade z_V is divided in two equal parts, the two half-blocks have grades $(1 - d)\, z_V$ and $(1 + d)\, z_V$, and this coefficient d does not depend on the size of V. When splitting the blocks further up to elementary blocks of size v, it is shown that the grades follow a logbinomial distribution and that their logarithms (rather than the grades themselves) have a variance of

---

[25] The accumulation along a vertical borehole is the product of the ore thickness and the average ore grade in the borehole; it represents a quantity of metal per unit surface.

the form (2.60). Therefore for a point support this model has close ties with the lognormal distribution (Matheron, 1955; 1962, pp. 308–311; or 1987b). Its principle of similarity allows it to be considered as a fractal model (Mandelbrot, 1982, pp. 376–377), or more precisely a multifractal model (Evertsz and Mandelbrot, 1989).

Hole Effect Models

Because the magnitude of a covariance $C(h)$ is bounded by its value at the origin $C(0)$, the maximum hole effect is reached when the covariance reaches $-C(0)$. The effect is obtained in $\mathbb{R}$ with the periodic covariance

$$C(h) = \cos(h/a) \qquad (h \in \mathbb{R}, \quad a > 0) \qquad (2.61)$$

This model is not strictly positive definite, since the variance of $Z(x + 2\pi a) - Z(x)$, equal to $2 (C(0) - C(2\pi a))$, is here equal to zero. Since this model has no damping factor, it is often combined with the exponential model, which gives

$$C(h) = \exp(-|h|/a_1) \cos(h/a_2) \qquad (h \in \mathbb{R}, \quad a_1, a_2 > 0) \qquad (2.62)$$

In higher dimensions an isotropic covariance cannot give such strong hole effect as (2.61). As a consequence of (2.24), the maximum hole effect in $\mathbb{R}^n$ is obtained with the *J-Bessel model*:

$$C_n(r) = \kappa_n(r/a) = 2^{n/2-1} \, \Gamma(n/2)(r/a)^{1-n/2} \, J_{n/2-1}(r/a) \qquad (a > 0) \qquad (2.63)$$

The ratio $C(r)/C(0)$ for an isotropic covariance cannot be less than -0.403 in $\mathbb{R}^2$, -0.218 in $\mathbb{R}^3$, and -0.133 in $\mathbb{R}^4$ (Matérn, 1960, p. 16).

These covariances are very regular: Their Taylor expansions include only even terms. Their behavior at infinity is a sine wave multiplied by $1/r^{(n-1)/2}$. Letting $n = 1$ in (2.63) gives back (2.61). For $n = 3$ we obtain the *cardinal-sine model*:

$$C(r) = (a/r) \sin(r/a) \qquad (h \in \mathbb{R}^3) \qquad (2.64)$$

The minimum value (-0.218) is reached for $r \approx 4.50 \, a$ (Figure 2.22e).

Since the model (2.62) is the product of an exponential model, which can be extended to $\mathbb{R}^n$, and a cosine model, which is valid only in $\mathbb{R}$ (as a function of $r = |h|$), we could imagine that it is valid only in $\mathbb{R}$. This model can nevertheless be extended to an isotropic model in $\mathbb{R}^n$ under a condition on the parameters. For example,

$$C(r) = \exp(-r/a_1) \cos(r/a_2) \qquad (2.65)$$

is a covariance in $\mathbb{R}^2$ if and only if $a_2 \geq a_1$, and it is a covariance in $\mathbb{R}^3$ if and only if $a_2 \geq a_1 \sqrt{3}$ (Yaglom, 1987, Vol. I, p. 366).

2.5.2 Anisotropic Models

Geometric Anisotropy

Geometric anisotropy is obtained by simple stretching of an isotropic model. By definition, a variogram in $\mathbb{R}^n$ displays a geometric anisotropy if it is of the form

$$\gamma(h) = \gamma_0(\sqrt{\mathbf{h}'\mathbf{Q}\mathbf{h}}) \tag{2.66}$$

where γ_0 is an isotropic model and $\mathbf{Q}$ a $n \times n$ positive definite matrix. For clarity the $n \times 1$ matrix $\mathbf{h}$ represents the components $\{h_i : i = 1, \ldots, n\}$ of vector h. The eigenvalues $\{b_i^2 : i = 1, \ldots, n\}$ of matrix $\mathbf{Q}$ and the eigenvectors define a new orthogonal coordinate system in which the quadratic form $\mathbf{h}'\mathbf{Q}\mathbf{h}$ can be expressed as a sum of squares $\sum_{i=1}^{n} b_i^2 \tilde{h}_i^2$ of the new components $\{\tilde{h}_i : i = 1, \ldots, n\}$ of vector $\mathbf{h}$. Combining a change of coordinate system with a scaling of $\tilde{h}_i$ into $\hat{h}_i = b_i \tilde{h}_i$ restores isotropy (b_i is chosen positive). The variogram can therefore be written as

$$\gamma(h) = \gamma_0(|\mathbf{A}\,\mathbf{h}|) \tag{2.67}$$

where the matrix $\mathbf{A}$ defines the transformation from the initial space to the isotropic space. The simplest case is when the anisotropy axes (i.e., the eigenvectors) coincide with the coordinate axes. No rotation is necessary, and thus (2.67) amounts to

$$\gamma(h) = \gamma_0\left(\sqrt{\sum_{i=1}^{n} b_i^2 h_i^2}\right)$$

or equivalently the matrix $\mathbf{A}$ is simply the diagonal matrix of the b_i.

The expression of $\mathbf{A}$ is more complex when it includes a rotation. In $\mathbb{R}^2$, denoting by θ_1 and $\theta_1 + \pi/2$ the main directions of anisotropy, a rotation θ_1 defines a new coordinate system with axes parallel to the anisotropy directions. The matrix $\mathbf{A}$ is then

$$\mathbf{A} = \begin{bmatrix} b_1 & 0 \\ 0 & b_2 \end{bmatrix} \begin{bmatrix} \cos\theta_1 & \sin\theta_1 \\ -\sin\theta_1 & \cos\theta_1 \end{bmatrix}$$

If $\gamma_0(r)$ is a linear variogram with unit slope, b_1 and b_2 are the slopes of $\gamma(h)$ along directions θ_1 and $\theta_1 + \pi/2$, and the graph of the reciprocal $1/b(\theta)$ of the slope as a function of θ describes an ellipse. Similarly, when $\gamma_0(r)$ is a transition model with unit range, $\gamma(h)$ has range $a_1 = 1/b_1$ in the direction θ_1 and range $a_2 = 1/b_2$ in the direction $\theta_1 + \pi/2$, and the graph of the range $a(\theta)$ describes an ellipse. The isovariogram curves are also concentric ellipses, so that this type of anisotropy is named elliptic anisotropy. Usually the anisotropy is described by the ranges a_1 and a_2 rather than by the parameters b_1 and b_2. θ_1 is often taken as the direction of maximum range, and a_1/a_2 is named the *anisotropy ratio*.

The generalization to $\mathbb{R}^3$ is straightforward when one of the main anisotropy axes is the vertical, which is often the case. In the general case, however, denoting the coordinate axes by Ox, Oy and Oz, the definition of a coordinate system involves three successive rotations: (1) a rotation θ_1 around the Oz axis, leading to new Ox' and Oy' axes; (2) a rotation θ_2 around the Ox' axis so that the Oz axis comes to its final position Oz''; and (3) a rotation θ_3 around this Oz''; axis to place the other two axes in their final position. In this final system, three anisotropy parameters b_1, b_2, and b_3 are associated with the main directions. The matrix $\mathbf{A}$ is therefore

$$\mathbf{A} = \begin{bmatrix} b_1 & 0 & 0 \\ 0 & b_2 & 0 \\ 0 & 0 & b_3 \end{bmatrix} \begin{bmatrix} \cos\theta_3 & \sin\theta_3 & 0 \\ -\sin\theta_3 & \cos\theta_3 & 0 \\ 0 & 0 & 1 \end{bmatrix} \begin{bmatrix} 1 & 0 & 0 \\ 0 & \cos\theta_2 & \sin\theta_2 \\ 0 & -\sin\theta_2 & \cos\theta_2 \end{bmatrix} \begin{bmatrix} \cos\theta_1 & \sin\theta_1 & 0 \\ -\sin\theta_1 & \cos\theta_1 & 0 \\ 0 & 0 & 1 \end{bmatrix}$$

Isovariogram surfaces are now ellipsoids, and likewise the range or the inverse of the slope as a function of direction is represented by an ellipsoid.

Generalizations of this anisotropy model can be defined, but there are consistency requirements to ensure that these are valid. For example, the slope of a linear variogram cannot vary arbitrarily with direction, even if continuously. For a power variogram, Matheron (1975a, p. 96), shows that $\varphi(h/|h|)\,|h|^\alpha$ is a variogram if and only if there exists a necessarily unique symmetric nonnegative measure μ on the unit sphere S_0 of $\mathbb{R}^n$ such that

$$\varphi(s) = \int_{S_0} |\langle s, s'\rangle|^\alpha \, \mu(ds') \qquad (s \in S_0)$$

This condition equivalently ensures that $\exp(-\varphi(h/|h|)\,|h|^\alpha)$ is a valid covariance, as a consequence of the theorem on the characterization of variograms.

For a linear variogram ($\alpha = 1$) in the 2D space ($n = 2$), $\varphi(s)$ is a symmetric function $b(\theta)$ representing the slope of the variogram in the direction θ, and the above condition is equivalent to the condition that the graph of the polar function $\rho(\theta) = 1/b(\theta)$ describes the boundary of a symmetric, closed, and convex set. This is obviously the case for a geometric anisotropy, where the graph of $1/b(\theta)$ describes an ellipse. For $n > 2$ the convexity property is no longer sufficient [refer to (Matheron, 1975a, pp. 96–98) for a precise characterization].

Zonal Anisotropy

In this model, also called *stratified anisotropy*, the variogram depends only on some components of the vector h (possibly after an appropriate change of coordinate system). In 3D the simplest case is when the variogram only depends on the vertical component h_z of vector h, namely is of the form $\gamma(h) = \gamma_0(h_z)$. The variogram in a direction θ with respect to the Oz axis is then $\gamma_\theta(r) = \gamma_0(r\cos\theta)$: If γ_θ has sill C and range a, γ_θ has sill C and range $a/\cos\theta$, except in any direction orthogonal to Oz where the variogram is identically zero. This is the variogram of a variable that remains constant in any horizontal plane and thus has a layered aspect. Another simple case is a variogram that only depends on the horizontal components h_x and h_y of h: the variable is constant along any vertical.

In practice a real phenomenon can very seldom be represented by a pure zonal model. But we often have several components $\gamma_j(h)$, at least one of which is zonal. In $\mathbb{R}^n$, when the elementary variogram $\gamma_j(h)$ is zonal and depends on only m components of the vector h, it must be a model admissible in $\mathbb{R}^m$, and not necessarily in $\mathbb{R}^n$. This model can of course have its own geometric anisotropy in $\mathbb{R}^m$. If the various components have sills C_j, $\gamma(h)$ has a sill $\Sigma\ C_j$, except in directions orthogonal to those of the components showing the zonal anisotropy. For example, if we have in $\mathbb{R}^3$

$$\gamma(h) = \gamma_1\left(\sqrt{h_x^2 + h_y^2 + h_z^2}\right) + \gamma_2\left(\sqrt{h_x^2 + h_y^2}\right) + \gamma_3(h_z)$$

the sill is $C_1 + C_2$ in the horizontal directions and is $C_1 + C_3$ in the vertical direction, whereas it is equal to $C_1 + C_2 + C_3$ in all other directions. This makes it easy to test the zonal character of an anisotropy.

One must beware of models that partition the coordinates, such as

$$\gamma(h) = \gamma_1(h_x) + \gamma_2(h_y) \qquad \text{in } \mathbb{R}^2$$

or

$$\gamma(h) = \gamma_1\left(\sqrt{h_x^2 + h_y^2}\right) + \gamma_2(h_z) \qquad \text{in } \mathbb{R}^2$$

The first model is obtained for an RF of the type $Z(x, y) = Z_1(x) + Z_2(y)$, where x and y are the coordinates of point x. For, such an RF, certain linear combinations can have a zero variance. Such is the case with

$$Z(x, y) - Z(x, y + v) - Z(x + u, y) + Z(x + u, y + v)$$

because for such an RF the linear combination is itself zero. Thus $-\gamma(h)$ is not a strictly conditionally positive definite function. Such a model is used only if it is imposed by the physics of the problem (here the sum of two 1D structures).

Other Anisotropies

Another variant is the *separable covariance*, also called *factorized covariance*, whose general form in $\mathbb{R}^n$ is

$$C(h) = \prod_{i=1}^{n} C_i(h_i) \tag{2.68}$$

where the h_i are the n components of vector h and the $C_i(h_i)$ are covariances in $\mathbb{R}$. This is the model obtained with the product of n one-dimensional independent zero-mean RFs along the various coordinate axes. It has nice conditional independence properties (cf. Sections 3.6.1 and 5.8.1).

No model allows the sill to vary continuously with the direction u of the vector h. If $\gamma_0(h)$ is an isotropic transition model, $\sigma^2(u)\,\gamma_0(h)$ is a variogram if and only if the symmetric function $\sigma^2(u)$ is constant almost everywhere (otherwise the stationary covariance $C(h) = C(0) - \sigma^2(u)\,\gamma_0(h)$ would not satisfy Slutsky's condition (1.12) and thus could not be the covariance of a stationary and ergodic random function). It may happen that a sample variogram does not satisfy this property. If the effect exceeds what can be explained by mere fluctuations, this does not mean that our anisotropy models are not rich enough but rather that the studied regionalized variable cannot derive from a stationary model.

2.5.3 Internal Consistency of Models

As soon as one departs from the Gaussian model, it is no longer legitimate to use just any covariance or variogram function. The proof that a positive definite function is a covariance is based on the construction of a Gaussian SRF with this covariance, but it does not establish that an SRF with any other type of spatial distribution can have this covariance. The same holds true for IRFs. Although a model is never the reality, it should at least be internally consistent. This question was studied by Matheron (1987a). We will simply present some examples often found in practical applications: indicators and lognormal SRFs. But let us first introduce a new concept that generalizes the variogram and is useful to characterize internal consistency: the variogram of order α.

Variogram of Order α

If instead of considering the squares of increments, we consider their magnitudes, or more generally their magnitudes raised to the power $\alpha > 0$, we obtain the variogram of order α (Matheron, 1987a) defined by

$$\gamma_\alpha(h) = \tfrac{1}{2}\mathrm{E}|Z(x+h) - Z(x)|^\alpha \qquad (2.69)$$

For $\alpha = 2$ we have the ordinary variogram, and for $\alpha = 1$ the *madogram*, namely the expected value of the magnitude of increments. Matheron (1987a) shows that any madogram, and more generally any variogram of order α, $0 < \alpha < 2$, is an ordinary variogram (i.e., $-\gamma_\alpha$ is a conditionally positive definite function). But this cannot be any model. For example, the inequality

$$|Z(x+h+h') - Z(x)| \le |Z(x+h+h') - Z(x+h)| + |Z(x+h) - Z(x)|$$

entails that the madogram satisfies the triangular inequality

$$\gamma_1(h+h') \le \gamma_1(h) + \gamma_1(h') \qquad (2.70)$$

More generally, a variogram of order α, $\alpha > 0$, should satisfy the necessary (but not sufficient) condition

$$[\gamma_\alpha(h + h')]^{1/\alpha} \leq [\gamma_\alpha(h)]^{1/\alpha} + [\gamma_\alpha(h')]^{1/\alpha}$$

A sufficient condition is the following: If $\gamma(h)$ is a variogram of order 2, then $\gamma(h)^{\alpha/2}$, $\alpha > 0$, is a variogram of order α. In particular, $\sqrt{\gamma(h)}$ is a madogram.

In practice, however, the variogram of order α is rarely studied for itself and is modeled with reference to the associated variogram of order 2, either for a robust estimation of the latter (see Section 2.2.5) or for modeling the bivariate distributions (see Section 6.4.4). For example, for a diffusive SRF (e.g., a Gaussian random function) the variogram of order 1 is, up to a multiplicative factor, the square root of the variogram of order 2, whereas for a mosaic SRF (e.g., i.i.d. random values assigned to the cells of a random partition) these two variograms (and all the variograms of order α, provided that they exist) are proportional.

Variogram of a Random Set

Let us consider a stationary random set, or equivalently its indicator which will be denoted by $I(x)$, to remind us that it is a binary SRF. Obviously it satisfies

$$[I(x + h) - I(x)]^2 = |I(x + h) - I(x)|$$

so that its variogram is identical to its madogram and must therefore satisfy the triangular inequality (2.70):

$$\gamma(h + h') \leq \gamma(h) + \gamma(h') \tag{2.71}$$

If this variogram behaves like $|h|^\alpha$ at short distances, then necessarily $0 < \alpha \leq 1$—Indeed, for small $|h|$ (2.71) entails $\gamma(2h) \leq 2\gamma(h)$, hence the result. In particular, an indicator cannot be m.s. differentiable. The case $\alpha = 1$ corresponds to a set whose boundary has a specific surface area of finite expectation, whereas for $\alpha < 1$ the boundary is of fractal type; that is, its Hausdorff dimension in $\mathbb{R}^n$ is greater than $n - 1$.

Condition (2.71) is necessary but not sufficient. A stricter set of conditions is proved by Matheron (1993) in a more general setting than the stationary case, the variogram being considered as a separate function of x and $x + h$ and not simply of h: A necessary condition for a function $\gamma(x, x')$ to be the variogram of an indicator is that for all $m \geq 2$ and for any configuration $x_1, \ldots, x_m$, the values $\gamma_{ij} = (x_i, x_j)$ satisfy

$$\sum_{i=1}^{m} \sum_{j=1}^{m} \varepsilon_i \varepsilon_j \gamma_{ij} \leq 0 \qquad \forall \varepsilon_i \in \{-1; 0; 1\} \text{ such that } \sum_{i=1}^{m} \varepsilon_i = 1$$

Matheron formulates the conjecture that this condition is also sufficient. These conditions are nevertheless not easy to use. In practice, the only covariances that are known to be admissible for indicator functions are those of known random sets.

In 1D a well-known example is obtained by assigning i.i.d. values 0 or 1 to the segments delimited by a Poisson point process, which yields an exponential covariance whose scale parameter is the inverse of the intensity of the Poisson point process. It is extended to the n-dimensional space by assigning i.i.d. values to the Poisson polyhedra delimited by a Poisson hyperplane process, which yields an isotropic exponential covariance (cf. Section 7.6.5).

A more subtle example in 1D is the covariance

$$C(h) = \sigma^2 \exp(-|h|/a_1) \cos(h/a_2) \tag{2.72}$$

It cannot be the covariance of a random set if $a_1 > a_2$, because in this case the corresponding variogram does not satisfy the triangular inequality (2.70). On the other hand, if $a_1 \leq a_2$, it is possible to construct a random set with this covariance. Indeed, if $a_1 = a_2 = a$ and $\sigma^2 = 1/4$, (2.72) is the covariance of an *alternating process* based on a gamma renewal process with shape parameter $\alpha = 2$ (renewal processes are described in Section 7.6.5).[26] Now, if we intersect this random set with an independent random set having an exponential covariance, the resulting covariance is the product of covariances and is of the form (2.72) with $a_1 < a_2$. Note that the graph of this model is a damped sine wave.

Another example of an unusual model is the covariance of a process used to model the concentration of minerals in the vicinity of germs and therefore called a *migration process* (Haas et al., 1967)[27]: Denoting by p the proportion of 1's and by $q = 1 - p$ the proportion of 0's, its covariance is

$$C(h) = \frac{pq}{q-p} \left[q \exp\left(-\frac{|h|}{pa}\right) - p \exp\left(-\frac{|h|}{qa}\right) \right] \tag{2.73}$$

namely the difference of two exponential functions. In the case $p = q = 1/2$, this expression is not valid, and a direct calculation (Lantuéjoul, 1994) gives

$$C(h) = \frac{1}{4} \left(1 - \frac{2|h|}{a} \right) \exp\left(-\frac{2|h|}{a}\right) \tag{2.74}$$

These models display a hole effect. A more general family of models, this time in $\mathbb{R}^n$, is provided by Boolean random sets (Section 7.7.1): their covariances are of the form

$$C(h) = e^{-2\lambda K(0)} \left(e^{\lambda K(h)} - 1 \right) \tag{2.75}$$

where λ is a positive intensity and $K(h)$ a geometric covariogram or a mixture of geometric covariograms. Conversely, a simple covariance model such as the spherical model cannot be put in that form and therefore cannot be the covariance of a Boolean random set of $\mathbb{R}^3$ (Emery, 2010b).

---

[26] Because a gamma random variable with shape parameter $\alpha = 2$ can be considered as the sum of two independent random variables with the same exponential distribution, this alternating process can be defined as follows: (1) Start from a Poisson point process with intensity $\lambda = 1/a$, (2) merge the segments delimited by these points two by two (choose randomly whether the segment containing the origin is merged with the preceding or the following segment), and (3) assign the values 0 and 1 alternately to each of these new segments (choose randomly whether the new segment containing the origin is valued 0 or 1). The covariance of this random set can be obtained by considering that the bivariate distribution of $Z(x)$ and $Z(x+h)$ depends on the number of points of the initial Poisson point process falling between x and $x + h$, which is a Poisson random variable with mean $\theta = |h|$.

[27] This migration process is defined as follows: (1) Consider a Poisson point process with intensity $\lambda = 1/a$, (2) subdivide each interval of this point process into two unequal parts, corresponding to the proportions p (left) and $q = 1 - p$ (right) of the initial interval, and (3) assign the value 1 to the left part and the value 0 to the right part.

Variogram of an Indicator Variable

Other random set models are obtained from an SRF with a continuous marginal distribution: for any variable $Z(x)$ we can define the z-level *excursion set*, or equivalently the indicator function associated with the threshold z, by

$$I(x; z) = 1_{Z(x) < z} = \begin{cases} 1 & \text{if } Z(x) < z, \\ 0 & \text{otherwise} \end{cases}$$

Indicator variograms have two types of application:

1. *Calculation of Robust Variograms.* Except in the case of a high threshold, the indicator is not sensitive to high values. Therefore the variogram of an indicator—for example, that associated with the median threshold—is sometimes used as robust variogram.

2. *Indicator Kriging.* Following Journel (1982), numerous authors use kriging estimates of the indicator $I(x; z)$ to approximate the probability that $Z(x_0) < z$ at some point x_0 conditionally on the neighboring data.

However, there are caveats:

1. In the case of a long-tailed distribution, the variogram of the median indicator may be effective at revealing the underlying structure. But for a distribution with little spread, the gain in robustness often comes with a loss of structure, as will be seen next in the Gaussian case.

2. The variogram of the indicator usually becomes destructured at extreme thresholds (Matheron, 1982b). The use of indicator kriging can therefore lack performance. We will return to this point in Section 6.3.3.

From the point of view of variographic analysis the variogram of an indicator is not the same as the variogram of Z. For example, as is proved in Section 6.4.3, the covariance $C(h; z)$ of the indicator $I(x; z)$ of a standard bi-Gaussian SRF $Z(x)$ is related to the correlogram $\rho(h)$ of $Z(x)$ by

$$C(h; z) = \frac{1}{2\pi} \int_0^{\rho(h)} \exp\left(-\frac{z^2}{1+u}\right) \frac{du}{\sqrt{1-u^2}} \tag{2.76}$$

In particular, for the median threshold $z = 0$ this gives the well-known formula

$$C(h; 0) = \frac{1}{2\pi} \arcsin \rho(h) \tag{2.77}$$

Note that the same *centered* covariances are obtained for the indicator $1 - I(x; z)$ of the event $Z(x) \geq z$.

The models defined by (2.76) are different from those commonly used. Figure 2.24 shows graphs of the correlation of indicators as functions of ρ for

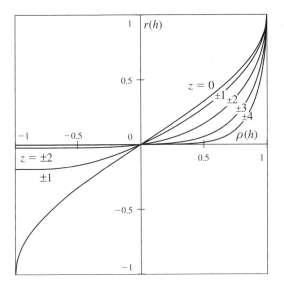

FIGURE 2.24 Graph of correlogram values $r(h) = C(h; z)/C(0; z)$ of the indicator $1_{Z(x)<z}$ (or equivalently of $1_{Z(x)\geq z}$) of a standard bi-Gaussian SRF as a function of correlogram values $\rho(h)$ of the bi-Gaussian SRF, for $z = 0, \pm1, \pm2, \pm3, \pm4$.

various thresholds (note the destructuring effect as $|z|$ increases). For a given threshold z, the indicator covariance $C(h; z)$ is an increasing function of ρ and can therefore be inverted. However, the function $\rho(h)$ obtained by applying (2.76) to an arbitrary indicator function is not necessarily a covariance because not all random sets can be obtained by truncation of a Gaussian SRF. Lantuéjoul (2002, p. 216) proved that the function

$$\rho(h) = \sin\left(\frac{\pi}{2}\exp\left(-\frac{|h|}{a}\right)\right)$$

is a valid correlogram in $\mathbb{R}^n$ for all n. Thus the exponential covariance $C(h) = \frac{1}{4}\exp(-|h|/a)$ can be regarded as the indicator covariance of the median-level excursion set of a Gaussian SRF. On the contrary, the spherical covariance cannot be regarded as the covariance of an excursion set of a Gaussian SRF, even in one-dimensional space, irrespective of the threshold z considered (Emery, 2010b). The same holds for the circular and triangular covariances.

Coming back to (2.76), we note that for ρ close to 1 (i.e., small h and no nugget effect) we have approximately

$$\gamma(h; z) = C(0; z) - C(h; z) \approx \frac{1}{\pi\sqrt{2}}\exp\left(-\frac{1}{2}z^2\right)\sqrt{1 - \rho(h)}$$

Near the origin the indicator variograms of a Gaussian SRF behave like $\sqrt{1 - \rho(h)}$ and are thus far less regular than the variogram of Z itself. Typically, if the variogram of Z is linear near the origin, indicator variograms behave like

$\sqrt{|h|}$ for all thresholds. This explains the apparent nugget effect often observed with indicator variograms. It would not be consistent, for example, to use the same variogram model for the Gaussian and for an indicator.

By contrast, for a mosaic RF all indicator variograms are proportional to the variogram of the RF itself (see Section 6.3.3).

Matheron (1982b) shows in the general case that the indicator variograms $\gamma(h; z)$ and the madogram (variogram of order 1) of $Z(x)$, $\gamma_1(h)$, are related by

$$\int_{-\infty}^{+\infty} \gamma(h; z) \, dz = \gamma_1(h)$$

Furthermore, knowledge of the direct and cross-covariances or variograms of the indicators at all thresholds is equivalent to the knowledge of the bivariate distributions of the SRF. Indeed

$$E[I(x; z_1)] = F(z_1),$$
$$E[I(x; z_1)I(x + h; z_2)] = F_h(z_1, z_2)$$

where F denotes the marginal distribution of $Z(x)$ and F_h represent the bivariate distribution of $Z(x)$ and $Z(x + h)$. It follows that the centered direct and cross-covariances for all possible thresholds are related to the centered covariance of Z by

$$\iint C(h; z_1, z_2) \, dz_1 dz_2 = C(h)$$

Thus in theory the study of indicators allows a determination of the bivariate distributions of an SRF. But, in practice, this is not so simple, and we will see that to get a consistent model it is preferable to model the bivariate distributions directly (see Section 6.4).

Variogram of a Lognormal Variable

When the marginal distribution is clearly non-Gaussian, we often try to make it Gaussian by a prior transformation of the variable. If, for example, $Z(x)$ has a lognormal marginal distribution, we take $Y(x) = \log Z(x)$, which has a Gaussian marginal distribution. If the bivariate distributions of the SRF $Y(\cdot)$ are Gaussian, the means m_Y and m_Z, the variances σ_Y^2 and σ_Z^2, and the covariances $C_Y(h)$ and $C_Z(h)$ are related by the equations

$$m_Z = \exp\left(m_Y + \tfrac{1}{2}\sigma_Y^2\right),$$

$$\sigma_Z^2 - m_Z^2[\exp(\sigma_Y^2) - 1], \tag{2.78}$$

$$C_Z(h) = m_Z^2[\exp(C_Y(h)) - 1]$$

as shown in the Appendix, Section A.9. Similarly, the variograms $\gamma_Y(h)$ and $\gamma_Z(h)$ are related by

$$\gamma_Z(h) = m_Z^2 \exp(\sigma_Y^2)[1 - \exp(-\gamma_Y(h))]$$

Because the variogram of a bilognormal SRF is difficult to estimate when the coefficient of variation of the lognormal distribution is not small, these equations make it possible to model the variogram of Z from that of Y. However, they are only valid if Z is bilognormal (i.e., its bivariate distributions and not only its marginal distribution are lognormal).

Conversely, if we decide to directly model the covariance of Z, it must be of the specific form (2.78), where the covariance C_Y can be any positive definite function since $Y(x) = \log Z(x)$ is a bi-Gaussian SRF. In other words, not just $C_Z(h)$ but also $\log(1 + C_Z(h)/m_Z^2)$ must be positive definite, which excludes certain models. Since C_Y satisfies $-C_Y(h) \le C_Y(0)$ according to the Schwarz inequality, it follows that C_Z must satisfy

$$-C_Z(h) \le \frac{m_Z^2 \, C_Z(0)}{m_Z^2 + C_Z(0)}$$

Matheron (1987a) proves for example that the geostatistician's best friend, the spherical model, is *not* compatible with *bilognormality* if the relative variance exceeds a finite (but unknown) threshold, and conjectures that it is not compatible with lognormality in any case (i.e., *multivariate* lognormality, not just a lognormal marginal distribution).

If now Y is a bi-Gaussian IRF with an unbounded variogram, Matheron (1974a) shows that the sample variogram of Z is, in expectation, proportional to $1 - \exp(-\gamma_Y(h))$, which extends the formula of the stationary case: γ_Z shows an apparent range, which is artificial. This calls again for analyzing $\log Z$ rather than Z itself.

2.5.4 Covariance Models in Special Spaces

Covariance on the Sphere

Most applications consider a small portion of the earth, so that we work in the 2D Euclidean space of a planar projection of the data points. Global models, however, require covariance functions defined on the sphere. Schoenberg (1942) characterizes continuous isotropic covariances on the sphere of $\mathbb{R}^n$, thus expressed as continuous functions of the form $C(\theta)$ of the spherical or central angle distance $\theta \in [0, \pi]$. Considering only the case $n = 3$, a function $C(\theta)$ is a covariance if and only if it is of the form

$$C(\theta) = \sum_{k=0}^{\infty} c_k \, P_k(\cos \theta)$$

where the coefficients $c_k \geq 0$ have a finite sum and P_k is the Legendre polynomial of order k. For example, $P_0(u) = 1$, $P_1(u) = u$, $P_2(u) = (3u^2 - 1)/2$. This formula gives the spectral representation of isotropic covariances on the sphere. When r is replaced by θ, the spherical model (2.47) and the exponential model (2.51) as well as its Radon transforms (2.52) and (2.54) are valid covariances on the sphere (Huang et al., 2009), whereas this is not the case for the Gaussian model (2.54) (Gneiting, 1999b).

A subclass of valid isotropic models on the sphere is obtained by restricting isotropic covariances in the 3D Euclidean space (Yaglom, 1987, Vol. I, Section 22.5): indeed, because the Euclidean distance between two points separated by a central angle θ on the sphere of unit radius is $2 \sin(\theta/2)$, if $C_E(r)$ is an isotropic covariance in $\mathbb{R}^3$, then the function

$$C(\theta) = C_E\left(2 \sin(\theta/2)\right)$$

is a valid covariance model on the sphere. In that case, it is equivalent to work with the covariance C and the central angle distance θ, or with the covariance C_E and the separation distance r in the Euclidean space. Geometric anisotropy has no equivalent on the sphere, but there is the need for models displaying nonstationarity in latitude and/or longitude. This is not a simple task and the reader is referred to Jun and Stein (2008), who present several approaches and references to other authors.

Covariance on a River Network

The concentration in nutriments along a river can be considered as a one-dimensional random function indexed by the curvilinear abscissa along the river instead of the 2D Euclidean coordinates. This is possible as long as the river receives no affluent. At the confluence of two rivers, the flow rates add and the resulting concentration is the average of the two input concentrations weighted by the flow rates (this is an ideal case where we neglect geochemical interactions). Special random function and covariance models have been developed for such networks. They are usually obtained by a weighted moving average, also called kernel convolution, of a Gaussian white noise defined along the rivers. A model with total independence between branches is obtained with a weighting function acting unilaterally upstream (Ver Hoef et al., 2006; Cressie et al., 2006). It only allows spatial dependence between two points of the river network that are upstream of each other. On the contrary, Bailly et al. (2006) and Monestiez et al. (2005b) propose a model based on conditional independence at branching points with respect to what happens downstream. It amounts to using a weighting function acting unilaterally downstream. De Fouquet and Bernard-Michel (2006) and Polus-Lefebvre et al. (2008) propose another point of view on these models and extend them to an intrinsic hypothesis. Ver Hoef and Peterson (2010) and Garreta et al. (2010) synthetize the above models in a hybrid model. These papers also propose special inference methods.

2.6 FITTING A VARIOGRAM MODEL

To use the geostatistical methods developed in the next chapters, we need the variogram at any lag value. Since $-\gamma(h)$ is a conditionally positive definite function, we cannot simply interpolate linearly between the lags of the sample variogram. We have to fit a model that is known to be a valid variogram function. Here we will only consider the key aspects of fitting. Numerous examples can be found in the geostatistical literature; for example, see Matheron (1962, 1968a), David (1977), and Journel and Huijbregts (1978).

2.6.1 Manual Fitting

Fitting the Behavior near the Origin

Let us consider the variogram of Figure 2.25. In current practice we extrapolate the linear behavior of the first few points of the variogram and obtain the fit *a* with a nugget effect equal to 1.2. But nothing prevents us from considering other fits near the origin, the two extremes being:

- a structure with a range smaller than the data interdistance and with no nugget effect (fit *b*),
- a very regular behavior near the origin combined with a strong nugget effect (fit *c*).

In such a situation the choice of a behavior near the origin is a vulnerable anticipation in the sense that it can always be confirmed or invalidated by

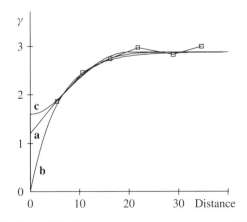

FIGURE 2.25 Sample variogram and three types of fit.

additional data on a tighter sampling grid. If this choice is not well-founded, kriging loses its optimality and the order of magnitude of the calculated kriging variances can be completely wrong. This point is illustrated by Armstrong and Wackernagel (1988) on a set of 52 topographic data. The consequences are even more serious for a conditional simulation, which actually claims to reproduce variability in detail.

Fortunately, the choice of a fit is usually guided by the physics of the problem or by subject matter knowledge. For example, in the case of Figure 2.25 we may consider the following situations:

1. If we are dealing with a geophysical potential field (gravity or magnetism) which is known to have a very regular spatial structure, fit c is called for, and the nugget effect reflects measurement errors.

2. If we are studying the top of a fairly continuous formation, and if the data come from boreholes and are free of errors, the apparent nugget effect is a short-range structure that can be modeled, as in b, for example, by using a spherical model (the choice of range is then subjective, or is taken from other data).

3. The fit a is used for variables, such as grades or porosities, that show strong variations and are measured on small supports, thus revealing the presence of microstructures.

Measurement errors, microstructures, and short-range components are often mixed. Sometimes the variance of the measurement errors is known by other means, which facilitates the fitting; for example, the stated precision of an equipment gives the standard deviation (or more commonly twice the standard deviation) of the measurement error. However, in order to accurately determine the short-range components, there is hardly any solution other than to take additional samples. In mining exploration, boreholes are first drilled in a fairly regular pattern. To improve the knowledge of the variogram at short distances, it is common practice to supplement these data with a cross of about 30 boreholes at a short spacing (about one-fourth of the grid size), located in an area reflecting the average spatial behavior. An alternative is to scatter the additional data across the entire study area so as to obtain close pairs giving a representative variogram at short lags [e.g., see Chilès et al. (1996)]. This can be measured by the variance of estimation of the variogram, which can be cal-culated, at least in the Gaussian case, provided that a tentative model has been selected (see Section 2.9.1). This approach has been systematized to design sampling schemes optimizing the inference of all variogram parameters. A reference on this subject is the book of Müller (2007, Chapter 6).

Modeling the Continuous Component

Once the type of behavior near the origin has been chosen (for distances less than the first lag), there remains to model the variogram at medium and large distances. Depending on the case, we can use a single basis model or combine several elementary components corresponding to different ranges (nested structures). Figures 2.26 and 2.27 show three examples of such fits. The points that need to be well reproduced are the slope at the origin, the range, and the sill (at least for a variogram showing this type of behavior).

The slope at the origin is evaluated from the first points of the variogram. The sill is placed at the level at which the variogram stabilizes (as we have seen in Section 2.1.2, the sample variance tends to underestimate the sill). The range is sometimes found less easily because the point at which it meets the sill is not always obvious. If a spherical model is considered (very common case in practice), the range can be derived from the fact that the tangent at the origin intersects the sill at two-thirds of the range.

In the absence of physical guidelines (as available for gravity or magnetism) or of internal consistency constraints, the geostatistician has a certain latitude in the choice of the basis models. Once the behavior near the origin has been set, the different possible fits will be graphically very similar at the scale of the study domain and will lead to nearly identical kriging results, as shown for example by Chilès (1974) with the topographic data from Noirétable (Section 2.7.3). Other comparisons have also been published; for example, see Diamond and Armstrong (1984). In other words, given the behavior near the origin, the fit is robust with regard to the analytical form of the selected model. For example, Journel and Huijbregts (1978) show that the graph of an exponential variogram can be modeled with excellent precision by the sum of two spherical models (p. 234) and that similarly the graph of a regularized logarithmic variogram can be modeled by the sum of two spherical models with a nugget effect (p. 168). However, the fit

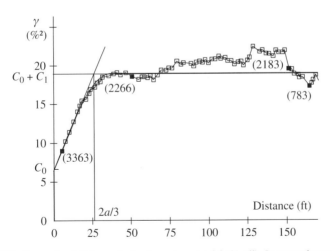

FIGURE 2.26 Example of fit by a single elementary model (% oil). In parentheses: number of pairs; a is the range. [From Dowd and Royle (1977), with permission of The Australasian Institute of Mining and Metallurgy; Journel and Huijbregts (1978), with permission of Academic Press Limited, London.]

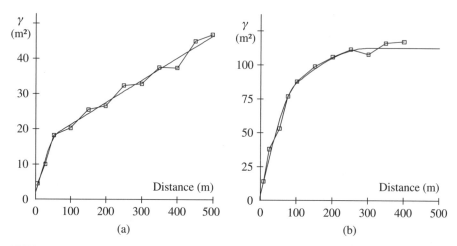

FIGURE 2.27 Examples of fit of nested structures: (a) thickness of nickel-bearing lateritic ore; (b) thickness of cover. Tiébaghi, New Caledonia. [From Chilès (1984), with kind permission from Kluwer Academic Publishers.]

can lack robustness when the behavior at short distances is parabolic; indeed the irregular terms of the variogram, which have the main role in variance calculations, may be masked by the parabolic part; in other terms, the variogram is not micro-ergodic (Section 2.9.2).

In the case of anisotropy, the variogram has to be fitted along the different directions by a global model, obviously including the anisotropies, so as to obtain a consistent model (and not, for example, as a sum of one-dimensional variograms along the main directions of anisotropy). The fits are more complex, but also richer, when a more global consistency is required, for example:

- consistency of variograms with regard to data differing in their measurement supports, measurement precision, and positioning precision;
- consistency of variograms with regard to variables linked by functional relationships, such as variables for different supports, a variable and one of its indicators, a differentiable variable and its partial derivatives, or a lognormal variable and its Gaussian transform.

These cases require a globally consistent fit including the theoretical relationships between the different models. This involves the following procedure:

1. Choice of a variogram model for the reference variable (point support variable, Gaussian variable, etc.).

2. Calculation of the theoretical form of the different sample variograms. This calculation takes into account the theoretical relationships between the

reference variogram and the variograms considered (obtained after regularization, differentiation, transformation, etc.).

3. Comparison of the theoretical variograms with the sample variograms, along with iteration of steps 1 to 3 until satisfactory fits are obtained.

Modeling a Proportional Effect

A proportional effect is usually considered as a departure from stationarity: The regionalized variable derives from a locally but not globally stationary RF model. Locally, namely in the neighborhood V_0 of point x_0, the variance of $Z(x+h) - Z(x)$ depends on h and not on x, but it varies from one neighborhood to the other. More precisely the variogram in V_0 is modeled as a function of h and V_0 of the form

$$\gamma_{V_0}(h) = b(m_{V_0})\, \gamma(h) \tag{2.79}$$

where b is some function of the local mean m_{V_0} and $\gamma(h)$ is a global model. However, Matheron (1974a) shows that a proportional effect is not always incompatible with global stationarity (in which case a global model suffices for global estimation). The local variogram may be regarded as the variogram of the random function conditioned on the data that belong to V_0. By studying the form of SRF variograms conditioned on the local mean m_{V_0}, Matheron shows that if the spatial distribution is lognormal, the local variogram γ_{V_0} takes the form (2.79) with $b(m_{V_0}) = m_{V_0}^2$ and $\gamma(h)$ proportional to the variogram of the nonconditioned SRF. In this case the value of modeling the proportional effect is to be able to provide local estimation variances conditioned on the local mean. These theoretical considerations are confirmed by the example presented by Clark (1979), which also shows that an improper modeling of a proportional effect can give worse results than a globally stationary model. An example of successful modeling of a clear proportional effect is given by Sans and Blaise (1987): The regression of the local variance on the local mean, in a bilogarithmic scale, is a perfect straight line with a slope 2.17 (uranium grades).

Expression (2.79) is valid locally, but these local models taken together do not constitute a valid global model. If the reference model $\gamma(h)$ is associated with a correlogram $\rho(h)$, a straighforward global nonstationary model for the covariance of $Z(x)$ and $Z(x')$ is

$$C(x, x') = \sigma(x)\, \sigma(x')\, \rho(x' - x)$$

where $\sigma^2(x) = b(m_{V_x})$ is the local variance of $Z(x)$.

Modeling a Nonstationary Covariance

In the above model the local shape of the covariance is always the same and we only modulate the local variance. We may have to also modulate the local range, anisotropy, or even covariance shape. It is easy to conceive a local model by spatially varying the variogram parameters. These can be given by auxiliary variables (e.g., seismic attributes may define the anisotropy direction) or by local variogram calculation and fitting (Machuca-Mory, 2010). Care must be taken to ensure smooth spatial variations of these parameters. A simple global covariance model can be derived by generalization of the covariogram (2.30), defined by autoconvolution of an integrable and square integrable function $w(u)$:

$$g(h) = \int w(u)\, w(u+h)\, dx$$

If we replace $w(u)$ by a dilution or kernel function $w(x; u)$ also depending on x, integrable and square integrable in u whatever x, and define

$$g(x, x') = \int w(x; u)\, w(x; u')\, du\, du'$$

then $g(x, x')$ is a nonstationary covariance [e.g., Higdon et al. (1999)]. Indeed a random function with that covariance is obtained by the dilution method presented in Section 7.5.2. (Higdon, 2002). Other models can be developed if we have replications in time (see Section 5.8).

2.6.2 Automatic Fitting

Least Squares with Predefined Weights

Least squares techniques can provide an automatic fit of the sample variogram. Generally we will look for the variogram within a family $\gamma(h; \mathbf{b})$, where $\mathbf{b}$ represents a vector of k parameters $b_1, \ldots, b_k$, belonging to a subset B of $\mathbb{R}^n$ (the parameters are, for example, the nugget effect and the anisotropy directions, ranges, and sills of several spherical components). Let us denote by $\{\hat{\gamma}(h_j): j = 1, \ldots, J\}$ the values taken by the sample variogram for J values of the vector h, each calculated from the $N(h_j)$ available pairs. The vector $\mathbf{b}$ is selected to minimize

$$Q(\mathbf{b}) = \sum_{j=1}^{J} w_j\, [\hat{\gamma}(h_j) - \gamma(h_j; \mathbf{b})]^2$$

The weights w_j are positive and typically equal to the ratio of the number of pairs $N(h_j)$ by the lag value $|h_j|$, in order to give more weight to well-estimated variogram lags and to short distances. If the variogram has been calculated along different directions, it is advisable to update these weights so that the various directions receive the same total weight.

When the variogram is linear in the parameters b_k, the function Q is a quadratic form of the b_k and its minimization is easy. Therefore, early implementations let the user fix anisotropies and ranges and minimized only on nugget effect and sills. Nonlinear minimization was sometimes carried out but limited to a few parameters (David, 1977, Section 6.3.2). Nowadays it is possible to efficiently optimize the value of several tens of linear and nonlinear parameters. In the implementation proposed by Desassis and Renard (2011) the geostatistician has only to give the types of the basis models (e.g., a nugget effect and four spherical components); ranges, sills, and anisotropies are automatically fitted.

A Newton-type algorithm is used to find the vector $\mathbf{b}^*$ minimizing Q (Madsen et al., 1999). This is an iterative procedure that starts from an initial guess $\mathbf{b}^{(0)}$ (e.g., in the case of a model consisting of three nested anisotropic spherical variograms, three ranges from a short distance up to the size of the domain, partial sills equal to one-third of the sample variance, anisotropy ratios set to 1). At step s, the objective function Q is locally approximated around the current vector $\mathbf{b}^{(s)}$ by a quadratic form, using a Taylor series expansion. Then $\mathbf{b}^{(s+1)}$ is defined as the minimum of the quadratic form. Newton methods are known to achieve fast convergence around local minima, but are not robust when starting far from $\mathbf{b}^*$. To circumvent this drawback, the minimum is searched in a trust region around $\mathbf{b}^{(s)}$. At each iteration, the size of the trust region is increased or decreased, depending on a quality assessment of the quadratic approximation at the previous iteration. Parameters with a negligible influence are eliminated. This results in nearly all cases in a fit close to an ideal manual fit.

This method can obviously include the fitting of a reference model related to several sample variograms corresponding, for example, to different supports, to several indicators of an assumed Gaussian variable, and so on.

Generalized Least Squares

In the above approach, the weights are chosen heuristically and the minimization does not take into account the correlations between the different lags of the sample variogram. Generalized least squares (Genton, 1998b) do not have that drawback: $\mathbf{b}$ is selected to minimize

$$Q(\mathbf{b}) = [\hat{\boldsymbol{\gamma}} - \boldsymbol{\gamma}(\mathbf{b})]' \mathbf{V}^{-1} [\hat{\boldsymbol{\gamma}} - \boldsymbol{\gamma}(\mathbf{b})]$$

where $\hat{\boldsymbol{\gamma}}$ is the vector of the $\hat{\gamma}(h_j)$, $\boldsymbol{\gamma}(\mathbf{b})$ is the vector of the $\gamma(h_j ; \mathbf{b})$, and $\mathbf{V}$ should be the covariance matrix of the vector $\hat{\boldsymbol{\gamma}}$ (if $\hat{\gamma}(h_j)$ is the traditional sample variogram (2.6) the variances can be calculated by using the covariance functions G_h defined in Section 2.9.1; the covariances can be calculated in a similar manner).

In this last case it is necessary, for the calculation of $\mathbf{V}$, to know $\gamma(h)$ as well as the fourth-order moments of the random function. In practice, we work within the framework of Gaussian RFs and proceed by iteration from an initial solution (e.g., obtained by weighted least squares): the vector $\mathbf{b}$ being fixed, we calculate $\mathbf{V}$, determine the new value of $\mathbf{b}$ minimizing Q, and start over again. The calculation of the matrix $\mathbf{V}$ is not a simple task, and the method is therefore limited to data on a regular grid.

A suboptimal but simpler algorithm consists in ignoring the correlations between lags of the sample variogram: The covariance matrix of the $\hat{\gamma}(h_j)$ is replaced by the diagonal matrix of the variances (Cressie, 1985). This amounts to weighted least squares with weights w_j equal to the reciprocals of $\text{Var}[\hat{\gamma}(h_j)]$ (these weights are updated during the iterative procedure). In the Gaussian case, $\frac{1}{2}[Z(x+h) - Z(x)]^2$ is distributed as $\gamma(h)\chi_1^2$, where χ_1^2 is a chi-square variable on one degree of freedom. If the correlations between the $N(h_j)$ increments taken into account by the calculation of $\hat{\gamma}(h_j)$ are negligible, we then have

$$\text{Var}[\hat{\gamma}(h_j)] \approx \frac{2\gamma(h_j; \mathbf{b})^2}{N(h_j)}$$

and the problem reduces to minimizing

$$Q(\mathbf{b}) = \frac{1}{2}\sum_{j=1}^{J} N(h_j)\left[\frac{\hat{\gamma}(h_j)}{\gamma(h_j; \mathbf{b})} - 1\right]^2$$

Even when based on a Gaussian assumption, weighted or generalized least squares are fairly robust: if this assumption is not really valid, the quadratic form to be minimized will not be optimal, but the result will not be biased.

Maximum Likelihood and Bayesian Method

Some methods work directly from sample data, without requiring the calculation of the sample variogram. Since these methods are blind, they tend to be used only when the presence of a strong drift causes the sample variogram to be hopelessly biased. Section 4.8 discusses the use of optimal quadratic estimators and regression. Let us examine here the maximum likelihood method and the Bayesian approach which have been applied to stationary and nonstationary fields.

The maximum likelihood method is widely used in statistics for parameter estimation and has been proposed as a means of estimating variogram parameters (Kitanidis and Lane, 1985)—some authors would say an *objective* means. Its principle is the following: Assume that the N observed data are from a multivariate Gaussian distribution with mean vector $\mathbf{m} = (\text{E}[Z(x_1)], \ldots, \text{E}[Z(x_N)])'$ and covariance matrix $\boldsymbol{\Sigma}$. The joint probability density of the sample $\mathbf{z} = (z(x_1), \ldots, z(x_N))'$ is

$$f(\mathbf{z}) = (2\pi)^{-N/2}|\boldsymbol{\Sigma}|^{-1/2}\exp\left(-\frac{1}{2}(\mathbf{z} - \mathbf{m})'\boldsymbol{\Sigma}^{-1}(\mathbf{z} - \mathbf{m})\right)$$

If the mean $\mathbf{m}$ and the covariance $\boldsymbol{\Sigma}$ are unknown and depend on parameter vectors $\boldsymbol{\beta}$ and $\mathbf{b}$, respectively, one can regard $\mathbf{z}$ as fixed and $f(\mathbf{z})$ as a function of $\boldsymbol{\beta}$ and $\mathbf{b}$, called the likelihood function $L(\boldsymbol{\beta}, \mathbf{b})$. Maximizing the likelihood, or equivalently minimizing the log-likelihood $-\log L(\boldsymbol{\beta}, \mathbf{b})$, yields estimates of the parameters.

In the case of an SRF with known mean, the vector $\boldsymbol{\beta}$ is absent. But this is not the usual case. When the mean of the SRF is not known, and more generally when the RF has a drift, the method has the advantage of allowing a joint determination of drift and covariance parameters, but it is then prone to biases that may be severe [e.g., see Matheron (1970, Chapter 4, exercise 16]. A variant of the method, known as *restricted maximum likelihood*, is to write the likelihood of the data in terms of the $N - 1$ increments $Z(x_{\alpha+1}) - Z(x_\alpha)$ or of more general increments so as to eliminate the mean or the drift from the set of unknowns, which has the effect of reducing the bias on the variogram parameter estimates (Kitanidis and Vomvoris, 1983; Kitanidis and Lane, 1985). In practice, this approach relies of course on a strict Gaussian assumption, which is questionable, to say the least. Also it lets the selection of parameters depend on sampling fluctuations.

Handcock and Wallis (1994) present an interesting study of 88 data of a meteorological field (average temperature over one winter) using a maximum likelihood method. The fitting was not done blindly but, instead, was done after a careful exploratory spatial data analysis. The variogram is assumed to be an isotropic Matérn model, and therefore **b** has three parameters: the shape parameter ν which controls the behavior $|h|^{2\nu}$ near the origin, the scale parameter, and the sill. They obtain $\nu = 0.55$, a value close to $1/2$ which would correspond to an exponential covariance. Indeed a fit of the sample variogram to an exponential model would seem natural.

The authors further use the Bayesian framework to include uncertainty in the variogram estimates and derive the joint posterior distribution of the parameters. The shape parameter ν is found to lie between 0.1 and 1.5 with a sharp mode at about 0.2. Accordingly, the variogram behaves like $|h|^{0.4}$ near the origin, namely with an infinite slope characterizing an extremely irregular field, even though meteorological arguments indicate that this average temperature field is continuous and may even be differentiable. This result is due to the choice of prior distribution: A probability density proportional to $1/(1+\nu)^2$ has been used for $\nu \in]0, \infty[$ because it is supposed to be "noninformative" in the region where the likelihood has mass and rules out very large values of ν. In reality, this model gives maximum prior probability to a pure nugget effect, which contradicts the physics of the problem. To produce sensible results, the method should rather use an *informative* prior distribution, but of course the results would then depend on the information introduced by the physicist or the geostatistician [see the example provided by Mostad et al. (1997)]. Tarantola (2005) introduces the concept of homogeneous distribution for representing the distribution of a parameter when only fundamental properties such as symmetries or positivity are known. In the absence of any other information, he suggests to choose the homogeneous probability distribution as a priori distribution, a choice that is as informative as any other, as he notes, and avoids the use of the "noninformative" terminology. On the Bayesian approach, the reader is referred to Chapter 7 of the book of Diggle and Ribeiro (2007), which considers uncertainty on all variogram parameters and shows its consequence on Bayesian kriging.

Another approach to the determination of the "belt" of the plausible models associated with a given sample variogram is proposed by Pilz et al. (1997). It assumes Gaussianity of the random field and is based on the spectral representation of the variogram. The same problem is addressed by Solow (1985) with the "*spatial bootstrap.*" The idea is to apply the bootstrap procedure of sampling from the empirical distribution function (Efron, 1979) to study the variability of covariance estimates. However, since the bootstrap procedure assumes independent samples, Solow starts by orthogonalizing the centered data $\mathbf{Z}$ using $\mathbf{X} = \mathbf{L}^{-1}\mathbf{Z}$, where $\mathbf{S} = \mathbf{L}\,\mathbf{L}'$ is the empirical covariance, possibly adjusted. The uncorrelated Xs are then bootstrapped independently and the reverse transformation $\tilde{\mathbf{Y}} = \mathbf{L}\,\tilde{\mathbf{X}}$ produces correlated realizations of the Ys, from which a new empirical covariance is computed. Repeating the procedure many times leads to approximate confidence intervals for covariances and estimation variances. This approach can be extended to the definition of prediction intervals of Z at target points (Schelin and Sjöstedt-de Luna, 2010).

Fuzzy Variogram Fitting

Fuzzy logic (Zadeh, 1965; Zimmermann, 2001) provides another means to represent the uncertainty on variogram parameters, as shown by Bardossy et al. (1990), and Loquin and Dubois (2010). It is helpful when the sample variogram cannot be easily interpreted and when a sensible a priori distribution for the variogram parameters cannot be specified but some experts' opinion can be formalized. The parameter values are not represented by crisp numbers or random variables but by fuzzy intervals. The reader interested in the link between possibility theory and probability theory is referred to Dubois (2006).

2.6.3 Validation

Statistical Tests

In order to validate the consistency of the data with an assumed model (a "hypothesis"), the standard statistical approach is to consider some function of the observations, a *test statistic*, and derive its probability distribution under the assumed model. If the observed test statistic is "too large," namely falls in the tails of the distribution, the model is rejected. There is a probability α that a correct hypothesis is rejected, depending on the definition of the tails (typically $\alpha = 0.05$). When the assumed model depends on some parameter, the set of values of this parameter for which the model is not rejected constitutes a confidence interval at level $1 - \alpha$.

It is very difficult to construct statistical tests in geostatistics because of the spatial dependence between observations. The successive lags of the sample variogram are correlated, as already noticed by Jowett (1955b). Moreover, the sample variogram, like the regional variogram, can exhibit large fuctuations at large distances compared to the theoretical model, as will be shown in Section 2.9.2. This limits the applicability of statistical tests to the main features of the variogram, mainly to its behavior near the origin. Switzer (1984) proposes a few tests for variogram scale and shape parameters and the nugget effect, as well as for their inversion into confidence regions. For example, he considers the family $\sigma^2\,\gamma(h;\,a)$, where γ is a fixed variogram shape, σ^2 is the sill, and a is the range. The idea is to linearly transform the data to uncorrelated quantities of constant variance and then consider certain rank orderings. A simple application of these ideas is a test for the range: Select a subset of data points whose interpoint distances all exceed the range; then test for randomness based on the rank correlation between $|Z(x_\beta) - Z(x_\alpha)|$ and $|x_\beta - x_\alpha|$, which is then a standard statistical problem [e.g., Spearman's coefficient; cf. Kendall (1970)].

Cross-Validation

A powerful model validation technique is to check the performance of the model for kriging. Here we have to anticipate on results presented in Chapter 3. Consider N data $Z(x_\alpha)$ and a variogram model fitted from the sample variogram calculated from these data. The principle of cross-validation, also called "leave-one-out method," is to estimate $Z(x)$ at each sample point x_α from neighboring data $Z(x_\beta)$, $\beta \neq \alpha$, as if $Z(x_\alpha)$ were unknown. Thus at every sample point x_α we get a kriging estimate $Z^*_{-\alpha}$ and the associated kriging variance $\sigma^2_{K\alpha}$. Because the true value $Z_\alpha = Z(x_\alpha)$ is known, we can compute the kriging error $E_\alpha = Z^*_{-\alpha} - Z_\alpha$ and the standardized error $e_\alpha = E_\alpha / \sigma_{K\alpha}$. If $\gamma(h)$ is the theoretical variogram, E_α is a random variable with mean zero and variance $\sigma^2_{K\alpha}$, and e_α is a zero-mean unit-variance random variable. Note that kriging errors are not independent. The following results are inspected (for example, see Figure 4.12):

- the posted standardized errors e_α;
- the histogram of standardized errors e_α;

- the $(Z^*_{-\alpha}, Z_\alpha)$ scatterplot;
- the $(Z^*_{-\alpha}, e_\alpha)$ scatterplot.

These plots should be examined in the context of the properties of the kriging estimator. In the case of simple kriging, we can check the smoothing relationship, the orthogonality of the estimate and the error, and the conditional unbiasedness in the Gaussian case. In the case of ordinary kriging, these properties do not necessarily hold. Nevertheless, scatterplots allow us to gauge how far we are from the ideal case of simple kriging. The histogram of standardized errors also shows if the kriging error can be considered Gaussian. Moreover, these plots are the basis for an interactive data analysis. They highlight data that are poorly "explained" by their neighbors, which can be meaningful anomalies, as well as erroneous data that ought to be corrected or set aside. Clustered anomalies can indicate a fault or a discontinuity in their neighborhood, which must be taken into account. A contour map of standardized errors shows if the magnitude of the error is homogeneous in space or, on the contrary, points to a lack of stationarity—for example, a proportional effect. Once these problems are fixed adequately, one ought to recompute the sample variogram and redo the cross-validation until acceptable results are obtained. Bradley and Haslett (1992) show an illustrative example of exploratory data analysis.

Comparing the results of two cross-validations performed under different conditions can help one decide between two candidate models. However, it is not a good idea to generalize this validation method to an automatic variogram fitting procedure—for example, a blind fit by minimization of $S = \frac{1}{N}\sum_{\alpha=1}^{N} E_\alpha^2$ under the constraint that $s = \frac{1}{N}\sum_{\alpha=1}^{N} e_\alpha^2$ is close to 1. Indeed, if a variogram $\gamma(h)$ leads to $s \neq 1$, replacing it by $\gamma'(h) = \gamma(h)/s$ leaves S unchanged ($S' = S$) and leads to $s' = 1$ even if the variogram shape has nothing to do with the sample variogram. Obviously the fact that s is close to 1 is an indicator of the quality of the fit only if this constraint was not included in the fitting procedure. Besides, the value of S is often essentially due to the contribution of a few points (anomalies or erroneous data). A blind fit by minimization of S would give an overwhelming weight to a few globally or locally extreme points, which is not desirable. Several robust techniques can attenuate this effect: considering the magnitude of errors rather than their squares, thresholding the errors (e.g., at 2.5 standard deviations), and comparing two options by their scores (counting how many times option 1 performs better than option 2 versus how many times option 2 is better than 1). Such techniques allow a refined comparison between two options.

In practice, the estimation of $Z(x_\alpha)$ from its neighbors is usually accomplished with a moving neighborhood. However, too close neighbors are eliminated if the data form a cluster or are located along profiles, in order to mimic the subsequent kriging conditions. If the number of points is not too large, a global neighborhood may also be used. This means solving N linear systems of size $N \times N$, which may be computationally expensive. A special technique has been developed to replace these calculations by the inversion of a single matrix of size $(N+1) \times (N+1)$ (see Section 3.6.6).

2.6.4 Spectral Modeling Approach

So far we have ignored spectral analysis methods, although they are widely used, especially for the study of time series. A fairly complete account is given by Yaglom (1987, Vol. I, Chapter 3 in $\mathbb{R}$, Section 22.2 of Chapter 4 in $\mathbb{R}^n$), along with many references [in 2D, also see Guyon (1993, Section 4.5)]. These methods require data sampled on a regular grid forming a parallelepiped in $\mathbb{R}^n$, which is rarely the case in the applications that we consider. Moreover, if the knowledge of the spectral measure and of the covariance are theoretically equivalent, passing from one to the other is not always easy. Signal-processing applications (e.g., the design of filters) essentially involve the spectrum, whereas geostatistical applications (kriging, change of support) involve the covariance, which explains why we favor variographic analysis. Still it is interesting to see how our signal-processing colleagues, who have given much thought to spectral estimation and modeling, proceed. For simplicity we will restrict the presentation to the one-dimensional case. The transposition to higher dimensions is fairly straightforward.

Because the data are sampled on a one-dimensional regular grid, we will index the data points by n instead of α. Consider a sequence of N sample values $Z_n = Z(n\Delta x)$, $n = 1, \ldots, N$, and to simplify notations, let us assume that $\Delta x = 1$. Furthermore, $Z(x)$ is an SRF with mean zero (in practice, the mean is subtracted initially), covariance $C(h)$, and spectral density $f(u)$. From the N data Z_n it is possible to calculate $2N - 1$ values of the sample covariance

$$\hat{C}_m = \frac{1}{N - |m|} \sum_{n=1}^{N-|m|} Z_n Z_{n+|m|}, \qquad m = 0, \pm 1, \ldots, \pm(N - 1)$$

$\hat{C}_m$ is an unbiased estimator of the theoretical covariance $C_m = C(m\Delta x)$, provided that $Z(x)$ has indeed a zero mean. Under certain conditions (e.g., met for a Gaussian RF), $\hat{C}_m$ is a consistent estimator of C_m in the sense that the quadratic mean of $\hat{C}_m - C_m$ tends to zero when $N \to \infty$. But for fixed N the quadratic mean of $\hat{C}_m - C_m$ is large for large values of m because $\hat{C}_m$ is then computed with few pairs of points. Moreover, contrary to the theoretical covariances C_m, the $2N-1$ sample covariances $\hat{C}_m$ do not necessarily constitute a discrete positive definite function. Their inverse Fourier transforms can take negative values. To avoid this, another estimator is considered, namely

$$\tilde{C}_m = \frac{1}{N} \sum_{n=1}^{N-|m|} Z_n Z_{n+|m|}, \qquad m = 0, \pm 1, \ldots, \pm(N - 1) \tag{2.80}$$

For fixed N, $\tilde{C}_m$ is a biased estimator of C_m (except if $m = 0$), but it is asymptotically unbiased when $N \to \infty$. Despite this bias, for fixed N, the quadratic mean of $\tilde{C}_m - C_m$ is generally less than or equal to that of $\hat{C}_m - C_m$ (this is obviously the case when $C_m = 0$, i.e., beyond the range). Furthermore, the $2N - 1$ terms $\tilde{C}_m$ form a discrete positive definite function. Hence the spectral density estimator

$$\tilde{f}(u) = \sum_{m=-N+1}^{N-1} e^{-2\pi i \, um} \tilde{C}_m$$

This (random) function is called the *periodogram*. It can be computed directly from the data as

$$\tilde{f}(u) = \frac{1}{N} \left| \sum_{n=1}^{N} e^{-2\pi i \, un} Z_n \right|^2$$

Since the data and the covariance are known only at discrete points, $\tilde{f}(u)$ is a (random) periodic function with period 1, and it will be considered only in the interval $u \in \,]-1/2, 1/2]$. For fixed N, $\tilde{f}(u)$ is a biased estimator of the spectral density $f_1(u) = \sum_{m=-\infty}^{+\infty} e^{-2\pi i \, um} C_m$ of the infinite random *sequence* Z_n, but it is asymptotically unbiased when $N \to \infty$. However, even when $N \to \infty$, the quadratic mean of $\tilde{f}(u) - f_1(u)$ does not tend to zero: $\tilde{f}(u)$ is not a consistent estimator of $f_1(u)$. Since $\tilde{f}(u)$ and $\tilde{f}(u')$ are uncorrelated (unless $u' = \pm u$), the periodogram exhibits erratic fluctuations. To attenuate this effect, the periodogram is convolved with a weighting function $A(u)$, also periodic with period 1. This leads to the *smoothed periodogram*

$$\varphi(u) = \int_{-1/2}^{1/2} A(u - u') \tilde{f}(u') \, du'$$

The weighting function $A(u)$ has for Fourier transform the sequence

$$a_m = \int_{-1/2}^{1/2} e^{2\pi i \, um} A(u) \, du, \qquad m = 0, \pm 1, \ldots, \pm(N-1)$$

The function $A(u)$ is called the *spectral window*, whereas the sequence a_m defines the *lag window*. As the Fourier transform exchanges convolution and multiplication, one has

$$\varphi(u) = \sum_{m=-N+1}^{N-1} e^{-2\pi i \, um} a_m \tilde{C}_m$$

To define $A(u)$ over $]-1/2, 1/2]$, one chooses a unit-sum function with a maximum at zero and a rapid fall off with increasing $|u|$ (of course there is a great diversity of such functions). The sequence a_m is then equal to 1 at zero and decays progressively as $|m|$ increases, which has the effect of reducing the values of the sample covariance at large lags (those which are in general poorly estimated).

The calculations can be accomplished by means of three successive discrete Fourier transforms (DFTs): (1) a DFT of the Z_n series, which by squaring gives the periodogram for $u = k/N$, (2) an inverse DFT of $\tilde{f}_k = \tilde{f}(k/N)$ to get the covariances $\tilde{C}_m$, and lastly (3) a DFT of the tapered covariances $a_m \tilde{C}_m$ to get the smoothed spectral density φ.

In principle, it is not necessary to model φ since any nonnegative integrable symmetric function can be a spectral density. However, parametric methods exist. The most common one is to start from the first $p + 1$ values of the covariance C_m $(m = 0, \ldots, p)$ and select the spectral density $f(u)$, $u \in]-1/2, 1/2]$, maximizing the entropy E defined by[28]

$$E = \int_{-1/2}^{1/2} \log f(u) \, du$$

while still matching the covariance C_m, namely satisfying the conditions

$$\int_{-1/2}^{1/2} e^{2\pi i \, um} f(u) \, du = C_m, \qquad m = 0, 1, \ldots, p \tag{2.81}$$

It can be shown that the solution is of the form

$$f(u) = \frac{\sigma^2}{|1 - \alpha_1 e^{-2\pi i \, u} - \cdots - \alpha_p e^{-2\pi i \, up}|^2}$$

[28] Note that this definition is slightly different from that of the entropy of a *probability* density function $f(z)$, which is defined by $E = -\int f(z) \log f(z) \, dz$.

This is the general form for the spectral density of an autoregressive process of order p (these will be presented in Section 7.5.1; the α_i are the coefficients of the autoregression and σ^2 is the variance of the innovations). In practice, the theoretical covariances C_m are not known and are replaced by the estimates $\tilde{C}_m$. Among all spectral densities satisfying conditions (2.81), the maximum entropy spectral density has the special property of corresponding to the linearly most unpredictable random sequence. Indeed the best linear predictor Z_{n+1}^* of Z_{n+1} from the present value Z_n and all past data $Z_{n-1}, Z_{n-2}, \ldots$ (the best estimator altogether in the Gaussian case) has the variance

$$\mathrm{Var}(Z_{n+1}^* - Z_{n+1}) = \exp\left(\int_{-1/2}^{1/2} \log f(u)\, du\right)$$

as shown, for example, by Doob (1953, Section XII 4). Maximizing entropy thus amounts to maximizing this variance. An illuminating derivation of this result is to note that the knowledge of the $(p+1)$ covariance values C_m allows the determination of the optimal linear predictor based on the last p values. The autoregressive sequence of order p is precisely that for which the p-value predictor coincides with the predictor based on the infinite past and therefore maximizes the innovation brought by the new value Z_{n+1} (Yaglom, 1987, Vol. II, p. 104).

The above theory is concerned with random *sequences* and requires data on a grid. These can be obtained by sampling a continuous signal. However, the spectrum f_1 of the sequence and the spectrum f of the continuous signal coincide only if frequencies higher than the Nyquist frequency $1/(2\,\Delta x)$ either do not exist in the signal, or have been filtered out before sampling. This is a problem for most geostatistical applications where sampling is very fragmentary and applying an anti-aliasing filter before sampling is simply impossible. Covariance estimation does not require the signal to be band-limited. The variogram has the additional advantage of not requiring a preliminary estimation of the mean, nor the stationarity of $Z(x)$, but only of its increments.

2.7 VARIOGRAPHY IN THE PRESENCE OF A DRIFT

When a drift is present one generally turns to the universal kriging (UK) model of Section 3.4. In this model the RF $Z(x)$ is considered as the sum of a deterministic drift $m(x)$ (usually a polynomial with unknown coefficients) and a zero-mean stationary or intrinsic random residual $Y(x)$. We will see that the presence of a drift poses difficult inference problems. Their definitive solution requires abandoning the UK model for the broader IRF–k model studied in Chapter 4, except in some simplified cases that will be presented here.

2.7.1 Impact of a Drift on the Raw Variogram

Let us consider in $\mathbb{R}^n$ an RF of the form

$$Z(x) = m(x) + Y(x)$$

where the drift $m(x)$ is a deterministic drift and the ("true") residual $Y(x)$ is an SRF or an IRF with zero mean and variogram $\gamma(h)$. The stochastic version $\Gamma_R(h)$ of the regional variogram (2.7) associated with the domain D,

$$\Gamma_R(h) = \frac{1}{2|D \cap D_{-h}|} \int_{D \cap D_{-h}} [Z(x+h) - Z(x)]^2 dx$$

has the expected value

$$E[\Gamma_R(h)] = \gamma(h) + \frac{1}{2|D \cap D_{-h}|} \int_{D \cap D_{-h}} [m(x+h) - m(x)]^2 dx \qquad (2.82)$$

Should the drift have some amplitude, the additional term can mask $\gamma(h)$. It is obviously the same for the sample variogram (2.6) calculated from the raw data.

This phenomenon is often reflected in an apparent anisotropy. In the case of a linear drift with gradient a, the additional term is equal to $|<a, h>|^2$. Whatever the direction considered, it is a parabola, but its influence is maximum in the direction of the gradient. This can be seen on the raw variogram of Figure 2.28, which was constructed from piezometric data of the Crau aquifer (Delhomme, 1976, 1978). The anisotropic behavior is very marked at large distances and shows a very rapid increase in the NE–SW direction. The large amplitude of the deviations in this direction corresponds to the general orientation of the aquifer in which the groundwater flows from NE to SW with an average hydraulic gradient of 3 to 4 m/km. The anisotropy is related to the variation of the average gradient of the aquifer with the direction. Different effects can be seen with other types of drift; for example, a dome-shaped drift gives a dome-shaped variogram.

In geostatistical calculations, however, it is the variogram $\gamma(h)$ of $Y(x)$, known as the underlying variogram, that is used. When dealing with a repetitive phenomenon, such as meteorology, we may have a large number of similar situations and thus be able to deduce the drift at a monitoring site by averaging the observations at this site. We can then subtract this drift from the data and obtain the "true" residuals directly. This is the approach developed by Gandin (1963, p. 27) in meteorology. But when the phenomenon considered is unique, as is generally the case here, we cannot directly separate the drift from the residual. We still have to find a way of determining the underlying variogram by neutralizing the auxiliary term responsible for the bias in the raw variogram.

2.7.2   Variogram of Residuals

Since the drift introduces a bias in the raw variogram, we subtract the drift and work on the residuals. Or rather, since the drift $m(x)$ is not known exactly, we take an unbiased estimate $\hat{m}(x)$ and subtract it from $Z(x)$ to obtain the (estimated) residuals $R(x_\alpha) = Z(x_\alpha) - \hat{m}(x_\alpha)$ at the N sample points. Let us compute the sample variogram of these residuals, more precisely its stochastic version, defined similarly to (2.6) by

$$\hat{\Gamma}_{Res}(h) = \frac{1}{2N_h} \sum_{x_\beta - x_\alpha \approx h} [R(x_\beta) - R(x_\alpha)]^2$$

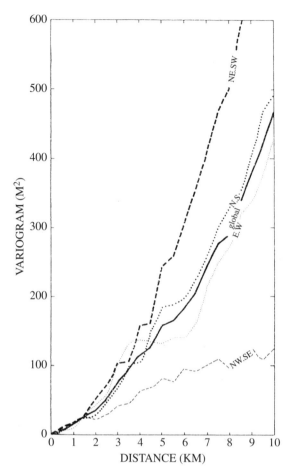

FIGURE 2.28 Raw variogram of the piezometry of the Crau aquifer, characteristic of a NE–SW linear drift. [Reprinted from Delhomme (1976, 1978), with permission from BRGM and Elsevier Science.]

Using the notation $Z_\alpha, \hat{m}_\alpha, R_\alpha$ instead of $Z(x_\alpha), \hat{m}(x_\alpha), R(x_\alpha)$ to simplify, the contribution of the sample points x_α and x_β to the sample variogram of residuals has for expected value

$$\gamma_{\mathrm{Res}}(x_\alpha, x_\beta) = \tfrac{1}{2}\mathrm{E}(R_\beta - R_\alpha)^2 = \gamma(x_\beta - x_\alpha) - \mathrm{Cov}(Z_\beta - Z_\alpha, \hat{m}_\beta - \hat{m}_\alpha)$$

$$+ \tfrac{1}{2}\mathrm{Var}(\hat{m}_\beta - \hat{m}_\alpha) \tag{2.83}$$

When the estimators $\hat{m}_\alpha$ and $\hat{m}_\beta$ are linear, (2.83) is expressed with the variogram $\gamma(h)$. It depends on x_α and x_β separately and not only on $x_\beta - x_\alpha$.

Therefore we define the variogram of residuals by $\gamma_{\text{Res}}(h) = \text{E}[\hat{\Gamma}_{\text{Res}}(h)]$, namely an average of terms $\gamma_{\text{Res}}(x_\alpha, x_\beta)$. There are obviously as many variograms of residuals as there are ways of estimating the drift. The expression of $\gamma_{\text{Res}}(x_\alpha, x_\beta)$ is only simplified when the *optimal* linear drift estimator $m^*(x)$ is used, as shown in Section 3.4.6 in the framework of the universal kriging model. The central term of the right-hand side of (2.83) is then twice the third term [cf. (3.30)], so that (2.83) becomes

$$\gamma_{\text{Res}}(x_\alpha, x_\beta) = \gamma(x_\beta - x_\alpha) - \tfrac{1}{2}\text{Var}(m_\beta^* - m_\alpha^*)$$

Thus, even if the estimator of the drift is the optimal one, which means that we already know the underlying variogram $\gamma(h)$ we are looking for, the variogram of the residuals is systematically biased downward. This bias is small at short distances but can be large at medium and large distances. It can lead to the erroneous conclusion that the residuals are uncorrelated.

To identify the underlying variogram $\gamma(h)$ an idea is to use relation (2.83) to calculate the theoretical variogram of residuals $\gamma_{\text{Res}}(x_\alpha, x_\beta)$ corresponding to a tentative $\gamma(h)$ and compare this with the sample variogram of the residuals. To this end, we can use for example the ordinary least squares estimator of the drift, which can be calculated easily (it is the optimal one if the underlying variogram is a pure nugget effect). This approach has been tried in the case of regularly spaced data: The variogram of residuals is computed in moving windows of 11 consecutive points, under the approximation of a locally linear drift. The average sample variogram of residuals is then compared to theoretical variogram of residuals "type curves" obtained by application of relation (2.83) (Huijbregts and Matheron, 1971; Sabourin, 1976). But the solution is not unique (Matheron, 1970, Section 4.6), and very different models $\gamma(h)$ can produce very similar variogram of residuals $\gamma_{\text{Res}}(h)$. This difficulty disappears with the generalized variogram (Section 4.7).

2.7.3   A Few Favorable Cases

Despite the presence of a drift, it is often possible to find a way of returning to the standard structural analysis of the stationary case.

Very Mild Drift

In the case of a very mild drift the term added to $\gamma(h)$ in (2.82) is negligible at short distances. This allows a good determination of the variogram at this scale, which is often sufficient for the requirements of kriging.

An example of this is found in a study of the Noirétable area topography (France) (Chilès and Delfiner, 1975). The aim of the study was to determine the precision with which one can reconstruct by kriging the topography of a 5-km^2 area from a pattern of 573 points. Figure 2.29a shows the sample variogram

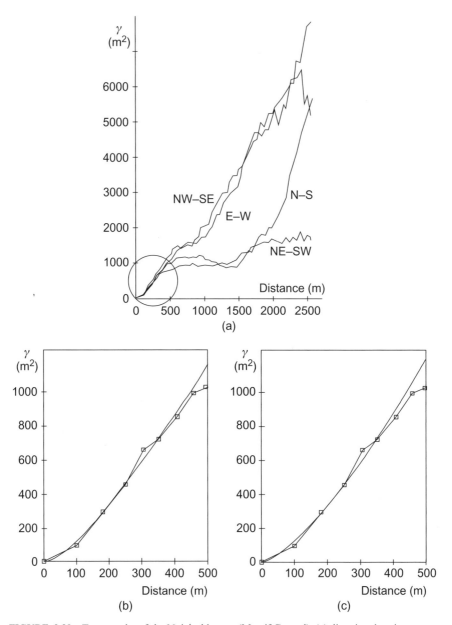

FIGURE 2.29 Topography of the Noirétable area (Massif Central): (a) directional variograms at large distances; (b) average variogram up to 500 m, and fit by a regularized linear model; (c) average variogram up to 500 m, and fit by a model of type $|h|^{\alpha}$. [From Chilès and Delfiner (1975), with permission of SFPT (a, b); Journel and Huijbregts (1978), with permission of Academic Press Limited, London (c).]

calculated in four directions. Beyond 500 m, the various curves diverge and exhibit a behavior characteristic of a complex-shaped drift becoming appreciable at this scale. But kriging involves neighborhoods that do not exceed 500 m in diameter. We can thus consider the phenomenon as being drift-free and isotropic at this scale. There remains to fit a model to the first 500 m of the average variogram. Several fits were proposed, all of which are suitable and give very close results—in particular, a regularized linear model given by (2.39) with $l = 500$ m (the equality of l and the maximum modeling distance of the variogram has no particular meaning) (op. cit.; Figure 2.29b) or an $|h|^\alpha$ model with $\alpha = 1.4$ (Journel and Huijbregts, 1978; Figure 2.29c).

Unidirectional Drift

Often the drift is not felt in all directions. If so and if the hypothesis of an isotropic variogram appears reasonable, we can use an isotropic model fitted to the sample curve obtained for the direction with no drift. We have already seen an example of this with the piezometry of the Crau aquifer (Figure 2.28).

  Another example is provided by Delfiner (1973). It concerns a study of the 500 mbar geopotential, roughly the altitude of the 500-mbar atmospheric pressure surface. Because of the rotation of the earth, the geopotential shows no E–W drift. There is, however, a very clear decrease from the equator to the pole; this is a well-known phenomenon reflecting the existence of a zonal wind (the wind is to a first approximation orthogonal to the gradient of the geopotential). The phenomenon is pictured in Figure 2.30a, which shows the average value of the data in sections of 5 and 10 degrees of latitude and longitude. It is no surprise that the sample variogram reaches a sill in the E–W direction and shows a parabolic behavior in the N–S direction (Figure 2.30b). Because of the physics of the problem, the basic drift functions are necessarily trigonometric functions depending only on the latitude θ. Here one can consider that, at least locally, it is of the form $a + b \cos \theta$. A model fitted to the E–W curve is taken as the variogram, namely a Gaussian model with a scale parameter $a = 1470$ km (the practical range being of the order of 2500 km), a sill $C = 25{,}000$ m^2, and a nugget effect $C_0 = 300$ m^2 (variance of the uncertainty of the radiosonde measurements, difficult to calibrate on the variogram but known by another way).

  Notice that in this particular case, the form of the drift is not invariant by translation, unlike usual drifts (cf. Sections 3.4.6 and 3.4.7). But the spherical form of the earth is a good reason to favor spherical coordinates.

Global Drift with a Simple Form

Sometimes the drift has a very simple form at the scale of the entire region of interest, such as linear or polynomial. In such cases we can subtract a global estimate of the drift from the data and work on the residuals. Generally, the least squares estimator is chosen because, regardless of the number of data, it only requires solving a small linear system. The variogram of residuals is

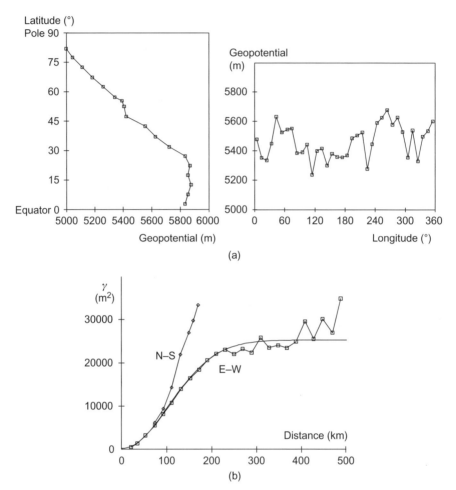

FIGURE 2.30 500-mb geopotential: (a) profiles of the mean versus longitude and latitude; (b) E–W and N–S variograms. [From Delfiner (1973), © SMF.]

certainly biased, but since this bias is negligible at short distances, we can correctly determine the variogram at the scale of the neighborhoods to be used for subsequent kriging.

To illustrate this, Figure 2.31 shows the raw variogram and variograms of residuals for civil engineering microgravimetric data (Bouguer anomaly sampled on a 15-m grid, locally refined) (Chilès, 1979b). The study zone is a 380-m × 350-m rectangle. The average raw variogram is clearly parabolic and reflects an essentially linear NW–SE global drift which is evident from the data. The directional raw variograms (not reproduced here) show, however, that no direction is entirely free of drift. Because the drift has a good overall shape, the sample variograms of residuals were calculated for polynomial drifts of degree 1, 2, and 3 determined by ordinary least squares. They show that in

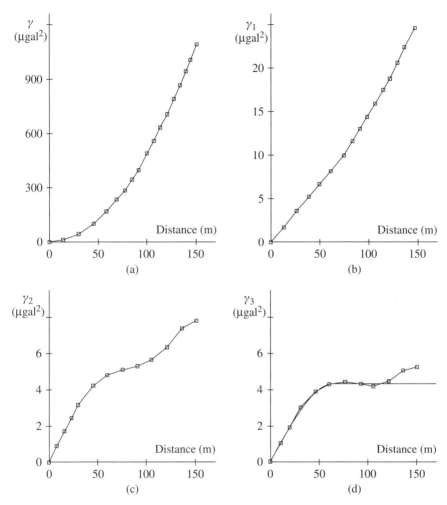

FIGURE 2.31   Microgravimetry in a quarry: (a) raw variogram; and variogram of residuals obtained after subtracting a global drift of degree k estimated by least squares: (b) $k = 1$; (c) $k = 2$; (d) $k = 3$. Notice the change in vertical scale between the raw variogram and the variograms of residuals. [From Chilès (1979b).]

order to eliminate the effect of the drift, we must consider a global drift of degree 2, if not 3. The variogram has a range of 60 m, or about one-sixth of the length of the domain. In this case the variogram of residuals is practically unbiased up to half the length of the domain (this is apparent when constructing simulations without drift using the techniques presented in Chapter 7, and then comparing, on the simulations, the raw variogram with the variogram of residuals). We obtain a good fit of γ_3 (variogram of residuals associated with a drift of degree 3) by a spherical model with range $a = 60$ m and sill $C = 430 \ \mu\text{gal}^2$. To this we must add a measurement error variance of 4 μgal^2, not perceptible on the sample variogram; geophysicists have told us, however, that the measurements are made with an error of ± 4 μgal (hence a standard deviation of 2 μgal and a variance of 4 μgal^2). To cross-validate the fit, 146 sample points were estimated from their neighbors and gave a mean standardized square error of 1.06, which is excellent.

2.8 SIMPLE APPLICATIONS OF THE VARIOGRAM

We conclude this chapter with simple applications of the variogram to variance calculations and sampling design. Here the form of the estimator is fixed, as when practically dictated by symmetries. We wish to compute (1) elementary estimation variances, such as the variance of estimation of a square or rectangular block by a central sample or by four samples located at the vertices, and (2) a global estimation variance when the data are more or less regularly located within the domain of study so that the sample mean can be taken as a sensible estimator of the regional mean. The variogram also enables the calculation of dispersion variances of small units partitioning larger domains. These results are very useful to design a sampling scheme. In mining, for example, one may wish to achieve a given accuracy for a global estimation or to optimize a pre-exploitation grid and control the local dispersion of the average grade of elementary mining units. It is assumed here that the phenomenon has no drift.

2.8.1 Estimation Variance

Let us begin with a very simple case, considering only two locations, x and $x + h$, and assuming that we know $Z(x)$ but not $Z(x + h)$; in the absence of any other information, we assign location $x + h$ the known value of location x. Since $2\,\gamma(h)$ is equal to the variance of $Z(x + h) - Z(x)$, $2\,\gamma(h)$ quantifies the error incurred when estimating $Z(x + h)$ by $Z(x)$. More generally, if we estimate the average value in a domain V

$$Z_V = \frac{1}{|V|} \int_V Z(x)\, dx$$

by the average of N sample points x_α,

$$\hat{Z} = \frac{1}{N} \sum_{\alpha=1}^{N} Z(x_\alpha)$$

we obtain, by extending (2.10) to the case of stochastic integrals,

$$\mathrm{Var}\left(\hat{Z} - Z_V\right) = -\frac{1}{N^2} \sum_{\alpha=1}^{N} \sum_{\beta=1}^{N} \gamma(x_\beta - x_\alpha) + \frac{2}{N|V|} \sum_{\alpha=1}^{N} \int_V \gamma(x_\alpha - x)\, dx$$
$$- \frac{1}{|V|^2} \int_V \int_V \gamma(x' - x)\, dx\, dx' \tag{2.84}$$

This result alternates exact and approximated expressions of the same integral, with the result that the estimation variance becomes smaller as:

- the network of sample points is tighter and more regular;
- the variogram is more regular, namely the variable itself is smoother in its spatial variation.

Such a variance is called an *estimation variance*. It is also referred to as an *extension variance* when one extends the value measured at one point, or on a small volume v, to the volume V. Denoting by

$$Z_v = \frac{1}{|v|} \int_v Z(x)\, dx$$

the average value of $Z(x)$ in v, (2.84) takes the form

$$\mathrm{Var}(Z_v - Z_V) = -\frac{1}{|v|^2} \int_v \int_v \gamma(x' - x)\, dx\, dx' + \frac{2}{|v||V|} \int_v \int_V \gamma(x' - x)\, dx\, dx'$$

$$-\frac{1}{|V|^2} \int_V \int_V \gamma(x' - x)\, dx\, dx' \tag{2.85}$$

The arithmetic mean $\hat{Z}$ is seldom used for local estimation. It is preferable to consider samples in V and its immediate vicinity and apply some weighting (kriging). For global estimation, however, if the data are evenly distributed so that they have approximately similar zones of influence, kriging will give approximately equal weight to all data points, and the estimator $\hat{Z}$ can be used. The estimation variance can then be calculated by formula (2.84). But if N is large and/or the domain V has a complex shape, which is often the case in real applications, this formula is quite cumbersome. A variety of approximations have been proposed to handle the most common cases encountered in the global estimation of an orebody, a forest, or an agricultural field. Some of them are presented in Section 2.8.3; they allow the calculation of the estimation variance of the resource in any 2D or 3D volume from regular or random sampling. Approximation formulas based on the irregular terms of the variogram are also available. Thanks to other formulas, it is also possible to deal with line sampling (along boreholes or transects) by cutting V into slices centered on the lines and combining errors incurred by estimating the lines from the samples and estimating the slices from the lines. The interested reader is referred to Matheron (1965, 1970), David (1977), and Journel and Huijbregts (1978) for a thorough description of the methods. Applications to the estimation of fish abundance are presented by Rivoirard et al. (2000).

2.8.2 Dispersion Variance

Let us consider a domain V that can be partitioned into N cells v_i which are identical to a cell v up to a translation. Let us also consider a particular realization $z(x)$ of the IRF $Z(x)$. If the mean values $z(v_i)$ of $z(x)$ in the cells v_i are all known, the mean of the N values $z(v_i)$ is simply the mean value $z(V)$ in the domain V. We therefore define the sample variance of v_i in V by

$$s^2(v \mid V) = \frac{1}{N} \sum_{i=1}^{N} \left[z(v_i) - z(V) \right]^2 \tag{2.86}$$

This variance measures the dispersion of the values in the cells v_i partitioning V. It can also be viewed as the variance of the estimation of $z(V)$ by $z(v_i)$, with the cell v_i being chosen at random among the N cells partitioning V.

Now, if the values $z(v_i)$ are not available, $s^2(v \mid V)$ cannot be calculated, but we can obtain its expected value $\sigma^2(v \mid V)$ by randomizing (2.86) with respect to the realization. This is equal to the expected value of the extension variance of $Z(v_i)$ to $Z(V)$, with the cell v_i being chosen at random, or equal to the mean of the N extension variances similar to (2.85) (with v_i instead of v). Because v_i is a partition of V, the cross term in (2.85) is, on the average, equal to the third term of the right-hand side, so that the variance of the $Z(v_i)$ in V becomes

$$\sigma^2(v \mid V) = \frac{1}{|V|^2} \int_V \int_V \gamma(x' - x) \, dx \, dx' - \frac{1}{|v|^2} \int_v \int_v \gamma(x' - x) \, dx \, dx' \tag{2.87}$$

$\sigma^2(v \mid V)$ is called the dispersion variance of v within V. For usual variogram models the second integral in (2.87) increases with the volume v of the support, hence the commonly observed decreasing-variance-with-volume relationship. In the case where v is limited to a single point, the preceding expression is reduced to

$$\sigma^2(0 \mid V) = \frac{1}{|V|^2} \int_V \int_V \gamma(x' - x) \, dx \, dx' \tag{2.88}$$

which generalizes the sample variance (2.5). If the variogram $\gamma(h)$ reaches a sill and if the domain V is very large with respect to the range, the variance of $Z(x)$ in V approaches the global point variance $\gamma(\infty) = C(0)$. This does not hold true for unbounded variograms.

Due to (2.88), the dispersion variance (2.87) can be written

$$\sigma^2(v \mid V) = \sigma^2(0 \mid V) - \sigma^2(0 \mid v)$$

When v is not a point and V cannot be partitioned into identical cells v_i, this formula is taken as a definition of the dispersion variance of v within V, but it no longer has a precise significance. It can even be negative, for example, one has $\sigma^2(V \mid v) = -\sigma^2(v \mid V)$.

Dispersion variances satisfy the following *additivity property*: If $v \subset V \subset \mathcal{V}$, then

$$\sigma^2(v \mid \mathcal{V}) = \sigma^2(v \mid V) + \sigma^2(V \mid \mathcal{V}) \tag{2.89}$$

Equation (2.89) is sometimes called "Krige's relationship" and is similar to an analysis of variance formula (note that this property is also valid among experimental dispersion variances).

From a practical point of view, it should be remembered that the double integrals over the same domain, which we find in expressions (2.84), (2.85), (2.87), and (2.88), can be expressed as simple integrals using the Cauchy algorithm. Analytical expressions for the various simple or double integrals can be obtained in 2D for rectangular cells for simple models such as the spherical one. In 3D a symbolic calculation of these integrals has been considered by Marbeau and Marbeau (1989). The value of these integrals can also be read from graphs [e.g., see Journel and Huijbregts (1978, pp. 125–147), for a spherical or exponential variogram in 2D and 3D].

2.8.3 Sampling Design

The preceding results enable us to design sampling patterns, namely to choose the locations of N points x_i that will be sampled for estimating a given domain V. This is possible because the estimation variance does not depend on the unknown values $Z(x_i)$ but only on their location and the variogram. Of course some data are necessary to identify the variogram. But once this is known, it is possible to predict the estimation variance for any proposed sampling pattern and thus optimize it to achieve a desired precision. We begin with an elementary example and then compare three main patterns: random sampling, stratified random sampling, and regular grid. Naturally in real applications the design of a sampling strategy is largely constrained by practical considerations. These can be dealt with, to some extent, by considering kriging estimators.

A Simple Example

Suppose that one is interested in the mean value of some parameter, say temperature, over a one-hour period. Which of these two estimates is better: the temperature at half the hour or the average of two consecutive measurements on the hour?

The problem amounts to the estimation of the average of $Z(x)$ over the segment $[0, L]$ either by the estimator $\hat{Z}_1 = Z(L/2)$ or by $\hat{Z}_2 = \frac{1}{2}[Z(0) + Z(L)]$. The corresponding estimation variances are given by formula (2.84). To express them in a simple manner, let us introduce two auxiliary functions of the variogram $\gamma(h)$:

$$\chi(h) = \frac{1}{h} \int_0^h \gamma(u) \, du$$

$$F(h) = \frac{1}{h^2} \int_0^h \int_0^h \gamma(x' - x) \, dx \, dx' = \frac{2}{h^2} \int_0^h u \, \chi(u) \, du \qquad (h > 0)$$

The two variances of estimation—σ_1^2 from a single central sample, σ_2^2 from two extreme samples—are then

$$\sigma_1^2 = 2\chi\left(\frac{L}{2}\right) - F(L), \qquad \sigma_2^2 = 2\chi(L) - F(L) - \frac{\gamma(L)}{2}$$

Let us assume a variogram of type $\gamma(h) = b\,|h|^\alpha$. The auxiliary functions take the form

$$\chi(h) = \frac{b\,h^\alpha}{\alpha + 1}, \qquad F(h) = \frac{2b\,h^\alpha}{(\alpha + 1)(\alpha + 2)}$$

which allows the calculation of σ_1^2 and σ_2^2. The results are given in Table 2.1 for typical values of α and $b = 1$.

We observe the following:

- For $\alpha = 0$, the limiting case of a pure nugget effect, only the number of samples matters so that σ_2^2 is half σ_1^2.
- This advantage decreases when α increases, and for $\alpha = 1$ the two patterns are equivalent.
- As α increases beyond 1, the variogram becomes more regular, and a single centrally located sample becomes better than two ill-placed ones.
- For $\alpha = 2$, which would correspond to a linear random function, the variances are zero.

Coming back to the initial problem, we conclude that in a turbulent environment ($\alpha = 2/3$) it is better to use the pattern with two measurements, whereas a centrally located sample is better in the case of smooth temperature variations ($\alpha > 1$).

A similar approach is used in mining to determine the final drill hole spacing ensuring that the average grade of blocks delimited by the drill holes is known with a prespecified estimation variance.

TABLE 2.1 Estimation Variance of an Interval of Length L from One Central Sample or Two Extreme Samples for the Variogram $\gamma(h) = |h|^\alpha$

α	σ_1^2	σ_2^2
0	1	0.5
0.5	$0.409L^{1/2}$	$0.300L^{1/2}$
1	$0.167L$	$0.167L$
1.5	$0.054L^{3/2}$	$0.071L^{3/2}$
2	0	0

Pure Random Sampling

Consider now the more general problem of the design of a pattern for estimating the average value Z_V of $Z(x)$ over a domain V. The survey will include N samples $\{Z(x_i) : i = 1, \ldots, N\}$, and Z_V will be estimated by $\hat{Z} = \frac{1}{N}\sum_{i=1}^{N} Z(x_i)$. How should this pattern be selected?

A first possibility is a random pattern where the samples are randomly scattered within V. More specifically, the x_i are independently located in V with uniform density $1/|V|$. An elegant way to derive the estimation variance is the following: For a given realization $z(x)$, the values $z(X_i)$ are random through X_i and are independent. Since X_i is uniformly distributed within V, $z(X_i)$, considered as an estimator of z_V, satisfies

$$E[z(X_i) - z_V] = 0$$
$$E[z(X_i) - z_V]^2 = s^2(0 \mid V)$$

where $s^2(0 \mid V)$ is the variance of $z(x)$ within V. Since the estimator $\hat{z}$ is the mean of the $z(X_i)$ and the X_i are independent, $\hat{z}$ is unbiased and its estimation variance is $\frac{1}{N}s^2(0 \mid V)$.

Randomizing the realization, we find that

$$\sigma_{\text{Rand}}^2 = E(\hat{Z} - Z_V)^2 = \frac{1}{N}\sigma^2(0 \mid V)$$

Note that this variance is that of the random pattern *before* the locations of the samples have been chosen. Once the pattern is fixed, the correct estimation variance is that of formula (2.84). The preceding result can then be used as an approximation, since it is equal to the average of the estimation variances associated with all possible realizations of the set of sampling points.

Stratified Random Sampling

This time, V is divided into N similar disjoint zones of influence v_i. Within each v_i a sample is placed at random with uniform density and independently of other samples. For a given realization $z(x)$, the error is

$$\hat{z} - z_V = \frac{1}{N}\sum_{i=1}^{N}\left[z(X_i) - z_{v_i}\right]$$

The partial errors $z(X_i) - z_{v_i}$ are independent. By a similar argument to that just seen, we find after randomization of the realization

$$\sigma_{\text{Strat}}^2 = E(\hat{Z} - Z_V)^2 = \frac{1}{N}\sigma^2(0 \mid v)$$

The variance of estimation has the same form as in the purely random case, except that $\sigma^2(0 \mid V)$ is replaced by $\sigma^2(0 \mid v)$, namely the variance of a point sample within its zone of influence v. Now by the additivity relationship (2.89) we have

$$\sigma^2(0 \mid V) - \sigma^2(0 \mid v) = \sigma^2(v \mid V) \geq 0$$

which proves that the stratified random pattern is always more efficient than the purely random pattern.

Square Grid

In this pattern the sampling is performed at the nodes of a regular grid with square or cubic cells if the variogram is isotropic, or on a grid with a direction and a ratio of elongation adapted to the geometric anisotropy of the variogram. Since in the latter case isotropy can be restored by means of a linear transformation of the coordinates, we will only consider the isotropic case. The domain V can therefore be partitioned into square or cubic cells v_i with a sample at the center of each cell. The estimation variance should be computed by application of formula (2.84), but this is usually too cumbersome. Since the estimation error is of the form

$$\hat{Z} - Z_V = \frac{1}{N} \sum_{i=1}^{N} \left[Z(x_i) - Z_{v_i} \right]$$

Matheron (1965, Section XII-3) proposes an approximation principle that consists in assuming that these partial errors are uncorrelated. This approximation has been shown to be quite good for usual isotropic variogram models provided that we use a square grid. Then

$$\sigma^2_{\text{Grid}} = \mathrm{E}(\hat{Z} - Z_V)^2 \approx \frac{1}{N} \sigma^2_{\mathrm{E}}(0, v)$$

where $\sigma^2_{\mathrm{E}}(0, v)$ is the extension variance of a central sample to a cell v. The estimation variance is simply computed by dividing the elementary extension variance of a sample to its zone of influence by the number of samples N.

For the usual variogram models, $\sigma^2_{\mathrm{E}}(0, v)$ is smaller than $\sigma^2(0 \mid v)$, so that we finally have

$$\sigma^2_{\text{Grid}} \leq \sigma^2_{\text{Strat}} \leq \sigma^2_{\text{Rand}}$$

Comparison of the Three Patterns

To get an idea of orders of magnitude, let us consider the 2D case where v is a square of side l. For the unit square $S_1 = [0, 1] \times [0, 1]$ we have

$$\int_{S_1} |x| \, dx = \frac{\sqrt{2} + \log(1 + \sqrt{2})}{3} = 0.765$$

$$\int_{S_1} \int_{S_1} |x' - x| \, dx \, dx' = \frac{2 + \sqrt{2} + 5 \log(1 + \sqrt{2})}{15} = 0.521$$

so that by application of (2.88) and (2.84) we obtain for the square of side l and the linear variogram $\gamma(h) = b\,|h|$,

$$\sigma^2(0\,|\,v) = 0.521b\,l, \quad \sigma_E^2(0, v) = 0.244\,b\,l$$

The ratio $\sigma^2(0\,|\,v)/\sigma_E^2(0, v) = 2.139$ shows that for a linear variogram the regular grid pattern is twice more efficient than the stratified random pattern, itself more efficient than the purely random scheme.

In order to have a complete comparison between the sampling patterns, let us further assume that the domain V is itself a square with side $l\sqrt{N}$. Then the variance of a sample within V is

$$\sigma^2(0\,|\,V) = 0.521b\,l\sqrt{N}$$

We can express the three estimation variances in terms of $\sigma^2 = \sigma^2(0\,|\,V)$ and the number of samples N, which are two independent parameters:

$$\sigma_{\text{Rand}}^2 = \frac{\sigma^2}{N}, \quad \sigma_{\text{Strat}}^2 = \frac{\sigma^2}{N^{3/2}}, \quad \sigma_{\text{Grid}}^2 \approx \frac{\sigma^2}{2.139N^{3/2}}$$

We can appreciate the benefit of a sampling pattern exploiting spatial correlations over crude random sampling: The variance is reduced by a factor of $N^{3/2}$ instead of N. A slight improvement over a regular square grid can be obtained with an hexagonal grid, as shown by Matérn (1960). The square grid remains usually preferred because it is in phase with the request of most applications to also deliver the local estimations of average values in square or cubic blocks.

The same type of comparison can be made in the case of a bounded variogram. But the results now depend on an additional parameter: the range a of the variogram. Let us focus on the comparison between the stratified random sampling and the regular grid for a spherical isotropic variogram. The ratio $\sigma^2(0\,|\,v)/\sigma_E^2(0, v)$ is always greater than 1, which proves that the regular grid performs always better than stratified random sampling. However, this advantage fades away as the size l of the grid cell becomes large with respect to the range a of the variogram: The ratio decreases from 2.14 for $l/a = 0$ to 1 for $l/a = \infty$, with an intermediate value 1.61 for $l/a = 1$. The reason is that for a large l/a ratio, the influence of a sample is purely local anyway, and the center of the square looses its strategic superiority.

The above considerations address an idealized sampling problem that came up at the early stages of the development of mining geostatistics, when statisticians, uninformed of Matérn's work in forestry (1960), were trying to force random sampling on mining engineers who preferred systematic grids. In reality the optimum data collection plan is very application-specific. It depends on the objective pursued, which may not just be the reduction of variance. Practical constraints are also important, such as cost or access (what if there is a large rock where a soil sample is supposed to be collected?). The optimal sampling further depends on the estimator selected. In the above comparison we have just used the plain mean, but we could also consider more sophisticated estimators, such as kriging estimators, and calculate estimation variances for different sample placements. Instead of minimizing the global estimation variance, we could also consider minimizing the average or the maximum of kriging variances of points or blocks discretizing the study domain. The interested reader may consult Rodríguez-Iturbe and Mejía (1974) for the design of rainfall networks, Bras and Rodríguez-Iturbe (1985) for sampling in hydrogeology, Gilbert (1987) for pollution monitoring, Rivoirard et al. (2000) for the survey of fish abundance, and Cressie (1991, Section 5.6) as well as Müller (2007) for an overview of approaches.

Designing the sampling to minimize a global on local estimation variance does not lead to the same scheme as optimizing the inference of the variogram (Section 2.6.1 on the behavior of the variogram at the origin; also see below). It is possible to work with a mixed objective in a Bayesian framework. Marchant and Lark (2007) consider a total variance that includes the kriging variance and the uncertainty on variogram parameters and use simulated annealing to optimize data placement. The reader is referred to the book of Diggle and Ribeiro (2007) for complementary information.

2.9 COMPLEMENTS: THEORY OF VARIOGRAM ESTIMATION AND FLUCTUATION

Estimating and modeling the variogram pose problems that pertain more to epistemology than to mathematical statistics. These problems have attracted the attention of geostatisticians since the beginning (Jowett, 1955b; Matheron, 1965, Chapter 13) and have fostered extensive studies (see Alfaro, 1979, 1984, and the important methodological work of Matheron, 1978). Without going into details that are beyond the scope of this book, we will outline the main points. We consider here random functions without drift. In the case of a drift, similar results are obtained with the generalized variogram.

2.9.1 Estimation of the Regional Variogram

Regional Variogram and Sample Variogram

In practice, we consider a regionalized variable $z(x)$ defined over a bounded domain D. By interpreting $z(x)$ as a realization of an RF $Z(x)$, we have provided a theoretical definition of the variogram $\gamma(h)$. Generally, however, the phenomenon under study is unique (e.g., an orebody), and it is primarily the variogram of the regionalized variable in D that is of interest. This variogram is the regional variogram defined by (2.7):

$$\gamma_R(h) = \frac{1}{2|D \cap D_{-h}|} \int_{D \cap D_{-h}} [z(x+h) - z(x)]^2 dx$$

where D_{-h} represents the translate of set D by vector $-h$, $D \cap D_{-h}$ is the set of points such that x and $x + h$ belong to D, and $|D \cap D_{-h}|$ is the measure of this set (i.e., the geometric covariogram of D for the distance h). The regional variogram is a purely deterministic and empirical quantity. If we know $z(x)$ at every point of D, γ_R is completely determined. It constitutes a summary of the structural characteristics of the regionalized variable and, in this sense, conveys a physical significance independently of the probabilistic interpretation that we can construct.

In practice, $z(x)$ is only known at a certain number of sample points $\{x_\alpha : \alpha = 1, \ldots, N\}$. As the regional variogram cannot be determined directly, we calculate the sample variogram (2.6),

$$\hat{\gamma}(h) = \frac{1}{2N_h} \sum_{x_\beta - x_\alpha \approx h} [z(x_\beta) - z(x_\alpha)]^2$$

where the sum is extended to the N_h pairs (x_α, x_β) of sample points separated (approximately) by the vector h.

Is this sample variogram a good approximation of the regional variogram? In rare cases it is possible to provide an experimental answer to this question. One such case is Narboni's (1979) exhaustive survey of the Ngolo tropical forest. The forest was subdivided into 50-m × 50-m plots, and the number of trees was counted in each plot. As a result the regional variogram of the variable "number of trees per plot" can be calculated exactly.

The study area contains about 100 transects, each with 200 plots so that, by considering only one transect in 10, one obtains a 10% sampling. By varying the position of the first transect, it is possible to simulate 10 different sampling choices. Figure 2.32a shows the ten corresponding sample variograms, whose average gives the regional variogram (the variograms are along the direction of the transects). These ten variograms are all similar, even

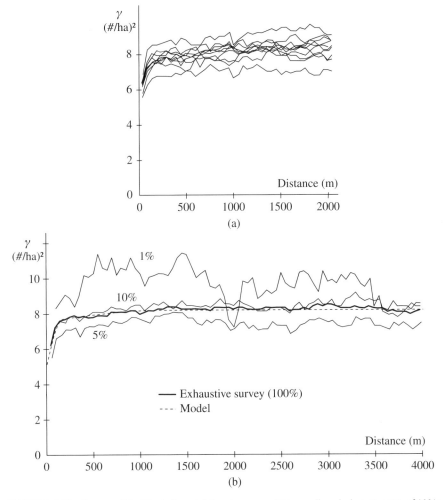

FIGURE 2.32 Survey of the Ngolo forest: (a) variograms of 10 sampling choices at a rate of 10%; (b) examples of variogram for sampling rates of 100% (average of the ten preceding variograms), 10%, 5%, and 1%. [From Narboni (1979), © Bois et Forêts des Tropiques.]

though their sills deviate by as much as 10% or 15% from the average sill. So with a 10% sampling one can get a good estimate of the regional variogram.

Figure 2.32b shows sample variograms for different sampling rates: 100% (the regional variogram, average of the curves from Figure 2.32a), 10%, 5%, and 1%. The deterioration is obvious at the 1% sampling rate (a single transect sampled). This is also seen from the variance of the samples: The variance obtained on sampling a single transect can vary, depending on which transect is selected, between 4.4 and 11.4. With a 10% rate, however, the variance varies only between 7.2 and 9.0 for the 10 possible sampling choices. Notice that the overall aspect of the sample variograms is correct, which is the most important: It is only the value of the sill that is poorly estimated. As we will see in Section 3.4.3, this inaccuracy only affects the kriging variance and not the estimate itself.

Estimation Variance of the Sample Variogram

Exhaustive sampling situations are exceptional, and we must consider the general case where the regional variogram γ_R cannot be determined experimentally. If we fix h and let

$$q_h(x) = \tfrac{1}{2}[z(x+h) - z(x)]^2$$

we see that $\gamma_R(h)$ is simply the average value of $q_h(x)$ over $D \cap D_{-h}$, and that $\hat{\gamma}(h)$ is the average value of the N_h data $q_h(x_i)$, where the x_i constitute a subset of the x_α. It is reasonable to expect that if sufficient data are available and fairly well distributed, $\hat{\gamma}(h)$ is close to $\gamma_R(h)$. If, for example, the data are on a regular grid, and if h is a multiple of the grid spacing, $\hat{\gamma}(h)$ is simply the discrete approximation of the integral defining $\gamma_R(h)$.

To go further, we have to determine the behavior of the regionalized variables $z(x)$ and $q_h(x)$. In our models we interpret $z(x)$ as the realization of an RF $Z(x)$, and then $q_h(x)$ is a realization of the RF

$$Q_h(x) = \tfrac{1}{2}[Z(x+h) - Z(x)]^2$$

We assume here that $Z(x)$ is an SRF or an IRF with variogram $\gamma(h)$ and that the RF $Q_h(x)$ has second-order moments and is stationary. Let $G_h(x' - x)$ denote the covariance of $Q_h(x)$ and $Q_h(x')$. To avoid confusion, Γ_R and $\hat{\Gamma}$ will now denote the random versions of the regional variogram and the sample variogram, or explicitly

$$\Gamma_R(h) = \frac{1}{2|D \cap D_{-h}|} \int_{D \cap D_{-h}} [Z(x+h) - Z(x)]^2 dx = \frac{1}{|D \cap D_{-h}|} \int_{D \cap D_{-h}} Q_h(x)\, dx \qquad (2.90)$$

$$\hat{\Gamma}(h) = \frac{1}{2N_h} \sum_{x_\beta - x_\alpha \approx h} [Z(x_\beta) - Z(x_\alpha)]^2 = \frac{1}{N_h} \sum_{x_\alpha - x_\alpha \approx h} Q_h(x_\alpha) \qquad (2.91)$$

Naturally,

$$E[\Gamma_R(h)] = E[\hat{\Gamma}(h)] = \gamma(h)$$

since $Q_h(x)$ has for expectation $\gamma(h)$. We can thus characterize the error incurred by taking the sample variogram for the regional variogram by the variance of $\hat{\Gamma}(h) - \Gamma_R(h)$. Considering their respective definitions (average of N_h values $Q_h(x_i)$ for one, average of $Q_h(x)$ over $D \cap D_{-h}$ for the other) brings us back to a standard calculation of estimation variance to be carried out with the covariance $G_h(x' - x)$ of the RF $Q_h(x)$.

$G_h(x' - x)$, however, is a fourth-order moment of the RF $Z(x)$. Determining the precision of the calculation of the second-order moment thus requires prior knowledge of the fourth-order moment. But the latter can generally only be evaluated with mediocre precision, related to the eighth-order moment, and so on. This way the problem can be displaced endlessly. We can nevertheless determine orders of magnitude by considering classic cases of spatial distribution. The Gaussian case will be examined first, and then indications will be given for random functions with skewed marginal distributions.

Gaussian Case

We know that if U_1 and U_2 are jointly Gaussian random variables with zero mean and covariance σ_{12}, the centered covariance of U_1^2 and U_2^2 is

$$\text{Cov}\left(U_1^2, U_2^2\right) = 2\,\sigma_{12}^2$$

From this we can deduce that for an IRF $Z(x)$ with bi-Gaussian increments, the covariance $G_h(x' - x)$ is given by

$$G_h(x' - x) = \tfrac{1}{2}[\gamma(x' - x + h) - 2\gamma(x' - x) + \gamma(x' - x - h)]^2 \tag{2.92}$$

This expression allows us to calculate the estimation variance of the variogram. Matheron (1965, pp. 229–230) shows that when the data are on a regular grid in n-dimensional space, we have to a first approximation

$$\text{Var}[\hat{\Gamma}(h) - \Gamma_R(h)] = 4\gamma(h)\sigma_h^2$$

where σ_h^2 denotes the variance of estimation of the mean value of $Z(x)$ in $D \cap D_{-h}$ from the N_h data $Z(x_\alpha)$ involved in the expression (2.91) of $\hat{\Gamma}(h)$. Thus, provided that there are enough pairs of points to calculate $\hat{\gamma}(h)$, we are sure of being able to evaluate $\gamma_R(h)$ with good precision: $\hat{\gamma}$ is a consistent estimator of γ_R. Figure 2.33 shows an example of the relative standard deviation (coefficient of variation)

$$\frac{\sqrt{\text{Var}[\hat{\Gamma}(h) - \Gamma_R(h)]}}{\gamma(h)}$$

in the case of 11×11 data on a regular 2D grid and for a spherical variogram or a variogram of type $|h|^\alpha$ (the figure results from theoretical calculations and not from simulations). It is seen that:

- The precision improves with the regularity of the variogram (large range or high α).
- The precision is practically of the same order of magnitude for all values of $h < L/2$.

General Case

To give a glimpse of the precision that one can expect with non-Gaussian RFs, let us consider three different RFs derived from independent standard Gaussian RFs $U(x)$ and $V(x)$ with the same correlogram $\rho(h)$:

1. $Z_1(x) = U(x)V(x)$. Its marginal distribution has the probability density function $f(z) = \frac{1}{\pi}K_0(|z|)$, where K_0 is the order-0 Bessel function of the second kind; its covariance is $C(h) = \rho(h)^2$.

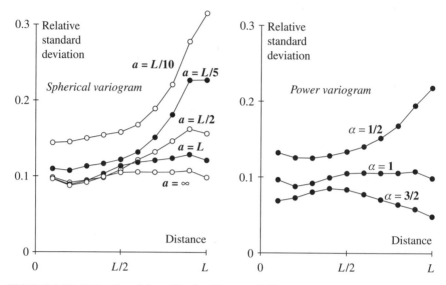

FIGURE 2.33 Estimation of the regional variogram: relative standard deviation in the Gaussian case for a spherical variogram with range a (on the left) and for a variogram of type $|h|^{\alpha}$ (on the right). The data are on a square $L \times L$ sampled by 11×11 points.

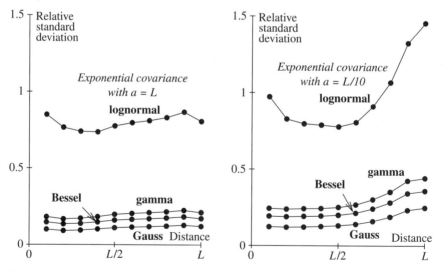

FIGURE 2.34 Estimation of the regional variogram: relative standard deviation for different spatial distributions, for an exponential variogram with scale parameter L (on the left) and $L/10$ (on the right). The data are on a square $L \times L$ sampled by 11×11 points.

2. $Z_2(x) = U(x)^2$. An RF with a gamma marginal distribution with parameters $1/2$ and $1/2$ (chi-square distribution on one degree of freedom); its covariance is $C(h) = 2 \, \rho(h)^2$ and its coefficient of variation is $\sqrt{2} = 1.414$.

3. $Z_3(x) = e^{U(x)}$. A lognormal RF with covariance $C(h) = e \, [e^{\rho(h)} - 1]$, thus with a fairly high coefficient of variation ($\sqrt{e - 1} = 1.311$).

These three RFs were selected because it is not too difficult to calculate the covariance $G_h(x' - x)$ from the results relative to the Gaussian RFs [see Alfaro (1979)]. For the first two cases it is very easy to obtain an exponential covariance ($\rho(h)$ must also be exponential with twice the scale parameter). As for the lognormal RF, it has been assigned a covariance as close as possible to the exponential model. Figure 2.34 shows the results for this type of model. Three conclusions can be drawn:

1. The variogram is less precise than in the Gaussian case.
2. It is acceptable for distributions such as those of Z_1 and Z_2 which are not too long tailed.
3. The sample variogram of a variable with a long-tailed distribution, like Z_3, bears only a distant relationship to the regional variogram.

This last result confirms the interest of analyzing the Gaussian-transformed data in complement to the raw data.

2.9.2 Modeling the Regional Variogram

Fluctuation Variance of the Regional Variogram

Even if we knew the value of the regionalized variable $z(x)$ at every point of the studied domain D and were capable of calculating the regional variogram $\gamma_R(h)$ for any vector h, this would exhibit so many variations of detail that we would have to simplify it to be able to express it in a usable form. This amounts to considering that two very similar regional variograms have the same parent variogram $\gamma(h)$. This simplification represents exactly a passage to the mathematical expectation: If one considers the studied regionalized variable as the realization of an IRF, namely as one realization among a set of similar realizations, the regional variogram of the regionalized variable is one among a family of regional variograms whose mean, or in probabilistic terms mathematical expectation, is a theoretical variogram $\gamma(h)$.

The passage to the IRF model enables us to define criteria for the precision required during modeling. In the framework of this model, the deviation $\Gamma_R(h) - \gamma(h)$ is a random variable. Its expectation is zero, and we can quantify the possible deviations by the fluctuation variance $\mathrm{Var}[\Gamma_R(h) - \gamma(h)]$. Let us therefore consider a given value of h and use the notations of the previous section. In view of definition (2.90) of $\Gamma_R(h)$, the variance of the fluctuation of $\Gamma_R(h)$ is simply the variance of the fluctuation of the mean of $Q_h(x)$ in $D \cap D_{-h}$. It can therefore be expressed in terms of the covariance $G_h(x' - x)$ of the RF $Q_h(x)$,

$$\mathrm{Var}[\Gamma_R(h) - \gamma(h)] = \frac{1}{|D \cap D_{-h}|^2} \int_{D \cap D_{-h}} \int_{D \cap D_{-h}} G_h(x' - x) \, dx \, dx' \qquad (2.93)$$

Calculating this variance again involves the fourth-order moments of the IRF $Z(x)$. Given the same remarks as in the preceding section, let us first examine the Gaussian case.

Gaussian Case

In view of the expression (2.92) for $G_h(x' - x)$, (2.93) is expressed as a function of $\gamma(h)$. The explicit calculation is complex, so for simplification we will consider the case where $\gamma(h)$ near the origin is equivalent to $b\,|h|^{\alpha}$. The relative fluctuation variance for small $|h|$ is, to a first approximation, of the form

$$\frac{\mathrm{Var}[\Gamma_R(h) - \gamma(h)]}{\gamma(h)^2} \approx A|h|^{4-2\alpha} + B|h|^n$$

where n is the dimension of the space (1, 2, or 3) [see Matheron (1978, p. 113) and the proof in $\mathbb{R}$ in Matheron (1970, Section 2.10.3)]. This relative variance tends to zero with $|h|$ provided that $\alpha < 2$.

Micro-Ergodicity

In the above situation the convergence of $\Gamma_R(h)/|h|^{\alpha}$ to a constant b when $h \to 0$ is ensured, provided that the RF is not m.s. differentiable. The parameter b then has an objective meaning: If we increase the number of sample points by refining the sampling grid, we can estimate it with precision. This is the concept of micro-ergodicity. It differs from conventional ergodicity, where one extends the data domain to infinity, which is of little interest to us because we always work in a bounded domain D. Micro-ergodicity refers to the case of a finite domain where we let the number of sample points tend to infinity by filling in the available space [Cressie (1991) calls this "infill asymptotics"].

The micro-ergodicity of the variogram in the neighborhood of the origin is therefore established, provided that it is not too regular (we find ourselves in an inverse situation to that of the estimation of the variogram). This is easily explained: If α is very close to 2, the realization $z(x)$ is a very regular function. When choosing a sufficiently small domain D, $z(x)$ can be assimilated to a linear function of x. But, depending on the location of D, the linear functions can have extremely variable slopes and thus impart almost any value of the coefficient of the term $|h|^{\alpha}$ of the regional variogram: The parameter b no longer has an objective meaning. In this case the probabilistic model is poorly adapted, at least at the scale of a small domain D.

Although it is generally possible to determine $\gamma(h)$ in the neighborhood of the origin, the fluctuation variance increases very rapidly with h (except if $\gamma(h)$ has a very small range with respect to the domain D). Thus in $\mathbb{R}$, for example, for an RF with a linear variogram known on [0, 1] [straightforward application of (2.93); see Matheron (1970), Chapter 2, exercise 16], we have

$$\frac{\mathrm{Var}[\Gamma_R(h) - \gamma(h)]}{\gamma(h)^2} = \begin{cases} \dfrac{4}{3}\dfrac{h}{L-h} - \dfrac{1}{3}\dfrac{h^2}{(L-h)^2} & \text{if } 0 \le h \le \dfrac{L}{2}, \\[4mm] 2 - \dfrac{4}{3}\dfrac{L-h}{h} + \dfrac{1}{3}\dfrac{(L-h)^2}{h^2} & \text{if } \dfrac{L}{2} \le h \le L \end{cases}$$

At $h = L/2$, the relative fluctuation variance already equals one, which is prohibitive.

Figure 2.35 shows the fluctuation standard deviation curves in $\mathbb{R}^2$ for the same cases as in Figure 2.33. Apart from the case of a small range, the theoretical variogram and the regional variogram may have only a distant relationship if $|h|$ is greater than half the length of the domain (if not sometimes less). This is, however, not serious insofar as geostatistical estimations depend much more on the behavior of the variogram at small distances than at large distances. From a practical point of view, we can accept that the variogram generally has no objective meaning at large distances and that it is pointless to try to refine the associated fit.

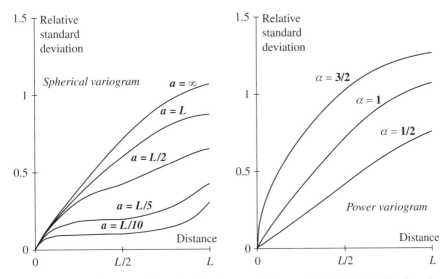

FIGURE 2.35 Fluctuation of the regional variogram: relative standard deviation in the Gaussian case for a spherical variogram with range a (on the left) and for a variogram of type $|h|^{\alpha}$ (on the right). The domain is a square $L \times L$.

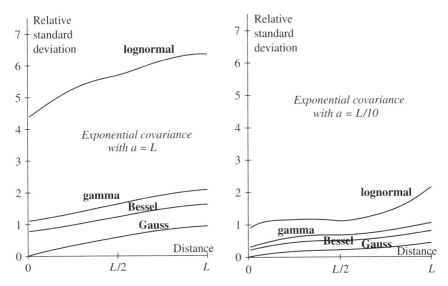

FIGURE 2.36 Fluctuation of the regional variogram: relative standard deviation for different spatial distributions, for an exponential variogram with scale parameter L (on the left) and $L/10$ (on the right). The domain is a square $L \times L$.

General Case

The conclusions are less encouraging for RFs that are clearly non-Gaussian, as has been shown by Alfaro (1979). We can see this from Figure 2.36, which shows the same RFs as Figure 2.34. Two conclusions can be drawn:

1. The fluctuation variance can be much larger than in the Gaussian case.
2. The relative variance no longer necessarily tends to zero when the lag h tends to zero: In other words, micro-ergodicity is no longer ensured, and the regional variogram does not even reproduce the behavior near the origin of the theoretical variogram.

In practice, this is frequently reflected in a proportional effect. If we treat it as such, which amounts to working on a variable conditioned on its local mean, we end up with a more satisfactory model. It is generally better to use methods based on prior transformation of the data into Gaussian variables (lognormal kriging, disjunctive kriging).

Micro-ergodicity is also no longer ensured for an indicator, or more generally for a mosaic RF (even if the marginal distribution is Gaussian). If the domain of study D is not large, a realization can very well be constant and thus give a regional variogram that is identically zero.

CHAPTER 3

Kriging

Once a map is drawn people tend to accept it as reality

—Bert Friesen

3.1 INTRODUCTION

A central problem in geostatistics is the estimation of a variable of interest over a domain on the basis of values observed at a limited number of points. Typically, one may want to build a grid model in view for example of drawing a contour map or of running some numerical simulator. Or the objective may be to make an inventory and compute amounts (of ore, trees, contaminants, etc.) within given areal or volumetric units. More generally, the desired quantity may be some function of the observed variable, but attention will be restricted here to linear functions.

From a deterministic viewpoint this is an interpolation problem. The variable of interest is approximated by a parametric function whose form is postulated in advance, either explicitly (e.g., polynomials) or implicitly (e.g., minimum curvature condition). The parameters are selected so as to optimize some criterion of best fit at the data points. Once the approximating function is determined, it is a simple matter to evaluate it wherever needed.

We will focus here on a probabilistic approach known as *kriging*, a term coined by G. Matheron in 1963 in honor of Danie Krige. It produces an interpolation function based on a covariance or variogram model derived from the data rather than an a priori model of the interpolating function. This is analogous to the prediction problem in time series: Given values of the past, usually at regular time intervals, predict the value of the signal at some time in the future. First the signal is analyzed, typically by computing and modeling the spectrum, and then a filter (= a predictor) is designed. Kriging follows a similar approach, but in a spatial setting where there is no general concept of past and future.

Geostatistics: Modeling Spatial Uncertainty, Second Edition. J.P. Chilès and P. Delfiner.
© 2012 John Wiley & Sons, Inc. Published 2012 by John Wiley & Sons, Inc.

TABLE 3.1   The Main Forms of Linear Kriging

Kriging Form	Mean	Drift Model	Prerequisite
Simple kriging (SK)	Known	None	Covariance
Ordinary kriging (OK)	Unknown	Constant	Variogram
Universal kriging (UK)	Unknown	Function of coordinates	Variogram
Kriging with external drift (KED)	Unknown	External variable	Variogram

A remark on terminology is in order here. The word *prediction* has become established as the determination of the value of a *random* quantity, whereas *estimation* refers to the inference of some *fixed* but unknown parameter of a model. This distinction originates from regression where we have the choice to either estimate the mean value of Y given X, or predict the value of an individual observation given X. The results are the same, but their variances are different and therefore it is necessary to distinguish them. Because spatial interpolation is closely related to regression, the terminology has followed, but the distinction is no longer critical. In standard English we predict an earthquake, or any particular event in the future, and we estimate mineral resources. In this book we will generally use the common language verb "estimate" in lieu of the technically focused "predict," and we will likewise use "confidence interval" for "prediction interval."[1]

Two other differences between the spatial processes considered in this book and time processes must be emphasized. First, the data are usually *irregularly spaced*, which rules out most of the methods commonly used in digital signal processing of either time signals or images. Second, there are *support effects* that require, at least in concept, a continuous rather than a discrete location indexing space.

The present chapter concentrates on solving the estimation problem by use of linear estimators. The theory is developed within the scope of a second-order statistical model involving only the mean $m(x)$ and the covariance function $\sigma(x, y)$ or the variogram $\gamma(x, y)$, *assumed known*. Note that the stationarity of σ or γ is not required to derive the estimators. In some cases such as space−time phenomena, it is possible to consider a nonstationary model. In most applications, however, a stationary covariance $C(h)$ or variogram $\gamma(h)$ is used. Table 3.1 summarizes the four main forms of kriging that we will discuss in this chapter and their underlying models.

The method of kriging has been extended in several directions. One extension deals with broader forms of nonstationarity than the UK model and shows how the minimal prerequisites for kriging can be reduced: It is the subject of Chapter 4. Another extension covers the multivariate case, namely when $Z(x)$ is vector-valued. The method, referred to as *cokriging*, is developed in Chapter 5, except for the special problem of filtering a random error in the data which is treated here. Finally, nonlinear estimators developed to evaluate nonlinear functions of $Z(x)$ are reviewed in Chapter 6.

---

[1] A confidence interval is a random interval around a deterministic parameter, whereas with a prediction interval both the predicted value and the interval are random.

3.2 NOTATIONS AND ASSUMPTIONS

- We denote by

$$\{Z(x): x \in D \subset \mathbb{R}^n\}$$

a random function used as a model for the regionalized variable of interest

$$\{z(x): x \in D \subset \mathbb{R}^n\}$$

representing reality. $z(x)$ is a realization of $Z(x)$. For simplicity we make no notational distinction between the (uppercase) parent random function $Z(x)$ and its particular (lowercase) realization $z(x)$, it being clear that all probabilistic calculations involve the random function and all numerical estimates involve the realization. The lowercase notation is used occasionally, when we want to emphasize the deterministic character of the expression (e.g., kriging as an interpolant).

- S denotes the set of points where $Z(x)$ has been sampled. In most cases, S is finite and consists of N data points denoted with Greek subscripts:

$$S = \{x_\alpha: \alpha = 1, \ldots, N\}$$

Occasionally, S can be infinite such as when data are continuously recorded along a profile.

- Values of functions at sample points are referenced by the subscripts of these points, such as

$$
\begin{aligned}
Z_\alpha &= Z(x_\alpha) & &\text{the data,} \\
m_\alpha &= m(x_\alpha) & &\text{mean value of } Z(x_\alpha)(= \mathrm{E}(Z_\alpha)), \\
\sigma_{\alpha\beta} &= \sigma(x_\alpha, x_\beta) & &\text{covariance between } Z(x_\alpha) \text{ and } Z(x_\beta)
\end{aligned}
$$

- The estimated quantity (the "objective") is of the general form

$$Z_0 = \int Z(x) p_0(dx)$$

for some integrable measure p_0. The case of point estimation $Z_0 = Z(x_0)$ corresponds to a Dirac measure $p_0(dx) = \delta(x - x_0)$ at the target point x_0, usually a grid node, whereas the estimation of a spatial average over a block v,

$$Z_0 = \frac{1}{|v|} \int_v Z(x) dx$$

corresponds to $p_0(dx) = (1/|v|) 1_v(x) dx$, where $1_v(\cdot)$ stands for the indicator function of the block v centered at the point x_0.

- Following established usage, kriging estimators are marked with an asterisk (*) superscript. In full explicit notations the kriging estimator of $Z(x_0)$ is of the form

$$Z^*(x_0) = \sum_{\alpha=1}^{N} \lambda_\alpha(x_0)Z(x_\alpha) + \lambda_0(x_0)$$

where $\lambda_\alpha(x_0)$ is a weight placed on $Z(x_\alpha)$ and $\lambda_0(x_0)$ is a constant that depends on x_0. For brevity this expression will be condensed to[2]

$$Z^* = \sum_{\alpha} \lambda_\alpha Z_\alpha + \lambda_0$$

with the understanding that the summation is extended over all α indexes in S. It must be kept in mind that *the weights λ_α depend on the location x_0* where the function is being estimated.

Support

We have said that the data were measured at points x_α. In reality the data are never collected at a single point but always involve a support of finite dimensions. This does not create any particular difficulty as long as all data have the same support and as long as this support is "compatible" with that of the objective. We speak of "point kriging" when the objective has exactly the same support as the samples (Sections 3.3 and 3.4). When the objective has a larger support, we speak of "block kriging" (Section 3.5). For compatibility the block must either be very large compared with the samples, which are then treated as points, or be a finite union of sampling units.

Neighborhoods

The theory is always derived as if all N data points were used in the estimation; this is the so-called global neighborhood case. In practice, N may be too large to allow computation and a "moving neighborhood" or "local neighborhood" has to be used, including only a subset of the data for the estimation of each grid node. This may alter the relationships between estimates at different grid nodes and introduce spurious discontinuities. Different methods have been proposed to handle large data sets and are discussed in Section 3.6.

3.3   KRIGING WITH A KNOWN MEAN

In this section we consider what could be called "the wonderful case of a known mean" which underlies the theory of *Simple Kriging* (SK). Indeed, knowing the mean of a stationary RF is already knowing very much. For some practical

---

[2] To make the presentation more accessible, the convenient and concise tensor notation $\lambda^\alpha Z_\alpha$ used by Matheron (1969a, 1970) will not be employed here.

purposes the mean alone may provide enough information. Knowing the mean also makes the theory very simple and endows the kriging estimator with all the nice properties. In case of a Gaussian RF, it coincides with the conditional expectation $E(Z_0|Z_1,\ldots,Z_N)$, which is the ideal estimator of Z_0 in the mean square sense. In all circumstances the error $Z^* - Z_0$ is uncorrelated with every Z_α and with Z^* itself.

In the real world the mean can be known only if there are repetitions of the phenomenon, as with space–time processes, or when the number of data becomes so large as to estimate the mean almost to perfection. The properties established under this condition may thus be regarded as *limit properties*.

3.3.1 Derivation of the Equations

For simplicity we consider here the case of point estimation. We want to estimate $z_0 = z(x_0)$ from N observations $z_1,\ldots,z_N$, using the affine estimator

$$z^* = \sum_\alpha \lambda_\alpha z_\alpha + \lambda_0$$

interpreted in the model as a realization of the random variable

$$Z^* = \sum_\alpha \lambda_\alpha Z_\alpha + \lambda_0$$

The constant λ_0 and the weights λ_α are selected so as to minimize in the model the expected mean square error $E(Z^* - Z_0)^2$.

First let us concentrate on λ_0. The mean square error (m.s.e.) can be written as

$$E(Z^* - Z_0)^2 = \mathrm{Var}(Z^* - Z_0) + [E(Z^* - Z_0)]^2$$

Since variances are insensitive to shifts, only the bias term on the right-hand side involves λ_0. To minimize the m.s.e., it is necessary to choose λ_0 so as to cancel the bias $E(Z^* - Z_0)$:

$$\lambda_0 = m_0 - \sum_\alpha \lambda_\alpha m_\alpha$$

The estimator Z^* becomes

$$Z^* = m_0 + \sum_\alpha \lambda_\alpha (Z_\alpha - m_\alpha) \tag{3.1}$$

This amounts to estimating the zero-mean variable $Y(x) = Z(x) - m(x)$ by the linear estimator

$$Y^* = \sum_\alpha \lambda_\alpha Y_\alpha$$

and adding the mean afterward. Thus we have established that the case of a known mean is equivalent to the case of a zero mean with $\lambda_0 = 0$, and from now on in this section we will consider that $Z(x)$ has a zero mean.

The m.s.e.—which now coincides with the variance—can be expanded in terms of the centered covariance $\sigma(x, y)$ of $Z(x)$,

$$E(Z^* - Z_0)^2 = \sum_\alpha \sum_\beta \lambda_\alpha \lambda_\beta \sigma_{\alpha\beta} - 2 \sum_\alpha \lambda_\alpha \sigma_{\alpha 0} + \sigma_{00}$$

The minimum of this quadratic function is obtained by canceling its partial derivatives with respect to the weights λ_α,

$$\frac{\partial}{\partial \lambda_\alpha} E(Z^* - Z_0)^2 = 2 \sum_\beta \lambda_\beta \sigma_{\alpha\beta} - 2\sigma_{\alpha 0} = 0$$

(That this is indeed a minimum is ensured by the positivity property of the covariance function—the m.s.e. is a convex function). The λ_α are solutions of the linear system of N equations

Simple Kriging System

$$\sum_\beta \lambda_\beta \, \sigma_{\alpha\beta} = \sigma_{\alpha 0}, \qquad \alpha = 1, 2, \ldots, N \qquad (3.2)$$

In matrix notation we have

$$\Sigma \lambda = \sigma_0$$

where $\Sigma = [\sigma_{\alpha\beta}]$ is the $N \times N$ matrix of data-to-data covariances, $\sigma_0 = [\sigma_{\alpha 0}]$ is the N-vector of covariances between the data and the target, and $\lambda = [\lambda_\alpha]$ is the N-vector of solutions.

These equations are the "best linear prediction" equations famous since the work of A. N. Kolmogorov (1941c) and N. Wiener (1942). They also appear as the "Yule-Walker" equations in time series [e.g., Box and Jenkins, (1976, p. 55)] and as the "normal equations" in linear regression [e.g., Rao, (1973, p. 266)].

The system (3.2), usually called the simple kriging system (SK), has a unique solution, provided that the matrix Σ is nonsingular.[3] This is always the case if the covariance function $\sigma(x, y)$ is strictly positive definite and if all sample points are distinct, which we will always assume (nonunique solutions may occur in case of a pure zonal anisotropy). Solving (3.2) is a routine computational problem.

The estimation variance σ_{SK}^2, called the *kriging variance*, associated with Z^* is obtained by substituting the solution of (3.2) in the m.s.e. Premultiplying (3.2) by λ_α and summing over all α gives

---

[3] If Σ is singular, the Singular Value Decomposition algorithm (SVD) can provide a particular solution minimizing the sum of squared weights [e.g., Press et al. (2007, Section 2.6)].

Simple Kriging Variance

$$\sigma_{SK}^2 = E(Z^* - Z_0)^2 = \sigma_{00} - \sum_\alpha \lambda_\alpha \sigma_{\alpha 0} \tag{3.3}$$

In matrix form we have

$$\sigma_{SK}^2 = \sigma_{00} - \boldsymbol{\lambda}' \boldsymbol{\sigma}_0$$

The kriging variance, or rather its square root, the kriging standard deviation, provides a measure of the error associated with the kriging estimator. Its use is discussed in Section 3.4.5. Notice that it does not depend on the values of the data but only on their locations. Notice also that if the covariance is multiplied by an arbitrary (positive) constant, the kriging weights do not change and the kriging variance is multiplied by that constant. Therefore in case of a stationary covariance function, one can scale the covariance by the common variance and write the system (3.2) in terms of correlation coefficients $\rho_{\alpha\beta}$ and $\rho_{\alpha 0}$, but the kriging variance is then also scaled. Finally, and still in the stationary case, it is clear that the kriging weights and variance are *shift invariant*: They do not change if the whole kriging configuration is shifted by an arbitrary vector h.

Examples

1. Consider the simplest case of only one sample point. Then (3.2) reduces to a single equation and the kriging estimator is

$$Z^*(x_0) = (\sigma_{10}/\sigma_{11})Z_1 = \rho_{10}(\sigma_{00}/\sigma_{11})^{1/2}Z_1$$

where ρ_{10} is the correlation coefficient between Z_1 and Z_0. One recognizes the standard linear regression of Z_0 on Z_1. As a function of x_0, $Z^*(x_0)$ is proportional to the covariance function σ_{10} and assumes the value Z_1 when $x_0 = x_1$. As x_0 moves away from x_1, the correlation ρ_{10} usually falls off and so does, logically, the influence of Z_1 on the estimation of Z_0. At large distances from x_1, $\rho_{10} = 0$ so that $Z^* = 0$; we can do no better than estimate $Z(x)$ by the mean.

From (3.3) the kriging variance is

$$\sigma_{SK}^2 = \sigma_{00} (1 - \rho_{10}^2)$$

2. We now complicate the problem and consider two points x_1 and x_2. Solving the 2×2 kriging system (3.2) for λ_1 and λ_2 and rearranging the terms lead to the estimator

$$Z^*(x) = \frac{1}{\Delta}[Z_1\, \sigma_{22} - Z_2\, \sigma_{12}]\, \sigma_{10} + \frac{1}{\Delta}[Z_2\, \sigma_{11} - Z_1\, \sigma_{21}]\, \sigma_{20}$$

with $\Delta = \sigma_{11}\,\sigma_{22} - \sigma_{12}^2$. This solution is represented graphically in Figure 3.1 (left), along the line joining x_1 and x_2; The stationary covariance function considered is the "cubic" model (2.49) with range a.

The estimator passes through Z_1 and Z_2 and tends to the mean zero as x_0 moves away from the data points, reaching zero exactly when the distance exceeds the range a. The three curves in Figure 3.1 illustrate the influence of the

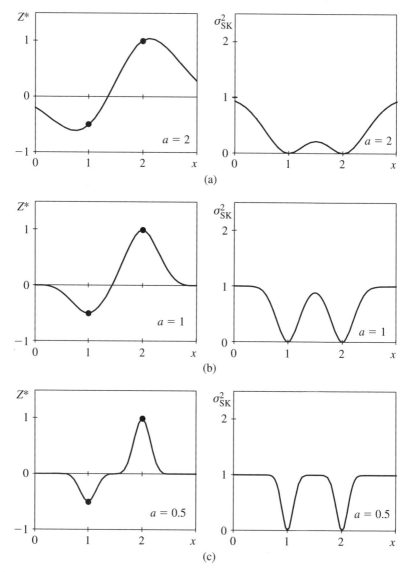

FIGURE 3.1 Simple kriging estimate and variance for the case of two data points and a cubic covariance model with $C(0) = 1$. From (a) to (c): the estimator becomes "wigglier" as the range a decreases.

range a on the solution: as a decreases, the covariance function falls off more rapidly and the information carried by the data Z_1 and Z_2 is considered more and more local.

The kriging variance is readily obtained from (3.3):

$$\sigma_{SK}^2 = \sigma_{00} - \frac{1}{\Delta} [\sigma_{22} \sigma_{10}^2 + \sigma_{11} \sigma_{20}^2 - 2\sigma_{12} \sigma_{10} \sigma_{20}]$$

This function is plotted in Figure 3.1 (right) along the line (x_1, x_2) for various values of a. It is seen that σ_{SK}^2 is zero at the sample points and reaches the maximum value of 1 under extrapolation. As could be expected, the variance reaches a local maximum at the midpoint between x_1 and x_2. When a is small, this maximum is equal to 1 because all sites are considered as extrapolated except in the immediate vicinity of the data.

3.3.2 Interpolation Properties of the Kriging Estimator

We can now generalize some of the properties observed in the above examples.

Consistency with Data Points

The kriging estimator is an exact interpolant. If x_0 coincides with a sample point, say x_1, then Z^* is equal to $Z(x_1)$. This can be verified by checking that the set of weights $\lambda_1 = 1$, $\lambda_\alpha = 0$ if $\alpha \neq 1$, satisfies the equations (3.2), and since the solution is unique, this is it. But it is simpler to note that $Z^* = Z(x_1)$ is certainly the best estimator of $Z(x_1)$ in the m.s.e. sense as it makes the error exactly zero. The kriging variance σ_{SK}^2 is naturally also zero.

Smoothing Relationship

Since kriging performs a linear averaging, we expect kriging estimates to be less dispersed than the data. This can be proved easily by considering the variance of $Z^*(x_0)$. From (3.2), we have

$$\text{Var } Z^* = \sum_\alpha \sum_\beta \lambda_\alpha \lambda_\beta \sigma_{\alpha\beta} = \sum_\alpha \lambda_\alpha \sigma_{\alpha 0}$$

and using (3.3) together with $\text{Var } Z_0 = \sigma_{00}$, we get the so-called *smoothing relationship*

$$\text{Var } Z^* = \text{Var } Z_0 - \sigma_{SK}^2 \tag{3.4}$$

$\text{Var } Z^*$ differs from $\text{Var } Z_0$ by an amount exactly equal to the kriging variance σ_{SK}^2 (which depends on x_0). This effect is illustrated in Figure 3.2: The kriging estimate wiggles near the data points, where σ_{SK}^2 is small, and merges gradually with the mean as data become sparser.

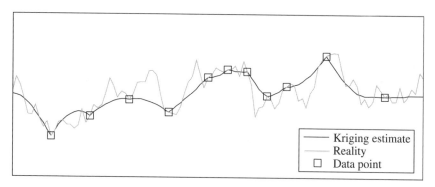

FIGURE 3.2 Illustration of the smoothing effect of kriging.

Note that despite its name the smoothing relationship does not, by itself, prove the existence of a spatial smoothing effect in the sense of a preferential attenuation of high frequencies. (In spectral analysis terms, we have only proved that the total power is reduced.) However, this spatial smoothing does usually take place when a global kriging neighborhood is used.

Incidentally we have also shown that $\sigma_{SK}^2 \le \text{Var } Z_0$. In a stationary model the estimation variance can never exceed the global variance, even if x_0 is very far from all data points (else the mean would become the estimator).

Kriging as an Interpolant

Though derived in a stochastic model, the function $Z^*(x) = z^*(x)$, once the data $Z_\alpha = z_\alpha$ are fixed, is a deterministic interpolant. To emphasize this fact, in this section we switch to lowercase notations for z, and to x instead of x_0 for the interpolated point. In order to identify the explicit form of the kriging interpolant we solve (3.2) for λ, and letting **z** be the N-vector of data, we get

$$z^* = \mathbf{z}' \boldsymbol{\Sigma}^{-1} \boldsymbol{\sigma}_x \tag{3.5}$$

In this expression, only $\boldsymbol{\sigma}_x$ depends on the location x. Defining $\mathbf{b} = \boldsymbol{\Sigma}^{-1} \mathbf{z}$, we have

$$z^*(x) = \mathbf{b}' \boldsymbol{\sigma}_x = \sum_\alpha b_\alpha \, \sigma(x_\alpha, x) \tag{3.6}$$

This is a linear combination of N covariance functions centered at the sample points x_α. The weights b_α do not depend on x, but they depend on the z_α (cf. Examples 1 and 2).

The interpolation formula (3.6) shows how the covariance function model of $Z(x)$ determines the continuity and regularity properties of the interpolant $z^*(x)$. If the covariance function is parabolic near the origin, $z^*(x)$ is differentiable; if it is linear near the origin, $z^*(x)$ is continuous but with cusps at the data points. If the covariance has a discontinuity at zero, there will be isolated jumps at the data points. These behaviors are illustrated in Figure 3.3. Notice in Figure 3.3b that kriging interpolation does not necessarily "look nice." Kriging is not designed to optimize the look of the interpolation, but instead its accuracy (e.g., as opposed to spline interpolation). The kriged map represents our real knowledge of reality; it shows details in densely sampled areas and is flat where data are scarce. By contrast, as we will see in Chapter 7, a simulation imagines how reality could look.

In fact the interpolation formula (3.5) can be derived directly by considering an interpolant of the form (3.6) and fitting the coefficients b_α so that $z^*(x)$

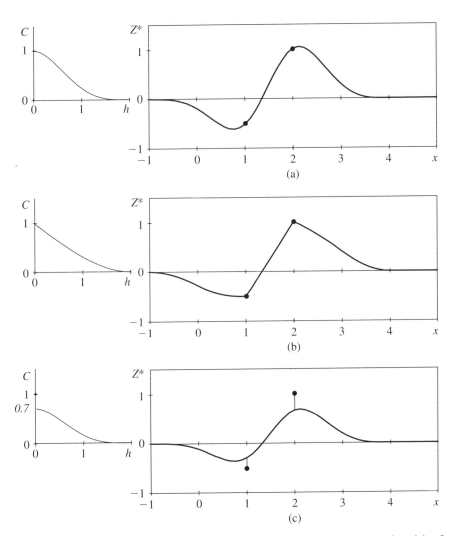

FIGURE 3.3 The dependence of the simple kriging estimate on the regularity near the origin of the covariance function: (a) parabolic behavior; (b) linear behavior; (c) discontinuity.

passes through all the data points. Indeed, the conditions $z^*(x) = z_\beta$ at each x_β entail that the b_α satisfy

$$\sum_\alpha b_\alpha \sigma_{\alpha\beta} = z_\beta \qquad \text{in matrix terms} \qquad \Sigma \mathbf{b} = \mathbf{z} \qquad (3.7)$$

which together with (3.6) entail (3.5). Depending on the order in which we multiply matrices in (3.5) we obtain the following two equivalent systems:

$$\begin{cases} \Sigma \lambda = \sigma_x \\ z^* = \lambda' \mathbf{z} \end{cases} \qquad \begin{cases} \Sigma \mathbf{b} = \mathbf{z} \\ z^* = \mathbf{b}' \sigma_x \end{cases} \qquad (3.8)$$

The second formulation has the advantage that $\mathbf{b}$ can be determined once for all and used to compute $z^*(x)$ by a simple scalar product with σ_x. This does not work for the kriging variance, however.

3.3.3 Kriging as a Projection

A simple geometric presentation of kriging can be made within the framework of Hilbert spaces of random variables (Section 1.1.2). In addition to providing intuitive insight, this approach permits the only rigorous proof for the case of continuously sampled data.

We consider here the finite-dimensional vector space $\mathcal{H}_{N+1}$ generated by all linear combinations of the $N+1$ zero mean, finite variance, random variables $Z_0 = Z(x_0)$ and $\{Z_\alpha : \alpha = 1, \ldots, N\}$, and all their limits in the mean square sense. A scalar product $<X, Y> = E(XY)$ is defined by the *noncentered* covariance, and the associated norm is $\|X\| = \sqrt{\langle X, X \rangle}$. The kriging problem can be reformulated as follows: Find X in the subspace $\mathcal{H}_N$ generated by the data $\{Z_\alpha : \alpha = 1, \ldots, N\}$ minimizing $\|Z_0 - X\|$. From Hilbert space theory we know that the minimum is achieved by a single element Z^*, called the "projection of Z_0 onto $\mathcal{H}_N$," which is the foot of the perpendicular dropped from Z_0 on $\mathcal{H}_N$. It is characterized by the orthogonality property (1.1),

$$\langle Z^* - Z_0, X \rangle = 0 \qquad \forall X \in \mathcal{H}_N \qquad (3.9)$$

As $\{Z_\alpha : \alpha = 1, \ldots, N\}$ forms a basis of $\mathcal{H}_N$, the above is equivalent to the N conditions

$$\langle Z^*, Z_\alpha \rangle = \langle Z_0, Z_\alpha \rangle, \qquad \alpha = 1, \ldots, N \qquad (3.10)$$

which by expansion give the system (3.2). This concise form reveals another characteristic property of the SK estimator: Z^* has the same covariance with each Z_α as Z_0 itself.

A particular case of (3.10) is the orthogonality of the kriging error and the kriging estimator

$$\langle Z^* - Z_0, Z^* \rangle = 0 \tag{3.11}$$

Furthermore, this orthogonality holds between the kriging error at x_0 and the kriging estimator $Z^*(x)$ at any other point, provided that $Z^*(x)$ is obtained from the *same* data as $Z^*(x_0)$. This property is the basis of the conditioning algorithm by kriging (Section 7.3.1).

The kriging variance is easily obtained from

$$\sigma_{SK}^2 = \|Z^* - Z_0\|^2 = \langle Z^* - Z_0, Z^* \rangle - \langle Z^* - Z_0, Z_0 \rangle$$

and the application of (3.11). We get

$$\sigma_{SK}^2 = \langle Z_0, Z_0 \rangle - \langle Z^*, Z_0 \rangle$$

which by expansion is seen to coincide with (3.3).

The smoothing relationship (3.4) has a simple geometric interpretation. From (1.2), or directly from the orthogonal decomposition $Z_0 = Z^* + (Z_0 - Z^*)$, we have

$$\|Z_0\|^2 = \|Z^*\|^2 + \|Z_0 - Z^*\|^2$$

In the case of zero-mean random variables, this is equivalent to

$$\text{Var } Z_0 = \text{Var } Z^* + \sigma_{SK}^2$$

The smoothing relationship is the Pythagorean theorem!

Notice that (3.11) entails $\text{Cov}(Z_0, Z^*) = \text{Var}(Z^*)$ so that

$$\rho^2 = \text{Corr}^2(Z_0, Z^*) = \text{Var}(Z^*)/\text{Var}(Z_0)$$

and the simple kriging variance takes the familiar form of a residual variance about a regression

$$\sigma_{SK}^2 = (1 - \rho^2)\text{Var } Z_0$$

These results are similar to (1.6) and (1.7) obtained in the case of the conditional expectation estimator—for a good reason explained in the next section. The case of a nonzero mean can be dealt with as above by adding an extra weight λ_0 and augmenting the space $\mathcal{H}_N$ by the constant random variable 1.

We now turn to the case of continuous sampling (Matheron, 1969a). The set S of data points is now infinite, and we look for an estimator of the form

$$Z^* = \int_S \lambda(dx)Z(x) \tag{3.12}$$

where $\lambda(dx)$ is a weighting function on S (mathematically a measure on S). The kriging system obtained is analogous to (3.2) but with a continuous index x instead of a discrete α,

$$\int_S \lambda(dy)\, \sigma(x, y) = \sigma(x, x_0) \qquad \forall x \in S \tag{3.13}$$

The kriging variance is

$$\sigma^2_{SK} = \left\| Z^* - Z_0 \right\|^2 = \sigma_{00} - \int_S \lambda(dx)\, \sigma(x, x_0)$$

However, here some mathematical complications arise due to the infinite-dimensional nature of the Hilbert spaces considered. The subspace $\mathcal{H}$ generated by the data $\{Z(x) : x \in S\}$ contains all finite linear combinations of elements of S as well as their limits in the mean square sense, but these limits are not necessarily measures (they can be "generalized functions" in L. Schwartz's sense; for example, the space may contain the derivatives of $Z(x)$). The projection Z^* of Z onto $\mathcal{H}$ still exists and is unique, but it is not necessarily of the form (3.12). The only valid statement is to say that if $\lambda(dx)$ satisfies (3.13), then (3.12) defines the optimal estimator. This difficulty only occurs when the covariance function is very regular, the classic example being $\sigma(x, y) = \exp(-(y-x)^2/a^2)$. In this case, knowledge of the process over an arbitrarily short interval allows *perfect* extrapolation anywhere into the future (Yaglom, 1962, p. 190). But this solution is based on a Taylor's series and not on the system (3.13). One could be tempted to ignore these mathematical difficulties and just solve the equations numerically by discretizing S, but the problem turns up as numerical instability.

3.3.4 Gaussian Regression Theory

When $Z(x)$ is a Gaussian RF, the simple kriging estimator Z^* coincides with the conditional expectation $E(Z_0 | Z_1, \ldots, Z_N)$. This follows immediately from the linearity of the regression function of the multivariate Gaussian distribution (see Appendix, Section A.8) and the characteristic property (1.3) of conditional expectation, since we have in particular

$$E[(\lambda_1 Z_1 + \cdots + \lambda_N Z_N - Z_0)\, Z_\alpha] = 0, \qquad \alpha = 1, \ldots, N$$

which are exactly the simple kriging equations (3.2).

The conditional distribution of Z_0 given $Z^* = z^*$ is also Gaussian and has for mean z^* and variance σ^2_{SK} (which does not depend on z^*). Indeed $Z_0 - Z^*$ and Z^* are jointly Gaussian (as linear combinations of Z values) and uncorrelated. Therefore they are *independent*, and the conditional expectation of $Z_0 - Z^*$ and of $(Z_0 - Z^*)^2$ given Z^* are equal to their unconditional expectations:

$$E[Z_0 - Z^* \mid Z^*] = E(Z_0 - Z^*) = 0$$
$$E[(Z_0 - Z^*)^2 \mid Z^*] = E(Z_0 - Z^*)^2 = \sigma_{SK}^2$$

It follows that

$$E[Z_0 \mid Z^*] = Z^*$$
$$\text{Var}[Z_0 \mid Z^*] = \sigma_{SK}^2 \qquad (3.14)$$

The property that $E(Z_0|Z^*) = Z^*$, called "conditional unbiasedness," is of great practical significance in the context of resource assessment problems.[4] For example, in selective mining the decision to process a block as ore or send it to waste is based on an estimate Z^* of the average grade of this block, but the actual ore recovery depends on Z_0. Conditional unbiasedness ensures that, on the average, we get what we expect. In mining, this property is considered more essential than minimum variance [Krige himself has always insisted on this point; e.g., Krige (1951, 1997), David (1977), David et al. (1984), Journel and Huijbregts (1978)]. We will come back to this question in Chapter 6.

The kriging procedure appears especially suited to Gaussian random functions. It does the best job in using the data since what remains, the kriging error, is totally unpredictable from these data. In non-Gaussian cases, kriging still provides the best linear estimator, but linear estimators may not be efficient if the regression function is highly nonlinear.

Another strong property of the Gaussian model is *homoscedasticity* of the conditional kriging error: its variance does not depend on the conditioning data values. As a consequence, confidence intervals based on the kriging standard deviation σ_{SK} also constitute *conditional confidence intervals*. This is very nice but unfortunately specific to the Gaussian model.

3.4 KRIGING WITH AN UNKNOWN MEAN

In most practical situations the mean $m(x)$ is not known. An obvious approach would be to estimate it and subtract it from the data and thus recover the zero-mean case. This approach is commonly used for processing time series, where it is known as "detrending." The problem is that estimated residuals are not the same as true residuals, they depend on how the mean is estimated, and the statistical properties of the whole procedure are difficult to analyze. The kriging approach presented now provides an optimal solution that involves only one estimation step.

The simplest case is when the mean is a constant $m(x) = m$ and leads to *ordinary kriging* (OK). It was developed by Matheron in the early 1960s and is the form of kriging used most because it works under simple stationarity assumptions and does not require knowledge of the mean.

---

[4] Conditional unbiasedness has been established here for point estimation but remains valid in the case of blocks.

The general model, which Matheron (1969a) named the *universal kriging* model for reasons explained below, assumes that the mean function can be represented as a response surface function

$$m(x) = \sum_{\ell=0}^{L} a_\ell f^\ell(x) \qquad (3.15)$$

where the $f^\ell(x)$ are known basis functions and a_ℓ are fixed but unknown coefficients. Usually the first basis function (case $\ell = 0$) is the constant function identically equal to 1, which guarantees that the constant-mean case is included in the model. The other functions are typically monomials of low degree in the coordinates of x (in practice, the degree does not exceed two). In the case of monomials, the superscript ℓ, which is an *index*, has the meaning of a power (in 1D, $f^\ell(x) = x^\ell$). Note that (3.15) may be regarded as a local approximation to $m(x)$; that is, the coefficients a_ℓ may vary in space but sufficiently slowly to be considered constant within estimation neighborhoods.

In some applications a simplified physical model can define drift basis functions that are particularly well suited to the problem. An example is the interpolation of hydraulic head in the presence of producing wells. Pumping creates a drawdown cone whose shape is described by the logarithm of the distance to the well, which can be introduced as a basis function. (Brochu and Marcotte, 2003). Similarly, in the assessment of noise sources for environmental purposes, an analytical approximation of the acoustic field can be computed and used as a drift function (Baume et al., 2009).

The universal kriging model is the decomposition of the variable $Z(x)$ into the sum

$$Z(x) = m(x) + Y(x)$$

of a smooth deterministic function $m(x)$, describing the systematic aspect of the phenomenon, and called the *drift*, and a zero-mean random function $Y(x)$, called the *residual* and capturing its erratic fluctuations.[5] Note that the drift refers to a technically precise notion (the mean of the RF Z), whereas *trend* is a generic term designating a general tendency, a systematic effect (besides, "trend" may imply an underlying driving force). Naturally the decomposition into drift and residual pertains to a certain *scale* of description. Seen from the road, a mountain appears as a drift while local accidents of the relief appear as residuals, but seen from an airplane the mountain itself is a fluctuation in the mountain range.

An important development of the universal kriging model is the use of external variables to model the drift function $m(x)$. For example, the depth $Z(x)$ of a geological horizon may be related to the travel time $T(x)$ of a seismic wave from the surface to that horizon by a model of the form

$$Z(x) = a_0 + a_1 T(x) + Y(x) \qquad (3.16)$$

---

[5] The term "drift" is standard for a stochastic process (e.g., a Brownian motion with a drift). The residual considered here is the *true* residual, by contrast with the calculated residual which is what remains after subtraction of a fit; *fluctuation* would be a more neutral term.

The first two terms represent the large-scale variations of depth, and the residual $Y(x)$ accounts for details that cannot be captured at the resolution of surface seismic. The coefficient a_1 can be interpreted as a velocity, and a_0 as a reference plane depth.[6] From a statistical point of view, this is a linear regression with correlated residuals. Alternative names found in the literature are *spatial regression* or *geo-regression*.

The variable $T(x)$ in (3.16) is treated as a deterministic function assumed known everywhere in the domain of interest—in practice, sampled densely enough to make interpolation errors negligible. Formally, it plays the same role as the $f^{\ell}(x)$ function in (3.15), but there are specific aspects to be discussed. One of them is smoothness. If $T(x)$ is rough, that roughness will be transferred to $Z^*(x)$, especially in sparsely sampled areas, which is probably not desirable. We should then perhaps smooth $T(x)$ before using it as an external drift function.

3.4.1 Ordinary Kriging

Let us first see how not knowing the mean affects the estimation problem in the case of a constant mean $m(x) = a_0$. Consider again the affine estimator $Z^* = \sum_{\alpha} \lambda_{\alpha} Z_{\alpha} + \lambda_0$. Its m.s.e. can be written as

$$E(Z^* - Z_0)^2 = \text{Var } (Z^* - Z_0) + \left[\lambda_0 + \left(\sum_{\alpha} \lambda_{\alpha} - 1 \right) a_0 \right]^2$$

Only the bias term on the right-hand side involves λ_0, but this time we cannot minimize it without knowledge of a_0. An intuitive solution would be to replace a_0 by an estimate $\hat{a}_0$ and solve for λ_0, but this estimate would necessarily depend on the data so that λ_0 would no longer be a constant. The only real solution is to set $\lambda_0 = 0$ and impose the condition $\sum \lambda_{\alpha} - 1 = 0$ on the weights λ_{α}. The bias $E(Z^* - Z_0)$ is then zero whatever the unknown constant a_0. The consequence for not knowing the mean is to restrict ourselves to a linear estimator with weights adding up to 1.

Subject to this condition the m.s.e. is equal to the variance of the error $Z^* - Z_0$ and depends only on covariances

$$\text{Var}(Z^* - Z_0) = \sum_{\alpha} \sum_{\beta} \lambda_{\alpha} \lambda_{\beta} \sigma_{\alpha\beta} - 2 \sum_{\alpha} \lambda_{\alpha} \sigma_{\alpha 0} + \sigma_{00}$$

Our problem can now be reformulated as follows: Find N weights λ_{α} summing to 1 and minimizing Var $(Z^* - Z_0)$. This is classically solved by the method of Lagrange multipliers. We consider the function

$$Q = \text{Var } (Z^* - Z_0) + 2\mu \left(\sum_{\alpha} \lambda_{\alpha} - 1 \right)$$

---

[6] To simplify, the presentation T in this equation is *one-way* time. In the usual case where T is two-way time in milliseconds, the time-to-depth function is $Z = z_0 + v \, T/2000$ with a velocity v in meters per second and a depth in meters.

where μ is an additional unknown, the Lagrange multiplier, and determine the unconstrained minimum of Q by equating its partial derivatives to zero:

$$\frac{\partial Q}{\partial \lambda_\alpha} = 2 \sum_\beta \lambda_\beta \sigma_{\alpha\beta} - 2\sigma_{\alpha0} + 2\mu = 0, \qquad \alpha = 1, \ldots, N,$$

$$\frac{\partial Q}{\partial \mu} = 2\left(\sum_\alpha \lambda_\alpha - 1\right) = 0$$

(That the extremum is indeed a minimum is again guaranteed by the convexity of Var $(Z^* - Z_0)$ as a function of the λ_α.) This leads to the following set of $N+1$ linear equations with $N+1$ unknowns

Ordinary Kriging System

$$\begin{cases} \sum_\beta \lambda_\beta \, \sigma_{\alpha\beta} + \mu = \sigma_{\alpha0}, & \alpha = 1, \ldots, N \\ \sum_\alpha \lambda_\alpha = 1 \end{cases} \tag{3.17}$$

The kriging variance is obtained by premultiplying the first N equations of (3.17) by λ_α, summing over α, and then using the last equation. The result is the OK variance:

$$\sigma_{OK}^2 = E(Z^* - Z_0)^2 = \sigma_{00} - \sum_\alpha \lambda_\alpha \sigma_{\alpha0} - \mu \tag{3.18}$$

The linear system (3.17) has a unique solution if and only if the covariance matrix $\Sigma = [\sigma_{\alpha\beta}]$ is strictly positive definite, which is the case if we use a strictly positive definite covariance function model and if all data points are distinct.

When Only the Variogram Is Known

The condition that the kriging weights add up to 1 entails that the kriging error $(Z^* - Z_0)$ is an allowable linear combination and therefore, according to Section 2.3.2, that its variance can be calculated with the variogram. The resulting OK equations are obtained by simply substituting $-\gamma$ for σ in (3.17).

$$\begin{cases} \sum_\beta \lambda_\beta \, \gamma_{\alpha\beta} - \mu = \gamma_{\alpha0}, & \alpha = 1, \ldots, N, \\ \sum_\alpha \lambda_\alpha = 1 \end{cases} \tag{3.19}$$

$$\sigma_{OK}^2 = \sum_\alpha \lambda_\alpha \gamma_{\alpha0} - \mu$$

Examples

3. Consider again the special case of only one sample point, but assume now an unknown constant mean. The solution of (3.19) is simply $\lambda_1 = 1$ and $\mu = -\lambda_{10}$ so that $Z^*(x_0) = Z_1$ and $\sigma^2_{OK}(x_0) = 2\gamma_{10}$. This is very different from the SK solution found in example 1.

4. In the case of two sample points at x_1 and x_2 the kriging equations yield

$$\lambda_1 = \frac{1}{2}\left[1 + \frac{\gamma_{20} - \gamma_{10}}{\gamma_{12}}\right], \qquad \lambda_2 = \frac{1}{2}\left[1 + \frac{\gamma_{10} - \gamma_{20}}{\gamma_{12}}\right], \qquad \mu = -\frac{1}{2}[\gamma_{10} + \gamma_{20} - \gamma_{12}]$$

so that

$$Z^*(x_0) = \frac{Z_1 + Z_2}{2} + \frac{(\gamma_{10} - \gamma_{20})(Z_2 - Z_1)}{2\gamma_{12}},$$

$$\sigma^2_{OK}(x_0) = \gamma_{10} + \gamma_{20} - \frac{(\gamma_{10} - \gamma_{20})^2}{2\gamma_{12}} - \frac{\gamma_{12}}{2}$$

These results are particularized in Figure 3.4 for three variogram models of type $\gamma(h) = |h|^{\alpha}$. The Z^* curve always goes through Z_1 and Z_2 and for large $|x_0|$ behaves like $|x_0|^{\alpha-1}$.

For $\alpha = 1$ kriging simply interpolates linearly between Z_1 and Z_2 in the interval $[x_1, x_2]$ (with $\mu = 0$) and outside assumes the value of the nearest end point. The optimal estimator of $Z(x)$ at $x_0 > x_2$ is just the last value observed. This property derives fundamentally from the Markov character of processes with independent increments with which the linear variogram is closely associated. To show this, we can use a standard (invariance principle) argument of probability theory which goes as follows: *If the solution of a problem only depends on certain characteristics (e.g., the first two moments) and if we can find the solution in an easy special case (e.g., Gaussian RF), then it is the general solution.* Here we consider the special case of a Brownian motion $X(t)$ without drift. It has independent and stationary increments, with a Gaussian distribution, and enjoys the Markov property of conditional independence: Once we know the value $X(t)$ reached by the process at time t, its future does not depend on the path it took to get there. To predict $X(t+\tau)$ on the basis of past values $\{X(t'): t' \leq t\}$, the only relevant information is $X(t)$. The derivation of the predictor is straightforward starting from the decomposition

$$X(t+\tau) = X(t) + [X(t+\tau) - X(t)]$$

Since the increment $X(t+\tau) - X(t)$ is independent of $X(t)$, and of any earlier value, we have

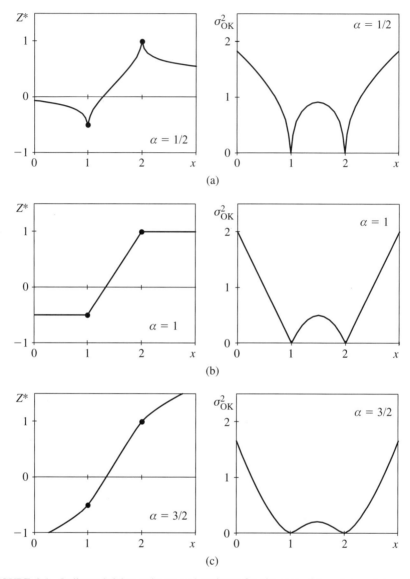

FIGURE 3.4 Ordinary kriging estimate and variance for the case of two data points and a variogram model $\gamma(h) = |h|^{\alpha}$: (a) $\alpha = 1/2$; (b) $\alpha = 1$; (c) $\alpha = 3/2$.

$$E[X(t + \tau) \mid X(t'), t' \leq t] = X(t) \qquad (\tau \geq 0)$$

This predictor is clearly unbiased and optimal, and it happens to be linear. This is the kriging solution associated with a linear variogram in 1D. Stock prices, for example, have been modeled by a process with independent increments [e.g., Box and Jenkins (1976, p. 150)] with the disappointing consequence that the best forecast of stock price at any time in the future is just the current value of the stock.[7]

Observe that for $\alpha = 1.5$ the estimator is not confined to the data range $[Z_1, Z_2]$. When $x_0 > x_2$, for example, $\lambda_2 > 1$ and $\lambda_1 < 0$. *Kriging weights can be negative or greater than* 1, even when the mean is constant. This effect is associated with high variogram regularity (power $\alpha > 1$).

Kriging variances are zero at $x_0 = x_1$ and $x_0 = x_2$ and increase rapidly without limits as x_0 departs from the $[x_1, x_2]$ interval. Extrapolation is risky!

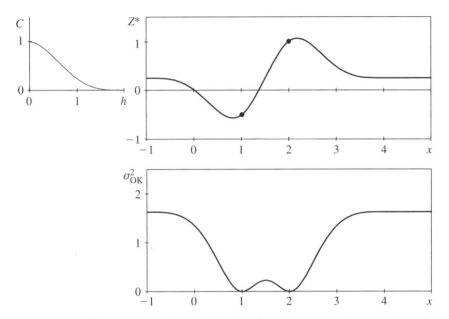

FIGURE 3.5 Ordinary kriging estimate (top) and variance (bottom) for the case of two data points and a variogram model with a sill 1. Compare with simple kriging (Figures 3.1a and 3.3a): At a distance from the data points the estimator is equal to the mean of the data (rather than the mean of the random function) and the kriging variance now exceeds the variance $C(0)$ of the RF.

Similar results are sketched in Figure 3.5 for a variogram with a finite sill of 1. In extrapolation the kriging estimator approaches $(Z_1 + Z_2)/2$, which is an unbiased estimator of the mean, while the kriging variance tends to 1.62 σ^2 and is thus larger than the global variance of $Z(x)$ itself. This is the penalty for not knowing the mean.

---

[7] The most popular model for the evolution of stock prices, which underlies the derivation of the famous Black–Scholes formula for options pricing, has the logarithm of stock price as a Brownian motion with a linear drift [e.g., Jarrow and Rudd (1983)]. But that does not make forecasting easier, as Example 5 shows.

3.4.2 Universal Kriging and Kriging with an External Drift

We want to estimate $Z_0 = Z(x_0)$ using a linear estimator $Z^* = \sum_\alpha \lambda_\alpha Z_\alpha$ and seek to minimize the m.s.e., which, as usual, can be decomposed as

$$E(Z^* - Z_0)^2 = \text{Var}(Z^* - Z_0) + [E(Z^* - Z_0)]^2$$

Now the mean is no longer constant but of the form (3.15), with the understanding that this representation also includes external drifts such as in (3.16). The bias can be expanded as

$$E(Z^* - Z_0) = \sum_\alpha \lambda_\alpha \sum_\ell a_\ell f_\alpha^\ell - \sum_\ell a_\ell f_0^\ell$$

using the notations

$$f_\alpha^\ell = f^\ell(x_\alpha), \qquad f_0^\ell = f^\ell(x_0)$$

and the convention that summation on ℓ extends over all possible values $\ell = 0, 1, \ldots, L$. By interchanging the order of summations on ℓ and α, we get

$$E(Z^* - Z_0) = \sum_\ell a_\ell \left(\sum_\alpha \lambda_\alpha f_\alpha^\ell - f_0^\ell \right)$$

In order to minimize $E(Z^* - Z_0)^2$, we have to make $[E(Z^* - Z_0)]^2$ zero whatever the unknown coefficients a_ℓ, which implies annihilating their factors in the above. This leads to the set of $L + 1$ conditions

$$\sum_\alpha \lambda_\alpha f_\alpha^\ell = f_0^\ell, \qquad \ell = 0, 1, \ldots, L \tag{3.20}$$

that Matheron (1969a) called *universality conditions*, hence the name universal kriging (UK). They express that the estimator Z^* is unbiased for *all* values of a_ℓ.

Subject to these conditions, the m.s.e. is equal to the variance of the error $Z^* - Z_0$:

$$\text{Var}(Z^* - Z_0) = \sum_\alpha \sum_\beta \lambda_\alpha \lambda_\beta \sigma_{\alpha\beta} - 2 \sum_\alpha \lambda_\alpha \sigma_{\alpha 0} + \sigma_{00}$$

Using Lagrange multipliers, we minimize

$$Q = \text{Var}(Z^* - Z_0) + 2 \sum_{\ell=0}^{L} \mu_\ell \left[\sum_\alpha \lambda_\alpha f_\alpha^\ell - f_0^\ell \right]$$

where $\mu_\ell, \ell = 0, \ldots, L$, are $L + 1$ additional unknowns, the Lagrange multipliers, and we determine the unconstrained minimum of Q by equating the partial derivatives of Q to zero:

$$\frac{\partial Q}{\partial \lambda_\alpha} = 2 \sum_\beta \lambda_\beta \, \sigma_{\alpha\beta} - 2\sigma_{\alpha 0} + 2 \sum_\ell \mu_\ell f_\alpha^\ell = 0, \qquad \alpha = 1, \ldots, N,$$

$$\frac{\partial Q}{\partial \mu_\ell} = 2 \left[\sum_\alpha \lambda_\alpha f_\alpha^\ell - f_0^\ell \right] = 0, \qquad\qquad \ell = 0, 1, \ldots, L$$

This leads to the following set of $N + L + 1$ linear equations with $N + L + 1$ unknowns.

Universal Kriging System

$$\begin{cases} \sum_\beta \lambda_\beta \, \sigma_{\alpha\beta} + \sum_\ell \mu_\ell f_\alpha^\ell = \sigma_{\alpha 0}, & \alpha = 1, \ldots, N \\[2mm] \sum_\alpha \lambda_\alpha f_\alpha^\ell = f_0^\ell, & \ell = 0, \ldots, L \end{cases} \tag{3.21}$$

In matrix notation the system (3.21) is of the form $\mathbf{A}\mathbf{w} = \mathbf{b}$ with the following structure:

$$\underbrace{\begin{bmatrix} \boldsymbol{\Sigma} & \mathbf{F} \\ \mathbf{F}' & \mathbf{0} \end{bmatrix}}_{\mathbf{A}} \underbrace{\begin{bmatrix} \boldsymbol{\lambda} \\ \boldsymbol{\mu} \end{bmatrix}}_{\mathbf{w}} = \underbrace{\begin{bmatrix} \boldsymbol{\sigma}_0 \\ \mathbf{f}_0 \end{bmatrix}}_{\mathbf{b}} \tag{3.22}$$

where $\boldsymbol{\Sigma}$, $\boldsymbol{\lambda}$, and $\boldsymbol{\sigma}_0$ are defined as for simple kriging and where

$$\mathbf{F} = \begin{bmatrix} 1 & f_1^1 & \cdot & f_1^L \\ 1 & f_2^1 & \cdot & f_2^L \\ \cdot & \cdot & \cdot & \cdot \\ \cdot & \cdot & \cdot & \cdot \\ \cdot & \cdot & \cdot & \cdot \\ 1 & f_N^1 & \cdot & f_N^L \end{bmatrix}, \quad \boldsymbol{\mu} = \begin{bmatrix} \mu_0 \\ \mu_1 \\ \cdot \\ \cdot \\ \cdot \\ \mu_L \end{bmatrix}, \quad \mathbf{f}_0 = \begin{bmatrix} 1 \\ f_0^1 \\ \cdot \\ \cdot \\ \cdot \\ f_0^L \end{bmatrix}$$

The kriging variance is obtained by premultiplying the first N equations of (3.21) by λ_α, summing over α, and then using the last $(L+1)$ equations. The result is the UK variance:

$$\sigma_{\text{UK}}^2 = \mathrm{E}(Z^* - Z_0)^2 = \sigma_{00} - \sum_\alpha \lambda_\alpha \sigma_{\alpha 0} - \sum_\ell \mu_\ell f_0^\ell \tag{3.23}$$

or in matrix form

$$\sigma_{\text{UK}}^2 = \sigma_{00} - \boldsymbol{\lambda}'\boldsymbol{\sigma}_0 - \boldsymbol{\mu}'\mathbf{f}_0 = \sigma_{00} - \mathbf{w}'\mathbf{b}$$

These equations were established independently by several authors, including Zadeh and Ragazzini (1950) as an extension of Wiener's prediction theory, Goldberger (1962) in the scope of a generalized linear regression model, and

Matheron (1969a) within the framework of infinite-dimensional Hilbert spaces (continuous sampling).

Conditions for Nonsingularity

The linear system (3.22) has a unique solution if and only if its matrix $\mathbf{A}$ is nonsingular. This holds under the following set of sufficient conditions: (1) that the submatrix Σ is strictly positive definite, (2) that the submatrix $\mathbf{F}$ is of full rank $L+1$ (equal to the number of columns). The proof follows from straightforward matrix algebra.

Strict positive definiteness of Σ is ensured by the use of a strictly positive definite covariance function and the elimination of duplicate data points. The condition on $\mathbf{F}$ expresses that the $L+1$ basis functions $f^\ell(x)$ are linearly independent on S:

$$\sum_\ell c_\ell f^\ell(x) = 0 \qquad \forall x \in S \qquad \Rightarrow \qquad c_\ell = 0, \qquad \ell = 0, \dots, L$$

This is a standard condition of "sampling design," encountered, for example, in the theory of least squares ($\mathbf{F'F}$ must be nonsingular). For one thing, there must be at least as many data points as there are basis functions (thus $N \geq L+1$). Moreover, the arrangement of the points must provide enough constraints to allow the determination of the coefficients a_ℓ in the linear model (3.15). A counterexample in 2D is when $m(x)$ is a plane and all sample points are aligned: Obviously the plane is not constrained by a single line. Likewise, when $m(x)$ is a quadratic function, the system is singular if all data points lie along two lines, a circle, an ellipse, a parabola, or a hyperbola. In view of these remarks, one must be careful, particularly when using moving neighborhoods, not to create singular systems by a bad selection of the data points.

Solving the Kriging Equations

When selecting a computer subroutine to solve the equations (3.21), it must be noted that the matrix $\mathbf{A}$ of the system is no longer positive definite as for the simple kriging system (3.2). This rules out direct use of the symmetric Cholesky decomposition, "which has all the virtues" (Wilkinson, 1965, p. 244). However, the problem can be reduced to the solution of two subsystems with positive definite matrices. The following procedure results directly from the matrix form of (3.21):

$$
\begin{array}{llll}
\text{1. Solve} & \Sigma \mathbf{w}_1 = \mathbf{F} & \text{and} \quad \Sigma \lambda_K = \sigma_0 & \text{for } \mathbf{w}_1 \text{ and } \lambda_K \\
\text{2. Compute} & \mathbf{Q} = \mathbf{F'w}_1 & \text{and} \quad \mathbf{R} = \mathbf{F'}\lambda_K - \mathbf{f}_0 & \\
\text{3. Solve} & \mathbf{Q}\mu = \mathbf{R} & (\mathbf{Q} \text{ is positive definite}) & \\
\text{4. Compute} & \lambda = \lambda_K - \mathbf{w}_1 \mu & &
\end{array}
\tag{3.24}
$$

Note that the first two systems can be solved in parallel and that the third is a small system of size $L+1$. Note also that λ_K is the solution of the simple kriging equations (3.2). The kriging estimator in the case of an unknown mean is thus equal to the simple kriging estimator Z_K^* computed as if the mean were zero plus a term that will be interpreted as a correction for the mean (cf. Section 3.4.7).

When Only the Variogram Is Known

If the constant function 1 is included in the set of basis drift functions $f^\ell(x)$, the universal kriging system (3.21) can be rewritten in variogram terms by replacing σ by $-\gamma$.

Universal Kriging System

$$\begin{cases} \sum_\beta \lambda_\beta \, \gamma_{\alpha\beta} - \sum_\ell \mu_\ell f_\alpha^\ell = \gamma_{\alpha 0}, & \alpha = 1, \ldots, N, \\ \sum_\alpha \lambda_\alpha f_\alpha^\ell = f_0^\ell, & \ell = 0, \ldots, L \end{cases} \tag{3.25}$$

UK Variance

$$\sigma_{UK}^2 = E(Z^* - Z_0)^2 = \sum_\alpha \lambda_\alpha \gamma_{\alpha 0} - \sum_\ell \mu_\ell f_0^\ell$$

with the usual notations

$$\gamma_{\alpha\beta} = \gamma(x_\alpha, x_\beta), \quad \gamma_{\alpha 0} = \gamma(x_\alpha, x_0), \quad f_\alpha^\ell = f^\ell(x_\alpha), \quad f_0^\ell = f^\ell(x_0)$$

The matrix **A** of the kriging system (3.25) is similar to (3.22) with a variogram matrix $\mathbf{\Gamma}$ in lieu of the covariance matrix $\mathbf{\Sigma}$. Use of a variogram function such that $-\gamma$ is strictly conditionally positive definite (which common models are) together with **F** of full rank and no duplicate data point ensure that **A** is nonsingular (see Section 4.6.1). The variogram matrix $\mathbf{\Gamma}$ itself is invertible but certainly not positive definite (0's on the diagonal), so that Cholesky's decomposition as in (3.24) does not apply. When the variogram is bounded, it is easier to express the kriging system in covariance terms. If not, the variogram usually admits a locally equivalent stationary covariance at the scale of the kriging neighborhood (see Section 4.6.2), which also enables use of the kriging system (3.21).

3.4.3 Comments on the Kriging Equations

The kriging equations (3.21) or (3.25) capture four aspects of the interpolation problem:

- The geometry of the sample points, through the $\sigma_{\alpha\beta}$ or $\gamma_{\alpha\beta}$ terms. These are functions of the interpoint distances and correct for the redundancy in the information.

- The position of the estimated point x_0 with respect to the data, through $\sigma_{\alpha 0}$ or $\gamma_{\alpha 0}$.
- The lateral continuity of the phenomenon, through the covariance or variogram model.
- The presence of a systematic location-dependent effect (trend), through the drift model.

The influence of the drift model depends on its complexity in relation with the data. In the vocabulary of regression analysis, $N-(L+1)$ would represent the number of degrees of freedom left in the residuals. This interpretation does not apply here because the data are correlated, but it helps to think in these terms. At one extreme, when $N = L + 1$, the solution is completely constrained by the unbiasedness conditions, and UK reduces to a purely deterministic fit. As L decreases, there are more and more degrees of freedom left, and the probabilistic nature of UK increases. Note that in all cases the unbiasedness conditions (3.20) ensure that the drift function $m(x)$ is interpolated exactly; that is, if $Z(x)$ coincides with the drift, interpolation is perfect.

The kriging estimator always coincides with the data at the sample points, and the kriging variance there is zero. This is true even in the presence of a nugget effect, but the estimate then has a discontinuity at each data point. If one is interested, rather, in the continuous component of the phenomenon, then slightly different equations should be used (see Section 3.7.1).

From equations (3.21) and (3.25) it is seen that the kriging weights, and therefore the kriging estimates, do not change if the covariance or variogram values are multiplied by a constant factor, while the μ_ℓ and the kriging variance are multiplied by that same factor.

When the variogram or the covariance is stationary, the kriging weights and variance are also invariant with respect to the origin of coordinates provided that the space generated by the basis drift functions is itself invariant under shifts (see Section 3.4.6). This is the case for a polynomial (or trigonometric) basis, but not for an external drift function. An external drift function is tied to a geographic location; that is, it is *localized*.

Examples

5. To generalize example 4, we now consider, still in 1D, a random function with a linear variogram $\gamma(h) = b|h|$ and a linear drift $m(x) = a_0 + a_1 x$. This time there are N sample points at arbitrary locations $x_1 < x_2 < \cdots < x_N$. The solutions of the kriging equations (3.25) are the following, with appropriate (x_0-dependent) values of μ_0 and μ_1:

$x_0 \leq x_1$ $\quad Z^*(x_0) = Z_1 - [(Z_N - Z_1)/(x_N - x_1)](x_1 - x_0),$

$x_i \leq x_0 \leq x_{i+1}, \quad Z^*(x_0) = [(x_{i+1} - x_0)/(x_{i+1} - x_i)]Z_i + [(x_0 - x_i)/(x_{i+1} - x_i)]Z_{i+1},$

$x_0 \geq x_N$ $\quad Z^*(x_0) = Z_N + [(Z_N - Z_1)/(x_N - x_1)](x_0 - x_N)$

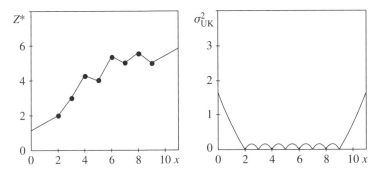

FIGURE 3.6 Universal kriging estimate and variance for the case of a linear variogram and a linear drift.

When $x_1 \leq x_0 \leq x_N$, the solution is simply linear interpolation in each subinterval. The same would hold true for a constant mean (OK). Notice again that the two adjacent sample points screen off the influence of all other data. In extrapolation the estimator is the straight line joining the first and the last data points (Figure 3.6). Considering $x_0 > x_N$, for example, the solution may be regarded as the sum of the OK estimator Z_N and a correction $\hat{a}_1 (x_0 - x_N)$ for the linear drift, in which $\hat{a}_1 = (Z_N - Z_1)/(x_N - x_1)$ is the estimator of the slope. Since $(Z_N - Z_1)$ and $(Z_0 - Z_N)$ are uncorrelated increments, the OK variance $2\gamma_{0N}$ is simply augmented by the variance of the drift correction, leading to a parabolic growth

$$\sigma_{UK}^2 = E(Z^* - Z_0)^2 = 2b(x_0 - x_N)[1 + (x_0 - x_N)/(x_N - x_1)] \qquad (x_0 > x_N)$$

Extrapolation is even more hazardous when there is a drift!

3.4.4 Kriging with Estimated Parameters

So far we have assumed that the variogram is known exactly. In reality it is *estimated*, and since this estimation is not perfect, two things happen: (1) The computed kriging estimates are different from the optimal kriging estimates, and (2) the computed kriging variances are different from the true error variances associated with the suboptimal estimates. How large can these differences be?

An answer to point 1 is that the quality of the kriging estimates obtained differs from the true optimum by an amount that is *second order* in the precision to which the optimal solution is determined. In other words, even a fairly crudely determined set of kriging weights can give excellent results when it is applied to data. A similar observation is made by Press et al. (2007, p. 651) regarding the determination of the optimal Wiener filter (analogous to kriging with random errors). The important factor is to make sure that the variogram behavior near the origin is correctly represented. For example, it matters to know that the behavior is linear, but the slope itself has no influence on the kriging weights. It takes a gross misspecification of the variogram model to have a dramatic impact on kriging estimates, such as using a continuous model when significant noise is present in the data, or the opposite. On the contrary, for point 2, the computed kriging variance is directly affected by the variogram fit.

Since the computed kriging variance is obtained by "plugging-in" an estimated variogram model assumed known without error, the reported kriging variance does not reflect the *total* uncertainty. This effect is ignored in standard geostatistical practice, which makes the kriging variance *optimistic*—some authors call it "naïve," some even call it "wrong." Assessing the sensitivity of kriging results to misspecification of the statistical model is a difficult problem. Cressie (1991, pp. 289–299) distinguishes *mathematical stability*, what happens when a (not too) different variogram or covariance function is used in place of the true one, and *statistical stability*, the effect of kriging with estimated parameters. Stein (1999, pp. 199–223) looks at the latter issue in detail and illustrates the dangers of using a Gaussian variogram without nugget effect to model a differentiable random field.

The Bayesian framework provides an elegant solution for taking into account the uncertainty on variogram or covariance parameters. Calling θ the vector of unknown covariance parameters and Z_0 the objective, the posterior probability density $f(Z_0 \mid \mathbf{Z})$ of Z_0 given the vector of observations $\mathbf{Z} = (Z_1, \ldots, Z_N)'$ is (Cressie, 1991, p. 171; Stein, 1999, p. 223)

$$f(Z_0 \mid \mathbf{Z}) = \int f(Z_0 \mid \mathbf{Z}, \theta) f(\theta \mid \mathbf{Z}) \, d\theta$$

where $f(\theta \mid \mathbf{Z})$ is the posterior density of θ given $\mathbf{Z}$ and is related to the prior density $f(\theta)$ by

$$f(\theta \mid \mathbf{Z}) = \frac{f(\mathbf{Z} \mid \theta) f(\theta)}{\int f(\mathbf{Z} \mid \theta') f(\theta') \, d\theta'}$$

Knowing $f(Z_0 \mid \mathbf{Z})$, estimates and confidence intervals can be derived.

While appealing in concept, this approach, known as Bayesian Kriging, may require much more information than is actually available. We need to postulate the analytical form of the conditional distribution of Z_0 given $\mathbf{Z}$ and θ, and of $\mathbf{Z}$ given θ, and choose a prior distribution for θ, all this simply to estimate one or two covariance parameters! Mathematical difficulties also arise with "improper priors" that do not give valid posterior distributions [Stein (1999) reports an example with the Matérn model], and extensive numerical integration is required [e.g., Handcock and Wallis (1994)]. For these reasons, Bayesian Kriging is rarely used in practice.

A practical way to account for model uncertainty is to compare the kriging results obtained under several variogram scenarios, generally differing by the range and the relative nugget effect. (The variogram sill does not impact kriging estimates, and the behavior at the origin should be set by physical considerations.) The sensitivity of the kriging results to the choice of the variant shows to what extent the uncertainty on the variogram is a problem and in some cases guides the design of additional sampling to improve variogram determination, usually short distance sampling. This scenario approach is similar in concept to sampling $f(\mathbf{Z} \mid \theta)$ for different values of θ.

An alternative approach to deal with model uncertainty is fuzzy logic. For example, Bardossy et al. (1990) use kriging with "fuzzy" variograms, which produces fuzzy estimates and fuzzy kriging variances. It separates the spatial uncertainty from the imprecision in the model parameters: The level of the fuzzy kriging variance reflects the probabilistic uncertainty of the interpolation, assuming exact variogram parameters, whereas the interval width of the fuzzy kriged values can be used to measure the effect of the imprecision in the variogram parameters. A review and discussion of the literature on fuzzy logic in geostatistics can be found in Loquin and Dubois (2010).

Finally, let us mention an approach proposed by Pilz et al. (1997) which consists in modeling a whole class of plausible variogram functions, rather than a single one assumed to be correct, and then using a new kriging method ("minimax kriging") to minimize the maximum possible kriging variance in that class.

3.4.5   Confidence Intervals

If the kriging error has a Gaussian distribution, this distribution is completely specified by its mean (zero) and its variance σ_K^2. Assuming a *known variogram*, the kriging variance is determined without error (i.e., is nonrandom), and it is possible to make a probabilistic statement such as

$$\Pr(|Z^* - Z_0| > 1.96\,\sigma_K) = 0.05$$

which leads to the traditional 95% confidence interval for Z_0 (approximating 1.96 by 2)

$$[Z^* - 2\sigma_K, Z^* + 2\sigma_K]$$

When the error is not Gaussian, this interval may not have a 95% coverage probability but is still used as a *nominal* (or *conventional*) confidence interval. A potential problem is that these bounds may not be consistent with the global constraints placed on the data. A negative lower bound may be found for an ore grade, or an upper bound greater than 1 for a variable Z defined as a proportion. A workaround is to perform a preliminary Gaussian transformation of the data, do the kriging, and back-transform the confidence interval. Since the transformation is order-preserving, the coverage probability remains the same.

Schelin and Sjöstedt-de Luna (2010) propose a semiparametric bootstrap method to deal with non-Gaussian behavior. Spatially correlated samples are simulated using the spatial bootstrap (Section 2.6.2), ordinary kriging estimates are formed, and the empirical distribution of kriging errors is obtained, from which percentiles are picked to construct a confidence interval. This method requires knowledge of the marginal distribution of the decorrelated variables.

Den Hertog et al. (2006) rely on the bootstrap for a different reason. They claim that "the kriging variance formula used in the literature is wrong because it neglects the fact that certain correlation parameters are *estimated*." A parametric bootstrapping of kriging variances is proposed instead, based on a simulation of kriging errors under a multivariate Gaussian model whose covariance model is specified but not its parameters. Simulation results, obtained in the context of DACE (Section 3.4.9) using Gaussian covariances and mainly artificial test functions, show that the bootstrapped kriging variance is generally larger than the classic kriging variance. However, the significance of these results is questionable. When the covariance model is selected arbitrarily, the computed kriging variance can only be a crude representation of the true error variance. Furthermore, for very smooth variables the covariance fit is critical, and using a pure Gaussian model is known to give unreliable results. Finally, sound geostatistical practice always includes a cross-validation step whose goal is precisely to calibrate the covariance on empirical kriging errors.

A Useful Inequality

Although the Gaussian character, or at least the symmetry, of the kriging error can be improved by transformation, its real distribution remains unknown. Then the following question is of interest: What is the real significance level (coverage probability) of this interval for plausible distributions of the error, not necessarily Gaussian? An inequality established by Vysochanskiĭ and Petunin (1980) and discussed by Pukelsheim (1994) gives us the answer. The only assumption, a very mild one, is that the error distribution is *continuous and unimodal*. Because of its general interest, we give the complete result here and particularize it to our problem.

The inequality states that if X is a random variable with a probability density f that is nondecreasing up to a mode ν and nonincreasing thereafter and if $d^2 = E(X - \alpha)^2$ is the expected squared deviation from an *arbitrary* point α, then

$$\Pr(|X - \alpha| \geq t\,d) \leq \frac{4}{9t^2} \qquad \text{for all } t \geq \sqrt{8/3}$$

$$\leq \frac{4}{3t^2} - \frac{1}{3} \qquad \text{for all } t \leq \sqrt{8/3}$$

When X is the kriging error and $\alpha = 0$, then $d^2 = \sigma_K^2$ and

$$\Pr(|Z^* - Z_0| \geq 2\sigma_K) \leq \frac{1}{9}$$

So under the stated assumptions, the nominal confidence interval has a significance level of 89%. In order to get a 95% interval, we need $t = 3$, since $4/81 = 0.049$. The penalty for not knowing the distribution is a loss of 6% confidence for the $\pm 2\sigma_K$ interval, or broadening the 95% interval to $\pm 3\sigma_K$ instead of $\pm 2\sigma_K$.

For $t > 1.63$ the Vysochanskiĭ–Petunin inequality coincides with the Gauss inequality for deviations from the mode (case $\alpha = \nu$) dating back to 1821 [see Cramér (1945, p. 183)]. Note that the bound is less than half the Bienaymé–Tchebycheff bound $1/t^2$. The usefulness of the Gauss inequality in the context of kriging was pointed out by Alfaro (1993).

Local and Regional Modulations

Another aspect of the kriging variance needs to be discussed. We have seen that the kriging variance does not directly depend on the data values used for the estimation: It is an *unconditional* variance. What this means is illustrated in Figure 3.7 from Ravenscroft [in Armstrong (1994)]. Two blocks are estimated from four samples, they have the same kriging variance (since the data layout is the same) and also the same kriging estimate of 10. Yet clearly the right-hand case carries a higher uncertainty. The kriging variance is an *average* of such cases. If the left-hand scenario is the most frequent one, the uncertainty will occasionally be underestimated; if the right-hand scenario is the rule, the

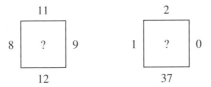

FIGURE 3.7 Which scenario is a safer bet? [From Armstrong (1994), with kind permission from Kluwer Academic Publishers.]

heterogeneity will translate into the variogram, and occasionally the uncertainty will be exaggerated. In this sense the kriging variance has the meaning of a spatial average as introduced in the discussion on the objectivity of probabilistic parameters for unique phenomena (Section 1.2).

Except in the case of a Gaussian RF with known mean, the kriging variance is not sufficient to model the conditional distribution of Z_0 given $Z_1, \ldots, Z_N$. This fact sometimes comes as a disappointment, but one should not ask too much from a simple linear approach. Modeling conditional distributions is a very ambitious goal, and in fact unrealistically ambitious without a specified theoretical model (see Section 6.3.2).

The kriging variance still has its merits as a precision indicator. Referring to the terminology used by Switzer (1993), the kriging error reflects two scales of variability. The first is a "local modulation" related to the spatial configuration of the data in the kriging neighborhood: Errors are smaller close to the data points, clustered samples carry less information than isolated ones, and so on. The other component is a "regional modulation" reflecting variability in the region of interest. When a global variogram is used over the whole domain, the kriging variance mainly reflects the local modulation. When data permit, however, the variogram parameters are adjusted regionally (proportional effects or other techniques), and the kriging variance, without being conditional, becomes an interesting indicator of uncertainty. Interpolated surfaces are too easily accepted as reality, especially when displayed and manipulated on powerful 3D modeling workstations. The kriging variance allows an intelligent use of these surfaces with consideration of possible errors.

3.4.6 Drift Estimation

We have seen that universal kriging provides an estimator of $Z(x_0)$ without having to estimate the mean. In fact, as intuition suggests, the mean is estimated anyhow, but *implicitly*. To see how this falls out, let us first consider the problem of drift estimation by itself.

Suppose that the objective is to estimate $m(x_0)$ at some point x_0. Following the approach of kriging, a linear estimator is formed,

$$m^*(x_0) = \sum_\alpha \lambda_\alpha Z_\alpha$$

with weights selected so that

$$E[m^*(x_0) - m(x_0)] = 0 \qquad \text{all } a_\ell$$
$$\text{Var}[m^*(x_0) - m(x_0)] \qquad \text{minimum}$$

In matrix notations where $\mathbf{Z}$ is the N-vector of Z_α data, $\boldsymbol{\lambda}$ the N-vector of weights, and $\boldsymbol{\mu}$ the $L+1$-vector of Lagrange multipliers, this leads to the following system:

Drift Estimation System

$$\begin{cases} \boldsymbol{\Sigma}\boldsymbol{\lambda} + \mathbf{F}\boldsymbol{\mu} = \mathbf{0}, \\ \mathbf{F}'\boldsymbol{\lambda} = \mathbf{f}_0 \end{cases} \qquad (3.26)$$

and thus

$$m^*(x_0) = \boldsymbol{\lambda}'\mathbf{Z} = \mathbf{f}_0'(\mathbf{F}'\boldsymbol{\Sigma}^{-1}\mathbf{F})^{-1}\mathbf{F}'\boldsymbol{\Sigma}^{-1}\mathbf{Z}$$

with

Drift Estimation Variance

$$E[m^*(x_0) - m(x_0)]^2 = -\boldsymbol{\mu}'\mathbf{f}_0 = \mathbf{f}_0'(\mathbf{F}'\boldsymbol{\Sigma}^{-1}\mathbf{F})^{-1}\mathbf{f}_0$$

Notice that the drift estimation system (3.26) coincides with a UK system (3.21) in which all covariances $\sigma_{\alpha 0}$ on the right-hand side are zero. The UK estimator merges into the drift estimator when the estimated point x_0 is at a distance from all data points greater than the range.

An alternative derivation of these results is as follows: In the finite sample case the UK model can be regarded as a linear regression model with correlated "residuals," whose expression in matrix terms is[8]

$$\mathbf{Z} = \mathbf{Fa} + \mathbf{Y}$$

The optimal estimator of the vector $\mathbf{a}$ of a_ℓ coefficients is classically obtained by generalized least squares (GLS), namely by minimizing

$$(\mathbf{Z} - \mathbf{F}\hat{\mathbf{a}})'\boldsymbol{\Sigma}^{-1}(\mathbf{Z} - \mathbf{F}\hat{\mathbf{a}})$$

over all choices of $\hat{\mathbf{a}}$ [e.g., Rao (1973)]. The GLS solution is

$$\mathbf{a}^* = (\mathbf{F}'\boldsymbol{\Sigma}^{-1}\mathbf{F})^{-1}\mathbf{F}'\boldsymbol{\Sigma}^{-1}\mathbf{Z} \qquad (3.27)$$

---

[8] In standard least squares notations, $\mathbf{F}$ would be $\mathbf{X}$, $\mathbf{a}$ would be $\boldsymbol{\beta}$, and $\mathbf{Y}$ would be $\boldsymbol{\varepsilon}$.

It is unbiased and has minimum variance. The variance–covariance matrix of this estimator is

$$\text{Cov}(\mathbf{a}^*, \mathbf{a}^*) = (\mathbf{F}'\boldsymbol{\Sigma}^{-1}\mathbf{F})^{-1} \tag{3.28}$$

Turning to residuals, for the optimal choice $\mathbf{a}^*$ the covariance of the estimated residuals is

$$E(\mathbf{Z} - \mathbf{F}\mathbf{a}^*)(\mathbf{Z} - \mathbf{F}\mathbf{a}^*)' = \boldsymbol{\Sigma} - \mathbf{F}\,(\mathbf{F}'\boldsymbol{\Sigma}^{-1}\mathbf{F})^{-1}\mathbf{F}'$$

or, explicitly

$$\text{Cov}(Z_\alpha - m_\alpha^*, Z_\beta - m_\beta^*) = \sigma_{\alpha\beta} - \text{Cov}(m_\alpha^*, m_\beta^*)$$

The covariance of the estimated residuals is a biased estimate of the covariance of the true residuals Y; in particular, the variances are systematically underestimated:

$$\text{Var}(Z_\alpha - m_\alpha^*) = \text{Var}(Z_\alpha) - \text{Var}(m_\alpha^*)$$

One property of the GLS regression solution (3.27) worth noticing is its invariance under a change of coordinates in the linear subspace generated by the columns of $\mathbf{F}$. If we use a new set of basis functions $\varphi^\ell(x)$ that are linearly related to the $f^\ell(x)$ by

$$\varphi^\ell(x) = \sum_s B_s^\ell f^s(x)$$

the drift coefficients estimates will of course be different but the value of $m^*(x_0)$ itself will not change. This is clear from (3.27): If $\mathbf{F}$ becomes $\mathbf{FB}$, with $\mathbf{B}$ invertible, then $\mathbf{a}^*$ becomes $\mathbf{B}^{-1}\mathbf{a}^*$ so that $\mathbf{F}\,\mathbf{a}^*$ remains invariant.

When Only the Variogram Is Known

A difficulty appears here. Since an intrinsic RF is defined only through its increments, the drift coefficient a_0 (associated with the constant function $f^0 \equiv 1$) is fundamentally indeterminate, and it is in principle impossible to estimate $m(x_0)$. In fact we are unable to calculate the variance of $m^*(x_0) - m(x_0)$ because it is not an allowable linear combination of the data: The weights of the Z data involved in this error add up to one instead of zero as they do for a kriging error.

We can work around this difficulty by considering that, over a bounded domain, the function $A - \gamma(h)$ is a covariance for some large positive constant A (this is true for all variogram models of practical interest). Rewriting (3.26), we get the following equations where $\mu_0' = \mu_0 + A$.

Estimation of the Drift Value $m(x_0)$

$$\sum_\beta \lambda_\beta \gamma_{\alpha\beta} - \mu'_0 - \sum_{\ell>0} \mu_\ell f^\ell_\alpha = 0, \quad \alpha = 1, \ldots, N$$

$$\sum_\alpha \lambda_\alpha f^\ell_\alpha = f^\ell_0, \qquad\qquad \ell = 0, \ldots, L$$

(3.29)

and the estimation variance is

$$E[m^*(x_0) - m(x_0)] = A - \mu'_0 - \sum_{\ell>0} \mu_\ell f^\ell(x_0)$$

The drift estimator does not depend on the constant A, but its variance does. In other words, we can derive a drift estimate, but its uncertainty is arbitrarily large! The problem arises from the fact that an intrinsic RF is defined up to a constant and this constant remains arbitrary. If we filter it by considering a drift *increment*, the problem goes away. Solving (3.29) at the points x_0 and y_0, we get

$$\text{Var}[m^*(y_0) - m^*(x_0)] = -\sum_{\ell>0} [\mu_\ell(y_0) - \mu_\ell(x_0)][f^\ell(y_0) - f^\ell(x_0)]$$

Applying this result to the data points $x_0 = x_\alpha$ and $y_0 = x_\beta$, we get the following interesting result:

$$\text{Cov}(Z_\beta - Z_\alpha, m^*_\beta - m^*_\alpha) = \text{Var}(m^*_\beta - m^*_\alpha) \tag{3.30}$$

which leads to the formula for the variogram of residuals, as indicated in Section 2.7.2:

$$\gamma_{\text{Res}}(x_\alpha, x_\beta) = \gamma(x_\beta - x_\alpha) - \tfrac{1}{2}\text{Var}(m^*_\beta - m^*_\alpha)$$

Example

6. We consider the same setup as in example 5—that is, linear variogram, linear drift, and $N+1$ data points in 1D at arbitrary locations $x_1 < x_2 < \ldots < x_N$. What is the optimal estimator of the drift?

It is a well-known property of the Brownian motion that it is the line joining the end points of the data interval. We can verify that indeed (3.29) admits the following solution:

$$\lambda_1 = \frac{x_N - x}{x_N - x_1}, \quad \lambda_N = \frac{x - x_1}{x_N - x_1}, \quad \lambda_i = 0, \quad \text{all } i \neq 1 \text{ or } N$$

We can calculate the variance of the slope estimator with the variogram $\gamma(h) = b|h|$,

$$\mathrm{Var}\left(\frac{Z_N - Z_1}{x_N - x_1}\right) = \frac{2b}{x_N - x_1}$$

and therefore calculate the variance of any drift increment, but not the variance of the drift estimator at any particular point.

Remarks for an External Drift

The invariance of kriging solutions under a change of coordinates was brought up on several occasions as a desirable property, which does not hold when using drift functions defined by a physical model or by external variables. What difference does it make in practice? To illustrate the point, consider a dome-shaped geological structure (an anticline). Using a quadratic surface model, least squares will find a dome *somewhere*, where it best matches the depth data. On the other hand, if the drift function is defined from external information, typically a seismic survey, the position of the dome will be pretty much fixed and least squares will find the best possible dome *at that place*. Localized information is superior to floating information, but only if it is reliable. Nothing bars us from using a seismic map from South America as an external drift for a depth map in Asia!

Because KED implicitly fits the drift by generalized least squares, we may wish to assess the goodness of fit by checking the coefficient of determination R^2. If we compute it with the standard formula

$$R^2 = 1 - \text{residual sum of squares/total sum of squares}$$

we obtain a lower R^2 than with ordinary least squares because these are designed to minimize the residual sum of squares. We may then incorrectly conclude that there is a loss of performance. For one thing, the interpretation of the above R^2 as the fraction of total variance explained only works for ordinary least squares and would need to be adapted to generalized least squares (Buse, 1973). More importantly, the quality of KED should not be gauged by the goodness-of-fit of the drift at the data points but by the accuracy of estimates at new points, which is measured by the kriging variance.

3.4.7 Additivity Relationship

Coming back to the implicit nature of drift estimation in universal kriging, consider the UK and the SK systems, in matrix form, the latter being distinguished by the subscript K:

$$\begin{cases} \boldsymbol{\Sigma}\boldsymbol{\lambda} + \mathbf{F}\boldsymbol{\mu} = \boldsymbol{\sigma}_0 \\ \mathbf{F}'\boldsymbol{\lambda} = \mathbf{f}_0 \end{cases} \qquad\qquad \boldsymbol{\Sigma}\boldsymbol{\lambda}_K = \boldsymbol{\sigma}_0$$

Subtracting the second system from the first and letting $\boldsymbol{\lambda}_D = \boldsymbol{\lambda} - \boldsymbol{\lambda}_K$ yields

$$\begin{cases} \Sigma \lambda_D + F \mu = 0, \\ F' \lambda_D = f_0 - F' \lambda_K \end{cases} \tag{3.31}$$

The UK estimator can be decomposed into the sum

$$Z^* = Z_K^* + Z_D^* \tag{3.32}$$

of the SK estimator Z_K^* calculated as if the mean were known and subtracted from the data ($\lambda_K = \Sigma^{-1} \sigma_0$ only involves covariances), and a corrective term Z_D^*. Solving (3.31) and taking into account (3.27), it is found that Z_D^* is of the form

$$Z_D^* = \sum_\ell a_\ell^* \left(f_0^\ell - \sum_\alpha \lambda_{K\alpha} f_\alpha^\ell \right) = m_0^* - \sum_\alpha \lambda_{K\alpha} m_\alpha^* \tag{3.33}$$

It is a *drift correction* involving the optimal drift estimates at point x_0 and at the x_α. Recombining this result with (3.32) gives

$$Z^*(x_0) = m_0^* + \sum_\alpha \lambda_{K\alpha}(Z_\alpha - m_\alpha^*) \tag{3.34}$$

Formula (3.34) is exactly the same as (3.1) for simple kriging except that the mean m is replaced by its optimal estimator m^*. In other words, universal kriging is equivalent to optimum drift estimation followed by simple kriging of the residuals from this drift estimate, as if the mean were estimated perfectly. This property only holds when the mean is estimated in a statistically consistent manner—that is, by generalized least squares and not by ordinary least squares.

The additivity relationship (3.32) extends to variances as well. The UK kriging error is

$$Z^* - Z_0 = (Z_K^* - Z_0) + Z_D^*$$

and by the characteristic orthogonality property of SK the error $(Z_K^* - Z_0)$ has zero covariance with all Z_α and thus with Z_D^*. Hence

$$\sigma_{UK}^2 = \sigma_{SK}^2 + \text{Var}(Z_D^*) \tag{3.35}$$

When only the variogram exists, the SK estimator is not defined. Additivity relations similar to (3.32) and (3.35) can be written with the OK estimator, except that the constant drift term a_0 cancels out from Z_D^*, since OK weights add up to one (see example 5). In all cases the drift correction variance is the price to pay for imposing unbiasedness constraints on the UK estimator. Though implicit, drift estimation is not free.

Special Case of a Constant Mean

In the case of a constant mean, these results provide insight into the workings of OK. From (3.33) the drift correction is

$$Z_D^* = \left(1 - \sum_\alpha \lambda_{K\alpha}\right) m^*$$

and the OK estimator is

$$Z_{OK}^* = Z_K^* + \lambda_m m^* \quad \text{with} \quad \lambda_m = 1 - \sum_\alpha \lambda_{K\alpha} \tag{3.36}$$

It differs from SK by the addition of an extra term carrying the influence of the overall mean. λ_m is often called *the weight on the mean*. The OK variance is

$$\sigma_{OK}^2 = \sigma_{SK}^2 + \lambda_m^2 \operatorname{Var}(m^*) \tag{3.37}$$

When the mean is known, or has a very small variance, OK reduces to SK. The optimal estimator m^* of the mean and its variance are given by (3.27) and (3.28):

$$m^* = (\mathbf{1}'\boldsymbol{\Sigma}^{-1}\mathbf{1})^{-1}\mathbf{1}'\boldsymbol{\Sigma}^{-1}\mathbf{Z}, \qquad \operatorname{Var}(m^*) = (\mathbf{1}'\boldsymbol{\Sigma}^{-1}\mathbf{1})^{-1} \tag{3.38}$$

Finally, solving (3.31) gives the Lagrange parameter μ

$$\mu = -\lambda_m \operatorname{Var}(m^*) \tag{3.39}$$

Since λ_m is generally positive, we conclude that μ is generally negative, a remark that will be useful for simulations.

Invariance under Linear Transformation of the f^ℓ

A direct consequence of the additivity relationship (3.32) is the invariance of the UK estimator and variance under a linear transformation of the basis drift functions. By definition, Z_K^* does not involve the drift at all, and Z_D^* is invariant under a linear transformation of the f^ℓ because the GLS estimator of the drift is itself invariant, as seen above. Now monomials of degree $\leq k$ satisfy a relationship of the form (binomial formula)

$$f^\ell(x+h) = \sum_s B_s^\ell(h) f^s(x)$$

showing that a shift of the points is equivalent to a linear transformation of the basis drift functions. Therefore, when the drift is a polynomial function of degree k, with all monomials included, the UK solution remains invariant under shifts. Notice, however, that the μ_ℓ do change.

The shift invariance property is attached to the limited class of functions that are closed under translations, namely the exponential polynomials, as will be seen in the context of IRF$-k$ theory (Section 4.3.4). These are good functions to use unless there is a reason to tie the drift to a specific geographic location, which is exactly what an external drift does.

3.4.8   Wonderful Properties Revisited

In the zero mean case, three wonderful properties were established: orthogonality of the error and the data, the smoothing relationship, and conditional unbiasedness in the Gaussian case. None of these properties holds anymore when the mean is unknown, but similar results can be stated.

Orthogonality Properties

When unbiasedness constraints are introduced, the kriging estimator is selected within a restricted class of linear combinations, and $Z^* - Z_0$ is no longer orthogonal to all Z_α. Indeed from equation (3.21) we have

$$< Z^* - Z_0, Z_\alpha > = - \sum_\ell \mu_\ell f_\alpha^\ell \tag{3.40}$$

This is "the curse of the μ's." The kriging error is no longer orthogonal to *all* linear combinations of Z_α but only to a *restricted class* of linear combinations, namely those which annihilate the basis drift functions f_α^ℓ. Specifically,

$$< Z^* - Z_0, \sum_\alpha \nu_\alpha Z_\alpha > = 0 \tag{3.41}$$

for any set of weights ν_α satisfying,

$$\sum_\alpha \nu_\alpha f_\alpha^\ell = 0, \qquad \ell = 0, 1, \dots, L \tag{3.42}$$

Formula (3.41) follows from a straightforward reformulation of the kriging equations (3.21) or (3.25). Like for simple kriging, these equations have a geometric interpretation in terms of projection in Hilbert spaces (Matheron, 1969a; Journel and Huijbregts, 1978).

The constraints (3.42) generalize the permissibility condition $\sum_\alpha \nu_\alpha = 0$ encountered with the variogram and play a central role in the theory of intrinsic random functions of order k. Suffice it to say here that any linear unbiased estimator of a residual $Z(y) - m(y)$ is of the form (3.42) so that, for any linear unbiased estimator $\hat{m}(x)$ of the mean, one has

$$< Z^* - Z_0, Z_\alpha - \hat{m}_\alpha > = 0, \qquad \alpha = 1, \dots, N$$

Smoothing Relationship

When the mean is unknown there is no guarantee that $\text{Var } Z^* \leq \text{Var } Z$ because $\text{Var } Z^*$ also carries the imprecision about the estimation of the mean. But a similar inequality holds between estimated residuals. By virtue of the above, the decomposition

$$Z_0 - \hat{m}_0 = (Z_0 - Z^*) + (Z^* - \hat{m}_0)$$

is an orthogonal one for *any* linear unbiased estimator $\hat{m}_0$ of the mean so that

$$\text{Var}(Z^* - \hat{m}_0) = \text{Var}(Z_0 - \hat{m}_0) - \sigma_{\text{UK}}^2$$

which generalizes (3.4). (Incidentally we find that $\sigma_{\text{UK}}^2 \leq \text{Var}(Z_0 - \hat{m}_0)$, which simply means that Z^* is a better estimator of Z_0 than $\hat{m}_0$.)

Toward Conditional Unbiasedness

Unfortunately, it is no longer true that $E(Z_0 \mid Z^*) = Z^*$ when the mean is estimated from the data, even in the Gaussian case. However, the following relation shows that any minimization of mean square error tends also to minimize the conditional bias:

$$E(Z_0 - Z^*)^2 = E\left[\text{Var}(Z_0 \mid Z^*)\right] + E\left[E(Z_0 \mid Z^*) - Z^*\right]^2 \qquad (3.43)$$

This formula is completely general (no Gaussian assumption) and follows directly from (1.5) and the fact that $E(Z_0 - Z^*) = 0$. Thus kriging, by design, tends to reduce conditional bias.

Due to mining applications, conditional unbiasedness has received attention in the case of OK. The criterion used most is the slope β of the linear regression of Z_0 on Z^*, which from (3.39) and (3.40) is found to be

$$\beta = \frac{\text{Cov}(Z_0, Z^*)}{\text{Var}(Z^*)} = 1 - \lambda_m \frac{\text{Var}(m^*)}{\text{Var}(Z^*)}$$

One solution to achieve $\beta \approx 1$ is to select a kriging neighborhood that is large enough for a good (implicit) estimation of the mean, thus making $\text{Var}(m^*)$ small. Now $\text{Var}(m^*) \leq \text{Var}(Z^*)$ since m^* is by design the minimum-variance linear combination of the data subject to the constraint that weights add up to 1. Hence we have the inequalities

$$1 - \lambda_m \leq \beta \leq 1 \qquad \text{if } \lambda_m \geq 0,$$

$$1 \leq \beta \leq 1 - \lambda_m \qquad \text{if } \lambda_m \leq 0$$

When the weight on the mean λ_m is small, the slope β is close to 1, even if the neighborhood is not large. Thus λ_m is a criterion for selecting the size of the kriging neighborhood (Rivoirard, 1987).

3.4.9 Dual Kriging and Radial Basis Functions

We now revisit the interpolation properties of kriging, this time with UK instead of SK. Using the matrix notations of (3.22) and switching to lowercase for z which is now considered deterministic, the UK estimator $z^*(x)$ at point x can be written as

$$z^*(x) = \boldsymbol{\lambda}'\mathbf{z} = [\mathbf{z}' \ \ 0]\begin{bmatrix}\boldsymbol{\lambda}\\ \boldsymbol{\mu}\end{bmatrix} = [\mathbf{z}' \ \ 0]\begin{bmatrix}\boldsymbol{\Sigma} & \mathbf{F}\\ \mathbf{F}' & 0\end{bmatrix}^{-1}\begin{bmatrix}\boldsymbol{\sigma}_x\\ \mathbf{f}_x\end{bmatrix} = [\mathbf{b}' \ \ \mathbf{c}']\begin{bmatrix}\boldsymbol{\sigma}_x\\ \mathbf{f}_x\end{bmatrix} \quad (3.44)$$

This leads to two equivalent formulations of kriging equations and estimates, depending on whether we start by multiplying the first two or last two matrices in the above expression.

$$\begin{bmatrix}\boldsymbol{\Sigma} & \mathbf{F}\\ \mathbf{F}' & 0\end{bmatrix}\begin{bmatrix}\boldsymbol{\lambda}\\ \boldsymbol{\mu}\end{bmatrix} = \begin{bmatrix}\boldsymbol{\sigma}_x\\ \mathbf{f}_x\end{bmatrix} \qquad \begin{bmatrix}\boldsymbol{\Sigma} & \mathbf{F}\\ \mathbf{F}' & 0\end{bmatrix}\begin{bmatrix}\mathbf{b}\\ \mathbf{c}\end{bmatrix} = \begin{bmatrix}\mathbf{z}\\ 0\end{bmatrix} \qquad (3.45)$$
$$z^* = \boldsymbol{\lambda}'\mathbf{z} \qquad\qquad\qquad z^* = \mathbf{b}'\boldsymbol{\sigma}_x + \mathbf{c}'\mathbf{f}_x$$

The second system in (3.45), called the *dual kriging* system, is obtained by multiplying the first two matrices in (3.44). It can also be written in terms of the variogram, provided that $f^0(x) \equiv 1$. The term "dual" originates from an alternative derivation of these equations by minimization in a functional space, similar to splines (Matheron, 1981a,c).

From (3.44) and (3.45) $z^*(x)$ is of the form

$$z^*(x) = \sum_\alpha b_\alpha\, \sigma(x_\alpha, x) + \sum_\ell c_\ell f^\ell(x) \quad (3.46)$$

with

$$\sum_\alpha b_\alpha f^\ell(x_\alpha) = 0, \qquad \ell = 0, 1, \ldots, L \quad (3.47)$$

The coefficients b_α and c_ℓ are linear functions of the data, and in the case of a global neighborhood they do not change with the location x. It suffices to compute them once for all, and the estimates $z^*(x)$ are obtained using (3.46). This does not work, however, for the kriging variance.

Formulas (3.46) and (3.47) have taken a new importance due to their connection with *radial basis functions* [e.g., Buhmann (2003)]. A radial basis function is an isotropic function whose value only depends on the distance, usually the Euclidean distance, from some fixed origin x_c, called a *center*:

$$\varphi(x, x_c) = \varphi(|x - x_c|)$$

The radial basis function approach takes the data points as centers and builds interpolants of the form

$$\Phi(x) = \sum_{\alpha} b_{\alpha} \, \varphi(|x - x_{\alpha}|) \qquad (3.48)$$

Constraining it to fit the N data points, gives N linear equations for the N coefficients b_{α}:

$$\sum_{\alpha} b_{\alpha} \, \varphi(|x_{\beta} - x_{\alpha}|) = z_{\beta}, \qquad \text{all} \quad \beta$$

These are exactly the equations (3.8) for the SK estimator, provided that $\varphi(r) = \sigma(r)$ is a stationary and isotropic covariance function ($r = |x|$). The attractive feature of a covariance is that it guarantees that the system can be solved.

Unfortunately, in applications the function φ is not necessarily a covariance. The following is a selection of commonly used radial basis functions in $\mathbb{R}^{n}$:

- Gaussian $\qquad\qquad\qquad \varphi(r) = \exp(-r^2/r_0^2)$
- Multiquadric $\qquad\qquad \varphi(r) = (r^2 + r_0^2)^{\frac{1}{2}}$
- Inverse multiquadric $\quad\; \varphi(r) = (r^2 + r_0^2)^{-\frac{1}{2}}$
- Thin plate spline in 2D $\;\; \varphi(r) = r^2 \log(r)$

The Gaussian is a covariance and so is the inverse multiquadric, as a special case of the Cauchy model (2.55). The multiquadric itself is not a covariance but a variogram (up to an additive constant). As for the thin plate spline, it is neither a covariance nor a variogram but a *generalized covariance*, as we will see in Section 4.6.4.

To guarantee that a unique solution exists, the interpolant (3.48) is augmented by a low-degree polynomial and becomes

$$\Phi(x) = \sum_{\alpha} b_{\alpha} \, \varphi(|x - x_{\alpha}|) + \sum_{\ell} c_{\ell} f^{\ell}(x) \qquad (3.49)$$

with coefficients subject to the constraints (3.47).

Replacing φ by σ in (3.49), writing that $\Phi(x_{\alpha}) = z_{\alpha}$ at all data points and appending the side conditions (3.47) leads exactly to the dual kriging system (3.45). Radial basis functions and dual kriging solutions coincide.

The side conditions (3.47) that appear somewhat artificially added to (3.49) turn up naturally if we start from standard kriging. When a variogram is used rather than a covariance, a condition of unit sum must be placed on the weights, which translates to the condition that the b_{α} sum to zero. Likewise, a thin plate spline model in 2D requires three polynomial side conditions. Radial basis functions are conditionally positive (or negative) definite functions of order k, and these come with side conditions that ensure the nonsingularity of the interpolation matrix. Micchelli (1986) is a major reference on the subject (see also Section 4.5.4).

Design and Analysis of Computer Experiments

Sacks et al. (1989) introduced the use of kriging for the design and analysis of computer experiments (DACE). This pioneering work opened up a completely new territory for kriging methods, solving engineering problems in a multi-dimensional space where the coordinates are no longer geographic but represent scalar design variables. The idea is to approximate an objective function, depending on design parameters, by an interpolating function (called *metamodel*) based on data from a limited number of experiments (costly computer runs) and then use the metamodel as a surrogate of the objective function to find an optimum. The advantage of kriging over traditional polynomial response surfaces is an exact fit at known design points and the possibility to tune the response surface through the covariance model smoothness, range, and nugget effect parameters. Given the fundamental anisotropy of the design space, the covariance model is often taken to be the product of one-dimensional covariances, typically exponentials or Gaussians. An alternative could be to split the parameters into two groups of sizes p and q and use nonseparable covariance models of the Gneiting class to define a covariance model in $\mathbb{R}^p \times \mathbb{R}^q$ [see Gneiting (2002) and Section 5.8.2]. In high-dimensional space with relatively few data points, the choice of the covariance has a significant influence on the relevance of the results.

These developments, combined with experimental design techniques to minimize the number of experiments, have generated a huge number of applications in such diverse areas as the design of aircrafts [e.g., Chung and Alonso (2002)], of automobiles, of computer chips, and so on. In the literature we can refer the reader to the works of Kleijnen, notably van Beers and Kleijnen (2004), a monograph (Kleijnen, 2008), and a review paper (Kleijnen, 2009).

3.4.10 A Bayesian Bridge Between Simple and Universal Kriging

This title of a paper by Omre and Halvorsen (1989) conveys the idea very clearly. Simple and universal kriging can be viewed as extreme cases in which we either know the drift perfectly or else know nothing about it. A Bayesian model introduced by Omre (1987) assumes that some prior knowledge is available about the drift and establishes a continuum between these two approaches. The model is specified at the order 2 and can be translated into prior and posterior distributions, assuming that all distributions are jointly Gaussian.

As a background it is interesting to first reconsider a variant of the UK model, the *random drift* model, proposed early on by Matheron (1970). Specifically,

$$Z(x) = Y(x) + M(x)$$

where E $Y(x) - 0$ and $M(x)$ is now a random function, not necessarily independent of $Y(x)$ but much smoother than $Y(x)$. So the dichotomy is physically meaningful, and $M(x)$ may be expanded in the usual way as

$$M(x) = \sum_\ell A_\ell f^\ell(x) \tag{3.50}$$

except that now the coefficients are *random*. Define the first two moments

$$E(A_\ell) = a_\ell \quad \text{and} \quad \text{Cov}(A_\ell, A_s) = K_{\ell s}$$

and the noncentered covariances

$$\begin{cases} E[Y(x)Y(y)] = \sigma(x, y) \\ E[Y(x)M(y)] = R(x, y) \\ E[M(x)M(y)] = K(x, y) + \sum_\ell \sum_s a_\ell \, a_s \, f^\ell(x) f^s(y) \end{cases}$$

The (nonstationary) covariance of $M(x)$ is simply

$$K(x, y) = \sum_\ell \sum_s K_{\ell s} f^\ell(x) f^s(y) \tag{3.51}$$

and the cross covariance between $Y(x)$ and $M(y)$ is of the form

$$R(x, y) = \sum_\ell E[Y(x)A_\ell] f^\ell(y) = \sum_\ell R_\ell(x) f^\ell(y)$$

If $Y(x)$ is a stationary RF then $R_\ell(x) = R_\ell$ is a constant, and assuming as usual that $f^0 \equiv 1$, we have the expansion

$$R(x, y) = \sum_\ell \sum_s R_{\ell s} f^\ell(x) f^s(y) \tag{3.52}$$

where $R_{\ell s} = 0$ for $s > 0$. If the mean is not assumed known, we consider an estimator of Z_0 of the form

$$Z^* = \sum_\alpha \lambda_\alpha Z_\alpha = \sum_\alpha \lambda_\alpha Y_\alpha + \sum_\alpha \lambda_\alpha M_\alpha$$

whose mean square error, using the foregoing relationships, can be written as

$$E(Z^* - Z_0)^2 = E\left(\sum_\alpha \lambda_\alpha Y_\alpha - Y_0 \right)^2 + \sum_\ell \sum_s (K_{\ell s} + a_\ell \, a_s + 2R_{\ell s})$$
$$\times \left(\sum_\alpha \lambda_\alpha f^\ell_\alpha - f^\ell_0 \right) \left(\sum_\beta \lambda_\beta f^s_\beta - f^s_0 \right)$$

Here Matheron argues that it is not possible to estimate the terms $K_{\ell s} + a_\ell a_s$ from the data Z_α because the noncentered covariance of $M(x)$ is not stationary, and even if it were, its inference would be very poor because of a too-high regularity, as was shown in Section 2.9.2. Therefore the only solution is to cancel the other terms and set

$$\sum_\alpha \lambda_\alpha f^\ell_\alpha - f^\ell_0 = 0 \qquad \ell = 0, 1, \ldots, L$$

Minimizing the mean square error under these conditions leads to the usual UK system.

Omre and Halvorsen (1989) start from a similar model except that they assume Y and M to be independent RFs (hence $R(x, y) \equiv 0$) and that the means a_ℓ and covariances $K_{\ell s}$ of the random drift coefficients are known from prior knowledge. Thus

$$
\begin{cases}
\mathrm{E}[Z(x)|A_\ell : \ell = 0, \ldots, L] = M(x) = \sum_\ell A_\ell f^\ell(x) \\[2mm]
\mathrm{Cov}[Z(x), Z(y)|A_\ell : \ell = 0, \ldots, L] = \sigma(x, y) \\[2mm]
\mathrm{E}[Z(x)] = \mathrm{E}[M(x)] = m(x) = \sum_\ell a_\ell f^\ell(x) \\[2mm]
\mathrm{Cov}[Z(x), Z(y)] = \sigma(x, y) + \sum_\ell \sum_s K_{\ell s} f^\ell(x) f^s(y)
\end{cases}
\tag{3.53}
$$

Consider the following unbiased estimator of Z_0 where prior means are involved:

$$Z^*_{\mathrm{B}} = \sum_\alpha \lambda_\alpha (Z_\alpha - m_\alpha) + m_0$$

It is optimized by unrestricted minimization of the variance

$$
\mathrm{E}(Z^*_{\mathrm{B}} - Z_0)^2 = \mathrm{E}\left(\sum_\alpha \lambda_\alpha Y_\alpha - Y_0\right)^2 \\
+ \sum_\ell \sum_s K_{\ell s} \left(\sum_\alpha \lambda_\alpha f^\ell_\alpha - f^\ell_0\right)\left(\sum_\beta \lambda_\beta f^s_\beta - f^s_0\right)
$$

This is the bridge between simple and universal kriging. At one end we have exact prior knowledge of the drift, so $K_{\ell s} = 0$; at the other end we have complete prior ignorance, so $K_{\ell s} \to \infty$ and the weights must satisfy the unbiasedness constraints of UK to keep the expression finite. In between, the "Bayesian kriging" equations are just those of SK but with the nonstationary covariance $\mathrm{Cov}[Z(x), Z(y)]$ defined by (3.53). As with standard SK, the solution is an exact interpolant, and no minimum number of points is required; in particular, we may have $N < L + 1$. Evidently the associated error variance is comprised between the SK variance and the UK variance.

Special Case of a Constant Random Mean

In the case of a constant random mean the model (3.53) simplifies to

$$E[Z(x)|M] = M, \qquad \text{Cov}[Z(x), Z(y)|M] = \sigma(x, y)$$
$$E[Z(x)] = E(M) = m, \qquad \text{Cov}[Z(x), Z(y)] = \sigma(x, y) + \text{Var}(M)$$

where m and $\text{Var}(M)$ are assumed known from prior knowledge. The Bayesian kriging equations are those of the SK system (3.3) written with the unconditional covariance of Z:

$$\sum_{\beta} \lambda_{\beta}[\sigma_{\alpha\beta} + \text{Var}(M)] = \sigma_{\alpha 0} + \text{Var}(M), \qquad \alpha = 1, 2, \ldots, N \qquad (3.54)$$

The optimal estimator M^* of the mean is of the form

$$M^* = m + \sum_{\alpha} \lambda_{M\alpha}(Z_{\alpha} - m)$$

with weights satisfying the equations

$$\sum_{\beta} \lambda_{M\beta}[\sigma_{\alpha\beta} + \text{Var}(M)] = \text{Var}(M), \qquad \alpha = 1, 2, \ldots, N \qquad (3.55)$$

These equations can be established directly or deduced from (3.54) by letting $\sigma_{\alpha 0} = 0$. The variance of the optimal estimator of the mean is

$$\text{Var}(M^*) = \left(1 - \sum_{\alpha} \lambda_{M\alpha}\right) \text{Var}(M) \qquad (3.56)$$

Since we introduce prior information, we expect this variance to be less than the variance $\text{Var}(m^*)$ of the OK estimator of the mean computed by (3.38) with the covariance $\Sigma = [\sigma_{\alpha\beta}]$. Indeed by solving (3.55) it can be shown that

$$\text{Var}(M^*) = \left(\sum_{\alpha} \lambda_{M\alpha}\right) \text{Var}(m^*) \qquad (3.57)$$

which proves the inequality since by (3.56) the sum of weights is less than 1. Combining (3.56), and (3.57), we can see that the sum of weights is between 0 and 1, the value 0 being reached when $\text{Var}(M) = 0$ and the value 1 when $\text{Var}(M) \to \infty$. Concerning kriging variances, similar to (3.36) it can be established that

$$Z_B^* = M^* + \sum_{\alpha} \lambda_{K\alpha}(Z_{\alpha} - M^*) = Z_K^* + \lambda_m M^*, \qquad \lambda_m = 1 - \sum_{\alpha} \lambda_{K\alpha}$$

where Z_K^* is the SK estimator and $\lambda_{K\alpha}$ is its weights, hence the additivity relationship

$$\sigma_B^2 = E(Z_B^* - Z_0)^2 = \sigma_{SK}^2 + \lambda_m^2 \text{Var}(M^*)$$

Combining the above with (3.37) and (3.57), we get

$$\sigma_{OK}^2 - \sigma_B^2 = \lambda_m^2 [\text{Var}(m^*) - \text{Var}(M^*)] = \lambda_m^2 \left(1 - \sum_\alpha \lambda_{M\alpha}\right) \text{Var}(m^*)$$

The Bayesian kriging variance is less than the OK variance, as expected. When $\text{Var}(M) \to \infty$, Bayesian kriging converges to OK; and when $\text{Var}(M) = 0$, it coincides with SK.

Prior Knowledge

The central question now is this: How can one make a "qualified guess" of the drift—that is, come up with an estimate and its associated uncertainty before any Z data become available? In the applications presented, this guess is always obtained from a different but related data set. Abrahamsen (1993), for example, builds a comprehensive multilayer reservoir model in which the Z variables are depths of geological surfaces measured in wellbores and the random drift is a function of seismic reflection time. The prior means and covariances of the coefficients of this function are derived from the analysis of velocity information at the wellbores. This technique uses seismic times and squared times as basis drift functions and is akin to the external drift method, except that some knowledge is assumed about the drift coefficients.

Pilz (1994) extends the theory to the case where only partial prior knowledge of the first two moments of the drift coefficients is available. This knowledge in effect restricts the possible prior distributions of the drift coefficients to a subfamily, and a "Bayes robust" estimator is derived by minimizing the maximum mean square estimation error over all possible prior distributions in this subfamily. For example, in the case of ordinary kriging we may know that the expected value of M, namely the true mean, lies in a given interval $[a, b]$ and that the variance of M is less than some value K_0. Then, using the notations of the additivity formula (3.34) for the UK estimator, the Bayes robust estimator derived by Pilz is

$$\hat{Z}_B = Z_K^* + \left(1 - \sum_\alpha \lambda_{K\alpha}\right)\hat{m}_B$$

where $\hat{m}_B$ is related to the optimum estimator of the mean m^* given by (3.27), the interval midpoint $m_0 = (a + b)/2$, the uncertainty variance K_0, and the maximum interval variance $K_1 = (b - a)^2/4$ by

$$\hat{m}_B = (m^* + \theta m_0)/(1 + \theta) \quad \text{with} \quad \theta = \text{Var}(m^*)/(K_0 + K_1)$$

(in the above m^* and $\mathrm{Var}\, m^* = (\mathbf{1}'\boldsymbol{\Sigma}^{-1}\mathbf{1})^{-1}$ are conditional on $M=m$). In general, $\hat{m}_B$ and $\hat{Z}_B$ are biased estimators, but $\hat{Z}_B$ achieves a smaller mean square error σ_B^2 than the OK estimator. Specifically,[9]

$$\sigma_{OK}^2 - \sigma_B^2 = \left(1 - \sum_\alpha \lambda_{K\alpha}\right)^2 (\mathrm{Var}\,(m^*))^2 / (\mathrm{Var}\,(m^*) + K_0 + K_1)$$

The Bayesian approach can also be used to study the effect of uncertainty on the covariance parameters [e.g., Handcock (1994)], but the theory becomes more complex because of nonlinearities. The noticeable difference is that incorporating the uncertainty on the covariance tends to increase the uncertainty on the final results, whereas modeling the uncertainty on the drift through a Bayesian analysis tends to *reduce* it.

Under the initial impulse of H. Omre, the Bayesian approach in geostatistics has seen a wave of applications to petroleum problems where data are initially scarce and replaced by explicit prior guesses (see Section 7.8.4).

3.4.11 Lognormal Kriging and Generalization

The linear estimators considered so far involved no assumption on the finite-dimensional distribution of the data other than on first and second moments. We saw that this worked very well in the case of a Gaussian RF with a known mean because then the simple kriging estimator coincides with the regression function of $Z(x)$ on the data. If, on the other hand, we know that the regression function is highly nonlinear, it would not be very smart to use linear estimators. The classic and important example is the lognormal case, namely when the logarithm of the random function $Z(x)$

$$Y(x) = \log Z(x)$$

is a Gaussian RF with a known mean or an unknown constant mean. Assume first that the mean is known. The conditional distribution of Y_0 given $Y_1, \ldots, Y_N$ is a Gaussian with mean the simple kriging estimator Y_{SK}^* and variance σ_{SK}^2. The simple lognormal kriging estimator (SLK) of $Z_0 = \exp(Y_0)$ is the conditional expectation (see Appendix, Section A.9)

$$Z_{SLK}^* = \mathrm{E}\big(e^{Y_0} \mid Y_1, \ldots, Y_N\big) = \exp\big(Y_{SK}^* + \tfrac{1}{2}\sigma_{SK}^2\big) \tag{3.58}$$

The conditional estimation variance is

$$\mathrm{Var}\big(Z_{SLK}^* - Z_0 \mid Z_1, \ldots, Z_N\big) = \mathrm{Var}\big(e^{Y_0} \mid Y_1, \ldots, Y_N\big) = (Z_{SLK}^*)^2 \big[\exp(\sigma_{SK}^2) - 1\big]$$

[9] A misprint in Pilz's article has been corrected in this formula.

This conditional variance depends on the data values, but the variance scaled by $(Z_{SLK}^*)^2$ does not.

SLK enjoys the same conditional unbiasedness property as simple linear kriging:

$$E(Z_0 \mid Z_{SLK}^*) = Z_{SLK}^*$$

This property results from the fact that Y_{SK}^* and $Y_0 - Y_{SK}^*$ are uncorrelated, and therefore independent, thanks to the Gaussian assumption

$$
\begin{aligned}
E(Z_0 \mid Z_{SLK}^*) &= E\left(\exp(Y_{SK}^* + Y_0 - Y_{SK}^* \mid Y_{SK}^*)\right) \\
&= \exp(Y_{SK}^*) \, E\left(\exp(Y_0 - Y_{SK}^*) \mid Y_{SK}^*\right) = \exp(Y_{SK}^*) \exp\left(\tfrac{1}{2}\sigma_{SK}^2\right)
\end{aligned}
$$

In case of an unknown mean, the problem becomes more complex. It is possible to construct optimal linear estimators in the logarithmic scale and also devise a reverse transformation that ensures unbiasedness, but the optimality properties of such procedures are unclear. Matheron (1974a) discussed this question in detail in an important report entitled "The return of the sea serpent," but no general solution emerges. Considerable simplification occurs if we accept to work with quantiles rather than moments because they simply follow the transformation. Since the distribution of the error $Y_{OK}^* - Y_0$ is symmetric (Gaussian), its median coincides with its mean and is thus zero; therefore $\exp(Y_{OK}^*)$ is a *median unbiased* estimator of Z_0:

$$\Pr\{\exp(Y_{OK}^*) \geq Z_0\} = \Pr\{\exp(Y_{OK}^*) \leq Z_0\} = 0.5$$

Likewise, any confidence interval on Y can be back-transformed. For example,

$$\Pr\left\{\exp(Y_{OK}^* - 2\,\sigma_{OK}) \leq Z_0 \leq \exp(Y_{OK}^* + 2\,\sigma_{OK})\right\} = 0.95$$

However, neither $\exp(Y_{Ok}^*)$ nor $\exp(Y_{OK}^* + \tfrac{1}{2}\sigma_{Ok}^2)$ is an unbiased estimator of Z_0. Using a constant multiplicative, bias correction, we obtain the ordinary lognormal kriging estimator

$$Z_{OLK}^* = \exp\left[Y^* + \tfrac{1}{2}\left(\mathrm{Var}(Y_0) - \mathrm{Var}(Y_{OK}^*)\right)\right] \tag{3.59}$$

The correction term $\mathrm{Var}\,Y_0 - \mathrm{Var}\,Y_{OK}^*$ is computed from the OK system (3.17) or (3.19) as

$$
\begin{aligned}
\mathrm{Var}(Y_0) - \mathrm{Var}(Y_{OK}^*) &= \sigma_{00} - \sum_\alpha \sum_\beta \lambda_\alpha \lambda_\beta \, \sigma_{\alpha\beta} = \sigma_{OK}^2 + 2\mu \\
&= \sum_\alpha \sum_\beta \lambda_\alpha \lambda_\beta \gamma_{\alpha\beta}
\end{aligned}
\tag{3.60}
$$

No general statement can be made on the sign of this correction other than noting that it is positive if all OK weights are positive.

The unbiased estimator (3.59) must be used with caution because it is nonrobust against departures from the lognormal model. Also, contrary to linear kriging, the variogram sill or slope directly affects the estimator itself, making the correction formula very sensitive to the variogram model. David (1988, p. 119) reports "horror stories" on this subject.[10] To avoid gross errors, various schemes are used to "calibrate" the estimates on the untransformed data. For example, Journel and Huijbregts (1978, p. 572) propose, in the stationary case, to scale the kriging estimates by a constant factor such that the arithmetic mean of these estimates equals the estimate of the mean obtained directly from the Z data. But the new estimates no longer honor the data points (the technique was suggested for block estimation). To preserve the interpolation property, a modified estimator can be defined as

$$Z^*_{OLK} = \exp\left(Y^*_{OK} + B\tfrac{1}{2}\left(\mathrm{Var}(Y_0) - \mathrm{Var}(Y^*_{OK})\right)\right)$$

with a calibration factor B to be determined by cross-validation (Delfiner, 1977). This procedure amounts to fine tuning the sill of the variogram of Y.

A common approach is to perform two studies in parallel, one on Z and one on $\log Z$, and compare the results in light of the properties of the lognormal model (see end of Section 2.5.3).

The lognormal model is a natural for positive skewed data, such as low geochemical concentrations, pollution levels, or permeability. In many cases it suffices to work on $\log Z$ instead of Z. In mining, the lognormal model is used in the context of block estimation (see Section 6.6.2 for change-of-support issues). Switzer and Parker (1976), Dowd (1982), David (1988), and Rivoirard (1990) are references on the use of the lognormal model in mining. In principle, this model requires stationary data as do all models involving distributional assumptions, since residuals tend to be severely biased. But highly skewed data may exhibit a strong drift even after logarithmic transformation. For example, after a nuclear test, plutonium concentrations in the soil range over almost 5 orders of magnitude, with a large maximum at the detonation point and a rapid falloff in all directions (Delfiner and Gilbert, 1978).

General Transformation

The lognormal kriging approach can be generalized to an arbitrary transformation $Z = \varphi(Y)$, where Y is a Gaussian RF. In the stationary case the function φ is determined from the histogram of the data; and if this is continuous (no accumulation of frequency at discrete values), φ has an inverse. We can then work on the Y values for which linear estimators are well-suited and transform the results back. Of course, assuming that the RF $Y(x)$ is Gaussian as a whole just because its marginal distribution is Gaussian takes a big leap of faith, but if we go for it, the wonderful properties of simple kriging provide a general formula for reducing the bias in the reverse transformation.

---

[10] Michel David once described lognormal kriging as "riding a wild horse" [quoted in Snowden (1994)].

Borrowing from a standard argument of bias reduction in statistical estimation [e.g., Cox and Hinkley (1974, p. 260)], suppose that φ is regular enough to have a Taylor expansion at order 2 and that the simple kriging error $Y^* - Y$ is relatively small; thus

$$\varphi(Y) \approx \varphi(Y^*) + (Y - Y^*)\varphi'(Y^*) + \tfrac{1}{2}(Y - Y^*)^2\varphi''(Y^*)$$

Taking the conditional expectation given Y^* and applying the results (3.14), we get

$$E[\varphi(Y)|Y^*] \approx \varphi(Y^*) + \tfrac{1}{2}\sigma_{SK}^2\varphi''(Y^*)$$

So the estimator

$$Z^* = \varphi(Y^*) + \tfrac{1}{2}\sigma_{SK}^2\varphi''(Y^*) \tag{3.61}$$

approximately satisfies the conditional unbiasedness relationship

$$E(Z^* \mid Y^*) = E(Z \mid Y^*)$$

which is stronger than just unbiasedness. The correction formula hinges on the assumption of small kriging variances and the correct determination of the variogram sill.

In the lognormal case, (3.61) translates into

$$Z^* = \exp(Y^*)\left[1 + \tfrac{1}{2}\sigma_{SK}^2\right]$$

and does recover the leading term in the series expansion of the exact solution $\exp(Y^* + \tfrac{1}{2}\sigma_{SK}^2)$.

3.5   ESTIMATION OF A SPATIAL AVERAGE

Kriging was originally developed not for estimation of point values but of average grades over mining panels. Its main goal was to avoid the systematic overestimation that takes place when high-grade panels are selected solely on the basis of internal samples. To understand this important problem, it is interesting to look at it in its original form and explain the approach proposed by Krige (1951), one of the pioneers of geostatistical methods.

3.5.1   Krige's Regression Effect

When a panel is selected because it contains high-grade samples, the samples surrounding it tend, by definition, to have lower grades. These data should also be included in the estimation of the panel or else the sampling is biased, and this is a cause of severe systematic error (here a constant mean is considered). On the average, high-grade panels are poorer than their internal samples suggest, and low-grade panels are richer.

Gold miners in South Africa were aware of this fact and used empirical correction factors. Krige provided a theoretical justification for these corrections in the scope of regression theory. He started from the assumption that the expected value of the mean grade $\overline{Z}_v$ of samples taken inside a panel is equal to the panel's true mean grade. This is an exact property if the samples are selected at random within the panel, or an approximate one if a subgrid is used. For the sake of simplicity, let us assume that grades are Gaussian (Krige considered the lognormal case). Then the regression line giving the conditional expectation of $\overline{Z}_v$ as a function of the panel grade is simply the first bisector (Figure 3.8). Necessarily, the other regression line

$$E(Z_V | \overline{Z}_v) = m + \beta(\overline{Z}_v - m) \tag{3.62}$$

relating the expected panel grade to the mean sample grade has a slope β less than one (the product of slopes is the square of the correlation coefficient). This is the correction formula used by Krige; it pulls the estimate toward the mean.

Now, in practice, the overall mean m is not known and is replaced by the mean $\overline{Z}$ of the N samples in the orebody so that (3.62) is evaluated by

$$\hat{Z}_V = \overline{Z} + \beta(\overline{Z}_v - \overline{Z})$$

This is a linear combination of data of the form

$$\hat{Z}_V = \sum_{\alpha} \lambda_{\alpha} Z_{\alpha}$$

that assigns the same weight to each sample inside the panel and assigns another constant weight $(1 - \beta)/N$ to each outside sample, the weights adding up to one.

Kriging generalizes this approach by personalizing the weight assigned to each sample.

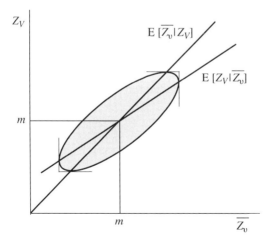

FIGURE 3.8 The regression effect. On the average the true panel grade is less than the sample mean grade for $\overline{Z}_v > m$ and greater for $\overline{Z}_v < m$.

3.5.2 Kriging Equations

We want to estimate the mean value of $Z(x)$ over a block v, using a linear combination of "point" data $Z(x_\alpha)$. We will consider directly the case of an unknown mean. The theory follows exactly the steps explained in Section 3.4.2, the only differences arising from the objective which is now

$$Z_0 = \frac{1}{|v|} \int_v Z(x)dx \quad (= Z_v)$$

The derivation involves the mean values of the drift functions over the block,

$$f_0^\ell = \frac{1}{|v|} \int_v f^\ell(x)dx \quad (= f_v^\ell)$$

the covariances $\sigma_{\alpha 0}$ between each sample Z_α and the block, and the variance σ_{00} of the block:

$$\sigma_{\alpha 0} = \frac{1}{|v|} \int_v \sigma(x_\alpha, x)dx \ (= \sigma_{\alpha v}) \quad \text{and} \quad \sigma_{00} = \frac{1}{|v|^2} \int_v \int_v \sigma(x, y)\, dx\, dy \quad (= \sigma_{vv})$$

With these values for $f_0^\ell, \sigma_{\alpha 0}$, and σ_{00} the kriging system remains the same as (3.21), and the kriging variance is (3.23). (In parentheses are the standard geostatistical notations for block averages.)

As before, these results can be reexpressed in terms of the variogram leading to the system (3.25) with the appropriate f_0^ℓ and $\gamma_{\alpha 0}(= \gamma_{\alpha v})$. For the kriging variance, however, the term $-\gamma_{00}$ corresponding to σ_{00} is no longer zero as in the case of point kriging but

$$\gamma_{00} = \frac{1}{|v|^2} \int_v \int_v \gamma(x, y)dx\, dy \quad (= \gamma_{vv})$$

and the kriging variance is

$$E(Z^* - Z_0)^2 = -\gamma_{00} + \sum_\alpha \lambda_\alpha \gamma_{\alpha 0} + \sum_\ell \mu_\ell f_0^\ell$$

Likewise, the theory can be developed for the estimation of an arbitrary moving average:

$$Z_0 = \int p_0(x)Z(x)dx \quad \text{with} \quad \int p_0(x)dx = 1$$

That just changes the expressions of $f_0^\ell, \sigma_{\alpha 0}$, and σ_{00}, which become

$$f_0^\ell = \int f^\ell(x)\, p_0(x)\, dx, \qquad \sigma_{\alpha 0} = \int \sigma(x_\alpha, x)\, p_0(x)\, dx,$$

$$\sigma_{00} = \iint \sigma(x, y)\, p_0(x)\, p_0(y)\, dx\, dy$$

A similar approach enables kriging from data with different supports.

3.5.3 Piecing Together Local Estimates

If the domain V to be estimated is the union of several nonoverlapping blocks v_i, the mean grade Z_0 of V is related to the mean grades Z_i of the v_i,

$$Z_0 = \frac{1}{|V|} \sum_i |v_i| Z_i \quad \text{with} \quad |V| = \sum_i |v_i|$$

If the estimation of Z_0 and the Z_i is carried out from the same data, it follows immediately from the linearity of the kriging system that the estimators are in the same relationship, namely

$$Z^* = \frac{1}{|V|} \sum_i |v_i| Z_i^*$$

This property allows a global estimation by piecing together local kriging estimates. In practice, the local estimates are not calculated from all the data but only from those of a neighborhood. If this is well-designed, the result is close to the optimum that would be obtained if all data were used. Note, however, that there is no similar relationship between kriging variances (the covariances between kriging errors would need to be introduced).

3.5.4 A Case Study in Forest Inventory

We consider again the tropical forest example introduced in Section 2.9.1 [from Narboni (1979)], in which an exhaustive survey of trees of the "gaboon" species (a variety of mahogany) was conducted over an area of 20,000 ha (1 hectare $= 10,000$ m$^2 = 2.471$ acres). This involved the analysis of 80,000 sampling units of 50 m $\times$ 50 m. Such considerable work was motivated by the desire to validate the geostatistical approach by comparison with the traditional estimation method, and most important, with *reality*.

To ensure stationarity, the area was divided into four approximately equal zones of 5000 ha (Ngolo I through IV). Inventory estimation was carried out on 500-m $\times$ 400-m rectangular tracts using 10% of the data. Tracts are centered on 50-m-wide strip transects further divided into the basic sampling units (Figure 3.9). The classic estimate $\hat{Z}_1$, in number of trees per hectare, is simply the mean of the sampling units inside the tract. Kriging, on the other hand, also uses information from outside the tract. Data that are close or symmetric with respect to the center of the estimated tract play a similar role in the kriging equations and receive the same weights. For computational simplicity they are aggregated within rings, and an overall weight is applied to the mean value within the ring (this is sometimes called *random kriging*). Using a pattern of five rings such as depicted in Figure 3.9, we have a loss of precision on variance of about 1%. In case of systematic sampling, the same kriging configuration can be used for all tracts, and the weights $\lambda_1, \ldots, \lambda_5$ assigned to each ring are calculated once for all.

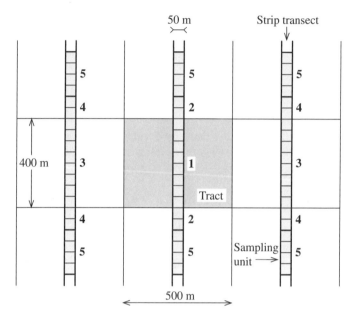

FIGURE 3.9 Kriging configuration for a tract (shaded area). The numbers define the rings. [From Narboni (1979), © Bois et Forêts des Tropiques.]

TABLE 3.2   Classic and Kriging Estimates Compared with Reality

NGOLO Zones	$\overline{Z}_v$	$\overline{\hat{Z}}_1$	Kriging Weights					$\overline{Z}^*$	σ_K^2	$2\sigma_K/\overline{Z}^*$
			λ_1	λ_2	λ_3	λ_4	λ_5			
I	1.18	1.25	0.38	0.10	0.18	0.08	0.26	1.25	0.20	72%
II	1.26	1.29	0.38	0.08	0.18	0.08	0.28	1.28	0.23	75%
III	1.53	1.58	0.44	0.10	0.16	0.06	0.24	1.53	0.34	76%
IV	1.69	1.61	0.41	0.10	0.17	0.07	0.25	1.62	0.36	74%

Note: $\overline{Z}_v$, true value; $\overline{\hat{Z}}_1$, mean of classic estimates; $\overline{Z}^*$, mean of kriging estimates; σ_K^2, kriging variance.

Source: Narboni (1979).

Table 3.2 compares the mean estimates over the four zones; multiplication by the areas gives the total number of trees. Notice that outside samples are weighted more than inside samples ($\lambda_1 < 0.5$). This is due to the presence of a large nugget effect (see Figure 2.32).

If we only compare the means of the classic and the kriging estimates, we find very little difference, except for the third zone where kriging provides a better estimate. The two methods are nearly equivalent for *global estimation*. But these aggregates average out a large disparity of *local* situations. Figures 3.10 and 3.11 show the difference in a striking manner. The scatterplot of true values (plotted

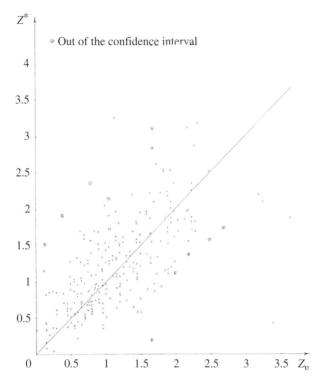

FIGURE 3.10   Kriging estimates (Y axis) and true values (X axis). Each point represents a 20-ha tract (10% sampling on Ngolo I). [From Narboni (1979), © Bois et Forêts des Tropiques.]

horizontally) versus kriging estimates is nicely centered about the unit slope line, with 11 points outside the 95% confidence interval (we expect about $250 \times 0.05 = 12.5$ points). The local kriging estimator satisfies the conditional unbiasedness property $E(Z_v|Z^*) = Z^*$, even though ordinary kriging rather than simple kriging is used. Compare this with the classic estimator $\hat{Z}_1$ in Figure 3.11: Tracts with $\hat{Z}_1 > m$ are overestimated, whereas tracts with $\hat{Z}_1 < m$ are underestimated. This is exactly the regression effect discussed earlier.

3.5.5   Estimation over a Variable Support

In a number of applications the property of interest is the product or the ratio of two variables. This is typically the case for studies carried out in 2D by using vertical integrals of point properties as operational variables. For example, one of the most influential properties for oil and gas resources computation is "net-to-gross," which is the ratio of the cumulative length of the intervals that contain producible hydrocarbons ('net pay thickness') to the gross thickness of the layer. Another example is the assessment of underground pollution, which involves the polluted *thickness* $H(x)$ ($x \in \mathbb{R}^2$) and the pollutant *accumulation* $A(x)$, representing the product of thickness by mean vertical pollutant concentration $Z(x)$ over this thickness. Denoting by H_S and A_S the spatial means of $H(x)$ and $A(x)$ in the domain, and S the area of the domain, the products $T = \rho S H_S$ and $Q = \rho S A_S$ represent respectively the tonnage of contaminated ground

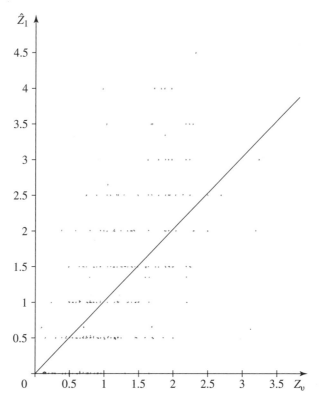

FIGURE 3.11 Classic estimates (*Y*-axis) and true values (*X*-axis). Each point represents a 20-ha tract (10% sampling on Ngolo I). [From Narboni (1979), © Bois et Forêts des Tropiques.]

that should be removed or treated and the quantity of pollutant it contains (ρ is mass density). The mean contaminant concentration is then $Q/T = A_S/H_S$.

H_S and A_S are additive quantities and can be estimated by kriging without special problems. Estimation of mean areal concentration, however, is tricky because $Z(x)$ values are defined over different supports (thicknesses). Ignoring this effect and kriging the Z's directly would produce volume inconsistent results. An alternative is to estimate A_S and H_S separately and take the ratio $Z^* = A^*/H^*$ as an estimate of A_S/H_S. However, this procedure carries a risk of bias which can be mitigated by using the same variogram, up to a multiplicative constant, for kriging A_S and H_S. If the same sample points are used for kriging A^* and H^*, the weights are then the same and

$$Z^* = \frac{A^*}{H^*} = \frac{\sum \lambda_\alpha A_\alpha}{\sum \lambda_\beta H_\beta} = \frac{\sum \lambda_\alpha H_\alpha Z_\alpha}{\sum \lambda_\beta H_\beta} = \sum \nu_\alpha Z_\alpha \quad \text{with} \quad \nu_\alpha = \frac{\lambda_\alpha H_\alpha}{\sum \lambda_\beta H_\beta}$$

The ν_α have a unit sum, and Z^* is therefore a weighted average, which is not the case if A_S and H_S are estimated using nonproportional variograms. When thickness varies significantly, it is preferable to use the above non-optimal estimates of H_S and A_S and ensure that Z^* is a weighted average. Such practice concerns positive variables and should not be used, for example, in the case of a variable with zero mean. The estimation variance of Z^* can be calculated exactly or approximately under certain conditions [e.g., see Journel and Huijbregts (1978), Section V.C.3].

3.5.6 Filling-in the Gaps

In applications such as mining or forestry where data are collected on a regular grid, considerable saving is achieved by always using the same pattern of data, since the kriging weights can then be computed once for all (i.e., if weights are shift invariant). But often the data grid has gaps, which ruins this plan. The three perpendiculars then come to the rescue: If the gaps are sparse, one can replace the missing data by their kriging estimates and proceed as if all data were present. Kriging variances, however, are underestimated by an amount equal to the variance of the difference between estimators with and without gaps.

The Three Perpendiculars

If

$$Z_{N+M}^* = \sum_{\alpha=1}^{N+M} \lambda_\alpha Z_\alpha$$

is the kriging estimator based on $N+M$ data of some quantity Z_0, then the kriging estimator of Z_0 based on the first N data points is

$$Z_N^* = \sum_{\alpha=1}^{N} \lambda_\alpha Z_\alpha + \sum_{\alpha=N+1}^{N+M} \lambda_\alpha Z_\alpha^*$$

where Z_α^* is the kriging estimator of Z_α from the N data. Kriging variances are related by

$$\text{Var}(Z_N^* - Z_0) = \text{Var}(Z_{N+M}^* - Z_0) + \text{Var}(Z_N^* - Z_{N+M}^*) \qquad (3.63)$$

In the case of a known mean, there is a simple geometric proof illustrated in Figure 3.12: The projection of Z_0 onto the Hilbert space $\mathcal{H}_N \subset \mathcal{H}_{N+M}$ can be accomplished by first projecting onto $\mathcal{H}_{N+M}$ and then projecting this projection onto $\mathcal{H}_N$. A similar proof can be given in the case of an unknown mean using the orthogonality relationship (3.41).

In our case the $N+M$ data points represent the complete kriging configuration and M the number of gaps. For the property to apply, the missing data should be reconstructed only from the N data present in the configuration. Intuitively though, the final results should be even better if the gaps are filled using the best possible neighborhoods.

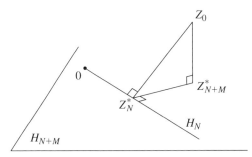

FIGURE 3.12 The three perpendiculars in the case of a known mean (simple kriging).

3.6 SELECTION OF A KRIGING NEIGHBORHOOD

The selection of data points to be included in the estimation is a key problem in the application of kriging. In theory, the minimum mean square error is achieved when all points are included, since any smaller neighborhood can be viewed as a constrained optimization with weights zero placed on the discarded points. But a global neighborhood may result in a kriging matrix that is too large to be inverted numerically. The maximum size of an invertible full matrix these days is about 5000, while large data sets can exceed one million points. Another important consideration is the geostatistical model itself, which may only have local validity.

The solution is to restrict the point selection to a subset of the data, changing with the estimated point, and is thus called a *moving neighborhood*. In doing so, however, we render the results dependent on the particular data selection algorithm and tend to create spurious discontinuities in regions where control points are scarce due to the sudden change of sample points from one neighborhood to the next. This difficulty is of course not specific to kriging but is a feature of all neighborhood methods.

We will first discuss special cases where the kriging weights vanish rapidly with distance away from the estimated point and give practical recommendations for neighborhood selection. Then we will present three approaches that ensure continuity. The first two retain a global neighborhood but modify the kriging matrix, or its inverse, to make it sparse and therefore numerically tractable. The third method builds continuity in moving neighborhood estimates. Other approaches not discussed here include fixed-rank kriging (Cressie and Johannesson, 2008) and Gaussian predictive process models (Banerjee et al., 2008).

3.6.1 Screening Effect and Relay Effect

Screening Effect

In a mathematical sense the "screening effect" describes a situation in which nonzero kriging weights are concentrated on a subset of samples in the immediate vicinity of the estimated point or block. These samples screen off the influence of all other data. In 1D we have seen that with a linear variogram the ordinary kriging estimator only depends on the two adjacent samples, or on the nearest end point (examples 4 and 5). For simple kriging the same circumstance occurs for the exponential covariance due to its associated Markov property: If $x_0 < x_1 < x_2$, then given Z_1, Z_0 and Z_2 are uncorrelated.

In higher dimensions there are some special covariance models for which any closed contour acts as a perfect screen between internal and external data. Let us consider for simplicity the isotropic case (a geometric anisotropy does not alter the conclusion). For SK these special models, expressed as functions of $r = |h|$, are in 2D the Matérn model (2.56) with $\nu = 1$, namely $C(r) = (r/a) \ K_1(r/a)$

(Whittle, 1954), and in 3D the covariance measure $C(r) = \exp(-r/a)/(r/a)$ [e.g., Arfken (1985)] ($a > 0$ in both cases). For OK the solutions are variogram measures: in 2D the de Wijs model $\gamma(r) = \log r$, and in n-D $\gamma(r) = 1/r^{n-2}$ (Matheron, 1965, p. 252).

In 2D a particular screening effect occurs for simple kriging when the covariance can be separated along the components h_x, h_y of the lag vector h as

$$C(h_x, h_y) = C_1(h_x) \, C_2(h_y)$$

Consider points $x_1, x_2, x_3, \ldots,$ aligned along a parallel to one coordinate axis and x_0 on the perpendicular to that line through x_1 (Figure 3.13a). Then the SK estimator of Z_0 from $Z_1, Z_2, Z_3, \ldots,$ assigns a zero weight to every point other than Z_1. Indeed, if $\lambda_i = 0$, $i > 1$, the SK equations reduce to

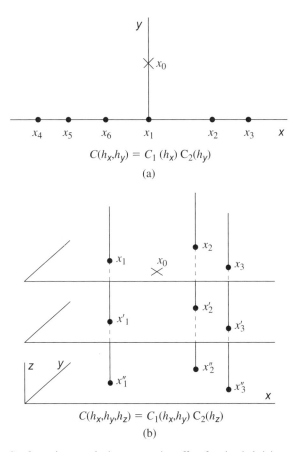

FIGURE 3.13 Configurations producing a screening effect for simple kriging under a separable covariance model: (a) points other than x_1 receive a zero weight; (b) points other than x_1, x_2, x_3 receive a zero weight.

$$\lambda_1 \, \rho_{1j} = \rho_{0j} = \rho_{01} \, \rho_{1j} \qquad \forall j$$

and are satisfied when $\lambda_1 = \rho_{01}$. Similar results hold in 3D. For example, if data lie in horizontal planes (Figure 3.13b) and if the covariance separates as

$$C(h_x, h_y, h_z) = C_1(h_x, h_y) \, C_2(h_z)$$

then the plane at the same elevation as the estimated point screens off the other data planes.

An example of separable model is the Gaussian covariance $C(h) = \exp(-|h|^2/a^2)$. Another model, widely used for digital image compression, is the separable exponential covariance

$$C(k, \ell) = \sigma^2 \rho_v^{|k|} \rho_h^{|\ell|}$$

where ρ_v and ρ_h, respectively, denote the vertical and horizontal correlation coefficients between adjacent pixels of an image, and k and ℓ are vertical and horizontal lags (Rabbani and Jones, 1991). Techniques of linear predictive coding exploit the high correlation between adjacent pixels of an image (ρ_v and ρ_h typically exceed 0.9) to reduce the quantity of bits transmitted over communication lines. Figure 3.14 illustrates a configuration used in forming the optimum linear predictor (i.e., the SK estimator) of a pixel X_m based on four previous pixels A through D. For the above separable model the weights are

$$\lambda_A = \rho_h, \qquad \lambda_B = -\rho_v \, \rho_h, \qquad \lambda_C = \rho_v, \qquad \lambda_D = 0$$

Pixel D receives a weight zero: Its influence is screened off by pixel C.[11] Note that in the absence of pixel A we would be in the situation described in Figure 3.13a, and only pixel C would receive a nonzero weight equal to ρ_v.

Aside from special mathematical cases, the screening effect represents a physical approximation: In the presence of strong spatial autocorrelation, the data of the first "ring" matter most, and very little additional information is

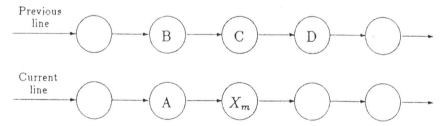

FIGURE 3.14 Differential pulse code modulation predictor configuration. [From Rabbani and Jones (1991).]

---

[11] On the subject of digital image processing, see the article by Yfantis et al. (1994) and the comment by P. Delfiner (same reference).

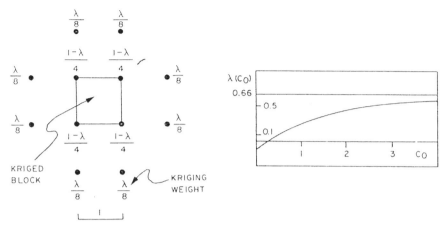

FIGURE 3.15 Variation of kriging weights with C_0, computed with the variogram model $\gamma(h) = C_o + |h|$ ($h \neq 0$). As C_o increases, all weights tend to $1/12$. [From David (1977).]

gained by considering sample points beyond that. To take a specific example, consider the problem depicted in Figure 3.15: the estimation by ordinary kriging of a square panel from its 12 nearest samples in a square grid, using a linear variogram with a nugget affect [from David (1977, p. 258)]. The symmetry of the pattern reduces the problem to the determination of a single weight λ. The variation of λ as C_o is progressively increased is shown in the adjacent chart. The eight samples of the outer ring share this weight λ which is small when C_o is small. In this case the kriging neighborhood can be limited to the first ring. However, when the nugget effect increases all weights tend to become equal to $1/12$. The nugget effect destroys autocorrelation and in the limit makes all samples equivalent. Hence the celebrated geostatistical saying: *The nugget effect lifts the screening effect* (Matheron, 1968a).

In practice, the screening effect should not be gauged by the weights but rather by the variance. In a study of the behavior of kriging weights, Rivoirard (1984) showed that when the neighborhood is enlarged, negative weights appear, which are not small even for the most common variogram models, but do not improve precision significantly. Figure 3.16 displays the SK and OK weights for the spherical model with range twice the grid spacing and the weight on the mean $\lambda_m = 1 - \sum \lambda_{K\alpha}$ (see Section 3.4.7). Observe that all variances are comparable: Very little is gained by extending the kriging neighborhood beyond the four nearest neighbors. Haslett (1989) proposes a stepwise algorithm for optimizing subset selection.

Relay Effect

It is often believed that when the variogram has a finite range it is not necessary to include sample points beyond that range in the kriging neighborhood. This assertion is certainly false when the mean is unknown, since the mean is not

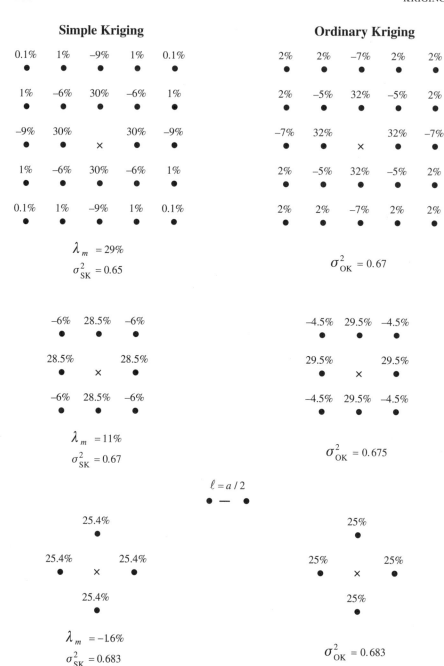

FIGURE 3.16 Simple and ordinary kriging weights for a spherical covariance with range $a = 2\ell$. [From Rivoirard (1984).]

local information, but it is also false in the case of simple kriging, as noted earlier. The reason is that points beyond the range may exert an influence through their correlation with points within the range by virtue of the so-called *relay effect*. In Figure 3.16, for example, all 16 points of the second ring have zero correlation with the estimated point, but the first ring acts as a relay. Similarly in Figure 3.14, pixel B would receive a weight zero in the absence of pixel A. Notice that relays tend to produce small or negative weights.

In a different setting, the autoregressive and moving average models (ARMA) of Box and Jenkins (1976) provide another angle at this. Autoregressive models, which by design depend on a finite number of past values, have correlation functions that extend to infinity, whereas the prediction of moving average models, which have a finite range, involves an infinite number of past values.

3.6.2 Practical Implementations of Neighborhood Selection

The additivity relationship (3.34) portrayed universal kriging as a two-stage procedure: One is the optimum estimation of the drift, and the other is simple kriging of the residuals. Unfortunately, these two stages raise conflicting demands on the neighborhood: If simple kriging can benefit from the screening effect, drift estimation, on the contrary, as we know from least squares theory, is best accomplished with sample points as far apart and well-scattered as possible. The obvious solution of estimating the drift (optimally) with a global neighborhood and subtracting it from the data is often impractical for the same reasons that preclude the use of a global neighborhood in the first place: too many data points and/or, more important, the poor validity of a simple drift model over the whole domain.

Sophisticated neighborhood selection algorithms have been devised to reach a compromise between near and far sample points. They usually include all points of the first ring and then more distant points, following a strategy that attempts to sample all directions as uniformly as possible while keeping the number of points as low as possible (octant search). Obviously, good azimuthal coverage is important to avoid the risk of bias. Typically, 16 to 32 points are retained, from at least five octants or four noncontiguous octants; the search is accelerated by initially classing the data points in the cells of a coarse grid. For contour mapping purposes, where continuity is important, larger neighborhoods may be considered to provide more overlap. Another approach is to divide the gridded domain into overlapping sections, estimate each of them with a global neighborhood, and splice the resulting subgrids.

A useful technique to improve continuity without spoiling the accuracy of the estimates is selective low-pass filtering. The idea is to run a smoothing filter on the kriged grid but modify the estimates only within a confidence interval defined by the kriging standard deviation so as to modulate the degree of smoothing by the uncertainty about the estimates. The following low-pass filter has good spectral properties: a moving average with weights $(1/4, 1/2, 1/4)$ along the rows and then along the columns of the grid, followed by a moving

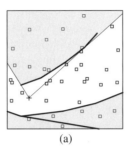

(a)

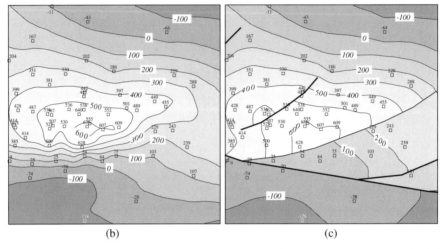

(b)   (c)

FIGURE 3.17 Kriging with faults: (a) the fault as a screen: the data points located in the shaded area cannot be seen from the target point marked by a cross; (b) result of kriging without fault; (c) result of kriging with faults.

average with weights $(-1/4, 3/2, -1/4)$ to restore some power in the high frequencies.

Another ad hoc technique is employed to deal with discontinuities such as geological faults. It is illustrated in Figure 3.17: The fault is considered as a screen and only the sample points directly "seen" from the estimated point are included in the kriging neighborhood. This technique is a stopgap solution. It assumes that fault displacement is only vertical, and it is limited by the availability of sample points in the "fault blocks." The correct approach is to model faults using 3D geomodeling techniques [Mallet (2008) and Section 2.2.6]. (See also the external drift method).[12]

[12] Structural geologists have developed the general concept of *retrodeformability*. Let us quote Suppe (1985, p. 57): "The fact that the rocks were originally deformed may seem trivial to mention, but it is actually an important key to the solution of many structural problems. It must be geometrically possible to undeform any valid cross section to an earlier less deformed or undeformed state; the cross section must be *retrodeformable*." Retrodeformable cross sections are popular in the petroleum industry under the name of *balanced cross sections*.

3.6.3 Covariance Tapering

Furrer et al. (2006) propose a global neighborhood approach that they applied successfully to a large irregularly spaced dataset involving over 5900 observations and a kriged grid of more than 660,000 points. The principle is to taper the covariance function to zero beyond a certain range so that the kriging matrix is sparse and the kriging system can be solved efficiently. Let K_θ be a covariance function that is identically zero outside a particular range described by θ. The tapered covariance function is defined by

$$\sigma_{\text{tap}}(x, y) = \sigma(x, y) K_\theta(x, y)$$

As the product of two covariance functions, this is indeed a covariance. The choice of the taper covariance K_θ is essential and is governed by two requirements: It should be identically zero beyond a certain distance (and not just asymptotically), and it should preserve the behavior near the origin of the true covariance $\sigma(x,y)$, because this behavior controls the lateral continuity of the interpolant.

The second condition implies that the taper K_θ should be more regular than the original covariance σ. This can be seen heuristically by considering principal irregular terms (Section 2.3.1). Suppose that σ and K_θ are stationary isotropic covariance functions $C(h)$ and $K_\theta(h)$ and admit near the origin an expansion of the form

$$C(h) \approx \sigma^2(1 - a|h|^\alpha) \qquad \text{and} \qquad K_\theta(h) \approx 1 - b|h|^\beta$$

where $\alpha > 0$ and $\beta > 0$ are the lowest non-even powers. The tapered covariance is then

$$C_{\text{tap}}(h) \approx \sigma^2(1 - a|h|^\alpha - b|h|^\beta)$$

In order to preserve the principal irregular term and its coefficient, it is necessary that $\alpha < \beta$.

Furrer et al. (2006) develop the theory for the Matérn covariance model $C_{\alpha,\nu}$ (2.56). Its principal irregular term is $|h|^{2\nu}$ if the smoothness parameter ν is noninteger and $|h|^{2\nu} \log|h|$ if ν is an integer. For the tapers they use polynomial covariance functions that turn out to be members of the spherical models family. Specifically, as a function of the power $\alpha\ (= 2\nu)$ the possible taper covariance models are the spherical (2.47) for $\alpha < 1$, the cubic (2.49) for $\alpha < 3$, and the penta (2.50) for $\alpha < 5$.

3.6.4 Gaussian Markov Random Field Approximation

The approach of Gaussian Markov random fields may be seen as the opposite of that of covariance tapering in the sense that it seeks to make the inverse of the covariance matrix—and not the covariance matrix itself—sparse. It was first used to generate simulations (Besag, 1974, 1975) but offers a new approach to kriging (Rue and Held, 2005). Consider a Gaussian random vector $\mathbf{Z} = \{Z_i : i = 1, \ldots, N\}$ with known mean $\mathbf{m}$ and covariance matrix $\mathbf{C}$. The conditional distribution of Z_i given the other components $\{Z_j : j \neq i\}$ is Gaussian, with mean and variance being the kriging estimate Z^*_{-i} of Z_i and the associated kriging variance. Denoting by $\mathbf{B}$ the inverse of $\mathbf{C}$, the kriging weights are found to be equal to $\lambda_j(i) = -B_{ij}/B_{ii}$ so that we have

$$Z^*_{-i} = m_i - \frac{1}{B_{ii}} \sum_{j \neq i} B_{ij} (Z_j - m_j), \qquad \sigma^2_{Ki} = \frac{1}{B_{ii}} \tag{3.64}$$

Since B_{ii} is the inverse of the conditional variance of Z_i given $\{Z_j : j \neq i\}$ (all except the ith), $\mathbf{B}$ is known as the precision matrix. Its off-diagonal elements are related to the conditional correlations of Z_i and Z_j given $\{Z_k : k \neq i, j\}$ by

$$\mathrm{Corr}(Z_i, Z_j \mid \{Z_k : k \neq i, j\}) = -\frac{B_{ij}}{\sqrt{B_{ii} B_{jj}}}$$

$\mathbf{B}$ is a symmetric positive-definite matrix. The pattern of zeroes of $\mathbf{B}$ can be used to define an undirected graph structure in which two nodes are connected by an edge when $B_{ij} \neq 0$. Let ne(i) denote the neighborhood of node i—that is, the set of nodes connected to i by an edge. The vector $\mathbf{Z}$ has the Markov property that Z_i is conditionally independent of $\{Z_k : k \notin \mathrm{ne}(i)\}$ given $\{Z_j : j \in \mathrm{ne}(i)\}$. The discretely indexed Gaussian $\mathbf{Z}$ is called a Gaussian Markov random field (GMRF).

Let us divide the N nodes into two groups, $\{i = 1, \ldots N_1\}$ and $\{j = N_1 + 1, \ldots, N\}$, the latter with $N_2 = N - N_1$ nodes. $\mathbf{Z}$ can be split accordingly in two vectors $\mathbf{Z}_1$ and $\mathbf{Z}_2$. The vector $\mathbf{Z}_2$ represents data, whereas $\mathbf{Z}_1$ is unknown and is to be estimated. The multivariate probability density function of $\mathbf{Z}$, usually given as function of $\mathbf{m}$ and $\mathbf{C}$, can also be expressed as a function of $\mathbf{m}$ and $\mathbf{B}$:

$$g(\mathbf{z}) = (2\pi)^{-N/2} |\mathbf{B}|^{1/2} \exp\left[-\tfrac{1}{2}(\mathbf{z} - \mathbf{m})' \mathbf{B} (\mathbf{z} - \mathbf{m})\right]$$

The conditional distribution of $\mathbf{Z}_1$ given $\mathbf{Z}_2$ is multivariate Gaussian. It is given in Section A.8 of the Appendix in terms of covariance matrices. The same can be done in terms of precision matrices: When we partition the matrix $\mathbf{B}$ into the four submatrices $\{\mathbf{B}_{pq} : p, q = 1, 2\}$ associated with the two groups, the conditional mean takes the form

$$\mathbf{m}_{1|2} = \mathrm{E}[\mathbf{Z}_1 \mid \mathbf{Z}_2 = \mathbf{z}_2] = \mathbf{m}_1 - \mathbf{B}_{11}^{-1} \mathbf{B}_{12}(\mathbf{z}_2 - \mathbf{m}_2) \tag{3.65}$$

where $\mathbf{B}_{11}$ is the conditional precision matrix of $\mathbf{Z}_1$ given $\mathbf{Z}_2$. This result generalizes (3.64). To compute the conditional mean (i.e., perform kriging), we do not need to invert $\mathbf{B}_{11}$ but only to solve the linear system

$$\mathbf{B}_{11}(\mathbf{m}_{1|2} - \mathbf{m}_1) = -\mathbf{B}_{12}(\mathbf{z}_2 - \mathbf{m}_2)$$

which can be done efficiently with sparse matrix algorithms. The GMRF model is very attractive for modeling random variables at the vertices of an undirected graph.

Suppose now that the Z_i represent the values of a Gaussian SRF $Z(x)$ at discrete points x_i over a domain of $\mathbb{R}^n$. Under what conditions can $\mathbf{Z}$ be a GMRF? Rozanov (1982, Sec. 3.1) shows that a Gaussian SRF of $\mathbb{R}^n$ has a Markov property if and only if its spectrum is of the form $f(u) \propto 1/P(u)$ where $P(u)$ is a nonnegative symmetric polynomial. A case of special interest is the Matérn covariance (2.56) whose spectrum is (Whittle, 1963)

$$f(u) = (2\pi)^{-n} \left(\frac{1}{a^2} - |u|^2 \right)^{-(\nu+n/2)}$$

It satisfies the criterion if and only if $\nu + n/2$ is an integer. This is a severe limitation, especially in $\mathbb{R}^2$ where no model with a linear behavior at the origin is possible. To overcome this, Rue and Held (2005) provide approximations of SRFs by GMRFs on a grid. The GMRF is parameterized by the nonzero B_{ij} of (3.64). The number of distinct parameters is reduced by accounting for symmetries, and border effects are avoided through periodic embedding. The parameters are chosen so that $\mathbf{B}^{-1}$ is close to $\mathbf{C}$ in some norm. A difficulty is to ensure that $\mathbf{B}$ is positive definite. A sufficient condition often used in practice is diagonal dominance, that is, $|B_{ii}| > \sum_{j \neq i} |B_{ij}|$ for all i, which amounts to $\sum_{j \neq i} |\lambda_j(i)| < 1$. Hartman and Hössjer (2008) give a detailed account of the implementation of GMRFs for kriging, and they propose a way to deal with data points not on the grid without displacing them to the nearest grid node. Rue and Held report that a 5×5 neighborhood is sufficient to reproduce an exponential covariance, whereas a neighborhood of at least 7×7 is required for a Gaussian covariance. Results also depend on the size of the grid mesh relative to the covariance range.

This approach has several extensions, in particular to models with a polynomial drift and to intrinsic random fields of order k, with prototypes of such Gaussian Markov fields being the random walk and its integrals.

We have seen in Section 2.5.1 that a Gaussian SRF with a Matérn covariance is the solution of a stochastic partial differential equation. In the case where it is a GMRF, Lindgren et al. (2011) solve the equation numerically using a finite element method with Delaunay triangles whose vertices are the nodes x_i. These nodes must include all data locations but otherwise may be placed at will, typically with a higher density in regions of interest. The solution provides an interpolation everywhere in the domain, although within triangles it boils down to a plain linear interpolation of the vertices. In the special case where all the vertices are points on a regular 2D grid and ν is an integer, the neighborhood ne(i) is a lozenge inscribed in a square of size $(2\nu+3) \times (2\nu+3)$ nodes only (including the node i), which drastically reduces the computational cost. It remains that this approach deals with very special cases not necessarily compatible with the application considered.

3.6.5 Continuous Moving Neighborhoods

Gribov and Krivoruchko (2004) developed an original method to ensure continuity with moving neighborhoods. The idea is to modify the kriging system so that data beyond a specified distance from the estimated point receive weights gradually approaching zero. This way, no discontinuity occurs when data points enter or exit the kriging neighborhood. Rivoirard and Romary (2011) propose an equivalent approach from a different perspective. The idea is to introduce a penalty on the kriging weights by minimizing an objective function of the form

$$Q = E(Z^* - Z_0)^2 + \sum_\alpha V_\alpha(\lambda_\alpha)^2 + 2\mu\left(\sum_\alpha \lambda_\alpha - 1\right) \tag{3.66}$$

In this expression (written for OK because of the last term) the penalty V_α plays the role of a noise variance and varies with the target point x_0. It is typically equal to 0 for data points x_α within a distance r of the estimated point x_0 (no penalty applied near the target point), and it increases continuously to infinity as x_α approaches the outer boundary of the kriging neighborhood, located at a distance R. An example of V_α profile is the following:

$$V_\alpha = \sigma^2\left(\frac{h-r}{R-h}\right)^2 \quad \text{for } r \leq h \leq R \quad \text{where } h = |x_\alpha - x_0| \quad \text{and} \quad \sigma^2 = \text{Var } Z(x)$$

By convention, V_α may be set to infinity when $h > R$ so that, to keep Q finite, all data points x_α on or beyond the neighborhood boundary necessarily receive a weight $\lambda_\alpha = 0$. Likewise, when the penalty is large, the weight must be small.

When the initial structure has no nugget effect, a refinement of the method is to introduce some correlation between added noises V_α according to a continuous, short-ranged, spatial covariance. This preserves the continuity of the kriging estimate at the target point when two data points merge. The resulting kriging system is given by (3.68) in the case of universal kriging.

Figure 3.18 compares the surfaces obtained with and without the use of continuous moving neighborhoods. The results speak for themselves. The beauty of the method lies in its simplicity and flexibility. Because it is solely based on the addition of a noise that increases with distance, the method works for all versions of kriging algorithms: OK, UK, and even IRF$-k$. Generalizations to more sophisticated neighborhoods (e.g., octant search) is a possibility.

Concerning the computed variance, continuous moving neighborhood outperforms straight kriging from data within the inner circle only, but underperforms kriging from all data inside the outer circle due to the constraints placed on the weights.

3.6.6 Global Neighborhood Cross-Validation

A special form of kriging neighborhood is involved in model cross-validation (Section 2.6.3): Each sample point x_α is estimated from all others, excluding x_α

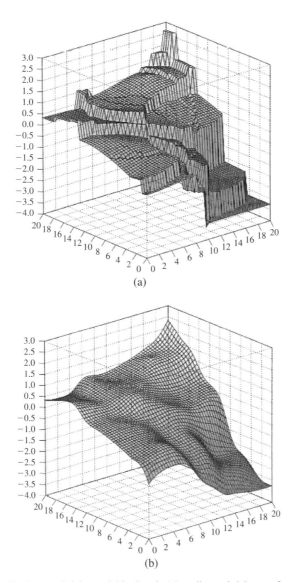

FIGURE 3.18 Continuous kriging neighborhood: (a) ordinary kriging surface obtained using a standard moving neighborhood with radius $R = 10$; (b) ordinary kriging surface obtained using a continuous moving neighborhood with radii $r = 7.5$ and $R = 12.5$. [From Rivoirard and Romary (2011), with kind permission of the International Association for Mathematical Geosciences.] (See color insert)

itself. Essentially the same global neighborhood is considered, but not quite. In a configuration with N data points and $L + 1$ drift functions, this involves, in principle, the resolution of N linear systems of size $(N + L) \times (N + L)$, which may be prohibitive. Dubrule (1983b) shows that all solutions can in fact be obtained by inversion of a single matrix, the matrix $\mathbf{A}$ of system (3.22). If

$\mathbf{B} = [B_{\alpha\beta}]$ denotes the inverse of $\mathbf{A}$ and if one considers the kriging estimator of $Z(x_\alpha)$ from $\{Z(x_\beta) : \beta \neq \alpha\}$, the kriging weights $\lambda_\beta(\alpha)$ and the kriging variance $\sigma^2_{K\alpha}$ are given by

$$\lambda_\beta(\alpha) = -B_{\alpha\beta}/B_{\alpha\alpha} \quad (\beta \neq \alpha), \qquad \sigma^2_{K\alpha} = 1/B_{\alpha\alpha} \qquad (3.67)$$

These are the same results as in (3.64), but their validity is established for UK while (3.64) and (3.65) are for SK only. If the kriging matrix is written in variogram terms, the sign must be changed on the right-hand side of the second formula. A generalization of the above results allows an estimation of $Z(x_\alpha)$ excluding more than a single observation.

3.7 MEASUREMENT ERRORS AND OUTLIERS

3.7.1 Filtering Nonsystematic Errors

Three types of error can affect the data: uncertainty on the exact positions of the measurements, systematic errors, and nonsystematic, also called *random*, errors. Positioning uncertainty is typically associated with marine and aerial surveys and can be modeled statistically (Section 2.4.4). The advent of satellite positioning (GPS) has made this uncertainty very small but perhaps still significant in applications where extreme precision is required, such as marine 3D seismic.

Systematic errors are the most dangerous. They usually go unnoticed, do not cancel out, and can ruin a whole analysis. They may have several origins: a drift of the instrument (as in gravimetric surveys), acquisition problems (tool malfunction, insufficient dynamic range, quantization error), model inadequacies with *computed data* (parameters, e.g., porosity or saturation are computed from wireline logs by inversion of a petrophysical model), and so on. Systematic errors, if suspected, can be corrected by multivariate methods and are discussed with them. We will focus here on random errors.

The simplest case is that of uncorrelated errors with the same amplitude. They turn up in the variogram as an additional nugget effect equal to the error variance. But errors may also be unequal (e.g., data from different sources) or even correlated. For example, in a bathymetric survey, data are acquired along profiles, and it can be assumed that errors are the same along a profile and independent across two different profiles. Furthermore, the standard deviation of the error is proportional to the local average depth (Chilès, 1977). If Z_ε denotes the measurements and Z represents the underlying RF, a comprehensive model is

$$Z_\varepsilon(x_\alpha) = Z(x_\alpha) + \varepsilon_\alpha$$

where ε_α are random errors defined only at the sample points and are subject to the following assumptions:

- Errors are nonsystematic, $\quad\quad E[\varepsilon_\alpha] = 0 \quad\quad\quad \alpha = 1, \ldots, N$
- Errors are uncorrelated with
  the studied RF $\quad\quad\quad\quad E[\varepsilon_\alpha \, Z(x)] = 0 \quad\quad \forall x, \, \alpha = 1, \ldots, N$
- Errors may be correlated
  among themselves $\quad\quad\quad E[\varepsilon_\alpha \, \varepsilon_\beta] = S_{\alpha\beta} \quad\quad \alpha, \, \beta = 1, \ldots, N$

We want to estimate the error-free value $Z_0 = Z(x_0)$ from observations $Z_\varepsilon(x_\alpha)$ "corrupted" by noise. In time series this is a standard problem known as *filtering* (the signal from the noise), and its solution in the frequency domain relies on the possibility of separating the spectral characteristics of the signal and of the noise. In the geostatistical terminology we regard this problem as a special case of *cokriging*—that is, estimating values of one variable on the basis of another. But the correlation structure is so simple that the result is only a slight modification of the standard kriging system. Our estimator is now

$$Z^* = \sum_\alpha \lambda_\alpha (Z_\alpha + \varepsilon_\alpha)$$

Since errors have zero mean, the unbiasedness conditions do not change. The m.s.e. becomes

$$E(Z^* - Z_0)^2 = E\left(\sum_\alpha \lambda_\alpha Z_\alpha - Z_0\right)^2 + \sum_\alpha \sum_\beta \lambda_\alpha \lambda_\beta S_{\alpha\beta}$$

Minimizing the complete expression leads to the cokriging system

$$\begin{cases} \sum_\beta \lambda_\beta (\sigma_{\alpha\beta} + S_{\alpha\beta}) + \sum_\ell \mu_\ell f_\alpha^\ell = \sigma_{\alpha 0}, & \alpha = 1, \ldots, N \\ \sum_\alpha \lambda_\alpha f_\alpha^\ell = f_0^\ell, & \ell = 0, \ldots, L \end{cases} \quad\quad (3.68)$$

and the cokriging variance

$$\sigma_{CK}^2 = E(Z^* - Z_0)^2 = \sigma_{00} - \sum_\alpha \lambda_\alpha \sigma_{\alpha 0} - \sum_\ell \mu_\ell f_0^\ell$$

Equations in terms of the variogram remain the same as (3.25) but with $\gamma_{\alpha\beta} - S_{\alpha\beta}$ replacing $\gamma_{\alpha\beta}$ (so the first N diagonal terms are no longer zeros but $-S_{\alpha\alpha}$).

The cokriging system (3.68) differs from the UK system (3.21) by the presence of the error covariance terms $S_{\alpha\beta}$. It relies on the assumption that the covariances $\sigma_{\alpha\beta}$ and $S_{\alpha\beta}$ are known separately. The cokriging estimator is no longer an exact interpolant: it does its job of filtering measurement errors.

In the special case of uncorrelated errors, the system (3.68) is identical to the standard kriging system with the covariance $\sigma + S$ except at data points (where the kriging system has $\sigma_{\alpha 0} + S_{\alpha 0}$ on the right-hand side). This shows why

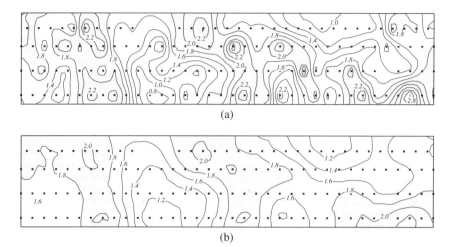

(a)

(b)

FIGURE 3.19 Filtering measurement errors (CO_2 concentration in soil): (a) contour map using a continuous variogram model; (b) contour map with noise filtered.

the kriging estimator also filters errors. It can also be seen directly by noting that at any point x_0 other than a sample point the kriging mean square error

$$E[Z^* - Z_\varepsilon(x_0)]^2 = E(Z^* - Z_0)^2 + E(\varepsilon_0^2)$$

differs from the cokriging error only by the addition of a constant term and thus has the same minimizer. At a sample point this relationship breaks down, and both the kriging estimate and its variance have a discontinuity (see Figure 3.3c). For contour mapping purposes it is preferable to grid the continuous component Z and filter measurement errors and the rest of the nugget effect. As for which variance should be reported, it makes sense to exclude genuine measurement error variances (since we are not interested in reconstructing them) but to include the nugget effect variance due to microstructures.

Figure 3.19 illustrates the effect of filtering error variances in mapping CO_2 concentration in soil. Straightforward estimation with a continuous variogram model produces the typical "fat in the soup" effect in which contour lines circle around data points because, due to errors, most of them are local extrema (Figure 3.19a). We know there are indeed uncertainties in the data because gas emanation is subject to very local variations of soil properties and to large measurement errors. By contrast, the filtered map shows a physically more meaningful picture of the CO_2 concentration distribution (Figure 3.19b).

Filtering Positioning Errors

We have seen in Section 2.4.4 the effect of positioning errors on the variogram in the stationary case. Looking now at kriging, the objective is to estimate $Z_0 = Z(x_0)$ from observation $Z(x_\alpha + U_\alpha)$, where U_α are positioning errors.

If the positioning errors are independent and with the same pdf $p(u)$, they will impact the ordinary kriging system only by transformation of covariances $\sigma_{\alpha\beta}$ from $C(x_\beta - x_\alpha)$ into $C_P(x_\beta - x_\alpha)$ and covariances $\sigma_{\alpha 0}$ from $C(x_0 - x_\alpha)$ into $C_p(x_0 - x_\alpha)$, where C_P is the regularized covariance $C*P$ and P the covariogram $p*\breve{p}$ of the pdf p.

In the case of a random function with a linear drift and a stationary or intrinsic residual, the positioning uncertainty will have no impact on the unbiasedness conditions provided that the positioning error has zero mean, but the variance to be minimized will include an additional term depending on the unknown slope of the drift. Chilès (1977) approximates this term by using an estimate of the slope, whereas Cressie and Kornak (2003) propose a Monte Carlo integration algorithm. As is the case for marine or aerial surveys, the generalization to correlated positioning errors is straightforward, provided that the bivariate distribution of the errors U_α and U_β is known for all data pairs (Chilès, 1977).

3.7.2 Poisson Kriging

The method presented here is formally equivalent to filtering a random error except that the added variability in the data is due to the observation process itself. The method was developed by Monestiez et al. (2005a; 2006) to estimate the abundance of fin whales in the Pelagos Sanctuary, a marine protected area in the western Mediterranean stretching between Corsica, the south of mainland France, and northern Italy. Knowledge of the spatial distribution of an animal population is essential to better understand the population's interaction with its environment. Making maps of animal abundance is difficult when animals are rare and the observation effort heterogeneous, because the raw maps reflect the effort as well as the abundance.

The observations are sighting counts recorded along random linear transects or onboard regular ferries between France and Corsica. Sighting event counts are preferred because the number of whales reported for a given sighting are often unreliable. Data from all available years are cumulated in cells of about 90 km^2, and the total time spent observing (in hours) is computed.

The number of sightings $Z(x)$ is modeled as a Poisson time process with intensity parameter $Y(x)$, which measures the sightings expectation at location x for a unit observation time. Thus $Z(x)$ is distributed as a Poisson with parameter $t(x)Y(x)$, where $t(x)$ is the observation time. $Y(x)$ is modeled as a positive SRF with constant unknown mean m and variogram $\gamma_Y(h)$. Conditionally on Y the random variables $Z(x)$ at different locations are mutually independent. In classical shorthand notations the properties of the Poisson distribution entail

$$E(Z_x|Y_x) = t_x Y_x, \qquad E(Z_x) = m\,t_x,$$

$$\text{Var}(Z_x|Y_x) = t_x Y_x, \qquad \text{Var}(Z_x) = t_x^2 \text{Var}(Y_x) + m\,t_x,$$

$$E(Z_x Z_y|Y) = t_x Y_x \delta_{xy} + t_x t_y Y_x Y_y, \qquad \delta_{xy} = 1 \text{ if } x = y, \quad \delta_{xy} = 0 \text{ only otherwise}$$

Using these relations an improved unbiased variogram estimator is obtained. For $h \neq 0$ this is

$$\gamma_Y^*(h) = \frac{1}{2N(h)} \sum_{x_\beta - x_\alpha \approx h} \left[\frac{t_\alpha t_\beta}{t_\alpha + t_\beta} (Z_\beta/t_\beta - Z_\alpha/t_\alpha)^2 - m^* \right],$$

$$N(h) = \sum_{x_\beta - x_\alpha \approx h} \frac{t_\alpha t_\beta}{t_\alpha + t_\beta}$$

where m^* is an estimate of the mean of Y. Compared to the standard variogram of Z/t, this improved estimator is smoother, reduces the sill to about one-fourth of its original value, and removes the nugget effect.

The intensity Y_0 at x_0 can be estimated by a linear combination $Y_0^* = \sum_\alpha \lambda_\alpha (Z_\alpha/t_\alpha)$, where the λ_αs are solution to a modified OK system that the authors name *Poisson Ordinary Kriging*:

$$\begin{cases} \sum_\beta \lambda_\beta C_{\alpha\beta} + \lambda_\alpha \dfrac{m}{t_\alpha} + \mu = C_{\alpha 0}, & \alpha = 1, \ldots, N \\ \sum_\alpha \lambda_\alpha = 1 \end{cases} \tag{3.69}$$

Here C is the covariance of Y and is derived from γ_Y. Note that (3.69) is formally the same as the error filtering system (3.68). The added diagonal term m/t_α represents the expected conditional variance of Z_α/t_α given Y_α:

$$\mathrm{E}\,\mathrm{Var}[(Z_\alpha/t_\alpha)|Y_\alpha] = \mathrm{E}(Y_\alpha/t_\alpha) = m/t_\alpha$$

The Poisson kriging variance has the same expression as the standard OK variance (3.18) but not the same value because the weights λ_α are different.

The maps obtained by Poisson kriging tend to be smoother than those obtained by OK of raw data, but overall are similar. The real difference lies in error variances. The authors report Poisson kriging variances lower by (a) a factor of 5 to 10 when an observation is present in the cell and (b) a factor of at least 2 when the cell is farther from available data.

Conversion of the sightings expectation maps, which represent whale sightings per hour, to whale abundance in animals per square kilometer requires accurate estimates of the mean area covered per hour of observation, and the mean size of observed groups of whales.

Poisson kriging has been applied to other rate data such as that found in epidemiology or criminology. An unpublished application deals with fertility rates from data aggregated by municipality.

3.7.3 Outliers

Kriging estimators being linear in the data are sensitive to the presence of a few unusually large (or sometimes small) values called "outliers". These were already a problem for variogram estimation and motivated the development of robust alternatives to the traditional variogram (Section 2.2.5). Robust versions of kriging have also been proposed (Hawkins and Cressie, 1984; Cressie 1993, pp. 144–150). The approach presented here is different. It relies on a fine tuning of the model of the spatial distribution of high values.

The high values considered here are not bad data. It is assumed, as should be the case in any geostatistical study, that bad data have been removed or corrected. The data are not extreme values either, in the sense of extreme value theory (natural disasters), which will be covered in Section 6.11. These high values typically come from heavy-tailed distributions, such as those of ore grades in gold or uranium deposits, where a small number of samples can be responsible for a large proportion of the metal content in the deposit. Parker (1991) reports a case where 4 out of 34 samples are responsible for 70% of the gold content, and 7 out of 34 for 90% of the gold content. Clearly this is a very uncomfortable situation because, as Parker notes, "even small changes in the grade near these assays or the proportion of high-grade material may be responsible for large differences between estimated and recovered reserves. If on the downside, these differences could have adverse consequences." On the other hand, discarding the high grades completely creates a bias and is an economic nonsense. The value of the mine may just come from high grades.

In spite of that, high grades may still be deliberately left out with the idea that it is more rewarding to revise estimates up than down. The problem is that this may lead to the abandonment of the project if it was a marginal one. The common practice is rather to exclude the high grades from the computation of the variogram and perform kriging on truncated data. The truncation logic is illustrated in Figure 3.20: Grades above a threshold z (called "top-cut" grade in

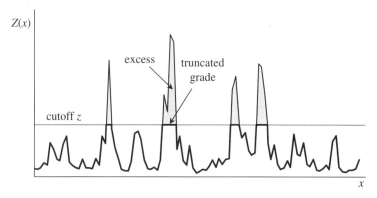

FIGURE 3.20 Truncated grades are defined by setting grades above the threshold z to the value z. They are considered as "normal" grades. The excess is treated separately and can be interpreted as a conventional income $B(z)$.

mining) are replaced by the value z, others are unchanged. The excess is either simply forgotten or kriged as a pure nugget effect and added to the estimates from the truncated data. This latter step amounts to scattering high grades all over the field, which may be adequate for global estimation but certainly not for local estimation, because high grades cannot be present where the truncated grades are low.

Truncation Model

Rivoirard et al. (2012) propose a relatively simple nonlinear model to deal with high values. The truncation described in Figure 3.20 gives the decomposition

$$Z(x) = \min(Z(x), z) + [Z(x) - z]1_{Z(x) \geq z}$$

The first term is the truncated grade; the second term, the *excess*, is what we lose by truncation. It is present only where $Z(x) \geq z$. These two quantities are positively correlated. The excess

$$B_z(x) = [Z(x) - z]1_{Z(x) \geq z}$$

can be interpreted as the conventional income (see Section 6.5.1) at the cutoff z, except that here z is not the true economic cutoff but usually a higher value. The expected values

$$b_z = E[B_z(x)] \quad \text{and} \quad T_z = E(1_{Z(x) \geq z}) = \Pr[Z(x) \geq z]$$

represent, respectively, the unconditional mean of the excess and the probability of exceeding the cutoff z, which in mining is interpreted as the ore tonnage above z (hence the notation T_z). In a stationary model these expected values do not depend on the location x. The conditional mean of the excess, computed only where it is positive, is equal to

$$E[(Z(x) - z)|Z(x) \geq z] = \frac{E[(Z(x) - z)1_{Z(x) \geq z}]}{P(Z(x) \geq z)} = \frac{b_z}{T_z}$$

Now the conditional expectation of $B_z(x)$ given $\min(Z(x), z)$ is 0 if this minimum equals $Z(x)$—that is, if $Z(x) < z$—and is equal to $E(Z(x) - z|Z(x) \geq z)$ otherwise, so that

$$E[B_z(x)|\min(Z(x), z)] = \frac{b_z}{T_z}1_{Z(x) \geq z} \qquad (3.70)$$

The truncated grade and the excess are related through the indicator of $Z(x) \geq z$. In fact the conditional expectation (3.70) can also be written as the regression of $B_z(x)$ on the indicator

$$E[B_z(x)|1_{Z(x) \geq z}] = \frac{b_z}{T_z} 1_{Z(x) \geq z}$$

The residual from this regression is

$$R_z(x) = B_z(x) - \frac{b_z}{T_z} 1_{Z(x) \geq z}$$

By construction it has mean zero and is uncorrelated with the indicator. It also turns out to be uncorrelated with the truncated grade. This lack of correlation applies *at the same point* x. It also applies spatially—that is, between the residual $R_z(x)$ at point x and the indicator $1_{Z(x+h) \geq z}$ at another point $x+h$—provided that there are no edge effects in the high value zone. This assumption can be tested by considering indicators at different cutoffs: In the absence of edge effects the indicator cross variogram $\gamma_{zz'}$ for two cutoffs $z' > z$ is proportional to the direct variogram γ_z of the lower indicator. This criterion guides the selection of the cutoff. There are also special models with lack of edge effect built in (mosaic model and model with orthogonal indicator residuals, Section 6.4.3).

In the end the grade is decomposed into a sum of three variables: the truncated grade, the weighted high-grade indicator, and an uncorrelated residual

$$Z(x) = \min(Z(x), z) + \frac{b_z}{T_z} 1_{Z(x) \geq z} + R_z(x)$$

For a sufficiently high cutoff the residual is nearly a nugget effect because it inherits the spatial destructuring of high grades, therefore its kriging estimate is its mean value zero. Replacing the residual by its mean value gives

$$Z(x) \approx \min(Z(x), z) + \frac{b_z}{T_z} 1_{Z(x) \geq z} \qquad (3.71)$$

The truncation model reduces to two spatially structured terms, the truncated grade and the indicator, which are free from high grades—this is the beauty of the decomposition.

These two terms can be estimated by cokriging, which involves modeling (a) the direct variograms of the truncated grade and of the indicator and (b) their cross-variogram.

When the residual is set to zero in formula (3.71), the only information retained from high grades is the geometry of high-grade zones and the mean of these high grades. This is quite interesting and can be used heuristically regardless of the underlying assumptions, at least for the estimate (the variance is more sensitive to the model). The improvement over a straight kriging of truncated grades is the addition of the mean excess grade b_z/T_z in proportion to its probability of presence.

Rivoirard et al. (2012) present an application of the truncation model to a gold deposit with vertical veins exploited in an open pit. The data consist of 8706 chemical assays on a 1-m support from 493 blast holes. Figure 3.21a shows the

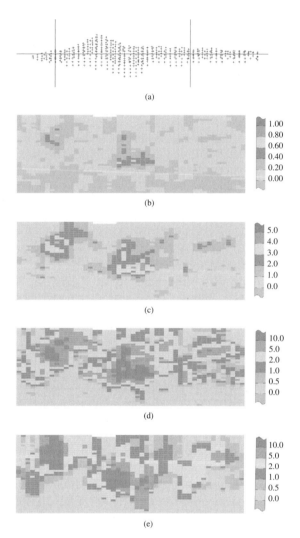

(a)

(b)

(c)

(d)

(e)

FIGURE 3.21 Vertical cross section through the block model along a gold vein. (a) Top view of
the deposit showing the trace of the cross-section and the position of the blast holes (drilling mesh
of 2.5 m perpendicular to the section and 5 m along the section); (b) Map of the indicator associated
with $Z(x) \geq 5$ g/t, estimated by ordinary cokriging from truncated and indicator data; (c) Map of
truncated grade below a 5 g/t cutoff, estimated by ordinary cokriging from truncated and indicator
data; (d) Map of the final cokriging estimates obtained by recombining the indicator and the
truncated grades by (3.71); (e) Map of direct ordinary kriging estimates of grades. Notice that the
scale of (c) ranges from 0 to 5 whereas that of (d) and (e) ranges from 0 to 10. [From Rivoirard et al.
(2012), with kind permission of the International Association for Mathematical Geosciences.]
(See color insert)

trace of the vertical vein and the position of the blast holes a few meters away on either side. The data have an extremely skewed distribution with a minimum of 0, a mean of 1.76, a maximum of 443 g/t, and a coefficient of variation σ/m of 7.74. A block model is built for short term planning, with block size $2.5 \times 5 \times 1$ m and at least one sample per block. A top-cut grade of 5 g/t is selected (much lower than the top-cut grade of 30 g/t commonly used in gold mines) to ensure lack of edge effects and a practically flat variogram of residual R_z.

Figure 3.21b shows the cokriging estimates of the indicator above 5 g/t within blocks. These represent estimated proportions of values above 5 g/t within each block—which explains why they remain fairly low. Figure 3.21c shows the cokriged truncated grade, with a color scale from 0 to 5 g/t, and Figure 3.21d is the final result obtained with formula (3.71). The result obtained by straight OK of the grade is shown in Figure 3.21e for comparison. Let us focus on the central and left-hand parts of the section. Both show areas with a relatively high truncated grade (between 3 and 4 g/t). However, high indicator values (between 0.4 and 0.6) are more abundant in the central part, which results in a more developed patch of high grades (> 10 g/t). The massive high-grade area (red) present on the left-hand side of the OK section and the large yellow patch on the right-hand side are likely artifacts caused by unwarranted interpolation of high grades.

3.8 CASE STUDY: THE CHANNEL TUNNEL

The Channel tunnel project may well be the ideal geostatistical case study. It is an important application, it is simple to understand, and—a rare event—it is possible to compare the geostatistical predictions with reality. Boring of the tunnel was completed one month ahead of the initial schedule despite the fact that most of the tunnel boring machines only came into service several months late. This is in contrast with the long delays usually encountered in tunnel boring around the world. It was the result of the excellent performance of men and machines, but also of a careful assessment of the geological risk through geostatistical analysis.

The first results of the study were published by Blanchin et al. (1989), and an overview of the Channel tunnel project with a historical perspective is given by Blanchin and Chilès (1993a). We will borrow heavily from these two papers, with an emphasis on methodological aspects.

3.8.1 Objectives and Methodology

The geological setting of the tunnel shown in Figure 3.22 can be summarized as follows: a favorable layer, the Cenomanian Chalk Marl, made of soft, generally impermeable and homogeneous rock, overlain by the Grey Chalk, a highly porous layer of typically fractured and altered rocks, and underlain by the Gault Clay, which cannot be penetrated without serious civil engineering

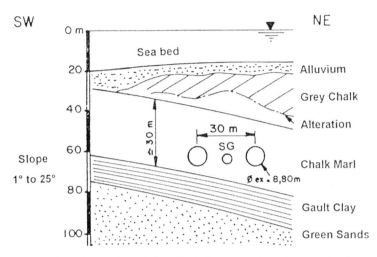

FIGURE 3.22 Channel tunnel: typical geological cross section showing the three tunnels. [Reprinted from Blanchin et al. (1989), with kind permission from Kluwer Academic Publishers; Blanchin and Chilès (1992, Figure 1), with kind permission from Springer-Verlag.]

problems. The Chalk Marl and the Gault Clay are in fact separated by a thin regular layer of Tourtia Chalk, but here it is lumped with the Chalk Marl for simplicity. Notice that what is called the tunnel comprises really three parallel tunnels 15 m apart, two for transport and a smaller one for servicing.

Although optimization of the tunnel alignment had to take into consideration the slope and curvature constraints imposed by a high-speed railway, the primary constraints were geotechnical and geological:

- No geophysical borehole could be intersected for fear of water inflow.
- The tunnel had to be at least 20 m deep below the seafloor to preserve the mechanical strength of the overlying formations.
- The tunnel could not be bored much deeper than 100 m below sea level because of the characteristics of the tunnel boring machines.
- The faults had to be intersected as orthogonally as possible.
- Most important, the tunnel had to be bored within the Chalk Marl formation.

The objective of the geostatistical study was to provide an accurate determination of the geometry of the Chalk Marl, which is only 30 m thick, and dipping, in order to prevent the risk of tunneling into the Gault Clay. It focused on the most critical variable, the top of the Gault Clay. A first estimation was made by kriging on the basis of the data available before the construction of the tunnel, with a careful evaluation of uncertainties. The results led the engineers to revise the initial layout. It was also realized that a better precision was needed in certain sections of the tunnel, and a complementary survey was designed by geostatistical analysis. Finally, data acquired during drilling of the service tunnel allowed a comparison with the geostatistical predictions.

3.8.2 Contour Mapping

In this phase the goal is to calculate a reliable digital model of the top of the Gault Clay and to produce meaningful contour maps and cross sections. The basis is 1500 km of bathymetric and reflection seismic surveys, recorded continuously (every 3 m). The data include (1) five longitudinal seismic profiles running parallel to the tunnel, 25 m apart on the French side and 250 m apart on the British side; (2) 83 transverse seismic profiles, at variable intervals between 250 m and 1000 m, and (3) 10 boreholes drilled in 1986 plus 90 old boreholes.

The depth $G(x)$ from sea level to the top of the Gault Clay is given by

$$G(x) = S(x) + V(x)T(x)/2$$

where $S(x)$ is the depth to the seafloor, computed from bathymetric data, $T(x)$ represents seismic two-way time from the seafloor to the top of the Gault Clay, computed from seismic profile data, and $V(x)$ is the average velocity obtained by various geophysical methods, such as sonic logs in some of the wells. The variables $S(x)$, $T(x)$, and $V(x)$ are estimated independently and combined through this equation to produce the final result $G(x)$.

Kriging variances are computed for each interpolated variable and because the errors are independent, the kriging variance attached to G is given by

$$\sigma_G^2 = \sigma_S^2 + (\sigma_V^2 T^2 + V^2 \sigma_T^2 + \sigma_V^2 \sigma_T^2)/4$$

where σ_S^2, σ_V^2, and σ_T^2 are the kriging variances associated with S, V, and T and all values depend on the location x.

Given the importance of risk assessment in this application and the geological heterogeneity from one part of the Channel to the other, a global structural analysis would be meaningless because not only the variogram parameters could change but also the variogram shape. The approach taken by Blanchin et al. (1989) is to divide the area into successive 1000-m-long units (37 units) and in each one compute (a) the histogram and statistical parameters and (b) the raw and residual variograms in the two profile directions (that coincide with the main directions of the anticline). To avoid sampling bias, some data in overrepresented areas are discarded. The outcome of this preliminary study is the definition of 16 homogeneous zones 1 to 5 km long, obtained by merging similar successive units.

To complete the analysis, the various sources of measurement errors are identified (e.g., tide correction, migration of seismic reflectors, velocity calculations) and included in the final structural model. The latter is validated in each zone and also globally, using standard cross-validation techniques.

The area of study is defined as a 1-km-wide and 40-km-long strip centered along the main axis of the first planned alignment. The variables are interpolated by kriging to the nodes of a 40-m × 20-m rectangular grid (20 m in the transverse direction), using for each variable a moving variogram model fitted zone by zone, with some smoothing between zones. Known faults are included as screens for the estimation of seismic times and velocities. Measurement errors that are locally constant along a profile are taken into account by adding a specific variance term to all covariances between two points on the profile.

Figures 3.23 and 3.24 show contour maps of depth and standard deviation for part of the French side of the tunnel layout. The general pattern of the top of the Gault Clay reflects the regional geological trend, an anticline whose axis is parallel to the alignment; the boundary of the Gault Clay outcrop is shown on the map as a dashed line. The standard deviation increases from south to north, reflecting the higher impact of velocity uncertainty with increasing seismic travel time. It shows well-marked minima in the vicinity of the profiles (good knowledge of seismic time) and the boreholes (consistent knowledge of seismic time and average velocity). Throughout the study area the standard deviation lies between 2 and 6 m, never exceeds 4 m along the underwater section of the tunnel route, and generally falls between 2 and 3 m.

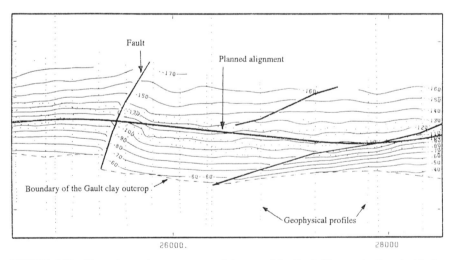

FIGURE 3.23 Channel tunnel: contour map of the top of the Gault Clay on the French side, in meters. [Reprinted from Blanchin and Chilès (1992, Figure 2), with kind permission from Springer-Verlag; Blanchin and Chilès (1993a), with kind permission of the International Association for Mathematical Geosciences; Blanchin and Chilès (1993b), with kind permission from Kluwer Academic Publishers.]

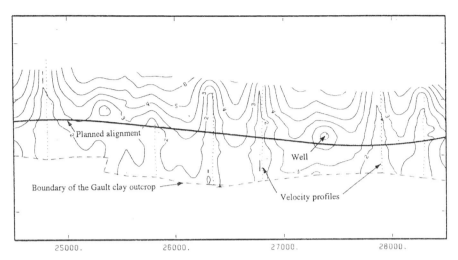

FIGURE 3.24   Channel tunnel: kriging standard deviation for the top of the Gault Clay on the French side, in meters. [Reprinted from Blanchin and Chilès (1992, Figure 4), with kind permission from Springer-Verlag; Blanchin and Chilès (1993a), with kind permission of the International Association for Mathematical Geosciences; Blanchin and Chilès (1993b), with kind permission from Kluwer Academic Publishers.]

3.8.3   Risk Assessment

We now have to answer the initial civil engineering question: Will the planned alignment intersect the Gault Clay? Or rather, given our incomplete knowledge, what is the *risk* that the planned alignment intersects the Gault Clay? This is where kriging standard deviation proves to be useful.

Cross sections are the tool of choice to visualize the geometry of the tunnel project in the vertical plane. They can be generated from the grid or directly by kriging at points along the three tunnel galleries. Figure 3.25 shows the results obtained with a spacing of 20 m along a section of the south tunnel (vertical scale exaggeration: 20). The seafloor and the top of the Gault Clay are represented with their nominal 68% confidence intervals (± 1 standard deviation). For bathymetry the estimation is so precise that the three lines are indistinguishable on the graph (the kriging standard deviation does not exceed 0.5 m). When considering the estimated top minus one standard deviation, one can see that the first alignment could intersect the Gault Clay in several places.

This led the engineers to revise the layout so as to maintain the tunnels nearly everywhere at least one standard deviation above the estimated top of the Gault, as shown in the cross section. The risk of penetrating the Gault Clay was thus reduced, but of course not entirely eliminated. Notice that the engineers chose to use σ rather than the statistician's sacred 2σ because, in sections where it mattered, they were ready to assume a 16% risk of hitting the Gault Clay (one-sided interval).

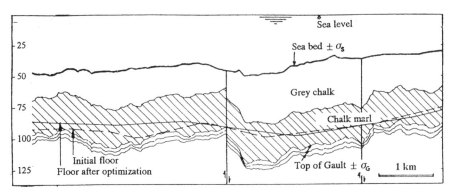

FIGURE 3.25   Channel tunnel: cross section of kriged results along the profile of the south tunnel. Vertical scale exaggeration: 20. [Reprinted from Blanchin et al. (1989), with kind permission from Kluwer Academic Publishers; Blanchin and Chilès (1992, Figure 1), with kind permission from Springer-Verlag.]

3.8.4   Optimum Design of a Complementary Survey

The tunnel project also included crossover excavations at two locations to enable the trains to pass from one tunnel to the other if necessary. Their construction required more accurate geological predictions than those used for the main tunnel, which required a complementary geophysical survey. How should the survey be designed to achieve a standard error of less than 1 m on the top of the Gault Clay?

Since kriging variances can be computed without knowing the values of the variables, it suffices to simulate the surveying process by adding fictitious data until the required precision is achieved. The result is a recommendation to place the transverse seismic profiles 25 m apart over the French crossover and 100 m apart over the British crossover and calibrate the seismic velocity by at least four boreholes at each crossover. With this new survey and a new pass of variogram analysis and kriging, maps and cross sections are redrawn with improved precision.

3.8.5   Geostatistical Predictions Versus Reality

As boring of the tunnel progressed, dual boreholes, one dipping to the north and the other to the south, were drilled downward from the central service tunnel to determine the actual depth and dip of the Gault Clay. Fifty-four dual boreholes were drilled along the first 13 km on the French side, and 31 were drilled along the first 15 km on the British side. The objective was to check the accuracy of the current estimates and, if needed, revise them to get a reliable geometric model for the continuation of the project.

The "reality–prediction" differences are plotted in Figure 3.26, and their statistical characteristics are summarized in Table 3.3, separately for the British side and the French side because their data patterns and spatial characteristics

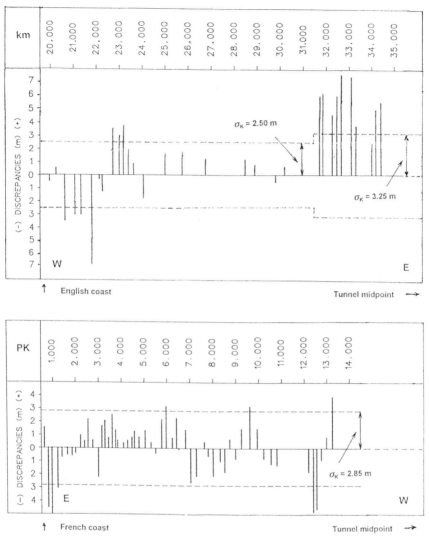

FIGURE 3.26 Channel tunnel: discrepancies observed in the service tunnel (reality−prediction): on the British side (above); on the French side (below). Both graphs go from the coast (left) toward the tunnel midpoint (right). The distance axes have opposite directions, and their origins differ. The two parts are separated by a zone of about 10 km without control borehole. [From Blanchin and Chilès (1993a,b), with kind permission of the International Association for Mathematical Geosciences and Kluwer Academic Publishers.]

are rather different. Typical of real case studies, the results are not as straight-forward as one would wish. On the British side the mean difference is 1.70 m and the standard deviation 3.4 m. The distribution of errors along the service gallery shows a good agreement between the actual and predicted depths for the first 20 points (dotted lines are $\pm\sigma_K$), but a systematic effect is apparent for the last

TABLE 3.3   "Reality–Prediction" Differences in British and French Sides

	French Side (13 km)	British Side (15 km)
Number of borehole pairs	54	31
Minimum difference	−5.00 m	−7.00 m
Maximum difference	+3.90 m	+8.00 m
Mean difference	+0.48 m	+1.70 m
Standard deviation from zero	2.02 m	3.40 m
Kriging standard deviation	2.85 m	2.55 m

Source: Blanchin and Chilès (1993a,b), with kind permission of the International Association for Mathematical geology and Kluwer Academic Publishers.

10 boreholes (average difference of 5.6 m). This prompted a review of the initial interpretation in light of the new data. A key factor is that these 10 discrepancies are clustered in a zone where the density of seismic profiles and borehole data is the lowest for the whole tunnel—that is, where the accuracy of the geostatistical mapping is low ($\sigma_K = 3.25$ m) compared with the average value for the British sector ($\sigma_K = 2.55$ m). A careful reinterpretation of the seismic data in that zone pinpointed two systematic errors: (1) An error in the time pick, the Tourtia Chalk was mistaken for the Gault Clay horizon lying in reality 3.5 m below; (2) there were errors in the calibration of the velocity data, due to poor positioning of old (1964–1965) geophysical boreholes in an area with a strong dip.

On the French side the positioning uncertainty on the geophysical boreholes was taken into account in the kriging process by a highly correlated error component in the velocity data. The observed discrepancies are small throughout, even apparently too small. But caution, the errors are far from independent! If they were, they would fluctuate back and forth around the zero line; on the contrary, they tend to stay on the same side of the line. Blanchin and Chilès (1993b) did a variogram analysis of the errors and concluded that on the French side the 54 observations are in fact equivalent to only 13 independent samples, while on the British side the 31 values are worth 21 independent samples.

In conclusion, the observations in the service tunnel were generally in good agreement with the geostatistical model and its predicted accuracy. When discrepancies occurred, they could be traced to systematic interpretation errors localized in sparsely sampled zones. This type of error is always a risk with geophysical (i.e., indirect) measurements, so it requires calibration data. The main objective of the geostatistical study, avoid penetrating the Gault Clay, was achieved; it never happened on the French side and happened twice on the British side, but where expected.

3.9   KRIGING UNDER INEQUALITY CONSTRAINTS

In natural phenomena one encounters inequalities of two types. The first are *global constraints* due to the very definition of the variables. For example, mineral grades, thicknesses of geological layers, are positive quantities, and rock properties such as porosity and fluid saturations vary between 0 and 1. The second type are *local constraints* providing bounds on the results, and these can be regarded as *data*. For example, if drilling was stopped at a depth z_0 without hitting a given geological surface, we know that the depth of the surface at that point is $Z(x) > z_0$. Naturally the information carried by interval data $a \leq Z(x) \leq b$ depends on the tightness of the bounds.

Simple ad hoc methods can sometimes solve the problem. For global constraints an adequate *transformation* of the data, followed by kriging and a back transform, can automatically produce estimates within the desired range. This is often useful even if the reverse transformation poses a bias problem. Another simple method is *censoring*: Estimation is carried out without consideration of the constraints, and the results are modified as needed to satisfy the constraints. For example, when a geological layer "pinches out," negative thickness estimates appear by continuity as the estimated point moves away from the layer boundary; it is legitimate to set these negative estimates to zero. Finally, adding *dummy points*, possibly with error variances, at critical locations may suffice to produce a quick result honoring the constraints.

In this section we review methods that incorporate the constraints explicitly in the estimation process by adaptations of linear kriging. Nonlinear methods based on conditional simulations are presented in Sections 7.6.1 and 7.6.3. We will focus here on three different approaches to handle inequalities: Place constraints on the kriging weights, constrain the estimates, and treat constraints as data. The advent of the Gibbs sampler has made the third approach the one of choice, but the other two still provide interesting methodological insight.

3.9.1 Nonnegative Kriging Weights

The reason why known global constraints may be violated by kriging estimates is the possibility of weights < 0 or > 1 (cf. Example 4 with a power $\alpha = 1.5$; Figures 3.14 and 3.16). The coincidence of a negative weight with a large sample value can produce negative estimates, or simply puzzling ones such as an estimated grade lower than every sample in the neighborhood. A sufficient but not necessary condition to preclude this is to constrain all kriging weights to be nonnegative. If, in addition, weights add up to one, the kriging estimates automatically lie within the minimum and maximum of the estimating data.

The ideal solution would be to use covariance or variogram models that automatically generate positive weights. Matheron (1986) investigated whether such models exist and stated the following conclusions.

Consider a set $I = \{1, 2, \ldots, N\}$ of indexes that we arbitrarily partition into sample points and estimated points. The random variables Z_i have a covariance matrix Σ, assumed to be strictly positive definite. We want to find Σ such that kriging weights are ≥ 0 for all partitions of the set I, namely all kriging configurations. The results are as follows:

1. The problem may have a solution only for SK and OK.
2. SK and OK weights are ≥ 0 if and only if all off-diagonal terms of the inverse matrix Σ^{-1} are ≤ 0.
3. Only in 1D do we know covariance or variogram functions ensuring positive kriging weights.

In the special case of SK, formula (3.64) shows the direct link between the sign of the off-diagonal terms of Σ^{-1} and the kriging weights. In the case of OK, the covariance matrix Σ in statement 2 can be replaced by the generalized covariance matrix $\mathbf{K}$ where $K_{ij} = -\gamma_{ij}$, or $K_{ij} = C - a_i - a_j - \gamma_{ij}$, and $\mathbf{K}$ is assumed strictly conditionally positive definite.

In 1D we already know that the exponential covariance ensures positive SK weights and the linear variogram positive OK weights. In fact the exponential covariance also ensures positive OK weights, and this property extends to completely monotone covariances (2.28). The (simple) kriging weights satisfy $\sum \lambda_\alpha \leq 1$ for all finite configurations and the optimal weights for the estimation of the mean are also positive. Likewise, positive OK weights are obtained with any variogram of the form

$$\gamma(h) = A|h| + C(0) - C(h) \qquad (A \geq 0)$$

where $C(h)$ is a completely monotone covariance. Unfortunately, these results are only for 1D. In 2D we know that the geostatistician's best friends, the spherical covariance and the linear variogram, generate negative weights (see Figures 3.16 and 3.15). In dimension higher than 1D the conditions on Σ^{-1}, to be useful, would require modeling Σ^{-1}, as in the Gaussian Markov random field approach (Section 3.6.4) rather than Σ as in the standard geostatistical approach.

An alternative is to place positivity constraints on the weights by using constrained optimization. For example, Barnes and Johnson (1984) reformulate ordinary kriging as

$$\text{Minimize } E\left(\sum_\alpha \lambda_\alpha Z_\alpha - Z_0\right)^2 \quad \text{subject to} \quad \sum_\alpha \lambda_\alpha = 1 \text{ and } \lambda_\alpha \geq 0 \quad \forall \alpha$$

which can be solved by quadratic programming. However, forcing all weights to be positive is artificial, it confines kriging in a straitjacket.

3.9.2 Minimization under Inequality Constraints

The general problem is to find an estimator $Z^*(x)$ such that

$$Z^*(x_\alpha) = z_\alpha \qquad \text{for } \alpha = 1, \ldots, N$$
$$zinf_\alpha \leq Z^*(x_\alpha) \leq zsup_\alpha \qquad \text{for } \alpha = N+1, \ldots, N+M$$

where the first N data are exact data and the remaining M are inequality data.

In two papers, Dubrule and Kostov (1986) and Kostov and Dubrule (1986) approach this problem by first noting that a direct constrained minimization of the quadratic form $E(Z^* - Z_0)^2$ leads to an impasse because a constraint at the point x_α only affects estimation at x_α itself, without any lateral continuity, producing jumps just like kriging of a sample point in the presence of a nugget effect. They reformulate the problem under dual kriging as follows: Find an interpolating function of the form

$$z^*(x) = \sum_{\alpha=1}^{N+M} b_\alpha K(x - x_\alpha) + \sum_\ell c_\ell f^\ell(x)$$

where the coefficients b_α and c_ℓ satisfy the conditions

$$\begin{cases} z^*(x_\alpha) = z_\alpha & \text{for } \alpha = 1, \ldots, N, \\ zinf_\alpha \leq z^*(x_\alpha) \leq zsup_\alpha & \text{for } \alpha = N+1, \ldots, N+M, \\ \displaystyle\sum_{\alpha=1}^{N+M} b_\alpha f_\alpha^\ell = 0 & \text{for } \ell = 0, \ldots, L \end{cases} \qquad (3.72)$$

The first and third equalities are similar to the usual dual kriging equations (3.45) for exact data. The interval constraints can be broken down into two one-sided inequalities.

These conditions on the form of the interpolating function do not suffice to determine a unique solution. By analogy with spline theory, the authors propose to select the solution b_α and c_ℓ minimizing the quadratic form

$$Q = \sum_{\alpha=1}^{N+M} \sum_{\beta=1}^{N+M} b_\alpha b_\beta K_{\alpha\beta}$$

under the constraints (3.72). When $M = 0$ (no inequalities), the solution is the usual dual kriging interpolant; when $M > 0$ and the spline covariance model is used for $K(x, y)$, Q is interpreted as the mean curvature of the interpolated surface, and the solution is then a thin plate spline.

A technical difficulty appears here because the function $K(x, y)$ used is not a genuine covariance but a "generalized covariance" (see Chapter 4). The quadratic form is not positive definite (strictly or not), and therefore not a convex function, which is inconvenient for minimization. The difficulty is only apparent because in reality we only deal with vectors $\mathbf{b}$ that satisfy the conditions $\mathbf{F}'\mathbf{b} = \mathbf{0}$ (allowable linear combinations), and the restriction of the quadratic form to these vectors is a convex function. Considering the top left block $\mathbf{U}$ of the partitioned inverse kriging matrix in equation (4.31), Langlais (1989) casts the minimization of Q in the equivalent form:

$$\begin{cases} \text{Minimize} \quad \mathbf{Z}'\mathbf{U}\mathbf{Z} & \text{subject to} \\ Z_\alpha = z_\alpha & \text{for } \alpha = 1, \ldots, N \\ zinf_\alpha \leq Z_\alpha \leq zsup_\alpha & \text{for } \alpha = N+1, \ldots, N+M \end{cases}$$

The first N components of the vector $\mathbf{Z}$ are the exact data and the remaining M components are the values at the constraint points determined by quadratic programming. Once $\mathbf{Z}$ is determined, the dual kriging solutions are computed as $\mathbf{b} = \mathbf{U}\mathbf{Z}$ and $\mathbf{c} = \mathbf{V}'\mathbf{Z}$. The constrained minimization problem has a unique solution if the UK problem for the N exact data has itself a unique solution (i.e., there are at least $L+1$ exact data points over which the drift functions are linearly independent); this condition is sufficient but not necessary (Langlais, 1989).

Figure 3.27 from Dubrule and Kostov illustrates the method in 1D using cubic spline interpolation: First, only the 9 exact data are used; then 18 inequality constraints are added, consisting of $M_1 = 12$ lower bounds, $M_2 = 6$ upper bounds, and two points ($x = 4$, $x = 6$) having two-sided inequalities. In the first pass, 12 inequalities are violated; in the second pass they are all satisfied of course, but notice the "clamping effect" at the bounds: The function is exactly equal to some of the bounds. This is how quadratic programming works: It selects some of the constraints (the "active" ones) and satisfies them at the bounds; then all the other constraints are automatically satisfied as well. If we knew which constraints are the active ones, it would be possible to introduce them as equality data, ignore the others, and proceed with normal kriging. The problem is to determine the active set, because it does not coincide with the constraints that are violated by an initial execution of the kriging procedure based only on exact data.

Minimization under constraints requires special care when moving neighborhoods are used because the algorithm may assign different values at the same inequality point depending on the neighborhood considered. The recommended procedure is to select exact data first, using the standard neighborhood search algorithm, and then include the constraint points that would have been selected by the search algorithm operating without distinction between data and constraints. To summarize, inequality data are treated as secondary information.

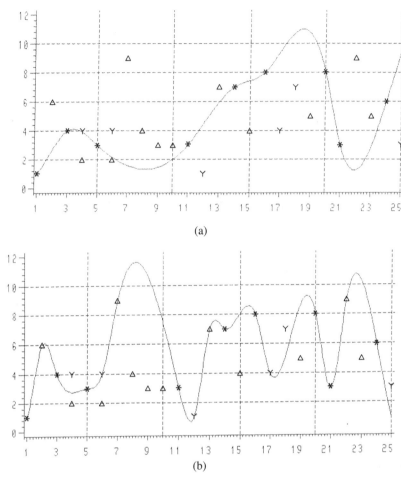

(a)

(b)

FIGURE 3.27 Interpolation (a) based on exact data only; (b) with constrained optimization. The function is a cubic spline obtained with $K(h) = |h|^3$ and a linear drift. Asterisks are exact data, triangles are lower bounds, and "Y" are upper bounds. [From Dubrule and Kostov (1986), with kind permission of the International Association for Mathematical Geosciences.]

3.9.3 Inequalities as Data

An alternative approach proposed by Langlais (1990) is to regard inequalities as data and replace them by exact values. The procedure is to simulate exact data satisfying the given inequalities, proceed to kriging from both actual and generated data, and finally average the results over several simulations of the inequality data. We will keep her approach but present a new implementation based on the use of the Gibbs sampler.

To simplify notations, we consider here that all data are of the form $Z_\alpha \in B_\alpha$, where B_α denotes an interval, or more generally a Borel set (e.g., a finite union of intervals), with the understanding that it may be reduced to a single point in the case of exact data. If Z is a Gaussian RF with known mean, consider the conditional expectation

$$E[Z(x)|Z_\alpha \in B_\alpha, \forall_\alpha]$$

In principle, this expectation can be calculated by integration of the Gaussian p.d.f., but in practice, this is an intractable problem. Instead, note that the simple kriging estimator is

$$E[Z(x)|Z_\alpha = z_\alpha, \forall \alpha] = \sum_\alpha \lambda_\alpha z_\alpha$$

so that

$$E[Z(x)|Z_\alpha \in B_\alpha, \forall \alpha] = \sum_\beta \lambda_\beta E(Z_\beta|Z_\alpha \in B_\alpha, \forall \alpha)$$

Thus it suffices to perform simple kriging after replacing the inequality data by their conditional expectations $E(Z_\beta|Z_\alpha \in B_\alpha, \forall_\alpha)$. We compute these means empirically by generating samples from the joint conditional distribution of the Z_α given all data $\{Z_\alpha \in B_\alpha\}$, using an algorithm known as the Gibbs sampler (see Section 7.6.2). This is implemented by repeating the following sequence:

1. Select an index α in the set of inequality data.
2. Simulate Z_α conditionally on $Z_\alpha \in B_\alpha$ and $Z_\beta = z_\beta$ for all $\beta \neq \alpha$ (β ranges over *all* data indexes except α).

The procedure can be initialized by generating each Z_α separately, conditionally on $Z_\alpha \in B_\alpha$. The index α may be scanned either periodically or using an irreducible Markov chain, the important point being that, in theory, each index is almost surely drawn infinitely often. It is possible to generalize the method to more complex constraints defined as a Borel set of $\mathbb{R}^N$ (e.g., ellipsoids instead of parallelepipeds; N is the number of data).

This approach finds its theoretical justification in the ideal case of a Gaussian RF with known mean. It can be used more generally, *as an algorithm*, by assuming that at each step the conditional distribution is Gaussian with mean the kriging estimate and variance the kriging variance. The algorithm ensures that the inequality data are accounted for in a consistent manner, but of course its optimality properties are unknown. The same approach is used effectively to generate conditional simulations constrained by inequality data (Section 7.6.3).

CHAPTER 4

Intrinsic Model of Order k

A trend is a trend is a trend
But the question is, will it bend?
Will it alter its course
Through some unforeseen force
and come to a premature end?

— Sir Alec Cairncross

4.1 INTRODUCTION

4.1.1 A Perspective

Intrinsic random functions of order k (IRF–k) were developed initially to work around the difficult problem of statistical inference of the variogram in the Universal Kriging model. Indeed, as we have seen in Chapter 2, the presence of a spatial trend creates a bias that affects both the raw variogram and the variogram of estimated residuals. The idea of IRF–k theory was to remove the trend from sight by using only linear combinations that filter out low order polynomials, and any trend is at least locally polynomial. As a side benefit, the new theory delivered a new class of covariance models, the polynomial generalized covariances, which are linear in their coefficients and thus make direct statistical inference possible without bias. In retrospect, however, it appears that the IRF–k model did not supersede the UK model, because the trend is often a physically meaningful component, especially when it can be estimated by regression on external variables.

The real benefit of IRF–k theory is that it brings unity and clarity to geostatistical theory. It singles out the *intrinsic properties* of a nonstationary random function, which are the only ones that really matter for kriging. It brings new insights into radial basis function interpolation and splines. And, most importantly, it builds a connection between geostatistics and physics by

Geostatistics: Modeling Spatial Uncertainty, Second Edition. J.P. Chilès and P. Delfiner.
© 2012 John Wiley & Sons, Inc. Published 2012 by John Wiley & Sons, Inc.

providing the correct conceptual model to represent the nonstationary solutions of stochastic partial differential equations.

4.1.2   From IRF–0 to IRF–k

The notion of intrinsic random function of order k (henceforth abbreviated as IRF–k) constitutes a natural generalization of the intrinsic random functions (i.e., with stationary increments) of traditional geostatistics. These correspond to the particular case $k = 0$ and will now be called IRF–0. In passing from stationary random functions (SRF) to IRF–0, the following changes take place:

- The basic working tool of the stationary case, the covariance $C(h)$, is replaced by the variogram $\gamma(h)$. This extends generality, since the class of valid variogram functions is broader than the class of covariance functions [$C(h)$ must be positive definite, whereas it is only required that $-\gamma(h)$ be conditionally positive definite]. In contrast to the covariance, the variogram may be unbounded, and this enables the description of phenomena with a potentially unlimited dispersion (theoretically infinite variance) such as the Brownian motion.

- In the stationary case there exists a mean value m about which the SRF fluctuates. The phenomenon remains "controlled" in the sense that the deviations from the mean are never too large nor last too long. In the case of an IRF–0 with an unbounded variogram, no such regulation exists. The Brownian particle has no memory. It bounces from its current position as if it were a new origin and shows no tendency to revert to its starting point. There is no constant mean value m. In effect such a process is defined up to an arbitrary constant and is generally studied only through its increments.

- In the case of an SRF, any linear combination

$$Z(\lambda) = \sum_i \lambda_i Z(x_i)$$

has the variance

$$\text{Var } Z(\lambda) = \sum_i \sum_j \lambda_i \lambda_j \, C(x_j - x_i)$$

where $C(h)$ is a centered covariance. In the case of an IRF–0, only special linear combinations have a finite variance, the allowable ones, satisfying the condition $\sum \lambda_i = 0$. The variance is then calculated using $C(h) = -\gamma(h)$ as if it were a covariance function:

$$\text{Var } Z(\lambda) = -\sum_i \sum_j \lambda_i \lambda_j \, \gamma(x_j - x_i)$$

Other combinations do not, in general, have a finite variance. Thus, at the cost of a relatively minor operating restriction (only use linear combinations summing up to zero), we gain the possibility of dealing with a large class of phenomena that, like the Brownian motion, cannot be represented by a stationary model.

Is it possible to go even further in the same direction? In other words, can we get access to broader classes of nonstationary phenomena at the cost of restrictions more severe than $\Sigma \lambda_i = 0$? The theory of IRF–k provides a positive answer to this question. The idea is to define models through increments of a sufficiently high order for stationarity to be reached, an approach generalizing the ARIMA models of time series analysis (Box and Jenkins, 1976). We discover that these models are characterized by a new structure function that is completely free of the influence of polynomial drifts and turns out to be the real minimum prerequisite for universal kriging.

4.2 A SECOND LOOK AT THE MODEL OF UNIVERSAL KRIGING

The basic model of universal kriging is the dichotomy

$$Z(x) = m(x) + Y(x) \tag{4.1}$$

where $Z(x)$ is the variable under study, $m(x)$ is the drift, and $Y(x)$ is the fluctuation, or residual, about this drift. From a mathematical point of view, the drift is well-defined as the expected value $m(x) = \mathrm{E}\,[Z(x)]$. But $m(x)$ is not an observable, except if there are repetitions allowing us to actually compute $m(x)$ as an average across several realizations of the same phenomenon.

When the phenomenon is unique, as it is in geological applications, this $m(x)$ is a purely theoretical construct. Its modeling is inspired by observations of "trends," namely systematic patterns of variations in the data. But in reality the drift is an elusive concept, sometimes unclear or very complex, sometimes clear but spurious.

4.2.1 Questioning the Dichotomy into Drift and Fluctuation

Except for replications, the most meaningful case for the dichotomy (4.1) is that of a phenomenon showing small local fluctuations about a clear overall trend liable to be modeled by a simple and smooth mathematical function (Figure 4.1). In the model the trend is treated deterministically because it is simple, whereas the

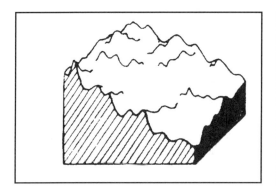

FIGURE 4.1 A favorable case for universal kriging. (Author: J. P. Delhomme.)

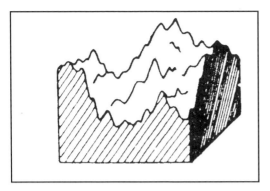

FIGURE 4.2   A puzzling case for drift modeling. (Author: J. P. Delhomme.)

fluctuations are too complex to be described in detail and are captured by probabilistic means. Specifically, $Y(x)$ is a stationary random function with a zero mean and a covariance whose range is short at the scale of the study. The drift may be estimated reliably and subtracted from the data to restore stationarity. The underlying covariance can be determined from the variogram of residuals which, at short distances, differs little from the true variogram. The drift then possesses an objective meaning in the sense of Section 1.2.

The situation is very different when the covariance of fluctuations does not have a short range. As we have seen in Section 2.7, the drift causes both the experimental variogram and the variogram of residuals to be considerably biased, which makes statistical inference of the covariance very difficult and thereby precludes the use of statistical tests aimed precisely at deciding whether there is a drift or not. Assuming that it can be defined, the drift is not necessarily simple enough to be modeled by an analytic expression valid over the whole studied domain (Figure 4.2). In such cases one can turn to a local model of the form

$$m(x) = \sum_{\ell} a_\ell(x_0) f^\ell(x) \tag{4.2}$$

valid only in a neighborhood $V(x_0)$ of each point x_0. As usual, the $f^\ell(x)$ are given functions (typically monomials), but the unknown coefficients a_ℓ vary with the neighborhood. Their estimation is even more difficult than before, since there are fewer data points in each neighborhood, and we are confronted with the problem of piecing together local estimates. The map obtained by moving neighborhood drift estimation often looks much less smooth than the kriging map, which is rather disturbing (Chilès, 1979a). In reality, in the absence of a clear separation of scales between $m(x)$ and $Y(x)$, the dichotomy is simply arbitrary, and the drift is not an objective parameter.

4.2.2   Examples of Zero-Mean Processes with Apparent Drifts

Chance fluctuations may produce clear trends. A well-known and striking example is shown in Figure 4.3 taken from Feller (1968, p. 87) displaying a record of a coin-tossing experiment. The function graphed is

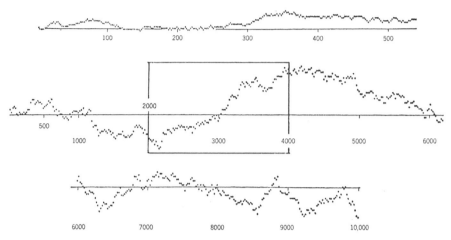

FIGURE 4.3 The record of 10,000 tosses of an ideal coin. [From Feller (1968, p. 87), with permission of John Wiley & Sons, Inc.]

$$S_n = X_1 + X_2 + \cdots + X_n$$

where $X_i = +1$ or $X_i = -1$ according to the outcome of the trial with a fair coin.

Given the string between $n = 2000$ and $n = 4000$ and told that these are daily oil prices, an "expert" would fit an upward trend and extrapolate it for the next decade. Alas, the trend turns and now points downward for the next 2000 samples! Thus, the next time the expert will try to forecast the "turning point." These trends are mere fluctuations of the random walk which has mean zero for each n.

Large fluctuations such as these can occur due to the lack of any regulating mechanism in the process. By contrast, a stationary process is subject, so to speak, to an elastic force pulling it back to its mean. For example, consecutive increments $Z(x_3) - Z(x_2)$ and $Z(x_2) - Z(x_1)$ at three aligned points x_1, x_2, x_3 have a correlation coefficient of $-1/2$ when the point interdistance is larger than the range: A move up tends to be compensated by a move down. However, the variogram of S_n does not have a range since it is unbounded ($\text{Var}(S_n - S_m) = |n - m|$).

Examples like this are not limited to one dimension. We can generate surfaces that show clear systematic patterns even though, by design of the simulation algorithm, $E[Z(x)] = 0$ for all x. Figure 4.4a shows a simulation of a Brownian RF in 2D. The process has a mean zero and a linear variogram; it was simulated by turning bands (see Section 7.4). Figures 4.4b and 4.4c were obtained similarly by radial integration of 4.4a (Section 4.5.8); they exhibit long-range patterns that could be interpreted as drifts, though more complicated ones than planes or quadratics. But we know that no drift was incorporated in the generation of the process.

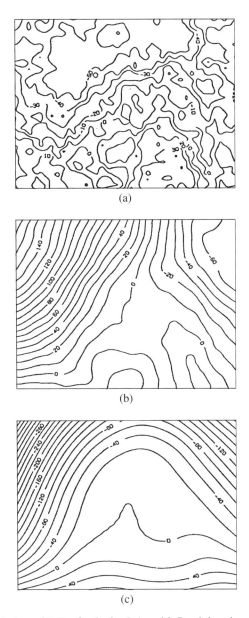

FIGURE 4.4   Simulations of IRFs of order $k = 0$, 1, and 2. Panels b and c suggest drifts that were not present in the construction of the models. [From Orfeuil (1972).]

These examples illustrate the existence of zero-mean processes whose fluctuations have an aspect usually attributed to the presence of a trend. There is an *apparent* drift but no genetic drift. Any interpretation of these pseudodrifts as systematic effects is totally spurious, and dangerous as well.

4.2.3 Toward Stationarity

From the grid of values of Figure 4.4b we construct a new grid by differencing as follows:

$$I(i, j) = Z(i - 1, j) + Z(i + 1, j) + Z(i, j - 1) + Z(i, j + 1) - 4 Z(i, j)$$

This new grid is displayed in Figure 4.5c. The "systematic" effects have been removed, the field appears stationary. The raw variograms which in Figure 4.5a

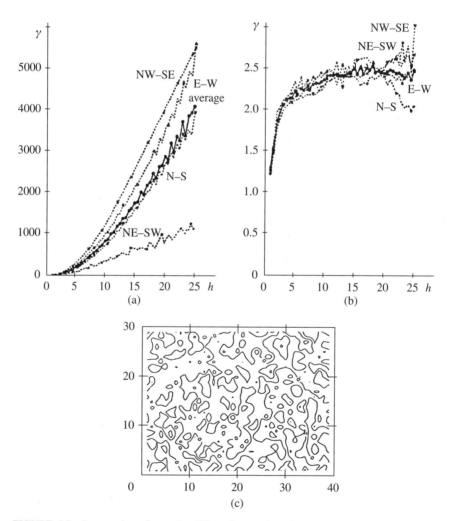

FIGURE 4.5 A case where first-order differencing renders data stationary: (a) raw directional variograms of grid displayed in Figure 4.4b; (b) directional variograms of first-order differences; (c) map of first-order differences.

exhibited a parabolic increase typical of a linear drift, and also a violent anisotropy, now stabilize around a clear sill in Figure 4.5b. The anisotropy has disappeared, which indicates that it was due to a (local) polynomial component that differencing has eliminated. Differencing turns nonstationary data into stationary ones.

Starting now from the highly nonstationary Figure 4.4c we apply the same differencing operation, but this time the raw variograms still exhibit a parabolic behavior (Figure 4.6b). Trying differencing again, which amounts to taking second-order differences of the initial data, establishes stationarity (Figures 4.6c and 4.6d). Again this is evidenced by a clear sill on the variograms and the restoration of isotropy. Note that by construction of the simulations, the noise level is zero so that the low correlation second-order differences, indicated by the flat variogram, is a genuine feature of the variable $Z(x)$ and not a mere reflection of a deteriorated signal-to-noise ratio.

This example shows that differencing can be an effective move toward stationarity. This technique can be extended to the more common case of scattered data, leading to the concept of Generalized Increment of order k.

4.3   ALLOWABLE LINEAR COMBINATIONS OF ORDER k

4.3.1   Allowable Measures and Generalized Increments of Order k

A set of weights λ_i applied to m points x_i of $\mathbb{R}^n$ defines a discrete measure λ of the form

$$\lambda = \sum_{i=1}^{m} \lambda_i \, \delta_{x_i} \tag{4.3}$$

where δ_{x_i} is the Dirac measure at the point x_i (Dirac delta function). The action of λ on a function $f(x)$ defines a linear combination that will be synthetically denoted by $f(\lambda)$:

$$f(\lambda) = \int f(x) \, \lambda(dx) = \sum_{i=1}^{m} \lambda_i f(x_i)$$

Definition. A discrete measure λ is allowable at the order k if it annihilates all polynomials P on $\mathbb{R}^n$ of degree less than or equal to k:

$$P(\lambda) = \sum_{i=1}^{m} \lambda_i \, P(x_i) = 0 \qquad \text{whenever degree } P \leq k \tag{4.4}$$

We call Λ_k the class of such allowable measures.

It is clear that (4.4) is achieved if and only if λ annihilates separately all monomials of degree up to k. In one-dimensional space there are $k+1$ conditions, one for each power of x. In n-dimensional space there are $k_n = \binom{k+n}{k}$

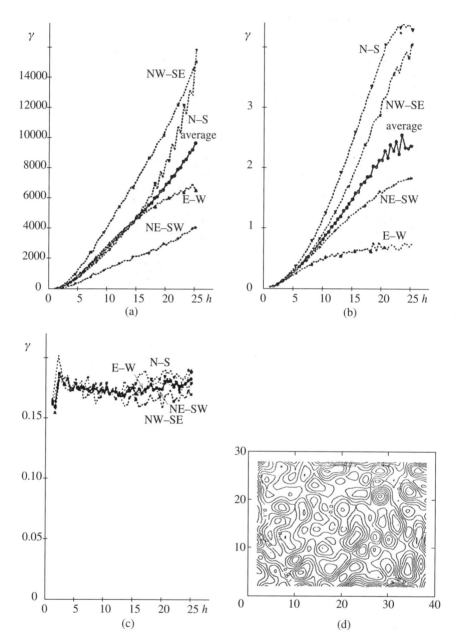

FIGURE 4.6 A case where second-order differencing is needed to make data stationary: (a) raw directional variograms of grid displayed in Figure 4.4c; (b) directional variograms of first-order differences; (c) directional variograms of second-order differences; (d) map of second-order differences.

monomials of degree less than or equal to k and thus as many conditions. To avoid quadruple indexes, it is convenient to use the following condensed notations:

$x_i = (x_{i1}, \ldots, x_{in})$ for a point in $\mathbb{R}^n$ (but in 2D we use (x_i, y_i) as the coordinates)

$\ell = (\ell_1, \ldots, \ell_n)$ for a set of nonnegative integers

$x_i^\ell = x_{i1}^{\ell_1} x_{i2}^{\ell_2} \cdots x_{in}^{\ell_n}$ for a monomial

$|\ell| = \ell_1 + \cdots + \ell_n$ for the degree of x^ℓ

Then (4.4) is equivalent to the set of conditions

$$\sum_{i=1}^{m} \lambda_i \, x_i^\ell = 0, \qquad |\ell| = 0, 1, \ldots, k \tag{4.5}$$

that is, all moments of order up to k, inclusive, are zero. Obviously we have $\Lambda_{k+1} \subset \Lambda_k$.

An allowable measure $\lambda \in \Lambda_k$ defines an *allowable linear combination of order k* (abbreviated as ALC–k), also called *authorized linear combination of order k*, or *generalized increment of order k*,

$$Z(\lambda) = \sum_i \lambda_i Z(x_i) \tag{4.6}$$

Link with Error Contrasts

The concept of ALC–k is of algebraic nature and can be presented in the framework of linear models and "error contrasts" familiar to statisticians. Suppose that in the standard notations of linear models

$$\mathbf{Y} = \mathbf{X}\boldsymbol{\beta} + \mathbf{U} \tag{4.7}$$

where $\mathbf{Y} = (Y_1, \ldots, Y_N)'$ is a vector of observations ($N > k_n$), $\mathbf{X} = (x_i^\ell)$ is the $N \times k_n$ matrix of all monomials evaluated at points $x_1, \ldots, x_N$, $\boldsymbol{\beta} = (\beta_0, \ldots, \beta_{k_{n-1}})'$ is a vector of coefficients, and $\mathbf{U} = (U_1, \ldots, U_N)'$ a vector of residuals. Formula (4.7) is exactly the discrete formulation of the universal kriging model with a polynomial drift of degree k. Then a linear combination $\sum_i \lambda_i Y_i$ is an ALC–k if the relation (4.5) is satisfied, namely if the vector of weights $\boldsymbol{\lambda} = (\lambda_1, \ldots, \lambda_N)'$ satisfies

$$\boldsymbol{\lambda}' \mathbf{X} = 0 \tag{4.8}$$

Combining (4.7) and (4.8) gives

$$\boldsymbol{\lambda}' \mathbf{Y} = \boldsymbol{\lambda}' \mathbf{U}$$

and it is seen that $\boldsymbol{\beta}$ (the drift coefficients) has been completely eliminated. Generalized increments are linear functions of $\mathbf{U}$ only; they are *error contrasts*. Conversely, if a function $\Phi(\mathbf{Y})$ of the data does not depend on $\boldsymbol{\beta}$ at all, namely satisfies the invariance property

$$\Phi(\mathbf{Y} + \mathbf{X}\mathbf{a}) = \Phi(\mathbf{Y}) \qquad \forall \mathbf{a}$$

then $\Phi(\mathbf{Y})$ depends on $\mathbf{Y}$ only through $N - k_n$ error contrasts (Delfiner, 1977). This justifies why in a theory where the drift is to be bypassed, it is necessary to allow some linear combinations and forbid others.

Before turning to examples, it is worth mentioning that the notion of ALC–k can be extended to the continuous case by considering

$$f(\mu) = \int \mu(dx) f(x)$$

for measures μ in the larger class $\mathcal{M}_k$ of measures vanishing outside a compact set (measures with compact support) and satisfying

$$\int \mu(dx) x^\ell = 0, \qquad |\ell| = 0, ..., k$$

4.3.2   Examples

Finite Differences on the Line

The forward finite difference of order $(k+1)$ of a function $f(x)$ is defined by

$$\Delta_a^1 f(x) = f(x + a) - f(x),$$
$$\Delta_a^{k+1} f(x) = \Delta_a^1 \Delta_a^k f(x) = (-1)^{k+1} \sum_{p=0}^{k+1} (-1)^p \binom{k+1}{p} f(x + pa)$$

As Δ_a^1 decreases by one the degree of any polynomial, it is easy to see by induction that finite differences of order $k+1$ annihilate polynomials of degree k. Thus they are ALC–k.

Five-Point Laplacian Approximation

In 2D consider 5 points $x_0, x_1, ..., x_4$ arranged as in Figure 4.7. Let $\lambda_1 = \lambda_2 = \lambda_3 = \lambda_4 = 1$ and $\lambda_0 = -4$, and let x_i and y_i denote the two coordinates of the point x_i. The measure λ is allowable at order 1, since

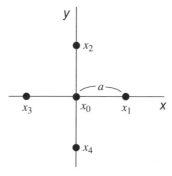

FIGURE 4.7 $f(\lambda) = f(x_1) + f(x_2) + f(x_3) + f(x_4) - 4 f(x_0)$ is an allowable linear combination of order 1 but not of order 2.

$$\sum_i \lambda_i = \sum_i \lambda_i x_i = \sum_i \lambda_i y_i = 0$$

Since $x_i y_i = 0$ for all i, λ also annihilates the monomial xy. But it does not annihilate x^2 nor y^2:

$$\sum_i \lambda_i x_i^2 = \sum_i \lambda_i y_i^2 = 2a^2$$

Therefore λ is not allowable at the order 2. (As a matter of fact, $f(\lambda)/a^2$ is a finite difference approximation to the Laplacian Δf of f at the point x_0.)

Increments on a Circle

$2k+2$ points placed on a circle at regular angular intervals $\pi/(k+1)$ and with alternating weights $+1$ and -1 define an allowable measure of order k. The proof follows from the orthogonality of the complex exponentials over the set of points and De Moivre's theorem $(\cos\theta + i\sin\theta)^n = \cos n\theta + i\sin n\theta$ (Chilès, 1977). Note, for example, that for $k=3$ there are 8 points for 10 monomials to annihilate.

Errors from Unbiased Estimation

A simple way of constructing an ALC–k is to take the difference between a value and a linear unbiased estimate of this value calculated under the assumption that the mean is a full polynomial of degree k. Indeed

$$\hat{Z}_0 = \sum_i \lambda_i Z_i$$

is an unbiased estimator of $Z_0 = Z(x_0)$ if and only if

$$\sum_i \lambda_i x_i^\ell = x_0^\ell, \qquad |\ell| = 0, \ldots, k$$

and therefore $\sum_i \lambda_i Z_i - \hat{Z}_0$ is an ALC–k.

The manner in which $\hat{Z}_0$ is obtained does not matter: it can be kriging (whatever the variogram) or least squares. For this purpose, least squares have the advantage of being the simplest, and they are used most. Note that several ALC–k may be obtained from the same least squares fit: all the residuals at the points used for the fit plus estimation errors at other arbitrary points not used for the fit.

4.3.3 Minimum Number of Points

According to equation (4.8), an ALC–k based on N points has weights λ_i which are a nontrivial solution of $\lambda' X = 0$. If the columns of X are linearly independent, necessarily $N \geq \text{rank}(X) + 1$. When all monomials of $\mathbb{R}^n$ are used, the rank of $\mathbf{X}$ is $k_n = \binom{k+n}{k}$ so that $k_n + 1$ is the minimum number of points required. Minimum values of N according to the dimensionality n of the space are

$$
\begin{array}{ll}
2 & \text{points for } k = 0 \\
n + 2 & \text{points for } k = 1 \\
(n+1)(n+2)/2 + 1 & \text{points for } k = 2
\end{array}
$$

However, if the points are such that there is linear dependence between the columns of $\mathbf{X}$, it is possible to construct increments with fewer points because some of the constraints are automatically satisfied. For example, on a circle, the fact that $x^2 + y^2 = 1$ implies that there are only $2k+1$ linearly independent monomials of degree up to k instead of $(k+1)(k+2)/2$ for the whole plane; so $2k+2$ points suffice to define an ALC–k as was indicated above. Geometrically, this means that all points lie on a curve or surface defined by an algebraic equation of the form $\mathbf{X}\mathbf{a} = 0$, such as a line or plane if $k = 1$, a conic or quadric if $k = 2$.

In practice, this possibility of taking advantage of the location of the points is interesting when data are along lines: increments of order k on the line, such as finite differences, are also valid increments in the plane or space. Indeed a change of coordinate system transforms a polynomial of degree k into another polynomial of degree k. Taking the line as one of the axes makes all other coordinates zero and shows that the conditions involving these coordinates are automatically satisfied.

4.3.4   Why Polynomials Are Used

We define the translate $\tau_h \lambda$ of the measure λ by the vector h as the measure with the same weights as λ but applied to a point configuration shifted by h, namely

$$
f(\tau_h \lambda) = f\left(\sum_i \lambda_i \delta_{x_i + h} \right) = \sum_i \lambda_i f(x_i + h)
$$

Does $\tau_h \lambda$ remain allowable when λ is allowable? In other words, is Λ_k closed under translations? The answer is in the affirmative owing to the binomial formula

$$
(x + h)^\ell = \sum_{s=0}^{\ell} \binom{\ell}{s} x^s h^{\ell - s}
$$

since

$$
\sum_i \lambda_i (x_i + h)^\ell = \sum_{s=0}^{\ell} \binom{\ell}{s} h^{\ell - s} \sum_i \lambda_i x_i^s = 0
$$

for $|\ell| \le k$ and $\lambda \in \Lambda_k$. In $\mathbb{R}^n$, $s = 0, \ldots, \ell$ is short for $s_j = 0, \ldots, \ell_j \; \forall j = 1, \ldots, n$, and $\binom{\ell}{s} = \ell!/[s!(\ell - s)!]$ where $\ell! = \ell_1! \ldots \ell_n!$. This result ensures that the translate $Z(\tau_h \lambda)$ of an ALC–k $Z(\lambda)$ is also an ALC–k, a property without which stationarity assumptions on generalized increments would not make sense. It is a direct consequence of the property of polynomials to be themselves closed under translations. Are polynomials special, or can other functions be used to construct generalized increments?

Mathematically, the problem is the following: find p functions $f^\ell(x)$ such that the finite-dimensional vector space $\mathcal{F}$ generated by the $f^\ell(x)$ is closed under translations. In other words, we want that

$$f(x) = \sum_{\ell=1}^{p} a_\ell f^\ell(x) \quad \Rightarrow \quad f(x+h) = \sum_{\ell=1}^{p} a_\ell(h) f^\ell(x)$$

Obviously this will hold if and only if it holds for each $f^\ell(x)$. So the $f^\ell(x)$ are solutions of the functional equation

$$f^\ell(x+h) = \sum_{s=1}^{p} B_s^\ell(h) f^s(x) \tag{4.9}$$

A general theorem (Matheron, 1979a) states that the only continuous (and even the only measurable) solutions of (4.9) are finite sums of functions of the form

$$f(x) = P(x) \exp(c_1 x_1 + \cdots + c_n x_n)$$

where $P(x)$ is a polynomial in $x = (x_1, \ldots, x_n)$ and $c = (c_1, \ldots, c_n)$ are real or complex coefficients.

For $n = 1$, $\mathcal{F}$ is generated by families of exponential monomials

$$\{x^\ell e^{cx} : \ell = 0, \ldots, k\} \tag{4.10}$$

Since we only consider real functions, if $\omega \neq 0$ the complex coefficient $c = \alpha + i\omega$ must have its conjugate counterpart, and $\mathcal{F}$ is generated by the functions

$$\{x^\ell e^{\alpha x} \cos \omega x, x^\ell e^{\alpha x} \sin \omega x : \ell = 0, \ldots, k\}$$

There are three remarkable subsets of this family:

1. Pure polynomials $P(x)$ of degree $\leq k$
2. Pure trigonometric functions $\{\cos \omega x, \sin \omega x\}$
3. Pure exponentials $\{\exp(\alpha x)\}$

Note that for exponentials or trigonometric functions, the rate α or the frequency $\omega/2\pi$ must be selected, whereas for polynomials there is no scaling parameter.

The same results hold formally when $n > 1$. It suffices to regard ωx as a scalar product $<\omega, x> = \omega_1 x_1 + \omega_2 x_2 + \cdots + \omega_n x_n$. However, when $n > 1$, $\mathcal{F}$ does not necessarily contain all terms of (4.10). For example, for $n = 2$ and $x = (x, y)$, the space $\mathcal{F}$ generated by the functions: 1; x; y; $x^2 + y^2$ is invariant under shifts but does not contain x^2 and y^2, nor xy.

From a theoretical point of view, it is possible to develop a theory of intrinsic random functions based on all the solutions of (4.9), including the exponential terms (Matheron, 1979a). This extension will not be presented here. Indeed in more than one dimension there is a genuine difficulty to use functions including an exponential component because they are scale- and orientation-dependent. Except perhaps in very specific cases it is desirable that allowable measures remain so under scaling or rotations. If this is also required, the polynomial solutions of (4.9) are the only possible ones.

In conclusion, polynomials are not used just for the sake of convenience but because they satisfy fundamental geometric invariance requirements.

4.4   INTRINSIC RANDOM FUNCTIONS OF ORDER k

Two mathematical definitions of IRF–k will be given: One is Matheron's original definition (1973a) which is abstract but mathematically more profound, and the other is a simpler definition, easier to understand.

4.4.1   Ordinary IRF–k

Definition 1. A random function $Z(x)$ is intrinsic of order k if for any allowable measure $\lambda \in \Lambda_k$ the random function

$$Z_\lambda(x) = Z(\tau_x \lambda) = \sum_i \lambda_i Z(x_i + x)$$

is second-order stationary in $x \in \mathbb{R}^n$ and has a zero mean.

This is equivalent to

$$\begin{cases} \mathrm{E}[Z_\lambda(x)] = 0, \\ \mathrm{E}[Z_\lambda(x)Z_\lambda(y)] = K_\lambda(y - x) \end{cases} \quad \forall x, y \in \mathbb{R}^n, \lambda \in \Lambda_k$$

An IRF–k is simply a random function with stationary increments of order k. The usual intrinsic model of geostatistics corresponds to $k=0$. Clearly an IRF–k is also an IRF–$(k+1)$ and of any higher order, since $\Lambda_{k+1} \subset \Lambda_k$. For example, an SRF is intrinsic at all orders; formally it would correspond to the case $k=-1$.

The condition that increments of order k have a zero mean is introduced for a simpler presentation and does not restrict generality. If these increments are stationary, their mean is necessarily a polynomial of degree $k+1$ at most (Matheron, 1973a)[1], which is eliminated by regarding $Z(x)$ as an IRF–$(k+1)$.

For example, we have seen in Section 2.1.2 that the mean of stationary increments is necessarily of the form $\mathrm{E}[Z(x+h) - Z(x)] = \langle a, h \rangle$; if $a \neq 0$ we should regard $Z(x)$ as an IRF–1 rather than an IRF–0.

By definition, an IRF satisfies $\mathrm{E}[Z_\lambda(x)^2] < \infty$ for any $\lambda \in \Lambda_k$. But $\mathrm{E}[Z(x)^2]$ may be infinite, or may at least depend on x; in the introductory coin tossing example we had $\mathrm{E}[S_n^2] = n$. Stationarity of increments allows for both nonstationarity in the mean and in the variance.

As usual with random functions, it will be assumed that $Z(x)$ is continuous in the mean square sense. This property is mathematically essential to extend the theory from the space Λ_k of discrete measures to the space $\mathcal{M}_k$ of measures with compact supports. If a discontinuity were present (nugget effect), it should be handled separately.

---

[1] Matheron names *intrinsic drift* of the IRF–k the part of degree exactly $k+1$ of this polynomial.

Examples of IRF–k

1. For $n = 1$ the integral of a zero-mean SRF is an IRF–0. Indeed let $X(x)$ be the SRF in $\mathbb{R}$ and $Y_0(x)$ the integral

$$Y_0(x) = \int_0^x X(t)\, dt$$

For any h the increment

$$Y_0(x + h) - Y_0(x) = \int_x^{x+h} X(t)\, dt$$

is a moving average of the stationary process $X(t)$ and is therefore stationary.

2. The $(k+1)$th integral of a zero-mean SRF is an IRF–k. We show this by induction. Assume that Y_k is an IRF–k, and consider the integral

$$Y_{k+1}(x) = \int_0^x Y_k(t)\, dt$$

Introducing the indicator function of the summation interval, we can write equivalently

$$Y_{k+1}(x) = \int Y_k(t) 1_{0 \leq t \leq x}\, dt$$

Thus for any discrete measure λ we have

$$Y_{k+1}(\lambda) = \sum_i \lambda_i Y_{k+1}(x_i) = \int Y_k(t)\, \mu(dt) = Y_k(\mu)$$

where

$$\mu(dt) = \left(\sum_i \lambda_i 1_{0 \leq t \leq x_i} \right) dt$$

Now $\lambda \in \Lambda_{k+1}$ implies that $\mu \in \Lambda_k$ (or more precisely $\in \mathcal{M}_k$), since

$$\int t^\ell \mu(dt) = \sum_i \lambda_i \int_0^{x_i} t^\ell\, dt = \frac{1}{\ell+1} \sum_i \lambda_i x_i^{\ell+1} = 0, \qquad \ell = 0, ..., k$$

For any x the translate $\tau_x \mu$ of the measure μ is defined by

$$Y_k(\tau_x \mu) = \sum_i \lambda_i \int_x^{x_i + x} Y_k(t)\, dt = \sum_i \lambda_i Y_{k+1}(x_i + x)$$

the last equality being a consequence of $\sum \lambda_i = 0$. Finally we have

$$Y_k(\tau_x \mu) = Y_{k+1}(\tau_x \lambda)$$

Since Y_k is an IRF–k and $\mu \in \Lambda_k$, $Y_k(\tau_x \mu)$ is stationary in x. Therefore $Y_{k+1}(\tau_x \lambda)$ is also stationary in x, and this is true for any $\lambda \in \Lambda_{k+1}$. Consequently Y_{k+1} is an IRF–$(k + 1)$.

Conversely, if an IRF–k is differentiable $(k + 1)$ times, its $(k + 1)$th derivative is stationary, being the limit of an ALC–k.

3. The same results hold if we start with an IRF–0. By integrating k times a Brownian motion $W_0(x)$, we obtain the IRF–k:

$$W_k(x) = \int_0^x \frac{(x - t)^{k-1}}{(k - 1)!} W_0(t)dt$$

where $W_k(x)$ vanishes at $x = 0$ as well as its first $(k - 1)$ derivatives (just like Y_k above).

4. Conversely, now in $\mathbb{R}^n$, if a random function $Z(x)$ is differentiable $(k + 1)$ times and if all its partial derivatives of order $(k + 1)$ are stationary and with zero mean, $Z(x)$ is an IRF–k. This property characterizes differentiable IRF–k's. Of course there exist nondifferentiable IRF–k's (e.g., an IRF–0 with a linear variogram), but it is shown below that any continuous IRF–k is the sum of an SRF and a differentiable IRF–k.

5. An ARIMA process (autoregressive integrated moving average process) is defined as a process whose finite difference of order d is a stationary ARMA process (Box and Jenkins, 1976; see Section 7.5.1 for a definition of ARMA models). Since a finite difference of order d is an ALC–$(d - 1)$, an ARIMA process is an IRF–$(d - 1)$. However, the ARIMA and IRF–k approaches differ in the following aspects:

- ARIMA models are completely specified, whereas IRFs are only second-order models.
- ARIMA models are one-dimensional, whereas IRFs are defined in $\mathbb{R}^n$.
- ARIMA models are essentially discrete, whereas IRFs are continuous or discrete.

4.4.2 Abstract IRF and Its Representations

If $Z(x)$ is an IRF–k and A_ℓ are random variables—independent or not of $Z(x)$—the new random function

$$Z_1(x) = Z(x) + \sum_{|\ell| \leq k} A_\ell x^\ell$$

is also an IRF–k: by definition, $\lambda \in \Lambda_k$ cancels all monomials x^ℓ, and thus

$$Z_1(\lambda) = Z(\lambda)$$

$Z_1(x)$ and $Z(x)$ are indistinguishable on the basis of ALC–k only.

In reality the concept of IRF–k relates to an *equivalence class* rather than a single function, namely the class of all random functions generating the same increments of order k. This motivated Matheron's definition of the IRF–k as a family of increments. To avoid any confusion, we will call this *an abstract IRF–k* and denote it with a tilde.

Definition 2. An abstract intrinsic random function $\widetilde{Z}$ is a linear mapping of Λ_k into a Hilbert space $\mathcal{H}$ of zero mean, finite variance random variables, such that for any $\lambda \in \Lambda_k$ the random function $\widetilde{Z}(\tau_x \lambda)$ is second-order stationary in x:

$$\widetilde{Z} : \Lambda_k \to \mathcal{H} \quad \text{such that} \quad \forall \lambda \in \Lambda_k \quad Z_\lambda(x) = \widetilde{Z}(\tau_x \lambda) \text{ is an SRF}$$

An ordinary IRF–k is a random function $Z(x)$ in the usual sense, whereas the abstract IRF $\widetilde{Z}$ is not a function of x but of λ. Any ordinary IRF–k $Y(x)$ generating the same increments as $\widetilde{Z}$, namely satisfying

$$Y(\lambda) = \widetilde{Z}(\lambda) \qquad \forall \lambda \in \Lambda_k$$

is called a *representation* of $\widetilde{Z}$. So, from this new perspective, what we defined as an ordinary IRF–k was in fact a representation of the abstract IRF $\widetilde{Z}$. But it is simpler to reason with representations because they "materialize" the equivalence class defined by the abstract IRF. The following two properties make it possible to identify $\widetilde{Z}$ with the class of all its representations:

1. Any abstract IRF–k $\widetilde{Z}$ has representations.
2. If one representation is known, all the others are deduced by addition of a polynomial of degree k with random coefficients.

Proof. It was shown in Section 4.3.2 that errors from linear unbiased estimation are ALC–k. We will use this property to construct a representation that has the structure of a residual. To this end, consider a collection of measures $\lambda_\ell(dx)$ satisfying for all $|\ell| \le k$ the conditions

$$\int \lambda_\ell(dx) x^s = \delta_\ell^s$$

($\delta_\ell^s = 1$ if $\ell = s$ and $= 0$ otherwise). One may think of these measures as defining unbiased estimators of the coefficients of a polynomial of degree k in the space $\mathbb{R}^n$ considered. For any $x \in \mathbb{R}^n$ the measure

$$\varepsilon_x(dt) = \delta_x(dt) - \sum_\ell x^\ell \lambda_\ell(dt) \tag{4.11}$$

belongs to Λ_k, since

$$\int \varepsilon_x(dt) t^s = \int \delta_x(dt) t^s - \sum_\ell x^\ell \int \lambda_\ell(dt) t^s = x^s - \sum_\ell x^\ell \delta_\ell^s = 0$$

Now we claim that the RF $Y(x)$ defined by

$$Y(x) = \widetilde{Z}(\varepsilon_x)$$

is a representation of $\widetilde{Z}$. Indeed, for any $\lambda = \sum \lambda_i \delta_{x_i} \in \Lambda_k$ the linearity of $\widetilde{Z}$ entails

$$Y(\lambda) = \sum_i \lambda_i Y(x_i) = \sum_i \lambda_i \widetilde{Z}(\varepsilon_{x_i}) = \widetilde{Z}\left(\sum_i \lambda_i \varepsilon_{x_i}\right)$$

But

$$\sum_i \lambda_i \varepsilon_{x_i} = \sum_i \lambda_i \delta_{x_i} - \sum_\ell \left(\sum_i \lambda_i x_i^\ell\right) \lambda_\ell = \lambda$$

the last equality being a consequence of $\lambda \in \Lambda_k$ (the coefficients of λ_ℓ are all zeros). Thus we have

$$Y(\lambda) = \widetilde{Z}(\lambda) \qquad \forall \lambda \in \Lambda_k$$

which proves point 1. In particular, for $\lambda = \varepsilon_x$ we obtain $Y(\varepsilon_x) = \widetilde{Z}(\varepsilon_x) = Y(x)$.

For point 2 suppose that $Z(x)$ is another representation of $\widetilde{Z}$. We have

$$Z(\lambda) - Y(\lambda) = 0 \qquad \forall \lambda \in \Lambda_k$$

So $Z_0(x) = Z(x) - Y(x)$ is a representation of the identically nil abstract IRF–k ($\widetilde{Z}_0(\lambda) = 0 \quad \forall \lambda \in \Lambda_k$). In particular, with the measure ε_x defined in (4.11),

$$Z_0(\varepsilon_x) = Z_0(x) - \sum_\ell Z_0(\lambda_\ell)\, x^\ell = 0$$

and thus $Z_0(x)$ is of the form

$$Z_0(x) = \sum_\ell A_\ell\, x^\ell$$

With this presentation we have three levels of abstraction instead of two as usual: the abstract IRF–k $\widetilde{Z}$, its representations $Y(x)$, which are random functions, and numerical realizations of $Y(x)$. We regard our regionalized variable as a realization of a representation of an abstract IRF $\widetilde{Z}$. This allows us to distinguish two kinds of properties:

- Those that do not depend on the representation: they are *intrinsic properties*.
- Those that depend on the representation.

The estimation of intrinsic properties only requires the specification of the abstract IRF–k model. To estimate other properties, it is necessary to qualify the representation being considered.

Internal Representations

The particular representation $Y(x) = Y(\varepsilon_x)$ that we have just constructed has the remarkable property of being itself an ALC–k. Therefore $Y(x)$ satisfies $E[Y(x)] = 0$ and $E[Y(x)^2] < \infty$.

To appreciate the specificity of $Y(x)$, consider another representation $Y_1(x)$. We have

$$Y(x) = Y_1(\varepsilon_x) = Y_1(x) - \sum_\ell x^\ell \int \lambda_\ell(dy)\, Y_1(y)$$

We see that $Y(x)$ is an *additive renormalization* of $Y_1(x)$ obtained by subtracting a polynomial of degree k whose coefficients $A_\ell = \int \lambda_\ell(dy)\, Y_1(y)$ are linear functionals[2] of $Y_1(x)$. A simple and standard example for an IRF–0 is the representation $Y(x) = Y_1(x) - Y_1(0)$. By construction, $Y(x)$ has a finite variance, while $Y_1(x)$ may not. Now, considering a compact domain D (with a nonempty interior), if all measures λ_ℓ have their support included in D, the representation $Y(x)$ only depends on values of $Y_1(x)$ within D. We say that $Y(x)$ is an *internal representation* over D. This notion is useful for the solution of partial differential equations of the type $\Delta Z(x) = Y(x)$ (see Dong, 1990).

---

[2] For true generality we must include the limits of all $Y_1(\varepsilon_x)$, where $\varepsilon_x \in \Lambda_k$ and with support in D.

An internal representation is generally not stationary, although it is sometimes possible to make it locally stationary by an astute choice of the measures λ_ℓ. We will revisit this subject later.

General Form of Representations

From the above we can conclude that all representations are the sum of an internal representation and an arbitrary polynomial of degree k at most

$$Y(x) = Y\left(\delta_x - \sum_\ell x^\ell \lambda_\ell\right) + \sum_\ell A_\ell x^\ell \qquad (4.12)$$

From now on we will leave the abstract IRF–k $\widetilde{Z}$ in the background and, following the simpler definition, will refer to $Z(x)$ and an IRF–k (instead of a representation of an abstract IRF–k).

4.5 GENERALIZED COVARIANCE FUNCTIONS

The correlation structure of a stationary random function $Z(x)$ is defined by its ordinary covariance function $C(h)$. If only the (ordinary) increments $Z(x+h) - Z(x)$ of the function are assumed stationary, the variogram $\gamma(h)$ is the structural tool. We saw that $\gamma(h)$ only allows the calculation of the variances of linear combinations whose sum of weights is zero. In the same manner, when the stationarity assumptions are limited to generalized increments of order k, what characterizes the correlation structure of $Z(x)$ is a new function called a generalized covariance function, abbreviated as GC and denoted by $K(h)$. Just as there were more models for variograms than for covariances, there are more models for generalized covariances than for both ordinary covariances and variograms. But there are also more restrictions attached to their use in variance calculations: They only work on allowable linear combinations. Figure 4.8 shows a family picture of C, γ, and K taken by J. P. Delhomme.

FIGURE 4.8 Covariance, variogram, and generalized covariance: a family picture. (Author: J. P. Delhomme.)

4.5.1 Existence and Uniqueness

Definition. Let Z be an IRF–k. A symmetric function $K(h)$ defined on $\mathbb{R}^n$ is called a generalized covariance (GC) of Z if

$$E[Z(\lambda)Z(\mu)] = \sum_i \sum_j \lambda_i \mu_j K(y_j - x_i) \qquad (4.13)$$

for any pair of measures λ, $\mu \in \Lambda_k$.

In fact it is sufficient that the above condition be satisfied in the case $\lambda = \mu$ (it is seen easily by expanding $[Z(\lambda + \mu)]^2$). In other words, it suffices to verify that

$$E[Z(\lambda)^2] = \sum_i \sum_j \lambda_i \lambda_j K(x_j - x_i) \quad (\lambda \in \Lambda_k) \qquad (4.14)$$

Formally $K(h)$ is used just as an ordinary covariance $C(h)$, but (4.14) only holds for $\lambda \in \Lambda_k$.

There is an existence and uniqueness theorem for any continuous IRF–k (i.e., whose representations are continuous in the mean square sense).

Theorem. Any continuous IRF–k has a continuous GC $K(h)$. This GC $K(h)$ is unique as an equivalence class, in the sense that any other GC is of the form $K(h) + Q(h)$, where $Q(h)$ is an even polynomial of degree $2k$ or less.

This theorem holds in $\mathbb{R}^n$ whatever the dimension n, and for any IRF–k. But the proof in the general case encounters technical difficulties which obscure the almost intuitive aspect of the result (Gel'fand and Vilenkin, 1961; Matheron, 1973a). Therefore we will restrict ourselves here to the simple case $n = 1$ and Z differentiable.

Proof of the Theorem for a Differentiable Z on $\mathbb{R}$

- *Existence.* Let $Z(x)$ be an IRF–k on $\mathbb{R}$, differentiable $k + 1$ times (in the mean square). Its $(k+1)$th derivative $Z^{(k+1)}(x)$ is stationary and has a stationary covariance $C(h)$. If we denote by $\sigma(x,y)$ the nonstationary covariance of $Z(x)$, it is related to $C(h)$ by

$$\frac{\partial^{2(k+1)}}{\partial x^{k+1} \partial y^{k+1}} \sigma(x,y) = C(y - x)$$

Integrating $(k + 1)$ times in x gives

$$\frac{\partial^{k+1}}{\partial y^{k+1}} \sigma(x,y) = (-1)^{k+1} \int_0^{y-x} \frac{(y - x - u)^k}{k!} C(u)du + \sum_{\ell=0}^{k} a_\ell(y)x^\ell$$

Integrating now $(k + 1)$ times in y for fixed x gives

$$\sigma(x,y) = (-1)^{k+1} \int_0^{y-x} \frac{(y - x - u)^{2k+1}}{(2k + 1)!} C(u)\, du + \sum_{\ell=0}^{k} b_\ell(y)x^\ell + \sum_{\ell=0}^{k} c_\ell(x)y^\ell$$

where $b_\ell(y)$ is a definite integral of order $k+1$ of $a_\ell(y)$. Let

$$K(h) = (-1)^{k+1} \int_0^h \frac{(h-u)^{2k+1}}{(2k+1)!} C(u)\,du$$

Verify that $K(h) = K(-h)$ so that $K(h)$ is a symmetric function just as $C(h)$; and since $\sigma(x, y)$ is symmetric in x and y, the $b_\ell(.)$ and $c_\ell(.)$ functions are identical. Thus the general form of the covariance $\sigma(x, y)$ is

$$\sigma(x, y) = K(y - x) + \sum_{\ell=0}^{k} c_\ell(y)x^\ell + \sum_{\ell=0}^{k} c_\ell(x)y^\ell \tag{4.15}$$

If λ is an allowable linear combination, then we have

$$\sum_i \lambda_i x_i^\ell = 0, \qquad \ell = 0, ..., k$$

and from (4.15) we obtain

$$\sum_i \sum_j \lambda_i \lambda_j \sigma(x_i, x_j) = \sum_i \sum_j \lambda_i \lambda_j K(x_j - x_i)$$

which proves that $K(h)$ is a generalized covariance.

• *Uniqueness.* Let us assume that $Z(x)$ has two distinct GCs, $K_1(h)$ and $K_2(h)$. From (4.15) the difference $K_0 = K_1 - K_2$ is of the form

$$K_0(y - x) = \sum_{\ell=0}^{k} c_\ell(y)x^\ell + \sum_{\ell=0}^{k} c_\ell(x)y^\ell$$

Because $Z(x)$ is differentiable $(k+1)$ times, $K_0(h)$ is differentiable $2(k+1)$ times. Letting $h = y - x$, we have

$$\frac{d^{2k+2}}{dh^{2k+2}} K_0(h) = (-1)^{k+1} \frac{\partial^{k+1}}{\partial x^{k+1}} \frac{\partial^{k+1}}{\partial y^{k+1}} K_0(y - x)$$

But this is zero identically, since x^ℓ and y^ℓ are monomials of degree strictly less than $(k+1)$. So $K_0(h)$ is a polynomial of degree $2k+1$ at most. Because $K_0(h)$ is a symmetric function, it is necessarily an even polynomial of degree $\leq 2k$ (i.e., with even powers only).

Conversely, for any even polynomial $Q(h)$ of degree $\leq 2k$, $K(h) + Q(h)$ is indeed a GC. It suffices to note that $Q(h)$ can be written just as $K_0(y - x)$, with the $c_\ell(\cdot)$ being polynomials of degree up to $2k$. □

Examples of GCs

1. If $Z(x)$ is an SRF with ordinary stationary covariance $C(h)$, then $C(h)$ is clearly also a GC. If $Z(x)$ is an IRF–0, then

$$K(h) = -\gamma(h) + \text{constant}$$

The GC is equal to $-\gamma$ up to an arbitrary constant (which is an even polynomial of degree $k=0$).

2. We have seen earlier that if $X(t)$ is a zero-mean SRF on $\mathbb{R}$ its $(k+1)$th integral

$$Y_k(x) = \int_0^x \frac{(x-t)^k}{k!} X(t)dt$$

is an IRF–k. The proof of the existence theorem given above shows that the GC of $Y_k(x)$ is

$$K_k(h) = (-1)^{k+1} \int_0^h \frac{(h-u)^{2k+1}}{(2k+1)!} C(u)du \qquad (h>0)$$

where $C(h)$ is the stationary covariance of $X(t)$.

4.5.2 Link Between Generalized and Ordinary Covariance Functions

The covariance and the variogram are easy to understand because they are directly related to the variable $Z(x)$. The GC is more abstract. To "materialize" it, let us see how it relates to the ordinary covariance. To this end, we consider the general form of representations (4.12),

$$Z(x) = Z(\delta_x - \sum_\ell x^\ell \lambda_\ell) + \sum_\ell A_\ell x^\ell$$

and calculate the centered covariance $\sigma(x, y)$. Because the first term is of the form $Z(\varepsilon_x)$ with $\varepsilon_x \in \Lambda_k$, its covariance can be calculated with the GC $K(h)$ using the general formula (4.13):

$$\begin{aligned}
E[Z(\varepsilon_x)Z(\varepsilon_y)] &= \iint \left(\delta_x(dt) - \sum_\ell x^\ell \lambda_\ell(dt)\right) K(t'-t) \left(\delta_y(dt') - \sum_s y^s \lambda_s(dt')\right) \\
&= K(y-x) - \sum_\ell x^\ell \int K(y-t)\lambda_\ell(dt) - \sum_\ell y^\ell \int K(t'-x)\lambda_\ell(dt') \\
&\quad + \sum_\ell \sum_s x^\ell y^s \iint \lambda_\ell(dt)K(t'-t)\lambda_s(dt') \qquad (4.16)
\end{aligned}$$

Now the covariance of $Z(x)$ is

$$\begin{aligned}
\mathrm{Cov}(Z(x), Z(y)) &= E\left(Z(\varepsilon_x) Z(\varepsilon_y)\right) + \sum_\ell x^\ell \mathrm{Cov}(A_\ell, Z(\varepsilon_y)) \\
&\quad + \sum_\ell y^\ell \mathrm{Cov}(Z(\varepsilon_x), A_\ell) + \sum_\ell \sum_s x^\ell y^s \mathrm{Cov}(A_\ell, A_s)
\end{aligned}$$

The covariance $\text{Cov}(A_\ell, .)$ cannot be calculated with the GC because A_ℓ is not an ALC–k. Anyhow, collecting the coefficients of x^ℓ and y^s with those in (4.16), we get a covariance of the form

$$\sigma(x,y) = K(y-x) + \sum_{|\ell|=0}^{k} c_\ell(y)x^\ell + \sum_{|\ell|=0}^{k} c_\ell(x)y^\ell + \sum_{|\ell|=0}^{k}\sum_{|s|=0}^{k} T_{\ell s}\, x^\ell y^s \qquad (4.17)$$

This is the same formula as (4.15) obtained by integration of a differentiable GC (the terms $x^\ell y^s$ can be distributed equally between the x^ℓ and y^s terms). It exposes the impact of the stationarity of increments of order k, an attenuated form of stationarity, on the covariance of $Z(x)$. There is a stationary part $K(y-x)$ and a nonstationary part involving polynomial terms separately in x and y. In general, the functions $c_\ell(x)$ are not polynomials, nor can they be determined from the data on the basis of a single realization because they involve nonstationary features of $Z(x)$. If we consider an arbitrary linear combination $Z(\lambda)$, its variance depends on the $c_\ell(x)$ and cannot be evaluated. But, if $\lambda \in \Lambda_k$, then these coefficients are filtered out.

Examples

1. The simplest illustration is with a GC–0 $K(h) = -\gamma(h)$. Then

$$\sigma(x,y) = -\gamma(y-x) + [\sigma(x,x) + \sigma(y,y)]/2$$

which is of the form (4.17) with $k=0$ and $c_0(x) = \sigma(x,x)/2$.

To take a specific example, consider in 1D a fractional Brownian motion $X(t)$ without drift and the representation $Y(t) = X(t) - X(0)$. This is a nonstationary RF with mean zero and a variance proportional to $|t|^\alpha$. We assume a scaling such that $\gamma(1) = 1$. Then the covariance of $Y(t)$ is

$$\sigma(t,t') = |t'-t|^\alpha + |t|^\alpha + |t'|^\alpha \qquad (0 < \alpha < 2)$$

When $\alpha = 1$ and both t and $t' > 0$, this takes the form $\sigma(t,t') = 2 \min(t,t')$.

2. Now suppose that $X(t)$ has a linear drift. We model it as an IRF–1 and consider the representation defined by

$$Y(t) = X(t) - X(0) - \frac{1}{2R}[X(R) - X(-R)]\, t$$

$Y(t)$ has the structure of an estimated residual and satisfies $E[Y(t)] = 0$. It is therefore an ALC–1 of X and has a finite variance that can be calculated with the GC $K(h) = -|h|^\alpha$. The nonstationary covariance of $Y(t)$ is found to be

$$\sigma(t, t') = -|t' - t|^\alpha + |t|^\alpha + |t'|^\alpha + (|R - t'|^\alpha - |R + t'|^\alpha)\frac{t}{2R}$$
$$+ (|R - t|^\alpha - |R + t|^\alpha)\frac{t'}{2R} + \frac{2^{\alpha-1}}{R^{2-\alpha}}tt'$$

which is of the form (4.17). It is interesting to note that for $\alpha = 1$ and t and $t' \in [-R, +R]$, the variogram of $Y(t)$ is

$$\frac{1}{2}\,\mathrm{E}[Y(t') - Y(t)]^2 = |t' - t| - \frac{1}{2R}(t' - t)^2$$

and depends only on $|t' - t|$. The variogram of estimated residuals is stationary (but biased).

4.5.3 Spectral Theory

The existence and uniqueness theorem for GCs is a direct consequence of Gel'fand and Vilenkin's theory of generalized random fields (1961, Chapter 3, Section 5.2). The result gives the general form of the correlation functional $B(\varphi, \psi)$ of a generalized random field acting on functions φ and ψ of the class of infinitely differentiable functions vanishing outside a compact set. This correlation functional is characterized by a "slowly growing measure" satisfying certain requirements. Matheron (1971b) established this result directly for the case of random functions (as opposed to random generalized functions).

The spectral theory of ordinary processes with stationary increments of arbitrary order (case $n = 1$) was established by Yaglom and Pinsker (1953). A comprehensive presentation of the theory is also given by Yaglom (1987, Vol. I, Chapter 4).

The class of GCs coincides with the class of continuous and symmetric functions $K(h)$ on $\mathbb{R}^n$ satisfying

$$\sum_i \sum_j \lambda_i \lambda_j K(x_j - x_i) \geq 0 \qquad (4.18)$$

for any real allowable measure $\lambda \in \Lambda_k$. This condition ensures that $\mathrm{E}\,[Z(\lambda)^2] \geq 0$. A real function satisfying (4.18) is said to be k-*conditionally positive definite*. Such functions are characterized by a certain spectral representation, just as ordinary covariances in the Bochner–Khinchin theory. The formula looks awesome, but it provides insight into the physical significance of IRFs.

Theorem. A continuous and symmetric function $K(h)$ on $\mathbb{R}^n$ is a GC of an IRF–k if and only if it is of the form

$$K(h) = \int \frac{\cos(2\pi <u, h>) - 1_B(u)P_k(2\pi <u, h>)}{(4\pi^2|u|^2)^{k+1}}\chi(du) + Q(h) \qquad (4.19)$$

where $P_k(x) = 1 - x^2/2 + \cdots + (-1)^k x^{2k}/(2k)!$, $1_B(u)$ is the indicator function of an arbitrary neighborhood of $u = 0$, and $\chi(du)$ is a unique positive symmetric measure, with no atom at the origin and satisfying

$$\int \frac{\chi(du)}{(1 + 4\pi^2|u|^2)^{k+1}} < \infty \qquad (4.20)$$

$Q(h)$ is an arbitrary even polynomial of degree $\leq 2k$.

$$u = (u_1, \ldots, u_n) \qquad \text{denotes a frequency vector}$$
$$|u|^2 = u_1^2 + \cdots + u_n^2 \qquad \text{denotes its squared modulus}$$
$$<u,h> = u_1\, h_1 + \cdots + u_n\, h_n \qquad \text{denotes a scalar product}$$

To understand this formula, first note that the term $1_B(u)\, P_k(2\pi <u, h>)$ under the integral represents the expansion at the order k of $\cos (2\pi <u, h>)$ in the neighborhood B of $u = 0$. It is an even polynomial of degree $2k$ in the argument h and represents exactly what must be subtracted from the cosine to make the integral converge at $u = 0$ given (4.20) (the difference is of the order $|u|^{2k+2}$). The value of the integrand is not defined at $u = 0$, but since $\chi(du)$ has no atom there, the integral converges.

The neighborhood B is arbitrary: If B_1 and B_2 are two different neighborhoods of $u = 0$, then the difference between the associated $K_1(h)$ and $K_2(h)$ is an even polynomial of degree $2k$ in h and may thus be incorporated in the arbitrary $Q(h)$ polynomial.[3]

When $K(h)$ is differentiable $2k$ times, the term $1_B(u)$ in formula (4.19) can be dropped (i.e., replaced by 1). Thus for $k = 0$ this term is never needed and formula (4.19) is seen to coincide with the spectral representation (2.21) of $-\gamma(h)$.

Formula (4.19) associates to each $K(h)$ the spectral measure

$$F(du) = \chi(du)/(4\pi^2 |u|^2)^{k+1}$$

defined on the space $\mathbb{R}^n - \{0\}$. This measure is the same for all GCs in the equivalence class and thus appears as the fundamental information on the correlation structure of an IRF–k. The synthetic condition (4.20) on the measure $\chi(du)$ is equivalent to the following two conditions on $F(du)$:

$$\int_{|u| < \varepsilon} |u|^{2k+2} F(du) < \infty, \qquad \int_{|u| > \varepsilon} F(du) < \infty \qquad (4.21)$$

where $\varepsilon > 0$ is arbitrary. Unlike for ordinary covariances it is not required that the integral of the spectrum converge but only that near-zero frequency $\int_{|u| < \varepsilon} |u|^{2k+2} F(du)$ converge while $\int_{|u| > \varepsilon} F(du)$ must converge at infinity.

In the case of stationary processes, $F(du)$ is interpreted as the power in the frequency interval $(u, u + du)$ and the integral $\int F(du) = C(0)$ as the total power of the process. With IRF–k the integral $\int F(du)$ may become infinite because the first condition (4.21) allows $F(du)$ to tend rapidly to infinity as $u \to 0$: There may be an infinite power at low frequencies, a phenomenon referred to as an *infrared catastrophe*. Such effect is observed with Brownian and fractional Brownian motions in one dimension (Mandelbrot, 1967, 1982, p. 389)—the spectral measure is then proportional to $du/u^{1+\alpha}$ $(0 < \alpha < 2)$.

It is this high power at low frequencies that is responsible for the apparent long-term "trends." It also explains why the restriction to allowable measures is necessary. Indeed consider $\lambda \in \Lambda_k$ and denote by

$$\widetilde{\lambda}(u) = \int \exp(-2\pi i <u, x>)\lambda(dx)$$

---

[3] If the assumption E $[Z(\lambda)] = 0$ were relaxed, $Z(\lambda)$ remaining of course stationary, the polynomial $Q(h)$ would be of degree $\leq 2k + 2$ instead of $2k$, the coefficient of h^{2k+2} not depending on 1_B. For example, $\gamma(h) + a\,|h|^2/2$ is a valid variogram if E $[Z(x+h) - Z(x)] = <a, h>$. But we exclude such possibility by requiring $a = 0$.

the Fourier transform of λ (i is the unit pure imaginary number). From (4.14) and (4.19) we obtain

$$E[Z(\lambda)]^2 = \int \frac{|\widetilde{\lambda}(u)|^2}{(4\pi^2|u|^2)^{k+1}} \, \chi(du)$$

Since λ has a compact support, the function $\widetilde{\lambda}(u)$ is infinitely differentiable, and $\lambda \in \Lambda_k$ implies that $\widetilde{\lambda}(u)$ and its first k derivatives vanish at $u = 0$. This ensures the convergence of the above integral given the fact that $\chi(du)$ itself is integrable near 0 (first condition 4.21). In other words, the Fourier transform $\widetilde{\lambda}(u)$ neutralizes catastrophes.

In the case that $\int F(du)$ converges, the integral in (4.19) may be written as the difference between two convergent integrals, and since the second integral is a polynomial of degree $2k$ in h, $K(h)$ finally takes the form

$$K(h) = C(h) + Q(h)$$

where $C(h)$ is an ordinary covariance function. This is sufficient to assert that the IRF–k $Z(x)$ possesses a stationary representation $Y_{St}(x)$ whose covariance is $C(h)$. From Section 4.4.2 we then know that $Z(x)$ differs from $Y_{St}(x)$ by a random polynomial of degree k: This is the universal kriging model.

Another special case is when $\int \chi(du) < \infty$. Then $K(h)$ is differentiable $(2k+2)$ times, which means that $Z(x)$ is differentiable $(k+1)$ times in the sense that all its partial derivatives of order $k+1$,

$$\frac{\partial^{k+1} Z(x)}{\partial^{\ell_1} x_1 \ldots \partial^{\ell_n} x_n}, \qquad \ell_1 + \cdots + \ell_n = k+1$$

exist and are stationary. Differentiating (4.19) under the integral sign (which is valid) yields the equation

$$\Delta^{k+1} K(h) = (-1)^{k+1} C(h) \tag{4.22}$$

where Δ^{k+1} is the iterated Laplacian operator and $C(h)$ is the stationary covariance associated with the spectral measure $\chi(du)$.

Finally let us note the following decomposition of the integral in (4.19),

$$K(h) = \int\limits_{|u| \leq u_0} \frac{\cos(2\pi <u, h>) - P_k(2\pi <u, h>)}{(4\pi^2|u|^2)^{k+1}} \chi(du) + \int\limits_{|u| > u_0} \cos(2\pi <u, h>) \frac{\chi(du)}{(4\pi^2|u|^2)^{k+1}}$$

The integral over $|u| \leq u_0$ corresponds to an IRF-k with no high frequencies, an infinitely differentiable IRF-k, while the integral over $|u| > u_0$ is an ordinary stationary covariance function $C(h)$ — thanks to the second relation (4.21). We can write the phenomenological "equation":

continuous IRF$-k$ = infinitely differentiable IRF$-k$ + stationary random function

In principle, this dichotomy into low and high frequencies could be used as a definition of a "drift + residual" model. However, the arbitrariness of the cutoff frequency u_0 highlights once again the elusive character of the notion of drift. In practice it is also nearly impossible to estimate these two components. This decomposition is thus mainly of theoretical interest.

4.5.4 A Characterization of Isotropic GCs on $\mathbb{R}^n$

Micchelli's landmark paper (1986) provides a very useful characterization of isotropic conditionally positive definite functions of order k on $\mathbb{R}^n$ for all n. We recall that a continuous function F on $[0,\infty[$ is completely monotone if it possesses derivatives $F^{(p)}(t)$ of all orders and if $(-1)^p F^{(p)}(t) \geq 0$ for $p = 0, 1, 2, \ldots$, and $t > 0$. In our notations, Micchelli's Theorem 2.1 can be restated as follows:

Theorem. An isotropic function K is conditionally positive definite of order k on $\mathbb{R}^n$ for all n whenever K is continuous on $[0,\infty[$ and $(-1)^{k+1} K^{(k+1)}(\sqrt{r})$ is completely monotone on $]0,\infty[$.

One of the motivations for this theorem was to show that the function $(r^2 + r_0^2)^{1/2}$ is a variogram, thereby proving the conjecture that multiquadric interpolation is always solvable. Indeed consider $K(r) = -(r^2 + r_0^2)^{1/2}$ and $k = 0$. Then $K(\sqrt{r}) = -(r + r_0^2)^{1/2}$ and $(-1)^1 K^{(1)}(\sqrt{r}) = \frac{1}{2}(r + r_0^2)^{-1/2}$, which has derivatives of alternating signs. To further illustrate the theorem, consider now the power law variogram $\gamma(r) = r^\alpha$ and $k = 0$. We have $K(\sqrt{r}) = -r^{\alpha/2}$ and $(-1)^1 K^{(1)}(\sqrt{r}) = \frac{\alpha}{2} r^{\alpha/2-1}$, which is completely monotone if and only if $\alpha < 2$. As a last example, consider the function $K(r) = r^2 \log r$ and $k = 1$. Then

$$K(\sqrt{r}) = r \log\sqrt{r} \propto r \log r \qquad \text{and} \qquad (-1)^2 K^{(2)}(\sqrt{r}) = 1/r$$

The function $1/r$ is completely monotone and we can conclude that $r^2 \log r$ is a GC–1 on $\mathbb{R}^n$ for all n. It is obviously a GC of higher order but not a GC–0 since $-(\log r + 1)$ is not positive for all r.

Micchelli's theorem is much easier to use than the spectral characterization (4.19); but its application is restricted to isotropic covariances that are valid whatever the space dimensionality n, whereas (4.19) is completely general.

4.5.5 Majorization of Generalized Covariances

From the spectral formula (4.19) and the majorization $|\cos x - P_k(x)| \leq x^{2k+2}/(2k+2)!$ it is seen readily that a GC–k must satisfy the inequality

$$|K(0) - K(h)| \leq a + b|h|^{2k+2} \qquad \forall h \in \mathbb{R}^n \qquad (4.23)$$

for some positive constants a and b. Likewise an IRF–k is differentiable (once) if and only if its GCs satisfy an inequality of the form ($a', b' \geq 0$)

$$|K(0) - K(h)| \leq a'|h|^2 + b'|h|^{2k+2} \qquad \forall h \in \mathbb{R}^n$$

The following two results are proved in Matheron's 1973 paper[4]:

1. $$\lim_{|h| \to \infty} K(h)/|h|^{2k+2} = 0 \qquad (4.24)$$

2. An IRF–k is the restriction to Λ_k of a stationary random function if and only if one of its GCs is bounded on $\mathbb{R}^n$. When $k = 0$, this means that the variogram must be bounded.

4.5.6   The Power Law Class

The $|h|^\alpha$ Model

We know that

$$\gamma(h) = |h|^\alpha$$

is a variogram, that is, $-\gamma$ is a GC–0, whenever $0 < \alpha < 2$. Likewise, the function

$$K(h) = (-1)^{1+\lfloor \alpha/2 \rfloor} |h|^\alpha \qquad (4.25)$$

is a GC of order k on $\mathbb{R}^n$ for any n whenever $0 < \alpha < 2k + 2$, where α is not an even integer. This can be verified using Micchelli's theorem.

The sign of $K(h)$ alternates with the value of α: It is negative for $0 < \alpha < 2$, positive for $2 < \alpha < 4$, negative for $4 < \alpha < 6$, and so on. When α is an even integer, $K(h)$ is an even degree polynomial and thus equivalent (in the same GC class as) the identically nil covariance $K(h) \equiv 0$. Note that the class of GCs expands as the order k increases.

For completeness, the spectral measure associated with $K(h) = \Gamma(-\alpha/2)|h|^\alpha$ in (4.19) is

$$F(du) = \chi(du)/(4\pi^2|u|^2)^{k+1} = \pi^{-\alpha-n/2}\, \Gamma\!\left(\frac{\alpha+n}{2}\right)|u|^{-\alpha-n}$$

where $\Gamma(\cdot)$ is the gamma function. Note that $\Gamma(-\alpha/2)$ has the same sign as $(-1)^{1+\lfloor \alpha/2 \rfloor}$.

The $|h|^{2k} \log|h|$ Model

Consider

$$K(h) = (-1)^{k+1} \lim_{\varepsilon \to 0} \frac{1}{\varepsilon}\left(|h|^{2k+\varepsilon} - |h|^{2k}\right)$$

$(-1)^{k+1} |h|^{2k+\varepsilon}$ is a GC–k on $\mathbb{R}^n$ for any n if $0 < \varepsilon < 2$, and so is $(-1)^{k+1}$ $(|h|^{2k+\varepsilon} - |h|^{2k})$ since $|h|^{2k}$ is an even polynomial of degree $2k$. The limit of a GC–k is still a GC–k and therefore

$$K(h) = (-1)^{k+1}|h|^{2k} \log|h| \qquad (4.26)$$

[4] All inequalities in this section assume an IRF–k with no intrinsic drift ($\mathrm{E}[Z(\lambda)] = 0$ for $\lambda \in \Lambda_k$).

is a GC of order k on $\mathbb{R}^n$ for any n. It is not a GC of any lower order because it would violate the condition (4.24).

A remarkable member of the family (4.26) is the GC–1 $K(h) = |h|^2 \log|h|$, which, as we will see, is associated with biharmonic splines in 2D.

4.5.7 Polynomial Covariance Models

Interesting elements of the power law class are the terms with odd integer exponent $\alpha = 2p + 1$. Then $(-1)^{p+1} |h|^{2p+1}$ is a GC–k, provided that $p \le k$. More generally, the function

$$K(h) = \sum_{p=0}^{k} (-1)^{p+1} b_p |h|^{2p+1} \tag{4.27}$$

is a GC of order k under conditions on the coefficients b_p, which are obviously satisfied if $b_p \ge 0 \ \forall p$.

Covariances of this form are called polynomial GCs. This is a slight misnomer because the functions are polynomials with respect to the modulus $|h| = (h_1^2 + \cdots + h_n^2)^{1/2}$ of the vector h and not with respect to the components $h_1, \ldots, h_n$ of this vector. Since they depend only on the modulus of h, polynomial GCs are *isotropic* models. (However, the techniques presented to deal with variogram anisotropies can be used here as well.)

The physical meaning of the exponent α is the following: The higher α, the more regular the model at a local scale and the more fluctuating at a global scale. A phenomenon with a linear covariance $-b_0|h|$ is continuous but not differentiable, with $b_1|h|^3$ it is differentiable once but also has very large fluctuations, and with $-b_2|h|^5$ it is differentiable twice but fluctuates wildly at large distances. Figure 4.4 illustrates the aspects of phenomena associated with the pure terms $-|h|, |h|^3, -|h|^5$. The last one represents a slowly varying twice differentiable component that in the universal kriging terminology would be called a *drift*.

The exact conditions to be placed on the coefficients b_p in (4.27) for $K(h)$ to be a valid GC–k on $\mathbb{R}^n$ are obtained by requiring that the measure $\chi(du)$ in the general spectral representation (4.19) be positive, leading to

$$\sum_{p=0}^{k} \frac{(2p+1)!}{p!} \Gamma\left[p + \tfrac{1}{2}(n+1)\right] b_p x^{k-p} \ge 0 \qquad \text{for} \quad x \ge 0$$

For $k = 0, 1$, and 2 and according to the space dimensionality n, the conditions are

$$
\begin{array}{llll}
k = 0: & b_0 \ge 0, & b_1 = 0, & b_2 = 0, \\
k = 1: & b_0 \ge 0, & b_1 \ge 0, & b_2 = 0, \\
k = 2: & b_0 \ge 0, & b_2 \ge 0, & b_1 \ge -\sqrt{\dfrac{20}{3}\left(1 + \dfrac{2}{n+1}\right)}\sqrt{b_0 b_2}
\end{array}
$$

TABLE 4.1   Polynomial GC Models for $k \leq 2$ with Nugget Effect Added

Filtered Polynomial (Drift)	k	Polynomial Generalized Covariance Model + Nugget Effect
Constant	0	$K(h) = C_0 \, \delta(h) - b_0 \, |h|$
Linear	1	$K(h) = C_0 \, \delta(h) - b_0 \, |h| + b_1 \, |h|^3$
Quadratic	2	$K(h) = C_0 \, \delta(h) - b_0 \, |h| + b_1 \, |h|^3 - b_2 \, |h|^5$

Constraints

In $\mathbb{R}^1 : C_0 \geq 0, \ b_0 \geq 0, \ b_2 \geq 0, \ b_1 \geq -2\sqrt{10/3}\,\sqrt{b_0 b_2}$

In $\mathbb{R}^2 : C_0 \geq 0, \ b_0 \geq 0, \ b_2 \geq 0, \ b_1 \geq -(10/3)\,\sqrt{b_0 b_2}$

In $\mathbb{R}^3 : C_0 \geq 0, \ b_0 \geq 0, \ b_2 \geq 0, \ b_1 \geq -\sqrt{10}\,\sqrt{b_0 b_2}$

Note that for $k = 2$ the condition on b_1 depends on the space dimensionality n and becomes more severe as n increases (lower bound $= -3.651, -3.333, -3.162$ for $n = 1, 2, 3$, respectively).

Table 4.1 summarizes the polynomial models for $k = 0, 1$, and 2; the nugget effect has been added to take into account microstructures or measurement errors. The main advantage of these models for applications is that they depend linearly on their parameters, which facilitates their statistical inference.

It is possible to augment (4.27) with logarithmic terms of the form (4.26). In practice this is limited to the $|h|^2 \log |h|$ term, and then it is preferable to drop the $-|h|^5$ covariance so as to keep the number of parameters as low as possible for better statistical inference. (The $-|h|^5$ covariance leads to poorly conditioned kriging matrices.) The resultant covariance with terms arranged by increasing regularity is

$$K(h) = C_0 \, \delta(h) - b_0 \, |h| + b_S \, |h|^2 \log|h| + b_1 |h|^3$$

$K(h)$ is a valid GC of order $k = 1$ —and of course of higher order as well—if and only if the coefficients satisfy the following inequalities (from Dubrule, 1981):

$$\text{In } \mathbb{R}^1 \quad C_0 \geq 0, \quad b_0 \geq 0, \quad b_1 \geq 0, \quad b_S \geq -\frac{\sqrt{24}}{\pi} \sqrt{b_0 b_1},$$

$$\text{In } \mathbb{R}^2 \quad C_0 \geq 0, \quad b_0 \geq 0, \quad b_1 \geq 0, \quad b_S \geq -\frac{3}{2} \sqrt{b_0 b_1},$$

$$\text{In } \mathbb{R}^3 \quad C_0 \geq 0, \quad b_0 \geq 0, \quad b_1 \geq 0, \quad b_S \geq -\frac{8}{\pi\sqrt{3}} \sqrt{b_0 b_1}$$

4.5.8   Construction of an IRF–k with a Polynomial GC in $\mathbb{R}^1$

Successive integrations of a process $W_0(x)$ with GC $K_0(h) = -|h|$ (W for Wiener–Lévy process) lead to a process

$$W_k(x) = \int_0^x \frac{(x-t)^{k-1}}{(k-1)!} W_0(t)dt$$

that is an IRF–k, with the polynomial GC:

$$K_k(h) = (-1)^{k+1}|h|^{2k+1}/(2k+1)!$$

Now consider a process of the form

$$Y(x) = \sum_{p=0}^{k} c_p W_p(x) \tag{4.28}$$

where all $W_p(x)$ are integrals of the *same process* $W_0(x)$. Then $Y(x)$ can be expressed as a sum of derivatives of $W_k(x)$:

$$Y(x) = \sum_{p=0}^{k} c_p D^{k-p} W_k(x)$$

where D^p denotes derivation of order p, from which the GC of $Y(x)$ is found to be

$$K(h) = \left(\sum_{p=0}^{k} c_p D^{k-p} \right) \left(\sum_{q=0}^{k} (-1)^{k-q} c_q D^{k-q} \right) (-1)^{k+1}|h|^{2k+1}/(2k+1)!$$

It is a polynomial GC. Conversely, any IRF–k in 1D with a polynomial GC has a representation of the form (4.28), which means that for any given valid set of covariance coefficients b_p defined in (4.27), we can always find the matching set of coefficients c_p (Matheron, 1973a). For $k=2$ the results are

$$c_0 = \sqrt{b_0}, \qquad c_2 = \sqrt{120b_2}, \qquad c_1 = \sqrt{6b_1 + 2c_0c_2}$$

This provides a simple algorithm for simulating in 1D any IRF–k with a given polynomial GC: It suffices to simulate an IRF–0 with a linear variogram and integrate it k times. The turning bands algorithm can then take over to generate a simulation with polynomial GC on $\mathbb{R}^n$.

4.6   ESTIMATION IN THE IRF MODEL

4.6.1   Intrinsic Kriging

The rule of the game here is to derive the kriging equations using allowable linear combinations, which are the only ones to have computable variances. This is done below in a straightforward manner. A geometric derivation of the equations in terms of projections in Hilbert spaces can be found in Matheron (1981c).

In our model the variable under study $Z(x)$ is regarded as a realization of an IRF–k of known order k and known GC $K(h)$; the inference problem will be considered later. To keep things simple, let us suppose that we want to estimate the value $Z_0 = Z(x_0)$ at a point x_0 using a linear combination of observed data $Z(x_\alpha)$. By placing appropriate conditions on the weights, we can ensure that the *estimation error* $Z^* - Z_0$ is an allowable linear combination of order k. Then its variance can be expressed in terms of $K(h)$, and the kriging equations are derived exactly as in the universal kriging approach.

Specifically, if

$$\sum_\alpha \lambda_\alpha x_\alpha^\ell - x_0^\ell = 0, \qquad |\ell| = 0, \ldots, k \tag{4.29}$$

then $Z^* - Z_0 = Z\left(\sum_\alpha \lambda_\alpha \delta_{x_\alpha} - \delta_{x_0}\right)$ is an ALC–k, and by formula (4.14) its variance is

$$\mathrm{E}(Z_0^* - Z_0)^2 = \sum_\alpha \sum_\beta \lambda_\alpha \lambda_\beta K(x_\beta - x_\alpha) - 2 \sum_\alpha \lambda_\alpha K(x_0 - x_\alpha) + K(0)$$

Minimizing this subject to (4.29) leads to the system

Intrinsic Kriging System

$$\begin{cases} \sum_\beta \lambda_\beta K(x_\beta - x_\alpha) + \sum_\ell \mu_\ell x_\alpha^\ell = K(x_0 - x_\alpha), & \alpha = 1, \ldots, N, \\ \sum_\alpha \lambda_\alpha x_\alpha^\ell = x_0^\ell, & |\ell| = 0, \ldots, k \end{cases} \tag{4.30}$$

The kriging variance is as usual.

Intrinsic Kriging Variance

$$\sigma_K^2 = K(0) - \sum_\alpha \lambda_\alpha K(x_0 - x_\alpha) - \sum_\ell \mu_\ell x_0^\ell$$

This system, sometimes named "intrinsic kriging," is exactly the same as in the universal kriging model (3.21) except that $K(h)$ is substituted for $\sigma(h)$. The kriging estimator and the kriging variance only depend on the GC class and not on the particular version used—namely not on the particular representation. They are intrinsic properties.

Just as for UK, there is a dual form of intrinsic kriging and the interpolant is given by exactly the same equations as (3.46) and (3.47) with K in place of σ and the f^ℓ as monomials.

Note that in this approach the conditions (4.29) are not introduced as unbiasedness conditions but as constraints to make the estimation error an allowable measure. For these conditions to be realizable, it is necessary that the matrix $\mathbf{X} = [x_\alpha^\ell]$ has a rank equal to its number of columns (full rank)—that is, the same conditions as in UK.

The conditions (4.29) ensure *numerical invariance* in the sense that an arbitrary polynomial of degree k may be added to $Z(x)$ without changing *the value* of $Z^* - Z_0$ (which is stronger than not changing the expected value). If moving neighborhoods are used for the estimation, then (4.29) may be interpreted physically as *filtering conditions*. They eliminate the effect of a local "drift," if we mean by this a smooth component *locally* approximable by a polynomial of degree k.

Properties of the Intrinsic Kriging Matrix

The kriging matrix of (4.30) is of the form

$$\mathbf{A} = \begin{bmatrix} \mathbf{K} & \mathbf{X} \\ \mathbf{X}' & \mathbf{0} \end{bmatrix}, \qquad \text{where} \quad \mathbf{K} = [K_{\alpha\beta}] \quad \text{and} \quad \mathbf{X} = [x_\alpha^\ell]$$

If the generalized covariance $K(h)$ is strictly conditionally positive definite of order k, all data points are distinct, and the matrix $\mathbf{X}$ is of full rank, the kriging matrix $\mathbf{A}$ is invertible. To show this suppose $\mathbf{A}\mathbf{w} = \mathbf{0}$ for some vector $\mathbf{w}' = (\boldsymbol{\lambda}', \boldsymbol{\mu}')'$, then

$$\mathbf{K}\boldsymbol{\lambda} + \mathbf{X}\boldsymbol{\mu} = \mathbf{0} \qquad \text{and} \qquad \mathbf{X}'\boldsymbol{\lambda} = \mathbf{0}$$

Premultiplying the first equation by $\boldsymbol{\lambda}'$ gives $\boldsymbol{\lambda}'\mathbf{K}\boldsymbol{\lambda} = 0$, which, by the positiveness of $\mathbf{K}$, implies $\boldsymbol{\lambda} = 0$. The first equation then reduces to $\mathbf{X}\boldsymbol{\mu} = \mathbf{0}$ and since $\mathbf{X}$ is of full rank, necessarily $\boldsymbol{\mu} = \mathbf{0}$. Therefore $\mathbf{A}$ is nonsingular.

Note that the matrix $\mathbf{K}$ itself may be singular when $k > 0$ (for $k = -1$ and $k = 0$ it is always invertible). A classic example is given by the GC $K(r) = r^2 \log r$ with data points $x_2, \ldots, x_N$ located on the unit circle centered at x_1. Then the first row and column of the matrix $\mathbf{K}$ consist of zeroes, and $\mathbf{K}$ is singular. The presence of $\mathbf{X}$ in the $\mathbf{A}$ matrix—that is, the side conditions (4.29)—solves the problem. Note that in IRF-k theory these conditions derive from the essence of the random function model, whereas for radial basis functions presented in Section 3.4.9 they appear as a trick. This result guarantees that interpolation by radial basis functions is always possible if the basis function is a generalized covariance and the proper side conditions are imposed.

It is interesting to go a bit further and investigate the properties of $\mathbf{A}^{-1}$. We define this inverse in partitioned form using block matrices $\mathbf{U}, \mathbf{V}, \mathbf{W}$ such that

$$\begin{bmatrix} \mathbf{U} & \mathbf{V} \\ \mathbf{V}' & \mathbf{W} \end{bmatrix} \begin{bmatrix} \mathbf{K} & \mathbf{X} \\ \mathbf{X}' & \mathbf{0} \end{bmatrix} = \begin{bmatrix} \mathbf{U}\mathbf{K} + \mathbf{V}\mathbf{X}' & \mathbf{U}\mathbf{X} \\ \mathbf{V}'\mathbf{K} + \mathbf{W}\mathbf{X}' & \mathbf{V}'\mathbf{X} \end{bmatrix} = \begin{bmatrix} \mathbf{I}_N & \mathbf{0} \\ \mathbf{0} & \mathbf{I}_{k_n} \end{bmatrix} = \mathbf{I}_{N+k_n} \qquad (4.31)$$

Where $k_n = \binom{k+n}{k}$ is the number of monomials of degree less than or equal to k in $\mathbb{R}^n$.

As $\mathbf{A}$ is symmetric, so is its inverse and $\mathbf{U}$ and $\mathbf{W}$ are symmetric matrices. $\mathbf{U}$ is obviously singular (since $\mathbf{U}\mathbf{X} = \mathbf{0}$) and, less obviously, nonnegative definite.

Indeed, postmultiplying the top left equation by $\mathbf{U}$ and taking into account $\mathbf{UX} = \mathbf{0}$ leads to

$$\mathbf{U}\mathbf{K}\mathbf{U} = \mathbf{U} \tag{4.32}$$

Now consider an arbitrary vector $\mathbf{x}$ and the weights vector $\boldsymbol{\lambda} = \mathbf{Ux}$. It defines an ALC since $\boldsymbol{\lambda}'\mathbf{X} = 0$. Therefore $\boldsymbol{\lambda}'\mathbf{K}\boldsymbol{\lambda} \geq 0$ (equals 0 only if $\boldsymbol{\lambda} = \mathbf{0}$) and from (4.32)

$$\boldsymbol{\lambda}'\mathbf{K}\boldsymbol{\lambda} = \mathbf{x}'\mathbf{U}\mathbf{K}\mathbf{U}\mathbf{x} = \mathbf{x}'\mathbf{U}\mathbf{x} \geq 0$$

Multiplying $\mathbf{A}^{-1}$ by the vector $[\mathbf{z}'\ \mathbf{0}]'$ the solutions $\mathbf{b}$ and $\mathbf{c}$ of the dual kriging equations are simply $\mathbf{b} = \mathbf{Uz}$ and $\mathbf{c} = \mathbf{V}'\mathbf{z}$. In the special case $\mathbf{K} = \boldsymbol{\Sigma}$, the vector $\mathbf{c}$ holds the optimal drift coefficients estimates and $-\mathbf{W}$ is its covariance matrix. There is no such interpretation for an arbitrary GC. The matrix $\mathbf{U}$ is also useful for kriging under inequality constraints (Section 3.9.2).

4.6.2 Locally Equivalent Stationary Covariances

From a computational point of view, it is interesting to replace the $\mathbf{K} = [K_{\alpha\beta}]$ matrix, which is only k-conditionally positive definite, by a genuinely positive definite matrix $\mathbf{C} = [C_{\alpha\beta}]$.

In principle, this is always possible by picking an internal representation $Y(x) = Z(\varepsilon_x)$ and using the nonstationary covariance $\mathrm{Cov}(Y(x), Y(y))$ in the kriging system (4.30) in lieu of $K(y - x)$. Since $\varepsilon_x \in \Lambda_k$, this covariance can be computed from the GC $K(h)$ using (4.16).

This method, however, is cumbersome and computationally slow. It would be much nicer if the IRF–k had a stationary representation: The stationary covariance $C(h)$ of this representation could then be used as a version of the GC $K(h)$. But in Section 4.5.5 we have seen that only an IRF–k with a bounded GC can have a stationary representation. This excludes the important case of IRFs with polynomial covariances.

Fortunately, we can be less demanding because in fact only *local stationarity* is needed. For any practical purpose we work in a restricted domain D—the studied area, or a moving neighborhood—and it is enough if the IRF–k $Z(x)$ has a representation $Y(x)$ that coincides within D with an SRF $Y_1(x)$ (being understood that $Y(x)$ may differ from $Y_1(x)$ outside the working domain D). This leads to the following definition:

Definition. An IRF–k is locally stationary over a bounded domain D if it has a representation $Y(x)$ that coincides on D with a stationary random function $Y_1(x)$.

The relationship between $Y_1(x)$, $Y(x)$, and $Z(x)$ is the following:

$$\begin{aligned} Y_1(x) &= Y(x) & \forall x \in D \\ Y(\lambda) &= Z(\lambda) & \forall \lambda \in \Lambda_k \end{aligned}$$

Hence

$$\sum_i \lambda_i\, Y_1(x_i) = \sum_i \lambda_i\, Z(x_i)$$

for any λ in the space $\Lambda_k(D)$ of allowable measures with support in D. The ordinary covariance $C(h)$ of $Y_1(x)$ is equivalent to the GC $K(h)$ over D in the sense that for any $\lambda \in \Lambda_k(D)$ we have

$$\sum_i \sum_j \lambda_i \lambda_j C(x_j - x_i) = \sum_i \sum_j \lambda_i \lambda_j K(x_j - x_i)$$

$C(h)$ is called a *locally equivalent stationary covariance*.

Not every IRF–k is locally stationary,[5] but the IRF–k's of practical interest are. In particular, any IRF–k with a polynomial GC is locally stationary on any bounded open set (Matheron, 1973a).

A locally equivalent stationary covariance $C(h)$ differs from its parent GC–k $K(h)$ by an even polynomial of degree $2k$ that depends on the dimensions of the domain D (it must since the IRF is not globally stationary)

$$C(h) = K(h) + Q(h)$$

Even for a fixed D, the covariance $C(h)$ is not unique. However, Matheron (1974b) shows that an IRF–k has at most one stationary *internal* representation over D (i.e., involving only values within D), but it is usually difficult to find!

Examples

1. Consider a Brownian motion without drift as in Example 1 of Section 4.5.2 with $\alpha = 1$, and restrict our attention to the interval $[-R, +R]$. If instead of $X(t) - X(0)$ we consider the representation

$$Y(t) = X(t) - \frac{1}{2}[X(-R) + X(R)]$$

(which satisfies $Y(-R) = - Y(R)$), we obtain the covariance

$$\sigma(t, t') = R - |t' - t|$$

$C(h) = R - |h|$ is therefore locally equivalent to the GC–0 $K(h) = -|h|$ for $|h| \leq 2R$. (It corresponds to the stationary internal representation on $[-R, +R]$.) An arbitrary constant $A > 0$ may be added so that the class of locally stationary covariances is a one-parameter family.

2. Matheron (1974b) shows that an IRF–0 with variogram $\gamma(h) = b|h|^\alpha$, where $0 < \alpha < 2$ (e.g., fractional Brownian motion) possesses locally stationary representations on $[-R, R]$ with covariances of the form

$$C(h) = b(A - |h|^\alpha) \quad |h| \leq 2R, \quad A \geq A_\alpha = \frac{R^\alpha}{\sqrt{\pi}} \Gamma\left(\frac{1+\alpha}{2}\right)\Gamma\left(1 - \frac{\alpha}{2}\right) \tag{4.33}$$

In the special case $\alpha = 1$ we recover the covariance $R - |h|$. By application of turning bands, (4.33) is also a locally stationary covariance on $\mathbb{R}^n$, provided that

[5] An analytic IRF–0 with an unbounded variogram, such as a Brownian motion convolved with a Gaussian density, cannot be locally stationary because it would then be stationary on the whole space, which is impossible since its variogram is not bounded. Another example is the integral of a zero-mean Gaussian process with a Gaussian covariance (Gneiting et al., 2001).

$$A \geq \frac{A_\alpha}{B_{n\alpha}} \quad \text{where} \quad B_{n\alpha} = \frac{\Gamma\left(\frac{n}{2}\right) \Gamma\left(\frac{1+\alpha}{2}\right)}{\Gamma\left(\frac{1}{2}\right) \Gamma\left(\frac{n+\alpha}{2}\right)} \tag{4.34}$$

For $k > 0$, Stein (2001) proves that the generalized power law GC–k (4.25) or (4.26) on $\mathbb{R}^n$ admits a locally equivalent stationary covariance on any bounded domain if $\alpha \leq 2k + \nu_0$ with $\nu_0 \approx 1.6915$.

3. If $X(t)$ is a Brownian motion with a linear drift, it is advantageous to consider the representation

$$Y(t) = X(t) - \frac{1}{2R} \int_{-R}^{+R} X(s) ds - \frac{1}{2R} [X(R) - X(-R)] t$$

By construction, $E[Y(t)] = 0$ and the covariance is found to be

$$\sigma(t, t') = \frac{R}{3} - |t' - t| + \frac{(t' - t)^2}{2R}$$

This is in fact the covariance of the stationary internal representation on $[-R, +R]$. The complete class of locally equivalent stationary covariances (for the IRF–1) is of the form (Matheron, 1974b)

$$C(h) = A - |h| + B h^2 \quad \text{with} \quad B \leq 1/(2R) \quad \text{and} \quad A \geq R(1 - 2BR) + \frac{1}{3} R(2BR)^2$$

This class contains the covariance $(1/2R) (R - |h|)^2$ (*quadratic model*, valid in 3D).

4. An example of a locally stationary representation in 2D is obtained using the construction based on Poisson lines presented in Section 7.6.5: The convex set D is intersected by a network of Poisson lines. Each line cuts D in two parts; random values are assigned to each part and cumulated. The resulting RF has the stationary covariance of the form

$$\sigma(x, y) = A \left[\frac{1}{2} |\partial D| - |y - x| \right]$$

where $|\partial D|$ is the perimeter of D and A is a positive constant.

5. Table 4.2 gives a particular family of covariances locally equivalent to polynomial GCs for $k \leq 2$. C_2 and C_3 are deduced by the turning bands method from C_1 which satisfies $C_1(2R) = -C_1(0)$ when $A = 0$. The formulas are written in terms of the modulus $r = |h|$ and hold for $r \leq 2R$. Adding a strictly positive constant A makes the covariances strictly positive definite. Note that for ALC–k the results of variance calculations do not depend on R or A because R and A are only involved in even powers of r.

6. Table 4.3 gives the locally equivalent stationary covariances for the GC model $K(h) = |h|^2 \log |h|$. Note that this is a valid model in $\mathbb{R}^n$ for any n although the equivalence with splines is only true in $\mathbb{R}^2$. $C_2(r)$ is graphed in Figure 4.9. It becomes negative at $r = R$, indicating a "hole effect" which is not surprising when one thinks of the physics of a flexed plate. As long as $r \leq 2R$, the function $C_2(r)$ may be used just as an ordinary covariance even without restrictions on the weights, but it is equivalent to the spline GC only if the three conditions ensuring that $k = 1$ are imposed. For $r \geq 2R$ the function $C_2(r)$ loses its positive-definiteness property.

TABLE 4.2 Stationary Covariances Locally Equivalent to the GC $K(h) = -b_0|h| + b_1|h|^3 - b_2|h|^5$ for $r = |h| \le 2R$

In $\mathbb{R}^1$: $C_1(r) = A - b_0(r - R) + b_1(r^3 - 3r^2R + 2R^3)$
$\qquad\qquad - b_2(r^5 - 5r^4R + 20r^2R^3 - 16R^5)$

In $\mathbb{R}^2$: $C_2(r) = A - b_0(r - \dfrac{\pi}{2}R) + b_1(r^3 - \dfrac{9\pi}{8}r^2R + \dfrac{3\pi}{2}R^3)$
$\qquad\qquad - b_2(r^5 - \dfrac{225\pi}{128}r^4R + \dfrac{75\pi}{8}r^2R^3 - 15\pi R^5)$

In $\mathbb{R}^3$: $C_3(r) = A - b_0(r - 2R) + b_1(r^3 - 4r^2R + 8R^3)$
$\qquad\qquad - b_2(r^5 - 6r^4R + 40r^2R^3 - 96R^5)$

TABLE 4.3 Stationary Covariances Locally Equivalent to the GC $K(h) = |h|^2 \log|h|$ for $r = |h| \le 2R$

In $\mathbb{R}^1$: $C_1(r) = \dfrac{1}{2}R^2 - \left(\dfrac{3}{2} - \log 2\right)r^2 + r^2 \log(r/R)$

In $\mathbb{R}^2$: $C_2(r) = R^2 - r^2 + r^2 \log(r/R)$

In $\mathbb{R}^3$: $C_3(r) = \dfrac{3}{2}R^2 - \left(\dfrac{11}{6} - \log 2\right)r^2 + r^2 \log(r/R)$

Source: Matheron (1981a).

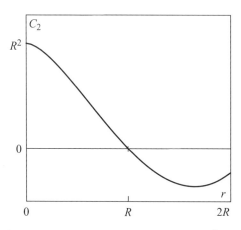

FIGURE 4.9 Stationary covariance locally equivalent to $K(r) = r^2 \log r$ in 2D for $r \le 2R$.

The r^2 and r^4 terms in the covariances are transparent to the kriging equations: In theory, they have no influence on either the kriging weights or the kriging variance. Numerically, however, these terms will dominate the other terms if R is large. These numerical considerations suggest to choose for R the smallest possible value. R is the radius of a circle (or a sphere) large enough to include all points involved in the estimation problem considered— that is, both the observations and the points to be estimated. For example, R may be the radius of the largest moving neighborhood or of the current neighborhood (hence the maximum distance involved in covariance calculations is $2R$).

4.6.3   Status of the Drift in IRF–k Theory

Since the concept of IRF–k encompasses a class of random functions (the representations) that are defined up to a polynomial, one could argue that this polynomial implicitly captures the concept of drift—whatever that means. But the drift itself can only be defined in an exceptional case: When the class of representations of the abstract IRF–k $\tilde{Z}$ contains a special random function $Y(x)$ that is distinguishable from all others by some remarkable property and can thus serve as an absolute reference. Then any other representation $Z(x)$ would differ from $Y(x)$ by a polynomial drift.

A case when this circumstance occurs is that of an IRF–k with a stationary representation $Y_{St}(x)$. Any other representation is then of the form

$$Z(x) = Y_{St}(x) + \sum_\ell A_\ell x^\ell$$

This is nothing but the universal kriging model (drift + stationary residual) with random coefficients. This case, however, is an exceptional one requiring that one of the GCs be bounded. In general, an IRF–k has no stationary representation and therefore no uniquely definable drift (IRF–k with polynomial GCs never have a stationary representation). Far from being an objection against the IRF model, this circumstance is consistent with our initial goal, which was precisely to deal with phenomena for which the dichotomy into drift and residual is meaningless, or at least arbitrary. The important point is that spatial estimation (kriging) remains possible within this model.

A Fallacy

At the early times of the development of the IRF theory, it was believed that locally equivalent stationary covariances could be used to perform drift estimation. But practical examples soon showed that it did not work. The drift estimates obtained did not reflect at all the general behavior of the phenomenon. Furthermore, these estimates differed with the radius R of the domain. The reason is now well understood. Locally equivalent stationary covariances are not unique even over a fixed domain D. As a consequence, the estimated drift is arbitrary and depends on the particular choice of the covariance. To some extent the terminology "locally equivalent stationary representation" is misleading, because one expects the equivalent $Y(x)$ to really look stationary over D. But that cannot be the case. If $Y_1(x)$ and $Y_2(x)$ are two distinct locally equivalent stationary representations, their difference

$$Y_1(x) - Y_2(x) = \sum_\ell A_\ell x^\ell$$

is necessarily a polynomial of degree k, and at least one of these functions has locally the aspect of a drift.

The locally equivalent stationary $Y(x)$ is no more than a mathematical construct, useful to derive covariance models but unable to give a meaning to the drift.

Estimation of a Regional Trend

For interpretation purposes it may be useful to estimate a spatial trend. Since our information on the phenomenon is most of the time very fragmentary, we must carefully distinguish two problems: that of the conceptual definition of the trend and that of the estimation of the trend so defined.

In light of the discussion on the objectivity of probabilistic parameters (see Section 1.5), we are led to ask the interpreter the following question: "*Assuming that instead of knowing the variable at a limited number of points you knew the values $z(x)$ everywhere in the domain of interest, which algorithm would you use to calculate your trend function?*" If the notion of trend as portrayed in the interpreter's mind has an objective meaning, such an algorithm must exist.

Typically the interpreter will respond with a conventional definition of the trend that translates into a linear algorithm—for example, fit a least squares polynomial, or apply a low-pass filter. We will then be able, on the basis of the available data, to estimate *that* specific trend using a kriging estimator.

Suppose, for example, that the user defines the trend as the polynomial of degree k that best fits $z(x)$ in the least squares sense over a specified domain D. The coefficients β_ℓ of that polynomial are linear functions of $\{z(x) : x \in D\}$ and can therefore be estimated from the available data $z_1, \ldots, z_N$. By the linearity of kriging, the estimators are simply the same linear functions of the kriging estimate $z^*(x)$. The procedure is equivalent to minimizing the integral

$$\int_D \left(z^*(x) - \sum_\ell \beta_\ell\, x^\ell \right)^2 dx$$

in which the true $z(x)$ are replaced by their kriging estimates $z^*(x)$.

It must be emphasized that this technique differs from a direct polynomial fit to the observations. We want the fit to be representative of the domain D and not only of the data points. The closest we can come to this objective is by reconstructing the values of $z(x)$ as accurately as possible over the whole domain D, or a discretized form of it, and then fit a polynomial. In the process we make use of the structure information contained in the GC, which the plain least squares fit at the observations does not.

4.6.4 Kriging and Splines

Interpolating Splines

Consider in 1D the IRF–1 $Z(x)$ with GC $K(h) = |h|^3$, and let us examine the behavior of the dual kriging interpolant $z^*(x)$ based on N consecutive points

$x_1, \ldots, x_N$. From (3.46) and (3.47) we obtain

$$\begin{cases} z^*(x) = \sum_{\alpha=1}^{N} b_\alpha |x - x_\alpha|^3 + c_0 + c_1 x, \\ z^*(x_\alpha) = z_\alpha, \qquad \alpha = 1, \ldots, N, \\ \sum_{\alpha=1}^{N} b_\alpha = 0 \quad \text{and} \quad \sum_{\alpha=1}^{N} b_\alpha x_\alpha = 0 \end{cases}$$

$z^*(x)$ is a cubic polynomial within each interval $[x_\alpha, x_{\alpha+1}]$, assumes the values z_α and $z_{\alpha+1}$ at its boundaries, and is continuous in the first and second derivatives across boundaries, including at x_1 and x_N where the second derivative is zero. In the intervals $]-\infty, x_1]$ and $[x_N, +\infty[$, $z^*(x)$ is linear as can be seen by expansion of the polynomial $(x - x_\alpha)^3$ and the application of the two constraints on b_α. Therefore $z^*(x)$ coincides with the cubic interpolating spline going through $z_1, \ldots, z_N$, which is the function $f(x)$ minimizing

$$J(f) = \int_{-\infty}^{+\infty} [f''(x)]^2 dx$$

subject to the N constraints $f(x_\alpha) = z_\alpha$. Note that the linear behavior of $f(x)$ on the outer intervals cancels $f''(x)$ and ensures that the integral is finite.

A physical interpretation of the problem is to consider a metal strip clamped at the N points $(x_1, z_1), \ldots, (x_N, z_N)$. The strip will adopt the shape that minimizes its flexing energy, which is proportional to the square of the curvature of the strip, a quantity approximated by $J(f)$.

The analog of this problem in 2D is to minimize the flexing energy of a thin metal plate, which is proportional to

$$J(f) = \iint \left[(\partial^2 f / \partial x^2)^2 + 2(\partial^2 f / \partial x \partial y)^2 + (\partial^2 f / \partial y^2)^2 \right] dx\, dy$$

under the N constraints $f(x_\alpha, y_\alpha) = z_\alpha$ [e.g., Gonzalez-Casanova and Alvarez (1985)]. (As usual in 2D, we use (x, y) as the coordinates of a point.) Duchon (1975) derived the explicit solution of this minimization problem

$$f(x, y) = \sum_\alpha b_\alpha K(r_\alpha) + c_0 + c_1 x + c_2 y \tag{4.35}$$

where

$$K(r) = r^2 \log r \quad \text{and} \quad r_\alpha^2 = (x - x_\alpha)^2 + (y - y_\alpha)^2$$

with coefficients satisfying

$$\sum_\alpha b_\alpha = 0 \qquad \sum_\alpha b_\alpha X_\alpha = 0 \qquad \sum_\alpha b_u Y_\alpha = 0$$

The spline interpolating function (4.35) turns out to have the same form as the kriging interpolant (3.46) with $k=1$ and a generalized covariance function $K(h) = |h|^2 \log |h|$.

The functions f presented above are known as *biharmonic splines* because each in its space of definition satisfies the equation $\Delta^2 f(x) = 0$ at every point $x \neq x_\alpha$ (Δ^2: iterated Laplacian). The basis function K itself satisfies $\Delta^2 K = \delta$. In 3D the solution is $K(r) = -r$. Higher-order splines can be defined by considering higher-order derivatives, but biharmonic splines are used most.[6] The reader is referred to Wahba (1990) for further information.

It is remarkable that biharmonic splines and kriging lead to the same results, considering that they proceed from two very different approaches. With splines one postulates a property of the interpolating surface, whereas with kriging one focuses on modeling the underlying random function itself. A common ground is the property of invariance under translations that is shared by IRF theory and by the differential operators with constant coefficients generally used to define splines.

The notion that spline interpolation is in certain cases equivalent to minimum variance linear prediction was shown by Kimeldorf and Wahba (1970) within the scope of a Bayesian analysis, and by Duchon (1976) who related splines with conditional expectations of random fields. Matheron (1981a) proved that the converse is also true: any kriging problem can be cast into a spline problem defined as the minimization of a norm associated with a linear operator. Such operator, however, is not necessarily with constant coefficients. Dubrule (1981) gives the following example: In 1D, kriging an IRF-1 with GC $K(h) = |h|^3$ is equivalent to splines with the differential operator $T = \partial^2/\partial x^2$. But if the covariance is $K(h) = -|h| + |h|^3$, the operator T cannot be differential with constant coefficients (the proof is not obvious). Likewise for a fixed order k, if we consider polynomial GCs

$$K(h) = \sum_{p=0}^{k} (-1)^{p+1} b_p |h|^{2p+1}$$

only one of them $(-1)^{k+1} |h|^{2k+1}$ corresponds to the minimization of the norm of a simple differential operator ($T = \partial^{k+1}/\partial x^{k+1}$).

Spline and kriging interpolation are thus equivalent in a formal way but not so in a practical way. Given the operator defining the spline, it is a tractable problem to find the equivalent kriging formulation, whereas it can be extremely difficult

---

[6] A fallacious argument sometimes heard to justify the use of splines for mapping geological structures goes roughly like this: Since geological materials have elastic properties and since thin-plate splines are a solution to the minimum stress problem, they are clearly physically correct. In reality, even if nature did create minimum stress surfaces, it did not force them to go through boreholes drilled millions of years later; that is, the boundary conditions were completely different. The fact that beds are necessarily minimum stress surfaces is also dubious, geologically. As shown by Suppe (1985), and simplifying to the extreme, the shape of a fold is largely a function of its thickness, with thicker layers having a longer wavelength. On the other hand, structural geologists have developed algorithms that take into account the geometric constraints imposed by the physics of deformation, most notably *balanced cross sections* (see footnote 12 of Section 3.6.2).

to identify the minimization problem associated with a given kriging solution. In this perspective, biharmonic splines constitute a very special and isolated case.

Smoothing Splines

When the data are subject to measurement errors—or when interpolating splines produce nonsensical results—one relaxes the constraints of exact fit at the sample points to request only that the fitted surface pass not too far from the data. Assuming equal error variances, the solution minimizes

$$\sum_\alpha [f(x_\alpha) - z_\alpha]^2 + \rho J(f) \qquad (\rho > 0) \tag{4.36}$$

within a class of functions f with continuous derivatives up to the appropriate order. The parameter ρ controls the trade-off between the smoothness of the curve and the fit at the data point. As ρ tends to zero, the solution tends to an interpolating spline; as ρ increases, the solution approaches a line or plane fit by least squares. Such interpolating functions are known as *smoothing splines.*

The formal equivalence with kriging extends to smoothing splines but in a slightly modified setup. The model is now

$$Z_\alpha = Y_\alpha + \varepsilon_\alpha$$

where the underlying $Y(x)$ is a *smooth* random function with covariance $K(h)$ and the ε_α are errors, uncorrelated with the RF $Y(x)$ and satisfying

$$E[\varepsilon_\alpha] = 0, \qquad E[\varepsilon_\alpha \varepsilon_\beta] = S_{\alpha\beta}$$

The estimation of the smooth component $Y(x)$ from noisy data Z_α is a cokriging problem analogous to equations (3.68) whose solution, in dual kriging terms, is of the form

$$\begin{cases} y^*(x) = \sum_\alpha b_\alpha K(x - x_\alpha) + \sum_\ell c_\ell x^\ell, \\ \sum_\beta b_\beta (K_{\alpha\beta} + S_{\alpha\beta}) + \sum_\ell c_\ell x_\alpha^\ell = z_\alpha, & \alpha = 1, \ldots, N, \\ \sum_\alpha b_\alpha x_\alpha^\ell = 0, & |\ell| = 0, \ldots, k \end{cases} \tag{4.37}$$

Biharmonic smoothing splines are a particular case of such cokriging, with a diagonal error covariance matrix $S_{\alpha\beta} = C_0\, \delta_{\alpha\beta}$ and a covariance $K(h)$ proportional to $|h|^3$ in 1D and to $|h|^2 \log |h|$ in 2D (Matheron, 1981a; Dubrule, 1983a). A simple parallel is established in 1D by Watson (1984).

A key problem in the application of smoothing splines is the choice of the smoothing parameter ρ. It is achieved by cross-validation techniques—in particular, the generalized cross-validation method of Craven and Wahba (1978).

This is where splines and kriging really meet because the endeavor to determine ρ from the data is a form of structure identification, a step central to kriging. Along these lines Dubrule (1983a) proposed to model the generalized covariance of $Z(x)$ with precisely the function that leads to smoothing splines in 2D:

$$K(h) = C_0\,\delta(h) + b_S\,|h|^2 \log|h|$$

The problem of selecting ρ subsumes into the determination of the structural parameters C_0 and b_S. In fact, by identification with the solution of (4.36) [e.g., Wahba (1990, p. 12)], we find simply $\rho = C_0/b_S$, which makes sense when interpreted as a "noise-to-signal ratio."

In conclusion of this section we note that the identification of splines with a special form of kriging is a tangible benefit of the IRF–k theory. It allows kriging to borrow techniques developed for splines, such as estimation under linear inequality constraints. From an operational perspective, however, the two approaches remain different because kriging leaves open the possibility of selecting a covariance model other than the standard spline type if the data so suggest. An empirical comparison of the predictive performance of kriging and splines, with a discussion, can be found in Laslett (1994). We give an excerpt from his conclusion: "It is when data are not sampled on a grid that kriging has the potential to outpredict splines, because it involves a translation of information on covariation of data values from intensely sampled regions to sparsely sampled regions; splines and some other nonparametric regression procedures do not appear to exploit such information."

4.7 GENERALIZED VARIOGRAM

Now that the theory is laid out, we are left with the crucial problem of structure identification: to determine the order k and the generalized covariance function $K(h)$. We will start with the special case of regularly spaced data along lines and introduce the generalized variogram, a new structural tool that inherits the advantages of the ordinary variogram. It can be calculated "nonparametrically" (i.e., without presuming a parametric model) and can be displayed and fitted graphically. The first definition of the generalized variogram can be found in Matheron (1972b), and it was first used by Orfeuil (1972) to check simulations of IRF–k. Applications include the analysis of microgravimetry data (Chilès, 1979b), geothermal data (Chilès and Gable, 1984), and fracture data (Chilès and Gentier, 1993). A similar tool, though more difficult to interpret, has been proposed by Cressie (1987).

4.7.1 Definition

The generalization of the simple increment or forward finite difference

$$\Delta_h Z(x) = Z(x+h) - Z(x)$$

is the increment or forward finite difference of order $k + 1$, which is the simplest ALC–k in 1D:

$$\Delta_h^{k+1} Z(x) = (-1)^{k+1} \sum_{p=0}^{k+1} (-1)^p \binom{k+1}{p} Z(x + ph) \qquad (4.38)$$

By definition, the generalized variogram of order k, denoted by $\Gamma(h)$ and abbreviated as GV, is the appropriately scaled variance of the increment of order $k + 1$:

$$\Gamma(h) = \frac{1}{M_k} \text{Var}[\Delta_h^{k+1} Z(x)] \qquad (4.39)$$

The scaling factor $M_k = \binom{2k+2}{k+1}$ is introduced to ensure that in the case of a pure nugget effect, $\Gamma(h) = C_0 \delta(h)$ as for an ordinary variogram.[7] Explicitly we have

$$k = 1 : \Gamma(h) = \frac{1}{6} \text{Var}[Z(x + 2h) - 2\ Z(x + h) + Z(x)],$$

$$k = 2 : \Gamma(h) = \frac{1}{20} \text{Var}[Z(x + 3h) - 3\ Z(x + 2h) + 3\ Z(x + h) - Z(x)],$$

$$k = 3 : \Gamma(h) = \frac{1}{70} \text{Var}[Z(x + 4h) - 4\ Z(x + 3h) + 6\ Z(x + 2h) - 4Z(x + h) + Z(x)]$$

4.7.2 Relationship Between Generalized Variogram and Covariance

Since the increment of order $k + 1$ is an ALC–k, its variance can be expressed in terms of the generalized covariance. Applying the definition (4.14) of a generalized covariance to (4.39) gives

$$\Gamma(h) = \frac{1}{M_k} \sum_{p=-(k+1)}^{k+1} (-1)^p \binom{2k+2}{k+1+p} K(ph) \qquad (4.40)$$

Explicitly

$$k = 1 : \Gamma(h) = K(0) - \tfrac{4}{3} K(h) + \tfrac{1}{3} K(2h)$$

$$k = 2 : \Gamma(h) = K(0) - \tfrac{3}{2} K(h) + \tfrac{3}{5} K(2h) - \tfrac{1}{10} K(3h)$$

$$k = 3 : \Gamma(h) = K(0) - \tfrac{8}{5} K(h) + \tfrac{4}{5} K(2h) - \tfrac{8}{35} K(3h) + \tfrac{1}{35} K(4h)$$

---

[7] Here we find further justification for naming $\gamma(h)$ a variogram rather than a *semi*-variogram; otherwise, $\Gamma(h)$ would not be the generalized variogram, even less the generalized semi-variogram, but the M_kth of a generalized variogram!

While being simple, the relationship between $\Gamma(h)$ and $K(h)$ is more complex than for the ordinary variogram. In particular, whether the GV determines the GC (up to the even polynomial) is still an unsolved problem (Chauvet, 1987, pp. 215–231). Indeed $\Gamma(h)$ only represents the variance of the increments associated with h, while what determines the GC is the set of all covariances of these increments for all h.

However, there are two important cases where the GV does determine the GC: When the GC is bounded and when the GC is of polynomial form. This is readily seen from (4.40) by observing that if $K(h)$ behaves like $|h|^{\alpha}$, so does $\Gamma(h)$, only the coefficient changes. On the other hand, if $K(h)$ is an ordinary covariance such that $K(h) = 0$ for $|h| \geq a$, then by (4.40) we obtain $\Gamma(h) = K(0)$ for $|h| \geq a$. The GV has the same range and sill as the ordinary variogram $K(0) - K(h)$; the nugget effect remains the same. All common covariance models have a polynomial behavior near the origin, and the GV inherits this but with changes in the coefficients; in particular, the even terms of degree $\geq 2k$ disappear. Figure 4.10 displays the GVs associated with the spherical model (2.47).

From a practical point of view, it should be kept in mind that the objective is to identify the GC function $K(h)$, since it is the quantity involved in estimation problems. So, after computing an experimental GV, one will try to fit a linear combination of GV models associated with known GC models. The above-mentioned models are the only ones used in practice. Since the GV models remain close to the GC models, fitting a model to an experimental GV is not more difficult than fitting a model to an ordinary variogram.

The statistical inference problems associated with the GV are essentially the same as for the ordinary variogram. The conclusions are also similar except that the GV can only be computed at shorter distances (the length of the region

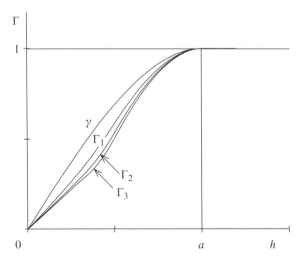

FIGURE 4.10 Generalized variograms associated with the spherical variogram model for $k = 0$, 1, 2, 3.

of interest divided by $k+1$) and fluctuation variances are higher. In the Gaussian case the regional GV $\Gamma_R(h)$ still converges to $\Gamma(h)$ when $h \to 0$ (micro-ergodicity), provided that $\Gamma(h)$ is not too regular near the origin ($\Gamma(h) \sim |h|^\alpha$ with $0 < \alpha < 2k+2$). More on statistical properties of the GV can be found in Chilès (1979b).

Polynomial Covariances

If $K(h) = (-1)^{1+\lfloor \alpha/2 \rfloor} b_\alpha |h|^\alpha$, then

$$\Gamma(h) = (-1)^{1+\lfloor \alpha/2 \rfloor} B_\alpha \, b_\alpha |h|^\alpha \qquad \text{with} \qquad B_\alpha = \frac{1}{M_k} \sum_{p=-k-1}^{k+1} (-1)^p \binom{2k+2}{k+1+p} |p|^\alpha$$

Explicit results are given below for the polynomial/logarithmic family model (where h stands for $|h|$; notice that b_p is here the coefficient of the monomial or logarithmic-monomial of degree p)

$$k=1 \begin{cases} K(h) = C_0 \, \delta(h) - b_1 h + b_2 h^2 \log h + b_3 h^3 \\ \Gamma(h) = C_0 \left[1 - \delta(h)\right] + \tfrac{2}{3} b_1 h + \tfrac{4}{3}(\log 2) b_2 h^2 + \tfrac{4}{3} b_3 h^3 \end{cases}$$

$$k=2 \begin{cases} K(h) = C_0 \delta(h) - b_1 h + b_2 h^2 \log h + b_3 h^3 - b_4 h^4 \log h - b_5 h^5 \\ \Gamma(h) = C_0 [1 - \delta(h)] + \tfrac{3}{5} b_1 h + \tfrac{3}{10}(\log 256 - \log 27) b_2 h^2 + \tfrac{3}{5} b_3 h^3 \\ \qquad\quad + \tfrac{3}{10}(27 \log 3 - 32 \log 2) b_4 h^4 + \tfrac{33}{5} b_5 h^5 \end{cases}$$

$$k=3 \begin{cases} K(h) = C_0 \, \delta(h) - b_1 h + b_2 h^2 \log h + b_3 h^3 - b_4 h^4 \log h - b_5 h^5 \\ \qquad\quad + b_6 h^6 \log h + b_7 h^7 \\ \Gamma(h) = C_0 \left[1 - \delta(h)\right] + \tfrac{4}{7} b_1 h + \tfrac{72}{35}(\log 4 - \log 3) b_2 h^2 + \tfrac{16}{35} b_3 h^3 \\ \qquad\quad + \tfrac{24}{35}(27 \log 3 - 40 \log 2) b_4 h^4 + \tfrac{16}{7} b_5 h^5 \\ \qquad\quad + \tfrac{8}{35}(1248 \log 2 - 729 \log 3) b_6 h^6 + \tfrac{2416}{35} b_7 h^7 \end{cases}$$

$\Gamma(h)$ is a polynomial of degree $2k+1$ in h, where all the logarithmic terms of $K(h)$ have turned into even degree terms.

4.7.3 An Application to the Topography of Fractures

Chilès and Gentier (1993) describe an application of generalized variograms to the study of the morphology of natural rock fractures. A precise determination of the topography of the two surfaces bordering a fracture is performed in the laboratory on cores by sampling the surfaces along profiles, using a special tool called a "profilometer." The goal is to understand the roughness of the fracture surfaces and the spatial variations of the aperture, which control the mechanical behavior of fractures under stress and the flow and transport within the fractures.

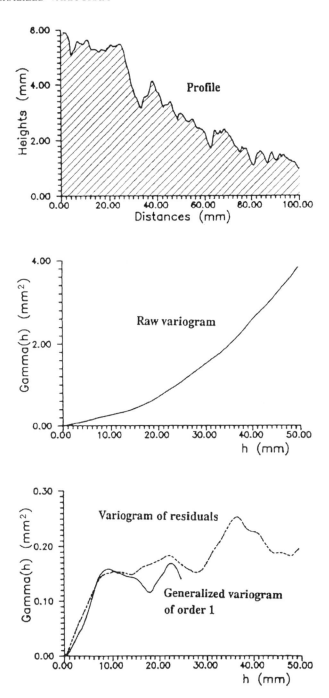

FIGURE 4.11 Fracture profile shown on cross-section along core axis, and associated variograms. [From Chilès and Gentier (1993), with kind permission from Kluwer Academic Publishers.]

Figure 4.11 shows a profile recorded across a fracture surface, its raw vario-gram, its variogram of residuals, and its GV for $k = 1$. The raw variogram has a parabolic shape that mainly reflects the bias due to the presence of a strong linear drift. By contrast the GV, displayed with a vertical exaggeration of about 10, exhibits a clear sill reached at a range of about 10 mm. It agrees with the variogram of residuals $\gamma_R(h)$ in the sense that they have the same range and sill, and the slope at the origin of $\Gamma(h)$ is two-thirds of the slope of $\gamma_R(h)$, as predicted by theory.

The excellent agreement of the two variograms in this example is due to the simplicity of the drift, which allows a global linear fit. In more complex cases a simple global fit of the drift would still leave bias in the variogram of residuals, whereas the GV remains unbiased at short distances because it filters the drift locally. In addition, $\Gamma(h)$ can reflect the unbounded behavior of the underlying variogram, while by construction a variogram of residuals is always bounded.

4.8   AUTOMATIC STRUCTURE IDENTIFICATION

In the general case of scattered data, it is not possible to compute a structure function such as the generalized variogram and graph it to suggest a model. We have to operate "blindly" and rely on an algorithm to pick the best set of parameters within a prespecified family of models. This is a parametric approach, whereas the generalized variogram is nonparametric. The design of that automatic procedure is the critical and also the most difficult part of the application of the IRF theory. We present here a method that claims no optimality properties but has the advantage of being completely general since it does not require the data points to be evenly distributed nor the underlying RF to be Gaussian. However, it critically depends on a good design of the measures λ used for the estimation, which requires careful tuning.

Since this method was introduced (Delfiner, 1976), numerous attempts have been made to improve on it and find estimators endowed with optimality prop-erties. One approach (Delfiner, 1977; Kitanidis, 1983, 1985; Marshall and Mardia, 1985; Stein, 1986a) consists of treating the covariance identification problem as one of estimation of variance components using Rao's minimum norm quadratic unbiased estimator (MINQUE) theory (Rao, 1970, 1971a,b, 1973, p. 303). Another approach is restricted maximum likelihood (REML) estimation based on maximizing the loglikelihood associated with a vector of linearly independent ALC–k (Kitanidis, 1983). The latter method relies heavily on the Gaussian assumption and is computationally intensive. Zimmerman (1989) proposes an efficient implementation for data on a lattice—but then, why not use generalized variograms?

4.8.1   General Principles

The method is based on the assumption that the covariance $K(h)$ is a linear combination of elementary models $K_p(h)$:

$$K(h) = \sum_p b_p K_p(h) \qquad (4.41)$$

For a polynomial GC model of order k, we take the elementary models $K_p(h) = (-1)^{p+1} |h|^{2p+1}$, $p = 0, \ldots, k$, plus the unit nugget effect $\delta(h)$ that we will associate with $p = -1$ (but we will maintain the usual notation C_0 rather than b_{-1} for the nugget effect value). It is also possible to include the spline model $|h|^2 \log |h|$, but then $-|h|^5$ should be dropped to keep the number of terms as low as possible. We could also consider for $K_p(h)$ models with given ranges and estimate the sills b_p.

Let us define the synthetic notation

$$K(\lambda) = \sum_\alpha \sum_\beta \lambda_\alpha \lambda_\beta \; K(x_\beta - x_\alpha) = \mathrm{E}[Z(\lambda)^2]$$

It results from (4.41) that

$$\mathrm{E}[Z(\lambda)^2] = K(\lambda) = \sum_p b_p K_p(\lambda) \qquad (4.42)$$

This is a *linear regression* of $Z(\lambda)^2$ on the predictor variables $K_p(\lambda)$. To determine the coefficients of this regression, we construct, based on the data points, a large number of ALC–k $Z(\lambda_i)$ and minimize

$$Q(\mathbf{b}) = \sum_i w_i^2 \, [Z(\lambda_i)^2 - \sum_p b_p K_p(\lambda_i)]^2 \qquad (4.43)$$

where $\mathbf{b} = [b_p] = (C_0, b_0, \ldots, b_k)'$ is the vector of the unknown coefficients. The weights w_i^2 are introduced to equalize the variances of $Z(\lambda_i)^2$ and therefore should, in theory, be equal to the reciprocals of $\mathrm{Var}[Z(\lambda_i)^2]$. But these variances are unknown. They involve the fourth-order moments of $Z(\lambda_i)$ or at least, in the Gaussian case, the second-order moments, which just depend on the covariance to be estimated. An iterative procedure could be considered, but it is not worth pursuing too far in this direction, since another basic assumption of least squares, the noncorrelation of the $Z(\lambda_i)^2$, is violated anyway. Thus one simply uses empirical weights. For example, one can take the weights associated with the elementary models $\delta(h)$, $-|h|$, or $|h|^3$ in the Gaussian case where

$$\mathrm{Var}[Z(\lambda_i)^2] = 2K(\lambda_i)^2$$

Just as a variogram is better known at short distances than at large ones, it is preferable to assign a larger weight to increments that are more "packed." Therefore the weighting derived from $|h|^3$ is used whenever possible. For $k = 0$ a weighting based on $-|h|$ is used.

In fact the problem departs from a plain least squares regression when we require the fit to constitute a valid GC at the order k considered. The coefficients are not free but must satisfy the inequalities of a polynomial GC. This would call for a constrained minimization of (4.43) using the techniques of quadratic programming. An easier way is to simply fit all possible regressions based on all subsets of the predictor variables $K_p(\lambda)$ and discard those that do not satisfy the constraints. Valid models are guaranteed to exist: The solutions associated with a single elementary model lead to a positive coefficient. If the search is limited to models for $k \leq 2$, there are $2^4 - 1 = 15$ possible regression equations to consider.

At this stage, two questions remain:

- How to select the order k?
- What is the best valid regression equation (4.42)?

The answers are unfortunately largely empirical due to the absence of solid goodness-of-fit criteria. For the second question least squares suggest that one ought to consider the residual sum of squares $Q(\mathbf{b})$ or its normalized version:

$$\frac{Q(\mathbf{b})}{Q(\mathbf{0})} = \frac{Q(C_0, b_0, \ldots, b_k)}{Q(0, 0, \ldots, 0)} = \frac{\text{residual sum of squares}}{\text{total sum of squares}}$$

This ratio is always less than 1. A value close to 1 indicates a bad fit, and a value close to 2/3 indicates a good fit. This number 2/3 is the theoretical value $E[Q(\mathbf{b})]/E[Q(\mathbf{0})]$ in the Gaussian case with the correct model and a perfect fit. The problem is that no significance test is available such as the F-test of standard least squares theory, and we must therefore base our selection on mere sample fluctuations. In practice, the quality of the fit is gauged not from the simple criterion Q but from several criteria considered simultaneously. The same holds true for the determination of the order k.

To illustrate the above, we now sketch the procedure implemented in a program named BLUEPACK developed at the Center for Geostatistics, Fontainebleau.[8]

4.8.2   An Implementation Example

BLUEPACK fits a polynomial GC model plus a nugget effect. (There is a variant, not presented here, including a spline term). The procedure comprises three steps:

1. Determine the order k (0, 1 or 2).
2. Compute all possible regressions (3, 7, or 15 according to the value of k), eliminate those that are not valid, and perform a first selection.
3. Compare the best regressions of step 2 to make a final selection.

---

[8] The predecessor of the current geostatistical package ISATIS®.

***Step* 1.** To determine the order, k the idea is to cross out known points and estimate them from neighboring data points, while varying only the number of unbiasedness conditions (1, 3, or 6)—that is, the order k of the IRF.

For each estimated data point Z_α the errors $\hat{Z}_\alpha - Z_\alpha$ are ALC–k of order $k = 0, 1,$ or 2. These errors are ranked by ascending magnitude, and the corresponding orders k are thus ranked by performance. The mean rank is computed for each k over all estimated Z_α, and the value selected is that with minimum mean rank. The advantage of this method is its robustness against outliers.

The mean square error criterion is also output by the program but only as a secondary check. In ambiguous cases it is advisable to select the lowest possible k.

Since at this stage of structure identification the covariance is not yet known, the above estimators $\hat{Z}_\alpha$ are simply determined by least squares. One design rule is to avoid using configurations of data points that have a high symmetry about the estimated point. Indeed a symmetric neighborhood tends to filter polynomials by itself, which leads to underestimate k. To introduce asymmetry, the data points are split into two concentric rings: data from the inner ring are used to estimate points from the outer ring, and vice versa.

***Step* 2.** Once k is determined, the possible regressions are computed. Those that are not valid are discarded (i.e., for polynomial GCs, those that do not fulfill the constraints expressed in Table 4.1). A selection is then made among the remaining regressions. The operational criterion involves a ratio of mean square errors. For any ALC–k $Z(\lambda_i)$ we have

$$E[Z(\lambda_i)^2] = K(\lambda_i) = \sum_p b_p K_p(\lambda_i)$$

Let

$$\hat{K}(\lambda_i) = \sum_p \hat{b}_p K_p(\lambda_i)$$

be its estimator. Since

$$E\left[\sum_i Z(\lambda_i)^2\right] = \sum_i K(\lambda_i)$$

the ratio

$$\rho = E\left[\sum_i Z(\lambda_i)^2\right] \Big/ E\left[\sum_i \hat{K}(\lambda_i)\right]$$

should be close to 1. The estimator

$$r = \sum_i Z(\lambda_i)^2 \Big/ \sum_i \hat{K}(\lambda_i) \tag{4.44}$$

is a biased estimator of ρ. To reduce the bias, we can split the sample and use the jackknife estimator

$$\hat{\rho} = 2r - \frac{n_1 r_1 + n_2 r_2}{n_1 + n_2}$$

where r_1 and r_2 are the ratios (4.44) computed separately in ring 1 and ring 2, comprising n_1 and n_2 measures, and r the ratio computed with all $n_1 + n_2$ measures (this eliminates a bias of the form b/n). The regression whose $\hat{\rho}$ is closest to 1 is selected.

***Step* 3 (*Optional Cross-Validation*).** The method is similar to the leave-one-out described in Section 2.6.3 for the stationary case, and it consists of evaluating candidate models through their performance in actual kriging situations. This is to avoid basing the selection on marginal differences in the jackknife statistics. The sample points, or a subset of them, are estimated from their neighbors as if they were unknown, using the order k determined at Step 1 and the best covariance models from Step 2 (ranked by increasing distance of $\hat{\rho}$ from 1). If the model is correct, the error $Z^*_{-\alpha} - Z_\alpha$ obtained when leaving the point x_α out has a variance equal to $\sigma^2_{K\alpha}$ so that the mean square standardized error

$$\text{m.s.s.e.} = \frac{1}{N} \sum_\alpha (Z^*_{-\alpha} - Z_\alpha)^2 / \sigma^2_{K\alpha}$$

should be close to 1. As a reference, note that the variance of this ratio is equal to $2/N$ if the variables $Z^*_{-\alpha} - Z_\alpha$ are independent and Gaussian.

The operational criterion is to minimize the mean square error

$$\text{m.s.e.} = \frac{1}{N} \sum_\alpha (Z^*_{-\alpha} - Z_\alpha)^2$$

under the constraint that the mean standardized square error defined above is not too far from 1. (Typically, a tolerance of $\pm 3\sqrt{2/N}$ is used.)

More sophisticated procedures may be used, such as pairwise ranking of models. Better yet, an analysis of the kriging errors $Z^*_{-\alpha} - Z_\alpha$ or their standardized version $(Z^*_{-\alpha} - Z_\alpha)/\sigma_{K\alpha}$ can be performed using graphical statistical procedures. This procedure also allows one to spot suspect data points in the context of their surrounding neighbors.

An Example

Figure 4.12 shows an output of the BLUEPACK Automatic Structure Identification options illustrating the first two steps. The data are the 573 topographic elevations of the Noirétable area presented in Section 2.7.3. In the first section each line indicates the performance of an order k. Results relative to measures based on polynomials fitted in rings 1 are reported under the heading "Ring 1"; they correspond to estimation of points of rings 2, and vice versa for "Ring 2." The means of the results obtained in Ring 1 and Ring 2 are given in the last column "Total." Naturally these means are weighted by the count of measures in each ring which are printed below the table ($n_1 = 805$ and $n_2 = 877$). A cutoff is applied on the sum of squares of the weights of the measures to avoid strong ill-conditioning; this explains why n_1 and n_2 may be different. The "mean neighborhood radius" (257 m) is also printed and tells us the scale at which the degree selection is made.

In the present case the degree $k = 1$ is the best with an average rank of 1.83. Note that the sum of the ranks is $1 + 2 + 3 = 6$. The mean square error criterion also indicates $k = 1$ as the best choice, which increases our confidence. Such agreement of criteria for drift identification is not always achieved, but it is not uncommon.

Only five out of the seven possible fits were found valid and are listed in the second section of Figure 4.12. The discarded models are $C_0\delta(h) - b_0|h|$ and $C_0 \delta(h) - b_0|h| + b_1|h|^3$.

The right-hand part of Figure 4.12 displays several criteria for selecting a model among the five candidates. The Q column shows the ratio $Q(\mathbf{b})/Q(\mathbf{0})$, while r_1, r_2, r, and $\hat{\rho}$ are given in the other columns. Here the model

$$C_0, = 0 \qquad b_0 = 0.230 \qquad b_1 = 0.960 \times 10^{-5}$$

stands out with a jackknife of 0.994, which turns out to also have the smallest ratio $Q(\mathbf{b})/Q(\mathbf{0}) = 0.761$. Figure 4.13 shows the results of cross-validation of the best four covariance models using all data points. These tests confirm the above selection with m.s.e. $= 21.12$ m^2

AUTOMATIC STRUCTURE IDENTIFICATION

1) IDENTIFICATION OF THE ORDER k

Degree	MEAN SQUARE ERRORS			RANKING OF THE TRIALS AVERAGE RANKS		
	Ring 1	Ring 2	Total	Ring 1	Ring 2	Total
2	342.3	349.1	345.8	1.92	1.85	1.88
1	119.7	137.1	128.8	1.82	1.85	1.83
0	298.2	293.3	295.7	2.26	2.31	2.28

Degree of the drift = 1
Counts: Ring 1 = 805   Ring 2 = 877   Total = 1682
Mean neighborhood radius = 256.7

2) IDENTIFICATION OF THE COVARIANCE

COVARIANCE COEFFICIENTS			EXPL/THEOR VARIANCE RATIOS				
C_0	$-b_0$	b_1	Q	r_1	r_2	r	$\hat{\rho}$
3.060	0	0	0.8461	13.5092	12.4363	12.9867	13.0047
0	−0.3392	0	0.7637	1.8649	1.7122	1.7904	1.7929
0	0	0.3904 E-4	0.9241	0.4692	0.4348	0.4525	0.4531
0	−0.2296	0.9595 E-5	0.7609	1.0321	0.9518	0.9931	0.9944
2.715	0	0.2798 E-4	0.8091	0.6276	0.5814	0.6052	0.6061

Provisional covariance fit
Order $k = 1$ $C_0 = 0$ $b_0 = 0.2296$ $b_1 = 0.9595E-5$

FIGURE 4.12 An example of automatic structure identification by BLUEPACK.

and m.s.s.e. $= 0.83$. Note that the model selected is not the one with the minimum m.s.e. because a good estimation of error variances is also required.[9]

Additional insight into the behavior of the selected model can be gained by inspection of the diagnostic plots shown in Figure 4.14. These plots are meant to detect the presence of residual structure not accounted for by the selected model and should be used *qualitatively* (e.g., the independence of Z^* and $Z - Z^*$ is established only for simple kriging and for a Gaussian SRF). Here we verify that posted standardized errors exhibit no particular clustering of large errors in specific areas; the scatterplot of Z versus Z^* shows no systematic effect hinting at a bias; the scatterplot of Z^* and standardized error indicates no major dependency of the error on the Z^* value (except for a decrease of the apparent scatter). Finally the histogram of standardized errors is symmetric about zero, with a normal shape but more squeezed in the middle and more tail-stretched than the Gaussian (this shape is very frequent).

[9] The stationary covariance locally equivalent to the selected GC for $|h| \leq 525$ m (approximately the neighborhood diameter, cf. Figure 4.12) allows an excellent fit of the sample variogram shown in Figure 2.29 b,c. This reconciles the variographic and IRF–k approaches.

	MODEL CROSS-VALIDATION

COVARIANCE COEFFICIENTS			JACKKNIFE	ERROR STATISTICS	
C_0	$-b_0$	b_1	$\hat{\rho}$	*M.S.E.*	*M.S.S.E.*
0	−0.2296	0.9595 E-5	0.9944	21.12	0.8276
2.715	0	0.2798 E-4	0.6061	20.50	1.3450
0	0	0.3904 E 4	0.4531	20.22	2.4520
0	−0.3392	0	1.7929	24.41	0.9697

Final-selection criteria:
1. M.S.E. close to its minimum value; and
2. M.S.S.E. close to 1, possibly in the interval 1 ± 3 Sqrt(2/N), i.e., [0.82, 1.18].

Final covariance fit
Order $k = 1$ $C_0 = 0$ $b_0 = 0.2296$ $b_1 = 0.9595E\text{-}5$

FIGURE 4.13 Cross-validation of the best four covariance models (ranked by increasing deviation of the jackknife from 1). Every data point has been estimated from 12 neighbors. m.s.e. (mean square error) shows that the first three fits outperform the fourth one. m.s.s.e. (mean square standardized error) is the closest to 1 for fit 1, and it lies in the tolerance interval for this fit only. Fit 1 is selected.

Discussion

One of the limitations of the above procedure is the isotropy of polynomial GCs. The inference of anisotropic models was considered by Chilès (1978) but was not used at the time because it requires too many parameters; this question could be revisited now with faster computers. While in the stationary case it is not acceptable to ignore anisotropy, in the nonstationary case one can postulate that the anisotropies are captured by the filtered polynomials (the "drift"). For $k = 0$ the only filtered terms are constants, which is a good reason, among others, to revert to a standard structural analysis. A stronger reason yet is that a model with a range may be more appropriate than a linear variogram.

The weakness of the automatic structure identification procedure lies in its lack of robustness against variables that do not fit the intrinsic hypotheses well: presence of a few isolated peaks, mixture of flat and jagged areas, and erroneous or "outlying" data. These heterogeneities have a strong influence on the criterion $Q(C_0, \dots, b_k)$ to be minimized. As we have seen, there are methods to check the model, but only after the fact.

Starks and Fang (1982) pointed out that in the case of the basic polynomial covariances $K_0(h) = -|h|$, $K_1(h) = |h|^3$, and $K_2(h) = -|h|^5$, there is a great deal of linear dependency between the predictor variables $K_0(\lambda)$, $K_1(\lambda)$, and $K_2(\lambda)$. This can cause instability in the estimation of the regression coefficients b_p.

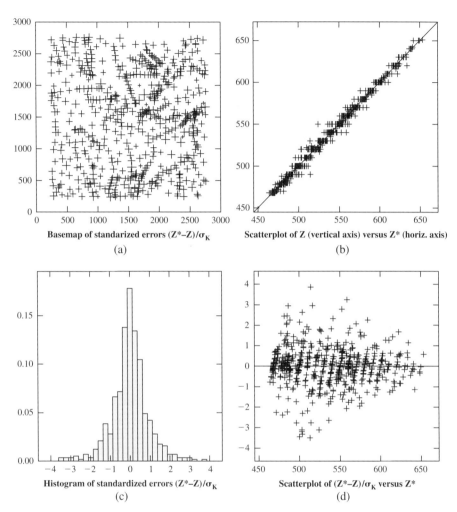

FIGURE 4.14 Diagnostic plots (topographic elevation, Noirétable): (a) posted standardized errors (the symbol size is proportional to the absolute value of the standardized error); (b) scatterplot of Z versus Z^*; (c) histogram of standardized errors; (d) scatterplot of standardized error versus Z^*.

An empirical study was made in one dimension with equispaced data (Chilès, 1978). Its conclusions cannot a priori be extended to scattered two-dimensional data, but they are consistent with the experience accumulated on automatic structure identification. The conclusions of this study are the following:

- If the covariance is predominantly linear with a nugget effect, all models can be fitted with fair precision.
- If the covariance is predominantly of type $|h|^3$, only models with a single other term may be fitted with acceptable precision.
- If the covariance model is predominantly of type $-|h|^5$, estimation of anything else than the coefficient of $|h|^5$ is meaningless.

4.9   STOCHASTIC DIFFERENTIAL EQUATIONS

Physical laws often take the form of differential or partial differential equations. Their transposition to the stochastic domain is fairly natural but can present specific difficulties, especially when white noise is involved. The questions asked for stochastic differential systems are the same as for deterministic ones, namely the existence and uniqueness of a solution, the analytical properties of the solutions, and how they depend on parameters and initial conditions. However, the introduction of stochastic elements brings new questions, such as the determination of the spatial distribution of the solution and the existence of stationary solutions. The study of stochastic differential equations is by itself a vast subject. The interested reader is referred to the works of Gikhman and Skorokhod (1972), Arnold (1973), Friedman (1975–1976), and Sobczyk (1991), to quote only a few, and of Gel'fand and Shilov (1958) for generalized random functions. We specialize our discussion to just one equation, but with broad utility, the Poisson equation $\Delta Z = Y$.

4.9.1   The Poisson Equation $\Delta Z = Y$

The stochastic *Poisson* equation is defined by

$$\Delta Z = \sum_{i=1}^{n} \frac{\partial^2}{\partial x_i^2} Z(x) = Y(x) \tag{4.45}$$

where Δ is the Laplace operator and Y and Z are RFs on $\mathbb{R}^n$. In this equation, Y represents a source term and is given. When Y is equal to zero, the equation is named the *Laplace* equation and its solutions are *harmonic functions*. The Poisson equation represents fluid flow in a porous medium with a constant permeability equal to one, where Z is the piezometric head and Y is the inflow/ outflow (due to rainfall, or injection, or pumping counted negatively). It can also represent the flow of heat in a homogeneous medium with thermal conductivity equal to one, where Z is the temperature and Y the heat production.[10] It is also used to find the electric potential for a given charge distribution.

Stochastic Model

The first question that arises is the existence of a stochastic model compatible with this equation. Let us first consider the 1D case. When Y is deterministic, the differential equation has for solution the second integral of Y, plus an arbitrary polynomial of degree 1. When Y is an SRF with zero mean, we have seen in Section 4.4.1 that the second integral of Y is a representation of an IRF–1. Since the various representations of an IRF–1 differ only by an

---

[10] The name *heat equation* usually refers to the equation $\partial Z/\partial t = a^2 \Delta Z$ which defines the temperature Z of an isotropic homogeneous medium as a function of point x and time t (the Laplacian being taken with respect to space coordinates) in the absence of heat production.

arbitrary polynomial of degree 1, the solution of the equation is an IRF–1 in the sense of the definition of Section 4.4.2. Likewise, if Y is an IRF–0, Z is an IRF–2. In both cases the solution is an IRF, that is, a class of ordinary RFs. Like in the deterministic case, boundary or initial conditions can restrict the generality of the solution to a particular representation. These results can be generalized to higher-dimensional spaces and to IRFs of a higher order, as shown by the following theorem from Matheron (1971b):

Theorem. If Y is a continuous IRF–k on $\mathbb{R}^n$, there exists a unique twice differentiable IRF–$(k+2)$ Z satisfying the differential equation $\Delta Z = Y$.

Uniqueness of course pertains to the IRF–$(k+2)$ regarded as a class of RFs.

Covariance Models

From this we can derive the relationship between the generalized covariances of Z and of Y, as well as the (generalized) cross-covariance of Y and Z (Matheron, 1971b; Dong, 1990, pp. 55–59). To keep things simple, let us consider only the case where Y is a zero-mean SRF. Let $x = (x_1, \dots x_n)$ and $y = (y_1, \dots, y_n)$ be two points of $\mathbb{R}^n$. Denoting by $C(h)$ the stationary covariance of Y, we get

$$C(y - x) = \mathrm{E}[Y(x)Y(y)] = \mathrm{E}[\Delta Z(x)\Delta Z(y)] = \mathrm{E}\left[\sum_{i=1}^{n}\sum_{j=1}^{n}\frac{\partial^2 Z(x)}{\partial x_i^2}\frac{\partial^2 Z(y)}{\partial y_j^2}\right]$$

With Z being twice differentiable, its nonstationary covariance $\sigma(x,y)$ is differentiable four times, and therefore we have

$$C(y - x) = \sum_{i=1}^{n}\sum_{j=1}^{n}\frac{\partial^2}{\partial x_i^2}\frac{\partial^2}{\partial y_j^2}\mathrm{E}[Z(x)Z(y)] = \Delta_x\Delta_y\sigma(x,y)$$

where Δx is the Laplace operator applied with respect to x. Because the ordinary covariance σ of Z and its generalized covariance K are related by (4.17), we get

$$\Delta_x\Delta_y\sigma(x,y) = \Delta_x\Delta_y K(y-x) = \Delta^2 K(h) \qquad \text{for} \quad h = y - x$$

The covariances C of Y and K of Z are therefore related by the equation

$$\Delta^2 K(h) = C(h) \tag{4.46}$$

Likewise for the cross-covariance we get

$$\mathrm{E}[Y(x)Z(y)] = \mathrm{E}[\Delta Z(x)Z(y)] = \mathrm{E}\left[\sum_{i=1}^{n}\frac{\partial^2 Z(x)}{\partial x_i^2}Z(y)\right] = \sum_{i=1}^{n}\frac{\partial^2}{\partial x_i^2}\mathrm{E}[Z(x)Z(y)]$$

$$= \Delta_x\sigma(x,y)$$

Given (4.17), for $k = 1$ this is of the form

$$E[Y(x)Z(y)] = \Delta K(h) + \Delta c_0(x) + \sum_{i=1}^{n} \Delta c_i(x)y_i \quad \text{for} \quad h = y - x \quad \text{and}$$

$$y = (y_1, \ldots, y_n) \tag{4.47}$$

In practice, we do not know which representation of the IRF–1 we are dealing with, and we only work on linear combinations allowable at order 1. The terms involving c_0 and c_i then disappear, and the covariance between $Y(x)$ and an ALC–1 $Z(\lambda)$ can be calculated using the generalized cross-covariance $\Delta K(h)$.

Case Where Y Is White Noise

If Y is not a continuous IRF–k but a random noise with covariance $C_0 \delta$ it should be interpreted as a generalized random function. The covariance K of Z is then a generalized function satisfying

$$\Delta^2 K = C_0 \, \delta$$

As seen in Section 4.6.4, this is the equation satisfied by the generalized covariance associated with biharmonic spline interpolation. Consequently the function K is, up to a multiplicative factor, the isotropic generalized covariance r^3 in $\mathbb{R}$, $r^2 \log r$ in $\mathbb{R}^2$, $-r$ in $\mathbb{R}^3$, $-\log r$ in $\mathbb{R}^4$, and $r^{-(n-4)}$ in $\mathbb{R}^n$, $n > 4$. Note that the irregularity of the RF Z increases with the space dimensionality.

Existence of a Stationary Solution

As an IRF–1, can Z have a stationary representation? Its covariance $K(h)$ has a spectral representation of the form (4.19) with $k = 1$. It can coincide with a spectral representation of a stationary covariance only if $\chi(du)/(4\pi^2|u|^2)^2$ has a finite integral. Now, in view of (4.46), the spectral measure $F(du)$ associated with the covariance $C(h)$ of Y is equal to $\chi(du)$.

Therefore Z has a stationary representation only if $F(du)/(4\pi^2|u|^2)^2$ is a spectral measure (with a finite integral)—that is, if Y has very little low frequency content. This is not true in general.

Note that (4.46) can be solved directly in simple cases, notably in 1D or when the covariance is isotropic [cf. Dong (1990, pp. 56–57)].

Solution under Boundary Conditions

Let us restrict ourselves to a domain D with boundary Γ. Let Z and Y be two ordinary functions satisfying

$$\begin{cases} \Delta Z(x) = Y(x) & \forall x \in D, \\ Z(x) = z(x) & \forall x \in \Gamma \end{cases}$$

The Green function (which only depends on the shape of the boundary[11]) makes it possible to express the real solution in the integral form

$$Z(x) = \int_D G(x, x')\, Y(x')\, dx' + \int_\Gamma \frac{d}{d\nu} G(x, x')\, z(x')\, dx' \qquad (4.48)$$

where ν is the outward unit vector at the point x' of Γ. This amounts to decomposing Z into a sum of two functions Z_1 and Z_2 such that

$$\begin{cases} \Delta Z_1 = Y & \text{over } D \\ Z_1 = 0 & \text{over } \Gamma \end{cases} \quad \text{and} \quad \begin{cases} \Delta Z_2 = 0 & \text{over } D \\ Z_2 = z & \text{over } \Gamma \end{cases}$$

where Z_1 depends on the shape of the boundary but not the boundary values. These determine Z_2. In view of the preceding stochastic framework, the solution of (4.45) is an IRF, but knowledge of Z on Γ characterizes a specific representation.

Cokriging of Y and Z

The solution of the above systems is generally nontrivial. Moreover, in usual applications Y is not known at every point of D nor Z at every point of Γ. The stochastic model then achieves its full value. Once the generalized covariance $K(h)$ of Z is known, as well as its Laplacian and its iterated Laplacian, we have all the elements to perform a cokriging of Z (or of Y) from data of both variables. Because Z is an IRF–1, we must of course impose conditions ensuring that in the expression of the cokriging error the terms involving Z constitute an ALC–1. This requires that we have enough Z data (at least $n + 2$ in n-dimensional space). If we have no Z data at all, we can still use cokriging, provided that we work on an internal representation of the IRF–1 (recall that an internal representation is an ALC–1 at every point x; see Section 4.4.2), but the result will depend on the selected internal representation, which is arbitrary.

Given the linearity of equation (4.45), the cokriged fields Y^{**} and Z^{**} satisfy the corresponding partial differential equation

$$\Delta Z^{**} = Y^{**}$$

---

[11] $G(x, x')$ is zero on the boundary Γ. For fixed $x' = x_0$, $G(., x_0)$ corresponds to the solution of the Poisson equation for a point source located at x_0

$$\Delta G(x, x_0) = \delta(x - x_0)$$

If the boundary Γ is at infinity, G is an isotropic (generalized) function of $r = |x' - x|$ that depends on the space dimension n: $G(r) = r/2$ if $n = 1$, $= (\log r)/(2\pi)$ if $n = 2$, $= -1/(4\pi r)$ if $n = 3$, and in general for $n \neq 2$, $G(r) = -\Gamma(n/2)/[2\,(n-2)\,\pi^{n/2}\,r^{n-2}]$ where the last Γ stands for the gamma function.

In particular, at points x_α where measurements of Y are available, we have $\Delta Z^{**}(x_\alpha) = Y(x_\alpha)$. In the same manner we can reconstruct conditional co-simulations of Y and Z (at least in the Gaussian case) satisfying the Poisson equation.

4.9.2 Stochastic Hydrogeology

Hydrogeology is the earliest and most prolific field of application of geostatistics to variables whose physical behavior is governed by partial differential equations. In an aquifer, permeability and the piezometric head are related by the flow equation: When permeability and boundary conditions are given, the head is determined. Therefore consideration of the flow equation is the key for developing physically consistent stochastic models of permeability and head.

Permeability observations are usually scarce, because they require setting up monitoring wells to measure the water level and carry out pumping tests, from which permeabilities are deduced. Piezometry data are easier to get and since the water level is directly related to the permeability field, piezometry data are used to reduce the uncertainty on permeability. This leads to the *inverse problem*: Find a permeability field that matches the permeability data and is consistent with the piezometry measurements.

When variables are treated as deterministic, the partial differential system has an infinite number of solutions that can be vastly different, many of them devoid of physical sense. A geostatistical approach makes it possible to capture spatial heterogeneity and thereby restrict the space of possible solutions. It also deals with the uncertainty due to the scarcity of data. In the end the set of solutions is represented as a family of conditional simulations.

The long list of contributors to stochastic hydrogeology includes Matheron himself with his 1967 monograph on porous media, de Marsily, Delhomme, Freeze, Dagan, Neuman, Gelhar, Gutjahr, Kitanidis, Rubin, Gómez-Hernández, Renard, to name only a few. A 2005 special issue of the *Hydrogeology Journal*, entitled "The Future of Hydrogeology," contains a number of papers related to geostatistics—in particular, those of de Marsily et al. dealing with spatial heterogeneity, Noetinger et al. on upscaling (further discussed in Section 7.10.2), and Carrera et al. on the inverse problem.

CHAPTER 5

Multivariate Methods

There are mountains here and hairy gooseberries don't grow on mountains

—D. H. McLain

5.1 INTRODUCTION

With kriging we estimate unknown values of a variable of interest Z from known values of the same variable Z. However, in applications the data often include values of other variables, possibly at other locations. These *secondary variables* may provide useful information on Z and should be taken into account. This calls for a multivariate generalization of kriging, which is developed in this chapter under the name of *cokriging*.

Notations and the geometry of sample sets are the main difficulties with cokriging theory, otherwise it is essentially the same as that of kriging. When drifts are present, one must consider the possible functional dependencies between them, but this is rather straightforward once the mechanism for deriving the equations is understood. We will not attempt to be exhaustive but will explain the general principles and concentrate on a few interesting problems. The real theoretical challenge is this: How can we model the relationships between the different variables in a globally coherent manner? And the practical challenge is: How can we infer this model from an often undersampled data set?

Due to the difficulty of building a complete multivariate model, simplified implementations are often used. The most popular one is known as *collocated cokriging*; it is optimal under a specific covariance model with screening properties. The common alternative to collocated cokriging is the external drift

Geostatistics: Modeling Spatial Uncertainty, Second Edition. J.P. Chilès and P. Delfiner.
© 2012 John Wiley & Sons, Inc. Published 2012 by John Wiley & Sons, Inc.

model presented in Chapter 3, based on linear regression, which treats secondary variables as deterministic functions known everywhere.

A number of applications where a variable is estimated from other variables are discussed in this chapter. Of particular interest are (a) the use of gradient information to constrain the shape of potential lines (for example, impose boundary conditions on a flow model) and (b) how the method can be developed toward potential function mapping.

In the atmospheric sciences and elsewhere, data are localized in both space and time. While the general cokriging framework would be formally applicable, specific space–time models have been developed to account for the specific nature of the time coordinate. Considerable progress has been achieved in the last 10 years or so to move away from the previous separability assumptions of time and space and allow space–time interaction either through specific covariance models or by physical modeling of the time evolution of dynamic systems. We will give a short account of the state of the art in these areas.

5.2 NOTATIONS AND ASSUMPTIONS

With cokriging theory, one quickly becomes short of space and names for indexes. We have tried to emphasize clarity by using a mix of Latin and Greek subscripts and matrix notations.

1. We are considering p simultaneous random functions $Z_i(x)$, indexed by Latin subscripts i ranging in the set $I = \{1, \ldots, p\}$ and defined over a domain D of $\mathbb{R}^n$:

$$\{Z_i(x) : x \in D \subset \mathbb{R}^n\}$$

Each $Z_i(x)$ is sampled over a set S_i of $N_i \geq 0$ points, referenced as usual with Greek indexes—which makes the distinction with random function indexes very clear:

$$S_i = \{x_\alpha \in D : Z_i(x_\alpha) \text{ known}\}$$

For simplicity we use the same notation x_α for generic data points, but *the sample sets S_i are in general different for the different indexes.* They may be pairwise disjoint, and some may even be empty. In fact the relationships between the different sets constitute one of the key aspects of a multivariate estimation problem.

2. As with universal kriging, we assume that each function $Z_i(x)$ may have a drift $m_i(x)$ that can be represented as a linear combination with unknown coefficients of known basis drift functions $f_i^\ell(x), \ell = 0, \ldots, L_i$, which in general may be different for the different RFs:

$$E[Z_i(x)] = m_i(x) = \sum_\ell a_{i\ell} f_i^\ell(x)$$

Cross-covariances between $Z_i(x_\alpha)$ and $Z_j(x_\beta)$ will be denoted by $\sigma_{ij}(x_\alpha, x_\beta)$:

$$\sigma_{ij}(x_\alpha, x_\beta) = \text{Cov}[Z_i(x_\alpha), Z_j(x_\beta)] = \text{E}[Z_i(x_\alpha)\,Z_j(x_\beta)] - m_i(x_\alpha)\,m_j(x_\beta)$$

At this point we make no assumption on covariances $\sigma_{ij}(x, y)$ other than the fact that they exist and are mathematically consistent. In the stationary case these covariances are of the form $C_{ij}(h)$ with $h = y - x$.

3. To avoid carrying quadruple indexes such as those above, it is useful to define vector notations. There are two ways of doing that. The natural one is to consider a vector-valued random function $\mathbf{Z}(x)$ (denoted with a bold $\mathbf{Z}$) whose components are the p RFs, namely

$$\mathbf{Z}(x) = \left(Z_1(x), Z_2(x), \ldots, Z_p(x)\right)'$$

However, this formulation implicitly assumes an equal sampling of all components, which is usually not the case. The other way, which we prefer to use here, is to pool all observations relative to the ith variable into an N_i-vector:

$$\mathbf{Z}_i = (Z_i(x_1), Z_i(x_2), \ldots, Z_i(x_{N_i}))'$$

Accordingly, an N_i-vector of weights is defined by $\boldsymbol{\lambda}_i = (\lambda_{i1}, \lambda_{i2}, \ldots, \lambda_{iN_i})'$ so that

$$\boldsymbol{\lambda}'_i\,\mathbf{Z}_i = \sum_\alpha \lambda_{i\alpha} Z_i(x_\alpha)$$

where the summation in α is short for "sum over all points $x_\alpha \in S_i$".

The covariance matrix between $\mathbf{Z}_i$ and $\mathbf{Z}_j$ data is the $N_i \times N_j$ matrix with general term $\sigma_{ij}(x_\alpha, x_\beta)$ (or $C_{ij}(x_\beta - x_\alpha)$ in the stationary case), $x_\alpha \in S_i$, $x_\beta \in S_j$. The letter $\mathbf{C}$ will be used for covariance matrices instead of Σ to avoid confusion with the summation sign:

$$\mathbf{C}_{ij} = \text{Cov}(\mathbf{Z}_i, \mathbf{Z}_j) = \text{E}(\mathbf{Z}_i\,\mathbf{Z}'_j) - \text{E}(\mathbf{Z}_i)\text{E}(\mathbf{Z}'_j)$$

In general, $\mathbf{C}_{ij}$ is not a symmetric matrix (even when it is a square matrix), however, from the above we have $\mathbf{C}_{ji} = \mathbf{C}'_{ij}$.

The vector of mean values can be written as

$$\text{E}(\mathbf{Z}_i) = \mathbf{F}_i\,\mathbf{a}_i$$

where $\mathbf{F}_i = \left[f_i^\ell(x_\alpha)\right]$ is the matrix of drift functions arranged by columns ℓ and rows α ($x_\alpha \in S_i$) and $\mathbf{a}_i = (a_{i\ell})$ is the column vector of drift coefficients for the RF Z_i, just as in standard UK theory.

4. The estimated variable corresponds to the value $i = 1$ of the subscript and will be called the *primary variable*, by opposition to the *secondary variables* with indexes $i \neq 1$. We will focus here on the estimation of the primary variable at a single target point x_0, knowing that any linear functional can be linearly deduced from such point estimates. Our objective is thus

$$Z_0 = Z_1(x_0)$$

Its mean value is

$$E(Z_0) = \sum_{\ell} a_{1\ell} f_1^{\ell}(x_0) = \mathbf{a}_1' \mathbf{f}_{10}$$

where $\mathbf{f}_{10} = \left(f_1^{\ell}(x_0) \right)$ is the vector of drift functions values at the point x_0 and $\mathbf{a}_1 = (a_{1\ell})$ is the vector of drift coefficients for the RF Z_1.

The covariance between the data vector $\mathbf{Z}_i$ and the objective Z_0 is the N_i-vector (lowercase)

$$\mathbf{c}_{i0} = \text{Cov}(\mathbf{Z}_i, Z_0) = E(\mathbf{Z}_i Z_0) - E(\mathbf{Z}_i) E(Z_0)$$

and the variance of the objective is

$$c_{00} = \text{Cov}(Z_0, Z_0) = \text{Var}(Z_0)$$

5. Finally the cokriging estimator of Z_0 is an affine function of *all* available data of the form

$$Z^{**} = \sum_{i \in I} \sum_{\alpha \in S_i} \lambda_{i\alpha} Z_i(x_\alpha) + \lambda_0 = \sum_{i \in I} \boldsymbol{\lambda}_i' \mathbf{Z}_i + \lambda_0$$

The double asterisk (**) is introduced to emphasize the difference with the kriging estimator; and the subscript CK will be used for the variance.

5.3   SIMPLE COKRIGING

Simple cokriging corresponds to the case where the means of the estimating and estimated RFs are known. Just as for simple kriging, elimination of the bias leads to

$$\lambda_0 = m_1(x_0) - \sum_i \sum_\alpha \lambda_{i\alpha} m_i(x_\alpha)$$

This amounts to subtracting the means and working with the zero-mean RFs:

$$Y_i(x) = Z_i(x) - m_i(x)$$

In the rest of this section we will consider that all RFs have zero means.

5.3.1 Derivation of the Equations

In order to estimate $Z_0 = Z_1(x_0)$ from the RFs $Z_i(x)$ known over the sample sets S_i, $i \in I$, we consider a linear estimator of the form

$$Z^{**} = \sum_i \sum_\alpha \lambda_{i\alpha} Z_i(x_\alpha) = \sum_i \lambda_i' \mathbf{Z}_i$$

Since the mean is zero, the mean square error is equal to the error variance:

$$E\left(Z^{**} - Z_0\right)^2 = E\left[\sum_i \sum_j \lambda_i' \mathbf{Z}_i \mathbf{Z}_i' \lambda_j - 2 \sum_i \lambda_i' \mathbf{Z}_i Z_0 + Z_0 Z_0\right]$$

$$= \sum_i \sum_j \lambda_i' \mathbf{C}_{ij} \lambda_j - 2 \sum_i \lambda_i' \mathbf{c}_{i0} + c_{00}$$

Canceling the partial derivatives with respect to λ_i yields the simple cokriging system and variance.

Simple Cokriging System

$$\sum_j \mathbf{C}_{ij} \lambda_j = \mathbf{c}_{i0}, \qquad i = 1, \ldots, p \tag{5.1}$$

$$\sigma_{CK}^2 = E\left(Z^{**} - Z_0\right)^2 = c_{00} - \sum_i \lambda_i' \mathbf{c}_{i0} \tag{5.2}$$

5.3.2 Simple Cokriging as a Projection

Just like its univariate brother, simple cokriging admits of a geometric interpretation in terms of projection onto a Hilbert space. Following Matheron (1970), we consider RFs defined over the product space

$$E = \mathbb{R}^n \times I$$

of $\mathbb{R}^n$ by the set $I = \{1, \ldots, p\}$ of indexes. An element of E is a pair (x, i), where $x \in \mathbb{R}^n$ and $i \in I$. The p random functions $Z_i(x)$ can be regarded as a single RF $Z(x, i)$ on this product space. Similarly, the sampling set of the RF $Z(x, i)$ is the collection of pairs (x, i) such that the point x belongs to S_i:

$$S = \{(x, i) : x \in \mathbb{R}^n, i \in I, x \in S_i\}$$

Now consider the Hilbert subspace $\mathcal{H}_S$ generated by linear combinations of zero-mean finite variance random variables $\{Z(x, i) : (x, i) \in S\}$ (i.e., the sampled ones) and all their L^2 limits, with the usual scalar product $\langle X, Y \rangle = E(XY)$. In explicit form, equations (5.1) can be written as

$$\text{Cov}\left[Z^{**}, Z_i(x_\alpha)\right] = \text{Cov}[Z_0, Z_i(x_\alpha)] \qquad \forall i \in I, x_\alpha \in S_i$$

which translates into

$$\langle Z^{**} - Z_0, X \rangle = 0 \qquad \forall X \in \mathcal{H}_S \tag{5.3}$$

showing that the cokriging estimator is the projection of Z_0 onto $\mathcal{H}_S$. Replacing X in (5.3) by all the $Z_i(x_\alpha)$, which form a basis of $\mathcal{H}_S$, leads to the system (5.1). Furthermore, since $Z^{**} \in \mathcal{H}_S$, the cokriging error is orthogonal to the cokriging estimator

$$\langle Z^{**} - Z_0, Z^{**} \rangle = 0$$

and the smoothing relationship holds:

$$\mathrm{Var}\, Z_0 = \mathrm{Var}\, Z^{**} + \sigma^2_{\mathrm{CK}}$$

This geometric formulation provides a synthetic expression of cokriging equations in terms of orthogonality relations that proves useful to shortcut cumbersome algebra. To take a specific example, consider, following Myers (1983), the problem of estimating a given linear combination of variables:

$$W(x) = \sum_{i=1}^{p} w_i\, Z_i(x)$$

(e.g., the sum over several geological horizons or the sum of values for different minerals). When all variables are equally sampled, we have the choice of either direct estimation of $W_0 = W(x_0)$ by kriging, or estimation by cokriging. How do they compare? For simplicity let us assume here that all means are zero. For every variable i the cokriging estimator of $Z_i(x_0)$ satisfies (5.3), so by linearity the linear combination

$$W^{**} = \sum_{i=1}^{p} w_i\, Z_i^{**}(x_0)$$

also satisfies (5.3) and is therefore *the* cokriging estimator of W_0. Consider now the straight kriging estimator W^*. The difference $W^* - W^{**} \in \mathcal{H}_S$ so that

$$\langle W^{**} - W_0, W^* - W^{**} \rangle = 0$$

and

$$\| W_0 - W^* \|^2 = \| W_0 - W^{**} \|^2 + \| W^{**} - W^* \|^2$$

The cokriging estimator has a smaller mean square error than the straight kriging estimator. Note, in passing, that cokriging ensures consistency: The estimate of the sum is the sum of the estimates, which would not be true if each variable Z_i were estimated separately by kriging.

5.4 UNIVERSAL COKRIGING

The simplicity and generality of simple cokriging is marred by the introduction of drifts because, to derive the equations, we have to specify if and how the means $m_i(x)$ of the different variables are related. Three cases will be considered: (1) algebraically independent drifts, (2) linearly dependent drifts, or (3) mixed case. The fourth possible case, nonlinear functional relationships between drifts, cannot be handled by universal cokriging; transformations and reparameterizations are required that are not discussed here.

Given the diversity of multivariate situations our intent here is not to be exhaustive but to lay out the general principles that will enable the user to develop the cokriging equations for a particular problem. We will focus on point estimation. The equations for block estimation and for drift estimation can be established in a similar straightforward manner.

5.4.1 Theory for Algebraically Independent Drifts

Here we assume that each RF $Z_i(x)$ has a drift of its own and that the drift coefficients of the different variables are not related. As before, we wish to estimate $Z_0 = Z_1(x_0)$ using an estimator of the form

$$Z^{**} = \sum_i \lambda_i' Z_i$$

We have

$$E(Z^{**} - Z_0) = \sum_i \lambda_i' F_i a_i - f_{10}' a_1$$

and want to make this zero for all drift coefficient vectors a_i and a_1. There are two cases to consider:

1. $S_1 = \{\emptyset\}$. Then the drifts of Z^{**} and Z_0 have no coefficient in common and the above can hold as an identity only if $f_{10} = 0$. But, in general, $f_{10} \neq 0$ (typically there is at least a 1) so that, except for particular cases, estimation is impossible if there is no observation of the primary variable.
2. $S_1 \neq \{\emptyset\}$. Then universal unbiasedness is achieved if and only if

$$\lambda_i' F_i = 0 \quad \text{for} \quad i \neq 1 \qquad \text{and} \qquad \lambda_i' F_1 = f_{10}'$$

These conditions can be rewritten in the synthetic form:

$$\lambda_i' F_i = \delta_{i1} f_{10}', \qquad \text{where} \quad \delta_{i1} = \begin{cases} 1 & \text{if } i = 1, \\ 0 & \text{if } i \neq 1 \end{cases}$$

Now

$$\mathrm{Var}\left(Z^{**} - Z_0\right) = \sum_i \sum_j \lambda_i' C_{ij} \lambda_j - 2 \sum_i \lambda_i' c_{i0} + c_{00}$$

Minimizing this variance subject to the unbiasedness constraints leads to the system with Lagrange parameters vector $\mu_i = (\mu_{i\ell})'$, $\ell = 0, \ldots, L_i$.

Universal Cokriging System

$$\begin{cases} \displaystyle\sum_j C_{ij} \lambda_j + F_i \mu_i = c_{i0} & i = 1, \ldots, p \\[2mm] F_i' \lambda_i = f_{10} \delta_{i1} & i = 1, \ldots, p \end{cases} \tag{5.4}$$

Cokriging Variance

$$\sigma_{\mathrm{CK}}^2 = \mathrm{E}\left(Z^{**} - Z_0\right)^2 = c_{00} - \sum_i \lambda_i' c_{i0} - \mu_1' f_{10} \tag{5.5}$$

(Note: If $\mathrm{E}[Z_i] = 0$ the corresponding matrix F_i is absent from the system.)

To take a specific example, if $p = 2$, the structure of the cokriging system is as follows:

$$\underbrace{\begin{bmatrix} C_{11} & C_{12} & F_1 & 0 \\ C_{21} & C_{22} & 0 & F_2 \\ \hdotsfor{4} \\ F_1' & 0 & 0 & 0 \\ 0 & F_2' & 0 & 0 \end{bmatrix}}_{A} \underbrace{\begin{bmatrix} \lambda_1 \\ \lambda_2 \\ \cdots \\ \mu_1 \\ \mu_2 \end{bmatrix}}_{X} = \underbrace{\begin{bmatrix} c_{10} \\ c_{20} \\ \cdots \\ f_{10} \\ 0 \end{bmatrix}}_{B} \tag{5.6}$$

where on the right-hand side

$$c_{10} = \left(\mathrm{Cov}[Z_1(x_\alpha), Z_1(x_0)], x_\alpha \in S_1\right)'$$
$$c_{20} = \left(\mathrm{Cov}[Z_2(x_\beta), Z_1(x_0)], x_\beta \in S_2\right)'$$

are the vectors of covariances between Z_1 and Z_0, and Z_2 and Z_0. Also,

$$\sigma_{\mathrm{CK}}^2 = c_{00} - X'B \tag{5.7}$$

Unlike kriging weights which do not depend on the variance level, cokriging weights depend on the *relative* dispersions of the different variables. For example, given auto- and cross-correlations between Z_1 and Z_2, the cokriging weights on the secondary variable Z_2 are proportional to the ratio of standard deviations σ_1/σ_2. This shows the importance of an accurate fit of covariances for use in cokriging.

Existence and Uniqueness of a Solution

With obvious notations the cokriging matrix $\mathbf{A}$ in (5.6) is of the form

$$\mathbf{A} = \begin{bmatrix} \mathbf{C} & \mathbf{F} \\ \mathbf{F}' & \mathbf{0} \end{bmatrix}$$

Notice that $\mathbf{C}$ is symmetric owing to the symmetry property $\mathbf{C}_{ji} = \mathbf{C}'_{ij}$. In fact the cokriging system has the same structure as the standard UK system.

If the multivariate covariance model is strictly positive definite (discussed later) and if there is no data duplication, the matrix $\mathbf{C}$ is nonsingular. Then, just as in the UK theory, the matrix $\mathbf{A}$ is nonsingular if and only if the columns of $\mathbf{F}$ are linearly independent—that is, if

$$\mathbf{F}\mathbf{a} = \mathbf{0} \quad \Rightarrow \quad \mathbf{a} = \mathbf{0}$$

Given the block diagonal structure of $\mathbf{F}$, this is equivalent to

$$\begin{cases} \mathbf{F}_1 \mathbf{a}_1 = \mathbf{0} & \Rightarrow \quad \mathbf{a}_1 = \mathbf{0} \\ \quad \vdots & \qquad \vdots \\ \mathbf{F}_p \mathbf{a}_p = \mathbf{0} & \Rightarrow \quad \mathbf{a}_p = \mathbf{0} \end{cases}$$

that is, every $\mathbf{F}_i$ is of full rank equal to its number of columns.

Each matrix $\mathbf{F}_i$ has N_i rows (the number of data points in S_i) and $L_i + 1$ columns (the number of basis drift functions). For the rank of $\mathbf{F}_i$ to be $L_i + 1$, it is required that $N_i \geq L_i + 1$. For a secondary variable the practical minimum is in fact $L_i + 2$, or else the variable plays no role. Indeed, if $N_i = L_i + 1$, the matrix $\mathbf{F}_i$ is then a square nonsingular matrix, and the constraint $\mathbf{F}'_i \boldsymbol{\lambda}_i = \mathbf{0}$ entails $\boldsymbol{\lambda}_i = \mathbf{0}$.

Discussion

1. It is intuitively obvious that if all variables are mutually uncorrelated (orthogonal) and if their means are algebraically independent, the secondary variables are of no help for the estimation of the primary variable. Indeed in this case the cokriging solution coincides with the UK estimator based on the primary variable alone. This is easily verified on (5.6) by letting $\mathbf{C}_{12} = \mathbf{C}_{21} = \mathbf{0}$ and $\mathbf{c}_{20} = \mathbf{0}$.

2. At the same time, it is difficult to conceive of an actual situation in which two variables are allowed to drift away independently and yet carry useful information about each other: The variations of the unknown drifts would dominate those of the residuals. The only case of practical interest for algebraically independent drifts is when all the means are constants, namely *ordinary cokriging*.

3. There must be at least enough data points of the primary (i.e., estimated) variable to be able to estimate its mean. However, if the primary variable has a zero mean, or equivalently if its mean is known, there is no minimum number

of points. In this particular case, estimation of the primary variable is possible on the basis of secondary variables only.

4. Being constrained by $\mathbf{F}_i' \boldsymbol{\lambda}_i = \mathbf{0}$, the secondary variables contribute to the estimator only as corrections (they only involve the true residuals of Z_i). Estimation will be improved only if the secondary variables are strongly correlated with the objective. Also, as noted above, the weights on secondary variables are nonzero only if their number exceeds the number of basis drift functions. For example, the weight assigned to a single secondary data point with an unknown mean is necessarily zero (but not so if the mean is known).

5. Instead of estimating a single variable, we may want to estimate p of them simultaneously. The efficient way to do this is to perform all calculations at once and solve one linear system with p right-hand sides rather than p linear systems with one right-hand side.

The cokriging estimator of a linear combination of variables is simply the same linear combination of the individual cokriging estimators of these variables. This quantity can be obtained directly by solving the system (5.4) for a right-hand side which is the same linear combination of the individual right-hand sides. The converse, however, is not true: It may be possible to estimate a linear combination (e.g., one that filters the drift) without being able to estimate its components individually.

5.4.2 A Worked-Out Example of Ordinary Cokriging: Incompletely Drilled Wells

In order to demonstrate the use of ordinary cokriging equations, we will work out a synthetic example that is also of practical interest. We consider two variables Z_1 and Z_2 with constant but unknown means m_1 and m_2—for example, the depths (counted positively downward) of

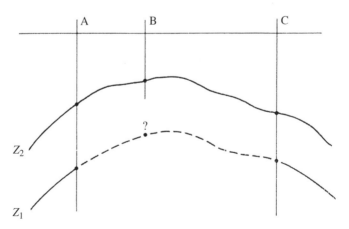

FIGURE 5.1 Estimation of the depth of Z_1 at well B from known depths of Z_1 and Z_2. Wells A and C have been drilled through both surfaces, but well B is incomplete.

two geological surfaces, where $Z_1 \geq Z_2$. If depths are determined from vertical wells, the coordinates of sample points are the same for the two surfaces—except that in some of the wells drilling may have been interrupted before hitting the deeper surface. The sample sets are related by $S_1 \subset S_2$. If wells are deviated, the sampling points are different for the two surfaces, and we may have $S_1 \cap S_2 = \emptyset$. For the example we consider the simple situation of Figure 5.1 with three vertical wells located at points A, B, C and well B drilled only down to the shallower surface. We wish to estimate the depth Z_1 at location x_B using all available depth values.

The cokriging estimator of $Z_1(x_B)$ is of the form

$$Z_1^{**}(x_B) = \lambda_{1A}Z_1(x_A) + \lambda_{1C}Z_1(x_C) + \lambda_{2A}Z_2(x_A) + \lambda_{2B}Z_2(x_B) + \lambda_{2C}Z_2(x_C)$$

Letting $p = 2$, and $x_0 = x_B$ the general cokriging equations (5.4) are

$$
\begin{bmatrix}
C_{11}(0) & C_{11}(AC) & C_{12}(0) & C_{12}(AB) & C_{12}(AC) & 1 & 0 \\
C_{11}(CA) & C_{11}(0) & C_{12}(CA) & C_{12}(CB) & C_{12}(0) & 1 & 0 \\
C_{21}(0) & C_{21}(AC) & C_{22}(0) & C_{22}(AB) & C_{22}(AC) & 0 & 1 \\
C_{21}(BA) & C_{21}(BC) & C_{22}(BA) & C_{22}(0) & C_{22}(BC) & 0 & 1 \\
C_{21}(CA) & C_{21}(0) & C_{22}(CA) & C_{22}(CB) & C_{22}(0) & 0 & 1 \\
1 & 1 & 0 & 0 & 0 & 0 & 0 \\
0 & 0 & 1 & 1 & 1 & 0 & 0
\end{bmatrix}
\begin{bmatrix}
\lambda_{1A} \\
\lambda_{1C} \\
\lambda_{2A} \\
\lambda_{2B} \\
\lambda_{2C} \\
\mu_1 \\
\mu_2
\end{bmatrix}
=
\begin{bmatrix}
C_{11}(AB) \\
C_{11}(CB) \\
C_{21}(AB) \\
C_{21}(0) \\
C_{21}(CB) \\
1 \\
0
\end{bmatrix}
\quad (5.8)
$$

where AB is shorthand for $x_B - x_A$.

Direct covariances $C_{11}(h)$ and $C_{22}(h)$ are symmetric, whereas cross-covariances are not but satisfy $C_{12}(AB) = C_{21}(BA)$ so that the matrix of the cokriging system is indeed symmetric.

To go further and get interpretable results, we must now particularize the covariance models. We will consider two cases: the additive model and the proportional covariance model.

Additive Model

In this model the thickness $H(x)$ of the layer delimited by the two surfaces is assumed uncorrelated with the top depth $Z_2(x)$:

$$Z_1(x) = Z_2(x) + H(x) \quad \text{with} \quad \text{Cov}[Z_2(x), H(y)] = 0 \quad \forall x, y$$

The covariance structure becomes

$$C_{12}(h) = C_{21}(h) = C_{22}(h) = C(h),$$
$$C_{11}(h) = C(h) + K(h)$$

where $K(h)$ is the covariance of $H(x)$. Substituting into (5.8), we find that the solution satisfies

$$\lambda_{2A} = -\lambda_{1A} \qquad \lambda_{2B} = 1 \qquad \lambda_{2C} = -\lambda_{1C} \qquad \mu_2 = 0$$

so that

$$Z_1^{**}(x_B) = Z_2(x_B) + \lambda_{1A}[Z_1(x_A) - Z_2(x_A)] + \lambda_{1C}[Z_1(x_C) - Z_2(x_C)]$$

where λ_{1A} and λ_{1C} are solutions of the system

$$
\begin{cases}
K_{AA}\lambda_{1A} + K_{AC}\lambda_{1C} + \mu_1 = K_{AB} \\
K_{CA}\lambda_{1A} + K_{CC}\lambda_{1C} + \mu_1 = K_{CB} \\
\lambda_{1A} + \lambda_{1C} = 1
\end{cases}
$$

which coincides with the ordinary kriging system of $H(x_B)$ from $H(x_A)$ and $H(x_C)$.

In this particular case, cokriging amounts to the obvious solution of estimating the thickness $H(x_B)$ by univariate kriging from $H(x_A)$ and $H(x_C)$ and adding the estimate to $Z_2(x_B)$:

$$
Z_1^{**}(x_B) = Z_2(x_B) + H^*(x_B)
$$

The cokriging variance computed from (5.5) coincides with the OK variance of $H(x_B)$:

$$
\sigma_{CK}^2 = K_{00} - \lambda_{1A} K_{AB} - \lambda_{1C} K_{CB} - \mu_1
$$

Proportional Covariance Model

In this model, all direct and cross-covariances are proportional to the same correlogram $C(h)$:

$$
\begin{aligned}
C_{11}(h) &= \sigma_1^2 C(h), \\
C_{22}(h) &= \sigma_2^2 C(h), \\
C_{12}(h) &= C_{21}(h) = \rho\,\sigma_1\sigma_2 C(h), \qquad -1 < \rho < 1
\end{aligned}
$$

That this is a valid multivariate covariance model is shown in Section 5.6.1, Example 3. Substituting into (5.8), we see that the last three covariance equations are satisfied identically with $\mu_2 = 0$ whenever

$$
\lambda_{2A} = -\rho\frac{\sigma_1}{\sigma_2}\lambda_{1A} \qquad \lambda_{2B} = \rho\frac{\sigma_1}{\sigma_2} \qquad \lambda_{2C} = -\rho\frac{\sigma_1}{\sigma_2}\lambda_{1C}
$$

The remaining equations then become

$$
\begin{cases}
C_{AA}\lambda_{1A} + C_{AC}\lambda_{1C} + \mu_1/[(1-\rho^2)\sigma_1^2] = C_{AB} \\
C_{CA}\lambda_{1A} + C_{CC}\lambda_{1C} + \mu_1/[(1-\rho^2)\sigma_1^2] = C_{CB} \\
\lambda_{1A} + \lambda_{1C} = 1
\end{cases}
$$

One recognizes the OK system for estimating $Z_1(x_B)$ from $Z_1(x_A)$ and $Z_1(x_C)$ and also for estimating $Z_2(x_B)$ from $Z_2(x_A)$ and $Z_2(x_C)$, without using $Z_2(x_B)$ of course. Finally the cokriging estimator can be written in the form

$$
Z_1^{**}(x_B) = Z_1^*(x_B) + \rho\frac{\sigma_1}{\sigma_2}\left[Z_2(x_B) - Z_2^*(x_B)\right] \tag{5.9}
$$

where $Z_1^*(x_B)$ and $Z_2^*(x_B)$ are the OK estimators of $Z_1(x_B)$ and $Z_2(x_B)$ based on the observations of Z_1 alone and Z_2 alone at points A and C.

The result is strikingly simple and intuitive: Cokriging improves ordinary kriging estimation of Z_1 by adding a correction equal to $\rho \sigma_1 / \sigma_2$ times the ordinary kriging error observed for Z_2. The cokriging variance is found to be

$$\sigma_{CK}^2 = \sigma_K^2 \left(1 - \rho^2 \right)$$

and is indeed smaller than the OK variance σ_K^2 of Z_1.

Under this model, cokriging estimation can be decoupled into two identical kriging problems. The same results are found if simple, instead of ordinary, kriging and cokriging are used.

For future reference it will be useful to rewrite (5.9) as

$$Z_1^{**}(x_B) = \rho \frac{\sigma_1}{\sigma_2} Z_2(x_B) + \lambda_{1A} \left[Z_1(x_A) - \rho \frac{\sigma_1}{\sigma_2} Z_2(x_A) \right] + \lambda_{1C} \left[Z_1(x_C) - \rho \frac{\sigma_1}{\sigma_2} Z_2(x_C) \right] \quad (5.10)$$

5.4.3 Collocated Cokriging

In the previous example, both variables are sampled equally (are collocated) except that the secondary variable is also available at the estimated point—where the primary variable of course is not. A simplified implementation of cokriging for two variables, introduced by Xu et al. (1992) under the name of *collocated cokriging*, is to retain only the value of the secondary variable that is co-located with the estimated point, thus using a total of $N+1$ values. Another implementation also includes the secondary data collocated with the primary data points—that is, $2N+1$ values in total. We propose to give this latter implementation the short name of collocated cokriging and name the less used single-point implementation *strictly collocated cokriging*. Both methods have become popular as techniques to simplify cokriging.

MM1 Model

Strictly collocated cokriging is implemented as simple cokriging, because otherwise the weight on the secondary data point would be zero (although the primary variable could have an unknown mean). The method is usually applied in conjunction with the special correlation structure:

$$\rho_{12}(h) = \rho_{12}(0) \, \rho_{11}(h) \quad (5.11)$$

where $\rho_{11}(h)$ is the correlogram of Z_1 and $\rho_{12}(h)$ is the cross-correlogram of Z_1 and Z_2. No assumption is made for the correlogram of Z_2 because it is not needed when a single secondary value is used.

Under this model the strictly collocated cokriging estimator is a variance-weighted linear combination of the simple kriging estimator of $Z_1(x_0)$ from Z_1 data and the linear regression estimator of $Z_1(x_0)$ from $Z_2(x_0)$:

$$Z_1^{**}(x_0) = \frac{\left(1 - \rho^2\right) Z_1^*(x_0) + \sigma_{SK}^2 \, \rho \, Z_2(x_0)}{\left(1 - \rho^2\right) + \rho^2 \, \sigma_{SK}^2}, \qquad \sigma_{CK}^2 = \sigma_{SK}^2 \, \frac{\left(1 - \rho^2\right)}{\left(1 - \rho^2\right) + \rho^2 \, \sigma_{SK}^2}$$

$$(5.12)$$

Here $\rho = \rho_{12}(0)$ and Z_1 and Z_2 have zero mean and unit variance. Necessarily $\sigma_{SK}^2 \leq 1$, with equality when the kriging estimate is zero (the mean), in which case the cokriging estimate reduces to the linear regression estimate $\rho Z_2(x_0)$. Formulas are derived from (5.1) and (5.2) and can also be established by Bayesian analysis (Doyen et al., 1996).

Model (5.11), introduced by Almeida and Journel (1994) as a "Markov-type" model and named MM1 (Journel, 1999), has the following conditional independence property in the Gaussian case (see Appendix, Section A.9): when $Z_1(x_0 + h)$ is known $Z_1(x_0)$ and $Z_2(x_0 + h)$ are independent. The primary value $Z_1(x_0 + h)$ "screens" $Z_1(x_0)$ from the influence of any further-away value $Z_2(x_0 + h)$. In other words, under (5.11) the simple cokriging estimator of $Z_1(x_0)$ from $Z_1(x_0 + h)$ and $Z_2(x_0 + h)$ assigns a zero weight to $Z_2(x_0 + h)$.

The model $Z_2(x) = Z_1(x) + \varepsilon(x)$, where $\varepsilon(x)$ is a pure nugget effect, provides an easy illustration of the screening effect: At locations where Z_1 is available, Z_2 brings no additional information since ε is independent of everything else, and with simple cokriging these Z_2 values are screened out. This is not necessarily the case with ordinary cokriging because the auxiliary data contribute to the implicit estimation of the means (screening occurs when $S_2 \subset S_1$).

Surprisingly, however, adding a value of Z_2 at a place where Z_1 is not known, and in particular at x_0, breaks the screening effect of Z_1 on Z_2 where they are both present. We can verify this, for example, in (5.9): All secondary data contribute to the estimation, and this remains true with simple cokriging (see also Section 5.6.4). Therefore the MM1 model (5.11) cannot be a justification for strictly collocated cokriging. Reducing the influence of the secondary variable to a single collocated value is a choice of neighborhood, not an optimality property.

MM2 Model

The question then arises: Which covariance model supports collocated cokriging without loss of information? This question has been studied in great detail by Rivoirard (2001, 2004) under various configurations of sampling sets. We will just examine the case of collocated cokriging of two variables with constant unknown means. We assume that $S_1 \subset S_2$, $x_0 \in S_2$, and of course $x_0 \notin S_1$, and instead of (5.11) we consider the alternative model:

$$\rho_{12}(h) = \rho_{12}(0)\,\rho_{22}(h) \tag{5.13}$$

This model, introduced as MM2 (Journel, 1999), has the following conditional independence property in the Gaussian case: When $Z_2(x_0)$ is known, $Z_1(x_0)$ and $Z_2(x_0 + h)$ are independent. The collocated $Z_2(x_0)$ screens $Z_1(x_0)$ from the influence of any further-away datum $Z_2(x_0 + h)$. Said differently, if (5.13) holds, the simple cokriging estimator of $Z_1(x_0)$ from $Z_2(x_0)$ and $Z_2(x_0 + h)$ assigns a zero weight to $Z_2(x_0 + h)$. The converse is also true so that (5.13) is a necessary condition for cokriging to be collocated.

Let σ_1 and σ_2 be the standard deviations of Z_1 and Z_2, let $\rho = \rho_{12}(0)$ and let a be the slope of the linear regression of Z_1 on Z_2; then (5.13) is equivalent to $C_{12}(h) = a\, C_{22}(h)$, that is,

$$\text{Cov}[Z_1(x) - aZ_2(x), Z_2(x+h)] = 0 \qquad \text{with} \quad a = \rho\frac{\sigma_1}{\sigma_2}$$

Therefore Z_1 admits the following orthogonal decomposition:

$$Z_1(x) = a\, Z_2(x) + R(x) \tag{5.14}$$

where $R(x)$ is *spatially* uncorrelated with Z_2, that is, $\text{Cov}[R(x), Z_2(y)] = 0$ for any points x and y. If the means of Z_1 and Z_2 are known and subtracted out, $R(x)$ is the residual from the regression of Z_1 on Z_2 and has mean zero; otherwise it has an unknown mean. The cokriging estimator of $R(x_0)$ is the simple or ordinary kriging of $R(x_0)$ computed with the covariance $C_{RR}(h) = C_{11}(h) - a^2 C_{22}(h)$. Since Z_2 is known at x_0, the cokriging estimator and cokriging variance of $Z_1(x_0)$ are

$$Z_1^{**}(x_0) = a\, Z_2(x_0) + R^*(x_0), \qquad \sigma_{CK}^2 = \text{E}\big[R^*(x_0) - R(x_0)\big]^2 \tag{5.15}$$

The covariance of residuals $C_{RR}(h)$ is the only one needed for cokriging. In Section 5.6.1 (Example 4) it is shown that $C_{RR}(h)$ is a valid covariance if the multivariate model is correct. This model also implies that Z_2 is at least as smooth as Z_1 or smoother, which is typically the case if it is defined over a larger support than Z_1. The results (5.10) of the incompletely drilled wells example illustrates the general formulas (5.15).

If Z_2 is not known at x_0, then in (5.15) $Z_2(x_0)$ is replaced by its SK or OK estimator based on all Z_2 data, and the cokriging variance is augmented by $a^2 \times$ kriging variance of $Z_2(x_0)$.

In conclusion, collocated cokriging is optimal if and only if the cross-covariance is proportional to the covariance of the secondary variable, which is the correct model if the secondary variable is smoother. We arrive at the following paradox noted by Rivoirard (2001): When collocated cokriging is fully justified, we don't need it since it reduces to kriging of a residual. In other cases, however, using a collocated neighborhood remains a convenient simplification.

5.4.4 Cokriging to Reduce the Impact of a Nugget Effect

An early application of cokriging dealt with the estimation of uranium reserves on the basis of chemical analyses from drill-hole cores and of radioactivity measurements (Guarascio, 1976). Although the relationship between radioactivity and uranium is not simple, in particular due to the fact that radioactivity integrates a much larger volume than cores, these data are still useful because they are sampled much more densely. Also, the nugget effect in the uranium grades is integrated out by radioactivity measurements.

In order to evaluate the benefit of cokriging as a function of the relative importance of the nugget effect, let us consider a controlled configuration in which the uranium grade (variable 1) of a block v is estimated from a central drill-hole and four drill-holes symmetrically located far from the block—that is, at a distance greater than the range. Radioactivity measurements (variable 2) are also assumed known at these locations. The two means are constant but unknown, and to simplify calculations, we assume that variable 2 and the continuous component of variable 1 conform to a proportional covariance model. Our complete model is therefore

$$C_{11}(h) = \sigma_0^2 \, \delta(h) + \sigma_1^2 \, \rho(h),$$

$$C_{22}(h) = \sigma_2^2 \, \rho(h),$$

$$C_{12}(h) = \sigma_{12} \, \rho(h)$$

where $\rho(h)$ is an isotropic correlogram. The covariance between the central drill-hole and the mean block value is $\sigma_1^2 \, \rho_{0v}$ while the covariance between the other four samples and that block is zero. Likewise the covariance between the central radiometry sample and the block is $\sigma_{12} \, \rho_{0v}$ and the other four covariances are zero. Finally the variance of the block itself is $\sigma_1^2 \, \rho_{vv}$, assuming that the nugget effect is averaged out. For the example we use the values $\rho_{0v} = 0.718$ and $\rho_{vv} = 0.624$ obtained by integration of a spherical model of unit sill over a square of length half the range. These numbers can also be read from graphs [e.g., Journel and Huijbregts (1978, pp. 127–128)], though with less precision.

Due to the symmetry of the configuration, it is possible to reduce the problem to the determination of the two weights λ_1 and λ_2 assigned to the central sample of Z_1 and Z_2, respectively. The weights assigned to the other four samples are the same and equal to $(1 - \lambda_1)/4$ and $-\lambda_2/4$ respectively. The cokriging system (5.6) comprises four equations:

$$\begin{cases} \lambda_1\left(\sigma_0^2 + \sigma_1^2\right) + \lambda_2 \, \sigma_{12} + \mu_1 = \sigma_1^2 \, \rho_{0v} \\ ((1 - \lambda_1)/4)\left(\sigma_0^2 + \sigma_1^2\right) - (\lambda_2/4)\sigma_{12} + \mu_1 = 0 \\ \lambda_1 \sigma_{12} + \lambda_2 \, \sigma_2^2 + \mu_2 = \sigma_{12} \, \rho_{0v} \\ ((1 - \lambda_1)/4)\sigma_{12} - (\lambda_2/4)\sigma_2^2 + \mu_2 = 0 \end{cases} \tag{5.16}$$

Although the four distant drill-holes have no correlation with the block, they play a role in its estimation: They help define the mean of the primary variable, and they permit λ_2 to be nonzero. Letting $\sigma_{12} = \rho_{12} \, \sigma_1 \sigma_2$ in (5.16), the weights are given by

$$\lambda_1 = \frac{1}{5} + \frac{4}{5} \frac{(1 - \rho_{12}^2)\sigma_1^2}{\sigma_0^2 + (1 - \rho_{12}^2)\sigma_1^2} \, \rho_{0v} \qquad \lambda_2 = \frac{4}{5} \frac{\sigma_0^2 \, \rho_{12}}{\sigma_0^2 + (1 - \rho_{12}^2)\sigma_1^2} \frac{\sigma_1}{\sigma_2} \, \rho_{0v} \tag{5.17}$$

After rearrangement of the terms, the cokriging variance is

$$\sigma_{CK}^2 = \lambda_1\left[\sigma_0^2 + (1 - \rho_{0v})\sigma_1^2\right] + \lambda_2 \sigma_{12}(1 - \rho_{0v}) - \sigma_1^2(\rho_{0v} - \rho_{vv}) \tag{5.18}$$

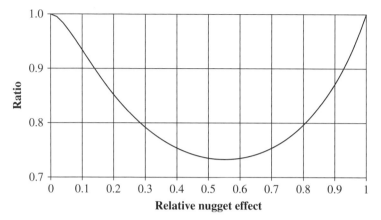

FIGURE 5.2 Cokriging-to-kriging variance ratio for a square panel as a function of the relative nugget effect. Vertical axis is σ_{CK}^2/σ_K^2 and horizontal axis is $\sigma_0^2/(\sigma_0^2 + \sigma_1^2)$.

The solutions for ordinary kriging are obtained by letting $\rho_{12} = 0$ in (5.17) and (5.18). Figure 5.2 displays the cokriging to kriging variance ratio as a function of the relative nugget effect $\sigma_0^2/(\sigma_0^2 + \sigma_1^2)$ for $\rho_{12} = 0.9$. It reaches a minimum of 0.73, representing a variance reduction of 27%, for a relative nugget effect of 0.55. This figure gives an idea of the maximum advantage that can be obtained by cokriging compared to kriging. The fact that $\lambda_2 = 0$ when the nugget effect vanishes is due to the screening effect of the proportional covariance model with equal sampling of all locations (Section 5.6.4). It may be noted that for a given relative nugget effect the cokriging-to-kriging variance ratio does not depend on either σ_1 or σ_2.

This example shows that the degradation of the "signal" due to noise on the primary variable can be partially offset by cokriging, if we have available a smooth secondary variable well correlated with the primary. However, when the noise variance becomes too large, in our example when it exceeds the signal variance, the advantage of cokriging declines.

When both primary and secondary variables have a nugget effect, it is usually the continuous components that contribute to cross-correlations. Marbeau (1976) made an extensive application of ordinary cokriging in forestry, where nugget effects are strong, and found a reduction in error standard deviation of about 10% with a maximum of 15%. Pan et al. (1993) compared cokriging and OK by cross-validation in a case study of a polymetallic deposit and report a standard deviation reduction of about 20%.

5.4.5 Filtering a Systematic Error

Suppose that we have two sets of data for a variable of interest Z_1, where the second set comprises a *systematic* error $\varepsilon(x)$ that is uncorrelated with Z_1 but spatially structured. Can the biased data help estimate Z_1?

Our model is

$$Z_2(x) = Z_1(x) + \varepsilon(x),$$
$$E[\varepsilon(x)] = m_\varepsilon \neq 0 \quad \text{and} \quad \text{Cov}(Z_1, \varepsilon) = 0$$

In the absence of indications on the bias m_ε, the means of Z_1 and Z_2 are algebraically independent and the cokriging equations (5.6) apply. We can therefore use the information in the second data set, provided that the decomposition $C_{22}(h) = C_{11}(h) + C_{\varepsilon\varepsilon}(h)$ is known and, most important, that we can tell which data are biased and which are not.

As usual, this requires the availability of at least enough data of Z_1 to estimate the mean—that is, evaluate the bias; clearly, biased data alone would be worthless unless the bias is very small. At the other extreme, if Z_1 and Z_2 are simultaneously present at most of the points, the practical solution is to fill in the gaps by interpolation of the error field and then base the estimation on the Z_1 data only, both actual and derived from Z_2. In intermediate sample configurations, cokriging can provide a useful solution.

Journel and Huijbregts (1978, p. 342) compared kriging with cokriging in a mining case of block estimation where cuttings are used in addition to a central core analysis, and they report a gain in standard deviation of about 20%.[1]

5.4.6 Universal Cokriging and Linear Regression

Cokriging exploits both correlations due to proximity in space and correlations among variables. We have already noted that when considering independent drifts, cokriging reduces to kriging if correlations among variables are zero, and we will now show that cokriging coincides with linear regression when means are known and all point-to-point correlations are zero. In this sense, cokriging constitutes a generalization of both approaches.

In linear regression theory, one considers independent samples of a random vector $\mathbf{Z} = (Z_1, \ldots, Z_p)'$ drawn from a multivariate distribution with mean vector $\mathbf{m} = (m_1, \ldots, m_p)'$ and a covariance matrix defined by $\sigma_{ij} = \text{Cov}(Z_i, Z_j)$. The linear regression of Z_1 on $Z_2, \ldots, Z_p$ is the function of the form

$$\hat{Z}_1 = \lambda_1 + \lambda_2 Z_2 + \cdots + \lambda_p Z_p$$

minimizing the expected mean square error. The solution is easily found to be

$$\hat{Z}_1 = m_1 + \sum_{i=2}^{p} \lambda_i (Z_i - m_i) \tag{5.19}$$

---

[1] The authors normalize the gain by the cokriging variance, whereas here we normalize it by the kriging variance so that the relative gain is always less than one.

with coefficients λ_i that satisfy the linear system

$$\sum_{j=2}^{p} \sigma_{ij} \lambda_j = \sigma_{i1} \qquad i = 2, \ldots, p$$

These equations coincide with the simple cokriging equations (5.1) for estimating Z_1 from $Z_2, \ldots, Z_p$, where vectors are replaced by scalars (one sample per variable) and $i_0 = 1$, with the set S_1 being empty. In the Gaussian case the estimator also coincides with the regression equation (conditional expectation). Note that the means are assumed constant and known and that no spatial aspect is involved so far.

In linear regression theory the multivariate samples are regarded as *independent* realizations of a parent random vector so that the estimation of a vector component Z_1 only involves the other components of the *same* vector. The equivalent geostatistical model where vectors $\mathbf{Z}$ are indexed by location x is to assume a multivariate nugget effect

$$\text{Cov}[Z_i(x), Z_j(y)] = \sigma_{ij} \, \delta(y - x)$$

or that all ranges are small compared with distances between samples (large grids). The simple cokriging estimator of Z_1 at a point x_0 where the other variables are known, only involves the values at this point and is similar to (5.19) except that the means are possibly location-dependent:

$$Z_1^{**}(x_0) = m_1(x_0) + \sum_{i=2}^{p} \lambda_i \, [Z_i(x_0) - m_i(x_0)] \qquad (5.20)$$

This remains the simple cokriging estimator of Z_1 when ranges are not small but x_0 is so far from the other data points that correlations have vanished (isolated point).

Universal cokriging generalizes this by allowing for unknown means defined in linear parametric form. The result is obtained by simply replacing all means in (5.20) by their cokriging estimates, by virtue of an additivity relationship between simple and universal cokriging similar to that between SK and UK:

$$Z_1^{**}(x_0) = m_1^{**}(x_0) + \sum_{i=2}^{p} \lambda_i [Z_i(x_0) - m_i^{**}(x_0)] \qquad (5.21)$$

Note that, in general, $m_i^{**}(x_0)$ is *not* the least squares estimator based on the values of the ith variables only but may involve all variables at all points. For example, if the data consist of N complete vectors in addition to the $(p-1)$ values at x_0, the means for $i > 1$ coincide with the least squares estimators based on the $N+1$ values of the ith variable, but for $i = 1$ the optimum estimator of the mean turns out to be the least squares estimator from the $N+1$ values $Z_1^{**}(x_0), Z_1(x_1), \ldots, Z_1(x_N)$, where the unknown $Z_1(x_0)$ is replaced by its optimal estimator (5.21). (This looks circular but it isn't!) In practice, the means and the coefficients of the regression are established once and for all by least squares from a subset of complete vectors, and a fixed equation is used thereafter.

The next generalization is to introduce spatial correlation among the vectors $\mathbf{Z}(x)$. But the price to pay for that extra power is the burden of statistical inference of $p(p+1)/2$ covariance functions.

5.4.7 Simultaneous Estimation of Several Variables

Instead of estimating a single variable, we may want to estimate a few or all of them simultaneously. Typical examples are (a) the evaluation of different elements, such as Pb, Zn, and Ag in a polymetallic mineral deposit, or (b) the joint reconstruction of geological surfaces to define a consistent three-dimensional model. This can be achieved by cokriging each variable from all the others, one at a time, but it would be a very inefficient algorithm. The better way is to perform all calculations at once and solve one linear system with p right-hand sides rather than p linear systems with one right-hand side. Denote the cokriging estimator of the kth variable at the point x_0 by

$$Z_k^{**}(x_0) = \sum_i \lambda_{ik}' \mathbf{Z}_i$$

where the second subscript k refers to the kth variable; then the joint cokriging equations for the vector

$$\mathbf{Z}^{**}(x_0) = \left(Z_1^{**}(x_0), Z_2^{**}(x_0), \dots, Z_p^{**}(x_0) \right)'$$

is obtained by writing the equations (5.4) for the different values of the index of the estimated variable:

$$\begin{cases} \sum_j \mathbf{C}_{ij} \lambda_{jk} + \mathbf{F}_i \, \mu_{ik} = \mathbf{c}_{ik} & i = 1, \dots, p; \quad k = 1, \dots, p \\ \mathbf{F}_i' \lambda_{ik} = \mathbf{f}_{k0} \, \delta_{ik} & i = 1, \dots, p; \quad k = 1, \dots, p \end{cases} \tag{5.22}$$

μ_{ik} is a matrix of Language parameters and $\mathbf{f}_{k0}$ is the vector of drift functions associated with the k-th variable and calculated at x_0. The vector $\mathbf{c}_{ik}$ is covariance between $\mathbf{Z}_i$ and $Z_k(x_0)$; it depends on x_0 but that dependence does not show in the notation. In (5.22) the index k ranges from 1 to p, but it could as well vary within a subset if one wanted to estimate a subset of the p variables.

The covariance of cokriging errors is given by

$$E\left[Z_k^{**}(x_0) - Z_k(x_0) \right]\left[Z_l^{**}(x_0) - Z_l(x_0) \right] = c_{kl} - \sum_i \lambda_{il}' \mathbf{c}_{ik} - \mu_{kl}' \mathbf{f}_{k0} \tag{5.23}$$

where c_{kl} is the covariance of $Z_k(x_0)$ and $Z_l(x_0)$. In the simple case $p = 2$, the awesome system (5.22) is simply written as

$$\begin{bmatrix} \mathbf{C}_{11} & \mathbf{C}_{12} & \mathbf{F}_1 & \mathbf{0} \\ \mathbf{C}_{21} & \mathbf{C}_{22} & \mathbf{0} & \mathbf{F}_2 \\ \mathbf{F}_1' & \mathbf{0} & \mathbf{0} & \mathbf{0} \\ \mathbf{0} & \mathbf{F}_2' & \mathbf{0} & \mathbf{0} \end{bmatrix} \begin{bmatrix} \lambda_{11} & \lambda_{12} \\ \lambda_{21} & \lambda_{22} \\ \mu_{11} & \mu_{12} \\ \mu_{21} & \mu_{22} \end{bmatrix} = \begin{bmatrix} \mathbf{c}_{11} & \mathbf{c}_{12} \\ \mathbf{c}_{21} & \mathbf{c}_{22} \\ \mathbf{f}_{10} & \mathbf{0} \\ \mathbf{0} & \mathbf{f}_{20} \end{bmatrix}$$

generalizing (5.6).

The problem of simultaneous cokriging of several variables was first considered by Myers (1982), who solved it by minimizing

$$\sum_{k=1}^{p} E\left[Z_k^{**}(x_0) - Z_k(x_0)\right]^2 = \text{Tr}(\mathbf{M_B})$$

where $\mathbf{M_B}$ is the covariance matrix of cokriging errors for a given matrix of weights $\mathbf{B}$ ensuring unbiasedness. Carr et al. (1985) published a computer program implementing this approach. Alternatively, Ver Hoef and Cressie (1993) proposed to find the matrix $\mathbf{B}$ such that for any other matrix of weights $\mathbf{A}$ ensuring unbiasedness, the matrix $\mathbf{M_A} - \mathbf{M_B}$ is positive definite, implying that any linear combination of variables $Z_i(x_0)$ is estimated better using $\mathbf{B}$ than $\mathbf{A}$. The equations turn out to be the same and coincide with (5.22). The error covariance matrix $\mathbf{M_B}$ is given by (5.23) and can be used to define confidence ellipsoids for the vector $\mathbf{Z}$ $(x_0) = (Z_1(x_0), \ldots, Z_p(x_0))'$.

5.4.8 Algebraic Dependence Between Drifts

We now turn to the case where the drift coefficients of the different variables are related. Each case is different but the principle is the same: The unbiasedness conditions must be modified to reflect the algebraic dependencies. In this section we will just examine the case of two variables with the same mean. Other cases will be encountered later.

Consider the cokriging estimator of $Z_1(x_0)$ based on Z_1 and Z_2:

$$Z_0^{**} = \lambda_1' \mathbf{Z}_1 + \lambda_2' \mathbf{Z}_2$$

If $Z_1(x)$ and $Z_2(x)$ have the same mean, the unbiasedness conditions are of the form

$$EZ_0^{**} = (\lambda_1' \mathbf{F}_1 + \lambda_2' \mathbf{F}_2)\, \mathbf{a} = \mathbf{f}_{10}'\, \mathbf{a} \qquad \forall \mathbf{a}$$

that is,

$$\mathbf{F}_1' \lambda_1 + \mathbf{F}_2' \lambda_2 = \mathbf{f}_{10}$$

and the cokriging system (5.6) becomes

$$\begin{bmatrix} \mathbf{C}_{11} & \mathbf{C}_{12} & \mathbf{F}_1 \\ \mathbf{C}_{21} & \mathbf{C}_{22} & \mathbf{F}_2 \\ \mathbf{F}_1' & \mathbf{F}_2' & 0 \end{bmatrix} \begin{bmatrix} \lambda_1 \\ \lambda_2 \\ \mu \end{bmatrix} = \begin{bmatrix} \mathbf{c}_{11} \\ \mathbf{c}_{21} \\ \mathbf{f}_{10} \end{bmatrix}$$

Here there is a single set of unbiasedness conditions and therefore a single μ. An application of this system is the estimation of point or block values from data with different supports. If the mean is constant, all points and blocks have the same mean, and the result applies directly. If it isn't, the system is still valid, simply the basis drift functions for Z_1 or Z_2 are replaced by their block average equivalents.

The structure of the above system allows for the estimation of Z_1 from Z_2 alone using

$$\begin{bmatrix} \mathbf{C}_{22} & \mathbf{F}_2 \\ \mathbf{F}_2' & \mathbf{0} \end{bmatrix} \begin{bmatrix} \boldsymbol{\lambda}_2 \\ \boldsymbol{\mu} \end{bmatrix} = \begin{bmatrix} \mathbf{c}_{21} \\ \mathbf{f}_{10} \end{bmatrix}$$

which is essentially a UK system in which the right-hand side has been modified to include the cross-covariances between the estimating and the estimated variables instead of direct covariances. We have already encountered one application of this: filtering a nonsystematic error in (3.68). Another could be the estimation of block values from point values or the reverse, provided that the point-to-point covariance is known.

Instead of postulating that all variables have the same mean, one could let the drifts differ by a free constant as when mapping subparallel horizons. This would be a *mixed case*. The resulting changes in the equations are left to the reader.

5.5 DERIVATIVE INFORMATION

A case of algebraic dependence that is particularly interesting for applications is when the auxiliary variables are derivatives of the primary variable. Typically we may know the depth of a geological surface and also its dip magnitude and azimuth. In the petroleum industry, those measurements are performed by dipmeter logging tools, which are resistivity devices that record several curves along the borehole wall or even a complete image. The dipping plane is determined from the displacements of these curves or by tracking the trace of the plane along the image [e.g., Hepp and Dumestre (1975), Antoine and Delhomme (1993)]. Another familiar example of gradient information is atmospheric pressure or geopotential and wind. In the so-called geostrophic approximation, wind is the gradient of atmospheric pressure or geopotential, up to a multiplicative factor and a 90° rotation due to the Coriolis force.

Two estimation problems may be considered, one is estimation *of* gradients, and the other is estimation *with* gradients. For a general treatment it is useful to introduce the notion of *directional derivative* of the RF $Z(x)$ with respect to the unit vector $u \in \mathbb{R}^n$ [e.g., Rockafellar (1970)]. This is a random variable defined as the following limit in the mean square sense:

$$\frac{\partial Z}{\partial u}(x) = \lim_{\rho \downarrow 0} \frac{Z(x + \rho u) - Z(x)}{\rho} \tag{5.24}$$

If $Z(x)$ is (m.s.) differentiable, the directional derivative in the direction u is the projection of the gradient $\nabla Z(x)$ onto the unit vector u with direction cosines $\cos \theta_i$:

$$\frac{\partial Z}{\partial u}(x) = \langle \nabla Z(x), u \rangle = \sum_{i=1}^{n} \frac{\partial Z(x)}{\partial x_i} \cos \theta_i$$

The mathematical conditions for the existence of directional derivatives in all directions u are the same as the differentiability of $Z(x)$, namely that the (generalized) covariance of $Z(x)$ be twice differentiable at the origin. Unfortunately, the variogram functions most used in practice—namely the spherical, the exponential, or the linear models—are not differentiable. The derivatives of $Z(x)$ are then simply not defined mathematically (infinite variance) and can in principle neither be estimated nor be used for estimation. Does it mean that slope information is essentially useless? We will discuss this question later; for now let us assume that derivatives exist.

5.5.1 Estimation of Derivatives

This part is straightforward. By linearity of the kriging system, the optimal estimator of the difference $Z(x + \rho u) - Z(x)$ is $Z^*(x + \rho u) - Z^*(x)$. Passing to the limit as in (5.24), we get

$$(\partial Z/\partial u)^* = \partial Z^*/\partial u \tag{5.25}$$

That is, the cokriging estimator of the derivative from Z alone is equal to the derivative of the kriging estimator. Hence the system for cokriging the derivative from Z is simply obtained by differentiating the right-hand side of the UK system (3.21), using the formula in the first line of (5.28). For the variance the term σ_{00} becomes $\mathrm{Var}[(\partial Z/\partial u)(x_0)]$. The variogram equations (3.25) can be modified in the same manner, but the above variance term must be added. (*Note*: A derivative is a linear combination whose sum of weights is zero.)

The relation (5.25) assumes that both the derivatives of Z and Z^* exist, but in fact only Z^* may have a derivative; for example, if the variogram of Z is linear near the origin, Z^* is differentiable nevertheless, except at the sample points. By extension we may accept $\partial Z^*/\partial u$ as the derivative estimator, but of course no estimation variance can be attached. Note that differentiating the kriging estimator does not lead to noise enhancement in the derivatives as would the differentiation of the Z data themselves because Z^* is a *smoothed* version of Z, except at the sample points.[2]

5.5.2 Estimation with Derivatives

Now we turn to the more difficult and also the more interesting problem. For the sake of simplicity we develop the theory in 2D but the results are basically the same in 3D or higher. Our cokriging estimator is of the form

$$Z^{**}(x_0) = \sum_{x_\alpha \in S_1} \lambda_{1\alpha} Z(x_\alpha) + \sum_{x_\beta \in S_2} \lambda_{2\beta} \frac{\partial Z}{\partial u_\beta}(x_\beta) + \sum_{x_\beta \in S_3} \lambda_{3\beta} \frac{\partial Z}{\partial v_\beta}(x_\beta) \tag{5.26}$$

S_1 is the sampling set of Z data, and the directional derivatives are taken along directions u and v which are allowed to vary with the point in the sampling sets S_2 and S_3 of derivatives. Typically, u and v are simply the directions of the

---

[2] In frequency terms, differentiation amounts to a multiplication of the signal frequency response $X(u)$ by the frequency u, thereby enhancing the high frequencies present in the signal. The smoothing nature of kriging attenuates high frequencies, at least away from the data points.

coordinate axes, but they can also be the gradient direction (e.g., downdip) and the orthogonal direction (strike). This general form of the estimator allows for the case where only one of the components is known, such as only the strike direction (the horizontal line of the bedding plane).

The functional relationship between $Z(x)$ and its derivatives carries over to the drift and covariances. From (5.26) the unbiasedness conditions are immediately found to be

$$\sum_{x_\alpha \in S_1} \lambda_{1\alpha} f^\ell(x_\alpha) + \sum_{x_\beta \in S_2} \lambda_{2\beta} \frac{\partial f^\ell}{\partial u_\beta}(x_\beta) + \sum_{x_\beta \in S_3} \lambda_{3\beta} \frac{\partial f^\ell}{\partial v_\beta}(x_\beta) = f^\ell(x_0) \qquad \text{for all } \ell$$

The cokriging system is then of the form

$$\begin{bmatrix} \mathbf{C}_{11} & \mathbf{C}_{12} & \mathbf{C}_{13} & \mathbf{F}_1 \\ \mathbf{C}_{21} & \mathbf{C}_{22} & \mathbf{C}_{23} & \mathbf{F}_2 \\ \mathbf{C}_{31} & \mathbf{C}_{32} & \mathbf{C}_{33} & \mathbf{F}_3 \\ \mathbf{F}_1' & \mathbf{F}_2' & \mathbf{F}_3' & 0 \end{bmatrix} \begin{bmatrix} \lambda_1 \\ \lambda_2 \\ \lambda_3 \\ \mu \end{bmatrix} = \begin{bmatrix} \mathbf{c}_{11} \\ \mathbf{c}_{21} \\ \mathbf{c}_{31} \\ \mathbf{f}_0 \end{bmatrix} \qquad (5.27)$$

and the cokriging variance is as in (5.7). On the left-hand side, $\mathbf{F}_1$ holds the drift function values over S_1, while $\mathbf{F}_2$ and $\mathbf{F}_3$ hold their derivatives in u and v over S_2 and S_3. All $\mathbf{F}_i$ have the same number of columns—that is, the number of basis drift functions—and a number of rows equal to the number of data points of each variable (the first column of $\mathbf{F}_1$ is usually composed of 1's, and those of $\mathbf{F}_2$ and $\mathbf{F}_3$ consist of 0's). Assuming the covariance model to be positive definite and to have no duplication of points, a sufficient condition for the cokriging system to have a unique solution is that $\mathbf{F}_1$ be of full rank, namely that kriging based on the Z values alone is possible. That condition is not necessary, however, since knowledge of derivative information may remove indeterminacy (e.g., specifying the normal to a plane in addition to one line in that plane).

The $\mathbf{C}_{ij}$ terms are the direct and cross-covariances between the Z data and the derivatives, and the $\mathbf{c}_{i1}$ terms are the covariances between, respectively, the vectors of Z, $\partial Z/\partial u$ and $\partial Z/\partial v$ data, and $Z(x_0)$. All these can be obtained by differentiation of the covariance $\sigma(x, y)$ of $Z(\cdot)$, provided that all derivatives involved exist and that differentiation and expectation may be interchanged. We get

$$\text{Cov}\left[Z(x), \frac{\partial Z}{\partial u}(y)\right] = \lim_{\rho \downarrow 0} \frac{\sigma(x, y + \rho u) - \sigma(x, y)}{\rho},$$

$$\text{Cov}\left[\frac{\partial Z}{\partial u}(x), \frac{\partial Z}{\partial v}(y)\right]$$

$$= \lim_{\rho, \rho' \downarrow 0} \frac{\sigma(x + \rho u, y + \rho' v) - \sigma(x, y + \rho' v) - \sigma(x + \rho u, y) + \sigma(x, y)}{\rho \rho'}$$

$$(5.28)$$

These formulas are general and do not require stationarity of the covariance. Explicit results can be obtained if the covariance is stationary and isotropic, that is, of the form

$$\sigma(x, x + h) = C(r) \qquad \text{with} \qquad r = |h|$$

Defining the projections of h on the unit vectors u and v

$$h_u = \langle h, u \rangle, \qquad h_v = \langle h, v \rangle$$

we have

$$\text{Cov}\left[Z(x), \frac{\partial Z}{\partial u}(x + h) \right] = h_u \frac{C'(r)}{r},$$

$$\text{Cov}\left[\frac{\partial Z}{\partial u}(x), \frac{\partial Z}{\partial v}(x + h) \right] = -h_u h_v \frac{1}{r^2}\left[C''(r) - \frac{C'(r)}{r} \right] - \langle u, v \rangle \frac{C'(r)}{r}$$

$$(5.29)$$

Here u and v are arbitrary unit vectors not assumed orthogonal. This generality is useful when only the gradient direction is known. There are two special cases of interest that emphasize the fundamental anisotropy of the derivative field: when the covariance is evaluated along the same direction as the directional derivative and when it is evaluated in the perpendicular direction. Letting $u = h/r$ and v orthogonal to u, we get from (5.29)

$$C_{\text{L}}(r) = \text{Cov}\left[\frac{\partial Z}{\partial u}(x), \frac{\partial Z}{\partial u}(x + ru) \right] = -C''(r),$$

$$C_{\text{N}}(r) = \text{Cov}\left[\frac{\partial Z}{\partial u}(x), \frac{\partial Z}{\partial u}(x + rv) \right] = -\frac{C'(r)}{r}$$

$$(5.30)$$

$C_{\text{L}}(r)$ and $C_{\text{N}}(r)$ are called the *longitudinal* and *transverse* (or *lateral*) covariance functions and play an important role in the theory of vector random fields.[3] See Daly (2001) for an application to strain modeling. These equations as well as (5.28) remain valid when the covariance $C(h)$ is replaced by $-\gamma(h)$, and the

---

[3] The covariance of isotropic vector random fields (Yaglom, 1987, p. 374) is of the form

$$C_{ij}(r) = [C_{\text{L}}(r) - C_{\text{N}}(r)]\frac{r_i r_j}{r^2} + C_{\text{N}}(r)\delta_{ij}$$

where i and j are vector component indexes, δ_{ij} is the Kronecker delta, and r_i and r_j are the components of the separation vector h along orthogonal axes i and j. The longitudinal and transverse covariance functions $C_{\text{L}}(r) = -C''(r)$ and $C_{\text{N}}(r) = -C'(r)/r$ verify the general relation for potential fields $C_{\text{N}}(r) = \frac{1}{r}\int_0^r C_{\text{L}}(x)dx$. This remains true for gradients in $\mathbb{R}^n$ since the longitudinal covariance remains $-C''(r)$. Furthermore, the general theory tells us that $C(r)$ is the covariance of a (scalar) differentiable isotropic random field in $\mathbb{R}^n$ if and only if $-C''(r)$ and $-C'(r)/r$ are covariances in $\mathbb{R}^{n+2}$.

cokriging system (5.27) can be written in terms of the variogram, provided that the constant function 1 is included in the set of drift basis functions.

The foregoing approach has been applied to meteorological problems in the scope of a broader study involving a model of four variables (geopotential, wind components, and temperature) over 10 layers. (Chilès, 1976; Chauvet et al., 1976). The example shown in Figure 5.3 is a 500-millibar geopotential map over Western Europe and the Atlantic Ocean obtained by using both geopotential and wind observations. The variogram of geopotential, similar to that shown in Figure 2.30b, was modeled as the sum of two Gaussian variograms, one with a scale parameter of about 1400 km and the other, representing the main component of the variogram of derivatives, with a scale parameter of 500 km. With a grid spacing of about 200 km the short-scale variogram allows winds to influence nearby grid nodes, but this influence decreases rapidly with distance. The map shows wind data in sparsely sampled areas. Notice that due to the Coriolis force, these winds are approximately *tangent* to geopotential contours.

5.5.3 Physical Validity of Derivatives

The singularity of common variogram models at the origin forces us to question the physical validity of representing gradient information by derivatives. In a mathematical sense a derivative is punctual information, but in the real world this is usually not what is measured. Consider, for example, geological dips. With a dipmeter tool there is a minimum distance between dip computation points corresponding to the diameter of the borehole, typically 20 to 25 cm. Even so there can be considerable variability in the magnitude and azimuth of these dips due to local sedimentary features such as ripples or cross-bedding. Since we are not interested in representing these local fluctuations, we consider only the dips that are likely to have significant lateral extension. Typically, this is what geologists call the *structural dip*. It is determined by statistical averaging within depth intervals with approximately constant dip magnitude and azimuth and is supposed to represent beds that were deposited on a nearly horizontal surface, with their present dips being the result of tectonic stresses [e.g., Serra (1985)]. Thus the slope information to be used for geological mapping is not a point gradient but rather an average over some support actually difficult to specify.[4] Renard and Ruffo (1993) modeled this dip as the derivative of the depth convolved by a Gaussian weighting function with main axes lengths (a_1, a_2), which were assigned arbitrary but plausible values ($a_1 = a_2 = 50$ m for a depth variogram range of 1400 m).

Another solution is to model derivative information by finite increments, which is where we started from in the definition (5.24), and this is also a form of averaging. In 2D, when both value and gradient are known at the

---

[4] The depth of investigation of the dipmeter tool should also be taken into account, and that in turns depends on the resistivity constrasts and also the geometry of the beds and the inclination of the wellbore.

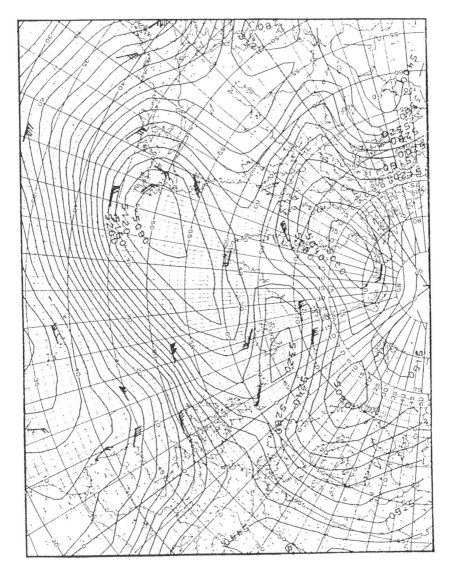

FIGURE 5.3 Cokriging of the 500-millibar geopotential field on 15/03/74—12:00. Unit: gpm. The symbol F represents wind data (direction and intensity). A guess field was used. [From Chilès (1976), with kind permission from Kluwer Academic Publishers.]

same data point P_0, a simple procedure is to create *dummy points* P_1 and P_2 that, together with P_0, recreate the correct tangent plane. For example, P_1 and P_2 are placed at a distance ρ from P_0, where P_1 is along the direction of steepest slope (e.g., downdip) with the value $Z_1 = Z_0 + \rho \|\nabla Z\|$ and P_2 is normal to this direction (e.g., along strike) with the value $Z_2 = Z_0$. The value of ρ depends on the lateral validity of the dip and is an *interpretation decision*. The main advantage of the method is that it can be implemented with a standard kriging program. When only the gradient is known, a similar method can be applied, except that the dummy points are involved only through the differences $Z_1 - Z_0$ and $Z_2 - Z_0$, which requires a genuine cokriging program.

When the RF $Z(x)$ is differentiable, the question of the lateral extent (the distance ρ) seems irrelevant, at first glance. But we know from experience that the influence of punctual derivatives may be so local as to have virtually no effect on the interpolated grid. In reality there is a difference between a *mathematical derivative* and a *physical derivative*. The mathematical derivative at the point x in the direction u is the limit of the ratio $[Z(x + \rho u) - Z(x)]/\rho$ as ρ becomes infinitely small. It suffices to modify a function only very slightly, for example, by convolving it with a smooth window of arbitrarily small width, to make it differentiable and even infinitely differentiable. The mathematical derivative has a physical existence only if this ratio remains approximately constant over a distance commensurate at least with the scale of observation and, preferably, the grid spacing. That distance represents the lateral extent of the gradient. Phrased differently, derivative data have an impact only if they exhibit enough spatial correlation at the scale of the study; otherwise, they appear like a nugget effect.

A final remark. The presence of a nugget effect on the variogram of $Z(x)$, which is the most adverse case of singularity at the origin, does not exclude the possibility of useful derivative information. If the derivative is computed from measurements of Z recorded by the same process, the measurement errors are the same and cancel out in derivatives. A case in point is again the dipmeter, where the uncertainty on the absolute depth of a single resistivity curve can reach several feet, but the displacement between curves can be determined within a fraction of an inch.

5.5.4 Kriging under Boundary Conditions

The example is taken from a work that J. P. Delhomme (1979) presented at a conference but never published, where he used gradient information as a means of constraining estimates to honor known physical boundary conditions. The study concerns an aquifer in which water, for geological reasons, cannot flow through the boundaries marked with a thick solid line in Figure 5.4a and is released through the western outlet. For the purpose of numerical modeling of the aquifer, it is desired to map the hydraulic head on the basis of 49 available piezometric measurements. Figure 5.4b shows the map produced by usual kriging; it is not acceptable for a hydrogeologist. First, the no-flow constraints are violated: Since the flow is orthogonal to hydraulic head contour lines, these lines should be perpendicular to the no-flow boundary, but in the southern part they are not. Second, to the west the water release is not in front of the actual outlet. Boundary conditions are now introduced by specifying that the gradient component normal to the no-flow boundary is zero. In principle, since the no-flow contour is

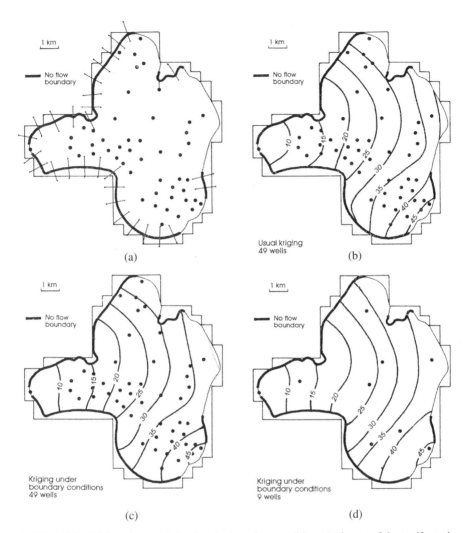

FIGURE 5.4 Kriging piezometric head under boundary conditions: (a) layout of the aquifer and locations of the wells and the dummy wells (thick lines are no-flow boundaries); (b) usual kriging from the 49 wells does not honor the constraints; (c) kriging under boundary conditions and the 49 wells honors the constraints; (d) kriging under boundary conditions using 9 wells only does practically as well as in (c). [From Delhomme (1979).]

continuous, one should consider a continuous cokriging estimator, but in practice, it suffices to discretize the problem and replace the gradient component by the differences between pairs of dummy points as shown in Figure 5.4a, considering

$$Z_0^{**} = \sum_\alpha \lambda_{1\alpha} Z(x_\alpha) + \sum_\beta \lambda_{2\beta} \left[Z\left(x_\beta + \rho v_\beta\right) - Z\left(x_\beta - \rho v_\beta\right) \right]$$

The vectors v_β are perpendicular to the boundary and the differences are zeros. One may wonder why these differences are included at all since their contribution to the estimator is zero. Because, and that is key, the weights $\lambda_{1\alpha}$ are different from kriging weights based on the Z_α alone. The resulting map in Figure 5.4c is the "ideal" one in the sense that it is consistent with everything that is already known. In order to evaluate the strength of the additional boundary conditions, the map was redrawn using only 9 wells plus these constraints and the result is pictured in Figure 5.4d, which also shows the locations of the dummy wells. The map is not very different from the ideal one, indicating that boundary conditions constitute very strong constraints.

5.5.5 Potential-Field Interpolation

Use of orientation data is at the core of an original method for building a 3D model of geological interfaces. The principle is to define a *potential field*, namely a scalar function $T(x)$ of location $x = (x, y, z)$ in 3D space, such that the geological interface corresponds to an isopotential surface—that is, the set of points x that satisfy $T(x) = c$, a given constant value (a *level set*). In the case of sedimentary deposits, T could be interpreted as geological time, or an increasing function of time, and an interface can be interpreted as an isochron surface, analogous to Mallet (2004).

The intriguing feature of this approach is that it permits an interpolation of potential differences using only potential differences equal to zero and gradients. The available data are as follows:

1. *Interface Points*: Locations of points $x_1, \ldots, x_m$ known to belong to the same interface. Two interface points x_1 and x_2 satisfy $T(x_2) - T(x_1) = 0$.
2. *Orientation Data*: The dip and strike of layers at selected points not necessarily on an interface. These define a local tangent plane where the potential remains constant. The unit vector normal to this plane and pointing toward increasing T is taken as the gradient of the potential field (this assumes the geologist is able to tell the direction of time). The gradient vector has for components three partial derivatives $\partial T/\partial u$, $\partial T/\partial v$, $\partial T/\partial w$ in directions u, v, w, which in practice usually coincide with the axes directions in 3D space.

T itself is defined up to an arbitrary constant taken to be the potential at an arbitrary reference point x_0. We estimate a potential difference using a cokriging estimator of the form

$$T^{**}(x) - T^{**}(x_0) = \sum_{\alpha=2}^{m} \lambda_\alpha [T(x_\alpha) - T(x_{\alpha-1})] + \sum_{x_\beta \in S_G} \sum_{i=1}^{3} \nu_{i\beta} \frac{\partial T}{\partial u_i}(x_\beta)$$

where S_G is the sample set of gradient points, and the u_i stand for the unit vectors u, v, w. Just as in the general case of gradients it is not neccessary to assume that all gradient components are known: If the i-th one is missing it suffices to set $\nu_{i\beta}=0$. The potential increments have a zero value, but their presence in the estimator causes the weights on the gradient to be different from what they would be based on the gradient data alone. Conversely, in the absence of gradient data the cokriging estimator would be identically zero. Note that this formula uses data from a single interface; if there are multiple interfaces, other terms similar to the first sum must be added.

The cokriging model includes a polynomial drift in 3D. For example, if interfaces are subparallel subhorizontal surfaces, it makes sense to use a linear drift in the vertical direction. For an ellipsoid-shaped body a quadratic drift in 3D (10 coefficients) may be considered. In all cases the constant function $f^0(x) \equiv 1$ should be omitted since the potential increments as well as the gradients filter it out. Concerning covariances, the method consists of postulating a twice differentiable covariance model for T—for example, a cubic model—and fitting its parameters so as to match the longitudinal and transverse variograms of derivatives given by (5.30). Figure 5.5 shows an example of such variograms for a dataset in central France.

It is advantageous to implement the cokriging algorithm in dual form because it allows an easy computation of the potential at any point x, and consequently the use of an efficient algorithm such as the *marching cubes*, to visualize iso-surfaces in 3D, without the need of a fine 3D grid.

Figure 5.6 illustrates the results obtained with the potential-field approach. It is especially useful to model the 3D geometry of geological bodies described by outcrop data and possibly a few drill-holes. Additional information on the

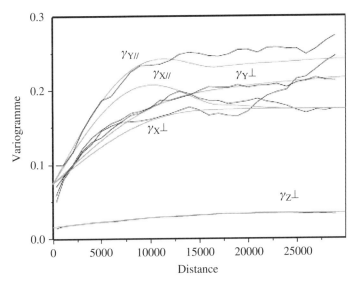

FIGURE 5.5 Longitudinal and transverse variograms of potential field derivatives for the Limousin dataset, Massif Central, France. Notations: $\gamma_{X//}$ is the longitudinal variogram of $\partial T/\partial x$ and $\gamma_X\perp$ the transverse. [From Aug (2004).] (See color insert)

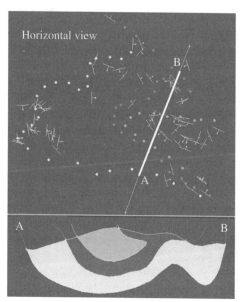

Vertical cross section

FIGURE 5.6   Potential field interpolation. Top: points at interfaces and structural data, sampled on the topographic surface; bottom: vertical cross section through the 3D model. [From Courrioux et al. (1998).] (See color insert)

method and its applications can be found in Lajaunie et al. (1997), Aug (2004), Chilès et al. (2005), and Calcagno et al. (2008).

5.6   MULTIVARIATE RANDOM FUNCTIONS

5.6.1   Cross-Covariances

The cross-covariance functions of a multidimensional stationary RF $\mathbf{Z}(x) = (Z_1(x), \ldots, Z_p(x))'$ with mean vector $\mathbf{m}(x) = (m_1(x), \ldots, m_p(x))'$ are defined by

$$C_{ij}(h) = \mathrm{E}[Z_i(x) - m_i(x)][Z_j(x+h) - m_j(x+h)] \qquad (5.31)$$

and only depend on the separation h between the two points. For simplicity and without loss of generality, we will assume here that all variables are centered to zero means.

The order in which variables are considered matters: Z_i is at the origin x, and Z_j is at the end point $x + h$. If we swap the variables, we get another cross-covariance:

$$\mathrm{E}[Z_j(x)Z_i(x+h)] = C_{ji}(h)$$

Trivially, we obtain

$$C_{ij}(h) = \mathrm{E}\big[Z_i(x)Z_j(x+h)\big] = \mathrm{E}\big[Z_j(x+h)Z_i(x)\big] = C_{ji}(-h)$$

but in general we get

$$C_{ij}(-h) \neq C_{ij}(h)$$

By Cauchy–Schwarz's inequality we have

$$|C_{ij}(h)| \leq \sqrt{C_{ii}(0)C_{jj}(0)}$$

so that $C_{ij}(h)$ is bounded. Applying the same inequality to one variable and an increment of another shows that $C_{ij}(h)$ is continuous at $h = 0$ whenever $Z_i(x)$ or $Z_j(x)$ is continuous in the mean square. A cross-covariance behaves very differently than a direct covariance function $C_{ii}(h)$. It is not necessarily symmetric, and its maximum may lie at $h = \tau \neq 0$. Such lagged correlation is common in time series due to a delay between the input and the output signals. It is less frequent in spatial phenomena, perhaps because it is incompatible with isotropy. But there are examples: offsets caused by faulting, by selective migration of a chemical element (in uranium mines, maximum radioactivity does not necessarily coincide with maximum uranium grades), or by "man-made" offsets between two images.

Example: The Pure Offset Model
The topography of the lower and upper surfaces of an horizontal fracture are described by two RFs $Z_1(x)$ and $Z_2(x)$, where x is the position of a point in the horizontal plane. If the two surfaces simply differ by a displacement of horizontal and vertical components τ and m (idealized case of a shear fracture), we obtain

$$Z_2(x) = Z_1(x+\tau) + m$$

so that

$$C_{12}(h) = C_{11}(h+\tau)$$

and is a maximum for $h = -\tau$, allowing the detection of τ (Chilès and Gentier, 1993). The same principle is used for locating patterns in an image using templates.

Characterization of Covariance Function Matrices

Since a positive definite function has its maximum at zero, a cross-covariance is generally not positive definite. Under which conditions is $C_{ij}(h)$ a valid cross-covariance function, or, rather, under which conditions does the set of $C_{ij}(h)$ constitute a valid model? The criterion is in the frequency domain and is due to Cramér (1940), who mentions in a footnote of his paper that the same result was found independently by Kolmogorov.

A continuous cross-covariance function $C_{ij}(h)$ has the spectral representation

$$C_{ij}(h) = \int e^{2\pi i \langle u, h \rangle} F_{ij}(du) \qquad (5.32)$$

where $u = (u_1, u_2, \ldots, u_n)$ is frequency in $\mathbb{R}^n$ and $\langle u, h \rangle = u_1 h_1 + \cdots + u_n h_n$ (of course the "i" in the exponential is the pure imagining number and has nothing to do with the index i in C_{ij}). The cross-spectral measure $F_{ij}(du)$ represents the common power of Z_i and Z_j in the infinitesimal spectral interval du. It satisfies the following symmetry relations for any Borel set B:

$$F_{ij}(-B) = \overline{F_{ij}(B)} = F_{ji}(B)$$

Criterion. (Cramér, 1940; Yaglom, 1987, Vol. I, p. 314)
The continuous functions $C_{ij}(h)$ are the elements of the covariance matrix of a multidimensional stationary RF of order 2 if and only if the cross-spectral matrix $\mathbf{M}(B) = [F_{ij}(B)]$ is positive definite for any (Borel) set B of $\mathbb{R}^n$, namely $\sum_i \sum_j \lambda_i \overline{\lambda_j} F_{ij}(B) \geq 0$ for any set of complex coefficients $\lambda_1, \ldots, \lambda_p$.
   This concise criterion ensures that given any linear combination with complex coefficients

$$Y(x) = \lambda_1 Z_1(x) + \cdots + \lambda_p Z_p(x)$$

its covariance function

$$\mathrm{E}\left[Y(x) \overline{Y(x+h)} \right] = \sum_{i=1}^{p} \sum_{j=1}^{p} \lambda_i \overline{\lambda_j} C_{ij}(h) = \int e^{2\pi i \langle u, h \rangle} \left[\sum_{i=1}^{p} \sum_{i=1}^{p} \lambda_i \overline{\lambda_j} F_{ij}(du) \right]$$

is the Fourier transform of a nonnegative measure and is therefore positive definite.[5] More generally, all variances of linear combinations calculated with the model are nonnegative. Note, however, that the criterion does *not* exclude the possibility of singular cokriging matrices due to linearly dependent variables. But the cokriging estimator is always unique (being a projection).
   If $C_{ij}(h)$ decreases fast enough for $C_{ij}(h)^2$ to be integrable, there exists a spectral density function $f_{ij}(u)$ such that $F_{ij}(du) = f_{ij}(u)\, du$. If in addition $|C_{ij}(h)|$ is integrable, $f_{ij}(u)$ is continuous and bounded and can be computed by inversion of the Fourier transformation (5.32). If all cross-covariances have a spectral density function, the validity criterion is a positive definite spectral density matrix:

$$\mathbf{M}(u) = \begin{bmatrix} f_{11}(u) & \cdots & f_{1p}(u) \\ \vdots & \cdots & \vdots \\ f_{p1}(u) & \cdots & f_{pp}(u) \end{bmatrix}$$

---

[5] As in Chapter 2, positive definite is taken synonymously to *nonnegative definite*; and to exclude the value zero, we refer to *strict* positive definiteness.

In general, $f_{ij}(u)$ is a complex function with the Hermitian symmetry (since $C_{ij}(h)$ is real)

$$f_{ij}(-u) = \overline{f_{ij}(u)} = f_{ji}(u)$$

The positive definiteness property entails in particular that for any pair (i, j) we obtain

$$|f_{ij}(u)| \leq \sqrt{f_{ii}(u) f_{jj}(u)} \tag{5.33}$$

Examples

1. *Independent RFs.* In this case $\mathbf{M}(u)$ is diagonal. Since all $f_{ii}(u)$ are real and nonnegative, being spectral densities of ordinary covariances $C_{ii}(h)$, $\mathbf{M}(u)$ is positive definite.

2. *Derivative.* In $\mathbb{R}$ consider a RF $Z(x)$ and its derivative $Z'(x)$ and let $f(u)$ be the spectral density associated with the covariance of $Z(x)$. The spectral density matrix is

$$\mathbf{M}(u) = \begin{bmatrix} 1 & 2\pi i u \\ -2\pi i u & 4\pi^2 u^2 \end{bmatrix} f(u)$$

We have $f(u) \geq 0$ and det $\mathbf{M}(u) = 0$ for all u so that $\mathbf{M}(u)$ is nonnegative definite. The same reasoning applies to the pure offset model, with

$$\mathbf{M}(u) = \begin{bmatrix} 1 & e^{2\pi i u \tau} \\ e^{-2\pi i u \tau} & 1 \end{bmatrix} f(u)$$

3. *Proportional Covariances.* Let $C_{ij}(h) = \sigma_{ij}\,\rho(h)$ and $f(u)$ the spectral density associated with $\rho(h)$. Then $M(u) = [\sigma_{ij}]\,f(u)$, and since $f(u) \geq 0$, $\mathbf{M}(u)$ is positive definite if and only if the matrix $[\sigma_{ij}]$ is also positive definite. Note that if $\rho(h)$ assumes negative values, the matrix $[C_{ij}(h)]$ is *negative* definite for those values of h.

4. *"Markov-Type" Models.* Consider two RFs, Z_1 and Z_2, with unit variance and correlation coefficient ρ. Assume that their cross-covariance is proportional to the covariance of Z_2 (MM2 model):

$$C_{12}(h) = \rho\, C_{22}(h)$$

If C_{11} and C_{22} (which are correlograms) have spectral densities f_1 and f_2, we have

$$\mathbf{M}(u) = \begin{bmatrix} f_1(u) & \rho f_2(u) \\ \rho f_2(u) & f_2(u) \end{bmatrix} \quad \Rightarrow \quad \det \mathbf{M}(u) = f_2(u)\left[f_1(u) - \rho^2 f_2(u)\right]$$

The multivariate model is valid if and only if $f_1(u) \geq \rho^2 f_2(u)$ for every frequency u. In particular, $f_1(u)$ cannot tend to zero faster than $f_2(u)$ as $u \to \infty$, showing that Z_1 *cannot be smoother than* Z_2. Equivalently, $C_{11}(h)$ is of the form

$$C_{11}(h) = \rho^2 C_{22}(h) + (1 - \rho^2) C_{RR}(h)$$

for some correlogram $C_{RR}(h)$. If Z_1 and Z_2 have zero mean, Z_1 can be decomposed as

$$Z_1(x) = \rho Z_2(x) + \sqrt{1 - \rho^2} R(x)$$

where $R(x)$ is a zero mean unit variance residual spatially uncorrelated with Z_2. Similar results are obtained if we start from the MM1 model $C_{12}(h) = \rho C_{11}(h)$, which may be more appropriate if Z_1 is smoother than Z_2.

5. *Nugget Effects.* In the multivariate case, it is not sufficient (nor necessary) for nugget effect constants to be positive; there are consistency constraints. We consider here the case where discontinuities are at the origin only, keeping in mind that cross-covariances can have discontinuities elsewhere (e.g., the pure offset model). If we admit that the phenomenon can be decomposed into two uncorrelated multivariate components, a pure nugget effect and a continuous component, each must be a valid model, and this condition is also clearly sufficient for the sum to be valid. A little technical difficulty appears if we want to apply the Cramér criterion to the nugget effect because it is applicable only to *continuous* covariances. A simple way to work around this difficulty is to consider the pure nugget effect as a case of proportional covariance model with a continuous correlogram $\rho(h)$ of arbitrarily small range:

$$C_{ij}(h) = \sigma_{ij} \rho(h)$$

We conclude immediately that the nugget effect matrix $[\sigma_{ij}]$ must be positive definite. For example, in a bivariate model we must have $\sigma_{11} \geq 0$, $\sigma_{22} \geq 0$, and $\sigma_{12}^2 \leq \sigma_{11} \sigma_{22}$. The magnitude of the nugget effect on the cross-covariance cannot exceed the geometric mean of the nugget effects on the two direct covariances. Again, if one of the variables is continuous, the cross-covariance must be continuous too.

Coherency and Phase Spectrum

The spectral density function $f_{ij}(u)$ is a very informative tool to study the relationship between two stationary RFs $Z_i(x)$ and $Z_j(x)$ and is used extensively for the analysis of time signals [e.g., Jenkins and Watts (1968), Koopmans (1974)]. We do not know of such use in a spatial geostatistical context, but it is interesting to briefly mention the approach. We will use the indexes 1 and 2 to denote the two generic variables.

Due to the absence of symmetry of $C_{12}(h)$, the spectral density $f_{12}(u)$ is a complex function

$$f_{12}(u) = c_{12}(u) - iq_{12}(u)$$

The real part $c_{12}(u)$ (not be confused with the covariance $C_{12}(h)$) is the Fourier transform of the even part of the cross-covariance and is usually called the *cospectrum*, while the imaginary part $q_{12}(u)$ is the Fourier transform of the odd part of the cross-covariance and is called the *quadrature spectrum*.

The Hermitian symmetry of $f_{12}(u)$ entails

$$c_{12}(-u) = c_{12}(u) \quad \text{and} \quad q_{12}(-u) = -q_{12}(u)$$

The polar representation

$$f_{12}(u) = |f_{12}(u)| e^{i\varphi_{12}(u)}$$

yields another set of spectral parameters that are perhaps the most useful because they can be interpreted quantitatively. Owing to the Cauchy–Schwarz inequality (5.33), the ratio

$$\rho_{12}(u) = |f_{12}(u)| / \sqrt{f_{11}(u) f_{22}(u)}$$

when defined, is always between 0 and 1. It is named the *coherency spectrum* and provides a nondimensional measure of the correlation between two RFs as a function of frequency. The term "coherency" (or "coherence") is borrowed from the study of light, and an interesting explanation of this idea can be found in Koopmans (1974, pp. 138ff.). Coherency remains invariant under scale changes in the frequency domain, namely under the application of independent linear filters to $Z_1(x)$ and $Z_2(x)$, which is a very important property when the data have been passed through linear filters with possibly unknown characteristics. The *phase spectrum* is defined by

$$\varphi_{12}(u) = -\arctan(q_{12}(u)/c_{12}(u))$$

It is interpreted as the average phase difference between $Z_1(x)$ and $Z_2(x)$ at frequency u. When there is no phase difference at any frequency, the cross-spectral density $f_{12}(u)$ is real and the cross-covariance is an even function.

5.6.2 Cross-Variograms

The cross-variogram was introduced by Matheron (1965) as the natural generalization of the variogram

$$\gamma_{ij}(h) = \tfrac{1}{2} E\big[Z_i(x+h) - Z_i(x)\big]\big[Z_j(x+h) - Z_j(x)\big] \tag{5.34}$$

for multivariate intrinsic random functions (of order 0), namely satisfying

$E[Z_i(x+h) - Z_i(x)] = 0$ for $i = 1, \ldots, p$

$\mathrm{Cov}\big[Z_i(x+h) - Z_i(x), Z_j(x+h) - Z_j(x)\big] = 2\gamma_{ij}(h)$ exists and depends only on h

The cross-variogram has two advantages over the cross-covariance: (1) It does not assume finite variances, and (2) the estimation of the cross-variogram is not corrupted by the estimation of the means. The relationship between the cross-variogram and the cross-covariance, when it exists, is the following:

$$\gamma_{ij}(h) = C_{ij}(0) - \tfrac{1}{2}\left[C_{ij}(h) + C_{ij}(-h)\right] \tag{5.35}$$

The cross-variogram satisfies $\gamma_{ij}(0) = 0$ and is an *even* function of h, whereas the cross-covariance in general is not. Here lies the shortcoming of the cross-variogram: There is a potential loss of information. The decomposition of the cross-covariance into even and odd parts

$$C_{ij}(h) = \tfrac{1}{2}\left[C_{ij}(h) + C_{ij}(-h)\right] + \tfrac{1}{2}\left[C_{ij}(h) - C_{ij}(-h)\right]$$

shows exactly what is lost, namely the odd part of the cross-covariance.

Example: The Pure Offset Model

Let $Z_1(x)$ and $Z_2(x)$ be two intrinsic random functions where $Z_2(x) = Z_1(x + \tau) + m$. Their cross-variogram is

$$2\gamma_{12}(h) = \gamma_{11}(h + \tau) + \gamma_{11}(h - \tau) - 2\gamma_{11}(\tau)$$

Unlike the covariance it cannot distinguish an offset τ from an offset $-\tau$.

In the extreme case where the cross-covariance is odd the cross-variogram is identically zero. For example, the cross-covariance between Z and Z' is the odd function $C'(h)$; as a consequence, $Z(y) - Z(x)$ and $Z'(y) - Z'(x)$ are always uncorrelated. Note in passing that the variogram derivative $\gamma'(h)$ is not a cross-variogram (because it is an odd function of h).

As a covariance of increments, $\gamma_{ij}(h)$ is subject to the Cauchy–Schwarz inequality

$$|\gamma_{ij}(h)| \le \sqrt{\gamma_{ii}(h)\,\gamma_{jj}(h)} \qquad \forall h \tag{5.36}$$

This inequality ensures the continuity of $\gamma_{ij}(h)$ at $h = 0$, provided that $Z_i(x)$ or $Z_j(x)$ is continuous in the mean square, and it also majorizes the growth of $\gamma_{ij}(h)$. Furthermore, the *coefficient of codispersion*

$$R_{ij}(h) = \frac{\gamma_{ij}(h)}{\sqrt{\gamma_{ii}(h)\,\gamma_{jj}(h)}} \tag{5.37}$$

introduced by Matheron (1965) provides an interpretive tool to analyze the correlation between the variations of $Z_i(x)$ and those of $Z_j(x)$.

Expansion of $\mathrm{Var}(\sum \lambda_i[Z_i(x + h) - Z_i(x)])$ shows that $[\gamma_{ij}(h)]$ is a positive definite matrix for every h (unlike $[C_{ij}(h)]$). The Cauchy–Schwarz inequality

(5.36) is a particular consequence of this, which is by no means sufficient to ensure the validity of the multivariate model, as the following example demonstrates. The true validity criterion is again in the frequency domain.

Example: An Invalid Cross-Variogram Model Satisfying the Cauchy–Schwarz Inequality

Consider the exponential cross-variogram model

$$\gamma_{11}(h) = \gamma_{22}(h) = 1 - e^{-b|h|} \qquad \gamma_{12}(h) = -\left(1 - e^{-c|h|}\right) \qquad (b, c > 0)$$

When $b > c$, these variogram functions satisfy the Cauchy–Schwarz inequality

$$\gamma_{11}(h)\gamma_{22}(h) - \gamma_{12}(h)^2 = \left(2 - e^{-b|h|} - e^{-c|h|}\right)\left(e^{-c|h|} - e^{-b|h|}\right) \geq 0$$

and yet the model is not valid. Indeed under a valid multivariate model the ordinary variogram of any linear combination of $Z_1(x)$ and $Z_2(x)$, expressed in terms of the model, must be a valid variogram function. Consider the sum $Y(x) = Z_1(x) + Z_2(x)$; its variogram is

$$\gamma_{YY}(h) = \gamma_{11}(h) + \gamma_{22}(h) + 2\gamma_{12}(h) = 2\left[e^{-c|h|} - e^{-b|h|}\right]$$

Since b is greater than c, this function is positive, but it is a suspicious-looking variogram that tends to 0 as $h \to \infty$. A direct proof (not by Fourier) is to consider that the covariance of $Y(x + \tau) - Y(x)$ for an arbitrary constant τ,

$$\gamma_{YY}(h + \tau) + \gamma_{YY}(h - \tau) - 2\gamma_{YY}(h)$$

must be a positive definite function of h and therefore bounded in absolute value by $2\gamma_{YY}(\tau)$. But for fixed h and large τ, the first two terms tend to zero so that this function is equivalent to $-2\gamma_{YY}(h)$, which can be greater in magnitude than $2\gamma_{YY}(\tau)$. So $\gamma_{YY}(h)$ is not a variogram and the multivariate model is invalid. The only case[6] where the model can be valid is $b = c$, which implies $\gamma_{YY}(h) \equiv 0$ and $Z_1(x) + Z_2(x) = $ constant.

Cokriging Equations in Terms of Cross-Variograms

In the special case where the RFs satisfy the symmetry condition

$$E\left[Z_i(x) - Z_i(x')\right]\left[Z_j(y) - Z_j(y')\right] = E\left[Z_j(x) - Z_j(x')\right]\left[Z_i(y) - Z_i(y')\right] \quad (5.38)$$

the covariance of any two increments is given by

$$E\left[Z_i(x) - Z_i(x')\right]\left[Z_j(y) - Z_j(y')\right]$$
$$= \gamma_{ij}(y' - x) + \gamma_{ij}(y - x') - \gamma_{ij}(y - x) - \gamma_{ij}(y' - x')$$

(Matheron, 1965, p. 146) so that in particular we have

$$E\left[Z_i(x) - Z_i(x_0)\right]\left[Z_j(y) - Z_j(x_0)\right] = \gamma_{ij}(x - x_0) + \gamma_{ij}(y - x_0) - \gamma_{ij}(y - x)$$

---

[6] A characterization of cross-covariances in $\mathbb{R}$ with exponential autocovariances is given by Yaglom (1987, Vol. 1, p. 315). In the example we have postulated that the cross-covariance is exponential.

From this is it easy to see that the variance of any linear combination of increments

$$\sum_i \sum_\alpha \lambda_{i\alpha} Z_i(x_\alpha) \quad \text{with} \quad \sum_\alpha \lambda_{i\alpha} = 0 \quad \forall i$$

(and, in particular, error variances) can be expressed in terms of the cross-variogram by simply substituting $-\gamma_{ij}(h)$ for $C_{ij}(h)$, the same rule as used in the univariate case. Thus ordinary and universal cokriging systems can be written in terms of cross-variograms, provided that for each variable the sum of weights is explicitly constrained to zero (i.e., even if the variable has mean zero) and that the symmetry condition (5.38) is satisfied.

Generalized Cross-Covariance

What would be the completely general equivalent of the variogram for a cross-structure function? In the case of an intrinsic model (of order 0), this would be a function $K_{ij}(h)$ enabling us to calculate the covariance of any pair of increments of the variables $Z_i(x)$ and $Z_j(x)$. Using the notations of IRF$-k$ theory, let us define such increments in the concise integral form

$$Z_i(\lambda) = \int Z_i(x)\lambda(dx), \qquad \text{where} \quad \int \lambda(dx) = 0$$
$$Z_j(\mu) = \int Z_j(y)\mu(dy), \qquad \text{where} \quad \int \mu(dy) = 0$$

Then the structure function should satisfy

$$E[Z_i(\lambda)Z_j(\mu)] = \iint \lambda(dx)\, K_{ij}(y - x)\, \mu(dy)$$

When the ordinary cross-covariance $C_{ij}(h)$ exists, it is obviously a solution. But so is the family

$$K_{ij}(h) = C_{ij}(h) - C_{ij}(0) - <c_1, h>$$

for any constant vector c_1. This leads to a multivariate generalization of the IRF-0 theory, in which the structural tool is a class of generalized cross-covariances $K_{ij}(h)$ with the spectral representations (Matheron, personal communication)

$$K_{ij}(h) = \int \frac{e^{2\pi i <u,h>} - 1 - i<u,h> 1_B(u)}{4\pi^2 |u|^2} \chi_{ij}(du) + c_0 + <c_1, h> \qquad (5.39)$$

where B is an arbitrary symmetric neighborhood of the origin, $\chi_{ij}(du)$ is a complex measure with the Hermitian symmetry, no atom at the origin, satisfying $\int \chi_{ij}(du)/(1 + 4\pi^2 |u|^2) < \infty$, and the matrix $[\chi_{ij}(B)]$ is positive definite for all Borel sets $B \subset \mathbb{R}^n$. The constant c_0 and vector c_1 are arbitrary.

Separating the real and imaginary parts of χ_{ij} in (5.39) yields a decomposition of $K_{ij}(h)$ into even and odd parts

$$K_{ij}(h) = \left[c_0 - \gamma_{ij}(h)\right] + \left[\langle c_1, h \rangle - \int \frac{\sin(2\pi\langle u, h \rangle) - \langle u, h \rangle 1_B(u)}{4\pi^2 |u|^2} Q_{ij}(du)\right]$$

where Q_{ij} is the imaginary part of χ_{ij}. When this is zero, $K_{ij}(h)$ is reduced to $c_0 - \gamma_{ij}(h)$.

In the case of the pure offset model with variogram $\gamma(h)$ and spectral measure $\chi(du)$, the generalized cross-covariance is of the form $-\gamma(h+\tau)$ for a fixed offset τ, and its spectral measure is $e^{2\pi i u \tau} \chi(du)$.

It does not seem possible to express $K_{ij}(h)$ as the expected value of increments and compute an experimental version of it. One must resort to postulating a parametric model and estimate its parameters like for generalized covariances. Dowd (1989) approaches this in the scope of a linear model of coregionalization. The definition of interesting models of generalized cross-covariances deserves further research.

5.6.3 An Example of Structural Analysis with Cross-Variograms

The following example illustrates the use of cross-variograms and codispersion graphs as analysis tools, independently of cokriging. It is taken from a study of the hydrothermal behavior of the ocean crust in a subduction zone located in the eastern Pacific Ocean (Chilès et al., 1991). Simply put, the objective of the work is to determine if the heat transfer in the ocean crust proceeds from a conductive or a convective regime. A conductive regime, because of thermal refraction, is associated with high heat transfer in deep water and low heat transfer in shallow water—that is, a positive correlation between variations of ocean depth and of heat flux. A convective regime (influx of cold water in deep permeable seabeds, followed by percolation and heating through the overlaying sedimentary layers up to shallower seabeds) results in the opposite effect. Although the site under consideration has been extensively studied, in particular in the scope of the ODP (Ocean Drilling Program), its hydrothermal system is still not well understood.

The most basic statistical analysis consists of a scatterplot of depth (in m) versus heat flux (in $\mathrm{mW/m^2}$), displayed in Figure 5.7a for the western part of the site. It shows a clear negative correlation, but the correlation coefficient is only -0.68. Geostatistical tools enable us to improve this correlation and also to analyze it more subtly.

The experimental cross-variogram between two variables is computed from all pairs of data points where both variables are known using the unbiased estimator

$$\hat{\gamma}_{ij}(h) = \frac{1}{2N(h)} \sum_{x_\beta - x_\alpha \approx h} \left[Z_i(x_\beta) - Z_i(x_\alpha) \right] \left[Z_j(x_\beta) - Z_j(x_\alpha) \right]$$

where $N(h)$ is the number of pairs (x_α, x_β) with separation h. Data points where only one of the variables is present are simply ignored. If the variables have no or too few samples in common to compute a cross-variogram, one may turn to the cross-covariance, but it is risky. Remember that for a single variable the covariance is biased by the estimation of the mean; for two variables, things may be worse as we subtract different means computed from data at different places. To ensure numerical consistency and avoid correlation coefficients greater than one, it is also recommended to compute the cross and direct variograms from the same set of points.

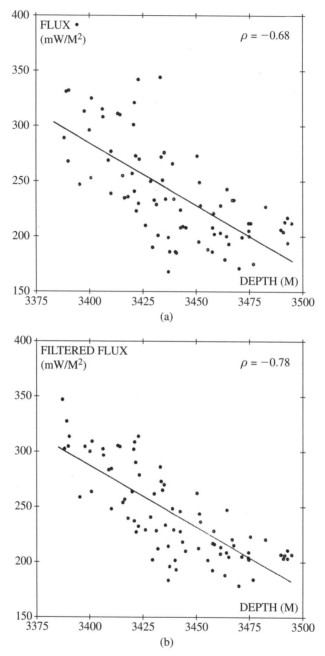

FIGURE 5.7 Scatterplots of heat flux versus depth and regression lines, western zone: (a) unfiltered flux; (b) filtered flux. [From Chilès et al. (1991).]

In the present case study the sampling requirement is fulfilled with the help of a little trick, a preliminary kriging interpolation to determine the depth at every point where the flux is known. This estimation having a good precision the kriging error may be neglected. Figures 5.8a to 5.8c show the direct and cross-variograms for depth (bathymetry) and heat flux in the western zone. These variograms are strongly anisotropic: The variations are more important in the N–S direction than in the E–W direction. This apparent anisotropy can also be interpreted as the effect of an N–S drift, but for our purpose we can leave the choice of an interpretation open.

The depth variogram is continuous at the origin. The heat flux variogram has the same shape but exhibits a nugget effect due to uncertainties on flux measurements. The flux measurement error variance has a value of about 300–400 $(mW/m^2)^2$, which corresponds to an error standard deviation of $17–20\ mW/m^2$, for a mean flux value of about $220\ mW/m^2$. The cross-variogram between depth and flux is negative as expected. The interesting new feature is that the correlation is weak, if not negligible, along the E–W direction. At the same time, the correlation in the N–S direction is stronger than what the correlation coefficient indicates, as can be seen from the coefficient of codispersion $R(h)$, defined by (5.37), graphed in Figure 5.8d. At medium to large distances the good correlation between depth variations and flux variations in the N–S direction shows up clearly.

Improvement of Correlation by Filtering of Measurement Error

The measurement errors on heat flux do not affect the covariance between flux and depth nor the cross-variogram. But they increase the scatter in the cross-plots and inflate the variance and variogram of the flux and therefore reduce the correlation and codispersion coefficients, especially at short distances. There are two possibilities to correct for measurement errors:

- Subtract the measurement error variance from the variance and variogram of the flux and recompute the correlation coefficient and the codispersion function.
- Filter the flux data themselves by the method of kriging with filtering of the nugget effect, and recompute the statistical parameters and variograms.

In the second approach the effect of measurement errors is considerably reduced, without being completely eliminated because the neighboring points used for kriging are only "pseudoreplicates" and they involve some lateral variations. But the first approach is very dependent on the fit of the nugget effect and so the second method is preferred.

Figure 5.7b displays the new depth–flux scatterplot for the western zone. Observe that it is less scattered along the regression line. The correlation coefficient improves from -0.68 to -0.78. The nugget effect is practically eliminated from the variogram of the filtered flux (not shown in the figure). Thus the new codispersion function $R(h)$ can be considered unaffected by measurement errors,

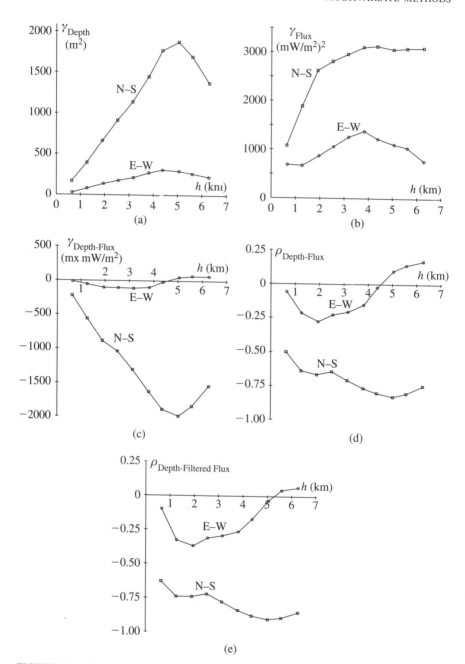

FIGURE 5.8 Direct and cross-variograms of depth and heat flux, western zone: (a) depth variogram; (b) flux variogram; (c) cross-variogram of depth and flux; (d) codispersion graph; (e) codispersion graph of depth and filtered heat flux. [From Chilès et al. (1991).]

at least at large distances; at short distances, where variations remain small, even small residual errors can produce an instability. The graph of this new $R(h)$, displayed in Figure 5.8e, reinforces the preceding analysis: low correlation in the E−W direction and high correlation in the N−S direction where $R(h)$ stabilizes at -0.86. This high correlation between depth and flux variations confirms the convective behavior in the western zone, at least along the N−S direction.

In the example presented, only one of the variables was subject to noise. However, the technique of spatial filtering provides a solution to a more challenging statistical problem: Determine the relationship between two variables when both are subject to measurement errors. The raw data scatterplot may be completely blurred by noise while the plot of filtered data may be focused enough to be interpretable.

5.6.4 Proportional Covariance Model and Screening Property

The proportional covariance model is the simplest multivariate model used in geostatistics. All covariances (variograms) are proportional to the same covariance (variogram) function

$$C_{ij}(h) = b_{ij} C(h) \qquad \text{or} \qquad \gamma_{ij}(h) = b_{ij}\gamma(h) \qquad (5.40)$$

with symmetric coefficients b_{ij} that define a positive definite matrix $\mathbf{B} = [b_{ij}]$. The conditions on $\mathbf{B}$ result directly from the symmetry $C_{ij}(h) = C_{ij}(-h)$ built in this model (hence $C_{ij}(h) = C_{ji}(h)$) and the Cramér criterion of Section 5.6.1. It is convenient to scale the basic structure function to $C(0) = 1$ so that $\mathbf{B}$ is the matrix of covariances of the variables at the same point, or of correlation coefficients if all variables are scaled to unit variance. If the matrix $\mathbf{B}$ is *strictly positive definite* ($\boldsymbol{\lambda}'\mathbf{B}\boldsymbol{\lambda} > 0$ for $\boldsymbol{\lambda} \neq 0$), it has an inverse and the cokriging system is always nonsingular. If $\mathbf{B}$ were singular, it would suffice to consider only a subset of linearly independent variables and deduce the others through the linear relationships. This model reduces the determination of $p\,(p+1)/2$ covariance (variogram) *functions* to the determination of one covariance (variogram) function and $p\,(p+1)/2$ *numbers*.

Matheron (1965, p. 149) introduced the proportional covariance model (which he named the "intrinsic correlation" model) not for the purpose of cokriging but to validate the usual statistical correlation coefficient in a geostatistical context. Why is there a problem? Because, when variables are spatially correlated, their variance or covariance within a finite domain V depends on V and on the support on which they are defined. We saw, for example, that the empirical variance s^2 is an estimate of $\sigma^2(0|V)$, the variance of a point within V, which is smaller than the variance of $Z(x)$ if it exists (Section 2.8.2). The general formula for the covariance within V of the variables $Z_i(v)$ and $Z_j(v)$ with support v, including the case $i=j$, is given by

$$\sigma_{ij}(v|V) = \frac{1}{V^2}\int_V\int_V \gamma_{ij}(y-x)dx\,dy - \frac{1}{v^2}\int_v\int_v \gamma_{ij}(y-x)dx\,dy$$

and their correlation coefficient within V is

$$\rho_{ij} = \frac{\sigma_{ij}(v|V)}{\sqrt{\sigma_{ii}(v|V)\,\sigma_{jj}(v|V)}} \tag{5.41}$$

This coefficient depends on v and V. It may tend to a limit when V becomes infinite, but this limit in general depends on the particular way V tends to infinity. Now, when all variograms are proportional, (5.41) becomes

$$\rho_{ij} = \frac{b_{ij}}{\sqrt{b_{ii}\,b_{jj}}}$$

This coefficient reflects the relationship between the two variables independently of the domain V and the support v. It is equal to the coefficient of codispersion $R_{ij}(h)$, which in this case does not vary with h.

A proportional covariance model can be generated by a linear combination of p mutually orthogonal zero-mean RFs $Y_j(x)$ with the *same covariance* $C(h)$, or the *same variogram* $\gamma(h)$:

$$Z_i(x) = m_i(x) + \sum_{j=1}^{p} A_{ij} Y_j(x)$$

where the matrix $\mathbf{A} = [A_{ij}]$ is nonsingular, satisfies $\mathbf{B} = \mathbf{AA}'$ and is computed, for example, by the Cholesky decomposition. In this expression the drifts $m_i(x)$ are assumed algebraically independent and modeled using the same set of basis drift functions. Equations (5.4) apply with similar matrices $\mathbf{F}$. This decomposition is often used to co-simulate correlated Z_i by simulating p independent RFs Y_j.

The proportional covariance model has the following screening property: *If all variables are measured at the same points, the cokriging estimator of a variable coincides with the kriging estimator based on that variable alone.* The secondary variables are "hidden" by the primary variable and receive zero weights.

This can be checked by letting $\lambda_i = 0$ for $i > 1$ in the general cokriging equations (5.4) written with $\mathbf{C}_{ij} = b_{ij}\,\mathbf{C}_{11}$ and $\mathbf{F}_i = \mathbf{F}$. In fact the screening property holds under the weaker conditions

$$\mathbf{C}_{i1}(h) = b_{i1}\mathbf{C}_{11}(h) \qquad \text{or} \qquad \gamma_{i1}(h) = b_{i1}\gamma_{11}(h)$$

which generalize the MM1 model (5.11). An interesting direct proof can be derived along the lines proposed by Rivoirard (2004) using a model with residuals similar to (5.14):

$$Z_i(x) = b_{i1}Z_1(x) + R_i(x), \quad i > 1; \qquad \text{Cov}[R_i(x), Z_1(y)] = 0 \quad \forall x, y$$

All residuals R_i are assumed spatially uncorrelated with Z_1 but the correlation structure between secondary variables is immaterial. If all variables are sampled at the same points the knowledge of Z_1 and Z_i is equivalent to the knowledge of Z_1 and R_i and cokriging estimation of $Z_1(x_0)$ involves the minimization of

$$Q = \mathrm{Var}\left[\sum_\alpha \lambda_{1\alpha} Z_1(x_\alpha) - Z_1(x_0)\right] + \sum_\ell \mu_{1\ell}\left[\sum_\alpha \lambda_{1\alpha} f^\ell(x_\alpha) - f^\ell(x_0)\right]$$

$$+ \mathrm{Var}\left[\sum_{i>1}\sum_\alpha \lambda_{i\alpha} R_i(x_\alpha)\right] + \sum_{i>1}\sum_\ell \mu_{i\ell}\sum_\alpha \lambda_{i\alpha} f^\ell(x_\alpha) \tag{5.42}$$

with the usual Lagrange parameters. This expression is of the form $Q_1 + Q_2$, where Q_2 simply cancels out by setting all weights $\lambda_{i\alpha} = 0$. The minimization of Q simplifies to that of Q_1, which gives the UK estimator of $Z_1(x_0)$ from Z_1 data alone, the stated screening property.

The result is the same if Z_1 is known at points where some Z_i are missing: The corresponding residuals are missing in (5.42) but still $Q_2 = 0$ and the screening effect persists. Another explanation is to remark that whenever a secondary variable is missing, we can always by thought give it an arbitrary value; once equal sampling is achieved, the screening property applies, making values of secondary variables irrelevant.

Now suppose on the contrary that Z_1 is missing at a place where Z_i is available. Then our observation is a composite of Z_1 and R_i, which ruins the separability in (5.42) and destroys the screening effect.

The considerable simplifications brought by the proportional covariance model should not hide the fact that it is a very special model. For example, it is practically incompatible with the presence of a nugget effect, because then it must be present also in cross-covariances, while microstructures usually do not cross-correlate. Another limitation of the model is its behavior under a change of support. Since the correlation between variables does not vary with the support, it cannot be improved by averaging over larger volumes, while such improvement may be observed. In view of this, the proportional covariance model is often used as part of another model presented next.

5.6.5 Linear Model of Coregionalization

The linear model of coregionalization is a sum of proportional covariance models. In matrix notations where $\mathbf{C}(h) = [C_{ij}(h)]$ is the $p \times p$ covariance matrix and similarly $\mathbf{\Gamma}(h) = [\gamma_{ij}(h)]$, this model takes the simple form

$$\mathbf{C}(h) = \sum_{k=1}^{s} \mathbf{B}_k C_k(h) \quad \text{or} \quad \mathbf{\Gamma}(h) = \sum_{k=1}^{s} \mathbf{B}_k \gamma_k(h) \tag{5.43}$$

where each $\mathbf{B}_k$ is called a *coregionalization matrix*. The explicit form is somewhat clumsy because triple indexes are involved, but hopefully the following will be clear:

$$C_{ij}(h) = \sum_{k=1}^{s} b_k(i,j)C_k(h) \quad \text{or} \quad \gamma_{ij}(h) = \sum_{k=1}^{s} b_k(i,j)\gamma_k(h)$$

In this model all covariances (variograms) are linear combinations of the same basic structures, indexed by k. If a particular elementary model is not present, its coefficient is set to zero; with this convention all covariances (variograms) comprise the same number s of structures. A sufficient condition for the model to be valid is that all matrices $\mathbf{B}_k$ are positive definite.

The name "linear model" originates from the fact that the above covariance structure can be obtained by linearly combining multivariate random functions $\mathbf{Y}_k(x) = (Y_{1k}(x), \ldots, Y_{pk}(x))'$ (the coregionalizations), whose components have a common structure function $C_k(h)$ or $\gamma_k(h)$ and where all RFs with different indexes are orthogonal. Indeed

$$\mathbf{Z}(x) = \sum_{k=1}^{s} \mathbf{A}_k \mathbf{Y}_k(x) \quad \Rightarrow \quad (5.43) \quad \text{with} \quad \mathbf{B_k} = \mathbf{A_k A_k'}$$

The functions $C_k(h)$ or $\gamma_k(h)$ are assumed to be known, and the $p \times p$ matrices $\mathbf{B}_k$ are to be estimated. Each component $C_k(h)$ or $\gamma_k(h)$ is associated with a certain structure scale, typically a nugget effect, a short- or medium-range model, and a long-range model. In practice, the number of structures does not exceed three because of the difficulty of clearly separating scales from an empirical variogram. When the elementary structures are scaled to unity ($C_k(0) = 1$), the sum of all $\mathbf{B}_k$ represents the covariance matrix of the variables at the same point.

By construction of the model, all cross-covariances are symmetric, and cross-variograms are therefore the correct structural tool. Due to the positivity assumption, for every (i, j) and every k, we obtain

$$|b_k(i,j)| \le \sqrt{b_k(i,i)\, b_k(j,j)} \tag{5.44}$$

In words, every basic structure present in the cross-covariance (variogram) of the variables i and j must also be present in the two direct covariances (variograms) of i and j, but the converse is not true.

Due to its simplicity and relative versatility, the linear model of coregionalization has received the most attention. Many applications can be found in the geostatistical literature. Journel and Huijbregts (1978, pp. 256ff.), after a study by Dowd (1971), show how the cross-variograms of lead, zinc, and silver in the Broken Hill mine can be nicely modeled by two basic structures, a nugget effect, and a spherical variogram with a range of 60 feet (so here $p = 3$ and $s = 2$). Wackernagel (1985, 1988) was able to fit the 120 variograms from 15 geochemical variables using just two structures (nugget effect and spherical model with a 5-km range). Daly et al. (1989) model microprobe X-ray images of six chemical elements with mixtures of spherical and cubic variograms and use this to optimize linear filtering of the noisy images. But the most exotic of this incomplete list must be the study by Steffens (1993) of animal abundance in the Kruger National Park in South Africa, where he analyzes variograms and cross-variograms of giraffe, impala, kudu, warthog, blue wildebeest, and zebra counts. We learn that zebra variograms have a very high nugget effect, while warthogs and kudus are negatively correlated. A spherical model (range 4 km) with a nugget effect fits these wildlife variograms reasonably well.

5.6.6 Fitting a Linear Model of Coregionalization

The fitting of a linear model of coregionalization can be done by the least squares method presented in Section 2.6.2 for a single variogram. Indeed, the sole difference with the univariate case is that the sill or slope of each elementary structure is replaced by a matrix of sills or slopes. An efficient implementation is proposed by Desassis and Renard (2011) who use the iterating algorithm proposed by Goulard (1989) and Goulard and Voltz (1992) to ensure the positivity of the coefficient matrices $\mathbf{B}_k$. Defining the empirical and model cross-variogram matrices

$$\hat{\mathbf{\Gamma}}(h) = \left[\hat{\gamma}_{ij}(h)\right] \quad \text{and} \quad \mathbf{\Gamma}(h) = \left[\gamma_{ij}(h)\right] = \sum_k \mathbf{B}_k \gamma_k(h)$$

the goodness-of-fit criterion is a weighted sum of squares (WSS) of all terms of the error matrix $\hat{\mathbf{\Gamma}}(h) - \mathbf{\Gamma}(h)$ summed over the set of lags J used for the fit. Specifically, it is the Euclidean norm

$$\text{WSS} = \sum_{h \in J} w(h)\text{Trace}\left\{\left[\mathbf{V}\left(\hat{\mathbf{\Gamma}}(h) - \mathbf{\Gamma}(h)\right)\right]^2\right\}$$

The weights $w(h)$ are positive and typically equal to the number of pairs used for variogram estimation at lag h. The matrix $\mathbf{V}$ is a positive definite matrix designed to equalize the influence of variables, typically the diagonal matrix of inverse variances—or the identity. The idea is to minimize the criterion by optimizing one $\mathbf{B}_k$ at a time and to repeat this until no improvement is possible. The residual for the current fit less the kth term is

$$d\mathbf{\Gamma}_k(h) = \hat{\mathbf{\Gamma}}(h) - \sum_{u \neq k} \mathbf{B}_u \gamma_u(h)$$

In the absence of positivity constraint, the optimal fit of $d\mathbf{\Gamma}_k$ by $\mathbf{B}_k \, \gamma_k(h)$ is obtained by canceling the derivative of WSS with respect to $\mathbf{B}_k$:

$$\frac{\partial \text{WSS}}{\partial \mathbf{B}_k} = -2\mathbf{V}\left[\sum_{h \in J} w(h)\gamma_k(h)\{d\mathbf{\Gamma}_k(h) - \gamma_k(h)\mathbf{B}_k\}\right]\mathbf{V} = 0$$

As $\mathbf{V}$ is nonsingular this gives

$$\mathbf{B}_k = (1/\alpha_k) \sum_{h \in J} w(h)\gamma_k(h)d\mathbf{\Gamma}_k(h), \quad \text{where} \quad \alpha_k = \sum_{h \in J} w(h)\gamma_k(h)^2$$

The constrained solution $\mathbf{B}_k^+ \geq 0$ is the positive definite matrix nearest to $\mathbf{B}_k$ according to the norm defined by $\mathbf{V}$. Being symmetric, the matrix $\mathbf{B}_k$ has a spectral decomposition of the form

$$\mathbf{B}_k = \mathbf{U}_k \mathbf{\Lambda}_k \mathbf{U}_k' \quad \text{with} \quad \mathbf{U}_k' \mathbf{V} \mathbf{U}_k = \mathbf{I}_p$$

where $\mathbf{U}_k$ is a matrix of eigenvectors of $\mathbf{B}_k \mathbf{V}$ and $\mathbf{\Lambda}_k$ is the diagonal matrix of its eigenvalues. The constrained solution is then

$$\mathbf{B}_k^+ = \mathbf{U}_k \mathbf{\Lambda}_k^+ \mathbf{U}_k'$$

where $\mathbf{\Lambda}_k^+$ is the matrix $\mathbf{\Lambda}_k$ in which all negative eigenvalues are replaced by zeros. The authors notice that this iterative algorithm always converges and that the solution does not depend on the starting point. Pelletier et al. (2004) propose several modifications of the above fitting procedure and compare them by Monte Carlo.

5.6.7 Factorial Kriging and Cokriging

We have seen that a linear model of coregionalization is associated with an orthogonal decomposition of the form $\mathbf{Z}(x) = \sum_k \mathbf{A}_k \mathbf{Y}_k(x)$. A method known as *Factorial Kriging Analysis*, developed by Matheron (1982a), permits the determination of the components $\mathbf{Y}_k$ on the basis of $\mathbf{Z}$, with the idea that they represent different genetic structures. The approach may be used for kriging or cokriging of a particular component on the basis of one or several RFs Z, or for *analysis* of a p-dimensional spatial vector $\mathbf{Z}$ in the same spirit as Principal Component Analysis. The latter is presented in the next section.

Consider the model

$$Z(x) = m(x) + \sum_{k=1}^{s} a_k Y_k(x) \tag{5.45}$$

where the a_k are known coefficients[7] and the $Y_k(x)$ mutually orthogonal RFs with zero mean and covariances $C_k(h)$. None of the $Y_k(x)$ is directly observable but can be estimated by cokriging from observations of $Z(x)$, provided that we know the cross-covariance between $Z(x)$ and $Y_k(x)$, which we do: It is $a_k C_k(h)$. The cokriging system for the estimation of $Y_k(x_0)$ is the following:

$$\begin{cases} \sum_{\beta} \lambda_\beta C(x_\beta - x_\alpha) + \sum_\ell \mu_\ell f_\alpha^\ell = a_k C_k(x_0 - x_\alpha) & \alpha = 1, \ldots, N \\ \sum_\alpha \lambda_\alpha f_\alpha^\ell = 0 & \ell = 0, \ldots, L \end{cases}$$

If the mean of Z were known and subtracted out, there would of course be no unbiasedness condition at all and no Lagrange parameter. Constraining the weights to zero has the effect of filtering out this mean without requiring an explicit estimation. Note that the left-hand side of the system remains the same for the estimation of all components Y_k, since the estimators are based on the same data, so that the work may be done in parallel. A standard kriging

---

[7] They are just scaling factors and can be set to one if the variances are included in the $Y_k(x)$ terms. Of course they have nothing to do with drift coefficients usually denoted by a_ℓ.

program may be used with a modified right-hand side—just as for filtering a nonsystematic error, which is a particular case of factorial kriging.

Given that the covariance of Z satisfies $C(x_0 - x_\alpha) = \sum a_k^2 C_k(x_0 - x_\alpha)$, the cokriging estimates automatically satisfy the consistency relation

$$Z_{UK}^*(x_0) = m^*(x_0) + \sum_{k=1}^{s} a_k Y_k^{**}(x_0)$$

where $m^*(x_0)$ is the optimal drift estimate computed from the system (3.26).

Applications of this approach include geochemical prospecting (Sandjivy, 1984), geophysical prospecting (Galli et al., 1984; Chilès and Guillen, 1984), remote sensing (Ma and Royer, 1988), and petroleum exploration (Jaquet, 1989; Yao and Mukerji, 1997). Two examples will illustrate the method. The first is from the study of gravity data by Chilès and Guillen (1984).

Extraction of a Deep Gravity Source

In magnetic or gravimetric surveys, one wishes to distinguish long wavelengths, associated with deep sources, from short wavelengths, reflecting shallower sources. This separation is traditionally performed on the basis of the two-dimensional power spectrum, but a direct spatial determination is possible and even advantageous. The experimental variogram of gravity data in this basin is shown in Figure 5.9. An excellent fit is obtained with a sum of two Cauchy models (2.55) with shape parameter $1/2$, especially suited to represent gravimetry data: a structure with scale parameter 5 km and sill 5 mgal^2 and another with scale parameter 11.4 km and sill 53 mgal^2, plus a very small measurement error nugget effect. In this variogram model the scale parameter

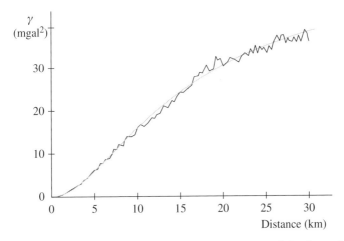

FIGURE 5.9 Sample variogram of gravity data and fit with two nested Cauchy models. [From Chilès and Guillen (1984).]

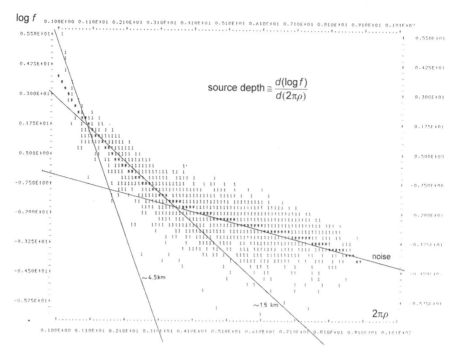

FIGURE 5.10 Logarithm of the isotropic power spectrum of gravity data. [From Chilès and Guillen (1984).]

represents twice the average depth of the sources, which are therefore located at 2.5 km and 5.7 km, in agreement with local geological knowledge. Figure 5.10 shows the logarithm of the isotropic power spectrum and the fit made by the geophysicist—don't ask how! The slopes of the lines indicate the depths of the sources: 1.5 and 4.5 km, consistent with the variogram fit. In order to compute this spectrum, the data had first to be interpolated to a regular grid, which was achieved by kriging using the fitted variogram. The kriged map and standard deviation are displayed in Figure 5.11. The decomposition of the kriged map into deep and shallow components is shown in Figures 5.12 and 5.13, on the right. Observe in Figure 5.13 that the anomaly in the upper right-hand corner coincides with a maximum of the kriging standard deviation and is therefore dubious. In performing the cokriging estimation, the neighborhood search algorithm must be customized to include near and far data so as to allow the separation of scales.

The results obtained with the spectral method are shown on the left-hand sides of Figures 5.12 and 5.13. The spectral and geostatistical maps are equivalent. The deep-field spectral map is smoother because of the global character of the spectral approach. Note that the separation between deep and shallow components, as well as the interpretation of gravimetric anomalies, is up to a constant. But the two methods do not correspond to the same constant.

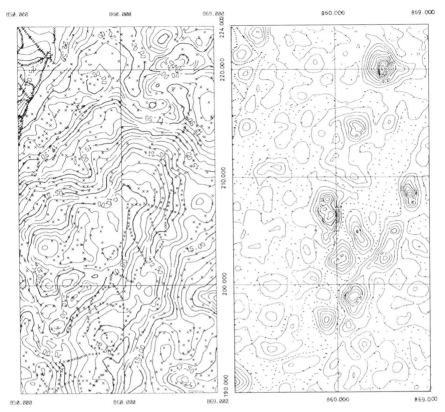

FIGURE 5.11 Kriged gravity field (left) and error standard deviation (right). [From Chilès and Guillen (1984).]

For geostatistical filtering, the mean of the shallow component has been set to zero, while it is a positive value for the spectral method.

The spectral method has some practical limitations that factorial kriging does not have:

1. It requires a complete rectangular grid of data; if these data are not acquired on a grid, or if the rectangle is not complete, a preliminary interpolation, or worse, extrapolation, is needed which already involves kriging (or similar) and alters the spectral characteristics.
2. It requires tedious tapering or padding to make the input grid periodic (wrap around in both directions).
3. It smears a local anomaly over all frequencies.

On the other hand, some phenomena, such as acoustic waves, are better understood in the frequency domain. Regardless of the method used, the separation of a field into spatial components must be supported by a physical model.

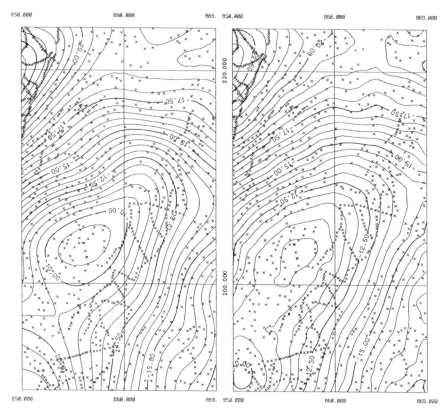

FIGURE 5.12   Estimates of deep gravity field: spectral (left), geostatistical (right). [From Chilès and Guillen (1984).]

Extraction of the Common Part of Two Surveys

In order to monitor fluid movements in a producing oil reservoir, seismic surveys are acquired at different times over the same area (4D seismic). However, due to many acquisition and processing differences, surveys need to be equalized before they can be compared. This is achieved by calibrating the surveys against a common reference established in an interval not affected by production (in practice in the overburden, ignoring geomechanical effects).

To define the reference, Coléou (2002) developed an automatic factorial cokriging procedure (AFACK) that extracts the common part between two surveys. The seismic signals Z_1 and Z_2 are modeled as sums:

$$Z_1(x) = S(x) + R_1(x), \qquad Z_2(x) = S(x) + R_2(x)$$

of a common part S representing the repeatable component (the geology) and residuals R_1 and R_2 representing acquisition effects (stripes + white noise).

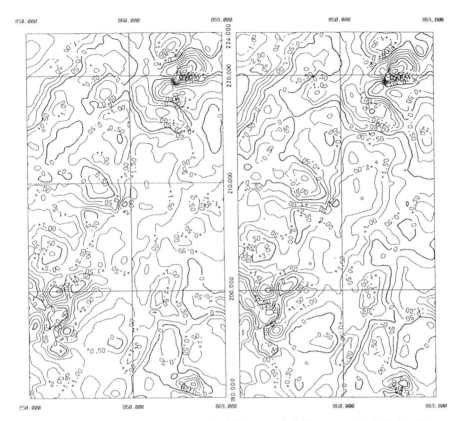

FIGURE 5.13 Estimates of shallow gravity field: spectral (left), geostatistical (right). [From Chilès and Guillen (1984).]

Residuals are assumed independent of S and also mutually independent. A cokriging estimator of S is formed:

$$S^{**}(x) = \sum_\alpha \lambda_{1\alpha} Z_1(x_\alpha) + \sum_\alpha \lambda_{2\alpha} Z_2(x_\alpha)$$

This can be computed using the covariances C_{11}, C_{22}, and C_{12}. Because the estimated points are on the same regular grid as the data points, the values of empirical covariances are available for all lags involved in the cokriging equations and there is no need to fit models. The positivity of C_{11} and C_{22} can be assured by taking estimators of the form (2.80) on a centered signal using a constant N as the denominator, and similarly for C_{12}. The regular grid geometry also permits an implementation of cokriging as a moving filter with constant weights computed once for all.

Under the independence assumptions, the complete covariance structure is

$$C_{11}(h) = C_S(h) + C_{R_1}(h), \qquad C_{22}(h) = C_S(h) + C_{R_2}(h), \qquad C_{12}(h) = C_S(h)$$

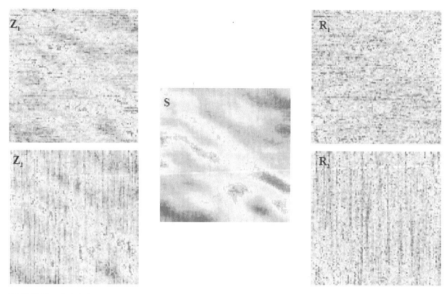

FIGURE 5.14 The common part S (center) of the two maps Z_1 and Z_2 (left) is extracted by cokriging. Residuals (right) display noise and stripes due to acquisition footprints. [From Coléou (2002).] (See color insert)

If the model is correct, $C_{12} = C_S$ is also a covariance and so are $C_{11} - C_S$ and $C_{22} - C_S$. This provides a quality check of the model. On the other hand, the model may serve only as a heuristics to derive the filtering algorithm, which is then judged by its results.

In Figure 5.14 the common part between two maps is extracted from very noisy synthetic data. The estimated common part S is a better reference to be used for the processing than any of the initial surveys. The stripes seen on the residuals are acquisition artifacts that can also be filtered by factorial cokriging.

The same approach is used for 3D cubes and between consecutive offset cubes, in a "movie" fashion. Angerer et al. (2003) extract an azimuth-dependent component in addition to the common part and the noise. Finally, factorial cokriging can be applied to extract the common part between more than two surveys.

5.6.8 Factorial Kriging Analysis

We now consider a p-variate spatial vector $\mathbf{Z}(x)$ and write the model as $\mathbf{Z} = \mathbf{m} + \sum_k \mathbf{A}_k \mathbf{Y}_k$, or in explicit notations

$$Z_i(x) - m_i(x) + \sum_{k=1}^{s}\sum_{j=1}^{p} A_k(i,j)\, Y_{jk}(x) \tag{5.46}$$

The $Y_{jk}(x)$ are mutually uncorrelated RFs, called *factors*, with zero mean and covariance

$$E[Y_{jk}(x) Y_{jk}(x+h)] = C_k(h) \quad \forall_j$$

In reality, only $\mathbf{B}_k = \mathbf{A}_k \mathbf{A}_k'$ is directly estimable and $\mathbf{A}_k$ is not uniquely defined. Given a symmetric positive definite matrix $\mathbf{B}$, one can find an infinite number of matrices $\mathbf{A}$ such that $\mathbf{B} = \mathbf{AA}'$. Principal component analysis (PCA), however, provides a natural determination of the matrices $\mathbf{A}_k$. Then the individual components $Y_{jk}(x)$ can be estimated by cokriging. When the drifts $m_i(x)$ are algebraically independent, the general equations (5.4) apply with the index pair (j, k) in the role of the index $i = 1$. The vectors $\mathbf{c}_{i0}$ are the N_i-vectors of the covariances between $Z_i(x_\alpha)$ and $Y_{jk}(x_0)$ for $x_\alpha \in S_i$, namely the vectors $(A_k(i, j) C_k(x_0 - x_\alpha): x_\alpha \in S_i)$, and the unbiasedness conditions are $\mathbf{F}_i' \boldsymbol{\lambda}_i = \mathbf{0}$ for all i (special case of a primary variable with no observation of it but a zero mean).

Now we need to elaborate on the definition of these "factors" that we are attempting to estimate. To this end, let us focus on a single scale and drop the index k. We are now dealing with a pure proportional covariance model with common correlogram $C(h)$. Let $\lambda_1 \geq \cdots \geq \lambda_p > 0$ be the eigenvalues of the covariance matrix $\mathbf{B}$ and let $\mathbf{u}_1, \ldots, \mathbf{u}_p$ be the associated eigenvectors, where $\mathbf{u}_i' \mathbf{u}_j = \delta_{ij}$. The $\mathbf{u}_i$, called the *principal axes*, define a new orthonormal coordinate system diagonalizing $\mathbf{B}$:

$$\mathbf{B} = \sum_{j=1}^{p} \lambda_j \mathbf{u}_j \mathbf{u}_j', \qquad \mathbf{I} = \sum_{j=1}^{p} \mathbf{u}_j \mathbf{u}_j', \qquad \mathbf{u}_i' \mathbf{B} \mathbf{u}_j = \lambda_j \delta_{ij} \qquad (5.47)$$

The *principal components* are the coordinates of $\mathbf{Z}$ in this new system. They are uncorrelated linear combinations $\mathbf{u}_j' \mathbf{Z}$ with variances λ_j. The quantities

$$Y_j = \mathbf{u}_j' \mathbf{Z} / \sqrt{\lambda_j}$$

are uncorrelated random variables with unit variance. From (5.47), $\mathbf{Z}$ can be expressed as

$$\mathbf{Z} = \sum_{j=1}^{p} (\mathbf{u}_j \mathbf{u}_j') \mathbf{Z} = \sum_{j=1}^{p} (\mathbf{u}_j' \mathbf{Z}) \mathbf{u}_j = \sum_{j=1}^{p} \sqrt{\lambda_j} Y_j \mathbf{u}_j$$

or, explicitly reintroducing the dependence on x and letting u_{ij} be the ith coordinate of $\mathbf{u}_j$, we have

$$Z_i(x) = \sum_{j=1}^{p} A(i,j) Y_j(x) \qquad \text{with} \qquad A(i,j) = \sqrt{\lambda_j} u_{ij}$$

The interesting feature of this decomposition is the possibility to reduce it to the first $q < p$ terms without losing much information—that is, if the first q eigenvalues account for most of the dispersion in the data (represent a high fraction of the trace of $\mathbf{B}$). Typically the first two terms will account for 80% of the variance, and the first three will account for 90% or more. Thus the Y_j can be interpreted as the *common factors* underlying the different observed variables.

To summarize the approach, we first compute direct and cross-covariances (variograms), fit a linear model of coregionalization, perform a PCA on each matrix $\mathbf{B}_k$ and deduce the coefficients $A_k(i, j)$ in (5.46), and finally estimate the leading factors $Y_{jk}(x)$ by cokriging.

The factors defined by PCA of $\mathbf{B}$ orthogonalize $\mathbf{C}(0)$, the covariance matrix at lag zero. The factors are uncorrelated *at zero distance*, but we can say no more because in general the eigenvectors of $\mathbf{C}(0)$ do not remain orthogonal with respect to $\mathbf{C}(h)$. Under the proportional covariance model, however, they do. In $\mathbf{C}(h) = \mathbf{B}\, C(h)$ the scalar $C(h)$ is just a multiplier and the same eigenvectors orthogonalize the covariance matrix $\mathbf{C}(h)$ for any lag h. The factors remain spatially uncorrelated

$$\mathrm{Cov}\left[Y_i(x),\, Y_j(x+h)\right] = \mathbf{u}_i'\, \mathbf{C}(h)\, \mathbf{u}_j / \sqrt{\lambda_i \lambda_j} = C(h)\, \mathbf{u}_i'\, \mathbf{B}\, \mathbf{u}_j / \sqrt{\lambda_i \lambda_j} = C(h)\delta_{ij}$$

and achieve a spatially consistent decomposition of $\mathbf{Z}(x)$.

The novelty of factorial kriging analysis is to enable a *scale-dependent* analysis of the correlation structure among variables. A separate PCA is carried out for each scale component rather than the raw observations, namely for each coregionalization matrix $\mathbf{B}_k$ rather than the covariance matrix which mixes the scales. This approach enabled Rouhani and Wackernagel (1990) to sort 16 piezometric observation wells into two groups, one with a relatively strong 12-month cycle seasonal component and the other with only a climatic 12-year cycle. Goovaerts et al. (1993) analyzed the chemical composition of groundwater and identified two scales of variation, a small scale likely due to the presence of local contamination and a large scale reflecting regional geological changes of the aquifer. Dousset and Sandjivy (1987), Wackernagel (1988), Bourgault and Marcotte (1993), and Goovaerts and Sonnet (1993) present applications in geochemistry, and Wackernagel et al. (1988) present others in soil science.

5.6.9 Min/Max Autocorrelation Factors

The factors defined by PCA are linear combinations, ordered by decreasing variance, of the p components *at the same point* of a p-variate spatial vector $\mathbf{Z}(x)$. As mentioned by Switzer and Green (1984), all data points are treated as replicates of a p-variate observation and no spatial properties are used to define the factors; that is, if all data points were rearranged, the factor definition would not change. Switzer and Green propose an alternative PCA method suited to geostatistics and named MAF, standing for maximal autocorrelation factors. The new factors are designed to maximize spatial autocorrelation at a given lag Δ.

Consider a p-variate random field $\mathbf{Z}(x) = (Z_1(x), \ldots, Z_p(x))'$ with mean and covariance:

$$\mathrm{E}[\mathbf{Z}(x)] = 0, \qquad \mathrm{Cov}[\mathbf{Z}(x), \mathbf{Z}(x + \Delta)] = \mathrm{E}[\mathbf{Z}(x)\mathbf{Z}'(x + \Delta)] = \boldsymbol{\Sigma}_\Delta$$

The covariance of a linear combination $\mathbf{a}'\mathbf{Z}(x)$ is

$$\mathrm{Cov}[\mathbf{a}'\mathbf{Z}(x), \mathbf{a}'\mathbf{Z}(x + \Delta)] = \mathbf{a}'\boldsymbol{\Sigma}_\Delta\,\mathbf{a} = \mathbf{a}'\boldsymbol{\Sigma}'_\Delta\mathbf{a} = \mathbf{a}'\tilde{\boldsymbol{\Sigma}}_\Delta\mathbf{a}$$

where

$$\tilde{\boldsymbol{\Sigma}}_\Delta = \tfrac{1}{2}\left[\boldsymbol{\Sigma}_\Delta + \boldsymbol{\Sigma}'_\Delta\right] \tag{5.48}$$

The cross-variogram matrix at lag Δ is related to the symmetrized matrix (5.48) by

$$\boldsymbol{\Gamma}_\Delta = \tfrac{1}{2}\mathrm{E}\left[\mathbf{Z}(x) - \mathbf{Z}(x + \Delta)\right]\left[\mathbf{Z}(x) - \mathbf{Z}(x + \Delta)\right]' = \boldsymbol{\Sigma}_0 - \tilde{\boldsymbol{\Sigma}}_\Delta$$

The autocorrelation of $\mathbf{a}'\mathbf{Z}(x)$ at lag Δ is therefore

$$\mathrm{Corr}[\mathbf{a}'\mathbf{Z}(x), \mathbf{a}'\mathbf{Z}(x + \Delta)] = 1 - \frac{\mathbf{a}'\boldsymbol{\Gamma}_\Delta\mathbf{a}}{\mathbf{a}'\boldsymbol{\Sigma}_0\mathbf{a}} \tag{5.49}$$

Let $\lambda_1 \geq \cdots \geq \lambda_p$ be the roots of the determinantal equation $|\boldsymbol{\Gamma}_\Delta - \lambda\,\boldsymbol{\Sigma}_0| = 0$ and let $\boldsymbol{a}_i$ be a generalized eigenvector of $\boldsymbol{\Gamma}_\Delta$ and $\boldsymbol{\Sigma}_0$:

$$\boldsymbol{\Gamma}_\Delta\,\mathbf{a}_i = \lambda_i\,\boldsymbol{\Sigma}_0\,\mathbf{a}_i \tag{5.50}$$

The maximum autocorrelation $1 - \lambda_p$ is reached for $\mathbf{a} = \mathbf{a}_p$, and $\mathbf{a}_p'\,\mathbf{Z}(x)$ is the first MAF. Considering the eigenvectors associated with increasing λs, we define p MAF factors $\mathbf{a}_i'\,\mathbf{Z}(x)$ from maximal to minimal spatial autocorrelation $1 - \lambda_i$. These MAF factors satisfy the following orthogonality properties:

$$\mathbf{a}_i'\boldsymbol{\Gamma}_\Delta\mathbf{a}_j = \mathbf{a}_i'\boldsymbol{\Sigma}_0\mathbf{a}_j = 0, \qquad i \neq j, \lambda_i \neq \lambda_j$$

$\mathbf{a}_i'\mathbf{Z}(x)$ and $\mathbf{a}_j'\mathbf{Z}(x)$ are uncorrelated at zero distance and so are their increments at lag Δ. In other words, the cross-variograms of the two MAFs cancel at lag Δ (and of course also at lag zero). Note that this does not imply that the MAFs themselves are uncorrelated at lag Δ; this is true only if cross-covariances are symmetric.

A nice property worth noting is the invariance of the MAF factors under a nonsingular linear transformation of the data; that is, $\mathbf{W}'\mathbf{Z}(x)$ has the same MAF factors as $\mathbf{Z}(x)$. This property is not shared by standard PCA, which gives different factors when a correlation rather than a covariance matrix is used.

From (5.50), λ_i and $\mathbf{a}_i$ can be viewed as the standard eigenvalues and eigenvectors of the matrix $\Sigma_0^{-1}\Gamma_\Delta$, but this asymmetric form is not convenient for calculations. Instead, the following algorithm is used:

1. Transform $\mathbf{Z}(x)$ into $\mathbf{Y}(x) = \Sigma_0^{-1/2}\mathbf{Z}(x)$ with a covariance equal to the identity matrix $\mathbf{I}_p$.
2. Compute differences $\mathbf{Y}(x) - \mathbf{Y}(x+\Delta)$ and the corresponding covariance matrix.
3. Find the eigenvectors of this covariance matrix and project $\mathbf{Y}(x)$ on these vectors.

The MAF technique was originally proposed by Switzer and Green (1984) to filter the noise component in multichannel spatial imagery data. The model assumes additive noise independent of the signal and proportional covariance models for both the signal and the noise, with autocorrelation decaying at a higher rate for the noise than for the signal. Then the maximal MAF is equivalent to maximizing the signal-to-noise ratio over all projections. The difference in decay rates is essential, because if all covariances of $\mathbf{Z}(x)$ itself are proportional, there is no MAF and all directions are equivalent [any $\mathbf{a}$ is an eigenvector of (5.50)].

Shapiro and Switzer (1989) apply the MAF procedure to extract time trends from multiple monitoring sites, the idea being that the trend is smooth, therefore strongly autocorrelated, and should be captured by the first MAF factors while the weakly autocorrelated time series should collect the high-frequency or noise components of the original input series. Nielsen et al. (2000) analyze irregularly sampled stream sediment geochemical data and display a collection of 40 MAF variograms that gradually evolve toward smaller ranges and higher nugget effects. Desbarats and Dimitrakopoulos (2000) use the MAF decomposition for a reduction of dimensionality and decorrelation of variables in a conditional simulation of pore-size distributions. The MAF approach is also used in ecology (Fujiwara, 2008).

5.6.10 Miscellaneous Models

In this section we give a brief overview of two models that can be of interest. Another one, indicator cokriging, is covered in Section 6.3.3.

Multivariate Matérn Model

Gneiting et al. (2010) introduce a multivariate covariance model in which both the direct and the cross-covariance functions are of Matérn type. What appears as a *tour de force* when poring over simulated realizations is that individual components can be very dissimilar. In the bivariate model on which we will focus here, one component may be rough and the other smooth, and they may even have different ranges.

Recall the Matérn isotropic covariance model with smoothness parameter ν and scale parameter a:

$$M(r\,|\,\nu, a) = \frac{2^{1-\nu}}{\Gamma(\nu)}(ar)^\nu K_\nu(ar) \qquad (a>0, \nu>0)$$

Here K_ν is a modified Bessel function of the second kind, and $1/a$ represents a scale parameter [unlike (2.56), where that parameter is a].

The bivariate full Matérn model is defined as

$$C_{11}(r) = \sigma_1^2 M(r \,|\nu_1, a_1), \qquad C_{22}(r) = \sigma_2^2 M(r \,|\nu_2, a_2),$$
$$C_{12}(r) = C_{21}(r) = \rho_{12}\, \sigma_1 \sigma_2\, M(r \,|\nu_{12}, a_{12})$$

This model is valid if and only if the magnitude of the collocated correlation coefficient ρ_{12} is less than a complicated function of the smoothness and scale parameters. Simpler *sufficient* conditions are more convenient. Letting

$$A = a_{12}^2 - \tfrac{1}{2}\left(a_1^2 + a_2^2\right), \qquad B = \nu_{12} - \tfrac{1}{2}(\nu_1 + \nu_2)$$

the sufficient conditions are

$$A \geq 0, \qquad B \geq 0, \qquad |\rho_{12}| \leq \frac{a_1^{\nu_1} a_2^{\nu_2}}{a_{12}^{\nu_1 + \nu_2}} \frac{\Gamma(\nu_{12})}{\Gamma(\nu_1)^{1/2}\,\Gamma(\nu_2)^{1/2}} \left(\frac{A}{B}\right)^B e^B$$

This full bivariate model has 9 parameters plus 2 more if nugget effects are added on the direct covariances. In view of this the authors define a *parsimonious* bivariate Matérn model in which

$$a_1 = a_2 = a_{12} = a, \qquad \nu_{12} = \tfrac{1}{2}(\nu_1 + \nu_2)$$

This reduces the number of parameters to 6 plus 2 for nugget effects. This model is valid if and only if

$$|\rho_{12}| \leq \frac{\Gamma\!\left(\nu_1 + \frac{n}{2}\right)^{1/2}}{\Gamma(\nu_1)^{1/2}} \frac{\Gamma\!\left(\nu_2 + \frac{n}{2}\right)^{1/2}}{\Gamma(\nu_2)^{1/2}} \frac{\Gamma\!\left(\frac{1}{2}(\nu_1 + \nu_2)\right)}{\Gamma\!\left(\frac{1}{2}(\nu_1 + \nu_2) + \frac{n}{2}\right)}$$

where n is the space dimensionality. When $n = 2$ (i.e., in 2D) this simplifies to

$$|\rho_{12}| \leq \frac{(\nu_1 \nu_2)^{1/2}}{\frac{1}{2}(\nu_1 + \nu_2)}$$

A parsimonious multivariate Matérn model is proposed for more than two variables in which all ranges are equal and all smoothness parameters ν_{ij} are arithmetic means of ν_i and ν_j.

Cross-Covariances Derived from Complex Models

The linear model of coregionalization has the following limitations:

- Direct and cross-covariances are even functions.
- There is limited flexibility for modeling cross-covariances: By the Cauchy–Schwarz inequality (5.44), any term included in the cross-covariance must also be present in the two direct covariances.

An innovative method to extend the model has been proposed by Grzebyk (1993), based on work by Lajaunie and Béjaoui (1991) to adapt kriging to either the case of complex variables or directional data in $\mathbb{R}^2$ (complex kriging). The idea is to consider the real part of a complex covariance model. If $Z(x)$ is a complex RF, with mean zero to simplify the presentation, its covariance

$$C(h) = \mathrm{E}\, Z(x)\overline{Z(x + h)}$$

is a complex function with the Hermitian symmetry $C(-h) = \overline{C(h)}$, whose Fourier transform is real and nonnegative. Consider now a complex proportional covariance model:

$$C_{ij}(h) = b_{ij}\, C(h)$$

where $\mathbf{B} = [b_{ij}]$ is a complex positive definite matrix with the Hermitian symmetry $b_{ji} = \overline{b_{ij}}$. That this is a valid model follows directly from the characterization criterion in Section 5.6.1; for the same reason, the complex conjugate model is also valid and therefore also the sum

$$\tfrac{1}{2}\left[C_{ij}(h) + \overline{C_{ij}(h)} \right] = \mathrm{Re}\left[C_{ij}(h) \right]$$

So the real part of the complex covariance matrix $[C_{ij}(h)]$ is a valid covariance model. The expression of these covariances in terms of the real and imaginary parts of $C(h)$ and b_{ij} is given by

$$\mathrm{Re}\left[C_{ij}(h) \right] = \mathrm{Re}\left(b_{ij}\right)\mathrm{Re}[C(h)] - \mathrm{Im}\left(b_{ij}\right)\mathrm{Im}[C(h)]$$

To go further, some explicit model of complex covariance must be selected. The above authors propose the following:

$$\mathrm{Re}[C(h)] = \sigma(h) \qquad \mathrm{Im}[C(h)] = \frac{1}{2}\sum_\theta p_\theta\left[\sigma(h - \tau_\theta) - \sigma(h + \tau_\theta)\right]$$

where $\sigma(h)$ is a real covariance function, $\{\tau_\theta\}$ is a family of translation vectors in $\mathbb{R}^n$, and $\{p_\theta\}$ is a set of positive constants satisfying $\sum p_\theta \le 1$ (sufficient but not necessary condition). Dropping the Re sign and renaming the real and imaginary parts of $\mathbf{B}$ as $\mathbf{G}$ and $\mathbf{H}$, respectively, the cross-covariance model is of the form

$$C_{ij}(h) = g_{ij}\,\sigma(h) - h_{ij}\frac{1}{2}\sum_\theta p_\theta\left[\sigma(h - \tau_\theta) - \sigma(h + \tau_\theta)\right] \tag{5.51}$$

Diagonal terms h_{ii} equal zero by the Hermitian symmetry of $\mathbf{B}$. In this formula the parametric form of $\sigma(h)$ is selected and so are the vectors $\{\tau_\theta\}$, which may have different directions. The coefficients g_{ij}, h_{ij}, p_θ are fitted, for example, by a least squares procedure described in Grzebyk (1993).

The model can be further generalized by introducing several scales of structures and lends itself to a decomposition into orthogonal factors leading to a "bilinear model of coregionalization." The reader is referred to Grzebyk and Wackernagel (1994) and Wackernagel (2003, Section 30) for the theory. The theory of complex kriging itself is presented in Wackernagel (2003, Section 29). Basically this approach minimizes the sum of the errors on the real and imaginary components $(Z_{\mathrm{Re}}^* - Z_{\mathrm{Re}})^2 + (Z_{\mathrm{Im}}^* - Z_{\mathrm{Im}})^2$ and spares us the necessity of modeling the cross-covariance between Z_{Re} and Z_{Im}, which cokriging of the components would require. However, as shown by Lajaunie and Béjaoui (1991), it is only when that cross-covariance is asymmetric that complex kriging may outperform a separate kriging of each component.

5.7 SHORTCUTS

Collocated cokriging, especially under the MM2 model, provides us with simplified cokriging. In this section we consider other approaches that bypass cokriging altogether but retain a multivariate character. We will begin with approaches that aim at introducing external knowledge in the interpolation of nonstationary phenomena. McLain (1980) poses the problem in a striking manner with an example of mapping the growth rate of hairy gooseberries in Ireland. In an area where there are no data points, a blind run of a computer mapping package produces a high growth rate, which the user knows to be incorrect because gooseberries don't grow on mountains. McLain diagnoses three causes for the computer's lack of success: "the computer didn't know where the mountains are, the computer didn't know that hairy gooseberries don't grow on mountains, and the computer wouldn't know how to use such information in any case." Here is McLain's solution: Establish a relationship between gooseberry growth rate and altitude, use the regression equation to predict the growth rate anywhere without interpolation, and finally correct the results by adding interpolated residuals. This is the guess field or the external drift approach, depending on whether the regression equation is fixed or not.

5.7.1 Guess Field Model

This model is associated with methods used in weather forecasting and called "objective analysis" (Cressman, 1959), by contrast with hand analyses. They were early applications of what is now known as *Data Assimilation*, a name actually already present in Rutherford (1972). The principle was to use the results from a numerical weather prediction model as an initial *guess field*, compute prediction errors at observation locations, and interpolate a correction using Gandin's "Optimum Interpolation" (Gandin, 1963), which coincides with simple or ordinary kriging. The guess field is powerful, it integrates diverse data from the recent past as well as the laws of physics to produce plausible estimates over areas where sampling is sparse (oceans and deserts). In Figure 5.3, for example, the 12-hour-lead forecast of the 500-mb geopotential surface was used as a guess field; if it were not, the results would be similar over Europe where observations are dense, but nonsensical gradients would be obtained over the Southern Atlantic, because extrapolation is unstable especially in the presence of a drift. Cokriging of geopotential and wind data (gradients) was used to produce this map (Schlatter, 1975; Chauvet at al., 1976).

The statistical model underlying the guess field approach is of the form

$$Z_1(x) = f(x, Z_2) + R(x) \qquad (5.52)$$

Z_1 is the variable of interest, of which we have a set of observations S_1. On the basis of other data, denoted generically by Z_2, a prediction $f(x, Z_2)$ of $Z_1(x)$ can be formed: This is the guess field. $R(x)$ is a residual independent of $Z_2(y)$ and represents the prediction error.

The function f in (5.52) is assumed known, at least to a good approximation, so that prediction errors are true residuals, without bias. It is also assumed that the guess field, which is usually defined on a grid, can be interpolated accurately to the observations' locations.

When $R(x)$ has mean zero and a stationary covariance model with a finite range, interpolation by simple kriging ensures a graceful blending with the guess field over sparsely sampled regions. On the other hand, assuming a constant unknown mean allows for local adjustments of the estimates when moving neighborhoods are used. The residual map may reveal anomalies calling for a revision of the model. For example, in the petroleum application of Delfiner et al. (1983) a water saturation residual tends to be systematically positive in the northern part of the field but negative in the southern part, indicating a possible tilt in the oil–water contact toward the south.

The guess field may be a regression, but it should not be calibrated on the observations S_1. In the MM2 model (5.14) the component $a Z_2(x)$ plays the role of a guess field. Actually, (5.52) is especially interesting when f is a complicated nonlinear function of Z_2 incorporating known physical relationships.

5.7.2 External Drift Model

The external drift model was introduced in Chapter 3 as a variant of the universal kriging model with polynomial drift. It may also be regarded as a multivariate model akin to collocated cokriging, and we will discuss how they relate. Unlike with the guess field model, here the regression coefficients are not assumed known but are estimated from the observations.

For the presentation to be specific, we revisit the example of seismic time-to-depth conversion. $Z(x)$ is the depth of a geological horizon at point x, and $T(x)$ is the travel time of a seismic wave from the surface to that horizon.[8] Seismic surveys provide a fine-mesh grid of the geometry of the subsurface, but depths are measured with high vertical resolution and absolute accuracy in boreholes only. By and large, it may be considered that seismic measurements describe the *shape* of a geologic object, whereas borehole data provide *depth control*.

To convert seismic reflection times to depths, geophysicists establish a time–velocity curve that is a best fit on a scatterplot of average velocity v against travel time T. Both quantities are measured between the mapped seismic horizon and an arbitrary reference plane (the datum plane), to correct for anomalous effects induced by significant elevations and/or near surface velocity changes. As a result of higher compaction at greater depth, velocity tends to increase with time and is often modeled, within zones, by a linear relationship. Average velocity and seismic derived depth are therefore

$$v = a_1 + a_2 T$$
$$E(Z) = a_0 + vT = a_0 + a_1 T + a_2 T^2$$

[8] Again to simplify the presentation, we consider one-way time. If T is two-way time, the equations should be written with $T/2$, and even $T/2000$ to account for differences in units (T in milliseconds and velocity in meters per second).

The coefficients a_0, a_1, and a_2 have to be determined from depth, time, and velocity information; they may vary from one zone to another. Due to the many uncertainties involved in seismic computations, it is unreasonable to expect a perfect match between the seismic depth and the depth measurements in boreholes. There are misties at the wells.

In the geostatistical model a residual fluctuation $Y(x)$ is allowed, accounting for details that cannot be captured at the resolution of surface seismic and also for local errors on datum plane determination. The depth at a point x is finally expressed as

$$Z(x) = a_0 + a_1 T(x) + a_2 T^2(x) + Y(x) \tag{5.53}$$

where $Y(x)$ is random and $T(x)$ deterministic. Statistically, this is a regression model with correlated residuals. When the regression coefficients are unknown, Kriging with External Drift (KED) equations are UK equations (3.21) or (3.25) with unbiasedness conditions:

$$\sum_\alpha \lambda_\alpha = 1 \qquad \sum_\alpha \lambda_\alpha T(x_\alpha) = T(x_0) \qquad \sum_\alpha \lambda_\alpha T^2(x_\alpha) = T^2(x_0)$$

Physically the first condition eliminates the influence of a constant shift due to an error in picking the seismic reflector; the other two conditions ensure that our estimator interpolates $T(x)$ and $T^2(x)$ exactly. Thus $Z(x)$ is estimated directly without an *explicit* time-to-depth conversion (though it is done implicitly by KED). Besides, the use of moving neighborhoods allows for lateral variations of these conversion coefficients. Kriging interpolation being exact, there are no misties at well locations.

As usual with UK, the variogram to be used is the variogram of the residuals $Y(x)$. If there are not enough data points to compute it, its shape can be hypothesized and the parameters fitted by cross-validation.

An application of this model was first published by Delhomme et al. (1981). We consider here a different example that is extreme but real, where only seven wells are available, five of them on top of a dome-shaped structure and two on the flanks (Delfiner et al., 1983). To make the problem more fun, a fault is present to the NE. Naturally, it is not possible to do any statistical inference with so few data, but we can still use kriging equations as *algorithms* and gauge the results on their merit. Figure 5.15a shows the time contour map of a marker close to the top of the formation of interest; this marker does not coincide with the reservoir top, and there is a small but unknown shift. We do not know which method was used to establish that map, but it does not matter. We accept it as the geophysicist's best guess, and of course it incorporates the fault information. On the basis of the seven wells alone, the best we can do is shown in Figure 5.15b; it is obtained by kriging with a forced quadratic drift, to achieve closure, and a spline covariance model. A fault screen is used: Note that no estimation is performed behind the fault due to the absence of wells.

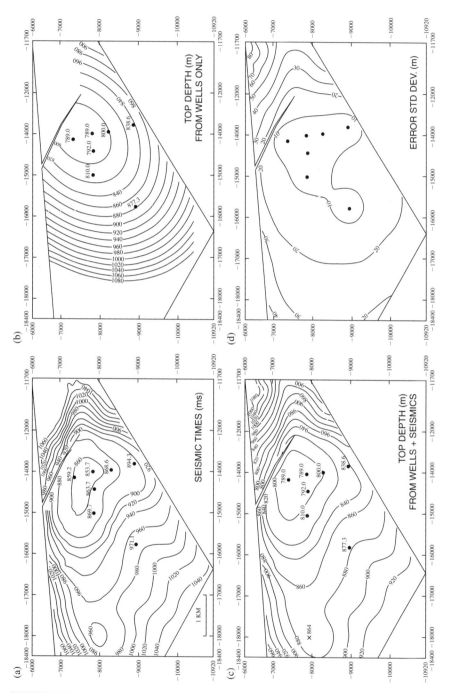

FIGURE 5.15 Use of seismic data in combination with well data to improve external structure definition: (a) seismic two-way time map; (b) top of reservoir from the wells alone; (c) map combining seismic times and well depths; (d) standard error map. [From Delfiner et al. (1983).]

Now the seismic times are introduced as an external drift and kriging is performed without a fault screen. The resulting map is shown in Figure 5.15c. It matches the well data exactly—see, for example, contour line 800 m—and its contours have the same shape as those of the seismic map. In particular, the influence of a secondary dome to the west marked by a seismic line at 960 m (Figure 5.15a) is reflected in the final depth contours, with some attenuation. Note that now there are contours behind the fault, thanks to the seismic information.

The method also allows the computation of a standard error map (Figure 5.15d), a conventional one in this case since no real inference was performed. As expected, the uncertainty is smaller near the wells (labeled less than 10 m), and one can see a sharp increase behind the NE fault. That increase takes place even though this time no fault was introduced in the estimation process. It seems that the method is intelligent and "knows" that a fault is there. Granted, we expect this to be the case for the map itself, since the fault information is present in the external drift, but standard errors do not depend on values. The explanation lies in the additivity relationship (3.35)

$$\sigma^2_{UK} = \sigma^2_{SK}$$
$$+ \sum_{\ell=0}^{2} \sum_{s=0}^{2} \text{Cov}(a^*_\ell, a^*_s) \sum_\alpha \sum_\beta \left[\lambda_{K\alpha} T^\ell(x_\alpha) - T^\ell(x_0) \right] \left[\lambda_{K\beta} T^s(x_\beta) - T^s(x_0) \right]$$

$$(5.54)$$

When $T(x_0)$ is very different from the calibration values $T(x_\alpha)$, the variance of the drift correction is large. In other words, the uncertainty increases as one moves away from the range of validity of the (implicit) velocity function—as a geophysicist would expect.

The external drift approach is easy to implement but requires the knowledge of $T(x)$ at every point involved in the kriging estimation, namely every Z sampling location and every grid node. In case $T(x)$ must be interpolated, the interpolation error is ignored.

Nothing in the theory requires $T(x)$ to be continuous. For example, Maréchal (1984) represents faulting along a profile with the discontinuous external drift function

$$m(x) = a_0 + a_1 x + a_2 H_c(x) + a_3 (x - c) H_c(x)$$

where $H_c(x)$ is the step function valued 1 if $x > c$ and 0 otherwise. Similar models are used in two dimensions and allow interpolation of faulted surfaces without the artificial recourse to fault screens.[9] Contouring the residuals enables visualization of the original unfaulted surface.

---

[9] The conditions of linear independence of the drift functions still entail that empty fault blocks cannot be interpolated.

KED and Collocated Cokriging: Do they Coincide?

Both methods use exactly the same data: the values of Z and T at estimating points and the value of T alone at the estimated point x_0. Are they the same? If not, which is better?

One obvious difference is that KED treats $T(x)$ as a deterministic drift whereas cokriging treats it as a random function. From an interpretation point of view, this difference may not be substantial. Indeed, considering the model

$$Z(x) = a_0 + a_1 T(x) + Y(x)$$

with a random $T(x)$ and conditioning on T

$$E[Z(x)|T(x)] = a_0 + a_1 T(x)$$

and recover the external drift model. However, while it is a simple matter to add a square term as in (5.53), the equivalent cokriging model would require the definition of three direct and three cross-covariances. External drift has more flexibility.

Let us then compare collocated cokriging under the MM2 model with KED under the drift model $E(Z) = a_0 + a_1 T$. By the additivity relationships of Section 3.4.7, KED is equivalent to an optimal estimation of the drift slope plus an OK estimation of the residual as if the slope was estimated perfectly. For collocated cokriging, the decomposition (5.15) is used and the residual is estimated by OK.

$$Z_{KED}^*(x_0) = a_1^* T(x_0) + \sum_\alpha \lambda_\alpha [Z(x_\alpha) - a_1^* T(x_\alpha)]$$

$$Z_{CK}^*(x_0) = a_1 T(x_0) + \sum_\alpha \lambda_\alpha [Z(x_\alpha) - a_1 T(x_\alpha)]$$

(5.55)

The difference between these two results is that cokriging assumes that the regression slope a_1 is known, while KED does not. In terms of variances we have

$$E\left[Z_{KED}^*(x_0) - Z(x_0)\right]^2$$

$$= E(a_1^* - a_1)^2 \left[T(x_0) - \sum_\alpha \lambda_\alpha T(x_\alpha)\right]^2 + E\left[R_{OK}^*(x_0) - R(x_0)\right]^2,$$

$$E\left[Z_{CK}^*(x_0) - Z(x_0)\right]^2 = E\left[R_{OK}^*(x_0) - R(x_0)^2\right]$$

Cokriging has a smaller variance and wins. Actually, not really, because KED incorporates the uncertainty on a_1 and is therefore more general. If the cokriging model is implemented with a regression slope a_1 equal to the optimal drift estimate a_1^*, the two estimators coincide.

This equivalence is surprising because, by design, cokriging is linear in T whereas KED is nonlinear in the drift functions (see Example 5 in Section 3.4.3). The contradiction is resolved by observing that when a_1 is replaced by a_1^*, the cokriging estimator ceases to be linear in T.

The equivalence between KED and the MM2 model, when it applies, justifies the requirement that the external drift be smooth. While it is intuitively clear that a rough T serving as a drift for a smooth Z does not make much sense, the MM2 model is simply not mathematically valid if Z, the primary variable, is smoother than (see Example 4 in Section 5.6.1). Further discussion on the structural link between variables in KED can be found in Rivoirard (2002).

5.7.3 Layer Cake Estimation

In sedimentary deposits the subsurface is often modeled as a stack of layers of thickness $Z_i(x)$, for x varying in the horizontal plane (layer cake model). Individual and total thicknesses are observed at well locations and we wish to interpolate them in a consistent manner—that is, ensure that the sum of individual thickness estimates is equal to the estimate of the sum. The complete cokriging solution satisfies this requirement but is impractical with a large number of layers. Furthermore, modeling errors on cross-covariances are likely to make the improved precision largely spurious. Hence the need for simpler estimators with the consistency property.

The simplest of all is to use the same kriging weights for all layers, but this disregards the spatial structure of individual layers. A more elaborate approach proposed by M. Riguidel quoted in Matheron (1979b) is to carry out independent kriging estimations of each layer and then apply a correction distributing the error on total thickness among individual layers. We present here a useful variant developed by Haas et al. (1998) which incorporates seismic information.

The first step is to estimate each layer independently from its own data by ordinary kriging. A collocated correction is then applied to take into account the total thickness obtained from seismic but with uncertainty. Defining

p number of layers

Z_i^* and σ_{Ki}^2 kriging estimator of ith layer thickness and kriging variance

$\hat{Z}_T$ and σ_T^2 total thickness from seismic thickness map and uncertainty variance

the corrected estimator is

$$\hat{Z}_i = Z_i^* + \lambda_i \left(\hat{Z}_T - \sum_{j=1}^{p} Z_j^* \right) \qquad \text{with} \qquad \lambda_i = \frac{\sigma_{Ki}^2}{\sum_{j=1}^{p} \sigma_{Kj}^2 + \sigma_T^2} \qquad (5.56)$$

The weight λ_i is derived by minimizing the estimation variance under the assumption that the kriging errors of individual layers are uncorrelated among themselves and with the seismic total thickness error.

The correction formula (5.56) has several nice properties. When the seismic total thickness variance σ_T^2 is zero, the weights λ_i add up to one and the total thickness is honored exactly. As the uncertainty on seismic total thickness increases, the constraint becomes looser. (Note that this uncertainty may vary with location.) The correction itself is proportional to the total thickness estimation error: The better the estimation, the smaller the correction. It is also weighted by the kriging variance of each layer, ensuring that no correction is applied at a point where the thickness is known, and therefore an exact match at the wells regardless of the seismic total thickness estimate. Error variances on individual and total thickness are

$$E\left(\hat{Z}_i - Z_i\right)^2 = \sigma_{Ki}^2 (1 - \lambda_i) \qquad E\left(\sum_i \hat{Z}_i - Z_T\right)^2 = \left(\sum_i \sigma_{Ki}^2\right) \sigma_T^2 / \left(\sum_i \sigma_{Ki}^2 + \sigma_T^2\right)$$

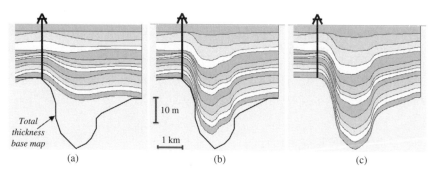

FIGURE 5.16 Kriging of layer thicknesses constrained by a seismic total thickness map. (a) Unconstrained ordinary kriging: total thickness is not reproduced; (b) constrained kriging with uncertainty on total thickness: total thickness is partly reproduced; (c) constrained kriging with no uncertainty on total thickness: total thickness is reproduced exactly. [From Haas et al. (1998).] (See color insert)

Both variances are smaller than what they would be in isolation. Errors between two different layers are negatively correlated with covariances:

$$E\big(\hat{Z}_i - Z_i\big)\big(\hat{Z}_j - Z_j\big) = -\sigma_{Ki}^2\, \sigma_{Kj}^2 \Big/ \bigg(\sum_n \sigma_{Kn}^2 + \sigma_T^2\bigg)$$

In the example shown in Figure 5.16, with a large vertical exaggeration, the total thickness base map and its uncertainty are derived from seismic. The interval is divided into 13 layers. In Figure 5.16a, each layer is kriged independently and total thickness is not honored. In Figure 5.16b, total thickness is introduced with an uncertainty variance: total thickness is partly reproduced. In Figure 5.16c, total thickness is known without uncertainty and is reproduced exactly. The method assumes implicitly that the wells are vertical; or nearly vertical, otherwise the seismic and the wells would not measure the same thicknesses.

The authors have considered applying the same correction procedure to a set of simulations. They also suggest for the first step to use the total thickness map as an external drift.

5.7.4 Compositional Data

These are vector RFs $\mathbf{Z}(x) = (Z_1(x), \ldots, Z_p(x))'$ whose components are proportions: They are positive and add up to one. We would like the estimates to have the same properties. Direct kriging or cokriging of the proportions cannot guarantee that. Furthermore, the implementation of cokriging is impractical for several reasons, the most important being that the covariances of the Zs must satisfy the closure relations $\sum_{j=1}^{p} C_{ij}(0) = 0 \;\; \forall i$ induced by the constant sum constraint, which result in spurious negative correlations and singular cokriging matrices. This leads to the use of preliminary transformations before applying kriging or cokriging.

Pawlowsky et al. (1994) use the *additive logratio transformation* of Aitchison (1986) and define

$$Y_i(x) = \log\big[Z_i(x)/Z_p(x)\big] \qquad i = 1, \ldots, p$$

where $Z_p(x)$ is the last component of $\mathbf{Z}(x)$. Note that all components $Z_i(x)$ must be strictly positive. The first $(p-1)$ transforms $Y_i(x)$ are estimated by cokriging, or kriging, while the last component $Y_p(x)$ is zero at all points and therefore can be ignored. The reverse transformation yields the estimates

$$Z_i^*(x) = \frac{\exp\left[Y_i^{**}(x)\right]}{1 + \sum_{j=1}^{p-1} \exp\left[Y_j^{**}(x)\right]} \qquad Z_p^*(x) = \frac{1}{1 + \sum_{j=1}^{p-1} \exp\left[Y_j^{**}(x)\right]}$$

By construction all estimates lie in the interval $[0, 1]$ and add up to one, as desired. On the negative side, the estimators are neither unbiased nor minimum variance, and the special role played by the last component appears arbitrary.

An alternative transformation is the *centered logratio transformation* of Aitchison (1986) defined as

$$Y_i(x) = \log\{Z_i(x)/g[\mathbf{Z}(x)]\}$$

where $g[\mathbf{Z}(x)] = \left[Z_1(x)Z_2(x) \cdots Z_p(x)\right]^{1/p}$ is the geometric mean of the components of $\mathbf{Z}(x)$, which must also be strictly positive. The symmetry with respect to the components is now preserved but the closure relation $\sum_{i=1}^{p} Y_i(x) = 0$ creates the same problems as with the initial proportions. Note that in both cases the transform vector is 0 when all proportions are equal. For clarity we now drop the dependence on location x.

The possible compositional vectors $\mathbf{u} = (u_1, u_2, \ldots, u_p)'$ form a set of points called a p-simplex and defined by

$$\mathbb{S}^p = \left\{\mathbf{u} = (u_1, u_2, \ldots, u_p)'; \quad u_i > 0, \quad i = 1, 2, \ldots, p; \quad \sum_{i=1}^{p} u_i = 1\right\}$$

$\mathbb{S}^3$, for example, is the portion of the plane $u_1 + u_2 + u_3 = 1$ in the first octant of $\mathbb{R}^3$, namely the interior of an equilateral triangle. In general the simplex $\mathbb{S}^p$ has the structure of a $(p-1)$-dimensional vector space. For any pair of compositions $\mathbf{u}$ and $\mathbf{v}$ in $\mathbb{S}^p$ the Aitchison scalar product is defined by

$$\langle \mathbf{u}, \mathbf{v} \rangle_a = \sum_{i=1}^{p} \log \frac{u_i}{g(\mathbf{u})} \log \frac{v_i}{g(\mathbf{v})}$$

This scalar product induces a distance which is useful to evaluate the difference between two compositions, and an Euclidean geometry on the simplex, called *the Aitchison geometry*. A correspondence between the Aitchison geometry and the standard Euclidean geometry is established by means of *isometric logratio transformations*, which associate to a point in the p-simplex its $(p-1)$ coordinates in an orthonormal basis of the simplex whose origin is the equal proportions composition. Egozcue et al. (2003) propose the following transformation:

$$y_i = \frac{1}{\sqrt{i(i+1)}} \log \frac{z_1 z_2 \ldots z_i}{(z_{i+1})^i}, \qquad i = 1, 2, \ldots, p-1$$

For example, for $p > 2$ the first two coordinates are

$$y_1 = \frac{1}{\sqrt{2}} \log \frac{z_1}{z_2}, \qquad y_2 = \frac{1}{\sqrt{6}} \log \frac{z_1 z_2}{z_3^2}$$

The obtained coordinates y_i are then treated by standard geostatistics and the transformation is reversed in the end using the constraint on the sum of proportions.

For further information and application examples on compositional data the reader is referred to the monograph by Pawlowski-Glahn and Olea (2004) and to the October 2005 issue of *Mathematical Geology* entirely devoted to this subject. More recently, this type of simplex approach has been proposed by Tolosana-Delgado et al. (2008) as an improvement over Indicator Kriging.

We conclude by mentioning the direct approach proposed by Walvoort and de Gruijter (2001), which consists in minimizing the sum of the kriging variances of the components subject to the positivity and unit sum constraints. One advantage of this method is that it can accommodate compositions with zero components.

5.8 SPACE–TIME MODELS

The treatment of variables $Z(x,t)$ distributed in both space and time has become so important that it would deserve a chapter of its own. However, rather than add pages to an already thick book, we prefer to outline the main approaches and refer the reader to recent review papers and books on the subject (Cressie and Wikle, 2011; Gneiting and Guttorp, 2010; Aanonsen et al., 2009; Evensen, 2009; Gneiting et al., 2007a).

Mathematically, $Z(x, t)$ may be regarded as a random function in the $(n + 1)$-dimensional space $\mathbb{R}^{n+1} = \mathbb{R}^n \times \mathbb{R}$—that is, n space dimensions plus one time dimension. If one can build a meaningful model of the covariance structure and of the drift in this space, then it is kriging as usual. This is true except that it misses a key point, the special role played by the time dimension. Isotropy may exist in space, but time flows only in one direction (unfortunately).

The main dividing line between space–time approaches is whether or not a physical model is available to describe the time evolution of the system. If such a model exists, *data assimilation* methods permit a merge of the numerical model with observations. These methods are applied in meteorology, oceanography, ecology, and petroleum engineering, to name the most frequent. Among them, sequential methods related to the *Kalman filter* and the *Ensemble Kalman Filter* (EnKF) will be discussed from the perspective of their connection with geostatistics.

If no dynamic model is available, then a purely probabilistic model is used, sometimes "inspired" by physics. Following Kyriakidis and Journel (1999), two conceptual viewpoints can be taken: Either regard $Z(x, t)$ as a single spatiotemporal RF, or else treat time or space location as an index and work in a multivariate setting in which space–time estimation is performed by cokriging.

In the multivariate approach a record of p time measurements at N monitoring stations may be regarded as N spatially correlated time series indexed by location, or as p serially correlated RFs indexed by time—in other words, "snapshots" of the $Z(x, t)$ process. The index is discrete and cannot be interpolated without further modeling. So, for example, the RF view does not allow time interpolation nor forecasting. However, given temporal stationarity, it allows the inference of nonstationary spatial models, which lead to better spatial estimates and standard errors than those obtained using a spatially averaged covariance structure.

Until about 10 years ago the only spatiotemporal models available were separable models, decomposing the space–time random function as a sum or a product of a purely spatial and a purely temporal component. Since then, families of nonseparable stationary covariance functions have been proposed, which allow space–time interaction and therefore a true spatio temporal approach.

5.8.1 Separable Space–Time Models

The straightforward extension of kriging is to consider that each point has space–time coordinates (x, t). For example, there may be two space coordinates, easting and northing measured in meters, and time which is measured, say, in hours. In view of statistical inference, it is assumed that the RF $Z(x, t)$, or its increments of some order, are second-order stationary in space and time. Thus, for example, the spatiotemporal variogram is a function of space and time separations:

$$\tfrac{1}{2}\mathrm{E}[Z(x', t') - Z(x, t)]^2 = \gamma(x' - x, t' - t), \qquad (x, t), \ (x', t') \in \mathbb{R}^n \times \mathbb{R}$$

When $t = t'$, the variogram reflects purely spatial variations; these are assumed statistically the same for all fixed times. When $x = x'$, the variogram captures the time variations only, and they are assumed statistically similar for all locations. Intermediate "directions" represent pairs separated both in space and time—for example, points 5 km and 3 hours apart; these directions are interesting for model discrimination and should not be overlooked. Naturally, there is a scaling problem between the time and the space dimensions, but it is not different, in principle, from the scaling problem encountered in 3D between vertical and horizontal variations. In case of "geometric anisotropy" a generalized distance such as $[(x' - x)^2/a^2 + (t' - t)^2/b^2]^{1/2}$ would correct for unequal spatial and temporal scale parameters a and b. The real difference is that phenomena are often periodic in time (diurnal period, yearly period, moon cycle, etc.), whereas they are not in space. This is the essential reason why going from $\mathbb{R}^2$ to $\mathbb{R}^2 \times \mathbb{R}$ is more difficult than from $\mathbb{R}^2$ to $\mathbb{R}^3$. A theory of space–time RFs has been proposed by Christakos (1992, 2000).

Great simplifications occur when the spatial and temporal variations can be completely separated. Two models share this nice property. The first one is a "zonal anisotropy" where the time direction plays the same role as the vertical direction in 3D problems (cf. Section 2.5.2)

$$\gamma(h, \tau) = \gamma_S(h) + \gamma_T(\tau), \qquad (h, \tau) \in \mathbb{R}^n \times \mathbb{R} \tag{5.57}$$

Structural analysis can be performed independently in space and time by considering pairs in the same spatial planes or in the same time series, possibly with tolerance limits. One way of checking this model is to look at the sill: It should be equal to the sill of γ_S in the space direction ($\tau = 0$), to the sill of γ_T in the time direction ($h = 0$), and to the sum of the two in all intermediate directions.

The additive model (5.57) can also be written in terms of generalized covariances, in which case, as noted by Rouhani and Hall (1989), who proposed it for groundwater data, the separation extends to the unbiasedness conditions as well; that is, there is no need for mixed space–time monomials. A known problem with model (5.57) is the possibility of singular kriging systems. More important, this model may be too crude. A more useful variant is the sum of an "isotropic" model in the space–time domain and a zonal term corresponding to the periodic time component of the phenomenon.

The second case of simplification occurs with the separable covariance model

$$C(h, \tau) = C_S(h)\, C_T(\tau), \qquad (h, \tau) \in \mathbb{R}^n \times \mathbb{R} \qquad (5.58)$$

Here the sill is the same in all directions and equal to $C_S(0)\, C_T(0)$. A quite useful property of this model is its associated screening effect pictured in Figure 3.13b: Given observations at time t, all other observations at the same space locations but different times are irrelevant for estimation at time t by simple kriging. In particular, *where the present is known, there is no information in the past.* However, this is only true in the case of a known mean.

The first use of the separable covariance (5.58), at least in the geosciences, seems to be traced back to a paper of Rodríguez-Iturbe and Mejía (1974) in which they optimized the design of a hydrologic network for the estimation of the long-term areal mean rainfall (an integral over space *and* time). Using a stationary model in time and space and a separable covariance, they showed that the time interval cannot be reduced too much: "No miracles can be expected in short times even from the most dense of all possible networks."

The above models are for the covariance structure. In general the spatial mean for fixed t will differ from the temporal mean for fixed x. There too a separable model is often used, decomposing the mean field $M(x, t)$ as the sum of a "space effect" $F(x)$ and a "time effect" $X(t)$, usually centered at zero, the baseline level of $Z(x, t)$ being included in the spatial component $F(x)$:

$$M(x, t) = F(x) + X(t)$$

For example, Séguret (1989) and Séguret and Huchon (1990) model the diurnal fluctuation of the magnetic field by a finite trigonometric expansion of the form

$$X(t) = \sum_i A_i \cos(\omega_i t) + \sum_i B_i \sin(\omega_i t)$$

where the ω_i are fixed angular frequencies (e.g., $2\pi/24$ for a daily cycle and t in hours) and A_i and B_i are unknown (possibly random) coefficients. This time effect is then removed by an adaptation of UK that filters the sine and cosine components (*trigonometric kriging*). General models for space–time trends can be found in Dimitrakopoulos and Luo (1997).

Once the covariance structure is determined and a drift model is postulated, kriging proceeds as usual by optimizing a linear estimator of the form

$$Z^* = \sum_{\alpha} \lambda_{\alpha} Z(x_{\alpha}, t_{\alpha})$$

Sample points are $\{(x_{\alpha}, t_{\alpha})\}$, and the kriging neighborhood is defined in both space and time.

5.8.2 Nonseparable Space–Time Covariance Models

Separable models are convenient but do not necessarily fit the data well. A simple example of a process with space–time interaction is the so-called *frozen field*. Imagine a purely spatial RF $Z_S(x)$, with stationary covariance C_S, moving in time with a constant velocity vector $v \in \mathbb{R}^n$. It defines a space–time process $Z(x, t) = Z_S(x - vt)$ with covariance

$$C(h, \tau) = C_S(h - v\tau)$$

This would apply to a mass of air under the influence of prevailing winds, or to a mass of water transported by ocean currents, over a short period of time. The velocity can also be a random vector V in which case the covariance becomes $C(h, \tau) = E[C_S(h - V\tau)]$ where the expectation is taken with respect to V (Gneiting and Guttorp, 2010).

Another classic example concerns covariances satisfying the so-called *Taylor hypothesis*, which states the equivalence of a time lag τ and a spatial lag $v\tau$ for some velocity vector $v \in \mathbb{R}^n$:

$$C(0, \tau) = C(v\tau, 0), \qquad \tau \in \mathbb{R}$$

This model was proposed by Taylor (1938) based on studies of turbulent flow and has been widely used because it allows the substitution of difficult space measurements by much easier time measurements at a fixed location. Gneiting et al. (2007a) provide a list of covariance functions that satisfy the Taylor hypothesis exactly, which includes the frozen field model.

Further in the vein of physically inspired models, partial differential equations are a natural source of nonseparable space–time covariances since they establish a relationship between time and space partial derivatives. Kolovos et al. (2004) provide a number of interesting examples.

To construct new models, Cressie and Huang (1999) establish that stationary space–time covariance functions admit the following representation:

$$C(h, \tau) = \int e^{2\pi i \langle u, h \rangle} \rho(u, \tau) \, du, \qquad h, u \in \mathbb{R}^n, \qquad \tau \in \mathbb{R}$$

where $\rho(u, \tau)$ is a continuous positive definite function of τ for all u, and derive covariances for specific choices of $\rho(u, \tau)$. Gneiting (2002) generalizes these results and proposes the following covariance family, which has become known as the *Gneiting class*:

$$C(h, \tau) = \frac{1}{\Psi(\tau^2)^{n/2}} \; \varphi\left(\frac{|h|^2}{\Psi(\tau^2)}\right), \qquad (h, \tau) \in \mathbb{R}^n \times \mathbb{R}$$

Here $\varphi(r)$, $r \geq 0$, is a completely monotone function and $\psi(r)$, $r \geq 0$, is a positive function with a completely monotone derivative (see definition in Section 4.5.4). The specific choices

$$\varphi(r) = \sigma^2 \exp(-cr^\gamma) \qquad \text{and} \qquad \psi(r) = (1 + ar^\alpha)^\beta$$

yield the parametric family

$$C(h, \tau) = \frac{\sigma^2}{\left(1 + a|\tau|^{2\alpha}\right)^{\beta n/2}} \; \exp\left(-\frac{c|h|^{2\gamma}}{\left(1 + a|\tau|^{2\alpha}\right)^{\beta\gamma}}\right) \qquad (5.59)$$

$a \geq 0$ and $c \geq 0$ are scale parameters of time and space, respectively, and σ^2 is the variance of the process. α and γ, both $\in [0,1]$, control the smoothness of the purely temporal and purely spatial covariances, and $\beta \in \,]0,1]$ is the space–time interaction. It is a bit surprising that the value $\beta = 0$ is excluded since we expect the model to accommodate the case of no space–time interaction. In fact it is a matter of parameterization: It suffices to multiply the covariance by the purely temporal covariance function $\left(1 + a|\tau|^{2\alpha}\right)^{-\delta}$ with $\delta > 0$ to get a separable model with $\beta = 0$ (Gneiting, 2002).

Notice that the purely temporal or purely spatial covariances defined in (5.59) display all types of behavior near the origin, with exponents 2α and 2γ between 0 and 2. Notice also that these covariances are spatially isotropic and fully symmetric in the sense that

$$C(h, \tau) = C(h, -\tau) = C(-h, \tau) = C(-h, -\tau)$$

Kent et al. (2011) draw our attention on the counterintuitive presence of a "dimple" in the covariance surface in certain cases. The effect is illustrated in Figure 5.17: For a fixed spatial lag $h > 0.7$ the temporal covariance is not a decreasing function of the temporal lag τ as one would normally expect, but increases from a minimum at zero lag before decreasing to zero, forming a valley or dimple in 3D. On the other hand, for spatial lags less than or equal to 0.7 the temporal covariance is well-behaved. So the real question for

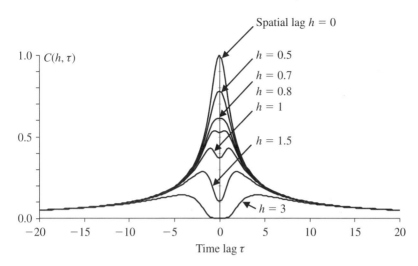

FIGURE 5.17 Temporal cross sections of the surface of the Gneiting class covariance $C(h, \tau)$ with parameters $a = c = 1$, $\sigma^2 = 1$, $\alpha = \beta = \gamma = 1$ and $n = 1$, for fixed values of spatial lag. From top to bottom: $h = 0, 0.5, 0.7, 0.8, 1, 1.5, 3$. A dimple is seen for $h > 0.7$.

applications is whether or not the dimple occurs at practically relevant spatial lags, which can be determined by time cross sections such as shown in Figure 5.17.

Stein (2005) gives a spectral characterization of continuous space–time covariance functions associated with Gaussian processes that are Markov in time, in the sense that the processes at times $t > 0$ and $s < 0$ are conditionally independent given the process everywhere at time 0. Unexpectedly, long-term dependence can take place despite the Markov property. From a theoretical perspective, Schlather (2010) unifies much of the existing literature, including Gneiting's class, in a unique class of normal scale mixtures [function φ of the form (2.27)].

5.8.3 Multivariate Approach

We now leave the view of the space–time $Z(x, t)$ as a random function in $\mathbb{R}^n \times \mathbb{R}$ to consider t or x as a discrete index. If $Z(x, t)$ is sampled synchronously at p time points, the data can be modeled as p serially correlated RFs $\{Z_i(x) = Z(x, t_i) : i = 1, \ldots, p\}$ indexed by time. This involves the inference of $p\,(p + 1)/2$ direct and cross-covariances between data at time t_i and t_j. Denoting by m_i and m_j the mean areal values at time t_i and t_j, we have

$$C_{ij}(h) = \mathrm{Cov}\left[Z_i(x), Z_j(x + h)\right] \approx \frac{1}{N} \sum_{\alpha=1}^{N} \left[Z(x_\alpha, t_i) - m_i\right]\left[Z(x_\alpha + h, t_j) - m_j\right]$$

This formula assumes spatial stationarity[10] but permits the estimation of a nonstationary temporal covariance by letting $h = 0$.

The alternative view is to treat the data as N time series $\{Z_\alpha(t) = Z(x_\alpha, t) : \alpha = 1, \ldots, N\}$ indexed by space location. This involves the inference of $N(N + 1)/2$ direct and cross-covariances between data at points x_α and x_β. Denoting by m_α and m_β the mean time values at locations x_α and x_β, we have

$$C_{\alpha\beta}(\tau) = \text{Cov}\left[Z_\alpha(t), Z_\beta(t + \tau)\right] \approx \frac{1}{p}\sum_{i=1}^{p}\left[Z(x_\alpha, t_i) - m_\alpha\right]\left[Z(x_\beta, t_i + \tau) - m_\beta\right]$$

This expression assumes temporal stationarity but permits the estimation of a nonstationary spatial covariance by letting $\tau = 0$.

Once the model is established, estimation can be carried out by cokriging, but this is rarely done. Rather, specific models are developed to take advantage of particular data configurations and relax stationarity assumptions. We will give three examples.

Estimation of Nonstationary Spatial Covariances

Consider a variable monitored at N sites x_α and p time points t_i so that the data are on a rectangular space–time lattice $\{Z(x_\alpha, t_i) : \alpha = 1, \ldots, N; i = 1, \ldots, p\}$. Typically, there are much fewer sites than time points. If we postulate that the process $Z(x, t)$ is temporally stationary, each time section $Z_i(x) = Z(x, t_i)$ may be regarded as a realization of the same parent random function $Z(x)$. Time plays no role except that the multiple realizations of $Z(x)$ may exhibit serial correlation if the lag between observation times is short. In the real world the assumption of temporal stationarity is never a trivial one and requires a careful selection of the data to ensure that they originate from the same underlying physical process, and in particular that the mean does not depend on t. For example, if we are interested in January temperatures, we may want to work only from January temperature data rather than a mix of all months.

The model is of the form

$$Z(x, t) = m(x) + Y(x, t)$$

where $Y(x, t)$ is a zero-mean residual process assumed stationary over time but nonstationary over space. In this model it is possible to estimate the mean at any monitoring site and to estimate the spatial covariance between two sites as time averages, without forcing spatial stationarity assumptions. For example,

$$\hat{\mathbf{m}} = \overline{\mathbf{Z}} = \frac{1}{p}\sum_{i=1}^{p}\mathbf{Z}_i, \qquad \hat{\mathbf{\Sigma}} = \frac{1}{p}\sum_{i=1}^{p}(\mathbf{Z}_i - \overline{\mathbf{Z}})(\mathbf{Z}_i - \overline{\mathbf{Z}})' \tag{5.60}$$

where as usual $\mathbf{Z}_i = (Z_i(x_1), \ldots, Z_i(x_N))'$.

$\hat{\mathbf{\Sigma}}$ is *numerically* positive definite, although possibly not strictly: It is singular if and only if there exist nonzero weights w_α such that $\sum_\alpha w_\alpha\left[Z_i(x_\alpha) - \overline{Z}(x_\alpha)\right] = 0$ for all i. Note that $\hat{\mathbf{\Sigma}}$ is a

---

[10] A space–time RF is spatially stationary in the wide sense if $C(x, t; x', t') = C(x' - x; t, t')$ and is time stationary in the wide sense if $C(x, t; x', t') = C(x, x'; t' - t)$ [from Christakos (1992, pp. 186–187)].

valid estimate of the true Σ even in the presence of serial correlation, but of course the estimate is more stable if the realizations are independent. (In this case, dividing by $p - 1$ eliminates the slight bias in the covariance estimate due to the estimation of the mean.)

Now suppose that we want to estimate $Z(x_0, t_0)$ at one of the observed time point $t = t_0$, using contemporaneous data only. We can subtract the means from the data and interpolate the residuals by simple kriging. It is possible to plug the covariance estimate (5.60) directly into the left-hand side of the simple kriging equations (3.2), which were established without stationarity assumptions. (This amounts to minimizing the time-averaged squared error.) By doing so, we introduce what Switzer calls *location-specific* information. However, we also need the covariances $\sigma_{\alpha 0}$ between the stations and the unobserved location x_0 and the variance σ_{00}.

First, in order to mitigate the dependence of the empirical covariance $\hat{\Sigma}$ between monitored locations on sampling variations, Loader and Switzer (1992) propose the alternative estimate

$$\tilde{\Sigma} = \nu \, \hat{\Sigma} + (1 - \nu)\mathbf{C} \qquad (5.61)$$

where $0 < \nu < 1$ is a constant and $\mathbf{C}$ is computed from a parametric model of the covariance function, typically obtained by forcing a stationary model. This is called a *shrinkage* estimate because it "shrinks" $\hat{\Sigma}$ toward $\mathbf{C}$. The value of the parameter ν is chosen from the data. A formal optimization formula (based on Wishart distributions) is provided but looks awesome; trial and error may be simpler, if not optimal.

Next the covariance vector $\boldsymbol{\sigma}_0 = (\sigma_{10}, \ldots, \sigma_{N0})'$ between the stations and an arbitrary unobserved point x_0 is estimated by interpolation of the columns of $\tilde{\Sigma}$, using as interpolation coefficients the simple kriging weights for estimation at the location x_0 calculated with the parametric covariance $\mathbf{C}$ (Switzer, 1989):

$$\tilde{\boldsymbol{\sigma}}_0 = \tilde{\Sigma} \, \mathbf{C}^{-1}\mathbf{c}_0 \qquad (5.62)$$

where $\mathbf{c}_0 = (c_{10}, \ldots, c_{N0})'$. With this choice the SK system (3.2) becomes

$$\tilde{\Sigma} \boldsymbol{\lambda} = \tilde{\Sigma} \, \mathbf{C}^{-1}\mathbf{c}_0$$

which when $\tilde{\Sigma}$ is nonsingular is equivalent to the SK system $\mathbf{C}\boldsymbol{\lambda} = \mathbf{c}_0$. The kriging estimator obtained using the location-specific $\tilde{\Sigma}$ is no different than that obtained using the parametric model $\mathbf{C}$. The improvement lies in the kriging variance, which reflects local covariance properties. This is given by

$$\sigma_{\mathrm{SK}}^2 = \operatorname{Var} Z_0 - \operatorname{Var} Z^* = \tilde{\sigma}_{00} - \boldsymbol{\lambda}_{\mathrm{K}}' \tilde{\Sigma} \boldsymbol{\lambda}_{\mathrm{K}} = \tilde{\sigma}_{00} - \tilde{\boldsymbol{\sigma}}_0'\big(\tilde{\Sigma}\big)^{-1}\tilde{\boldsymbol{\sigma}}_0 \qquad (5.63)$$

where $\tilde{\sigma}_{00}$ is an estimate of $\operatorname{Var} Z_0$ and $\boldsymbol{\lambda}_{\mathrm{K}}$ is the vector of simple kriging weights. The procedure suggested for estimating $\tilde{\sigma}_{00}$ is spatial smoothing of the diagonal elements of $\tilde{\Sigma}$. From (5.63) it is seen that the estimate is constrained by

$$\tilde{\sigma}_{00} \geq \tilde{\boldsymbol{\sigma}}_0'\big(\tilde{\Sigma}\big)^{-1}\tilde{\boldsymbol{\sigma}}_0$$

More generally, the consistency of the covariance model requires that the *augmented* covariance matrix obtained by adding rows and columns for q unobserved locations be positive definite. A necessary and sufficient condition is the positive definiteness of the matrix

$$\tilde{\Sigma}_q - \tilde{\boldsymbol{\sigma}}_q'\big(\tilde{\Sigma}\big)^{-1}\tilde{\boldsymbol{\sigma}}_q$$

where $\tilde{\Sigma}_q$ is the estimate of the $q \times q$ covariance matrix between the new sites and $\tilde{\sigma}_q = \tilde{\Sigma} \mathbf{C}^{-1} \mathbf{c}_q$ is the $N \times q$ covariance matrix between observed and unobserved sites. Loader and Switzer show that the estimation of $\tilde{\Sigma}_q$ can be carried out sequentially by adding one site at a time. The order in which points are added has no influence on the estimates of covariances between observed and unobserved locations but does influence the covariances between unobserved locations. The sequential algorithm only requires fresh estimation of the diagonal terms of the matrix, and the off-diagonal terms are obtained automatically from (5.62).

The limitation of the foregoing approach is the estimation of spatial covariances for pairs of unobserved locations. An alternative approach has been proposed by Sampson and Guttorp (1992) based on multidimensional scaling. In a nutshell, the empirical $\sqrt{\gamma_{\alpha\beta}}$ is a distance $d_{\alpha\beta}$ between the temporally stationary $Z(x_\alpha, t)$ and $Z(x_\beta, t)$. The idea is to represent the data as points in an image space such that the Euclidean interpoint distances match the order relationships among the $d_{\alpha\beta}$. The distances $|y_\beta - y_\alpha|$ in the image space and the corresponding $d_{\alpha\beta}$ are related by a monotone step function g such that $d_{\alpha\beta}^2 = g(|y_\beta - y_\alpha|)$. The function g is analogous to the usual variogram and is both stationary and isotropic. The extension to unobserved locations is achieved by fitting a parametric variogram model to this g function and modeling the mapping $y = f(x)$ between the geographic space and the image space. A discussion of this approach and an alternative rescaling procedure can be found in Monestiez and Switzer (1991).

A Model for Monitoring Data

The assumption of temporal stationarity may be unreasonable or may force us to ignore too much data. A milder form is to assume that this is only true for residuals. The above covariance estimation technique is then applied to residuals. There remains to define the model. A very detailed one has been proposed by Høst et al. (1995), based on the decomposition

$$Z(x, t) = M(x, t) + S(x, t) U(x, t)$$

where $M(x, t)$ is the mean, $S(x, t)$ the standard deviation or scale, and $U(x, t)$ a temporally stationary residual random function with zero mean and unit variance. The authors decide to also regard M and S as random functions and assume mutual independence among M, S, and U.

Next the mean field is modeled as the sum of a "space effect" $F(x)$ and a "time effect" $X(t)$:

$$M(x, t) = F(x) + X(t)$$

where $X(t)$ has zero mean over the discrete set of observation times. Similarly, the scale field is decomposed as

$$S(x, t) = H(x) + \kappa(t)$$

where $H(x)$ is a spatial scale field and $\kappa(t)$ is a temporal modulation such that $\kappa^2(t)$ has mean one. $F(x)$ and $H(x)$ are considered as second-order stationary random functions.

It is convenient to abbreviate $Z(x_\alpha, t_j)$ as $Z_{\alpha j}$ and write the model in the condensed notation

$$Z_{\alpha j} = F_\alpha + X_j + \kappa_j H_\alpha U_{\alpha j}$$

Taking means along rows and columns as is done for two-way tables yields the following estimates of the model components:

$$\hat{F}_\alpha = \frac{1}{p}\sum_{j=1}^{p} Z_{\alpha j} \qquad \hat{X}_j = \sum_\alpha \theta_\alpha (Z_{\alpha j} - \hat{F}_\alpha) \qquad \hat{H}_\alpha^2 = \frac{1}{p}\sum_{j=1}^{p}(Z_{\alpha j} - \hat{F}_\alpha - \hat{X}_j)^2$$

$$\hat{\kappa}_j^2 = \sum_\alpha \theta_\alpha (Z_{\alpha j} - \hat{F}_\alpha - \hat{X}_j)^2 / \sum_\alpha \theta_\alpha \hat{H}_\alpha^2 \qquad \hat{U}_{\alpha j} = (Z_{\alpha j} - \hat{F}_\alpha - \hat{X}_j)/\hat{H}_\alpha \hat{\kappa}_j$$

where θ_α are spatial weights adding up to one—for example, the optimal weights in the estimation of the constant mean of $F(x)$ (see Section 3.4.6). Once computed, these estimates are considered as exact. The estimate at an unobserved location x_0 and a monitoring time t_j is obtained by combining the separate kriging estimates

$$Z_{0j}^* = F_0^* + \hat{X}_j + \hat{\kappa}_j H_0^* U_{0j}^*$$

Here F_0^* and H_0^* are ordinary kriging estimates from the values $\hat{F}_\alpha$ and $\hat{H}_\alpha$ at monitoring stations, derived with fitted spatially stationary variogram models. U_{0j}^* is the simple kriging estimate from empirical residuals $\hat{U}_{\alpha j}$, based on a nonstationary covariance matrix obtained as indicated above. There is a difficulty here because the estimation of the residuals forces the linear constraint $\sum_\alpha \theta_\alpha \hat{H}_\alpha \hat{U}_{\alpha j} = 0$ for all j, making the empirical covariance matrix singular [cf. (5.60)]. To overcome this problem, Høst et al. suggest to replace the time effect estimate $\hat{X}_j$ by another time series $\tilde{X}_j$ which is the sum of annual and seasonal effects, and modify the residuals accordingly. Note that interpolation is only possible at monitored times, since no model is assumed for the temporal modulation $X(t)$ and $\kappa(t)$.

The interpolation variance is given by[11]

$$\text{Var}\left(Z_{0j}^* - Z_{0j}\right) = \text{Var}\left(F_0^* - F_0\right) + \kappa_j^2 \left(\nu^2 + \sigma_H^2\right) \text{Var}\left(U_{0j}^* - U_{0j}\right)$$
$$+ \kappa_j^2 \, \text{Var}\left(H_0^* - H_0\right) \text{Corr}^2\left(U_{0j}^*, U_{0j}\right)$$

where $\nu = E[H(x)]$ and $\nu^2 + \sigma_H^2$ is approximated by $\sum_\alpha \theta_\alpha \hat{H}_\alpha^2$. The squared correlation is equal to $\text{Var}(U_{0j}^*)$ because simple kriging is used and the U field has unit variance. This formula shows the contributions of the interpolation variance of the mean field F, the residual field U, and the scale component H.

The authors apply the approach to sulfur dioxide concentrations recorded as monthly averages over six years, and they compare the results with those obtained using standard kriging based only on data from the same time section. They conclude that the estimates are virtually identical but that the space–time model allows a better assessment of standard errors.

There are simpler versions of this very general model, for example, the scale can be made constant [e.g., Stein (1986b)].

Estimating a Time Trend

In all the preceding models the focus was on spatial estimation, and time was secondary. We used the time observations merely to strengthen spatial estimation. The priorities may be reversed, and the time evolution of the process may on the contrary be the primary focus of interest. A spectacular example is the discussion on global warming in relation with the

---

[11] This formula rectifies an error in formula (5) of the Høst et al. (1995) paper. Note that the parameter ν has nothing to do with the parameter of formula (5.61).

greenhouse effect. Is there a systematic increase of temperature liable to change the life conditions on earth within the next century? Here we have no interest in the spatial hetero-geneity among monitoring stations.

Sølna and Switzer (1996) propose the following temperature model:

$$Z(x, t) = a(x) + b(x)t + U(x, t)$$

where $a(x)$ is a deterministic baseline temperature field, $b(x)$ a spatially varying time-trend rate modeled as a second-order stationary RF, t is a year index, and $U(x, t)$ is a zero-mean residual random function, second-order stationary in both space and time. The random functions $b(x)$ and $U(x, t)$ are assumed uncorrelated. Two related quantities are of interest:

- Estimation of the time-trend rate averaged over a geographic region A:

$$m_A = \int_A b(x) d\mu(x)$$

- Estimation of an annual regional temperature change:

$$M_A(t_1, t_2) = \frac{1}{t_2 - t_1} \int_A [Z(x, t_2) - Z(x, t_1)] d\mu(x)$$

The idea is to estimate these quantities from the available records and compare them with the associated standard errors in order to test whether the rate of temperature change is signif-icantly different from zero. It is immediately apparent that these two quantities depend on the year-to-year differences

$$d(x, t) = Z(x, t) - Z(x, t - 1) = b(x) + [U(x, t) - U(x, t - 1)]$$

rather than on temperatures themselves. The derivation of the space–time variogram of $d(x, t)$ and of the kriging estimates and variances of m_A and M_A would take too long and can be found in the Sølna and Switzer paper. An application to temperature data in a climatically homogeneous region, the steppe of eastern Europe, shows a very strong spatial dependence allowing a precise estimation of regionally averaged temperature changes. On the other hand, the time series of these estimates show little temporal structure. On the basis of 40 years of temperature observations the authors concluded that "a warming trend cannot easily be discerned from a 'nonwarming' scenario," but it might with a record of 90 years. Naturally, these conclusions pertain to a relatively small geographic region and cannot be extended to the global scale.

The conclusions could also be different with the record of nearly 60 years that should be available now.

5.8.4   Sequential Data Assimilation

"*Data assimilation can be defined as the incorporation of observations into a dynamical model to improve forecasts*" (Bertino et al., 2003). There is an abun-dant literature on data assimilation (DA) and our purpose here is to highlight its links with geostatistics. We refer the reader to the monograph by Evensen (2009) for a complete treatment of DA, and we refer to Aanonsen et al. (2009) for a review focusing on applications to history matching of oil and gas reservoirs.

Kalman Filter

Consider a system evolving through time and sampled at regular time points. This system is characterized by a true state N-vector $\mathbf{x}_n$ where the subscript n denotes time. Typically, $\mathbf{x}_n$ holds the average values of the physical variables of interest over all active grid cells plus global variables. The dynamics of the system, which we assume here to be linear, is captured by an $N \times N$ state transition matrix $\mathbf{F}_n$. On the other hand, we have an M-vector of observations $\mathbf{y}_n$ with M generally much smaller than N. Observations are related to the true state space by an $M \times N$ observation matrix $\mathbf{H}$. Dynamics and observations are assumed imperfect, so model and observation errors are included in the model. In the notations of Bertino et al. (2003), which we generally follow here, the superscript "m" is used for model and "o" for observation. The equations of the system are then

$$\mathbf{x}_n = \mathbf{F}_n \, \mathbf{x}_{n-1} + \boldsymbol{\varepsilon}_n^m \qquad \boldsymbol{\varepsilon}_n^m \sim N(\mathbf{0}, \boldsymbol{\Sigma}^m)$$
$$\mathbf{y}_n = \mathbf{H} \, \mathbf{x}_n + \boldsymbol{\varepsilon}_n^o \qquad \boldsymbol{\varepsilon}_n^o \sim N(\mathbf{0}, \boldsymbol{\Sigma}^o) \tag{5.64}$$

$\boldsymbol{\varepsilon}_n^m$ and $\boldsymbol{\varepsilon}_n^o$ are mutually independent, zero mean Gaussian vectors, uncorrelated in time, with time-invariant covariance matrices $\boldsymbol{\Sigma}^m$ and $\boldsymbol{\Sigma}^o$. Evaluating $\boldsymbol{\Sigma}^o$ is simple if observation errors are assumed spatially independent: $\boldsymbol{\Sigma}^o$ is then diagonal and error variances can be estimated from the nugget effects of the measurements time series. Model errors are difficult to evaluate and require severe assumptions such as a stationary covariance (Bertino et al., 2002).

The Kalman filter may be described as a two-step process: prediction and update—or in DA terminology, forecast and analysis, denoted by the superscripts f and a. The forecast is made by taking the best state estimate at the previous time point and applying $\mathbf{F}_n$:

$$\mathbf{x}_n^f = \mathbf{F}_n \mathbf{x}_{n-1}^a$$

The forecast error is

$$\mathbf{x}_n^f - \mathbf{x}_n = \mathbf{F}_n \left(\mathbf{x}_{n-1}^a - \mathbf{x}_{n-1} \right) - \boldsymbol{\varepsilon}_n^m$$

By design, $\mathbf{x}_{n-1}^a$ is unbiased and therefore so is the forecast. The forecast error covariance is most important and can be computed by propagating the covariance $\mathbf{C}_{n-1}^a$ of the state vector estimation error at the previous time point

$$\mathbf{C}_{n-1}^a = \mathrm{E}\left(\mathbf{x}_{n-1}^a - \mathbf{x}_{n-1} \right)\left(\mathbf{x}_{n-1}^a - \mathbf{x}_{n-1} \right)'$$

using the formula

$$\mathbf{C}_n^f = \mathrm{E}\left(\mathbf{x}_n^f - \mathbf{x}_n \right)\left(\mathbf{x}_n^f - \mathbf{x}_n \right)' = \mathbf{F}_n \, \mathbf{C}_{n-1}^a \mathbf{F}_n' + \boldsymbol{\Sigma}^m \tag{5.65}$$

The analysis step improves the estimation of the state vector $\mathbf{x}_n$ by combining the forecast with the observations available at time n using an unbiased linear estimator of the form

$$\mathbf{x}_n^a = \mathbf{J}_n \mathbf{x}_n^f + \mathbf{K}_n \mathbf{y}_n$$

where $\mathbf{J}_n$ and $\mathbf{K}_n$ are matrices to be determined. Unbiasedness entails

$$\mathrm{E}\big(\mathbf{x}_n^a - \mathbf{x}_n\big) = (\mathbf{J}_n + \mathbf{K}_n \mathbf{H} - \mathbf{I})\, \mathrm{E}(\mathbf{x}_n) = 0 \quad \forall \mathrm{E}(\mathbf{x}_n) \qquad \Rightarrow \qquad \mathbf{J}_n = \mathbf{I} - \mathbf{K}_n \mathbf{H}$$

so that the updated estimate is of the form

$$\mathbf{x}_n^a = \mathbf{x}_n^f + \mathbf{K}_n\big(\mathbf{y}_n - \mathbf{H}\mathbf{x}_n^f\big) \tag{5.66}$$

A correction term proportional to the difference between actual and forecast observations is added to the forecast. The minimum variance $N \times M$ Kalman gain matrix $\mathbf{K}_n$ is determined by minimizing the trace of the error covariance matrix

$$\mathbf{C}_n^a = (\mathbf{I} - \mathbf{K}_n \mathbf{H})\, \mathbf{C}_n^f\, (\mathbf{I} - \mathbf{K}_n \mathbf{H})' + \mathbf{K}_n \mathbf{\Sigma}^o \mathbf{K}_n'$$

with respect to $\mathbf{K}_n$. This leads to the Kalman updating formulas

$$\begin{aligned} \mathbf{K}_n &= \mathbf{C}_n^f\, \mathbf{H}'\big(\mathbf{H}\, \mathbf{C}_n^f\, \mathbf{H}' + \mathbf{\Sigma}^o\big)^{-1} \\ \mathbf{C}_n^a &= (\mathbf{I} - \mathbf{K}_n \mathbf{H})\, \mathbf{C}_n^f \end{aligned} \tag{5.67}$$

These equations may also be derived in a Bayesian framework where the forecast plays the role of the prior and the analysis the role of the posterior.

In geostatistical terms the correction defined in (5.66) is the simple cokriging estimate of the forecast error $\mathbf{x}_n - \mathbf{x}_n^f$ from the measurement residuals $\mathbf{y}_n - \mathbf{H}\mathbf{x}_n^f$. Indeed the covariance of these residuals is

$$\mathrm{E}\big(\mathbf{y}_n - \mathbf{H}\mathbf{x}_n^f\big)\big(\mathbf{y}_n - \mathbf{H}\mathbf{x}_n^f\big)' = \mathbf{H}\mathbf{C}_n^f\mathbf{H}' + \mathbf{\Sigma}^o$$

while the covariance between these residuals and forecast errors is

$$\mathrm{E}\big(\mathbf{y}_n - \mathbf{H}\mathbf{x}_n^f\big)\big(\mathbf{x}_n - \mathbf{x}_n^f\big)' = \mathbf{H}\mathbf{C}_n^f$$

Writing the simple cokriging equations with these matrices yields the solution (5.67).

It is interesting to note that in the absence of observation errors the analysis step ensures an exact match of the observations. Since

$$\mathbf{H}\mathbf{K}_n = \mathbf{H}\mathbf{C}_n^f\mathbf{H}'\big(\mathbf{H}\mathbf{C}_n^f\mathbf{H}'\big)^{-1} = \mathbf{I}$$

we have

$$\mathbf{H}\,\mathbf{x}_n^a = \mathbf{H}\,\mathbf{x}_n^f + \mathbf{H}\,\mathbf{K}_n\big(\mathbf{y}_n - \mathbf{H}\,\mathbf{x}_n^f\big) = \mathbf{y}_n$$

In the case when observed variables are part of the state vector, the matrix $\mathbf{H}$ just operates a selection of these variables from the state vector, and the exact match property is equivalent to exact interpolation. In its simplest form, when considering a single state variable, the analysis step reduces to a simple kriging of the correction from the measurement residuals (Gandin's "Optimum Interpolation").

In general the Kalman filter differs in an essential way from a standard kriging or cokriging estimation in that it does not require stationary covariances in either space or time. The covariance matrix $\mathbf{C}_n^f$ is spatially nonstationary and is evolved in time by application of formula (5.65). This brings significant modeling flexibility.

Extended Kalman Filter

The Kalman filter was extended to nonlinear dynamics by the so-called *extended Kalman Filter* where the state transition and the observation models (5.64) are replaced by

$$\mathbf{x}_n = f\big(\mathbf{x}_{n-1}, \varepsilon_n^m\big), \qquad \mathbf{y}_n = h(\mathbf{x}_n) + \varepsilon_n^o$$

with nonlinear but differentiable functions f and h. At each time step the functions are linearized around the current state estimate and an approximate equation for the forecasting error covariance is derived. The extended Kalman filter gained an early popularity by being successfully applied for trajectory estimation in the Apollo space program (the first manned mission to the moon). However, it is of heuristic nature and may diverge in the case of highly nonlinear dynamics.

Ensemble Kalman Filter (EnKF)

While the Kalman filter is conceptually appealing, its implementation is simply not feasible computationally when the state dimension N is very large, typically from 10^5 to 10^6. Evensen (1994) introduced the *Ensemble Kalman Filter* (EnKF) to address this problem and also the possible divergence of the extended Kalman filter. The principle of EnKF is to propagate the uncertainty on initial conditions by generating multiple realizations of the state vector, called an *ensemble*, using successive forecast and analysis cycles. The ensemble may be thought of as a set of conditional simulations.

The workflow begins by generating an ensemble of realizations of the inputs to the dynamic system, typically $r = 100$ realizations that represent the prior knowledge and its uncertainty. This is achieved using geostatistical simulation techniques. Then for each realization, denoted by the subscript j, a forecast is computed by

$$\mathbf{x}_{n,j}^f = f_n\big(\mathbf{x}_{n-1,j}^a; \varepsilon_{n,j}^m\big) \qquad j = 1, \ldots, r$$

Unlike with the extended Kalman filter, no linearization of f_n is required. Furthermore, forecasts can be parallelized, which is an attractive feature of EnKFs.

Assuming a linear observation matrix $\mathbf{H}$, the analysis (conditioning) step is performed by

$$\mathbf{x}_{n,j}^a = \mathbf{x}_{n,j}^f + \mathbf{K}_n\left(\mathbf{y}_n - \mathbf{H}\,\mathbf{x}_{n,j}^f + \varepsilon_{n,j}^o\right) \tag{5.68}$$

where $\mathbf{K}_n$ is the classic Kalman gain matrix given by (5.67) except that the covariance matrix of forecasts is replaced by its ensemble estimate. Notice the inclusion of an observation error $\varepsilon_{n,j}^o$, not present in (5.66) and varying with the realization. This amounts to treating observations as random variables

$$\mathbf{y}_{n,j} = \mathbf{y}_n + \varepsilon_{n,j}^o$$

with means equal to the actual measurements $\mathbf{y}_n$. These randomized observations must also be simulated at every time step. As explained by Burgers et al. (1998), without this error term the analyzed ensemble covariance would have an underestimation bias equal to $\mathbf{K}_n\Sigma^o\mathbf{K}_n'$. If the mean of the analyzed ensemble is used as the best estimate, ensemble covariances may be interpreted as error covariances.

Let us now examine the consequences of replacing the forecast covariance by an ensemble covariance. We define an $N \times r$ matrix $\mathbf{X}$ whose columns are the ensemble members and an $M \times r$ matrix $\mathbf{Y}$ holding the randomized observations, i.e., including the $\varepsilon_{n,j}^0$ terms. With obvious notations and dropping the time index n the updating equation (5.68) becomes

$$\mathbf{X}^a = \mathbf{X}^f + \mathbf{K}\left(\mathbf{Y} - \mathbf{H}\mathbf{X}^f\right) \tag{5.69}$$

To compute $\mathbf{K}$ the forecast covariance matrix is replaced by the ensemble covariance

$$\mathbf{C}^f = \frac{1}{r-1}\left(\mathbf{X}^f - \overline{\mathbf{X}}^f\right)\left(\mathbf{X}^f - \overline{\mathbf{X}}^f\right)', \qquad \text{where} \quad \overline{\mathbf{X}}^f = \left(\frac{1}{r}\sum_j \mathbf{x}_j^f\right)\mathbf{1}_N' \tag{5.70}$$

Considering (5.67) and (5.70), the analysis equation (5.69) is then of the form

$$\mathbf{X}^a = \mathbf{X}^f + (\mathbf{X}^f - \overline{\mathbf{X}}^f)(\cdots)$$

The analyzed state vectors belong to the vector space spanned by the columns of $\mathbf{X}^f$, the forecast ensemble members. If the forecasting equations are linear, the analysis will be a linear combination of the initial ensemble. The small size of the ensemble compared to the number of state variables seriously limits the performance of the EnKF. A way around this is *localization*, which means local analysis using moving neighborhoods, possibly combined with covariance tapering to force the covariance to zero at a finite distance. Then the analysis is locally linear but globally it is not, and it lies in a much larger space than that spanned by the ensemble members (Evensen, 2009).

In this brief tour of EnKFs we encountered geostatistics at several places: for generating the initial ensemble, modeling spatial covariances, assimilating new data, and localizing the analysis. Other points of contact include Gaussian transform (Bertino et al., 2003; Simon and Bertino, 2009), change of support issues (Bertino et al., 2002), and facies modeling and simulation (Aanonsen et al., 2009).

Geostatistical Output Perturbation

For the sake of completeness we conclude by mentioning an approach proposed by Gel at al. (2004) to generate an ensemble of forecasts, the *geostatistical output perturbation* method. Contrary to EnKFs, which perturb the inputs to the numerical weather prediction model and then run the model many times, this method runs the model once and perturbs the output. These "perturbed" outputs are conditional simulations generated under an external drift model in which the forecast values define the regression function and the measurements are the conditioning data. This approach is intended for organizations that do not have the computing resources and data access required to perform a proper DA.

CHAPTER 6

Nonlinear Methods

We do not send estimates to the mill.

—Michel David

6.1 INTRODUCTION

Kriging permits the estimation of a variable Z at a point, or of its average value over a domain. But for a number of applications the real issue is whether $Z(x)$ exceeds a given threshold. The navigator gives special attention to the shallow parts of a bathymetric map. The environmentalist looks for places where contaminant concentration may exceed a given critical level. The meteorologist focuses on weather situations that may result in cooling below 0°C. The mining engineer attempts to predict which elementary blocks have a mean grade above the economic cutoff grade and the quantity of metal these blocks contain. When varying the threshold, it appears that the basic problem is the determination of a cumulative distribution function, or, more generally, the estimation of a nonlinear functional H of $Z(x)$. Now due to the smoothing property of kriging, one can see that $H(Z^*)$ is not a good estimator of $H(Z(x_0))$. Hence the need for a more powerful estimator than linear kriging. There are, however, specific spatial aspects that distinguish our problem from standard estimation of a frequency distribution:

1. *Domain*. The distribution is not necessarily estimated at a precise location but over a given domain. This may be the whole domain of interest D, using all available data (global estimation). Or it may be a *local* distribution based only on a subset of neighboring values (local estimation), in which case the distribution is also conditional on these values. The typical application is the determination of the grade distribution of

Geostatistics: Modeling Spatial Uncertainty, Second Edition. J.P. Chilès and P. Delfiner.
© 2012 John Wiley & Sons, Inc. Published 2012 by John Wiley & Sons, Inc.

small blocks in a mining panel. If the spacing between samples is large compared with the dimensions of a block, it is unrealistic to attempt predicting individual blocks—precision would be poor, and estimates would change very little from block to block. A more reasonable goal is to predict the *number* of blocks above the threshold. An equivalent way of expressing this, which may be meaningful for a variety of applications, is to say that rather than estimating the probability of exceeding the threshold at a given location x, we are estimating that probability at a *random* location $\underline{x}$ within the region of interest.

2. *Change of Support.* Although measurements are usually made on a point support—or at least small enough to be considered as such—one is often interested in larger supports. Thus in selective mining only the high-grade ore is processed for metal recovery. But a deposit is not mined with a teaspoon. In the absence of geological guidance, selection is performed on "small" blocks, which are in fact huge compared with the core samples. The distribution of these blocks is the quantity of interest and must be determined from the core samples, a challenging task that requires change-of-support models.

3. *Information Effect.* The classification of Z as above or below threshold is necessarily made on the basis of incomplete information, which results in a loss of efficiency. In selective mining, a selection of the blocks based on estimates Z^* will yield inferior economic results compared to a selection based on true values. This effect is named the *information effect*. Its evaluation requires modeling of the joint distribution of Z and Z^*.

In the first part of this chapter, we will examine the estimation of the point distribution. We will first review the solutions provided by simple methods whose implementation only requires modeling second-order moments. We will in particular discuss the capabilities and the limits of indicator kriging and its variants. We will then present a nonlinear estimator that is intermediate between simple or ordinary kriging and conditional expectation, *disjunctive kriging*. Since this requires modeling bivariate distributions, we will give a glimpse of the main models available and criteria for selecting one.

In the second part, we will turn to the change-of-support problem. There lies the full worth of these bivariate models, because they provide change-of-support models whose domain of validity far exceeds that of classic models, which will be presented too. The information effect will be discussed only briefly.

6.2   GLOBAL POINT DISTRIBUTION

Similar to SK requiring the knowledge of the mean m, estimating a local—or conditional—distribution requires the knowledge of the global—or marginal—distribution. We will seek to relax this prerequisite, like OK does for the mean. But first let us examine the problem of determining the marginal distribution of Z from the data.

6.2.1 Regional Distribution and Theoretical Distribution

Global estimation of the point distribution covers in fact two conceptually different problems:

1. Considering a regionalized variable $z(x)$, where x belongs to a domain D, one is interested in the regional histogram of the values of $z(x)$, $x \in D$, or in the regional c.d.f.

$$F_R(z) = \frac{1}{|D|} \int_D 1_{z(x) < z} \, dx$$

$F_R(z)$ is an ordinary function that exists independently of any probabilistic interpretation. However, since $z(x)$ is known only at a limited number of points $\{x_\alpha : \alpha = 1, \dots, N\}$, one is led to consider $z(x)$ as a realization of a RF $Z(x)$, and the problem becomes that of the estimation of

$$F_R(z) = \frac{1}{|D|} \int_D 1_{Z(x) < z} \, dx$$

for any value of z from the data $\{Z(x_\alpha) : \alpha = 1, \dots, N\}$. We will keep the same notation $F_R(z)$ for a c.d.f. that is now random (through Z). The problem is meaningful for general random function models, but in practice, it is only considered under an assumption of global or local stationarity.

2. Given that one turns to the model of a stationary and ergodic RF $Z(x)$, the global problem may be regarded as the estimation of the theoretical marginal distribution $F(z) = E[1_{Z(x) < z}]$.

This distinction generalizes the distinction between the theoretical mean, or mathematical expectation, of an SRF $Z(\cdot)$ and the regional mean of $Z(x)$ in the domain of interest D. In practice, we can at best determine the regional distribution, and most of the time only its approximation by an empirical distribution. We will often have to consider a framework in which the theoretical distribution is known. We will then assume that it can be identified with the empirical distribution or modeled from it, which amounts to assuming that we have a fairly large number of data and that the domain of interest is large with respect to the range (cf. the discussion in Section 2.3.5 on the integral range and the conditions under which the regional mean and the theoretical mean can be identified). Of course we will try to walk away as much as possible from the relatively strict stationarity that is implicit in this hypothesis.

Determination of the Empirical Distribution

Consider a stationary and ergodic RF $Z(x)$, with marginal distribution $F(dz)$. This distribution is generally unknown. We assume to be dealing with a realization $z(x)$ of the RF $Z(x)$, whose values $z_\alpha = z(x_\alpha)$ are known at N data points $\{x_\alpha : \alpha = 1, \ldots, N\}$. It is further assumed that we can construct an empirical distribution $\hat{F}$ that is a good approximation of the true F by assigning an appropriate weight to each sample. The empirical distribution $\hat{F}$ is of the form

$$\hat{F}(z) = \sum_{\alpha=1}^{N} w_\alpha 1_{z_\alpha < z} \tag{6.1}$$

where the w_α are nonnegative weights summing to one. The simplest is to give all weights the same value $1/N$. This solution is acceptable if the sampling pattern is nearly uniform. But data are often in clusters or aligned along profiles, or else the sampling grid is locally tighter, for example, in shallow areas, near pollution peaks, or in well-mineralized zones. We have seen in Section 2.2.6 that special care must be taken to get a sensible sample variogram from such data. Similarly, the raw data histogram provides a biased image of the mean characteristics. We have to seek a *declustering* technique to weigh each sample according to its location relative to the other samples and to the domain D of interest. Geometrical methods, such as cell declustering (Journel, 1983; Deutsch and Journel, 1992) and polygonal declustering (Isaaks and Srivastava, 1989; Deutsch, 2002), provide ad hoc answers to the presence of clusters but do not take spatial correlations into account.

A natural way of accounting for these correlations is to use kriging weights. Several solutions are possible according to the type of kriging considered. Switzer (1977) proposes to minimize a weighted sum of relative variances of the differences $\hat{F}(z) - F(z)$ for different values of z. If one is mainly interested in values near the median, he suggests to simply take the kriging weights of the estimation of the mean value in D of the indicator associated with the median. Another solution is to take the ordinary kriging weights of Z_D. This method has the merit of being consistent with the estimation of Z_D. Indeed the mean that can be deduced from the histogram thus estimated coincides by construction with the direct kriging of Z_D. The histogram obtained is a genuine histogram if all weights are positive, which is generally the case for global estimation. This histogram obviously depends on the domain D, so that the choice of D matters. It should not exceed too much the "envelope" of the sample points.

Evaluating the Tail of the Distribution

The empirical distribution (6.1) progresses by jumps at each value z that coincides with one of the data values z_α. This can be avoided by replacing $1_{z_\alpha < z}$ in (6.1) by a kernel $k(z_\alpha; z)$ that is an appropriate continuous function increasing from 0 to 1 in a neighborhood of z_α.

Another reason to do so is that a finite number of data does not sample the full range of the possible values. Even if the data are evenly located and are of good quality, the sample variance is on average smaller than the a priori variance of Z, because it is a variance of dispersion of point values in the domain defined by the union of all the data points.

Since $k(z_\alpha; \cdot)$ can be considered as a cumulative distribution function, examples of kernels are (a) the Gaussian c.d.f. for a variable with a symmetric distribution and (b) the lognormal c.d.f. for a positive variable with a positively skewed distribution. The mean of the distribution $k(z_\alpha; \cdot)$ is fixed at z_α so that the sample mean is not altered. Its variance is independent of z_α (Gaussian kernel), or proportional to z_α^2 (lognormal kernel), and is parameterized so as to obtain the desired tail or the desired variance for Z (for example, a variance equal to the sill of the variogram). This method, proposed by Deraisme and Rivoirard (2009), is very flexible.

Modeling of the Global Distribution

This empirical distribution may display peculiarities of the data set, reflecting the anecdote rather than the global behavior in the area of interest. So a smooth version of $\hat{F}$ may be retained for F (e.g., a piecewise linear function). In all cases it is important to pay attention to the modeling of F at both ends of the distribution, especially the tail of the distribution for a nonnegative variable. At that point, external knowledge can be incorporated. For example, a linear or negative exponential function can be used to model the end of the cumulative distribution up to what is considered as the maximum possible value for Z. We can control the fitting by comparing the empirical distribution and the final model, as well as the corresponding selectivity curves presented in Section 6.5.1.

6.2.2   Transform to a Gaussian Variable

In a nonlinear study the interest usually lies more in the estimation of a local conditional distribution than in the global distribution. Gaussian SRFs have nice properties for that. While it is rare to encounter a variable with a Gaussian spatial distribution, it is usually possible at least to transform it into one with a Gaussian *marginal*. If $Z(\cdot)$ has a continuous marginal distribution $F(dz)$ and if $G(dy)$ stands for the standard Gaussian distribution, the transformation $Y = G^{-1}(F(Z))$ transforms $Z(\cdot)$ into an SRF $Y(\cdot)$ with standard Gaussian marginal and is called the *normal score transform* (recall that $F(Z(x)) = G(Y(x))$ in distribution, being both uniform deviates). This *Gaussian transformation* can be defined "graphically" from the modeled c.d.f. F: Each point $(z, F(z))$ of the graph of the c.d.f. is associated with the point $(y, G(y))$ such that $F(z) = G(y)$ (Figure 6.1). Conversely, Z can be regarded as the transform of the Gaussian Y by $Z(x) = \varphi(Y(x))$, and this will turn out to be the useful formulation. The function $\varphi = F^{-1} \circ G$ is called the *transformation* (or *anamorphosis*) *function*. It can be represented by its expansion into Hermite polynomials (Section 6.4.4). The normal score transform is the function φ^{-1}.

The case where G is the standard normal c.d.f. is of special interest, but we will also consider other marginal distributions for Y—for example, a gamma distribution. The methodology is exactly the same as for a Gaussian

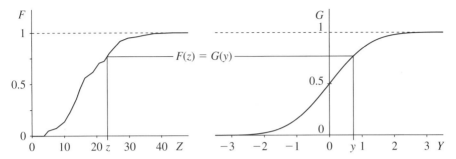

FIGURE 6.1   An example of graphical Gaussian transformation.

transformation, except that the polynomial expansion of φ is made with Laguerre polynomials (Section 6.4.4)

Improved Modeling of the Marginal Distribution in a Multi-Gaussian Framework

Emery and Ortiz (2005b) propose an original and elegant method for modeling the marginal distribution when $Z(\cdot)$ is regarded as the transform of a Gaussian random function $Y(\cdot)$—that is, an SRF with multivariate Gaussian distributions and not only a Gaussian marginal. They start from a first guess $\hat{F}(z)$ of the c.d.f. $F(z)$—for example, the empirical distribution defined by (6.1).

The N data $Z(x_\alpha) = z_\alpha$ are sorted by ascending values (assuming no ties to simplify the presentation). We denote the sorted data by z_i and denote the associated values of the c.d.f. by $\hat{F}_i = \hat{F}(z_i)$. The Gaussian value associated with z_i is $y_i = G^{-1}(\hat{F}_i)$. The c.d.f. $\hat{F}$, however, is an approximation to the true distribution F. The values y_i are therefore approximations to the true values $Y(x_i)$ at the locations x_i of the sorted data. These true values are unknown, but they honor the ranking

$$-\infty = Y(x_0) < Y(x_1) < Y(x_2) < \cdots < Y(x_{i-1}) < Y(x_i)$$
$$< Y(x_{x+1}) < \cdots < Y(x_N) < Y(x_{N+1}) = +\infty$$

Starting from the initial approximations y_i, a set of $\{Y(x_i) : i = 1, \ldots, N\}$ honoring these constraints and consistent with the spatial structure is simulated with the Gibbs sampler presented in Section 7.7.3: (i) An index i is chosen randomly and (ii) the value of $Y(x_i)$ is relaxed: $Y(x_i)$ is simulated conditionally on the other values $Y(x_j) = y_j, j \neq i$, under the constraint that the updated value $Y(x_i)$ belongs to the current interval $]y_{i-1}, y_{i+1}[$. The procedure is repeated for another index, and so on, until all the $Y(x_i)$ have been updated. The complete loop is repeated a large number of times. The average of the successive updates of $Y(x_i)$ for a fixed i converges to the conditional expectation of $Y(x_i)$.

The Gibbs sampler assumes that the variogram of Y is known, which is not the case. The method is therefore first applied with a model fitted to the sample variogram of the y_i derived from the first guess $\hat{F}$. At the end of each Gibbs sampler run, the variogram of the current simulation is calculated and taken as a new initial model to repeat the whole procedure. Emery and Ortiz show an example where convergence is obtained after about five iterations of the whole procedure.

This approach can be easily generalized to the presence of spikes, typically due to a zero effect (a significant proportion of zero values). It automatically takes the redundancy of clustered data into account, doing it in a fully consistent way. Moreover, it solves an important practical problem: With the standard normal score transform the Gaussian data $Y(x_j)$ have, by design, a variance equal or very close to 1, so that their variogram has a sill greater than 1, because the variance of the $Y(x_j)$ is a dispersion variance (the exception is when the range is much shorter than the size of the study domain). Then, contrary to expectations, $Y(x)$ is not a standard Gaussian SRF, which may result in unrealistically high back-transformed values. In order to get a sample variogram with unit sill and a sample variance smaller than 1, in conformity with the dispersion variance formula, Emery and Ortiz apply the Gibbs sampler with a variogram model rescaled to a unit sill: The successive sample variograms then progressively tend to have a unit sill.

Note that this methodology gives the transformation φ^{-1}, or equivalently the c.d.f. F, for $z = z_i$ only. Between these values the c.d.f. must be modeled, particularly at its tail, using one of the methods presented above.

6.3   LOCAL POINT DISTRIBUTION: SIMPLE METHODS

6.3.1   Local Estimation of the Point Distribution: Goals

Consider a subdomain V of the domain of interest D. The proportion of V occupied by the points where $Z(x)$ is below a given threshold z is

$$I(z) = \frac{1}{|V|} \int_V 1_{Z(x) < z} \, dx \tag{6.2}$$

Our goal will be its estimation for one or more values of z, or even for all possible values of z, from observations $Z(x_1), \ldots, Z(x_N)$ at data points $x_1, \ldots, x_N$. The optimum estimator is the conditional expectation $E[I(z) \mid \text{data}]$, where "data" represents the conditioning data $Z(x_1), \ldots, Z(x_N)$. The function $I(\cdot)$ is a random c.d.f. (through Z), and determining the conditional expectation of $I(z)$ amounts to determining the conditional c.d.f.

$$F_{\underline{x}}(z \mid \text{data}) = \Pr\{Z(\underline{x}) < z \mid \text{data}\} \tag{6.3}$$

where $\underline{x}$ represents a uniform random point of V. Indeed

$$\Pr\{Z(\underline{x}) < z \mid \text{data}\} = E[1_{Z(\underline{x}) < z} \mid \text{data}] = E[I(z) \mid \text{data}]$$

Any other estimator $I^*(z)$ can be considered as an approximation $\hat{F}_{\underline{x}}(z \mid \text{data})$ to the conditional distribution. The distinction between the conditional distribution and its approximation is important, especially when the objective is to define confidence limits. Indeed, if we know the conditional c.d.f. $F_{\underline{x}}(z \mid \text{data})$, a conditional confidence interval for $Z(\underline{x})$ at level α is

$$Z(\underline{x}) \in \left[F_{\underline{x}}^{-1}\left(\tfrac{\alpha}{2} \mid \text{data} \right), F_{\underline{x}}^{-1}\left(1 - \tfrac{\alpha}{2} \mid \text{data} \right) \right]$$

This interval covers a proportion $1 - \alpha$ of actual situations. But if the estimator $I^*(z)$ is only an approximation to the conditional probability, the interval $\left[I^{*^{-1}}\left(\tfrac{\alpha}{2} \right), I^{*^{-1}}\left(1 - \tfrac{\alpha}{2} \right) \right]$ does not necessarily cover a proportion $1 - \alpha$ of situations. Switzer and Xiao (1988) present a simple example (Gaussian SRF, four data points at the vertices of a square, $I^*(z)$ obtained by kriging of the indicator $1_{Z(x) < z}$) for which the interval $\left[I^{*^{-1}}(0.05), I^{*^{-1}}(0.95) \right]$, which could be expected to cover the true value in 90% of the cases, contains it only in 79% of the cases (results obtained by Monte Carlo simulation): Extreme situations are more likely than expected on the basis of $I^*(z)$. Additional results can be found in Lajaunie (1990), who started from this example.

Let us point out that our stated goal is different from the estimation of the probability that $Z(x)$ remains less than z at *every* point of V (i.e., that the *maximum* of $Z(x)$ in V remains less than z): That is a sensitive and difficult problem that will be considered in the case of random functions displaying extreme values (Section 6.11).[1] Our goal also differs from the estimation of the probability that the mean value of $Z(x)$ in V is less than z, a problem that will be examined later. However, formula (6.2) shows that the estimation of $I(z)$ derives directly from the estimation of $1_{Z(x) < z}$ at any point x of V. In terms of conditional distribution, (6.3) can be written as

$$F_{\underline{x}}(z \mid \text{data}) = \frac{1}{|V|} \int_V F_x(z \mid \text{data}) dx$$

where $F_x(\cdot \mid \text{data})$ is the conditional c.d.f. of $Z(x)$ at a fixed point x. Generally, the difference between the case where V is reduced to a point and that where V is really a volume will be of the same nature as point kriging and block kriging. To simplify the presentation, we will only consider here the case where the studied domain V is reduced to a point x_0. Rather than (6.2), the function $I(z)$ that we try to estimate is simply

---

[1] In the old days, air pollution standards were often defined by reference to peak values, a typical rule being that at each station where it is monitored, the pollution level should not exceed a prescribed threshold. Naturally, this is an incentive not to monitor. The statistician's recommendation to make the threshold dependent on the number of monitoring stations was regarded as unfair by politicians who contended that pollution standards should be the same for everyone. Nowadays, standards have become quite sophisticated.

$$I(z) = 1_{Z(x_0) < z}$$

Phrased differently, we will seek (an approximation to) the conditional distribution

$$F_{x_0}(z \mid \text{data}) = \Pr\{Z(x_0) < z \mid \text{data}\}$$

rather than the distribution (6.3). Before reviewing a few classic estimators I^* that constitute approximations to the conditional distribution, let us examine the conditional distribution itself. In its most general form, it is poles apart from the simple estimators that we consider here. But in practice it is applied only to Gaussian RFs (up to a transform), namely in a framework in which all multivariate distributions are entirely determined by their second-order moments—in other words, by the structural characteristics of linear geostatistics.

6.3.2   Conditional Expectation

General Case

Consider N sample points $\{x_\alpha : \alpha = 1, \ldots, N\}$ and the target point x_0, and denote by $F(z, z_1, \ldots, z_N)$ the $(N+1)$-dimensional distribution defined by

$$F(z, z_1, \ldots, z_N) = \Pr\{Z(x_0) < z, Z(x_1) < z_1, \ldots, Z(x_N) < z_N\}$$

From this distribution we can deduce

- the marginals $F_0(z) = \Pr\{Z(x_0) < z\}$, $F_1(z) = \Pr\{Z(x_1) < z\}$, ..., $F_N(z) = \Pr\{Z(x_N) < z\}$;
- the conditional distribution $F_0(z \mid z_1, \ldots, z_N)$ of $Z(x_0)$ when the $Z(x_\alpha)$ are set to values z_α.

The optimum estimator of $I(z)$ at the point x_0—that is, the conditional expectation of $1_{Z(x_0) < z}$—is precisely $F_0(z \mid z_1, \ldots, z_N)$.

This method is not practical in the general case because the inference of the multivariate distribution is beyond reach (if the interval of variation of Z is discretized in p classes, there are p^{N+1} probabilities to estimate, e.g., 10^{11} values for only 10 classes and 10 samples; the conditional expectation itself is an element of a functional space of dimension 10^{10}, and we cannot seriously claim to be optimizing in a space of such dimension!). Hence the absolute need to specify a model, despite the risks inherent to this.

The Multi-Gaussian Model

In practice the Gaussian model is sought for its good properties. Suppose that we transform the variables $Z(x_0)$, $Z(x_1)$, ..., $Z(x_N)$ to $N+1$ standard normal variates $Y(x_0)$, $Y(x_1)$, ..., $Y(x_N)$ by means of an appropriate transformation φ such that $Z(x_i) = \varphi(Y(x_i))$. The crucial hypothesis of the multi-Gaussian model is to assume that the $(N+1)$-dimensional vector $(Y(x_0), Y(x_1), \ldots, Y(x_N))'$ has a multivariate Gaussian distribution, and is therefore characterized by the matrix of covariances $\{\sigma_{ij} : i, j = 0, 1, \ldots, N\}$. The conditional distribution of $Y(x_0)$ is then Gaussian. More specifically,

$$Y(x_0) | \{ Y(x_\alpha) = y_\alpha : \alpha = 1, \cdots, N \} = y^*_{SK} + \sigma_{SK} U$$

where y^*_{SK} is the simple kriging estimate of $Y(x_0)$ and σ_{SK} the associated kriging standard deviation; U is an independent standard normal deviate (standardized kriging error).

We then have

$$\Pr\{Z(x_0) < z | \text{data}\} = G\left(\frac{y - y^*_{SK}}{\sigma_{SK}}\right) \quad \text{with} \quad y = G^{-1}(F(z)) \qquad (6.4)$$

where "data" represents the conditioning data $Z(x_\alpha) = z_\alpha$, or equivalently $\{Y(x_\alpha) = y_\alpha : \alpha = 1, \ldots, N\}$. More generally, the conditional expectation of any function of $Z(x_0)$, expressed as a function ψ of $Y(x_0)$, is obtained as

$$\text{E}[\psi(Y(x_0)) \,|\, \text{data}] = \int \psi\left(y^*_{SK} + \sigma_{SK} u\right) g(u) \, du$$

The conditional variance

$$\text{Var}[\psi(Y(x_0)) \,|\, \text{data}] = \text{E}\left[\psi^2(Y(x_0)) \,|\, \text{data}\right] - \left\{\text{E}[\psi(Y(x_0)) \,|\, \text{data}]\right\}^2$$

can be obtained in a similar manner. It depends on the data values and not only on the data configuration with respect to the target point; therefore, it is a more realistic measure of the dispersion of the true value around its estimate than the nonconditional variance. Confidence intervals for $Z(x_0)$, for example, are more realistic than those that would be obtained by a direct kriging of $Z(x_0)$ from data $Z(x_\alpha)$. Working with Gaussian transformed data thus provides a distinctive advantage over a direct study of $Z(x)$, provided that the multivariate Gaussian assumption is sensible.

All calculations can be conveniently carried out by working with expansions of the function ψ into Hermite polynomials, as shown by Emery (2005a) (also see useful formulas at the end of Section A.5.2).

All conditioning information is concentrated in the kriging estimate y^*_{SK}. The multi-Gaussian model strongly determines the form of the estimator, and it hardly allows the data "to speak" for themselves. The assumption on which it relies is practically unverifiable, but one can at least verify that bivariate distributions are Gaussian (see Section 6.4.4) and implement a validation procedure.

If $Z(x)$ is only locally stationary, with a slowly varying mean in the zone of interest, the same is generally true for $Y(x)$. So one can think of substituting ordinary kriging for simple kriging of $Y(x)$. But then the kriging estimate and variance no longer define the conditional distribution of $Y(x_0)$. Emery (2005a) shows, however, that an unbiased estimator of $Y(x_0)$ is obtained if we substitute y^*_{OK} for y^*_{SK} and $\sigma^2_{OK} + 2\mu$ for σ^2_{SK}, namely with

$$\psi^* = \int \psi\left(y^*_{OK} + \sqrt{\sigma^2_{OK} + 2\mu}\ u\right) g(u)\ du$$

where μ is the Lagrange parameter of the ordinary kriging system. The proof is similar to that presented in the framework of the discrete Gaussian model in Section 6.8.3.

The multi-Gaussian model is usually used under the assumption that the spatial distribution of the RF $Y(x)$ is Gaussian (i.e., *all* finite-dimensional distributions are Gaussian), so that we could simply call it "the Gaussian model." We will, however, use its usual denomination, which emphasizes that we are working under a stricter assumption than with the bi-Gaussian model presented later.

Case of a Lognormal SRF

If $Z(x)$ is an SRF with a lognormal spatial distribution,[2] let us consider $Y(x) = \log Z(x)$, which is a Gaussian SRF. With y^*_{SK} and σ^2_{SK} now denoting the simple kriging estimate and variance of this Gaussian, relation (6.4) becomes

$$\Pr\{Z(x_0) < z \mid \text{data}\} = G\left(\frac{\log z - y^*_{SK}}{\sigma_{SK}}\right)$$

In other words the conditional distribution of $Z(x_0)$ remains lognormal. As we have seen in Section 3.4.11, its arithmetic mean is the simple lognormal kriging estimator

$$Z^*_{SLK} = \exp\left(Y^*_{SK} + \tfrac{1}{2}\sigma^2_{SK}\right)$$

and its logarithmic variance is σ^2_{SK}.

If $Y(x)$ is only locally stationary and if the local mean is unknown, the conditional distribution is not known but we have seen in Section 3.4.11 that an unbiased estimator of $Z(x_0)$ is

$$Z^*_{OLK} = \exp\left(Y^*_{OK} + \tfrac{1}{2}\left(\sigma^2_{OK} + 2\mu\right)\right)$$

This result is similar to the preceding one, with the ordinary kriging estimator Y^*_{OK} replacing the simple kriging estimator Y^*_{SK} and the sum $\sigma^2_{OK} + 2\mu$ of the ordinary kriging variance and the Lagrange parameter replacing the simple kriging variance σ^2_{SK}. The ordinary lognormal kriging Z^*_{OLK} is thus equal to the mean of a lognormal distribution with logarithmic mean Y^*_{OK} and logarithmic variance $\sigma^2_{OK} + 2\mu$; nevertheless, that distribution is not the conditional distribution of $Z(x_0)$ and can only be considered as an approximation to it.

---

[2] A lognormal RF is a random function of the form $Z(x) = \exp(Y(x))$, where $Y(x)$ is a Gaussian SRF. Its marginal distribution has two parameters. In applications we sometimes have to add a translation parameter and to consider that $Z(x) = b + \exp(Y(x))$. Because the properties of the three-parameter lognormal distribution are easily deduced from those of the two-parameter distribution, we will only consider the latter.

6.3.3 Indicator Methods

Indicator Kriging

We still wish to estimate $I(z) = 1_{Z(x_0) < z}$ from data $\{Z(x_\alpha) : \alpha = 1, \ldots, N\}$. For a given threshold z, the indicator $1_{Z(x) < z}$, regarded as a function of x, is a random function, and our objective can be restated as the estimation of this RF at the point x_0 from the data $Z(x_\alpha)$. Indicator kriging, proposed by Journel (1982), consists of estimating $I(z) = 1_{Z(x_0) < z}$ by kriging the corresponding indicator RF $1_{Z(x) < z}$. We are thus back to a classic problem, one of simple kriging, given two successive simplifications:

- Replacement of the initial data $Z(x_\alpha)$ by indicator data $1_{Z(x_\alpha) < z}$.
- Replacement of conditional expectation by kriging.

While the second step may be regarded as an inevitable approximation to solve the problem within the scope of two-point statistics, the first step constitutes a clear loss of information.

In theory, if $Z(x)$ can be considered as an SRF with known marginal $F(dz)$, since $F(z)$ is by definition the mean of the RF $1_{Z(x) < z}$ one can use simple kriging with an additional term for the mean

$$I^*(z) = \left(1 - \sum_{\alpha=1}^{N} \lambda_\alpha\right) F(z) + \sum_{\alpha=1}^{N} \lambda_\alpha 1_{Z(x_\alpha) < z}$$

without introducing an unbiasedness constraint. In practice, however, the marginal is not always known, and one uses ordinary kriging

$$I^*(z) = \sum_{\alpha=1}^{N} \lambda_\alpha 1_{Z(x_\alpha) < z}$$

under the condition $\sum_{\alpha=1}^{N} \lambda_\alpha = 1$.

Note that if the objective is the c.d.f. of $Z(x)$ in a volume V rather than at a point x_0 (i.e., if $I(z)$ is defined by an expression such as (6.2)), it suffices to modify the right-hand sides of the kriging systems as indicated in Section 3.5.2. In both cases kriging must of course be done using the covariance of the indicator at the given threshold (which in general changes with z). Indicator kriging has a number of advantages:

- It takes into account the structure of each indicator $1_{Z(x) < z}$.
- It produces an estimation variance.
- It does not require prior modeling of the theoretical distribution F.
- It does not require global stationarity, but only local stationarity.

However, it has some drawbacks:

- There are as many variograms to model as there are levels z considered, and of course as many kriging systems to solve.
- Since kriging does not guarantee, except in special circumstances, that the weights λ_α are nonnegative, we are exposed to getting estimates $I^*(z)$ that are negative or greater than 1. Even if for each level z the weights λ_α are positive, there is no assurance that the estimates of $I(z)$ at the various levels satisfy the order relations of a c.d.f. ($z < z' \Rightarrow I(z) \le I(z')$). This problem is resolved, for example, by using a quadratic correction (Sullivan, 1984): From initial estimates $I_n^* = I^*(z_n)$ at the different levels z_n, define consistent estimators I_n^{**} by

$$\sum_n w_n [I_n^{**} - I_n^*]^2 \quad \text{minimum}$$

under the condition that monotonicity is achieved. The positive weighting coefficients w_n are selected according to the relative importance of the various levels z_n.

A simplification often made is to use the same variogram for all levels, for example, the variogram of $Z(x)$ or the variogram of the indicator of the median, as suggested by Journel (1983). In reality, this is justified only if all variograms are proportional. Matheron (1982b) showed that such property is true for all levels only in the very special case of the mosaic model.

Case of a Mosaic SRF

For such a random function, either $Z(x)$ and $Z(x')$ are equal with probability $\rho(x' - x)$, and the distribution of their common value is F, or else they are independent and have this same distribution F. In other words, the space is partitioned into random compartments, and the values assigned to the compartments are independent random variables with the same distribution F. It is immediately seen, denoting by m and σ^2 the mean and variance of F, that the centered covariances of $Z(x)$ and $1_{Z(x)<z}$ are

$$\text{Cov}(Z(x), Z(x+h)) = \sigma^2 \rho(h),$$
$$\text{Cov}(1_{Z(x)<z}, 1_{Z(x+h)<z}) = F(z)(1 - F(z))\,\rho(h)$$

In this very special model, indicator kriging is equivalent to kriging with the covariance of $Z(x)$. Furthermore, in this model, indicator cross-covariances for two different levels z and z' are also proportional to $\rho(h)$:

$$\text{Cov}(1_{Z(x)<z}, 1_{Z(x+h)<z'}) = F(\min(z, z'))[1 - F(\max(z, z'))]\,\rho(h)$$

The set of indicators associated with various threshold levels thus conforms to a proportional covariance model, and we have seen in Section 5.6.4 that kriging is then equivalent to cokriging.

But aside from the mosaic case, the shape of an indicator variogram usually changes with the level. In the Gaussian case, for example, a "destructuring" of the indicator variogram is observed as the threshold departs from the median (e.g., see Figure 2.24). The indicator variogram tends to the nugget effect model so that at extreme levels local estimation of an indicator by kriging becomes useless.

Estimation of a Conditional Variable

Once we have an approximation $I^*(z)$ to the conditional c.d.f. of $Z(x_0)$, it is possible to approximate any function $\psi(Z(x_0))$ by the expected value

$$\psi^* = \int \psi(z)\, I^*(dz)$$

Consider, for example, for a fixed threshold z_0, the variables $T_0 = 1_{Z(x_0) \geq z_0}$ and $Q_0 = Z(x_0)1_{Z(x_0) \geq z_0}$. In mining, if z_0 is the cutoff grade, these variables represent the indicator of ore at point x_0 and the corresponding quantity of metal (counted zero when the grade is less than z_0). They can be estimated by

$$T^* = 1 - I^*(z_0) \qquad \text{and} \qquad Q^* = \int_{z_0}^{\infty} z\, I^*(dz)$$

In the scope of a mosaic model, or simply if we are content to use the same variogram model (up to a multiplicative factor) for all thresholds z, it can be seen that these estimators are of the form

$$T^* = \sum_{\alpha=1}^{N} \lambda_\alpha 1_{Z(x_\alpha) \geq z_0} \qquad \text{and} \qquad Q^* = \sum_{\alpha=1}^{N} \lambda_\alpha Z(x_\alpha)\, 1_{Z(x_\alpha) \geq z_0}$$

where the λ_α are kriging weights. We can estimate T_0 and Q_0 by direct kriging of the variables $T(x) = 1_{Z(x) \geq z_0}$ and $Q(x) = Z(x)1_{Z(x) \geq z_0}$ and the corresponding ore grade (the conditional variable $Z(x_0) \mid Z(x_0) > z_0$) by the ratio $\hat{m} = Q^*/T^*$. Just as in Section 3.5.5, $\hat{m}$ is a weighted average (i.e., with weights adding up to one) of Z data above the threshold z_0. This would not be the case if different variograms were used for the estimation of T_0 and Q_0.

Indicator Cokriging

Indicator kriging takes into account the position of a value relative to the threshold but not its proximity: A value just slightly above the threshold is not distinguished from a very large value. A way to incorporate more information is to consider the value of the variables relative to a series of thresholds z_n. Reasoning in a multivariate context, it is logical to seek to estimate $I(z_n)$ by cokriging using not only the indicators $1_{Z(x_\alpha) < z_n}$ but also the indicators at all other levels $1_{Z(x_\alpha) < z_m}$. This is indicator cokriging. The estimator is then of the form

$$I^*(z_n) = \lambda_0 + \sum_{m}\sum_{\alpha=1}^{N} \lambda_{m\alpha}\, 1_{Z(x_\alpha) < z_m}$$

Again one can either consider cokriging with a known mean, that is, find the $\lambda_{m\alpha}$ by simple cokriging and let

$$\lambda_0 = F(z_n) - \sum_{m}\left[F(z_m)\sum_{\alpha=1}^{N} \lambda_{m\alpha} \right]$$

or use ordinary cokriging with the unbiasedness constraints

$$\lambda_0 = 0, \qquad \sum_{\alpha=1}^{N} \lambda_{n\alpha} = 1, \qquad \sum_{\alpha=1}^{N} \lambda_{m\alpha} = 0 \quad \text{if} \quad m \neq n$$

From a theoretical standpoint, this method is clearly more powerful than indicator kriging, since it incorporates the information that a sample value is slightly above the threshold z_n ($1_{Z(x_\alpha)<z_{n+1}} = 1$) or well above the threshold (e.g., $1_{Z(x_\alpha)<z_m} = 0$ for all levels considered), while in both cases $1_{Z(x_\alpha)<z_n} = 0$. But it has a downside: It requires estimation and modeling of covariances and cross-covariances of all the different levels, not to mention solving very large systems.

One might be tempted to take the same model for all these covariances, up to a scaling factor. But this leads to a proportional covariance model where, with all variables being sampled at the same points, cokriging is equivalent to kriging. It is therefore seen that the advantage of indicator cokriging over plain kriging relies on a fine modeling of all direct and cross-covariances. Unfortunately, we do not have theoretical models available for representing a set of indicators in a consistent manner. Journel and Posa (1990) give necessary but not sufficient conditions, and we have seen in Section 2.5.3 that the problem is already not simple when a single indicator is considered. The implementation of indicator cokriging thus remains difficult, which explains why the method has been presented many times since mentioned by Journel (1983), but apparently seldom applied to estimation problems, except in studies of a methodological nature, such as by Liao (1990); Goovaerts (1994, 1997), studying concentrations of chemical elements in the soil, obtains inferior results with indicator cokriging than with plain indicator kriging, which emphasizes this difficulty. This has motivated the development of *probability kriging* (Sullivan, 1984), which consists of cokriging each indicator with a single secondary variable: the rank obtained by simple sorting of the data or, equivalently, the variable $R(x) = F(Z(x))$ where F is the c.d.f. of Z. In fact, true cokriging of the whole set of indicators is achieved by disjunctive kriging, as will be shown in the next section. Indicator cokriging seems to be more used to build conditional simulations (see Section 7.7.1).

Elementary Comparisons

Let $Z(x)$ be an SRF whose marginal distribution is continuous and with median zero. We are interested in the indicator RF $1_{Z(x) \geq 0}$. More specifically, we want to evaluate the indicator of the event $Z(x_0) \geq 0$ at the point $x_0 = x_1 + h$ given the value $Z(x_1) = z$. Indicator kriging proposes to approximate this indicator by

$$I^* = \tfrac{1}{2} + r\left(1_{z \geq 0} - \tfrac{1}{2}\right)$$

where r is the correlation coefficient of the indicator values at x_0 and x_1. This problem can cover very different situations with the following extremes:

1. $Z(x)$ is a mosaic SRF. Indicator kriging coincides with conditional expectation and is the appropriate answer.

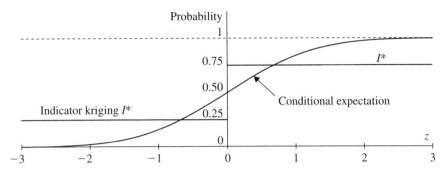

FIGURE 6.2 Estimates of the probability that $Z(x_0) \geq 0$ conditionally on $Z(x_1) = z$ as a function of z when the bivariate distribution of $Z(x_0)$ and $Z(x_1)$ is a standard normal with correlation coefficient 0.707: comparison between conditional probability and indicator kriging.

2. $Z(x)$ is a Gaussian SRF (and therefore with mean zero). The correlation coefficient ρ of $Z(x_0)$ and $Z(x_1)$ is related to the indicator correlation r by $\rho = \sin(\frac{\pi}{2}r)$ (see (2.77)). The distribution of $Z(x_0)$ conditionally on $Z(x_1) = z$ is Gaussian, with mean ρz and variance $(1 - \rho^2)\,\sigma^2$, where σ^2 denotes the variance of Z. Denoting by G the standard Gaussian c.d.f. we have

$$E[1_{Z(x_0) \geq 0} \mid Z(x_1) = z] = \Pr[Z(x_0) \geq 0 \mid Z(x_1) = z] = G\left(\frac{\rho z}{\sigma\sqrt{1 - \rho^2}}\right)$$

Figure 6.2 depicts the conditional expectation in the Gaussian case as a function of z and its approximation by indicator kriging for $\sigma = 1$, $\rho = 0.707$ and $r = 1/2$, which is seen to be rather crude, and hence the interest of indicator cokriging and more precisely of disjunctive kriging (which would here coincide with conditional expectation).

Note that neither the conditional expectation nor the kriged indicator assume the values 0 or 1. This is inherent to the mean square error minimization criterion. If we force the classification by using a decision rule based on the sign of the simple kriging estimator $Z^*(x_0) = \rho\, Z(x_1)$, the resulting probability of misclassification is given by

$$E\left[1_{Z^*(x_0) \geq 0} - 1_{Z(x_0) \geq 0}\right]^2 = \frac{1}{2} - \frac{1}{\pi}\arc \sin(|\rho|) = \frac{1}{\pi}\arc \cos(|\rho|)$$

In this formula ρ is the correlation between $Z(x_0)$ and $Z(x_1)$ and $|\rho|$ represents the correlation between $Z(x_0)$ and its simple kriging estimator $Z^*(x_0)$, which is the parameter of interest when applying (2.77). Since the kriging variance of $Z(x_0)$ is $\sigma_{SK}^2 = (1 - \rho^2)\,\sigma^2$, the formula can be written as

$$E[1_{Z^*(x_0) \geq 0} - 1_{Z(x_0) \geq 0}]^2 = \frac{1}{\pi}\arc \sin\left(\frac{\sigma_{SK}}{\sigma}\right)$$

This result applies for simple kriging estimation of $Z(x_0)$ from any number of points. As intuition suggests, the classification is perfect when $\sigma_{SK} = 0$ and worst when $\sigma_{SK} = \sigma$ (the maximum), where it becomes completely random and therefore wrong half of the time.

6.4 LOCAL ESTIMATION BY DISJUNCTIVE KRIGING

We have just seen that it is not easy to estimate the indicator $I(z) = 1_{Z(x_0) < z}$ by cokriging of indicators associated with a number of thresholds, if only because of the difficulty to represent direct and cross-covariances in a consistent way. But as mentioned in Section 2.5.3, the knowledge of direct and cross covariances of the indicators of an SRF for all possible levels is equivalent to the knowledge of the bivariate distributions of the SRF, since

$$\Pr\{Z(x) < z \quad \text{and} \quad Z(x+h) < z'\} = E[1_{Z(x) < z} 1_{Z(x+h) < z'}] = C_{zz'}(h)$$

where $C_{zz'}(h)$ is the noncentered cross-covariance between the indicators associated with the thresholds z and z'. We are now going to examine the approach based on a direct modeling of bivariate distributions, namely disjunctive kriging (Matheron, 1973b, 1976a).

6.4.1 Disjunctive Kriging

Objective

Let $Z(x)$ be a random function which for the time being we do not assume necessarily stationary. We have N sample points $\{x_\alpha : \alpha = 1, \ldots, N\}$ and therefore N data $Z_\alpha = Z(x_\alpha)$. We wish to estimate $I(z) = 1_{Z(x_0) < z}$, or more generally a quantity $Z_0 = f_0(Z(x_0))$. The approach is to carry out the estimation of Z_0 by cokriging of the indicators. If the domain of variation of $Z(x)$ is partitioned into classes B_m, we can associate to $Z(x)$ the indicators $1_{B_m}(Z(x))$. Cokriging Z_0 from indicators amounts to seeking an estimator of the form

$$Z^* = \lambda_0 + \sum_m \sum_\alpha \lambda_{m\alpha} 1_{B_m}(Z_\alpha) \tag{6.5}$$

If the line is partitioned into infinitely small intervals B_m, Z^* is in fact of the form

$$Z^* = \sum_{\alpha=1}^{N} f_\alpha(Z_\alpha) \tag{6.6}$$

where the f_α are measurable functions. The problem thus amounts to the estimation of Z_0 by a sum of univariate measurable functions of the data. This is an intermediate objective between kriging ($Z^* = \lambda_0 + \sum_{\alpha=1}^{N} \lambda_\alpha Z_\alpha$, linear functions f_α only) and conditional expectation ($Z^* = f(Z_1, \ldots, Z_N)$, a single function of all the data). This objective is called the *disjunctive kriging* of Z_0 (abbreviation DK), since in the expression (6.5) of Z^* the indicators $1_{B_m}(Z(x))$ achieve a disjunctive coding of information, in the sense that at a point x_α one and only one of the indicators is nonzero. Such type of coding is classic in correspondence analysis (Benzécri et al., 1973). As we will see, knowledge of the bivariate distributions of $\{Z_0, Z_1, \ldots, Z_N\}$ is sufficient to solve the problem. The determination of these distributions is not out of reach when we have enough data. Revisiting the example of Section 6.3.2, if the interval of variation is discretized into p classes and if disjunctive kriging is carried out from N data, we must determine $(N+1)(Np+2)p/2$ probabilities compared with p^{N+1} for conditional expectation and $(N+1)(N+2)/2$ for simple kriging. So, still with $p = 10$ and $N = 10$, we find that disjunctive kriging requires the inference of 5610 values, against 66 for simple kriging and 10^{11} for conditional expectation. With regularly spaced data, stationarity assumptions reduce the required numbers for kriging by a factor of the order of $N/2$, and this is further reduced to a few parameters by modeling the covariance functions.

General Equations of Disjunctive Kriging

We assume that the random variables $\{Z_\alpha : \alpha = 1, \ldots, N\}$ and Z_0 have second-order moments. Thus they belong to a Hilbert space equipped with the scalar product $\langle X, Y \rangle = E[XY]$. We denote by $\mathcal{H}_\alpha$ the subspace generated by random variables of the form $f_\alpha(Z_\alpha)$ with a finite second-order moment, where f_α is a measurable function. Let $\mathcal{H}$ be the subspace of random variables of the form $\sum_{\alpha=1}^{N} f_\alpha(Z_\alpha)$. The disjunctive kriging of Z_0—in other words, the optimal estimator of the form (6.6)—is simply the projection of Z_0 onto $\mathcal{H}$. By the projection theorem (cf. Section 1.1.2), the DK estimator Z^* is characterized by the orthogonality property

$$\langle Z^* - Z_0, X \rangle = 0 \qquad \forall X \in \mathcal{H} \tag{6.7}$$

Any random variable of the subspace $\mathcal{H}$ being of the form $X = \sum_{\alpha=1}^{N} f_\alpha(Z_\alpha)$ orthogonality must be satisfied in particular for any random variable belonging to one of the subspaces $\mathcal{H}_\alpha$, thus of the form $X = f_\alpha(Z_\alpha)$, and on the other hand, the linearity of the scalar product ensures that this condition is also sufficient to achieve (6.7).[3] Consequently, (6.7) is equivalent to

$$\langle Z^*, X \rangle = \langle Z_0, X \rangle \qquad \text{for all } X \text{ of the form } f_\alpha(Z_\alpha)$$

By the characteristic formula (1.3) for conditional expectation this condition is equivalent to

$$E[Z^* \mid Z_\alpha] = E[Z_0 \mid Z_\alpha], \qquad \alpha = 1, \ldots, N$$

Therefore the functions f_α corresponding to the DK $Z^* = \sum_{\alpha=1}^{N} f_\alpha(Z_\alpha)$ are characterized by

$$\sum_{\beta=1}^{N} E[f_\beta(Z_\beta) \mid Z_\alpha] = E[Z_0 \mid Z_\alpha], \qquad \alpha = 1, \ldots, N \tag{6.8}$$

This system involves the conditional distributions of Z_0 and of Z_β given Z_α and only requires knowledge of bivariate distributions.

It follows from (6.7) that the DK estimation variance is

$$\sigma_{DK}^2 = E[Z^* - Z_0]^2 = \text{Var}[Z_0] - \text{Var}[Z^*] = \text{Var}[Z_0] - \text{Cov}[Z_0, Z^*]$$

These results are interpreted like those of simple kriging (Section 3.3.2).

---

[3] Note that the $\mathcal{H}_\alpha$ are not disjoint since all contain the constant random variables. Also, the projection theorem assumes that the space $\mathcal{H}$ is closed, which is not proven. However, this question is not blocking since in the isofactorial models, used in practice, the projection can be found and is of the correct form.

Disjunctive Kriging Equations for a Transformed Isofactorial Model

There are SRFs for which the solution of the system (6.8) is greatly simplified, namely those whose bivariate distributions are isofactorial, the best known example being SRFs with Gaussian bivariate distributions. We will see shortly that this circumstance is far from being general. It can be extended, however, to the case where $Z(\cdot)$ is the *transform* of an SRF $Y(\cdot)$ with isofactorial bivariate distributions. We will present later the general properties of isofactorial models and the main models used in geostatistics.

Let us denote by $G(dy)$ the marginal distribution of the SRF $Y(x)$ and consider transforms $\psi(Y(x))$ where ψ is square-integrable for G. The main property of an isofactorial model is that there exists a set of functions $\{\chi_n(y):$ $n = 0, 1, \ldots\}$ (the *factors*) such that the transform $\psi(Y(x))$ can be decomposed into a sum $\sum_{n=0}^{\infty} \psi_n \chi_n(Y(x))$ whose coefficients ψ_n are given by

$$\psi_n = \int \psi(y)\, \chi_n(y)\, G(dy) \tag{6.9}$$

where the factors $\chi_n(Y(x))$ are uncorrelated SRFs satisfying

$$E[\chi_m(Y(x))\, \chi_n(Y(x+h))] = \delta_{mn}\, T_n(h) \tag{6.10}$$

and $T_n(h)$ is the covariance of the SRF $\chi_n(Y(x))$ with $T_n(0) = 1$. From the characteristic property (1.3) it follows that (6.10) is equivalent to

$$E[\chi_n(Y(x+h)) \mid Y(x)] = T_n(h)\, \chi_n(Y(x)) \tag{6.11}$$

One of the factors, ascribed the index 0, is the constant function $\chi_0(y) \equiv 1$. The corresponding (noncentered) covariance is $T_0(h) \equiv 1$.

Let us now reformulate our problem in the DK framework:

1. The SRF $Z(x)$ of interest is a transform $Z(x) = \varphi(Y(x))$ where $Y(x)$ is an SRF with isofactorial bivariate distributions.
2. We wish to estimate a function of $Z(x_0)$ that we consider of the form $Z(x_0) = \psi(Y(x_0))$. Typically, the objective is the estimation of $1_{Z(x_0)<z}$. It is equivalent to the estimation of $1_{Y(x_0)<y}$ where $y = \varphi^{-1}(z)$, since applying the threshold z on Z is the same as thresholding Y at $y = \varphi^{-1}(z)$.
3. We have data $\{Z(x_\alpha) : \alpha = 1, \ldots, N\}$ or, equivalently, data $\{Y(x_\alpha) : \alpha = 1, \ldots, N\}$.

To simplify notations, let $Y_0 = Y(x_0)$ and $Y_\alpha = Y(x_\alpha)$, and let us rewrite our objective as being the estimation of $Z_0 = \psi(Y_0)$ by an estimator of the form

$$\psi^* = \sum_{\alpha=1}^{N} f_\alpha(Y_\alpha)$$

$\psi(Y_0)$ and the $f_\alpha(Y_\alpha)$ can be decomposed on the basis of the factors χ_n as

$$\psi(Y_0) = \sum_{n=0}^{\infty} \psi_n \, \chi_n(Y_0),$$

$$f_\alpha(Y_\alpha) = \sum_{n=0}^{\infty} f_{\alpha n} \, \chi_n(Y_\alpha) \tag{6.12}$$

The coefficients ψ_n, defined by (6.9), are known, since the function ψ is given. The coefficients $f_{\alpha n}$ are to be determined. The disjunctive kriging system (6.8) is written here as

$$\sum_{\beta=1}^{N} E[f_\beta(Y_\beta) \mid Y_\alpha] = E[\psi(Y_0) \mid Y_\alpha], \qquad \alpha = 1, \ldots, N$$

As a consequence of (6.11) and (6.12) and by interchanging the summation order, we obtain

$$\sum_{n=0}^{\infty} \sum_{\beta=1}^{N} f_{\beta n} \, T_n(x_\beta - x_\alpha) \, \chi_n(Y_\alpha) = \sum_{n=0}^{\infty} \psi_n \, T_n(x_0 - x_\alpha) \, \chi_n(Y_\alpha), \qquad \alpha = 1, \ldots, N$$

Since the $\chi_n(Y(\cdot))$ are uncorrelated SRFs, these equations can be broken up into a distinct system for each value of n: The coefficients $\{f_{\alpha n} : \alpha = 1, \ldots, N\}$ are solutions of the system

$$\sum_{\beta=1}^{N} f_{\beta n} \, T_n(x_\beta - x_\alpha) = \psi_n \, T_n(x_0 - x_\alpha), \qquad \alpha = 1, \ldots, N \tag{6.13}$$

$T_n(h)$ being the covariance of the SRF $\chi_n(Y(x))$, this is the simple kriging system of $\psi_n \chi_n(Y_0)$ from the $\chi_n(Y_\alpha)$. For $n = 0$ this system is degenerate, and it suffices to take any $f_{\alpha 0}$ such that $\sum_{\alpha=1}^{N} f_{\alpha 0} = \psi_0$.

The advantage of the isofactorial model is that cokriging of factors reduces to kriging of each factor separately. We could say in an incorrect but illustrative manner that disjunctive kriging of $Z_0 = \psi(Y_0)$ is equivalent to a combination of simple kriging of Y_0 from the Y_α, of Y_0^2 from the Y_α^2, and so on (the image is incorrect because the factors are usually not monomials). Note that if the objective is the mean value of $\psi(Y(x))$ in a volume V and not at a point x_0 [e.g., if it is the c.d.f. $I(z)$ defined in (6.2)], it suffices in the systems (6.13) to replace the point covariance $T_n(x_0 - x_\alpha)$ on the right-hand side by the mean value of $T_n(x - x_\alpha)$ over V (cf. Section 3.5.2).

At this point let us recapitulate the DK procedure in the typical case of the estimation of $I(z_c) = 1_{Z(x_0) < z_c}$ where for clarity the index c is used to denote a fixed cutoff or threshold:

1. Transform the data $Z(x_\alpha)$ into isofactorial $Y(x_\alpha)$ defined by $Z(x_\alpha) = \varphi(Y(x_\alpha))$.

2. Translate the objective $I(z_c)$ into a function of $Y(x_0)$: $\psi(Y(x_0)) = 1_{Y(x_0) < y_c}$ with $y_c = \varphi^{-1}(z_c)$.

3. Compute the coefficients ψ_n of the expansion of the function ψ using (6.9); in this case this simplifies to $\psi_n = \int_{-\infty}^{y_c} \chi_n(y) G(dy)$.

4. Solve the systems (6.13) to obtain the weights $f_{n\alpha}$.

5. Compute the estimate as $I^*(z_c) = \sum_{\alpha=1}^{N} \sum_{n=0}^{\infty} f_{n\alpha} \chi_n(Y(x_\alpha))$.

Expression of Disjunctive Kriging from Estimates of the Factors

The simple kriging of the factor $\chi_n(Y_0)$ is

$$\chi_n^* = \sum_{\alpha=1}^{N} \lambda_{n\alpha} \chi_n(Y_\alpha) \tag{6.14}$$

with weights $\lambda_{n\alpha}$ solution of the system (6.13) written with $\psi_n = 1$ (in $\lambda_{n\alpha}$ we deliberately reverse the order of the indexes):

$$\sum_{\beta=1}^{N} \lambda_{n\beta} T_n(x_\beta - x_\alpha) = T_n(x_0 - x_\alpha), \qquad \alpha = 1, \ldots, N \tag{6.15}$$

By linearity, we can then form the DK estimate of *any* function $\psi(Y_0)$ using the formula

$$\psi^* = \psi_0 + \sum_{n=1}^{\infty} \psi_n \chi_n^* \tag{6.16}$$

Pseudo Conditional Distribution

Starting from (6.16) and replacing the coefficients ψ_n by their expression in (6.9), we get

$$\psi^* = \psi_0 + \int \psi(y) \left(\sum_{n=1}^{\infty} \chi_n(y) \chi_n^* \right) G(dy)$$

which amounts to the expected value of $\psi(Y_0)$ with respect to the pseudoprobability distribution

$$G^*(dy) = \left(1 + \sum_{n=1}^{\infty} \chi_n(y) \chi_n^* \right) G(dy)$$

This distribution is an approximation to the conditional distribution of Y_0 given the Y_α. It sums to one. However, there is no guarantee that $G^*(dy)$ is positive for any value of y.

Disjunctive Kriging Variance

The estimation variance of $\chi_n(Y_0)$ is the simple kriging variance

$$\sigma_n^2 = \text{Var}[\chi_n^* - \chi_n(Y_0)] = 1 - \sum_\alpha \lambda_{n\alpha} T_n(x_0 - x_\alpha) \tag{6.17}$$

It is of course zero for $n = 0$, since the factor $\chi_0(Y_0) \equiv 1$ is estimated without error. The DK variance of $Z_0 = \psi(Y_0)$ is the sum of the kriging variances of the factors $\psi_n \chi_n(Y_0)$:

$$\sigma_{\text{DK}}^2 = \text{Var}[\psi^* - \psi(Y_0)] = \sum_{n=1}^{\infty} \psi_n^2 \sigma_n^2 \tag{6.18}$$

6.4.2   Definition and General Properties of an Isofactorial Model

The isofactorial representation of the bivariate normal distribution has been known for a long time [e.g., Cramér (1945, p. 290)]. More generally, isofactorial bivariate models have first been introduced in quantum mechanics; they appeared in the field of stochastic processes with the study of Markov chains [e.g., Karlin and McGregor (1957, 1960)]; they are at the basis of correspondence analysis (Benzécri et al., 1973). Their introduction in geostatistics and the systematic study of a large number of models are due to Matheron (a presentation of most models, along with references to the original technical reports, can be found in Armstrong and Matheron, 1986a–c). Let us examine in more detail the properties of the isofactorial models that will be useful to us. Our presentation will concern the bivariate distributions of the $(Y(x), Y(x + h))$ pairs from an SRF (of order 2).

Hilbertian Basis

Let $Y(x)$ denote an SRF with marginal distribution G (not necessarily Gaussian). By definition, the system of functions $\{\chi_n(y) : n = 0, 1, \dots\}$ is a Hilbertian basis for the space $L^2(G)$ if

1. The functions $\chi_n(y)$ form a complete countable system for the marginal distribution G: Any measurable function ψ such that $\int \psi(y)^2 G(dy) < \infty$ can be represented by the series

$$\psi(y) = \sum_{n=0}^{\infty} \psi_n \chi_n(y) \tag{6.19}$$

with

$$\psi_n = \int \psi(y) \, \chi_n(y) \, G(dy) \qquad (6.20)$$

2. The functions $\chi_n(y)$ are orthonormal for the distribution G, namely

$$\int \chi_m(y) \, \chi_n(y) \, G(dy) = \delta_{mn} \qquad (6.21)$$

or, in probabilistic terms,

$$E[\chi_m(Y(x)) \, \chi_n(Y(x))] = \delta_{mn}$$

The functions χ_n are the factors. For most models in use, one of the factors, ascribed the index 0, is a constant function. Since its norm is 1, its value is either -1 or $+1$. The value $+1$ is usually selected:

$$\chi_0(y) \equiv 1$$

We will restrict ourselves to this case, because all the isofactorial models we will present satisfy this condition.[4] Letting $m=0$ in (6.21), it is seen that all the other factors have zero expectation:

$$\int \chi_n(y) \, G(dy) = 0 \qquad \text{or} \qquad E[\chi_n(Y(x))] = 0 \qquad (n>0)$$

Note that relation (6.21) simply expresses that *at the same point* x, $\chi_m(Y(x))$ and $\chi_n(Y(x))$ are uncorrelated $(m \neq n)$.

Definition of an Isofactorial Model

By definition, the bivariate distributions of an SRF $Y(x)$ with marginal distribution $G(dy)$ constitute an isofactorial model if the following property is satisfied:

- For any two points x and $x+h$, the bivariate distribution of the pair $(Y(x), Y(x+h))$ can be factorized in the form

$$G_h(dy, dy') = \sum_{n=0}^{\infty} T_n(h) \, \chi_n(y) \, \chi_n(y') \, G(dy) \, G(dy') \qquad (6.22)$$

for some Hilbertian basis $\{\chi_n : n=0, 1, \ldots\}$ of the space $L^2(G)$.

The factors χ_n are the same for all h, hence the denomination "isofactorial model."

---

[4] It is, however, easy from such a basis to derive a basis without constant function, such as by substituting $(\chi_0(\cdot) - \chi_1(\cdot))/\sqrt{2}$ and $(\chi_0(\cdot) + \chi_1(\cdot))/\sqrt{2}$ to χ_0 and χ_1.

Properties of an Isofactorial Model

The following properties can be verified:

- Formula (6.22) expresses that the conditional distribution of $Y(x+h)$ given $Y(x)=y$ is $\sum_{n=0}^{\infty} T_n(h)\,\chi_n(y)\,\chi_n(y')\,G(dy')$. As a consequence we recover (6.11)

$$E[\chi_n(Y(x+h)) \mid Y(x)] = T_n(h)\,\chi_n(Y(x))$$

which can be taken as an alternative definition of an isofactorial model. It is a little more general than definition (6.22), because it allows us, for example, to incorporate the case where $Y(x)$ and $Y(x+h)$ are equal and have a correlation of $+1$.

- From (6.22) we also have

$$\iint \chi_m(y)\,\chi_n(y')\,G_h(dy, dy') = \delta_{mn}\,T_n(h)$$

or, in probabilistic terms,

$$E[\chi_m(Y(x))\,\chi_n(Y(x+h))] = \delta_{mn}\,T_n(h)$$

which is (6.10). In other words, when $m \neq n$, $\chi_m(Y(x))$ and $\chi_n(Y(x))$ are not only orthogonal at the same point, as expressed by (6.21), but also spatially orthogonal; that is, $\chi_m(Y(x))$ and $\chi_n(Y(x+h))$ are orthogonal for all h. The *random functions* $\chi_n(Y(x))$ are therefore stationary (at order 2) and uncorrelated, and they have as respective covariances the $T_n(h)$ functions. These covariances in fact are correlograms, since the factors are normalized. In particular, $T_0(h) \equiv 1$, since $\chi_0(y) \equiv 1$ (the other factors have zero expectation, so T_0 is the only noncentered covariance).

- Given that $\chi_0(y) \equiv 1$, that the other factors have zero expectation, and that all these factors constitute an orthonormal system, it is easy to show that the mean and variance of the SRF $\psi(Y(\cdot))$ can be related to the ψ_n defined by (6.20) by

$$E[\psi(Y(x))] = \psi_0, \qquad \mathrm{Var}[\psi(Y(x))] = \sum_{n=1}^{\infty} \psi_n^2 \qquad (6.23)$$

and that its covariance takes the form

$$\mathrm{Cov}(\psi(Y(x)), \psi(Y(x+h))) = \sum_{n=1}^{\infty} \psi_n^2\,T_n(h) \qquad (6.24)$$

Likewise, if φ is another measurable function whose expansion has coefficients φ_n, the cross-covariance of $\varphi(Y(x))$ and $\psi(Y(x+h))$ can be written in the form

$$\mathrm{Cov}(\varphi(Y(x)), \psi(Y(x+h))) = \sum_{n=1}^{\infty} \varphi_n \psi_n T_n(h) \qquad (6.25)$$

Special Case of Polynomial Factors

Some models have polynomial factors and share the following additional properties:

- χ_n is a polynomial of degree n.
- Denoting the mean and standard deviation of $Y(x)$ by m_Y and σ_Y, the factor χ_1 is (with an arbitrary choice of sign)

$$\chi_1(y) = (y - m_Y)/\sigma_Y$$

- As a consequence, $T_1(h)$ coincides with the correlogram $\rho(h)$ of $Y(x)$.
- Since the above expression of χ_1 can be inverted as $y = m_Y + \sigma_Y \chi_1(y)$, the conditional expectation of $Y(x+h)$ is a linear regression:

$$E[Y(x+h) \,|\, Y(x)] = m_Y + \rho(h)[Y(x) - m_Y]$$

- The coefficients of the polynomial χ_n are related to the moments of order k of the distribution G, $k \le 2n$.
- The expansion (6.19) truncated to any order $n_{\max}$ is the best approximation of $\psi(y)$ by a polynomial of degree $n_{\max}$, in the least squares sense measured by $G(dy)$—that is, such that $\int [\psi(y) - \sum_{n=0}^{n_{\max}} \psi_n \chi_n(y)]^2 G(dy)$ is a minimum. However, if $\psi(y)$ is bounded, the expansion limited at the order $n_{\max}$ deviates indefinitely from $\psi(y)$ for $|y| \to \infty$: The approximation of $\psi((y))$ by a finite-degree polynomial can only be valid over a bounded domain of y.

For more information on families of orthogonal polynomials the reader is referred to Szegö (1939), Hochstrasser (1972), Beckman (1973), and Chihara (1978).

Additional Remarks

It seems easy to construct isofactorial models, since it is easy to construct a system of orthonormal polynomials χ_n over the distribution G, provided that this distribution has moments of all orders. This apparent simplicity is misleading:

1. If functions χ_n constitute an orthonormal system, nothing proves, even for fixed h, that there exist values T_n such that a bivariate distribution can be expanded as in (6.22).
2. Conversely, there can exist a system of nonpolynomial functions χ_n such that the bivariate distribution has the expansion (6.22). A simple example is that of a truncated bivariate Gaussian distribution: Starting from two

independent standard normals Y and Y', we keep only the part of the bivariate distribution where Y and Y' have the same sign. The marginals remain standard normal and the bivariate p.d.f. is

$$g_{\text{trunc}}(y, y') = (1 + \text{sign}(y)\,\text{sign}(y'))\,g(y)\,g(y')$$

where g is the univariate p.d.f. of the standard normal. The two factors of the bivariate density g_{trunc} are $\chi_0(y) \equiv 1$ and $\chi_1(y) = \text{sign}(y)$, which is not a polynomial.

3. Even if we find a system of functions χ_n enabling the expression of the bivariate distribution for fixed h as (6.22), the model is isofactorial only if the factors χ_n do not depend on h, which cannot be taken for granted [cf. discussion in Matheron (1989)].

6.4.3 Main Isofactorial Models

The class of bivariate distributions of an SRF coincides with the class of (noncentered) direct and cross-covariances of indicators obtained by thresholding the random function. Since a generalization of Bochner's theorem characterizing direct and cross-covariances of indicators does not exist, the same is true for bivariate distributions, and this remains the case in the limited scope of isofactorial models. The main source of models is derived from special random functions whose bivariate distributions can be written down in closed form: When these are isofactorial, the model obtained is isofactorial by construction. Unsurprisingly, there is a large number of special models that more or less overlap. It is not possible to give a complete inventory here (Matheron alone devoted over 600 pages to these models). We will present one model akin to the mosaic model and the classic models with polynomial factors, which have been used in most applications. Since in practice the isofactorial model does not directly concern the raw variable Z but a transform Y, we will assume, without loss of generality, that the transformed variable is standardized, and we will only consider this type of marginal distribution (standard normal, for example). Let us start with two simple examples that correspond to random functions we have already encountered, the bi-Gaussian and the mosaic models.

Bi-Gaussian Model

The normalized Hermite polynomials $\chi_n(y) = H_n(y)/\sqrt{n!}$ form an orthonormal basis with respect to the standard normal distribution. Hermite polynomials H_n are defined by Rodrigues's formula[5]

---

[5] In most textbooks the definition includes a $(-1)^n$ factor, which leads to odd-order Hermite polynomials of the opposite sign. We have kept the present definition for consistency with the geostatistical literature.

$$H_n(y)g(y) = \frac{d^n}{dy^n} g(y)$$

and are easily calculated by the recurrence relation

$$H_{n+1}(y) = -yH_n(y) - nH_{n-1}(y)$$

The bivariate normal distribution $G_\rho(dy, dy')$ with correlation coefficient ρ $(-1 < \rho < 1)$ has the bivariate p.d.f.

$$g_\rho(y, y') = \frac{1}{2\pi\sqrt{1-\rho^2}} \exp\left(-\frac{y^2 - 2\rho y y' + y'^2}{2(1-\rho^2)}\right)$$

It can be represented by the series

$$g_\rho(y, y') = \sum_{n=0}^{\infty} \rho^n \chi_n(y) \chi_n(y') g(y) g(y') \tag{6.26}$$

It is an isofactorial model, and the covariances T_n are equal to ρ^n. Therefore, if (Y, Y') is a pair of random variables with bivariate density function g_ρ, relation (6.11) takes the simple form

$$E[\chi_n(Y') \mid Y] = \rho^n \chi_n(Y) \tag{6.27}$$

Similarly, if $Y(x)$ is an SRF with correlogram $\rho(h)$, whose pairs $(Y(x), Y(x+h))$ are Gaussian, their bivariate distributions are of the form (6.22) with $T_n(h) = \rho(h)^n$ and

$$E[\chi_n(Y(x+h)) \mid Y(x)] = \rho(h)^n \chi_n(Y(x))$$

The possibility to develop the bivariate p.d.f.s as well as the functionals of $Y(\cdot)$ can be exploited for covariance calculations. An interesting example is the calculation of the covariance of indicators obtained by thresholding Y (Matheron, 1975b). From (A.10) the indicator $1_{Y(x)<y}$ can be represented by the series

$$1_{Y(x)<y} = G(y) + g(y) \sum_{n=1}^{\infty} \frac{\chi_{n-1}(y)}{\sqrt{n}} \chi_n(Y(x)) \tag{6.28}$$

By application of (6.25), the cross-covariance of the indicators $1_{Y(x)<y}$ and $1_{Y(x+h)<y'}$ associated with thresholds y and y' is

$$C_{yy'}(h) = g(y) g(y') \sum_{n=1}^{\infty} \frac{\chi_{n-1}(y) \chi_{n-1}(y')}{n} \rho(h)^n \tag{6.29}$$

This expression can be used directly for numerical computations. From an analytical point of view, for fixed y and y' this is a function of $\rho \equiv \rho(h)$, whose derivative is

$$\frac{dC_{yy'}}{d\rho} = g(y)g(y') \sum_{n=1}^{\infty} \chi_{n-1}(y)\, \chi_{n-1}(y')\, \rho^{n-1}$$

We recognize the isofactorial expansion of the bivariate density $g_\rho(y, y')$. By integration from 0 to $\rho(h)$, we obtain

$$C_{yy'}(h) = \int_0^{\rho(h)} g_u(y, y')\,du$$

In the special case $y = y'$, the direct covariance of the indicator is found to be

$$C_y(h) = \frac{1}{2\pi} \int_0^{\rho(h)} \exp\left(-\frac{y^2}{1+u}\right) \frac{du}{\sqrt{1-u^2}}$$

which establishes formula (2.76).

Mosaic Model

In the case of a mosaic random function, the random variables $Y(x)$ and $Y(x+h)$ are equal with probability $\rho(h)$, or uncorrelated with probability $1 - \rho(h)$. Their bivariate distribution is

$$G_h(dy, dy') = \rho(h)\, G(dy)\, \delta_y(dy') + [1 - \rho(h)]\, G(dy)\, G(dy')$$

where $G(y)$ is the marginal c.d.f. of $Y(\cdot)$ and $\rho(h)$ is its correlogram. If $\varphi(\cdot)$ is a measurable function, letting $m_\varphi = \int \varphi(y)\, G(dy)$ we have

$$E[\varphi(Y(x+h)) \mid Y(x)] = \rho(h)\, \varphi(Y(x)) + [1 - \rho(h)]\, m_\varphi$$

In particular, if $\{\chi_n(y) : n = 0, 1, \ldots\}$ constitutes an orthonormal set of factors for the distribution G, the relations (6.11) are satisfied with $T_n(h) = \rho(h)$ for $n > 0$ (and $T_0(h) \equiv 1$). The mosaic model is therefore an isofactorial model. However, its isofactorial representation brings nothing new, at least as long as it is not complemented by a change-of-support model.

Why Other Models?

The two preceding models correspond to extreme cases. For the same exponential covariance, the first one can represent a Gaussian diffusion process with continuous realizations, whereas the second one represents a random partition into Poisson polygons or polyhedra. One can think of describing other situations by introducing a transformation to an SRF with a Gaussian marginal. But experience has shown that a Gaussian transformation is not always desirable nor even possible. Besides, even if we can obtain an SRF with a Gaussian marginal, it does not follow that its bivariate distributions are Gaussian or mosaic. Therefore we need to broaden the class of possible models.

We will only give a glimpse of the two approaches that have received most attention (there are many others). The first one is to extend the mosaic model to less systematic effects, leading to the model with orthogonal indicator residuals, whose factors are not polynomials. On the contrary, the second approach is to develop models with polynomial factors. First, by a systematic study of diffusion processes we will identify all the classic marginal distributions that can be considered to construct isofactorial models with polynomial factors. Next, for each of these marginals we will characterize the class of isofactorial models that can be associated with it—in other words the general form of covariances $T_n(h)$.

Model with Orthogonal Indicator Residuals

A mosaic random function shows abrupt changes between low and high values. It has no *edge effect* in the sense that when a threshold is applied, the proximity to the boundaries of the threshold level set has no influence on the values assumed. It is therefore suitable to represent a deposit where we pass without transition from clear-cut waste to rich ore, or to model a pollution that remains concentrated near the contamination sources. However, even among models displaying abrupt changes, the mosaic model is an extreme case. The model with orthogonal indicator residuals, which belongs to the same class, has been developed by Rivoirard (1988, 1989) to capture more diverse spatial structures. We will give a glimpse of this model and refer the interested reader to either of Rivoirard's papers, as well as to the slightly different presentation given by Matheron (1989).

This model includes notably some Boolean integer-valued random functions like those presented in Figure 7.31 (Section 7.7.1). In these examples, $Y(x)$ is obtained by superimposition of independent random sets $A_1, \ldots, A_N$. Starting from a situation where all points are assigned a zero value, we set to 1 all points that belong to the random set A_1, and then we set to 2 all points that belong to A_2, and so on, until we set to the value N the points that belong to A_N (at each update the preceding values are overridden). Realizations of the random function are therefore piecewise constant and do not display edge effects when moving from high to low values (due to the construction, any lower value can be met when leaving the set of high values), but do show edge effects when moving from low to high values.

In this model the marginal distribution of $Y(x)$ is $F_n = \Pr\{Y(x) < n\} = \prod_{m=n}^{N} \Pr\{x \notin A_m\}$ and the regression of $1_{Y(x)<n}$ on $1_{Y(x)<n+1}$ is (F_n/F_{n+1}) $1_{Y(x)<n+1}$. The residuals from that regression are

$$R_n(x) = 1_{Y(x)<n} - \frac{F_n}{F_{n+1}} 1_{Y(x)<n+1} = F_n \left(\frac{1_{Y(x)<n}}{F_n} - \frac{1_{Y(x)<n+1}}{F_{n+1}} \right)$$

So letting

$$\chi_n(y) = \frac{1_{y<n}}{F_n} - \frac{1_{y<n+1}}{F_{n+1}}, \qquad n = 1, 2, \ldots, N-1,$$

$$\chi_N(y) = \frac{1_{y<N}}{F_N} - 1,$$

$$\chi_{N+1}(y) = 1$$

the normalized indicator $1_{Y(x)<n}/F_n$ takes the form

$$\frac{1_{Y(x)<n}}{F_n} = \chi_n(Y(x)) + \frac{1_{Y(x)<n+1}}{F_{n+1}} = \sum_{p=n}^{N+1} \chi_p(Y(x))$$

Any function of $Y(x)$ can therefore be expressed as a function of the $\chi_n(Y(x))$: The χ_n constitute an orthogonal basis for the marginal distribution defined by the F_n. Moreover, it can be shown that the random functions $\chi_n(Y(x))$ are uncorrelated. The functions $\chi_n(y)$ therefore constitute the factors of an iso-factorial representation of the bivariate distributions of pairs $(Y(x), Y(x+h))$.

The direct and cross-variograms of indicators are no longer all identical, up to a multiplicative factor, as is the case for a mosaic RF. However, the cross-variogram of two indicators is proportional to the direct variogram of the higher indicator, which allows an easy check of the suitability of the model for a given data set. This model captures a destructuring of low grades (the structure of the random set A_1 is altered by the superimposition of the other random sets). It is well-suited when high values constitute hot spots (local concentration of pollution or high-grade ore).

In applications where a destructuring of high grades is observed, the model is constructed in the reverse manner; that is, all points are initialized with the value N and random sets associated with decreasing values are superimposed. The cross-variogram of two indicators is then proportional to the direct var-iogram of the lower indicator. High values are found in areas not covered by the random sets, which can be seen as the pores when the random sets are made of grains. This model is well-suited to situations inheriting from this kind of geometry—for example, vein-type deposits and pollutions transferred by a circulation of fluids in the porous space.

Diffusion Models

Diffusion models are the prototype of models with *edge effect*, in the sense that to go from a value to another all intermediate values must be visited. Such model is appropriate, for example, in the case of a diffuse pollution or a deposit where there is no clear-cut discontinuity between the waste and the ore. From a theoretical standpoint, diffusion processes are symmetric continuous-time Markov processes and as such have isofactorial bivariate distributions (Matheron, 1989). In the continuous case, namely when they take their values in $\mathbb{R}$, they satisfy a diffusion equation similar to the heat equation (Feller, 1968, Section XIV.6; Feller, 1971, Section X.4 and X.5; Matheron, 1975b; Armstrong and Matheron, 1986a). Discrete diffusion processes, also named birth-and-death processes (Feller, 1971, Section XIV.6), take their values in $\mathbb{N}$ and only progress by unit jumps ($+1$ or -1). Their general study from the point of view of isofactorial properties is due to Matheron (1975c, 1984e, 1989). The system being in state i at time t (which means that $Y(t) = i$), in the interval $]t, t + dt]$ goes to state $i + 1$ with probability $a_i\, d_t$, goes to state $i - 1$ with probability $b_i\, dt$, and stays in state i with probability $1 - (a_i + b_i)\, dt$. The birth rates a_i and death rates b_i are called the diffusion coefficients of the process. The marginal distribution G of the stationary process $Y(t)$ associated with these diffusion coefficients, defined by the probabilities $p_i = \Pr\{Y(t) = i\}$, satisfies

$$a_i\, p_i = b_{i+1}\, p_{i+1}, \qquad i = 0, 1, 2, \ldots$$

It is shown in the general case that the covariances $T_n(h)$ of the factors of a diffusion process are of the form $\exp(-\lambda_n\, |\, h\,|\,)$. The process itself therefore has for covariance an exponential function (when the factors are polynomials) or a sum of exponentials (in the general case). The parameter λ_0 is zero, and the others are positive and increase with n. Their more or less rapid growth characterizes the destructuring of the factors with n. In the case where the marginal is Gaussian, $\lambda_n = n$, and we recover covariances $T_n(h)$ of the form $\rho(h)^n$; however, this circumstance is not general.

Diffusion processes are defined in one dimension. They are extended to higher dimensions by the substitution method. If $Y(t)$, $t \in \mathbb{R}$, is a diffusion process and if $T(x)$, $x \in \mathbb{R}^p$, is a random function with strictly stationary increments and an unbounded variogram, the RF $\widetilde{Y}(x) = Y(T(x))$ is an SRF whose bivariate distributions are isofactorial. It has the same marginal and same factors χ_n as the diffusion process $Y(t)$, but their covariances change: The covariances $\widetilde{T}_n(h)$ ($h \in \mathbb{R}^p$) of the factors of the RF $\widetilde{Y}(x)$ are

$$\widetilde{T}_n(h) = \mathrm{Cov}\big(\chi_n\big(\widetilde{Y}(x)\big), \chi_n\big(\widetilde{Y}(x+h)\big)\big) = \mathrm{E}[\exp(-\lambda_n\,|\,T(x+h) - T(x)\,|\,)]$$

In practice, the studied SRF is assumed to be of the form $\widetilde{Z}(x) = \varphi\big(\widetilde{Y}(x)\big)$. If the function $\varphi(y)$ can be expanded as $\sum_{n=0}^{\infty} \varphi_n\, \chi_n(y)$, the covariance of $\widetilde{Z}$ is

$$C_{\widetilde{Z}}(h) = \sum_{n=1}^{\infty} \varphi_n^2\, \widetilde{T}_n(h) = \mathrm{E}\left[\sum_{n=1}^{\infty} \varphi_n^2 \exp(-\lambda_n\,|\,T(x+h) - T(x)\,|\,)\right]$$
$$= \mathrm{E}[C_Z(|\,T(x+h) - T(x)\,|\,)]$$

where $C_Z(\cdot)$ is the covariance of the stationary process $Z(t) = \varphi(Y(t))$. Additional results on random functions obtained by substitution can be found in Section 7.7.6.

It can also be shown that any set of covariances of the form $\widetilde{T}_n(h) = \exp(-\lambda_n \gamma(h))$, where $\gamma(h)$ is any unbounded variogram in $\mathbb{R}^p$, leads to a valid model. This type of model can be used to define numerically the bivariate distributions of discrete diffusive SRFs. But the inference of the parameters of the underlying diffusion process and of the variogram $\gamma(h)$ requires specific tools and is very tricky (Lajaunie and Lantuéjoul, 1989).

For applications the preferred approach is to consider diffusion processes with polynomial factors known explicitly (Rodrigues formula definition, calculation of $\chi_{n+1}(y)$ from $\chi_n(y)$ and $\chi_{n-1}(y)$ using a recurrence relation whose parameters are known explicitly). Only eight classes of diffusion processes meet this condition. Three of them have a continuous marginal:

- The Gaussian model with Hermite polynomials.
- The gamma model with Laguerre polynomials. The gamma distribution, defined for positive variables, broadens the scope of the Gaussian model.

- The beta model with Jacobi polynomials, which has hardly ever been used in geostatistics so far (the beta distribution with parameters α and β is that of $X_1/(X_1 + X_2)$, where X_1 and X_2 are two independent gamma random variables, with the same scale and respective parameters α and β).

The five other classes of models are associated with a discrete marginal distribution:

- The binomial model with Krawtchouk polynomials, which is the discrete equivalent of the Gaussian model.
- The negative binomial model with Meixner polynomials, which is the discrete equivalent of the gamma model.
- The Poisson model with Charlier polynomials, which can be obtained as limit of the preceding ones.
- The discrete Jacobi-type model with discrete Jacobi polynomials, which is the discrete equivalent of the continuous model with a beta distribution.
- The discrete anti-Jacobi-type model which is related to the Jacobi one but has no continuous equivalent.

Other well-known models that are special cases of the above are of course to be included. Thus for $\alpha = \beta$ the Jacobi polynomials are ultraspherical polynomials and, in particular, the Chebyshev polynomials for $\alpha = \beta = 1/2$ and the Legendre polynomials for $\alpha = \beta = 1$; in this latter case the beta distribution is the uniform over [0, 1]. Diffusion models with a binomial, negative binomial, and Poisson distribution can be obtained as limits of the Jacobi-type model. Continuous diffusion models (Gaussian, gamma, beta) can also be obtained as limits of this model.

For the Gaussian, gamma, binomial, negative binomial, and Poisson models, λ_n is proportional to n, so that the $T_n(h)$ are of the form $\rho(h)^n$. For the beta and discrete Jacobi models the decrease of $T_n(h)$ with n is faster than that of $\rho(h)^n$, whereas it is slower for the anti-Jacobi model.

The main properties of the polynomials associated with the Gaussian, gamma, or negative binomial distributions, which are the most common target marginals, are summarized in the Appendix. The reader is referred to Beckman (1973) for additional results on most of these families of polynomials, as well as Armstrong and Matheron (1986b) and Matheron (1984a, 1984c, 1984e, 1989) for isofactorial models.

Pure Models with Isofactorial Factors

We call a pure model a model for which the $T_n(h)$ are of the form $\rho(h)^n$. We have seen that diffusion processes with a Gaussian marginal are of this type with an exponential correlogram $\rho(h)$. By substitution they provide SRFs that remain of this type with a broader choice of correlograms satisfying $\rho(h) > 0 \;\forall h$. We have likewise seen that random functions with a Gaussian spatial distribution, and thus Gaussian bivariate distributions, also have $T_n(h)$ of type $\rho(h)^n$ but without restriction on the correlogram $\rho(h)$. What for other marginal distributions? Grouping the results obtained by Beckman (1973, Section 6.2 and Appendix I) for continuous variables and by Matheron (1980) for discrete distributions, we obtain the following result.

Given a marginal distribution and its associated orthonormal polynomials, the coefficients $T_n = \rho^n$ do define a bivariate distribution under the following conditions:

- For any $\rho \in [-1, 1]$ if the marginal is Gaussian, binomial, or beta with $\alpha = \beta$.
- Provided that $\rho \in [0, 1]$ if the marginal is gamma, beta ($\alpha \neq \beta$), negative binomial, or Poisson.

For a random function a model $T_n(h) = \rho(h)^n$ is valid, provided that $\rho(h)$ satisfies the above condition for all h and is a correlogram (this being sufficient to ensure that $\rho(h)^n$ is a covariance for all n). Such a model shows a rapid destructuring of the factors as n increases, commensurate with the decrease of $\rho(h)^n$ with n.

Mixture of Pure Models: Hermitian, Laguerre, and Meixner Models

The univariate distribution of $Y(x)$ may be Gaussian and not its bivariate distributions. If the SRF $Y(x)$, with marginal $G(dy)$, has a system of orthonormal polynomials $\chi_n(y)$ and iso-factorial bivariate distributions, this does not imply that the covariances $T_n(h)$ of the factors are of the form $\rho(h)^n$ (i.e., assuming that such a model is valid, which we do not know in the discrete Jacobi or anti-Jacobi case). For a pair of variables, when the pure model associated with $T_n = \rho^n$ exists, it can be generalized by randomization of the correlation coefficient ρ, namely by taking T_n of the form

$$T_n = \int r^n \, \varpi(dr) \tag{6.30}$$

The support of the distribution ϖ must be included in the interval of permissible values of ρ—that is, the interval $[-1, 1]$ if the marginal is Gaussian or binomial and $[0, 1]$ if the marginal is gamma, negative binomial, or Poisson. Conversely, at least in the case of a Gaussian, gamma, or negative binomial marginal, any isofactorial distribution with polynomial factors can be expressed with coefficients T_n of the form (6.30) (Matheron, 1976a; Sarmanov, 1968; Matheron, 1984a, respectively). The corresponding isofactorial models are called by the name of their associated polynomials: *Hermitian model* (Gaussian marginal), *Laguerre model* (gamma marginal), and *Meixner model* (negative binomial marginal). It can be shown that if T_1 is a given positive value ρ, T_n is minimal and equal to ρ^n when the distribution ϖ is concentrated on the value ρ. In other words, pure models have the property of corresponding to the maximum destructuring.

Considering an SRF and not only a pair of random variables, the distribution ϖ associated with the $T_n(h)$ usually depends on h. We will denote it by ϖ_h. Restrictions must be placed on the family of distributions ϖ_h to guarantee that $T_n(h)$ is a covariance function for all values of n. In applications, in order to keep statistical inference simple, one limits the choice to distributions ϖ_h such that the $T_n(h)$ can be expressed as functions of the correlogram $\rho(h)$.

Example of Bi-Gamma Distributions

By analogy with the bi-Gaussian case, there are two ways of defining a bi-gamma distribution. But unlike in the bi-Gaussian case, these two definitions do not coincide. The first way is to define the bi-gamma distribution as a pure isofactorial model with T_n of the form ρ^n. We have seen that the bivariate distributions of gamma diffusion processes are of this type. The second way is to start from three independent unit-scale gamma random variables X_0, X_1, and X_2, with respective parameters $\rho\alpha$, $(1 - \rho)\alpha$, and $(1 - \rho)\alpha$, with $0 < \rho < 1$, $0 < \alpha < 1$, and define

$$Y_1 = X_0 + X_1,$$

$$Y_2 = X_0 + X_2$$

Y_1 and Y_2 are two unit-scale gamma random variables, with the same parameter α and correlation coefficient ρ (see Appendix, Section A.6.1). Matheron (1973b) shows that their bivariate distribution can be written in the isofactorial form (6.22) with normalized Laguerre polynomials and coefficients

$$T_n = \frac{\Gamma(\alpha)}{\Gamma(\alpha + n)} \frac{\Gamma(\alpha\rho + n)}{\Gamma(\alpha\rho)}$$

This quantity is the nth-order moment of the beta distribution with parameters $\rho\alpha$ and $(1 - \rho)\alpha$, thus with mean ρ, and is therefore of the form (6.30). This result can be immediately transposed to negative binomial random variables [e.g., see Feller (1968, p. 285)]. SRFs with this type of bivariate distribution can be obtained by regularization of a stationary orthogonal random measure with gamma or negative binomial distribution. When applied to random measures with a Gaussian or Poisson distribution, this procedure leads, however, to pure bivariate models (Matheron, 1973b; Armstrong and Matheron, 1986a).

Other Isofactorial Models

Let us recall that the factors of isofactorial models are not necessarily polynomials. For example, the uniform distribution over [0, 1] can be associated with various bivariate distributions with different factors (Matheron, 1975c):

- Legendre polynomials.
- Trigonometric factors $\cos(2\pi n\theta)$ and $\sin(2\pi n\theta)$ (e.g., Brownian motion on the circle).
- Walsh functions (nonpolynomial); this model exists in continuous and in discrete versions.
- Factors of the form $\chi_n\big(G^{-1}(z)\big)$, obtained by the change of variable $Z(\cdot) = G(Y(\cdot))$, starting from a model with factors $\chi_n(y)$ and an absolutely continuous marginal $G(dy)$. These factors are generally not polynomial functions.

6.4.4 Practice of Disjunctive Point Kriging

In the case of discrete variables, there have been attempts to check the adequacy of an isofactorial approach or to numerically define the isofactorial model from the data (Lajaunie and Lantuéjoul, 1989; Subramanyam and Pandalai, 2001). The task is more difficult when dealing with continuous variables. In practice, one uses theoretical models. Isofactorial models are very diverse, but in fact only a few have been used. The methodology has, naturally, first been developed for the Hermitian model (Maréchal, 1976). After a preliminary transformation to normal, this model can represent a large class of variables with a continuous, possibly mildly skewed, distribution. Nevertheless, we can only deplore that so many applications have used the bi-Gaussian model, which is quite special indeed, without questioning the validity of this choice. The application methodology for the Laguerre model is essentially due to Hu and Lantuéjoul (1988) and Hu (1988). This model extends the capabilities of the bi-Gaussian model to distributions that exhibit a cluster of values near the origin and a long distribution tail. As for the Meixner model, it makes it possible, at the price of a discretization, to properly account for a large proportion of zero values or to study a discrete variable directly (Demange et al., 1987; Kleingeld, 1987). The Hermitian, Laguerre, and Meixner models comprise a range of bivariate distributions that go from the mosaic to the pure diffusive model. Thus we have elected to limit our presentation of the practice of disjunctive kriging to these three models. Table 6.1 presents their main features: The left-hand columns give for each type of marginal distribution the model used and the corresponding family of polynomials, and the right-hand columns give the names of the bivariate distributions that can be associated with them. Other than that, the main model that has been applied, to

TABLE 6.1 Main Isofactorial Models with Polynomial Factors Classified by Type of Marginal and Bivariate Distributions

Marginal Distribution			Bivariate Distribution		
Type	Distribution	Polynomials	$T_n = \rho^n$	Intermediate	$T_n = \rho$
Continuous symmetric	Gauss	Hermite	Bi-Gaussian	Hermitian model	Mosaic Gaussian
Continuous skewed	Gamma	Laguerre	Pure gamma	Laguerre model	Mosaic gamma
Discrete	Negative binomial	Meixner	Pure negative binomial	Meixner model	Mosaic negative binomial

our knowledge, is the model with orthogonal indicator residuals [Bordessoule et al. (1988, 1989); Rivoirard, (1994), in particular the application of Chapter 13)]. In the case of discrete variables, Subramanyam and Pandalai (2001) give a characterization of isofactorial models; it can be used to check whether such a model is adequate, on the basis of the empirical distributions.

Choice of the Transformed Marginal Distribution

In applications, the data are seldom obliging enough to have a histogram matching one of the classic marginals. We have to use a transformation in which the variable of interest $Z(\cdot)$ is assumed of the form $Z(x) = \varphi(Y(x))$, and $Y(\cdot)$ is supposed to conform to an isofactorial model whose marginal is one of the classic distributions. As will be shown later, the transformation function φ is used through the coefficients φ_n of its expansion in the basis of orthonormal polynomials $\chi_n(y)$. If the marginal distributions G of $Y(\cdot)$ and F of $Z(\cdot)$ have similar shapes, few terms need to be retained in this expansion. This explains why a transformation to a uniform, which at first glance would seem appealing, is rarely advisable, since it would "pack" extreme data too much, with the consequence that a large number of Jacobi polynomials would be required to represent the transformation function.

When the data histogram reflects a continuous, not too skewed, distribution, a Gaussian transformation is appropriate. But if the data histogram is very skewed, with a large proportion of low values and a long tail, a Gaussian transformation will limit the amplitude of large values, which may be the goal, but also magnify unimportant differences between low values. To resolve this problem, it is better to transform to a skewed variable. The gamma distribution is particularly interesting for its diverse behaviors. A positive random variable Y follows a (standard) gamma distribution with parameter $\alpha > 0$ if its density is

$$g_\alpha(y) = \frac{1}{\Gamma(\alpha)} e^{-y} y^{\alpha-1} \qquad (y > 0)$$

where $\Gamma(\cdot)$ is the gamma function (A.1). The shape of the density g_α depends on α: If $\alpha < 1$, g_α is a decreasing function, unbounded at the origin; if $\alpha > 1$, g_α is a bell-shaped curve that tends to the Gaussian for large values of α; in the intermediate case $\alpha = 1$, g_1 is the density of the exponential distribution.

The data histogram can have an atom at the origin. For example, if Z represents the thickness of a geological formation that is sometimes absent, this is reflected in a proportion $p_0 > 0$ of zero values. If Z is a mineral grade, there may be a nonnegligible proportion p_0 of zero values (or at least considered as such because below the detection limit). This is called the *zero effect*. In the case of a Gaussian (or gamma) transformation, φ still exists but is not one to one, so we do not know which value of Y should be associated with a zero value of Z; we only know that Y belongs to the interval $]-\infty, G^{-1}(p_0)]$. This problem disappears if we choose for Y a discrete distribution such that $\Pr\{Y = 0\} = p_0$. On the other hand, we must discretize the continuous part of the distribution of Z, which is not a major hurdle if we can master this discretization. The negative binomial distribution is a good candidate, because it enables the representation of both a zero effect and a strong skewness of the rest of the distribution that generally comes with it. The negative binomial distribution with parameters $\alpha > 0$ and $p \in [0, 1]$ is defined by

$$p_i = (1 - p)^\alpha \frac{\Gamma(\alpha + i)}{\Gamma(\alpha)} \frac{p^i}{i!}, \qquad i \geq 0$$

The case $\alpha = 1$ corresponds to the geometric, or Pascal, distribution $p_i = (1 - p) p^i$, which is the discrete version of the exponential distribution. In general, the negative binomial distribution appears as a discrete version of the corresponding gamma distribution with the same parameter α.

In these last two cases, one problem is the choice of the parameter α (in the case of the negative binomial the choice of the parameter p results from the probability p_0 to be matched). It seems reasonable to choose a low value for α when the histogram is skewed and the relative variance high. Demange et al. (1987) choose for α the value that also preserves the ratio of the cutoff grade to the mean grade. A similar criterion is the preservation of the ratio of the median to the mean. Hu (1988) proposes to preserve a parameter that characterizes the relative dispersion of the distribution—for example, the relative variance or the selectivity index presented in Section 6.5.1.

Expansion of the Transformation Function into Factors

Having selected the marginal distribution $G(dy)$ of $Y(x)$, we have to make explicit the form of the transformation function φ such that $Z(x) = \varphi(Y(x))$. Suppose that the c.d.f. $F(z)$ of Z, modeled as seen in Section 6.2, is discretized by a series of points $\{(z_p, F_p) : p = 0, 1, \ldots, P\}$, where the $F_p = F(z_p)$ increase with p from $F_0 = 0$ to $F_P = 1$; z_0 and z_P are the minimum and maximum possible values for z. The value of Y associated with z_p is $y_p = G^{-1}(F_p)$. If $y_0 = -\infty$

(respectively, $y_P = +\infty$), this is replaced by a large negative (respectively, positive) value (e.g., -4 and $+4$ in the case of a Gaussian transformation). The transformation function φ satisfies $\varphi(y_p) = z_p$. Suppose that the discretization of F is fine enough for an approximation of φ by a piecewise linear function to be acceptable. φ is then defined by

$$\varphi(y) = a_p + b_p y \quad \text{with} \quad b_p = \frac{z_p - z_{p-1}}{y_p - y_{p-1}} \quad \text{and} \quad a_p = z_p - b_p y_p,$$
$$y \in \,]y_{p-1}, y_p], \quad p = 1, \ldots, P$$

We will need $\varphi(y)$ in the form of its expansion into factors $\chi_n(y)$:

$$\varphi(y) = \sum_{n=0}^{\infty} \varphi_n \chi_n(y)$$

According to (6.9) the coefficients φ_n are given by

$$\varphi_n = \int \varphi(y) \chi_n(y) G(dy) = \sum_{p=1}^{P} \int_{y_{p-1}}^{y_p} (a_p + b_p y) \chi_n(y) G(dy)$$

In the case of a Gaussian or gamma transformation, $a_p + b_p y$ can be expressed linearly as a function of the Hermite or Laguerre polynomials of degrees 0 and 1, so that the integral in the above expression can be calculated analytically using relations (A.9) and (A.11), or (A.17) and (A.18).

The polynomial expansion is truncated to an order n_{max}. It is usually not monotone and thus takes meaningless values for extreme values of y. We select n_{max} so that φ is monotone in the actual domain of variation of y and the sum of the φ_n^2, for n varying from 1 to n_{max}, is close to the variance of $Z(x)$. Usually one to a few dozen terms are sufficient.

Once the transformation has been defined, the data $Z(x_\alpha)$ are transformed into data $Y(x_\alpha)$ by application of the transformation φ^{-1}.

Choice of Isofactorial Model

Having defined the marginal distribution of Y and the transformation that leads to it, there remains to define the bivariate distributions. This is achieved by specifying the isofactorial model through the covariances $T_n(h)$ of the factors $\chi_n(Y(x))$ for all $n > 0$ (in all cases $T_0(h) \equiv 1$). Because the models considered here have polynomial factors, the covariance $T_1(h)$ is simply the correlogram $\rho(h)$ of the SRF $Y(x)$. This correlogram is related to the variogram $\gamma(h)$ of $Y(x)$ obtained by a classic structural analysis of the transformed data $Y(x_\alpha)$ by

$$\gamma(h) = C[1 - \rho(h)]$$

where the sill C is equal to 1 in the Gaussian case, to α in the gamma case, and to $\alpha p/(1-p)^2$ for a negative binomial transformation.

We have seen in Section 6.4.3 [relation (6.30)] that for the Hermitian, Laguerre, and Meixner models the covariances of the factors are of the form

$$T_n(h) = \int r^n \, \varpi_h(dr)$$

where ϖ_h is a probability distribution concentrated on the interval $[-1, 1]$ in the Hermitian case and on the interval $[0, 1]$ for Laguerre and Meixner models. In practice, one limits the choice to isofactorial models where the covariances $T_n(h)$ are functions of $\rho(h)$ and where $\rho(h)$ is nonnegative. Four isofactorial distribution models are used most (Figure 6.3):

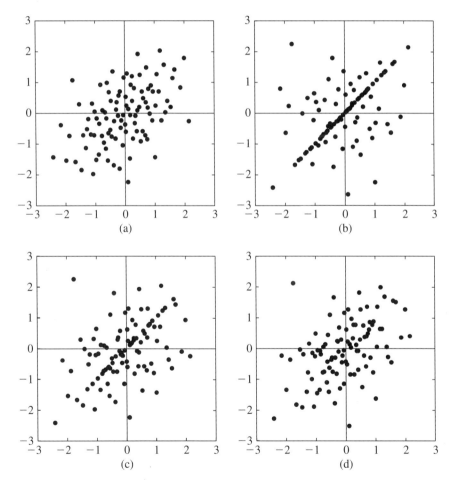

FIGURE 6.3 Scatterplot of 100 pairs $(Z(x), Z(x+h))$ with standard normal marginal distributions for a value h such that $\rho(h) = 0.5$: (a) Gaussian bivariate distribution; (b) mosaic model; (c) barycentric model with $\beta = 1/2$; (d) beta model with $\beta = 2$.

1. *The pure diffusive model*, with a distribution ϖ_h concentrated on the single value $\rho(h)$:

$$T_n(h) = \rho^n(h), \qquad n \geq 0$$

   The covariances of the factors tend to a pure nugget effect when n increases.

2. *The mosaic model*, which corresponds to a distribution ϖ_h concentrated on the values 0 and 1, with respective probabilities $1 - \rho(h)$ and $\rho(h)$:

$$T_n(h) = \rho(h), \qquad n > 0$$

   All factors have the same covariance.

3. *The barycentric model*, which is a mixture of the preceding two, in proportions β and $1 - \beta$: The distribution ϖ_h is concentrated on the values 0, $\rho(h)$ and 1, respectively, with probabilities $(1 - \beta)(1 - \rho(h))$, β and $(1 - \beta)\rho(h)$, where β is a parameter between 0 and 1, so that

$$T_n(h) = \beta \rho^n(h) + (1 - \beta)\rho(h), \qquad n > 0$$

4. *The beta model*, which is a mixture of all pure models associated with a positive correlation coefficient, since ϖ_h is a beta distribution with parameters $\beta \rho(h)$ and $\beta(1 - \rho(h))$, where β is a positive parameter; it leads to

$$T_n(h) = \frac{\Gamma(\beta)}{\Gamma(\beta + n)} \frac{\Gamma(\beta \rho(h) + n)}{\Gamma(\beta \rho(h))}, \qquad n \geq 0$$

   For an SRF with gamma marginal, this model generalizes the second type of bi-gamma distribution presented in Section 6.4.3. It has variants [see Hu (1988)].

In principle, the justification of the use of an isofactorial model and the choice of the model itself must proceed from the study of the regression curves $\chi_n(Y(x + h))$ as a function of $Y(x)$ or of $\chi_n(Y(x))$ [cf. relation (6.11)]. In practice, it would be tedious to examine lagged scatterplots (or h-scattergrams) for all values of n until n_{max} and for various classes of lag h, although it is advisable to inspect a few for validation [e.g., Goovaerts (1997)]. These models have the advantage of being distinguishable simply by inspection of the variogram of order 1 (or madogram) of $Y(x)$, defined as we have seen in Section 2.5.3 by

$$\gamma_1(h) = \tfrac{1}{2} \mathrm{E}[|Y(x + h) - Y(x)|]$$

Its sill C_1 is in all cases the dispersion indicator S of the marginal distribution of Y, as results from definition (6.32), but its shape depends on the model. Taking as structure functions the normalized variograms of order 1 $\tilde{\gamma}_1(h) = \gamma_1(h)/C_1$ and of order 2 (the usual variogram) $\tilde{\gamma}(h) = \gamma(h)/C = 1 - \rho(h)$, we have

$$\widetilde{\gamma}_1(h) = \begin{cases} \sqrt{\widetilde{\gamma}(h)} & \text{for the pure diffusive model,} \\[2mm] \widetilde{\gamma}(h) & \text{for the mosaic model,} \\[2mm] \beta\sqrt{\widetilde{\gamma}(h)} + (1-\beta)\widetilde{\gamma}(h) & \text{for the barycentric model with parameter } \beta, \\[2mm] \dfrac{\Gamma(\beta)}{\Gamma\left(\beta+\frac{1}{2}\right)}\dfrac{\Gamma\left(\beta\widetilde{\gamma}(h)+\frac{1}{2}\right)}{\Gamma(\beta\widetilde{\gamma}(h))} & \text{for the beta model with parameter } \beta \end{cases}$$

These relations express very simple links between the madogram and the variogram of $Y(x)$. In particular, by plotting $\widetilde{\gamma}$ as a function of $\widetilde{\gamma}_1$, we obtain a straight line for the mosaic model, a parabola for the pure diffusive model, and intermediate behaviors for the other models (Figure 6.4).

The final choice of the model can be validated by comparing the variogram of the initial variable $Z(x)$ with the theoretical variogram calculated from the covariances $T_n(h)$ using relation (6.24) with $\psi_n = \varphi_n$ (that is, if the variogram of $Z(x)$ is robust enough). It can also be validated with indicator variograms whose theoretical expression is also given by (6.24) where the ψ_n are now the coefficients of the expansion of the indicator (these are provided below).

The examination of indicator variograms may call into question the choice of the reference distribution. Indeed, in the case of random functions with bi-gamma distributions (pure diffusive version), Emery (2005b) shows that low grades are less structured than high grades. This means that the indicator

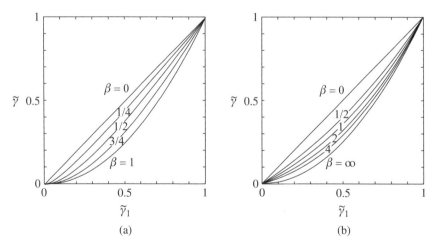

FIGURE 6.4 Relationship between the variogram and the madogram of an SRF with Gaussian, gamma, or negative binomial marginal for the main isofactorial models (both variograms are normalized): (a) barycentric model; (b) beta model. The extremes correspond to the pure diffusive model (parabola) and the mosaic model (straight line).

associated to the threshold $G_\alpha^{-1}(p)$, $p < 0.5$, has larger nugget effect or a shorter range than that related to the threshold $G_\alpha^{-1}(1 - p)$. This phenomenon is all the more pronounced as the marginal distribution is skewed (low α value). This is not really surprising, because in such a case low grades are very similar, so that the contour of the random set associated with a low threshold is fairly erratic. A different character may preclude the use of a gamma distribution with a low value of α (for a large α the gamma distribution tends to the normal, and then low and high grades are similarly destructured).

For completeness, let us mention another tool proposed by Journel and Deutsch (1993) to summarize bivariate distributions: the bivariate entropy function $E(h)$. It gives the entropy of the bivariate p.d.f. $g_h(y, y')$ of the $(Y(x), Y(x + h))$ pair as a function of h:

$$E(h) = - \int_{-\infty}^{+\infty} \int_{-\infty}^{+\infty} g_h(y, y') \log(g_h(y, y')) \, dy \, dy'$$

(the integral is in fact limited to the support of g_h). Among unbounded SRFs with the same covariance, Gaussian random functions correspond to maximum entropy, or "maximum disorder" [relation (6.30) has already shown that, for a given marginal, the maximum destructuring is obtained with the pure diffusive model]. But this entropy function cannot be determined experimentally, except if we have a complete image.

An Example of Structural Analysis

In a vein of the open-pit Salsigne gold mine (France), zones to be mined are delineated on the basis of boreholes on a 5-m × 5-m grid. The data from 1-m-long cores have been studied in detail (Liao, 1990; Chilès and Liao, 1993). Figures 6.5a and 6.5b display the histogram and the empirical c.d.f. of core gold grades. It has 30% of zero values and a long tail (grades spread out to 400 g/t, for a mean of 10 g/t). This suggests a gamma transformation. This choice is confirmed by the examination of the indicator variograms associated with two complementary proportions, namely 30% (threshold: 0.5 g/t) and 70% (threshold: 8.5 g/t). Figures 6.6a and 6.6b show that the indicator associated with the higher threshold has a much lower nugget effect than the other. This is not compatible with a Gaussian transformation but is consistent with a gamma transformation. Choosing the parameter α so that the percentage of waste is matched (59% of data below 5 g/t $\approx 0.5 \, m_Z$), we obtain $\alpha = 0.34$, which corresponds to a very skewed distribution. If we only kept the first term in the expansion of the transformation function, which amounts to assuming that grades follow a gamma distribution (up to a shift), we would already account for 89% of the variance ($\varphi_1^2 / \sigma_Z^2 = 0.89$). In practice, the expansion is truncated to $n_{max} = 16$ Laguerre polynomials so that over 99% of the variance is reproduced. Figure 6.5c shows the transformation function φ associated with the 16

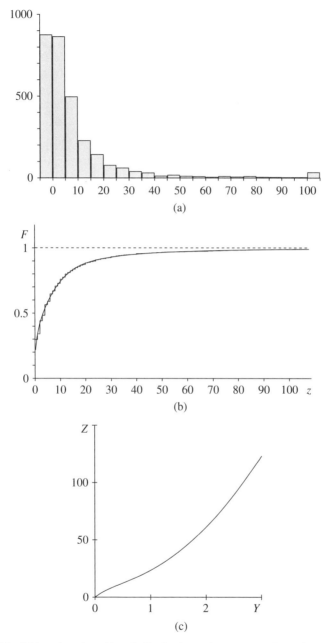

FIGURE 6.5 Salsigne deposit: gold grade distribution and gamma transformation: (a) histogram; (b) empirical c.d.f. and fit $G\circ\varphi^{-1}$; (c) transformation function φ obtained with 16 Laguerre polynomials. [From Liao (1990).]

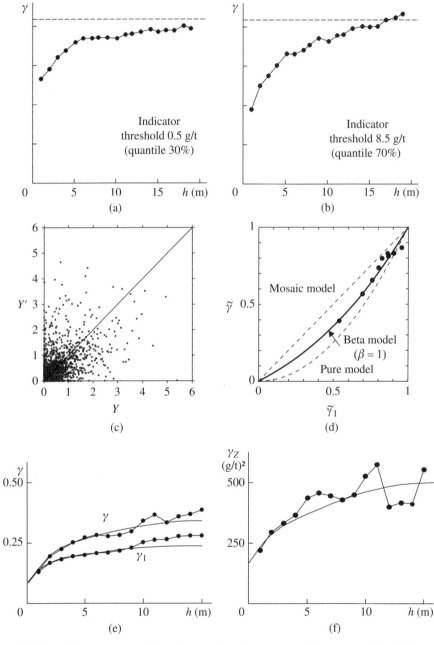

FIGURE 6.6 Variograms of gold grades and of transformed data: (a) variogram of the indicator associated with a threshold of 0.5 g/t (30% data below that value); (b) variogram of the indicator associated with a threshold of 8.5 g/t (70% data below that value) (c) scatterplot of gamma transformed pairs $Y(x)$ and $Y' = Y(x + h)$ for $h = 1$ m vertically; (d) relationship between normalized variogram and madogram of gamma transformed data; (e) variogram and madogram of gamma transformed data; (f) variogram of grades. [From Liao (1990) and Chilès and Liao (1993), with kind permission from BRGM and Kluwer Academic Publishers.]

Laguerre polynomials. Figure 6.5b shows that the theoretical c.d.f. it defines fits the empirical c.d.f. perfectly. As a comparison, to achieve an equivalent precision with a Gaussian transformation requires 27 Hermite polynomials, and the first term accounts for only 42% of the variance (i.e., the histogram of grades is far from Gaussian).

Figures 6.6d–f show the sample variogram $\hat{\gamma}$ and the sample madogram $\hat{\gamma}_1$ of the gamma variable Y associated with Z, the graph of the relation between the normalized variograms $\tilde{\gamma}_1$ and $\tilde{\gamma}$ (limited to zones where the variograms do not fluctuate too much), as well as the variogram $\hat{\gamma}_Z$ of grades. An empirical bivariate distribution of Y data is displayed in Figure 6.6c. It is not easy to interpret. By contrast, Figure 6.6d suggests an intermediate model between the diffusive and mosaic models. A beta model with $\beta = 1$ has been chosen. The theoretical variogram γ has then been selected to fit the three sample variograms (Figures 6.6e and 6.6f). Similar variograms can in fact be obtained starting from a Gaussian transformation. But a comparison of the performance of the two models (Hermitian and Laguerre) for predicting change-of-support effects demonstrated, as expected, the clear superiority of the model based on the gamma transformation.

Other examples can be found in Hu and Lantuéjoul (1988) (bi-Gaussian and Laguerre models applied to uranium grades) and Emery (2006a) (Laguerre model fitted to soil pollution data; negative binomial distribution modeling the count of caterpillar larvae infesting sugar cane).

Performing Kriging

The disjunctive kriging estimate of $Z_0 = \psi(Y(x_0))$ and the DK variance are derived by relations (6.16) and (6.18) from the simple kriging estimates and variances of the factors $\chi_n(Y(x_0))$. These are obtained by solving the systems (6.15) and applying (6.17). The generalization to the DK of the mean value of $\psi(Y(x))$ in a volume v is straightforward (cf. Section 3.5.2). The coefficients ψ_n are obtained by application of (6.20). Let us mention a few classic cases (for clarity the index c denotes a fixed cutoff or threshold):

- *Disjunctive kriging of* $Z(x_0)$. We simply have $\psi_n = \varphi_n$. However, disjunctive kriging is rarely used as a replacement of ordinary kriging because the improvement is often minimal (Puente and Bras, 1986).
- *Disjunctive kriging of* $I(z_c) = 1_{Z(x_0) < z_c}$. Since $1_{Z(x_0) < z_c} = 1_{Y(x_0) < y_c}$ with $y_c = \varphi^{-1}(z_c)$, we have

$$\psi_n = \int_{-\infty}^{y_c} \chi_n(y) \, G(dy)$$

The explicit form of the integral is deduced from (A.9) in the case of a Gaussian distribution and from (A.17) for a gamma distribution. For example, in the case of a Gaussian transformation we have

$$\psi_0 = G(y_c), \qquad \psi_n = \frac{1}{\sqrt{n}} \chi_{n-1}(y_c) g(y_c), \qquad n = 1, 2, \ldots$$

where G and g are, respectively, the c.d.f. and p.d.f. of the standard Gaussian and χ_n are the normalized Hermite polynomials.

- *Disjunctive kriging of point-support recovery functions.* By analogy with the (deterministic) recovery functions defined in Section 6.5.1, these are the (random) quantities

$$T(z_c) = 1_{Z(x_0) \geq z_c} \qquad \text{and} \qquad Q(z_c) = Z(x_0) 1_{Z(x_0) \geq z_c}$$

which can be rewritten as

$$T(z_c) = 1_{Y(x_0) \geq y_c} \qquad \text{and} \qquad Q(z_c) = \varphi(Y(x_0)) 1_{Y(x_0) \geq y_c}$$

with $y_c = \varphi^{-1}(z_c)$. Thus we have

$$\psi_n = \int_{y_c}^{\infty} \chi_n(y) G(dy) \qquad \text{for } T(z_c)$$

$$\psi_n = \int_{y_c}^{\infty} \varphi(y) \chi_n(y) G(dy) = \sum_{m=0}^{\infty} \varphi_m \int_{y_c}^{\infty} \chi_m(y) \chi_n(y) G(dy) \qquad \text{for } Q(z_c)$$

The Appendix provides elements for calculating the partial integrals [relations (A.9) and (A.11) in the Gaussian case, relations (A.17) and (A.18) in the gamma case].

The kriging systems (6.15) of the factors are constructed and solved iteratively, and the corresponding DK variances (6.17) are calculated; since two successive systems are very similar, the solution of system n can be taken as an initial solution of system $n + 1$, and an iterative improvement of the solution quickly converges to the exact solution.

Truncation of the Expansion

In the case where $T_n(h) = \rho^n(h)$, the off-diagonal terms of the left-hand side matrix of system (6.15) as well as the right-hand side terms tend rapidly to zero as n increases so that the estimator of $\chi_n(Y_0)$ also tends rapidly to zero with increasing n. This allows us to limit the expansion to a finite number of terms n_0. For the more complex models presented above, the left-hand side matrix tends toward a matrix that is no longer the identity matrix. Still simplifications can be implemented. Indeed, denote by $\hat{T}(h)$ the limit of $T_n(h)$ when $n \to \infty$, by $\{\hat{\lambda}_\alpha : \alpha = 1, \ldots, N\}$ the solution of the kriging system (6.15) written with $\hat{T}(h)$ instead of $T_n(h)$, and by

$$\hat{\chi}_n = \sum_{\alpha=1}^{N} \hat{\lambda}_\alpha \chi_n(Y_\alpha)$$

the nonoptimal estimates of the factors $\chi_n(Y_0)$ associated with these weights. Expressing the optimal estimates by reference to the nonoptimal ones, relation (6.16) can be written

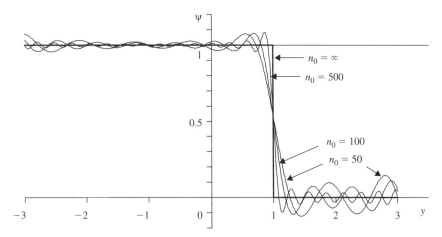

FIGURE 6.7 Approximation of the indicator function $\psi(y) = 1_{y<1}$ by a finite number n_0 of Hermite polynomials.

$$\psi^* = \psi_0 + \sum_{n=1}^{\infty} \psi_n \, \hat{\chi}_n + \sum_{n=1}^{\infty} \psi_n \left(\chi_n^* - \hat{\chi}_n \right)$$

Replacing $\hat{\chi}_n$ by its definition, we obtain

$$\psi^* = \left(1 - \sum_{\alpha=1}^{N} \hat{\lambda}_\alpha \right) \psi_0 + \sum_{\alpha=1}^{N} \hat{\lambda}_\alpha \, \psi(Y_\alpha) + \sum_{n=1}^{\infty} \psi_n \left(\chi_n^* - \hat{\chi}_n \right) \tag{6.31}$$

The first two terms are finite sums and the infinite sum can be truncated as soon as the difference $\chi_n^* - \hat{\chi}_n$ becomes negligible. This is a nice result because the objective $\psi(Y(x_0))$ is often a strongly nonlinear function of $Y(x_0)$—typically, an indicator function—so a satisfactory expansion of ψ would require a large number of factors (much more than the $n_{\max}$ factors of the expansion of the transformation function φ). Figure 6.7 shows the approximation of an indicator function by 50, 100, and 500 Hermite polynomials. The convergence to the indicator function is very slow. With 100 terms, this approximation is not acceptable at the extremes, and with 500 terms there remains a significant discrepancy (an indicator cannot be represented exactly by a finite number of polynomials). The application of relation (6.31) enables us to only consider a relatively small number of factors.

Locally Varying Mean

So far we have been assuming that $Z(x)$ is an SRF with a known mean, and even with a known marginal distribution. In practice, we only know an estimate of it, and it may also be that stationarity is only local. It is tempting, then, for calibration on the local characteristics of $Z(x)$, to (a) impose the unbiasedness condition of ordinary kriging to the kriging of factors and (b) substitute for all $n > 0$ the ordinary kriging system to the simple kriging system (6.15):

$$
\begin{cases}
\displaystyle\sum_{\beta=1}^{N}\lambda_{n\beta}T_n\left(x_\beta - x_\alpha\right) + \mu_n = T_n(x_0 - x_\alpha), & \alpha = 1, \ldots, N, \\
\displaystyle\sum_{\beta=1}^{N}\lambda_{n\beta} = 1
\end{cases}
$$

In this case beware that if the off-diagonal covariances on the left-hand side as well as those on the right-hand side tend to 0 when n increases, the estimator χ_n^* no longer tends to 0 but to the average value of the $\chi_n(Y_\alpha)$. Thus, in principle, it is necessary to compute all χ_n^*. Rivoirard (1994, Chapter 9) proposes an elegant trick to limit the expansion to a finite number of terms n_0. This trick is equivalent to (6.31) except that the estimators of the factors now correspond to ordinary kriging rather than simple kriging. As a consequence, the first term of the right-hand side disappears, so that the estimator no longer depends on ψ_0, which was our objective. However, the replacement of simple kriging of factors by ordinary kriging requires caution. Assuming that locally the $\chi_n(Y(x))$ are no longer with zero mean ($n > 0$) amounts to questioning the marginal distribution, thus the orthogonality of the factors and the bivariate distributions, and therefore the optimal character of disjunctive kriging.

6.4.5 Applications

The first implementations of disjunctive kriging were made in the scope of a bi-Gaussian isofactorial model (Jackson and Maréchal, 1979; Young, 1982). In mining, predictions obtained by disjunctive kriging from exploration surveys have been compared with the reality of exploitation as known from numerous blast-holes. The following conclusions can be drawn:

- Concerning the estimation of the grade at a point, or the mean grade of a panel, disjunctive kriging usually brings little improvement over simple or ordinary kriging. This is confirmed by various case studies, including electrical conductivity data (Yates et al., 1986), contamination of underground water (Yates and Yates, 1988), and soil geochemistry (Webster and Oliver, 1989). Yates et al. observe, for example, a reduction of estimation variance of about 5% for point kriging and 10% for block kriging. Therefore, disjunctive kriging is seldom used as a substitute for ordinary kriging to estimate the observed value Z itself.
- If we look for zones where Z has a chance greater than 50% to exceed a threshold z_c, it is better to estimate the indicator $1_{Z \geq z_c}$ by DK than select the zones where the kriging estimate of Z exceeds the threshold, especially if the distribution of Z is skewed [Webster and Oliver (1989) for grades in agricultural soils; Wood et al. (1990) for physical properties of soils].

- Point-support recoverable resources in a large panel corresponding to the zone of influence of a borehole can be predicted with an error of about 10%, which is considered as quite good compared to the results obtained using conventional methods (Jackson and Maréchal, 1979).
- Disjunctive kriging and indicator cokriging are theoretically equivalent, up to discretization effects (number of factors in one case, number of thresholds in the other). Liao (1990) verified this empirically on a data set from a gold deposit by applying (a) indicator cokriging with fits of all covariances and cross-covariances and (b) disjunctive kriging using a model fitted from the data. Let us mention that various studies have compared the performance of bi-Gaussian disjunctive kriging and indicator kriging by trying them on real or synthetic data sets. Some have concluded that indicator kriging could give better results than bi-Gaussian disjunctive kriging [e.g., Carr and Deng (1987) for earthquake ground motion data]. This should not be surprising, since none of these two methods is optimal. DK loses its optimality character if used with an arbitrary isofactorial model (in this case the bi-Gaussian model), and indicator kriging is poorer than indicator cokriging. Since these two methods are not equivalent, either one may outperform the other under given circumstances (e.g., diffusive or mosaic model).
- This shows the importance of an appropriate model selection. However, if the variable under study is poorly structured (strong nugget effect, short range compared with the data spacing), the choice of a model and a method (Gaussian or gamma transformation, disjunctive of indicator kriging) is of limited importance. Local estimations of the point distribution will lack sharpness [as defined by Gneiting et al. (2007b)], so that predictions will be meaningful only after aggregation of a large number of point results.

Disjunctive kriging can be easily generalized to disjunctive cokriging if a consistent set of bivariate distributions can be modeled. It is the case, for example, when all bivariate distributions are Gaussian. Muge and Cabeçadas (1989) developed such a model to study the short- and medium-term evolution of the pollution of a lake by eutrophication. The model takes into account the time delay of some variables and allows the estimation of the "probability" to exceed a threshold value in a given period of time.

6.5 SELECTIVITY AND SUPPORT EFFECT

We now turn to change-of-support problems. These are always tricky problems, especially when only point-support data, or those considered as such, are available, which will be our working assumption here. We will adopt the terminology used in mining, which historically has been the most important field of application. But selectivity problems may be posed in similar terms for the management of fish stock, agricultural land, or the remediation of contaminated areas. Before considering estimation problems, we will first define the concepts of selectivity, support effect and information effect. This presentation summarizes results presented by Matheron (1984b) and Lantuéjoul (1990), to which the reader is referred for more details.

6.5.1 Selectivity

Let us first review a number of tools characterizing probability distributions and their dispersion, which will be useful to study support and information effects.

Selectivity Curves

Consider a nonnegative random variable Z representing, for definiteness, the grade of a block v selected uniformly among identical blocks discretizing a panel V. The distribution of block grades is defined by the cumulative distribution function

$$F(z) = E[1_{Z < z}]$$

Let us assume that for mining we only select blocks whose mean grade exceeds a cutoff value z. $F(z)$ represents the proportion of blocks regarded as waste (i.e., whose mean grade is below the cutoff grade). In practice, one is more interested in the blocks selected and in the mineral they contain, and the following auxiliary functions named *recovery functions* or *selectivity curves* are defined (Figure 6.8):

- The *ore tonnage* at cutoff z, normalized by the total tonnage (i.e., with no cutoff)

$$T(z) = E[1_{Z \geq z}] = 1 - F(z)$$

- The *quantity of metal* residing in this ore, normalized by the total tonnage (of ore)

$$Q(z) = E[Z\, 1_{Z \geq z}] = \int_z^\infty u\, F(du)$$

- The *mean grade* of selected ore

$$\widetilde{m}(z) = E[Z \mid Z \geq z] = \frac{Q(z)}{T(z)}$$

- The *conventional income*

$$B(z) = E[(Z - z)1_{Z \geq z}] = Q(z) - z\, T(z)$$

This last function has an economic significance. Let us assume that the selection of blocks above the cutoff grade z is entirely free; that is, we can leave in place

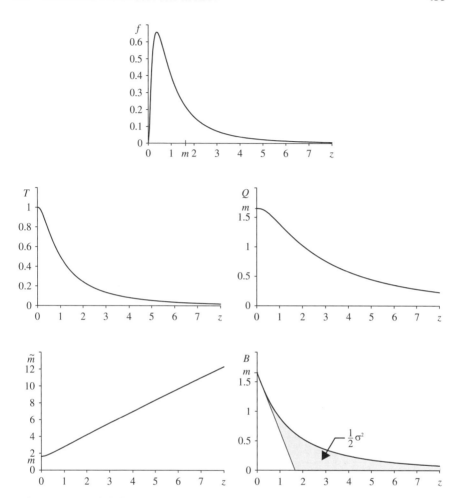

FIGURE 6.8 Probability density $f(z)$ and selectivity curves $T(z)$, $Q(z)$, $\tilde{m}(z)$, and $B(z)$. Example of a lognormal distribution with logarithmic mean 0 and logarithmic variance 1. σ^2 is the variance of the lognormal variable.

any block with a mean grade below the cutoff and mine any block with grade above the cutoff. Let us also assume that the cutoff grade z is chosen so that the quantity of metal recovered from a block of grade z pays for its marginal mining and processing costs. In other words, z is the marginal cutoff; it can be considered as a variable in the sense that it depends on the selling price of the recovered metal and on available technology. Denoting by c the selling price of metal (per unit weight) and by d the density of ore (assumed to be constant and thus independent of grade, making tonnage equivalent to volume), the marginal income from a block with mean grade Z is $cd\,|v|\,(Z - z)$ if Z is greater than or equal to z, and is zero otherwise (no stripping and removing of the block). Thus $B(z)$ represents, up to a normalization, the operating income before depreciation and fixed costs as a function of the marginal grade z. This function can also be written as

$$B(z) = \int_z^\infty T(u)\, du$$

The definitions of $Q(z)$ and $B(z)$ can be read on the graph of $T(\cdot)$, as shown in Figure 6.9: $Q(z)$ is the area of the domain under the tonnage curve defined by points with ordinates less than $T(z)$, whereas $B(z)$ is the area of the domain defined by points with abscissas greater than z.

We will present without proof the main properties of recovery functions. The functions T and Q each characterize the distribution F. They are not necessarily continuous, except if the distribution F has a density, and they are both nonincreasing. Their values at the origin are $T(0) = 1$ and $Q(0) = m$, mean of Z, and they vanish at infinity. The function $\tilde{m}$ is not necessarily continuous and is nondecreasing. Its value at the origin is $\tilde{m}(0) = m$.

B is a more interesting function because it is convex, and therefore continuous on $]0, +\infty[$, and nonincreasing. It assumes the value m at zero and vanishes at infinity. Its slope at the origin is equal to $-(1 - p_0)$, where $p_0 = F(0+)$ represents the proportion of zero values. B also characterizes the distribution F.

Because T is a nonincreasing function of z, z is a nonincreasing function of T. It is thus possible to eliminate z and reparameterize Q, $\tilde{m}$, and B as functions of the ore tonnage T (Figure 6.10). $Q(T)$ is a concave function, and thus continuous on $]0, m[$, and nondecreasing. $\tilde{m}(T)$ is a nonincreasing function. In particular, one has

$$\frac{dQ}{dT} = z(T) > 0, \qquad \frac{d\tilde{m}}{dT} = -\frac{\tilde{m}(T) - z(T)}{T} = -\frac{B(T)}{T^2} \leq 0$$

The functions $B(z)$ and $Q(T)$ are related by the duality formulas

$$B(z) = \sup_T \{\, Q(T) - zT \,\} \qquad Q(T) = \inf_z \{\, B(z) + zT \,\}$$

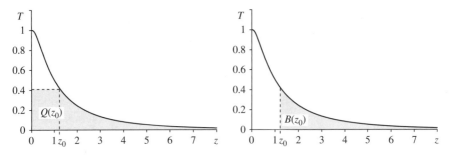

FIGURE 6.9 Graphical reading of $Q(z)$ and $B(z)$ from the graph of $T(\cdot)$.

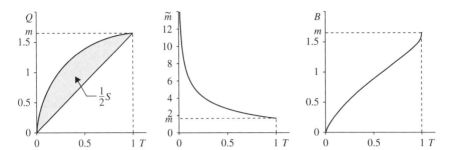

FIGURE 6.10 Selectivity curves $Q(T)$, $\tilde{m}(T)$ and $B(T)$. Example of a lognormal distribution with logarithmic mean 0 and logarithmic variance 1. S is the dispersion indicator.

As a consequence the function $Q(T)$, just as the function $B(z)$, characterizes the distribution F. Similar concepts are used in econometrics to study the concentration of income in the population [the fraction $x = F(z)$ of the people with income smaller than z earns a fraction $y = 1 - Q(z)/m$ of the global income: Lorenz curve; Atkinson (1970); Rothschild and Stiglitz (1973); Shorrocks (1983)].

Interpretation in Terms of Geochemistry

Recovery functions were introduced by Lasky (1950) to represent grade–tonnage relationships for porphyry copper deposits. Lasky used a law of the form $\tilde{m}(T) = \alpha(1 - \log T)$. It corresponds to an exponential distribution of grades, where $T(z) = \exp(-z/\alpha)$ and $\tilde{m}(z) = z + \alpha$. In other words, to get ore with mean grade above m_1, it suffices to mine at the cutoff grade $z = m_1 - \alpha$.

Recovery functions have been used in geochemistry at the global scale to model grades in the earth's crust. Ahrens (1954), on the basis of numerous examples, stated a "law" according to which the distribution of grades is lognormal. We have seen with the de Wisjian model that this type of distribution can be obtained by a process of enrichment of half of the ore and depletion of the other half, the phenomenon being repeated for each half over and over (multiplicative cascade); in the language of fractals, this is a multifractal model (Feder, 1988). Using a statistical approach, Vistelius (1960) reached much more circumstantiated conclusions, interpreting some long-tailed distributions as mixtures of Gaussians. Unsurprisingly, Ahrens' model has inspired a number of generalizations [e.g., de Wijs (1976)]. Turcotte (1986), working in the scope of fractal models, claimed that grades follow a Pareto distribution (power law) obtained by Rayleigh distillation where successive crystallizations enrich the still liquid fraction of the magma, and he proposed a multiplicative cascade model that is a variant of the de Wisjian model: After each subdivision, only the half with larger concentration is further subdivided into halves with enriched and depleted concentrations, respectively. Reexamining these different points of view in the light of geochemistry, Allègre and Lewin (1995) conjectured that distributions of fractal or multifractal type result from geochemical differentiation processes, whereas the other distributions can be explained by geochemical mixing processes. A synthesis of the various models is presented by Turcotte (1997) and Agterberg (2007). A nice example of fitting of a multiplicative cascade to arsenic data is detailed by Gonçalves (2001). Nature is of course more complex than these conceptual models would suggest. While such models are important for the understanding of the origin of deposits and thus mineral exploration, in geostatistics our objective is more modestly to fit the distribution actually observed in the area of interest.

Dispersion Indicator and Selectivity Index

The *dispersion indicator* of the distribution F of a random variable Z with values in $\mathbb{R}$ (not necessarily positive) is defined by

$$S = \frac{1}{2} \int_{-\infty}^{+\infty} \int_{-\infty}^{+\infty} F(dz) |z' - z| F(dz') = \int_{-\infty}^{+\infty} F(z)[1 - F(z)] \, dz \qquad (6.32)$$

This can be compared with the variance

$$\sigma^2 = \frac{1}{2} \int_{-\infty}^{+\infty} \int_{-\infty}^{+\infty} F(dz) (z' - z)^2 F(dz') = \int_{-\infty}^{+\infty} (z - m)^2 F(dz)$$

where $m = \int_{-\infty}^{+\infty} z F(dz)$ is the mean. S and σ^2 both summarize the dispersion of the distribution F, but S is more robust than σ^2: up to the factor $\frac{1}{2}$, S is the mean of the magnitude of $Z' - Z$ when Z and Z' are two independent random variables with the same distribution F, while σ^2 is the mean square value of $Z' - Z$. The dispersion indicator S has a finite value if and only if the mean m exists (i.e., $E[|Z|] < \infty$), whereas this condition does not guarantee that the variance σ^2 is finite. It can be shown that S and σ are related by the inequality $S \leq \sigma/\sqrt{3}$. Equality is obtained when F is uniform over a bounded interval. For a Gaussian distribution we have $S = \sigma/\sqrt{\pi}$, a value close to the maximum. The *normality index* $\nu = \sqrt{\pi} \, S/\sigma$ can also be defined, and it is a dimensionless parameter with value 1 for a Gaussian distribution. S is related to the interquantile distance associated with p and $1 - p$ by the following inequality which justifies the name dispersion indicator given to S:

$$z_{1-p} - z_p \leq \frac{S}{p(1 - p)} \qquad (p < \tfrac{1}{2})$$

In order to include the case of a distribution with discontinuities, it is understood that z_p is the smallest value such that $F(z_p) \geq p$ and z_{1-p} is the largest value such that $F(z_{1-p}) \leq 1 - p$.

When Z is a nonnegative variable, which we assume from now on, then S is strictly less than m. The *selectivity index* $s = S/m$, named Gini coefficient in econometrics (Gini, 1921), therefore belongs to the interval $[0, 1[$. In the case of a lognormal distribution, for example, this index is related to the logarithmic standard deviation σ' by $s = 2G(\sigma'/\sqrt{2}) - 1$ and can take any value in $[0, 1[$. The selectivity index s is more robust than the coefficient of variation σ/m, which can be infinite. In applications, however, the second-order moments need to be known—and therefore the coefficient of variation—since the variance is the only parameter whose evolution under a change of support can be predicted exactly. In this perspective, Matheron (1985) examines, for a number of distributions, to which extent the coefficient of variation can be related to the selectivity index.

The interest of the dispersion indicator and the selectivity index, aside from their robustness, appears clearly when considering the graph $Q(T)$. Selectivity is maximal if we can recover all the metal while mining only a minute fraction of the ore. This would be the case, for example, if we could pick the nuggets in a gold deposit. The graph of $Q(T)$ would reach its upper limit m already for a very low value of T. Conversely, no selectivity is possible if the ore has a uniform grade equal to the mean grade m. The function $Q(T)$ then coincides with the straight line of equation $Q = m\,T$. Generally, it is shown that the area between this line and the graph of $Q(T)$ is equal to $S/2$ (Figure 6.10):

$$\int_0^1 (Q(T) - mT)\, dT = \tfrac{1}{2} S$$

The variance satisfies a similar property, but its interpretation in terms of selectivity is not so clear (Figure 6.8): The area between the graph of $B(z)$ and the straight line $B = m - z$ (the tangent at the origin if there is no zero effect) is equal to $\sigma^2/2$:

$$\int_0^\infty (B(z) - (m - z)1_{z \le m})\, dz = \tfrac{1}{2}\sigma^2$$

6.5.2 Support Effect and Information Effect

Support Effect

Let us now consider an RF $Z(x)$, for example, representing a grade, and the grades of blocks with support v. The distribution of block grades

$$Z_v = \frac{1}{|v|} \int_v Z(x)\, dx$$

depends of course on their size v, with the general tendency being that the dispersion decreases as the size of the support increases: Very low as well as very high grades can be observed on cores, whereas the mean grades of large panels have a low contrast (see Figure 6.11). At the same time, selective mining is all the more difficult as the support v of the selection unit is large. In the limit, when v is very large, all blocks have grades very close to the mean m of Z, and no real selectivity is possible. From a probabilistic viewpoint, these facts can be explained by the three following properties:

1. The block mean grade (no cutoff) is independent of block size (it is the mean m of $Z(x)$).
2. The marginal distribution $F_v(\cdot)$ of block grades gets narrower around the mean as the support gets larger, according to the variance (2.33)

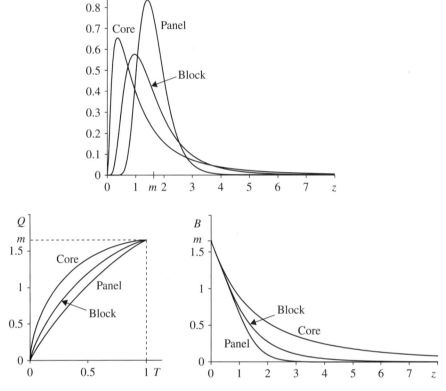

FIGURE 6.11 Evolution of the probability density $f(z)$ and the selectivity curves $Q(T)$ and $B(z)$ with the size of the support. From the cores to the blocks, and again from the blocks to the panels, the variance is reduced by a factor of 4.

$$\sigma_v^2 = \frac{1}{|v|^2} \int_v \int_v C(x' - x)\, dx\, dx' \tag{6.33}$$

where $C(h)$ is the covariance of the SRF $Z(x)$.

3. $F_v(\cdot)$ tends to become Gaussian when the support v tends to infinity in all directions, provided that the covariance of Z has a finite integral range (central limit theorem).

To improve the comparison of the two distributions, let us introduce a definition and a theorem.

Definition. If Z_1 and Z_2 have distributions F_1 and F_2 with the same mean m, we say that the distribution F_1 is more selective than F_2 if the associated recovery functions satisfy $Q_1(T) \geq Q_2(T)$ for any $T \in [0, 1]$, or $B_1(z) \geq B_2(z)$ for any $z \geq 0$ (the two definitions are equivalent given the duality between $B(z)$ and $Q(T)$).

To know whether a given distribution is more selective than another, we can use a general functional analysis theorem established by Cartier [see Alfsen (1971, Section 3)]:

Theorem. The c.d.f. F_1 is more selective than the c.d.f. F_2 if and only if there exists a bivariate distribution $F_{12}(dz_1, dz_2)$ with marginals F_1 and F_2 and such that

$$E(Z_1 \mid Z_2) = Z_2 \tag{6.34}$$

or equivalently if and only if

$$\int \varphi(z) F_1(dz) \geq \int \varphi(z) F_2(dz) \tag{6.35}$$

for any convex function φ.

Relation (6.34) is called *Cartier's relation*. By application of (1.4) and (1.5), this implies that F_1 and F_2 have the same mean and

$$\mathrm{Var}(Z_1) = \mathrm{Var}(Z_2) + \mathrm{Var}(Z_1 - Z_2) \tag{6.36}$$

Coming back to the support effect, if $\underline{x}$ is a uniform random point in a block v with grade Z_v, it is easy to show that

$$E[Z(\underline{x}) \mid Z_v] = Z_v \tag{6.37}$$

Likewise, if $\underline{v}$ is a block randomly selected among blocks with support v forming a partition of a panel V with mean grade Z_V, we have

$$E[Z(\underline{v}) \mid Z_V] = Z_V$$

By Cartier's theorem the distribution F_V of panels is less selective than the distribution F_v of blocks, which is itself less selective than the distribution F of cores (considered as points). This phenomenon, which is fundamental in geostatistics, is known as the support effect. Selectivity relations between the distributions F_V, F_v, and F also entail that the dispersion indicator and the selectivity index decrease as the support increases (at least for nested supports).

The first two properties (preservation of the mean and reduction of the variance according to (6.33)) and the fact that F_v is less selective than F are not enough to characterize the marginal distribution of block grades. Indeed, that distribution depends on the complete spatial distribution of the SRF $Z(x)$, and not only on its marginal and its covariance. In practice, the rigorous calculation of $F_v(z)$ is only possible in a few special cases, and we have to resort to change-of-support models based on assumptions that are only approximately verified. Several models are presented in Sections 6.6 to 6.9. All are consistent in that they honor the above properties (but not necessarily the convergence to a Gaussian distribution when v

gets larger and larger), which does not mean that they are always suitable. Any validation, even if partial, is useful; for example, one can construct composite samples (group metric cores into 10-m composites) and verify that their histogram is correctly predicted from the point-support histogram (core samples). It should be noted that these global change-of-support models determine the block distribution, and thus global recoverable resources, without associated uncertainty.

Information Effect

In selective mining, as elsewhere, people have to make decisions based on incomplete information. When only point-support data or regarded as such are at hand (core samples or blast holes), the decision to accept or reject a block v must be based on an estimated value since the block true grade is unknown. The best one can do is consider the conditional expectation $Z_v^* = E[Z_v|Z_\alpha : \alpha = 1, \ldots, N]$. It is a conditionally unbiased estimator:

$$E(Z_v \mid Z_v^*) = Z_v^* \tag{6.38}$$

It follows from Cartier's theorem that the distribution of Z_v^* is less selective than that of Z_v. In particular, by application of (6.36), estimated grades are less dispersed (smoother) than actual grades: For a high cutoff grade there will be far fewer blocks accepted on the basis of Z_v^* than on Z_v. Even if one considers a cutoff for which the bias on the ore tonnage is negligible, there will be a loss due to bad selection:

- Poor blocks are selected because estimated rich.
- Rich blocks are rejected because estimated poor.

In all cases this translates into a degradation of the value of the exploited ore, which increases with the dispersion of the scatterplot (Z_v, Z_v^*) around the first bisector. This degradation is called the *information effect*: Selection based on estimates rather than the true block values always results in a degradation of the conventional income.

In our reasoning we have overlooked a difficulty that deserves further explanation. Even though the selection is made on the basis of the estimated value Z_v^*, the grade of a mined block is the true grade Z_v and not Z_v^* ["We do not send estimated values to the mill" (David, 1977)]. Therefore the ore tonnage and the quantity of metal of the selected blocks are

$$T(z) = E\left(1_{Z_v^* \ge z}\right), \qquad Q(z) = E\left(Z_v \, 1_{Z_v^* \ge z}\right) \tag{6.39}$$

However, when the estimator Z_v^* is conditionally unbiased, we have that

$$E\left(Z_v^* \, 1_{Z_v^* \ge z}\right) = E\left(Z_v \, 1_{Z_v^* \ge z}\right) = Q(z)$$

This is an extremely desirable property: The true mean grade of the selected blocks is equal to the mean grade of their estimated values (in expectation). This justifies the calculation of selectivity curves directly from the distribution of Z_v^*. This simplification of course is no longer legitimate if we use an estimator that is not conditionally unbiased.

Selection on Kriged Grades

In practice, we cannot find the conditional expectation, and we do the selection on simpler estimators, such as the grade of a sample in the block or a kriging estimate from samples in the block and its surrounding. But kriging only guarantees unbiasedness, not conditional unbiasedness. This results in an additional loss. For example, if the regression $E[Z_v \mid Z_v^*]$ is linear, we have

$$E[Z_v \mid Z_v^*] = m + r[Z_v^* - m]$$

(i.e., actual increment $= r \times$ expected increment in view of the selection)

where

$$r = \rho \frac{\sigma_v}{\sigma_v^*}, \quad \sigma_v^2 = \mathrm{Var}[Z_v], \quad \sigma_v^{*2} = \mathrm{Var}[Z_v^*], \quad \rho = \mathrm{Corr}(Z_v, Z_v^*)$$

If Z_v^* is simply the grade of a sample taken in the block v, one has $\sigma_v^* > \sigma_v$, and thus $r < 1$. For $Z_v^* > m$ one tends to select an ore that is less rich than expected. At the other extreme, for very scattered grades and a small support v, when each panel is kriged with a very large neighborhood, it is possible to get $\sigma_v^* \ll \sigma_v$ and $r > 1$ and reach the wrong conclusion that the deposit is uneconomic.

To evaluate the quantity of metal that is actually recovered, it is convenient to introduce the regression between true and estimated grades:

$$H(z) = E(Z_v \mid Z_v^* = z)$$

The quantity of metal is then

$$Q(z) = E\left(Z_v \, 1_{Z_v^* \geq z}\right) = E\left[E\left(Z_v \, 1_{Z_v^* \geq z} \mid Z_v^*\right)\right] = E\left(H(Z_v^*) \, 1_{Z_v^* \geq z}\right)$$

If the function H is increasing, which is a reasonable assumption, the tonnage and the quantity of metal can be written as

$$T(z) = E\left[1_{H(Z_v^*) \geq H(z)}\right], \quad Q(z) = E\left[H(Z_v^*) 1_{H(Z_v^*) \geq H(z)}\right]$$

Therefore

$$T(z) = \widetilde{T}(H(z)), \quad Q(z) = \widetilde{Q}(H(z))$$

where $\widetilde{T}(z)$ and $\widetilde{Q}(z)$ stand for the recovery functions associated with $H(Z_v^*)$. Provided that we replace z by $H(z)$, the curves of the recovered tonnage and quantity of metal are identical to those of $H(Z_v^*)$. As a consequence, $Q(T) = \widetilde{Q}(T)$. In other words, to quote a figurative expression of G. Matheron (1984b), "we may say that we are mining conditional expectations rather than true grades." [An alternative point of view is to replace the conditionally biased estimate Z_v^* by the conditionally unbiased estimate $Z_v^{**} = H(Z_v^*)$ and select on the basis of Z_v^{**}. The result is unchanged if we replace the cutoff z by the cutoff $H(z)$.] The actual calculation of the recovery functions requires the knowledge of the conditional distribution of $H(Z_v^*) = \mathrm{E}[Z_v \mid Z_v^*]$, which amounts to knowing the bivariate distribution of Z_v and Z_v^*. This is not accessible from the data and requires a model. We will see that this is possible in the framework of the discrete Gaussian model and its extensions (provided of course that the model is applicable).

A Comparison of Support and Information Effects

To compare these two effects, Matheron (1984b) presents a simple example that can be solved explicitly: The decision to mine the block v is made in view of the grade of a single sample, randomly located in the block, and the bivariate distribution of the block and the sample is lognormal. We denote by σ_v^2 the logarithmic variance of the block grade Z_v, by σ^2 the logarithmic variance of the sample grade Z, and by m the common arithmetic mean of Z and Z_v. Cartier's relation $\mathrm{E}(Z \mid Z_v) = Z_v$ entails that the correlation coefficient of log Z and log Z_v is $\rho = \sigma_v/\sigma$. The conditional expectation $H = \mathrm{E}(Z_v \mid Z)$ is also lognormal with arithmetic mean m and logarithmic variance $\sigma_H^2 = \rho^2\sigma_v^2 = \sigma_v^4/\sigma^2$ (see Appendix, Section A.9). The different selectivity curves can then be calculated from the lognormal distributions. For $m = 1$, $\sigma^2 = 1$, and $\sigma_v^2 = 1/2$, for example, the selectivity index goes from 0.521 for point-support grades to 0.383 for block-support grades and to 0.276 for blocks selected on the basis of estimates. The first loss is due to the support effect—that is, selecting from true grades is more efficient at a point support than at a block support—and the second loss reflects the information effect. Notice that the two effects can be of comparable importance.

To summarize, one should not base the selection on single samples from the blocks and calculate the value of the deposit as if the grades of the selected blocks were equal to those of their samples—the implicit methodology behind an evaluation of recoverable resources by polygons of influence. Indexing the recovery functions by

- "pol" for an evaluation by polygons of influence which would identify the distribution of block grades with the distribution of the samples,
- "ide" for the ideal case of a selection based on true block grades,
- "opt" for a selection based on conditional expectation,
- "sub" for the actual selection based on suboptimal estimates,

we obtain the following ranking of conventional incomes:

$$B_{\text{sub}}(z) \leq B_{\text{opt}}(z) \leq B_{\text{ide}}(z) \leq \hat{B}_{\text{pol}}(z) \qquad \forall z$$

The hat over $\hat{B}_{\text{pol}}$ reminds us that it is a biased estimate (by excess) rather than an evaluation susceptible to be reached. The loss from B_{opt} to B_{sub} is due to the use of a suboptimal selection criterion; it could be reduced. The difference between B_{ide} and B_{opt} is due to the information effect, and the difference between $\hat{B}_{\text{pol}}$ and B_{ide} represents the support effect; these can only be reduced at the cost of new data and a reduction of the volume v of the selection unit. The loss $B_{\text{sub}} - \hat{B}_{\text{pol}}$ shows how illusory our expected income is when it is calculated with polygons of influence and the histogram of the blocks is confused with that of the samples.

6.6   MULTI-GAUSSIAN CHANGE-OF-SUPPORT MODEL

We have seen in Section 6.3.2 that the conditional point distribution is seldom rigorously accessible, with the noticeable exception of the multi-Gaussian model where $Z(x) = \varphi(Y(x))$ is a function of a Gaussian SRF $Y(x)$. So we will first consider the change of support in the multi-Gaussian model.

6.6.1   General Case

Let us first focus our interest on a given block v and seek the distribution of its mean grade, denoted by Z_v or $Z(v)$, conditioned on the data $Z(x_\alpha)$ or, equivalently, on the corresponding $\{Y(x_\alpha) : \alpha = 1, \ldots, N\}$ (in practice, only a subset of the data, inside the block or close to it, are taken into account). Due to the change of support, we know that distribution in special cases only (e.g., when the SRF Z is itself Gaussian). The multi-Gaussian approach can still be used if we approximate the block grade by the mean grade of M points x_i discretizing the block. Then

$$\begin{aligned}
F_v(z|\text{data}) &= \Pr\left\{ \frac{1}{M} \sum_{i=1}^{M} Z(x_i) < z \mid Z(x_\alpha) : \alpha = 1, \ldots, N \right\} \\
&= \Pr\left\{ \sum_{i=1}^{M} \varphi(Y(x_i)) < Mz \mid Y(x_\alpha) = y_\alpha : \alpha = 1, \ldots, N \right\}
\end{aligned} \tag{6.40}$$

Because the SRF $Y(x)$ is assumed Gaussian, the vector of conditional expectations of the $Y(x_i)$ given the $Y(x_\alpha)$ is Gaussian and is obtained by linear regression. Its distribution is entirely determined by the vector of conditional means and the conditional covariance matrix, which are the vector of the simple kriging estimates $y^*(x_i)$ and the covariance matrix of the simple kriging errors. The distribution (6.40) generally cannot be calculated analytically. It is calculated by the Monte Carlo method: We draw a large number of vectors $\{Y(x_i) : i = 1, \ldots, M\}$ from the conditional distribution and evaluate $F_v(z \mid \text{data})$ numerically as the proportion of simulated vectors satisfying the condition $\sum_{i=1}^{M} \varphi(Y(x_i)) < Mz$. The conditional mean of any functional of $Z(v)$ can be evaluated by the mean value of the functional on the simulations, or deduced from the evaluation of the conditional distribution [see Emery (2006b) for more details]. The conditional variance can be obtained in the same manner.

The multi-Gaussian approach is very powerful, because it can be extended straightforwardly to any functional of the $Z(x_i)$. For example, it allows the evaluation of *indirect* recoverable resources—that is, when the selection will be based on a *future estimate* of the block grade. To this end, it suffices to complement the set of points x_i discretizing the block with points where future information will be available.

In some situations, for example at the boundaries of an orebody, we want to relax the assumption that the Gaussian random function has a zero mean and switch to a model with a locally unknown mean. Emery (2006b) proposes an approximation based on ordinary kriging instead or simple kriging and providing an unbiased estimation of any functional of $Z(v)$.

The multi-Gaussian model has been proposed and applied by Verly (1983, 1984). Just like for point-support data, it is recommended, before using this method, to verify at least that the data are compatible with the hypothesis of stationarity of the RF $Y(x)$ and that the bivariate distributions of this SRF can be considered Gaussian. Verly (1986) gives details on an application to a uranium deposit; predictions made from limited information are validated by the real grades of blocks (approximated by the mean of blast holes). In an application to a copper deposit, Emery (2006b) compares the conditional expectation (known mean) to the approximation with a locally unknown mean. The latter seems to provide better results in border areas, but the differences remain small.

It should be noted that the nonconditional version of this model constitutes a global change-of-support model.

Though rigorous, up to the discretization of the block and the computation by the Monte Carlo method, the multi-Gaussian model is not used much because of it computational requirements. However, the calculations can be simplified if we are able to anticipate the shape of the nonconditional distribution F_v of $Z(v)$. This is possible, at the price of an approximation, in the lognormal case.

6.6.2 Permanence of Lognormality

In the lognormal case, it is often observed that the distribution of $Z(v)$ is still lognormal: There is *permanence* of lognormality. This permanence cannot be strictly true: The average of two lognormal random variables is not a lognormal random variable. However, numerical experiments show that the permanence of lognormality is a sensible approximation as long as v remains small [for definiteness, as long as the variance of $\log Z(x)$ is less than 1 and the size of v does not exceed two-thirds of the range; see discussion at the end of Section 6.8.1.

The characteristics of the lognormal random variable $Z(v)$ are defined by two relations: (i) $Z(x)$ and $Z(v)$ have the same mean, and (ii) the variance of $Z(v)$ derives from the covariance of the SRF Z according to (6.33). Since the mean, variance, and covariance of the lognormal SRF Z are related to those of the Gaussian SRF $Y(x) = \log Z(x)$ by the relations (2.78), let us take as main parameters the mean m, variance σ^2, and covariance $C_Y(h)$ of Y. Similarly, let us characterize the random variable $Z(v)$ by the mean m_v and variance σ_v^2 of $Y_v = \log Z(v)$. (Note that Y_v is not the mean value of $Y(x)$ in v.) Combining these relations, we get

$$m_v + \tfrac{1}{2}\sigma_v^2 = m + \tfrac{1}{2}\sigma^2,$$

$$\exp(\sigma_v^2) = \frac{1}{|v|^2} \int_v \int_v \exp(C_Y(x' - x)) \, dx \, dx' \tag{6.41}$$

from which σ_v^2 and m_v can be deduced immediately. The reduction of variance on the Gaussian variables can be characterized by $r^2 = \sigma_v^2/\sigma^2 < 1$.

We now turn to the local problem—that is, finding the distribution of the mean grade $Z(v)$ of a specific block v conditional on $\{Z(x_\alpha) : \alpha = 1, \ldots, N\}$. The vector of transformed data $Y(x_\alpha) = \log Z(x_\alpha)$ is Gaussian, and under the simplifying assumption of permanence of lognormality, $Y_v = \log Z(v)$ is a Gaussian random variable. The multi-Gaussian approach is greatly simplified by going one step further and assuming that the joint distribution of $Y_v, Y(x_1), \ldots, Y(x_N)$ is multivariate Gaussian. Indeed, in that case there is no need to discretize the block v: The conditional distribution of Y_v is Gaussian; its mean is the simple kriging estimate Y_{SK}^* from the data $Y(x_\alpha)$, and its variance is the simple kriging variance σ_{SK}^2. In particular, like in the point-support case, the conditional expectation of $Z(v)$ is the simple lognormal kriging estimator

$$Z_{SLK}^* = \exp\left(Y_{SK}^* + \tfrac{1}{2}\sigma_{SK}^2\right)$$

When carrying out this kriging, we should remember that Y_v and $Y(x)$ do not have the same mean. To implement the calculations, we need, in addition to the knowledge of the point covariance of the SRF Y, the covariances of Y_v and $Y(x_\alpha)$. With the same approach as for the determination of σ_v^2 we get

$$\exp(\mathrm{Cov}(Y_v, Y(x))) = \frac{1}{|v|} \int_v \exp(C_Y(x' - x)) \, dx'$$

If strict stationarity of $Y(x)$ is not assured, the conditional distribution is not known but, exactly like for a point support, the ordinary lognormal kriging

$$Z_{OLK}^* = \exp\left(Y_{OK}^* + \tfrac{1}{2}\left(\sigma_{OK}^2 + 2\mu\right)\right)$$

is an unbiased estimator of $Z(v)$. As explained for the point-support case, the lognormal distribution with logarithmic mean Y_{OK}^* and logarithmic variance

$\sigma_{OK}^2 + 2\mu$ can only be considered as an approximation to the conditional distribution of $Z(v)$.

When performing the ordinary kriging of Y_v, the relation (6.41) between the (local) means of $Y(x)$ and Y_v must be taken into account so that m only is considered unknown. Details of this method can be found in Journel (1980).[6] This procedure should be used with caution because it lacks robustness if the data depart from a lognormal distribution. Matheron (1974a) presents several variants of this approach; also see Dowd (1982), Rivoirard (1990), and Gessie (2006).

Parker et al. (1979) present an application to a uranium deposit, including a cross-validation of local point-support recovery functions. The uranium grade is not strictly stationary; the authors have thus used ordinary lognormal kriging and have carefully calibrated the sill of the variogram of $Y(x)$ (using a local calibration). Rendu (1979) presents a validation of lognormal kriging and of the calculation of the conditional distribution of blocks for a gold deposit in South Africa. The study shows an excellent agreement between the results predicted by geostatistics and reality. Because the mean is well known, the results obtained by simple kriging of Y_v are better than those obtained by ordinary kriging. However, forcing the lognormal formalism to data whose histogram is spread out but not lognormal can lead to real disasters (David et al., 1984).

6.7   AFFINE CORRECTION

Global Affine Correction

If $Z(x)$ is a Gaussian SRF, with mean m, variance σ^2, and covariance $C(h)$, then the mean value $Z(v)$ in the block v is Gaussian, with the same mean m, and variance σ_v^2 given by (6.33). Therefore the following equalities hold in distribution:

$$\frac{Z(x) - m}{\sigma} \overset{D}{=} \frac{Z(v) - m}{\sigma_v} \sim \mathcal{N}(0, 1) \tag{6.42}$$

The affine correction is to consider that even if the distribution of Z is not Gaussian, the relation (6.42) remains valid, at least to a first approximation. The model F_v of the block distribution is then derived from the model F of the point distribution by

$$F_v(z) = F(m + (z - m)/r) \qquad \text{with} \qquad r = \sigma_v/\sigma \tag{6.43}$$

The change-of-support coefficient r is less than 1 and decreases as the support v gets larger. This model has the advantage of simplicity: F_v can be immediately deduced from the sample histogram or any graphical fit of it without mathematical modeling. This change-of-support model is consistent but only valid, besides the Gaussian case, if the distribution of Z is continuous and if v is small

---

[6] In the case of an unknown mean, Journel assumes that the conditional variance of Y_v is equal to the simple kriging variance σ_{SK}^2; it must be changed to $\sigma_{OK}^2 + 2\mu$.

compared to the range (in particular, the variogram of Z should not have a nugget effect). Indeed:

1. It assumes a permanence of the distribution (F and F_v have the same type of distribution, only the scale parameter changes). This permanence is only true in the Gaussian case. We also know that if v becomes far larger than the range, F_v must tend to a Gaussian distribution by virtue of the central limit theorem, which is not the case in this model (except again if the SRF is Gaussian).

2. The interval of definition of $Z(v)$ narrows around the mean in the ratio r. If the distribution of Z has an atom at 0, the distribution of $Z(v)$ will have one at $m(1-r)$, which makes no physical sense.

Indicator Kriging and Local Affine Correction

Let us now consider a specific block v. If we cannot turn to the Gaussian case, the calculation of conditional expectation is generally impossible. So is kriging, or rather cokriging of the indicator $1_{Z(v)<z}$ from indicator data $1_{Z_\alpha<z'}$, since the v-support indicator is not the mean in v of point-support indicators. A possible solution is to apply a local affine correction to the distribution defined by kriging a series of indicators (for different cutoffs) of the point variable $Z(x_0)$, where x_0 is the center of the block v. Since this local distribution has for mean the kriging estimate $z^*_{x_0}$ of $Z(x_0)$, (6.43) becomes

$$\hat{F}_v(z) = \hat{F}_{x_0}\left(z^*_{x_0} + \left(z - z^*_{x_0}\right)/r_{x_0}\right)$$

where $\hat{F}_{x_0}$ is the c.d.f. derived from the kriged indicators and $r^2_{x_0}$ is the ratio of the conditional variance for support v to the variance of the conditional point-support distribution at x_0. Because these variances cannot be calculated in general, Journel (1984) proposes to approximate the conditional change-of-support coefficient r_{x_0} by the global change-of-support coefficient r.

Unfortunately, as shown by Emery (2008), the global coefficient is always greater than the local one, at least in expectation (see Section 6.8.3), so this approximation makes block grade distributions look more selective than they really are and creates biases in the assessment of recoverable resources. In fact this method very often produces results that are simply unacceptable (Liao, 1990; Rossi and Parker, 1994). Apart from exceptions, nonparametric methods such as indicator kriging cannot capture a change of support. This requires explicit modeling of bivariate distributions.

6.8 DISCRETE GAUSSIAN MODEL

Ideally we would like to have an isofactorial model in which all pairs of Gaussian variables associated to point and block values have Gaussian bivariate distributions, or even in which the multivariate distribution of these Gaussian variables is Gaussian. Indeed, this would make it easy to proceed further with the calculation of recovery functions after a change of support, including for indirect recovery functions (i.e., when the selection will be based on a future estimate of the true value rather than on the true value itself). Matheron (1976b) proposes two change-of-support models for RFs Z that are transforms of an SRF Y with a Gaussian marginal. The first one, the Hermitian model, is developed for an SRF Y whose bivariate point–point distributions are Hermitian. It includes the special case where the point–point distributions are Gaussian, but then the bivariate point–block distributions are not (they are Hermitian). This model can be used at the global scale but lacks flexibility for estimating local recovery functions. The other model, the discrete Gaussian model (DGM), satisfies our requirements at the cost of a few assumptions and, at least in its original version, of a discretization of the location of the data. We will only present that model. It is by far the most used model even if it is not applicable in all situations.

6.8.1 Global Discrete Gaussian Model

Let us consider an SRF $Z(x)$ that can be expressed as the transform of an SRF $Y(x)$ with standard normal marginal distribution. (*Note*: Y must be strictly stationary and not only of order 2; otherwise different blocks do not necessarily have the same distribution.) It is therefore of the form $Z(x) = \varphi(Y(x))$ with the transform $\varphi = F^{-1}G$, where F is the marginal c.d.f. of $Z(\cdot)$ and G the standard normal c.d.f. Consider a block v and a uniform random point $\underline{x}$ within v. The random variable $Z(\underline{x})$ has for c.d.f. the marginal distribution F of the SRF $Z(\cdot)$ and can be expressed as the transform $\varphi(Y(\underline{x}))$ of the random variable $Y(\underline{x})$. Similarly, we can consider that the mean grade $Z(v)$ of the block v is of the form $Z(v) = \varphi_v(Y_v)$, where Y_v is a standard normal and φ_v the block transformation function, which we want to determine. The crucial assumption of the discrete Gaussian model is that the bivariate distribution of the $(Y(\underline{x}), Y_v)$ pair is Gaussian.[7] It is characterized by a correlation coefficient r assumed positive. The block transformation function and its distribution are then derived.

Cartier's relation (6.37) is equivalent to

$$E[\varphi(Y(\underline{x})) \mid Y_v] = \varphi_v(Y_v) \tag{6.44}$$

Since the $(Y(\underline{x}), Y_v)$ pair is Gaussian, the conditional distribution of $Y(\underline{x})$ given $Y_v = y_v$ is Gaussian with mean $r\, y_v$ and variance $1 - r^2$. Hence

$$\varphi_v(y_v) = \int \varphi\!\left(r\, y_v + \sqrt{1 - r^2}\, u\right) g(u)\, du \tag{6.45}$$

---

[7] This assumption is usually an approximation: In the case of a Gaussian random function, the bivariate distribution of $Y(\underline{x})$ and Y_v is a mixture of bivariate Gaussian distributions with different correlation coefficients and therefore cannot be exactly Gaussian.

It is interesting to note that (6.45) has the form of a convolution product:

$$\varphi_v(y_v) = \varphi * g_{1-r^2}(ry_v)$$

where g_{1-r^2} is the p.d.f. of a zero-mean normal with variance $1 - r^2$. There are remarkable families of functions φ for which this convolution can be expressed in a simple and exact manner. An example is the exponential transformation (lognormal random functions): If φ is of the form $\varphi(y) = m_Z \exp(\sigma y - \sigma^2/2)$, the application of (6.45) leads to $\varphi_v(y_v) = m_Z \exp(r\sigma y_v - r^2\sigma^2/2)$ —there is permanence of lognormality. In usual cases the actual calculation of φ_v is carried out using the expansions of the transformation functions into normalized Hermite polynomials as

$$\varphi(y) = \sum_{n=0}^{\infty} \varphi_n \chi_n(y), \qquad \varphi_v(y) = \sum_{n=0}^{\infty} \varphi_{vn} \chi_n(y)$$

The coefficients φ_n are assumed known, and the φ_{vn} are to be determined. Applying relation (6.27), Cartier's relation (6.44) is expanded into

$$\sum_{n=0}^{\infty} r^n \varphi_n \chi_n(Y_v) = \sum_{n=0}^{\infty} \varphi_{vn} \chi_n(Y_v)$$

Since the χ_n constitute an orthonormal basis, we have $\varphi_{vn} = r^n \varphi_n$ for all n, so

$$\varphi_v(y) = \sum_{n=0}^{\infty} \varphi_n r^n \chi_n(y) \tag{6.46}$$

There remains to determine r. Two options are available. The first (referred to as DGM1), proposed by Matheron, is based on the fact that the variance defined by the block distribution must be consistent with the block variance σ_v^2 deriving from the covariance $C(h)$ of the SRF $Z(x)$. The former is the sum of the squared coefficients φ_{vn} for $n \geq 1$, while the latter is given by application of (6.33) with the covariance C of Z. Therefore, r is the solution of the equation

$$\sum_{n=1}^{\infty} \varphi_n^2 r^{2n} = \frac{1}{|v|^2} \int_v \int_v C(x' - x) \, dx \, dx' \tag{6.47}$$

The left-hand side is an increasing function of r, going from 0 to σ^2 as r increases from 0 to 1, where $\sigma^2 = C(0)$ is the variance of $Z(x)$; since the variance σ_v^2 is itself comprised between 0 and σ^2, this equation has indeed one and only one solution. The solution r is named the *change-of-support coefficient*. Its value tends to 0 when the support becomes very large (assuming a finite integral range for C) and is equal to 1 when the support reduces to a point. The change of support works all the better that v is small with respect to the range of C or, equivalently, that r is close to 1.

The second option (DGM2), proposed by Emery (2007a), is simpler but requires the additional assumption that the bivariate distribution of $Y(\underline{x})$ and $Y(\underline{x}')$ for two independent random points within the same block v is Gaussian. In that case, r^2 is the block variance of the SRF $Y(\cdot)$:

$$r^2 = \frac{1}{|v|^2} \int_v \int_v \rho(x' - x) \, dx \, dx' \qquad (6.48)$$

where ρ is the covariance of the SRF $Y(\cdot)$. Moreover, Y_v is simply the average $Y(v)$ of $Y(\cdot)$ in the block v, rescaled to a unit variance by the change-of-support coefficient r:

$$Y_v = Y(v)/r \qquad (6.49)$$

Proof. We need to be precise about notations. Concerning the original variable, we may indifferently denote by $Z(v)$ or Z_v the average of $Z(x)$ in v. But the Gaussian random variable Y_v we have associated with that average is not the average of $Y(x)$ in v. We must therefore carefully distinguish the average $Y(v)$ of $Y(x)$ in v, which is a random variable with mean zero and variance smaller than 1, and the standard normal RV Y_v associated with $Z(v)$. We recall that r is the correlation coefficient of the $(Y(\underline{x}), Y_v)$ pair and is also its covariance since Y and Y_v have unit variances.

Let us now consider a second uniform random point $\underline{x}'$ within v, independent of $\underline{x}$, and assume that the bivariate distribution of $Y(\underline{x})$ and $Y(\underline{x}')$ is also Gaussian. The covariances between $Z(\underline{x})$, $Z(\underline{x}')$, and $Z(v)$ can be obtained by applying the following transposition of relation (6.25) to the transforms $\varphi(Y)$ and $\psi(Y')$ of two standard normals Y and Y' whose bivariate distribution is Gaussian with correlation coefficient ρ:

$$\mathrm{Cov}(\varphi(Y), \psi(Y')) = \sum_{n=1}^{\infty} \varphi_n \psi_n \rho^n \qquad (6.50)$$

Using (6.46), this gives

$$\mathrm{Cov}(Z(\underline{x}), Z(v)) = \sum_{n=1}^{\infty} \varphi_n^2 \, r^n (\mathrm{Cov}(Y(\underline{x}), Y_v))^n,$$

$$\mathrm{Cov}(Z(\underline{x}), Z(\underline{x}')) = \sum_{n=1}^{\infty} \varphi_n^2 \, (\mathrm{Cov}(Y(\underline{x}), Y(\underline{x}')))^n$$

which entails that

$$\mathrm{Cov}(Y(\underline{x}), Y(\underline{x}')) = r \, \mathrm{Cov}(Y(\underline{x}), Y_v) \qquad (6.51)$$

Since r is defined as the covariance of $Y(\underline{x})$ and Y_v, we have

$$\mathrm{Cov}(Y(\underline{x}), Y_v) = r$$
$$\mathrm{Cov}(Y(\underline{x}), Y(\underline{x}')) = r^2$$

A random location has the same effect on covariances as an average value. The above relations thus imply

$$\mathrm{Cov}(Y(v), Y_v) = r,$$
$$\mathrm{Var}(Y(v)) = r^2$$

so that the correlation between the regularized variable $Y(v)$ and the Gaussian variable Y_v attached to the block is equal to 1. So

$$Y(v) = rY_v$$

and r^2 is the block variance (6.48) of $Y(v)$. $\qquad\qquad\square$

Discussion

The discrete Gaussian model is by design a consistent change-of-support model. Moreover, it is compatible with the central limit theorem: When r tends to zero, the coefficients $r^n \varphi_n$ of the expansion (6.46) for $n > 1$ become negligible in comparison with $r \varphi_1$, so that $Z(v)$ tends to be Gaussian. Nevertheless, DGM models are only approximations. Indeed, model DGM1 is based on a hypothesis (that the bivariate distribution of $Y(\underline{x})$ and Y_v is Gaussian) which is not fully true from a theoretical standpoint even though it is usually satisfied in practical applications when v is relatively small compared with the variogram range. Model DGM2 requires an additional assumption (that bivariate distributions of random points are Gaussian) so that its validity is more limited.

The validity of the discrete Gaussian model has been checked in the lognormal case by comparing the block transform φ_v predicted by the model with the "true" block transform obtained from point-support simulations of a fine grid discretizing the block: Matheron (1981b) for DGM1 in 1D, Emery (2007a) for DGM2 in 2D, Chilès (2012) for DGM1 and DGM2 in 2D. Both models give good results when the logarithmic variance is small, because then we remain close to the Gaussian case. Figure 6.12 illustrates the results for a fairly large logarithmic standard deviation of 1.5 (the coefficient of variation of $Z(x)$ is 2.91). They are based on 10 to 100 thousand simulations of a Gaussian SRF $Y(x)$ with a spherical variogram, transformed to a lognormal SRF $Z(x)$. When the block size is small with respect to the range—which is the usual situation in practical applications—DGM1 and DGM2 give the same answer, very close to the true one (Figure 6.12a for a block size of $1/10^{\mathrm{th}}$ the range). When the block size is equal to the range, DGM1 continues to perfectly model the true block transform whereas DGM2 deviates for high cutoff grades (Figure 6.12b). Finally, for very large blocks (10 times the range in Figure 6.12c) DGM1 again provides a very good answer except for extreme cutoffs, whereas DGM2 starts diverging as soon as the cutoff on Y moves away from zero. In this example we can conclude that DGM1 has a broader range of validity than DGM2. In both cases the presence of a nugget effect extends the validity range.

From a practical viewpoint, the DGM model does not explicitly require that the bivariate distributions of fixed-location point-support variables $Y(x)$ and $Y(x')$ be Gaussian. On the other hand, in model DGM1 the determination of the change-of-support coefficient r requires the knowledge of the covariance of

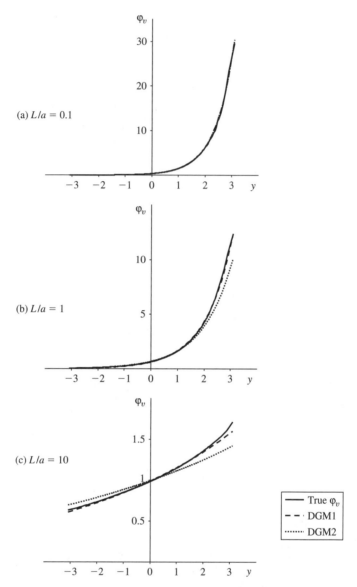

FIGURE 6.12 Comparison of models DGM1 and DGM2 in the lognormal case. Each figure shows the "true" Gaussian transformation φ_v of a block (determined from simulations) and its approximations by models DGM1 and DGM2. The support v is a square of side L. The variogram of the Gaussian SRF is spherical with variance σ^2 and range a. Comparison for $\sigma = 1.5$ (that is, $\sigma^2 = 2.25$) and: (a) $L/a = 0.1$, (b) $L/a = 1$, (c) $L/a = 10$. The "true" φ_v was determined from 10,000 simulations in (a), from 100,000 simulations in (b) and (c).

the SRF Z while model DGM2 requires the knowledge of the covariance of the SRF Y. Good practice is to model both the variogram of Z and the variogram of Y in a consistent manner by assuming that $Y(\cdot)$ has Gaussian bivariate distributions, so that the two covariances are related by $C(h) = \sum_{n=1}^{\infty} \varphi_n^2 (\rho(h))^n$. For both models, if the data originate from regularly spaced drill holes, it is recommended, as suggested by Maréchal (1976), to check the validity of the DGM model by applying it to the prediction of the distribution of the average grade of a group of four neighboring drill holes, which can be compared to the actual one.

6.8.2   Local Discrete Gaussian Model

A model in which all bivariate distributions are Gaussian can be obtained by generalizing the global discrete Gaussian model. To this end we must accept to discretize the deposit into a partition of small blocks of size v and forget the exact position of each sample in its block.

Domain Discretization

The deposit is regarded as the union of nonoverlapping blocks v_p which are identical up to a translation. To simplify, we denote by Z_{v_p} the mean grade of block v_p, and v the support of these blocks. The random variables Z_{v_p} thus constitute a discretized version of the deposit.

Concerning point-support grades, we will not attempt to consider their exact position inside the block, and will denote by $Z_p = Z(\underline{x}_p)$ the grade of a point $\underline{x}_p$ selected randomly in the block v_p (uniform density) and independently of any other point. Experimental data do not necessarily exist for all indexes p but for a subset $p_1, \ldots, p_N$.

Z_p and Z_{v_p} are transforms of standard normals Y_p and Y_{v_p} through the transformation functions φ and φ_v defined for the global model:

$$Z_p = \varphi(Y_p), \qquad Z_{v_p} = \varphi_v(Y_{v_p})$$

Specification of all Bivariate Distributions

The pairs (Y_p, Y_{v_p}) are assumed to follow the Gaussian bivariate distribution characterized by the correlation coefficient r determined for the global model. For the isofactorial model to be completely specified, two additional hypotheses are required:

- Given Y_{v_p}, Y_p is independent of the other Y_q, $q \neq p$, as well as the other Y_{v_q}, $q \neq p$.
- The pairs (Y_{v_p}, Y_{v_q}) are bivariate Gaussian.

All pairs of variables chosen among the random variables Y_p and Y_{v_p} then have Gaussian bivariate distributions. These are characterized by the corresponding correlation coefficients. If R_{pq} denotes the correlation coefficient between Y_{v_p} and Y_{v_q}, the discrete Gaussian model is defined as follows:

- Block – block distribution : $\text{Cov}(Y_{v_p}, Y_{v_q}) = R_{pq},$ (6.52)
- Point – block distribution : $\text{Cov}(Y_p, Y_{v_q}) = r\, R_{pq},$ (6.53)
- Point – point distribution : $\text{Cov}(Y_p, Y_q) = r^2\, R_{pq}, \quad p \neq q$ (6.54)

Relation (6.54) is also valid for $p = q$ if we consider two different random points in v_p. Otherwise the variance of Y_p is of course equal to 1. The proof of these relations is obtained by expressing the covariances of the corresponding Z variables in the form (6.50) and following lines similar to the proof establishing (6.51).

In this model, things are as though the main role was not held by the SRF $Z(x)$ but by a discretized version composed of the grades of the blocks v_p, or their Gaussian transforms Y_{v_p}.

This is at the origin of the name of the model, although it can now be used without discretization. The samples, to be regarded as randomly located in each block, have grades whose Gaussian transforms are of the form

$$Y_p = r Y_{v_p} + \sqrt{1 - r^2}\, \varepsilon_p \qquad (6.55)$$

where the ε_p are standard normal variables uncorrelated among themselves and with the Y_{v_q} (whether q is equal to p or not).

There remains to determine the correlation coefficients R_{pq}. In the approach of model DGM1, this is achieved by inverting numerically the relation

$$\text{Cov}(Z_{v_p}, Z_{v_q}) = \sum_{n=1}^{\infty} \varphi_n^2\, r^{2n}\, R_{pq}^n$$

since the left-hand side can be computed from the covariance $C(h)$ of the SRF $Z(x)$:

$$\text{Cov}(Z_{v_p}, Z_{v_q}) = \frac{1}{|v|^2} \int_{v_p} \int_{v_q} C(x' - x)\, dx\, dx'$$

There is no guarantee that the matrix $[R_{pq}]$ thus obtained is positive definite. If this is not the case, a covariance model is fitted to the set of these numerical covariances.

In the approach of model DGM2 the block variable Y_{v_p} is equal to the average value of Y in v_p, scaled by the change-of-support coefficient r defined by (6.48): $Y_{v_p} = Y(v_p)/r$. Thus the covariances R_{pq} derive directly from the covariance $\rho(h)$ of the Gaussian SRF $Y(x)$:

$$R_{pq} = \text{Cov}(Y_{v_p}, Y_{v_q}) = \frac{1}{r^2} \text{Cov}(Y(v_p), Y(v_q)) = \frac{1}{r^2 |v|^2} \int_{v_p} \int_{v_q} \rho(x' - x)\, dx\, dx'$$

The discrete Gaussian model is then consistent with respect to variances and covariances and provides all elements required by disjunctive kriging, which can be carried out by duplicating the method described in Section 6.4.1. In particular, this will provide the local recovery functions $T(z)$ and $Q(z)$, namely for a selection based on the true grades of the blocks. Matheron (1976b) presents an extension of the model enabling the incorporation of the information effect, even if the ultimate information is not yet available. This can be done more easily in the framework of the fully Gaussian model presented hereafter.

Conditional Distribution of a Block

Instead of using DK, the conditional distribution can be determined directly if we accept the framework of a full Gaussian model: Assuming that the vector of the Y_{v_p} is multivariate Gaussian, so is also the vector of the Y_{v_p} and Y_p. The conditional distribution of Y_{v_0} for a target block v_0 is then a Gaussian whose conditional expectation and variance coincide with the simple kriging estimate and variance of Y_{v_0} from the data Y_p of the informed blocks $p = p_1, \ldots, p_N$.

The assumption of multivariate rather than bivariate Gaussianity is of course more restrictive but has the advantage of providing a true conditional distribution rather than a pseudoconditional distribution as obtained by DK. The calculation can be done with model DGM1 or DGM2.

6.8.3 Local Model in a Multi-Gaussian Framework

A last step is to assume that the joint distribution of Y_v and the transformed data $Y(x_1), \ldots, Y(x_N)$ is multivariate Gaussian. It is not clear that this is far more restrictive than the assumption of a multivariate Gaussian distribution for the vector of the Y_{v_p} and Y_p, and it has the distinctive advantage that there is no more need to discretize the study domain and delocalize the data to get the covariances between the Gaussian variables. Since the covariances between the Gaussian data are given by the correlogram $\rho(h)$ of Y and the variance of Y_v is by definition equal to 1, we only need to specify the covariances $r_{\alpha v}$ between the $Y(x_\alpha)$ and Y_v.

In the framework of model DGM1, these covariances are derived by equating two formulas for the covariance of $Z(x_\alpha)$ and $Z(v)$: (i) the formula based on the bivariate distribution of $Y(x_\alpha)$ and Y_v and (ii) the average covariance of $Z(x_\alpha)$ and $Z(x)$ when x describes the block v, based on the bivariate distribution of $Y(x_\alpha)$ and $Y(x)$. Since the coefficients of the expansions of $\varphi(y)$ and $\varphi_v(y)$ into normalized Hermite polynomials are φ_n and $r^n \varphi_n$, respectively, equating these two formulas leads to the equation

$$\sum_{n=1}^{\infty} \varphi_n^2 \, r^n \, r_{\alpha v}^n = \frac{1}{|v|} \int_v \sum_{n=1}^{\infty} \varphi_n^2 \, [\rho(x - x_\alpha)]^n dx$$

where $r \in [0, 1]$ is the solution of equation (6.47). For fixed r the lefthand side is an increasing function of $r_{\alpha v} \in [0, 1]$; therefore the equation has a unique solution $r_{\alpha v}$ when the covariance $\rho(h)$ is nonnegative.

Working in the framework of model DGM2 is even simpler: Since $Y_v = Y(v)/r$, we have $r_{av} = \rho_{av}/r$, where ρ_{av} is the covariance of $Y(x_\alpha)$ and $Y(v)$.

The DGM1 approach corresponds exactly to the multi-Gaussian model developed in Section 6.6.2 for lognormal SRFs (the difference is that in the lognormal case we have an analytical correspondence between the covariances of the RFs Y and Z). This model has not been developed until now but is worth considering because the domain of validity of model DGM1 seems to be wider than that of model DGM2. For the rest of the presentation, we will adopt the latter since it is simpler and has already been used, but it should be remembered that similar results can be obtained with model DGM1, or even, as shown by Matheron (1976b) and Lantuéjoul (1990), by remaining in the truly discrete model seen at the end of Section 6.8.2 (i.e., multi-Gaussian after randomization of the location of the data).

The conditional distribution of Y_v is then a Gaussian whose conditional expectation and variance coincide with the simple kriging estimate and variance of Y_v. The simple kriging estimator of Y_v is

$$Y_v^* = \sum_{\alpha=1}^{N} \lambda_\alpha Y_\alpha$$

with λ_α solutions of the system

$$\sum_{\beta=1}^{N} \lambda_\beta \rho_{\alpha\beta} = \frac{\rho_{\alpha v}}{r}, \qquad \alpha = 1, \ldots, N$$

with

$$\rho_{\alpha\beta} = \rho(x_\beta - x_\alpha) \qquad \text{and} \qquad \rho_{\alpha v} = \frac{1}{|v|} \int_v \rho(x - x_\alpha) \, dx$$

and the kriging variance is

$$\sigma_{\mathrm{SK}v}^2 = 1 - \frac{1}{r} \sum_{\alpha=1}^{N} \lambda_\alpha \rho_{\alpha v}$$

The conditional distribution of Y_v is Gaussian and given by

$$\Pr\{Y_v < y \mid Y_\alpha = y_\alpha : \alpha = 1, \ldots, N\} = G\left(\frac{y - \sum_{\alpha=1}^{N} \lambda_\alpha y_\alpha}{\sigma_{\mathrm{SK}v}}\right)$$

This enables the estimation of local recovery functions for a future selection based on the true grade of the block. At the other extreme we may be interested in the recovery functions associated with a selection on the basis of the estimate Y_v^* (i.e., with the information currently available). They can be easily obtained as soon as the bivariate distribution of Y_v and Y_v^* is known. Since the estimator Y_v^* and the kriging error $Y_v^* - Y_v$ are orthogonal and the target Y_v has unit variance, the variance of the kriging estimator is

$$\sigma^{*^2} = \text{Var}[Y_v^*] = 1 - \sigma_{SKv}^2$$

and the covariance of Y_v and Y_v^* is also equal to σ^{*^2}. The pair $(Y_v, Y_v^*/\sigma^*)$ is therefore distributed as a standard bivariate Gaussian with correlation coefficient σ^*, and its isofactorial representation is given by (6.26) with $\rho = \sigma^*$. In practical applications, we are interested in the intermediate case of a selection based on a future estimate; this is presented below.

Local Change-of-Support Correction

Intuitively the conditional distribution of $Z(v)$ would be obtained by application of a local change-of-support to the conditional point-support distribution of $Z(\underline{x})$, or equivalently of $Y(\underline{x})$ for a random point $\underline{x}$ in the specific block v. The latter is Gaussian with mean the kriging estimate $Y^*(\underline{x})$ of $Y(\underline{x})$ and variance the corresponding kriging variance $\sigma_{SK}^2(\underline{x})$. The kriging estimator of $Y(\underline{x})$ is the average of the kriging estimator of $Y(x)$ when x describes the block v. By linearity of the kriging estimator, it coincides with the kriging estimator of $Y(v)$:

$$Y^*(\underline{x}) = Y^*(v)$$

While the kriging systems for $Y(\underline{x})$ and $Y(v)$ are identical, the kriging variances differ by the term representing the variance of the objective. $\sigma_{SK}^2(\underline{x})$ is therefore related to the kriging variance $\sigma_{SK}^2(v)$ of $Y(v)$ by

$$\sigma_{SK}^2(\underline{x}) = \sigma_{SK}^2(v) + 1 - r^2$$

According to (6.48), in model DGM2 the squared global change-of-support coefficient is equal to the ratio between the a priori variances of $Y(v)$ and $Y(\underline{x})$. The local change-of-support coefficient would therefore be defined by

$$\underline{r_x} = \frac{\sigma_{SK}(v)}{\sigma_{SK}(\underline{x})}$$

Since $Y_v = Y(v)/r$, this can also be expressed as

$$\underline{r_x} = \frac{r\,\sigma_{SKv}}{\sigma_{SK}(\underline{x})}$$

where σ_{SKv}^2 is the kriging variance of Y_v, and $\sigma_{SK}^2(\underline{x})$ can be written

$$\sigma_{SK}^2(\underline{x}) = 1 - r^2(\sigma_{SKv}^2 - 1)$$

This relation implies that $\sigma_{SK}^2(\underline{x})$ is larger than σ_{SKv}^2 and therefore that $r_x \leq r$: The local change-of-support coefficient is thus smaller than the global one. Equivalently, the reduction of variance is more pronounced in the local framework than in the global framework. Emery (2008) shows that this result is general and not specific to the discrete Gaussian model.

One may note that the assumptions that have been introduced along the way are on the whole not very different from those of the multi-Gaussian model of Section 6.6.1. The discrete Gaussian model can thus appear as an approximation to the multi-Gaussian model. However, this latter model involves calculations that are only approximate and much more cumbersome. Moreover, the estimation of recovery functions of a set of small blocks (in mining, the selective mining units) can be obtained in a single run by another discrete Gaussian method, uniform conditioning (see next Section), whereas the conditional expectation requires the combination of block by block calculations.

Locally Varying Mean

In order to apply the methodology of the discrete Gaussian model, we assume that $Z(x)$ is of the form $\varphi(Y(x))$ for some SRF $Y(x)$ with Gaussian bivariate (or even multivariate) distributions; we model φ from all the data, and we model the variogram of Z and/or Y from the original or Gaussian transformed data; finally, from these parameters we deduce the change-of-support coefficient r related to some block support v. But $Z(\cdot)$ and $Y(\cdot)$ may not be as stationary as requested by the model, so that we would like to relax the stationarity constraint. Let us consider the situation where we keep all the parameters as they have been fixed, except that we assume that the local mean of Y in the neighborhood of the block v we consider is not necessarily zero but some unknown value m, while the local variance of Y remains equal to one.

Since $Y_v = Y(v)/r$, Y_v now has mean $m_v = m/r$ instead of zero. Therefore it is easy to replace the simple kriging of Y_v by its ordinary kriging from the Y_α. Due to the relation between the means of Y and Y_v the sum of the weights must be equal to $1/r$. But contrarily to simple kriging, ordinary kriging does not define the conditional distribution of Y_v. However, as shown by Emery (2005a, 2006b), we can build unbiased estimators of local recovery functions. These estimators are identical to those which would be obtained if Y_v followed a normal distribution with mean Y_{OKv}^* and variance $\sigma_{OKv}^2 + 2\mu$, but this distribution is only an approximation to the conditional distribution of Y_v (Y_{OKv}^* denotes the ordinary kriging estimate and σ_{OKv}^2 represents the corresponding kriging variance, while μ is the Lagrange parameter of the kriging system).

Proof. We consider a recovery function expressed as a functional ψ of Y_v. Let us come back to the situation where $Y(\cdot)$ has a zero mean. The conditional expectation $\psi(Y_v)$ is

$$E[\psi(Y_v)|\text{data}] = \int \psi\left(Y_{SKv}^* + \sigma_{SKv} u\right) g(u) \, du$$

or equivalently

$$E[\psi(Y_v)|\text{data}] = E[\psi(Y_{\text{SK}v}^* + \sigma_{\text{SK}v} U) \mid Y_{\text{SK}v}^*] \tag{6.56}$$

where U is a standard normal random variable independent of $Y_{\text{SK}v}^*$. Since the SK estimator and the kriging error are orthogonal and the target Y_v has unit variance, the variance σ_{SK}^{*2} of the kriging estimator $Y_{\text{SK}v}^*$ and the kriging variance $\sigma_{\text{SK}v}^2$ add up to one, so that (6.56) can also be written

$$E[\psi(Y_v)|\text{data}] = E\left[\psi\left(Y_{\text{SK}v}^* + \sqrt{1 - \sigma_{\text{SK}v}^{*2}}\, U\right) \,\middle|\, Y_{\text{SK}v}^*\right]$$

In the case where the mean of $Y(\cdot)$ is unknown, let us consider the estimator obtained by replacing $Y_{\text{SK}v}^*$ by $Y_{\text{OK}v}^*$ and σ_{SK}^* by σ_{OK}^* (we assume that $\sigma_{\text{OK}}^{*2} \leq 1$ even if it cannot be taken for granted):

$$\psi^*(Y_v) = \int \psi\left(Y_{\text{OK}v}^* + \sqrt{1 - \sigma_{\text{OK}}^{*2}}\, u\right) g(u)\, du = E\left[\psi\left(Y_{\text{OK}v}^* + \sqrt{1 - \sigma_{\text{OK}}^{*2}}\, U\right) \,\middle|\, Y_{\text{OK}v}^*\right]$$

Its mean is

$$E[\psi^*(Y_v)] = E\left[\psi\left(Y_{\text{OK}v}^* + \sqrt{1 - \sigma_{\text{OK}}^{*2}}\, U\right)\right]$$

Since U is a standard normal random variable independent of $Y_{\text{OK}v}^*$, $Y_{\text{OK}v}^* + \sqrt{1 - \sigma_{\text{OK}}^{*2}}\, U$ has mean m_v and variance 1, thus the same distribution as Y_v. Therefore

$$E[\psi^*(Y_v)] = E[\psi(Y_v)]$$

and $\psi^*(Y_v)$ is an unbiased estimator of $\psi(Y_v)$.

The variance of the ordinary kriging estimator $Y_{\text{OK}v}^*$ is

$$\sigma_{\text{OK}}^{*2} = \text{Var}[Y_{\text{OK}v}^*] = 1 - \sigma_{\text{OK}v}^2 - 2\mu$$

where μ is the Lagrange parameter of the kriging system, so that the unbiased estimator can also be expressed in the form

$$\psi^*(Y_v) = \int \psi\left(Y_{\text{OK}v}^* + \sqrt{\sigma_{\text{OK}v}^2 + 2\mu}\, u\right) g(u)\, du$$

This result is identical to that which would be obtained if Y_v followed a normal distribution with mean $Y_{\text{OK}v}^*$ and variance $\sigma_{\text{OK}v}^2 + 2\mu$. Note that this result assumes that $\sigma_{\text{OK}v}^2 + 2\mu$ is positive, which cannot be taken for granted [see (3.60)]. $\qquad\square$

Selection on the Basis of Future Information

We have seen the case where the selection is based on the data currently available (exploration drill holes). When designing an exploitation project, the information which will be the basis for the final selection (blast hole data) is usually not yet available but we want to anticipate the evaluation of the recovery functions in order to choose technical parameters such as the size v of the selection units, the spacing of the blast holes, the cutoff grade, and so on. This is known as the problem of the indirect recovery functions.

We assume that the decision to send a block to the mill or to the waste will be taken on the basis on an estimator of the form

$$Z(w) = \sum_{i=1}^{M} w_i \, Z(x_i)$$

where the weights w_i are all positive and add up to 1. It is recommended to use ordinary kriging weights (provided that they are positive), but this is not mandatory. The M data points, denoted x_i, are those that will be available in the block and its neighborhood when the decision will be taken, and they must be distinguished from the data points x_α currently available.

The notation emphasizes the fact that the estimate is a weighted average of $Z(\cdot)$. When going from the point support of the individual data to the support of w (the set of the points x_i), we carry out a change of support. The results obtained for model DGM2 can be directly transposed to this change of support. To this aim, let us denote $Z(\bar{x})$ the value of $Z(\cdot)$ at a point $\bar{x}$ chosen randomly among the x_i according to the weights w_i. $Z(\bar{x})$ can be considered as deriving from $Y(\bar{x})$ through the transformation function φ, and $Z(w)$ can be regarded as deriving from a Gaussian random variable Y_w through some transformation function φ_w to be determined.

In order to apply model DGM2, let us assume that the distribution of the $(Y(\bar{x}), Y_w)$ pair is Gaussian with a positive correlation coefficient s and that the bivariate distribution of $Z(\cdot)$ at two independent random points among the data points x_i is Gaussian. We obtain that

$$Y_w = \frac{Y(w)}{s} \qquad \text{with} \qquad s^2 = \operatorname{Var} Y(\bar{x}) = \sum_{i=1}^{M} \sum_{j=1}^{M} w_i \, w_j \, \rho(x_j - x_i)$$

and that the transformation function φ_w is

$$\varphi_w(y) = \sum_{n=0}^{\infty} \varphi_n \, s^n \, \chi_n(y)$$

The pair (Y_v, Y_w) is therefore Gaussian, with a correlation coefficient ρ_{vw} equal to that of $Y(v)$ and $Y(w)$:

$$\rho_{vw} = \frac{\operatorname{Cov}(Y(v), Y(w))}{rs} = \frac{1}{rs} \sum_{i=1}^{M} w_i \, \frac{1}{v} \int_v \rho(x - x_i) \, dx$$

In order to determine the indirect recoverable resources, we need to know the bivariate distribution of the pair (Y_v, Y_w) conditional on the current data Y_α. Assuming that the multivariate distribution of Y_v, Y_w, and the current data Y_α is Gaussian, the conditional distribution of the pair (Y_v, Y_w) is Gaussian with mean (Y_v^*, Y_w^*), variance $(\sigma_{\mathrm{SK}v}^2, \sigma_{\mathrm{SK}w}^2)$, and covariance $\sigma_{\mathrm{SK}vw}$, where Y_v^* and Y_w^* are the simple kriging estimates of Y_v and Y_w from the Y_α, $\sigma_{\mathrm{SK}v}^2$ and $\sigma_{\mathrm{SK}w}^2$ the corresponding kriging variances, and $\sigma_{SK\,vw}$ the covariance of both kriging errors (the correlation of the conditional variables is therefore $\rho_{SK\,vw} = \sigma_{SK\,vw}/(\sigma_{SK\,v}\,\sigma_{SK\,w})$. Note that the covariance of the kriging errors is

$$\sigma_{SK\,vw} = E\left(Y_v^* - Y_v\right)\left(Y_w^* - Y_w\right)$$
$$= \mathrm{Cov}\left(Y_v^*, Y_w^*\right) - \mathrm{Cov}\left(Y_v^*, Y_w\right) - \mathrm{Cov}\left(Y_v, Y_w^*\right) + \mathrm{Cov}\left(Y_v, Y_w\right)$$

The first term is a double sum of covariances $\rho(x_\beta - x_\alpha)$. Due to the kriging equations, it is equal to the second as well as the third terms, which are simple sums and therefore it need not be calculated.

It is straightforward to deduce the indirect recoverable resources from that conditional bivariate distribution. By reasoning like for a selection based on the data currently available, these results can be extended to the case where the random function Y is considered as having a nonzero mean m in the vicinity of block v [see Emery (2006b), for a presentation of the method and an application to a copper deposit].

Estimation of Local Multivariate Recovery

In polymetallic deposits the selection of mined blocks is performed on the basis of the grade of the main valuable constituent, but one still wishes to know (a) the quantities of metal that will be recovered for the other constituents that can be valorized and (b) the average grades of the substances that have an impact on the ore processing or the value of the final product (presence of impurities). All methods based on a bivariate or multivariate Gaussian assumption can be extended to the case of several variables. This requires of course that the Gaussian-transformed variables associated with the various substances have bivariate (or multivariate) Gaussian distributions. With suitable assumptions the exercise presents no major difficulty, as shown by Maréchal (1982) for disjunctive kriging.

Discussion

We developed the local discrete Gaussian model as an approximation to conditional expectation. Which conditions are required for this approximation to be accurate? A prerequisite is that the global model correctly reproduces the true marginal block distribution; that question was discussed at the end of Section 6.8.1. Concerning local estimation specifically, the only piece of work, to our knowledge, is an experiment carried out by Cressie (2006) to study the estimation of the block value $Z(v)$ of a lognormal SRF. Cressie considers unbiased lognormal estimators that are exponentials of the simple or ordinary kriging estimators of $Y(v)$, which amounts to the assumptions of model DGM2. The experiment compares this ordinary lognormal kriging with conditional expectation obtained by the method of Section 6.6.1. The experiment is carried out with blocks composed of a finite number of points, so that the approximation due to the discretization of a true block in formula (6.40) is avoided.

The results indicate that ordinary lognormal kriging performs well in situations where the block size is small with respect to the range, the lognormal standard deviation is not too large, and the neighborhood is sparse. The first two conditions are required for the global model to be efficient. The third one expresses that conditional expectation makes better use of numerous data than an estimator whose form is limited to the exponential of a linear combination of the logarithms of the data.

To conclude, when Z can be regarded as the transform of a Gaussian SRF Y, it is worth applying the multi-Gaussian approach of Section 6.6.1 when the block and its neighborhood contain many data points. In usual applications however, such as the estimation of selection mining units or of small panels subject to a remediation process, we are not in this situation. Except in extreme cases (very skewed distribution), we can work in the framework of model DGM1 and even, when the distribution is moderately skewed and the range much larger than the block size, in the framework of the simpler model DGM2. These discrete Gaussian models provide a huge computational advantage over the multi-Gaussian approach if we are interested in millions of blocks, as is the case in many real-world applications.

6.8.4 Uniform Conditioning

In general, we are not interested in the distribution of the mean grade of a single block v but in the distribution of small blocks of size v making up a panel V. If we carry out the estimation by disjunctive kriging, the sum of the factors of the blocks can be kriged directly. If we prefer to use the conditional expectation, the result is obtained as a mixture of elementary distributions. If V is very large compared with v, the calculation can become prohibitive, so approximations are used, the most classic of which is uniform conditioning (Matheron, 1974c; Rivoirard, 1994, Chapter 11). The method consists in conditioning the grades of small blocks by the mean grade of the panel rather than by the samples in the blocks (working on the Gaussian transforms).

To be more specific, let us assume that the discrete Gaussian model DGM1 remains valid up to the size V of the panel and that all bivariate distributions are Gaussian. The mean grade $Z(V)$ of the panel can thus be considered as the transform of a Gaussian random variable Y_V through a function φ_V of the form

$$\varphi_V(y) = \sum_{n=0}^{\infty} \varphi_n R^n \chi_n(y) \qquad (6.57)$$

where the positive correlation coefficient R between Y at a random point in V and Y_V is given by the relation

$$R^2 = \frac{1}{|V|^2} \int_V \int_V \rho(x' - x) \, dx \, dx'$$

(Note that R is smaller than the block change-of-support coefficient r.)

The panel is made of identical blocks v_p, whose average grades are represented by random variables $Y_{v_p} = Y(v_p)/r$. Let us denote by $\underline{v}$ a random block in the panel. Its grade is represented by a standard normal random variable $Y_{\underline{v}}$ through the same transform φ_v as the fixed blocks v_p. $Z(\underline{v})$ and $Z(V)$ follow the Cartier's relation $E[Z(\underline{v})|Z(V)] = Z(V)$, which can be rewritten

$$E\left[\varphi_v\left(Y_{\underline{v}} \mid Y_V\right)\right] = \varphi_V(Y_V)$$

In the Hermite polynomial expansions of φ_v and φ_V, the coefficients of the nth-order term are thus identical up to a factor equal to the correlation coefficient of Y_v and Y_V raised to the power n. Because these expansions are given by (6.46) and (6.57), this correlation coefficient, which is the change-of-support coefficient from the block support to the panel support, is equal to R/r.

Conditioning on a Panel Estimate

In the uniform-conditioning approximation, we condition by $Y_{\underline{v}}{}^*$, or equivalently by Y_V^*, obtained by simple kriging, rather than by the $Y_{v_p}^*$ attached to each block v_p. Let us work within the multi-Gaussian assumption, where $Y_v = Y(\underline{v})/r$ and $Y_V = Y(V)/R$. Since $\underline{v}$ is a random block in the panel V, the simple kriging estimator of $Y(\underline{v})$ coincides with that of $Y(V)$, but the kriging variance is different: The variance of the target, which appears in the variance formula (the variance σ_{00} with the notations of Section 3.5.2), corresponds to the support of the block v rather than the panel V—that is, to r^2 instead of R^2. As a consequence, with a reasoning similar to what has been derived for a fixed block v, the simple kriging estimator of $Y_{\underline{v}}$ is

$$Y_{\underline{v}}^* = \sum_{\alpha=1}^{N} \lambda_\alpha Y_\alpha$$

with λ_α solutions of the system

$$\sum_{\beta=1}^{N} \lambda_\beta\, \rho_{\alpha\beta} = \frac{\rho_{\alpha V}}{r}, \qquad \alpha = 1,\dots,N$$

with

$$\rho_{\alpha\beta} = \rho\left(x_\beta - x_\alpha\right) \quad \text{and} \quad \rho_{\alpha V} = \frac{1}{|V|}\int_V \rho(x - x_\alpha)\,dx$$

and the kriging variance is

$$\sigma_{\mathrm{SK}\underline{v}}^2 = 1 - \frac{1}{r}\sum_{\alpha=1}^{N} \lambda_\alpha\, \rho_{\alpha V}$$

The conditional distribution of $Y_{\underline{v}}$ is Gaussian and given by

$$\Pr\{Y_{\underline{v}} < y \,|\, Y_{\underline{v}}^* = y_{\underline{v}}^*\} = G\left(\frac{y - y_{\underline{v}}^*}{\sigma_{SK\underline{v}}}\right)$$

The sole change with respect to the case of a fixed block v is thus the replacement of $\rho_{\alpha v}$ by $\rho_{\alpha V}$. The generalization to a lack of stationarity and to indirect recovery functions can be done easily by following the same lines as in the case of a fixed block v. The method is of course a crude approximation with respect to a mixture of block-by-block results but seems to produce sound and robust results (Guibal and Remacre, 1984).

Remacre (1984, 1987), Rivoirard (1994, Chapter 12), and Deraisme et al. (2008) present another uniform-conditioning approach, based on a conditioning on an estimate $Z^*(V)$ of $Z(V)$. That approach is taken in practical applications to get recoverable resources estimates at zero cutoff consistent with a previous ordinary kriging of the panel grades. It requires that $Z^*(V)$ is conditionally unbiased.

Since it is based on a multivariate Gaussian framework, uniform conditioning can also be extended to the recovery of several constituents, as shown by Deraisme et al. (2008). The application presented by these authors compares the multivariate Gaussian approach (Section 6.6.1) and the multivariate uniform conditioning; despite a notable dispersion in the results for the individual panels, the global results are largely similar.

6.9 NON-GAUSSIAN ISOFACTORIAL CHANGE-OF-SUPPORT MODELS

The discrete Gaussian model is very convenient but is not suited for all situations. Being developed for diffusive random functions, it is not adapted, for example, to mosaic random functions and to variables displaying a large proportion of zeroes. Even if the simplifications brought by multivariate assumptions will no longer be possible, there is the need for other isofactorial change-of-support models because they provide consistent solutions at the local and global scales. Local models provide all covariances needed for disjunctive kriging of any function of Y_{v_0} (and thus of Z_{v_0}) from data Y_α, by duplicating the method described in Section 6.4.1.

6.9.1 General Form of Isofactorial Change-of-Support Models

Let us first consider that the variable of interest is directly the SRF $Y(x)$ represented by an isofactorial model. The case where $Y(x)$ is a Gaussian SRF is a very special one, because then $Y(v)$ is also Gaussian. Such distributional permanence does not generally hold for other types of random functions. For example, if Y_1 and Y_2 are two i.i.d. gamma random variables with parameter α, their average follows a gamma distribution with parameter 2α and is thus more regular. Therefore in the general case we need point–point, point–block, and block–block isofactorial models for pairs of random variables whose marginal distributions may be of a different form. Considering also that we work on transformed data, the typical model is the following:

- Point-support grades are of the form $Z(x) = \varphi(Y(x))$, where $Y(x)$ is an SRF with marginal distribution $G(dy)$.
- Average grades over a support v are of the form $Z(v) = \varphi_v(Y_v)$, where Y_v has marginal $G_v(dy)$.
- The marginal distribution G has orthonormal factors $\chi_n(y)$, and $Z(x)$ can be expanded as

$$Z(x) = \sum_{n=0}^{\infty} \varphi_n \chi_n(Y(x))$$

- The marginal distribution G_v has orthonormal factors $\chi_n^v(y)$, and $Z(v)$ can be expanded as

$$Z(v) = \sum_{n=0}^{\infty} \varphi_{vn} \chi_n^v(Y_v)$$

- The bivariate point–point distribution of $Y(x)$ and $Y(x')$ is of the form

$$G_{xx'}(dy, dy') = \sum_{n=0}^{\infty} T_n(x, x') \chi_n(y) \chi_n(y') G(dy) G(dy')$$

- The bivariate point–block distribution of $Y(x)$ and Y_v is of the form

$$G_{xv'}(dy, dy') = \sum_{n=0}^{\infty} T_n(x, v) \chi_n(y) \chi_n^v(y') G(dy) G_v(dy')$$

- The bivariate block–block distribution of Y_v and $Y_{v'}$, where v and v' have same shape and dimensions, is of the form

$$G_{vv'}(dy, dy') = \sum_{n=0}^{\infty} T_n(v, v') \chi_n^v(y) \chi_n^v(y') G_v(dy) G_v(dy')$$

The transformation coefficients φ_n and φ_{vn} and the covariances T_n must verify consistency relationships. The first ones originate in that randomizing x in v has the same effect on covariances as averaging in v. Therefore:

$$\mathrm{Cov}(Z(\underline{x}), Z(\underline{x}')) = \mathrm{Cov}(Z(\underline{x}), Z(v')) = \mathrm{Cov}(Z(v), Z(v'))$$

This is true even if v and v' coincide, provided that the uniform random points $\underline{x}$ and $\underline{x}'$ are independent. Expanding $Z(\underline{x})$ and $Z(v)$ as functions of $Y(\underline{x})$ and Y_v through the transformation functions φ and φ_v, one shows by identification of the terms of order n that

$$\varphi_{vn} = \varphi_n T_n(\underline{x}, v), \tag{6.58}$$

$$\begin{aligned} T_n(\underline{x}, v') &= T_n(\underline{x}, v) T_n(v, v'), \\ T_n(\underline{x}, \underline{x}') &= T_n(\underline{x}, v)^2 T_n(v, v') \end{aligned} \tag{6.59}$$

In particular, when v and v' coincide (but the random points are chosen independently) we obtain

$$T_n(\underline{x}, \underline{x}') = T_n(\underline{x}, v)^2$$

In that model, the T_n of the point–point and point–block distributions derive from those of the block–block distributions.

Proof. The expansions of the above covariances are

$$\mathrm{Cov}(Z(\underline{x}), Z(\underline{x}')) = \mathrm{Cov}(\varphi(Y(\underline{x})), \varphi(Y(\underline{x}'))) = \sum_{n=1}^{\infty} \varphi_n^2 \, T_n(\underline{x}, \underline{x}'),$$

$$\mathrm{Cov}(Z(\underline{x}), Z(v')) = \mathrm{Cov}(\varphi(Y(\underline{x})), \varphi^v(Y_{v'})) = \sum_{n=1}^{\infty} \varphi_n \, \varphi_n^v \, T_n(\underline{x}, v'),$$

$$\mathrm{Cov}(Z(v), Z(v')) = \mathrm{Cov}(\varphi^v(Y_v), \varphi^v(Y_{v'})) = \sum_{n=1}^{\infty} (\varphi_n^v)^2 \, T_n(v, v')$$

If v and v' are identical, we obtain (6.58). Inserting this relation in the above relations, we obtain (6.59). $\qquad\qquad\qquad\qquad\qquad\qquad\qquad\qquad\qquad\qquad\qquad\qquad\qquad\square$

Another consistency relationship concerns the variance of $Z(v)$: It can be calculated either with (6.23) applied with $\psi_n = \varphi_n^v$ or with (6.33) applied to the covariance of Z, which results itself from the coefficients of φ_n and the $T_n(x, x')$ through relation (6.24) with $\psi_n = \varphi_n$. The two expressions must give identical results.

In the special case of isofactorial models whose first factor is equal to 1 and whose second factor is a linear function of y, Emery (2007a) shows that the random function Y_v is linearly related to the average $Y(v)$. However, this does not lead to the simplifications of the discrete Gaussian model (in its multivariate Gaussian framework) because the distribution of $Y(v)$ conditional on the data $Y(x_\alpha)$ cannot be deduced from its simple kriging.

6.9.2 Some Specific Isofactorial Change-of-Support Models

Consistent models exist, and even a large number of them, nearly all developed by Matheron. Their presentation is rather technical and would take too long, so we just refer the reader to the literature.

In the context of Gaussian marginals, Emery and Ortiz (2005a) propose an extension of the discrete Gaussian model to Hermitian bivariate distributions, provided that they are not of mosaic type.

The discrete approach can also be extended to marginal distributions other than the Gaussian. In the case of a gamma marginal, a gamma distribution with parameter α is chosen for Y, and a less selective gamma distribution is chosen for Y_v, namely with parameter $\alpha_v \geq \alpha$. Hu (1988) and Emery (2007a) develop the model for Laguerre bivariate distributions.

Matheron (1984d) proposes an elegant construction of a mosaic random function model which has simple isofactorial bivariate distributions (including for variables with different supports). Disjunctive kriging can be carried out very easily in this model, because it only requires the resolution of a single system. The solution is more elaborate than an indicator kriging.

For the model with orthogonal indicator residuals, which is very useful in the case of a nondiffusive variable, Matheron (1989) proposed a simple change-of-support procedure which is also presented by Rivoirard (1988, 1989). Suppose that the random function $Y(x)$ is obtained by superimposition of random sets, as explained in Section 6.4.3, and that these random sets are Boolean random sets. If we randomly select a proportion $r < 1$ of the primary points, we obtain another random function $Y_r(x)$ with orthogonal residuals. It can be shown that the bivariate distribution of $(Y(x), Y_r(x))$ is isofactorial. This is exploited to define a change-of-support model where $Y_r(x)$ and $Y(x)$ represent, up to a transformation, the grade of a block and the grade of a random point within the block, respectively.

Matheron (1984e, 1989) provides solutions to an important and difficult problem, the general theory of isofactorial change-of-support models for discrete variables. The special case of a negative binomial marginal distribution was already addressed by Matheron (1980). These models associate a discrete point-support variable with a block-support variable having the same distribution (but different parameters). It is also interesting to be able to represent the block characteristics by continuous variables. Mixed change-of-support models have been developed to this effect, namely the negative binomial/gamma model (Matheron, 1984a).

6.10 APPLICATIONS AND DISCUSSION

The first implementations of disjunctive kriging with change of support were made in the scope of the discrete Gaussian model. That model has gained in simplicity with the new insight of Emery (2007a). In mining, predictions obtained by disjunctive kriging from exploration boreholes have been compared with reality (the actual grade of mined panels being seldom known, each panel was discretized by the blast holes it contains; Jackson et al. 1979; Young, 1982). A number of authors have compared the performance of different estimation methods of recoverable resources on real or synthetic data sets. Liao (1990) and Chilès and Liao (1993) compared indicator cokriging (with affine correction for change of support) and disjunctive kriging with Gaussian or gamma transformation on data from three real deposits (including the Salsigne gold mine). These studies arrive at the following conclusions:

- Concerning global recoverable resources, the discrete model (Gaussian or gamma) generally produces acceptable results, as opposed to the affine correction which is to be banned except if the nugget effect is small and if the change of support remains moderate.
- Local recoverable resources can be predicted fairly well by disjunctive kriging, if applied carefully (e.g., straight elimination of isolated blocks and border areas if they are barren).
- Indicator kriging with affine correction can produce results close to those of disjunctive kriging in some limited cases (mosaic model, moderate support effect, and limited σ/m variability). It can even outperform disjunctive kriging if the model used for DK is selected arbitrarily and is ill-suited. It is not as good as disjunctive kriging with an appropriate isofactorial model. It often comes with a strong bias, especially regarding the estimation of the quantity of metal.
- Despite appearances the implementation of indicator kriging is no simpler than that of disjunctive kriging, except if the same variogram is used for all thresholds.

As emphasized earlier by Matheron (1985) and Lantuéjoul (1988), these works demonstrate the importance of a correct choice of the model. Indeed the advantages of nonlinear methods vanish if applied blindly. The value of modeling bivariate distributions lies in the availability of relatively varied models, of diagnostic tools for choosing among these models, and of methods for fitting their parameters and checking whether these models provide a reasonable representation of the data. The DK approach provides consistent change-of-support models. Of course it is ideal if both point-support data (samples) and block data (grades of mined blocks) are available to validate the model—and they can also be used for estimation.

6.11 CHANGE OF SUPPORT BY THE MAXIMUM (C. LANTUÉJOUL)

There are situations where the relevant quantity to estimate is not an average but a minimum or a maximum over a domain. Typical examples include the chance that a pollutant does not exceed some critical concentration at any point of a region, or that all parts of a mechanical structure can resist at a critical wind speed. This indicates that the work performed on averages should also be performed on extrema. This is a difficult task. If the statistics of extremes now benefits from a large body of theory, its extension to a spatial context is still in infancy. It is therefore not possible at this point to give a set of methods and practical tools meeting this objective. We simply give an account of the status of research in that field.

This section starts with a comparison of the extrema distribution associated with different supports. The case of large supports is then investigated. Like second-order stationary random functions tend to become multi-Gaussian when averaged over larger and larger supports and normalized, many stationary random functions tend to become max-stable when maximized over larger and larger supports and suitably normalized. From the general form of their spatial distribution a number of independence properties can be derived. Several constructions of max-stable random functions are also proposed.

For simplicity, technical issues are left aside. For more information, see Beirlant et al. (2004), de Haan and Ferreira (2006), and Resnick (2007).

6.11.1 Comparison of Extrema Distributions at Different Supports

Let Z be a stationary random function. For each bounded domain v, the infimum, the average, and the supremum of Z over v are considered:

$$\check{Z}(v) = \inf_{x \in v} Z(x), \qquad Z(v) = \frac{1}{v} \int_v Z(x)\, dx, \qquad \hat{Z}(v) = \sup_{x \in v} Z(x) \qquad (6.60)$$

Their distribution functions are respectively denoted by $\check{F}_v$, F_v, and $\hat{F}_v$.

Now, let v and V be two bounded domains. If v divides V, in the sense that V can be split into subdomains congruent to v, then $E[Z(\underline{v})|Z_V] = Z_V$ and the Cartier theorem applies (Section 6.5.2), which implies in turn

$$\int \varphi(z)\, dF_V(z) \le \int \varphi(z)\, dF_v(z) \qquad \forall \varphi \text{ convex} \qquad (6.61)$$

Conversely, if formula (6.61) holds, then there exists a bivariate distribution F with marginals F_v and F_V such that $E[X \mid Y] = Y$ for each pair of variables (X, Y) distributed like F.

Suppose now that V contains v. We clearly have $\hat{Z}(v) \le \hat{Z}(V)$, which implies $\hat{F}_V(z) \le \hat{F}_v(z)$ for each z, and more generally

$$\int \varphi(z)\, d\hat{F}_V(z) \le \int \varphi(z)\, d\hat{F}_v(z) \qquad \forall \varphi \text{ decreasing and bounded} \qquad (6.62)$$

Conversely, it has been established by Strassen (1965) that if formula (6.62) holds, then there exists a bivariate distribution F with marginals $\hat{F}_v$ and $\hat{F}_V$ such that $X \leq Y$ for each pair of variables (X, Y) distributed like F.

Applying these formulas to $-Z$, dual results are obtained for the infimum. In particular, we have

$$\int \varphi(z)\, d\check{F}_V(z) \leq \int \varphi(z)\, d\check{F}_v(z) \qquad \forall \varphi \text{ increasing and bounded}$$

In general, the distribution of the maximum (or the minimum) of a random function over a support is not mathematically tractable. The Boolean random function is a notable exception. As illustrated in Figure 7.30, this is the superior envelope of a family of primary functions f_i located at Poisson points s_i:

$$Z(x) = \sup_i f_i(x - s_i)$$

The supremum of Z over the support v does not exceed a cutoff z if and only if the excursion set $X(z)$ of Z above z—that is, $\{x : Z(x) > z\}$—avoids v. $X(z)$ is a Boolean model, the objects of which are the excursion sets of the primary functions above z and can be seen as independent copies of a random set $A(z)$. Then we have

$$\Pr\{\hat{Z}(v) < z\} = \Pr\{X(z) \cap v = \varnothing\} = \exp(-\lambda\, \mathrm{E}\,|A(z) \oplus \check{v}|)$$

where λ denotes the Poisson intensity of the Boolean random function and $\oplus$ represents dilation.[8]

6.11.2   Large Supports

In the case of SRFs with finite variance, the average of a large number of i.i.d. SRFs is close to Gaussian. Similarly, such an SRF tends to become Gaussian when averaged over larger and larger supports. The latter result is more difficult to prove because of spatial correlations and is subject to ergodicity conditions. The situation is similar when studying the maximum instead of the average. We will therefore examine the case of the maximum of a large number of random functions and admit that the situation is similar for a large support (which is indeed the case).

Let $(Z_n, n \geq 1)$ be a sequence of independent and identically distributed stationary random functions. Put

$$\check{Z}^{(n)} = \min(Z_1, \ldots, Z_n), \qquad Z^{(n)} = \frac{Z_1 + \cdots + Z_n}{n}, \qquad \hat{Z}^{(n)} = \max(Z_1, \ldots, Z_n)$$

If the Z_n's have finite mean m and finite variance σ^2, then the central limit theorem states that the spatial distribution of the normalized random function

$$Y^{(n)} = \frac{Z^{(n)} - m}{\sigma/\sqrt{n}}$$

tends to become standard Gaussian as n becomes very large. This means that all finite linear combinations of variables tend to be normally distributed, that pairwise uncorrelated groups of variables tend to be mutually independent, and so on. Moreover, all the statistical properties of $Y^{(\infty)}$ are specified by the covariance function of the Z_n's.

---

[8] The dilation $A \oplus B$ is the union of all translates of A by a vector of B [see Serra (1982)].

Things are a little more complicated when the maximum is considered instead of the average $Z^{(n)}$ (by duality, the case of the minimum can be treated exactly as the maximum and is not treated here).

Clearly, some normalization is necessary to avoid degeneracy. To this end, let us introduce a sequence (a_n) of real numbers and a sequence (b_n) of positive numbers. If those sequences have been suitably chosen, then the spatial distribution of $(\check{Z}^{(n)} - a_n)/b_n$ tends to the spatial distribution of a so-called *max-stable* random function. This terminology comes from the fact that, when replacing a max-stable random function by the maximum of finitely many independent copies of itself, its spatial distribution remains unchanged up to normalization.

By analogy, if suitable ergodicity conditions are met, the supremum of $Z(x)$ over a domain V tends to a max-stable random variable when the domain V tends to infinity in all directions. Max-stable RVs and RFs are attractors with respect to the maximum, like stable RVs and RFs are attractors with respect to the average (Gaussian RVs and RFs in the case of a finite variance).

6.11.3 Max-Stable Random Functions

We now examine the main properties of stationary max-stable random functions and present a structural tool playing a role similar to the variogram for SRFs. Let Z denote a max-stable SRF. All marginals are equal to a generalized extreme distribution (Fisher and Tippett, 1928; Gnedenko, 1943)

$$\Pr\{Z(x) < z\} = \exp\left(-\left[1 + \xi\left(\frac{z - \mu}{s}\right)^{-1/\xi}\right]_+\right)$$

where a_+ denotes $\max(a, 0)$, ξ and $\mu \in \mathbb{R}$, and $s > 0$. μ is a location parameter whereas s is a scale parameter and ξ is a shape parameter. Such a distribution includes the three standard extreme distributions, namely the Fréchet distribution

$$\Phi_\alpha(z) = \begin{cases} 0 & \text{if } z \leq 0 \\ \exp(-z^{-\alpha}) & \text{if } z \geq 0 \end{cases} \quad (\alpha > 0)$$

the Gumbel distribution

$$\Lambda(z) = \exp(-e^{-z})$$

and the Weibull distribution

$$\Psi_\alpha(z) = \begin{cases} \exp(-|z|^\alpha) & \text{if } z \leq 0 \\ 1 & \text{if } z \geq 0 \end{cases} \quad (\alpha > 0)$$

Note with Embrechts et al. (1997) that the three distributions are closely linked. Indeed, if $X > 0$, then

$$X \text{ has c.d.f. } \Phi_\alpha \Leftrightarrow \log X^\alpha \text{ has c.d.f. } \Lambda \Leftrightarrow -X^{-1} \text{ has c.d.f. } \Psi_\alpha \quad (6.63)$$

By using these formulas, Z can be assumed to have standard Fréchet marginals, that is, $\Pr\{Z(x) < z\} = \exp(1/z)$. This distribution has the inconvenience of having infinite mean and variance, but also has the advantage of providing a simple expression for the multivariate distribution $\Pr\{Z(x_1) < z_1, \ldots, Z(x_k) < z_k\}$, or more compactly $\Pr\left\{\max_{1 \leq i \leq k} \frac{Z(x_i)}{z_i} < 1\right\}$. Indeed, this distribution takes the form

$$\Pr\left\{\max_{1\le i\le k}\frac{Z(x_i)}{z_i}<1\right\}=\exp\left(-\int_{S_k}\max_{1\le i\le k}\frac{\omega_i}{z_i}\,dH(\omega)\right) \tag{6.64}$$

where the spectral measure H is a positive measure on the simplex S_k defined as

$$S_k=\{\omega=(\omega_1,\ldots,\omega_k):\omega_1,\ldots,\omega_k\ge 0;\omega_1+\cdots+\omega_k=1\}$$

Regarding the properties of H, apply (6.64) with z_i finite and the other z_j's infinite. This gives

$$\exp\left(-\frac{1}{z_i}\right)=\Pr\left\{\frac{Z(x_i)}{z_i}<1\right\}=\exp\left(-\int_{S_k}\frac{\omega_i}{z_i}\,dH(\omega)\right)$$

hence

$$\int_{S_k}\omega_i\,dH(\omega)=1 \tag{6.65}$$

Since $\sum\omega_i=1$, the total mass of H is obtained by summing formula (6.65) over all possible i values:

$$\int_{S_k}dH(\omega)=\int_{S_k}\sum_{i=1}^{k}\omega_i\,dH(\omega)=k$$

From formula (6.64) and its consequence (6.65), a number of statistical consequences can also be derived. Since we have

$$\int_{S_k}\max_{1\le i\le k}\frac{\omega_i}{z_i}\,dH(\omega)\le\int_{S_k}\sum_{i=1}^{k}\frac{\omega_i}{z_i}\,dH(\omega)=\sum_{i=1}^{k}\frac{1}{z_i}$$

it follows that

$$\Pr\left\{\max_{1\le i\le k}\frac{Z(x_i)}{z_i}<1\right\}\ge\prod_{i=1}^{k}\Pr\left\{\frac{Z(x_i)}{z_i}<1\right\} \tag{6.66}$$

In other words, the variables of a max-stable random function are *positively dependent*. It can be shown that the equality takes place when the $Z(x_i)$'s are pairwise independent (Resnick, 1987; Beirlant et al., 2004). Consequently, *pairwise independent* variables of a max-stable random function are *mutually independent*.

The particular case $k=2$ is of special interest. The bivariate distribution takes the simpler form

$$\Pr\{Z(x)<z,Z(x+h)<t\}=\exp\left(-\int_0^1\max\left(\frac{\omega}{z},\frac{1-\omega}{t}\right)dH_h(\omega)\right)$$

where H_h is a positive measure on $[0,1]$ with a total mass of 2. Owing to (6.65), the integral of ω is equal to 1. In the case where $z=t$, we can write

$$\Pr\{Z(x)<z,Z(x+h)<z\}=\exp\left(-\frac{\theta(h)}{z}\right)$$

with

$$\theta(h) = \int_0^1 \max(\omega, 1 - \omega) \, dH_h(\omega) \quad \tag{6.67}$$

Following Schlather and Tawn (2003), the function θ is called *extremal coefficient function*. It always satisfies $1 \le \theta(h) \le 2$. The extremal value $\theta(h) = 1$ corresponds to the case where $Z(x) = Z(x + h)$, whereas $\theta(h) = 2$ means that $Z(x)$ and $Z(x + h)$ are independent. Regarding its structural properties, Schlather and Tawn showed that $\theta(h) - 1$ is a madogram. A simple proof starts with

$$\tfrac{1}{2} E |1_{Z(x) < z} - 1_{Z(x+h) < z}| = \tfrac{1}{2} E\left(1_{Z(x) < z} - 1_{Z(x+h) < z}\right)^2$$

$$= \Pr\{Z(x) < z\} - \Pr\{Z(x) < z, Z(x + h) < z\}$$

$$= e^{-1/z} - e^{-\theta(h)/z}$$

from which it follows that

$$\lim_{z \to \infty} \frac{z}{2} E |1_{Z(x) < z} - 1_{Z(x+h) < z}| = \lim_{z \to \infty} z \left[e^{-1/z} - e^{-\theta(h)/z}\right] = \theta(h) - 1.$$

The extremal coefficient function determines the variograms of the indicators associated to all thresholds z. Conversely it can be fitted through any of these indicator variograms, which are experimentally accessible. This can be used to check the max-stable character of Z, because $\theta(h)$ should not depend on the threshold considered. The function $\theta(h)$ is thus a very useful structural tool. It should be mentioned that other tools have been developed for characterizing the dependence of extremes (Coles et al., 1999; Cooley et al., 2006; Falk and Michel, 2006; Naveau et al., 2009).

The distribution of the maximum of $Z(x)$ over several points $x_1, \dots, x_k$ derives from (6.64):

$$\Pr\left\{\max_{1 \le i \le k} Z(x_i) < z\right\} = \exp\left(-\frac{1}{z} \int_{S_k} \max_{1 \le i \le k} \omega_i \, dH(\boldsymbol{\omega})\right)$$

This is of the form

$$\Pr\left\{\max_{x \in K} Z(x) < z\right\} = \exp\left(-\frac{\theta(K)}{z}\right)$$

where K represents the configuration $\{x_1, \dots, x_k\}$ up to a translation. The function $\theta(K)$ generalizes the extremal coefficient function to more than two points. This result can be extended to compact subsets K (with sup replacing max) and thus give the solution of the global change of support by the supremum, provided however that $\theta(K)$ can be determined. This is the case for some random function models. See Lantuéjoul et al. (2011) for more details.

6.11.4   Models for Max-Stable Random Functions

A Gaussian SRF with standard normal marginal is fully characterized by its correlogram $\rho(h)$. This is not as simple for max-stable SRFs because there is no unique max-stable SRF model. This can be seen as either a drawback or an advantage (according to the application, we can choose among several models with different features). Numerous models for max-stable random functions can be found in the literature. Undoubtely, the most famous one is the *storm process* introduced by Smith (1990) and extended by Schlather (2002). Designed by Schlather (2002), the so-called *extremal Gaussian process* is complementary. This section focuses on these two models.

The Storm Process

Introduced for modeling local (convective) precipitations, this random function can be defined as

$$Z(x) = \sup_{(s,t) \in \Pi} \frac{1}{t} Y_{s,t}(x - s) \tag{6.68}$$

where Π is a homogeneous Poisson point process with intensity μ in $\mathbb{R}^n \times \mathbb{R}_+$ and ($Y_{s,t}, s \in \mathbb{R}^n$, $t \in \mathbb{R}_+$) is a family of independent copies of a positive random function Y on $\mathbb{R}^n$. In simple models, Y is a fixed deterministic function ($Y_{s,t}$ is thus independent of s and t) with local influence—for example, the indicator function of a ball or a Gaussian function of the form $Y(u) = \exp(-b|u|^2), b > 0$. In the general case, such function is randomized: ball of random diameter, Gaussian function with random b, Poisson polygons or polyhedra. Roughly speaking, the realizations of Y represent the possible shapes that a storm can take. s is a location parameter whereas $1/t$ specifies the magnitude of the storms. The value of the storm process at point x is the largest magnitude of all storms affecting x. Figure 6.13 shows a simple 2D example of a storm process where the storms have a Gaussian shape; in this representation the values of the storm process are transformed to a uniform marginal distribution (the three-dimensional perspective effect that can be perceived is a mere consequence of the fact that this function vanishes rapidly at infinity).

A remarkable feature with a storm process is that it is mathematically tractable. The spatial distribution of Z is

$$\Pr\left\{ \max_{1 \le i \le k} \frac{Z(x_i)}{z_i} < 1 \right\} = \exp\left(-\int_{\mathbb{R}^n} \mathrm{E}\left[\max_{1 \le i \le k} \frac{Y(x_i + s)}{z_i} \right] ds \right) \tag{6.69}$$

In particular, the margins of Z are either Fréchet or degenerate, depending on whether the expected integral of Y is finite or not:

$$\Pr\{Z(x) < z\} = \exp\left(-\frac{\mu}{z} \int_{\mathbb{R}^n} \mathrm{E}[Y(s)] ds \right)$$

FIGURE 6.13   Realization of a storm process with Gaussian storms. The simulation has been transformed to a uniform distribution. The largest values are in white.

When the margins are standard Fréchet, the extremal coefficient function depends only on μ and the transitive madogram of Y:

$$\theta(h) - 1 = \frac{\mu}{2} \int_{\mathbb{R}^n} \mathrm{E} |Y(s+h) - Y(s)| \, ds$$

More generally let us denote by

$$Z^K = \sup_{x \in K} Z(x)$$

the supremum of Z over a compact subset K of $\mathbb{R}^n$. By the definition of the storm process, it can also be written

$$Z^K = \sup_{(s,t) \in \Pi} \frac{1}{t} Y_{s,t}^{\tau_{-s}K}$$

If we assume that the function $s \to \mathrm{E}[Y^{\tau_{-s}K}]$ is finite and integrable for any K, the multivariate distribution of the suprema over different supports can be determined. In particular, the supremum of Z over K is also Fréchet distributed:

$$\Pr\{Z^K < z\} = \exp\left(-\frac{\mu}{z} \int_{\mathbb{R}^n} \mathrm{E}[Y^{\tau_{-s}K}] \, ds\right)$$

Extreme Gaussian Process

This model is defined as

$$Z(x) = \sup_{t \in \Pi} \frac{Y_t(x)}{t}, \qquad x \in R^n \tag{6.70}$$

where Π is a homogeneous Poisson point process with intensity μ on $\mathbb{R}_+$ and $(Y_t, t \in \mathbb{R}_+)$ are independent copies of a stationary Gaussian random function Y on $\mathbb{R}^n$. Since Y has a global rather local character, this model is more appropriate for regional (cyclonic) precipitations. Two examples are displayed in Figure 6.14.

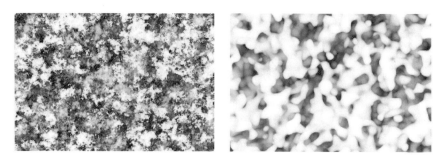

FIGURE 6.14 Realization of two extreme Gaussian rocesses. The underlying Gaussian random functions have exponential (left) and Gaussian (right) covariance functions. The simulation has been transformed to a uniform distribution. The largest valus are in white.

As stated by Schlather (2002), the Gaussian assumption is not strictly necessary and the definition makes perfect sense for a large variety of stationary random functions for Y. The spatial distribution of the associated process is given by the formula

$$\Pr\left\{\max_{1\leq i\leq k}\frac{Z(x_i)}{z_i}<1\right\}=\exp\left(-\mu\int_0^\infty\Pr\left\{\max_{1\leq i\leq k}\frac{Y(x_i)}{z_i}\geq t\right\}dt\right) \tag{6.71}$$

from which it can be readily seen that Z is stationary. Its univariate distribution

$$\Pr\{Z(x)<z\}=\exp\left(-\frac{\mu}{z}\int_0^\infty\Pr\{Y(x)\geq t\}dt\right) \tag{6.72}$$

bears only the positive part $Y_+=\max(Y,0)$ of Y. Indeed it is concentrated at 0 if $Y\leq 0$. It is also degenerate when the integral diverges. For that reason, Schlather (2002) assumes that the quantity $m=\mathrm{E}[Y_+(x)]$ satisfies $0<m<\infty$, in which case $Z(x)$ is Fréchet distributed with index $\alpha=\mu m$.

In the standard Fréchet case (i.e., $\mu m=1$), the extremal coefficient function is proportional to the madogram of Y_+:

$$\theta(h)-1=\frac{\mu}{2}\,\mathrm{E}|Y_+(s+h)-Y_+(s)|$$

Now, it should be noted that the madogram is strictly less than the mean. From this, $\theta(h)<2$ follows. Whatever the lag h considered, $Z(x)$ and $Z(x+h)$ cannot be independent.

In this model the supremum of Z over a compact set K is $Z^K=\sup_{t\in\Pi}\frac{Y_t^K}{t}$, so formula (6.72) can be generalized in

$$\Pr\{Z^K<z\}=\exp\left(-\frac{\mu}{z}\int_0^\infty\Pr\{Y^K\geq t\}dt\right)$$

Applications

In applications, it is possible to consider the random function under study as of the form $\varphi(Z(x))$, where φ is an anamorphosis function and Z a max-stable random function. The supremum of $\varphi(Z(x))$ over a compact K is then $\varphi(Z^K)$ so that the study of extreme values can be carried out on Z.

When the multivariate distribution of $\{Z^{K_i}:i=0,1,\ldots,N\}$ for different supports K_i has a mathematical expression, the conditional distribution of the supremum over K_0 given the suprema over the other K_i can be derived, at least in theory. When this direct approach is not feasible, this objective is achieved with simulations (simulation algorithms are available for most models; conditional simulation methods are in development).

These models have received applications to environmental problems such as precipitation and temperature, to materials corrosion, and to the analysis of financial series.

CHAPTER 7

Conditional Simulations

Je suis un mensonge qui dit toujours la vérité.

—Jean Cocteau

7.1 INTRODUCTION AND DEFINITIONS

7.1.1 Introductory Example

As an introduction, let us consider the somewhat simplified problem brought up by Alfaro (1979) and illustrated in Figure 7.1: A submarine cable is to be laid on the seafloor between two cities, and we want to predict the length of the cable. Measurements of the depth of the seafloor are available every 100 m (Figure 7.1a).

An easy solution is to estimate the depth by kriging at every point along the profile and to compute the length of this kriged curve (Figure 7.1b). As an example, the length obtained for the small portion of the profile shown in the figure is $L = 945$ m. But this result is not quite convincing, since the actual seafloor is far more complex than the kriged one (kriging has a smoothing effect). As a result, the length computed from the kriged profile certainly underestimates, perhaps quite severely, the actual length.

To get the exact solution, we need a continuous, or at least very dense, survey of depth (Figure 7.1c). In the present case a survey with samples taken every 10 m gives a length $L = 1182$ m. The previous estimate $L = 945$ m was thus indeed too low. Is there a way of avoiding the costs of the second survey? There is if we introduce a model.

The dependency between the length of a curve and the scale of investigation is a key issue in fractal theory (Mandelbrot, 1975b, 1977, 1982). It is easy to see that if the depth $Z(x)$ is a Gaussian IRF−0 with variogram $\gamma(h)$ and if the distance D is sampled at the interval d, the length L has expectation

Geostatistics: Modeling Spatial Uncertainty, Second Edition. J.P. Chilès and P. Delfiner.
© 2012 John Wiley & Sons, Inc. Published 2012 by John Wiley & Sons, Inc.

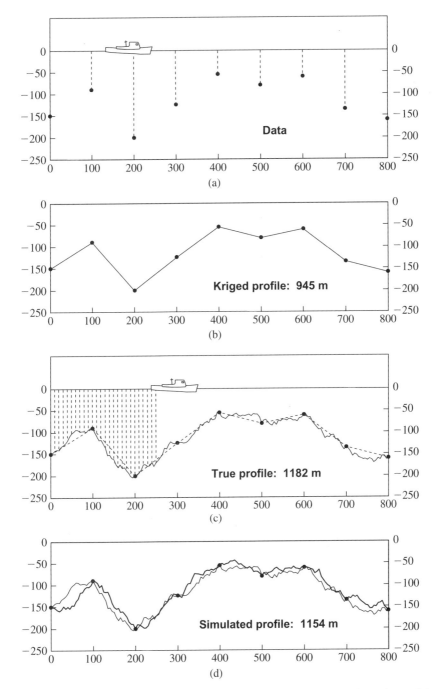

FIGURE 7.1 Submarine cable: (a) seafloor survey with 100-m spacing; (b) kriging estimate of the seafloor; (c) continuous survey (true profile); (d) conditional simulation based on the 100-m data, and true profile. [From Alfaro (1979).]

$$E(L) = D \int \sqrt{1 + \sigma^2 u^2} \, g(u) \, du$$

where $\sigma^2 = 2\,\gamma(d)/d^2$ and $g(u)$ denotes the standard Gaussian p.d.f. If the IRF is not differentiable, the length L increases as the sampling interval used to measure it becomes smaller, and it becomes so large as to be best considered infinite. If the variogram is of type $|h|^{\alpha}$, the relationship between the lengths $L(d)$ and $L(d_0)$ is

$$L(d) \approx (d/d_0)^{\frac{\alpha}{2} - 1} L(d_0)$$

provided that d and d_0 are small enough. In reality, even if the length of the seafloor along the profile were infinite, a cable has some rigidity. If we admit that it can hang above the sea bottom over distances of about 10 m, then the goal becomes an estimation of $L(d)$ for $d = 10$ m.

The idea of conditional simulations is to build a representation of the phenomenon that is consistent with the data observed at the 100-m sampling interval, as kriging is, and yet reproduces the local fluctuations at the scale of 10 m. Clearly a conditional simulation is not reality but only a possible version of it, among myriads of others. An example is shown in Figure 7.1d; the conditional simulation differs from the real seafloor, often more than kriging does, but has the same character. Not surprisingly, the cable measured on this simulation with a sampling interval of 10 m is very close to the true length: $L = 1154$ m instead of $L = 1182$ m. Other features of the seafloor can also be seen from the simulation, for example, the occurrence of large slopes. This is what makes simulations interesting.

"You are trying to entangle us in your mathematical tricks!" shouted a bewildered attendant at a presentation of conditional simulations. Likewise, the reader may at this point have at least three questions in mind:

1. How can one claim to reproduce a profile at a 10-m sampling interval from data at a 100-m sampling interval? By building a model at the 10-m scale, derived either from similar cases or from additional sampling at the 10-m interval. But this will not require a systematic survey of the seafloor every 10 m along the whole profile. A limited number of data at 10-m spacing will suffice to evaluate the variogram at distances of 10 m and more.

2. How confident can one be in the result of the simulation? Had another simulation been made, the cable length would have been different. Indeed. Therefore several simulations are generated so as to determine the probability distribution of the cable length.

3. Why not estimate the length L directly? The answer is that kriging—in its linear version—only allows estimation of quantities that are linear in the studied variable, such as the average depth along the profile. The length L is not a linear functional of depth but essentially depends on local fluctuations of the seafloor.

7.1.2 Definition and Use of Conditional Simulations

Nonconditional Simulations and Conditional Simulations

A (nonconditional) simulation of the random function $\{Z(x) : x \in \mathbb{R}^n\}$ is simply a realization of $Z(x)$, randomly selected in the set of all possible realizations. Its construction requires the knowledge of the spatial distribution of the RF $Z(x)$, at least implicitly. In practice this poses two problems:

1. The random function $Z(x)$ is usually defined so that some regionalized variable $\{z(x) : x \in D \subset \mathbb{R}^n\}$, observed at sample points $\{x_\alpha : \alpha = 1, \ldots, N\}$, can be considered as one of its realizations. Its finite-dimensional distributions can never be inferred from the data, even in a stationary framework. The situation is very different from image analysis where the complete image is available and allows, to some degree, an estimation of the multivariate distributions.[1] In geostatistics we can at best hope to obtain bivariate distributions, and possibly a few multivariate distributions, if we have data on a sufficiently regular grid.

2. Even if we knew the spatial distribution of $Z(x)$, we would in general still be unable to construct a simulation. The only truly general simulation procedure is the sequential method presented later in this chapter. However, this method has practical limitations and demands that we are able to calculate the conditional distributions, which is unusual. On the other hand, there are general methods for simulating Gaussian SRFs, for all covariance models. The Gaussian case aside, there is a vast number of random function models, each having particular spatial characteristics and corresponding to a class of possible covariance functions.

In view of this, we can consider a definition at order 2: A (nonconditional) simulation of the random function $Z(x)$ is a realization of an RF $S(x)$ chosen in a class of RFs with the same second-order moments as $Z(x)$, namely same covariance, variogram, or generalized covariance.

The choice of the RF class must be justified and driven by the problem and the data, although a certain measure of arbitrariness is inevitable. Whenever possible, we will attempt to increase specificity by restricting the choice of $S(x)$ to a narrower class than that defined at order 2. In the stationary case, for example, the marginal distribution of $Z(x)$ can be known, at least approximately, from the data, and we will require that $S(x)$ has the same marginal distribution. The type of the bivariate distributions can often be determined even from scattered data, and this can direct to specific RF models.

The random function $S(x)$ has an infinite number of realizations. Among them, some assume at the sample points the same values as those observed and

---

[1] This is the ideal case of image analysis. In practice, images often lack spatial homogeneity, so the multivariate distributions remain unknown.

thus can be considered to better represent the regionalized variable $z(x)$. They will be called conditional simulations. A conditional simulation is therefore a realization randomly selected from the subset of realizations that match the sample points. Equivalently, it is a realization of a random function with a conditional spatial distribution.

Use of Conditional Simulations

Conditional simulations are useful qualitatively, to obtain realistic pictures of spatial variability. Quantitatively, they are the tool of choice to evaluate the impact of spatial uncertainty on the results of complex procedures, such as numerical modeling of a dynamic system or economic optimization of the development of a natural resource. Conditional simulations fall in the scope of so-called Monte Carlo methods—they are in fact nothing but *spatially consistent* Monte Carlo simulations. Section 7.10 presents simulation case studies in mining and in petroleum.

The goal pursued here is not to reproduce the geological, physical or chemical processes that generated the observed phenomenon. More modestly it is simply to mimic its spatial variations as realistically as possible. The results of a simulation must always be considered with a critical eye and checked against the background of the application. This is very important when studying problems such as connectivity, fluid flow, or transport, which compel us to consider, explicitly or not, a model specified well beyond its second-order moments. This is illustrated in Figures 7.2 and 7.3, showing 1D and 2D views of nonconditional simulations of various SRFs with the same exponential covariance function. They look very different (notice simulations in Figures 7.2a, 7.3a, and 7.3b: they are nonergodic provocations!). If we built 100 simulations of each SRF instead of a single one to estimate a second-order moment numerically, each SRF would lead to the same result, up to fluctuations. But it is easy to imagine that the conclusions drawn with regards to connectivity or fluid flow would be very different. Chilès and Lantuéjoul (2005) present three different random set models with the same bivariate, trivariate and, for some cases, quadrivariate distributions. In practice it is impossible to discriminate these models from scattered data points and yet they have very different connectivity properties. Illustrative examples regarding fluid flow are given by Journel and Deutsch (1993), Zinn and Harvey (2003), and Emery and Ortiz (2011). In this respect, conditioning on the actual observations is a safeguard but does not exempt us from choosing an adequate SRF model.

How Many Simulations Should We Generate?

The answer depends on the objective and on the structure of the phenomenon. When modeling a stationary field over an area much larger than the range, a single simulation can give a view of a variety of possible local situations. This is often sufficient, for example, to assess the performance of a mining scenario that depends mainly on the local variability of ore grades or thicknesses. Conversely,

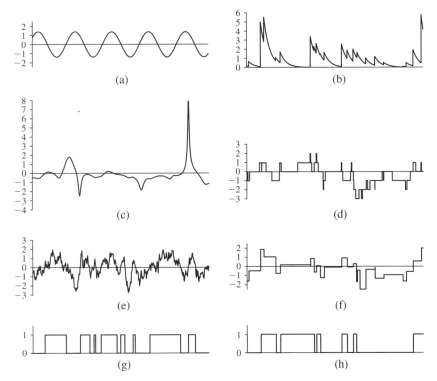

FIGURE 7.2 1D simulations of the same exponential covariance: (a) elementary spectral process; (b) Ambartzumian process (dilution of Poisson germs); (c) dilution by a symmetric function; (d) dilution by random segments; (e) Gaussian process; (f) mosaic process with Gaussian marginal (partition into Poisson segments); (g) truncated Gaussian process; (h) mosaic random set (Poisson segments set independently to 0 or 1). Note that the processes (a), (e), and (f) have the same Gaussian marginal distribution and that the random sets (g) and (h) have the same mean 1/2. The process (a) is not ergodic; and in practice, many elementary simulations are superimposed. Scale parameter: 1/30th the length.

when studying a nonstationary field such as a petroleum reservoir, a single simulation provides a single answer in terms of flow and production because these are global problems. It is therefore necessary to build several simulations if we want to assess the range of the possible results. Typically, 100 simulations may be needed, but this number depends largely on the distribution of the parameter of interest: If it is skewed and if we care even about extreme situations, a larger number of simulations is needed—but bear in mind that extreme value predictions usually lack robustness. It must be pointed out that conditional simulations are not equally probable; some are extreme and others are "average". Rather, they represent i.i.d. drawings from a (multivariate) conditional distribution that has no reason to be uniform. For some applications, simulations are ranked on the basis of a regional quantity computed from the realization (e.g., the volume above a reference level).

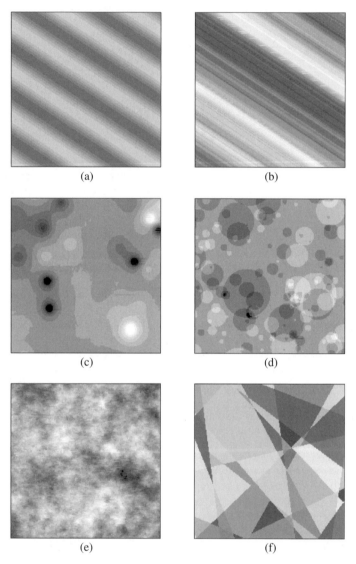

FIGURE 7.3 2D simulations showing the same exponential covariance: (a) elementary spectral RF; (b) elementary RF by turning bands with Gaussian marginal; (c) dilution by an isotropic function; (d) dilution by random disks; (e) Gaussian RF; (f) mosaic RF with Gaussian marginal (partition into Poisson polygons); (g) truncated Gaussian RF; (h) mosaic random set (Poisson polygons set independently to 0 or 1); (i) sequential indicator simulation; (j) nonmosaic random set (Poisson polygons set alternately to 0 or 1). Note that the RFs (a), (b), (e), and (f) have the same Gaussian marginal distribution and that the random sets (g), (h), (i), and (j) have the same mean 1/2. The RFs (a) and (b) are not ergodic, and in practice, many elementary simulations are superimposed. The irregularity of detail of (i) is an artifact of the simulation algorithm. Scale parameter: 1/10th the grid size.

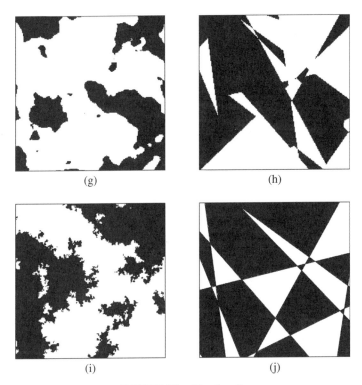

(g)

(h)

(i)

(j)

FIGURE 7.3 (Continued)

Should We Simulate the Uncertainty on the Parameters?

When we build several simulations, a question is whether or not to include the uncertainty on the parameters (mainly the histogram and the variogram). In our approach, only regional parameters have an objective meaning. Since we use models that are somehow stationary, we choose theoretical parameters close to the empirical ones. But the latter may differ from the regional parameters. This issue was addressed in Section 3.4.4 for kriging. Some solutions can be transposed to conditional simulations. It is the case for the Bayesian approach: If the prior density $f(\theta)$ of the parameter vector θ can be assessed, it is sensible to sample θ for each simulation from the posterior density $f(\theta \mid \mathbf{Z})$ of θ given the vector $\mathbf{Z}$ of the observations $\{Z(x_\alpha): \alpha = 1, \ldots, N\}$. It is not a simple task.

Sometimes there are not enough observations nor enough information on the physical laws governing the studied phenomenon for us to choose among candidate random function models or determine the behavior of the variogram at the origin. Then modeling our ignorance by an a priori distribution may not be an adequate solution. We prefer to generate simulations corresponding to several scenarios. The specification of these scenarios must proceed from an agreed methodology and cannot simply result from geostatistical considerations. In case of complex projects, methods of elicitation of expert opinions may be useful [e.g., Ayyub (2001)].

7.1.3 A Classification of the Methods

From the point of view of applications, three broad categories of quantities are to be simulated:

1. *Continuous Variables.* These are measurements of physical properties, such as a mineral grade, the thickness of a geological layer, or the velocity of sound. They are continuous variables in the sense of a continuous histogram, and also in a spatial sense, the nugget effect, if any, being treated separately. Continuous variables have a natural ranking and we can do maths on them (e.g., add, subtract, multiply, divide, take logarithms).

2. *Categorical Variables.* These are discrete variables representing elements of a classification. Typical examples are lithofacies types or indicators of large genetic units. Categorical variables may or may not be rank ordered, but in general we cannot do maths on them. The simulations define a partition of the space into cells with a constant value—within which physical variables can be simulated.

3. *Objects.* These represent entities of the application domain, whose morphology is captured as a whole and simulated in one shot. For example, a fracture is simulated as a disk with given dimensions and orientation, a meandering channel as a sinusoidal strip with given wavelength, amplitude, axis orientation, thickness, width, and optionally length. In contrast to *object-based* simulations, the other types of simulations are generated point by point and often referred to as *pixel-based* simulations.

The models used to represent these quantities do not quite follow this classification, mainly because categorical variables can be simulated either directly or by truncation of a continuous variable. We will distinguish four basic model types:

1. *Diffusive Model.* The simulations are continuous functions (almost surely). The name originates from physical diffusion, the slow process by which molecules spread out evenly. The prototypical example is the Brownian motion, whose transition probabilities satisfy the celebrated diffusion equation. In geology one can think of diffusive impregnation. The most used diffusive model is simply the Gaussian RF model. By analogy, in the case of an integer random function, a discrete diffusive model is one that varies by unit jumps.

2. *Jump Model.* The simulations are piecewise constant and progress by random jumps (usually noninteger). The prototype is the Poisson random walk and in a wider sense the dilution random functions.

3. *Mosaic Model.* The simulated space is partitioned randomly into cells within which the simulation is constant, and the values taken in the various cells are drawn independently from the same distribution. A typical example is a partition into Voronoi polygons or polyhedra based on a Poisson point process.

4. *Random Set Model.* The simulations assume the values 0 or 1 only, representing the indicator of a random set (binary simulation). The typical example is the Boolean model obtained by independent placement of random sets.

Table 7.1 summarizes the various methods presented in this chapter, in the order in which they are introduced, diffusive or jump models first, then mosaic and indicator models. The following characteristics are considered:

- *Covariance Model.* Some methods are very general (but limited to the Gaussian case), whereas others are limited to particular covariance models; some are limited to $\mathbb{R}$, whereas others are usable directly in $\mathbb{R}^n$.

- *Conditioning Method.* For diffusive and jump models, the simulation methods are mostly proven ones, and in the case of Gaussian RFs there are simple methods for conditioning on data, either directly or in a separate kriging step. For mosaic or random set models, some of the simulation methods are recent, and there is no general method of conditioning. However, most methods allow the construction of conditional simulations, generally by use of iterative algorithms.

- *Ergodicity and Reproduction of the Covariance.* We generally wish each simulation to have a sample mean close to the theoretical mean m and a sample covariance close to the theoretical covariance $C(h)$, especially if the simulated domain is sufficiently large. This implies that all simulations are drawn from the realizations of an RF that is ergodic in the mean value and the covariance (second-order ergodicity). Some simulation methods do not satisfy this condition (see the continuous spectral simulations of Figures 7.2a and 7.3a). Nonergodic methods are in fact utilized by adding a large number of independent simulations.

- *Conditions of Use.* Considerations are points on a systematic grid or not, maximum number of simulated points, exact or approximate reproduction of the covariance, and possibility of carrying out several simulations in parallel.

The methods for generating nonconditional simulations usually produce realizations of strictly stationary RFs with zero mean. It is interesting to note that the addition, properly scaled, of a large number of such independent simulations tends to a Gaussian RF. Specifically, if $S_1(\cdot), S_2(\cdot), \ldots,$ is a sequence of independent zero-mean, finite-variance simulations with the same spatial distribution, the spatial distribution of the random function $T_k(\cdot)$ defined by

$$T_k(x) = \frac{S_1(x) + \cdots + S_k(x)}{\sqrt{k}}$$

tends to a Gaussian RF as $k \to \infty$. Indeed the central limit theorem (Feller, 1971, Section VIII.4) ensures that the distribution of any linear combination of

TABLE 7.1 Main Characteristics of the Various Simulation Methods

Simulation Method	$\mathbb{R}^n$	RF Model	Covariance Model	Systematic Grid Required	Conditioning Method
Sequential Gaussian	All n	Gaussian	All C	No	Direct
Matrix decomposition	All n	Gaussian	All C	No	Direct
Turning bands	$n > 1$	Gaussian	All C, γ, K	No	Kriging
Autoregressive	$n = 1, 2$	Gaussian	Damped exponential	Yes	Kriging
Moving average	$n = 1, 2$ $n > 2$	Gaussian	Most C with short range	Yes	Direct Kriging
Dilution	$n = 1$ $n > 1$	Jump	Most C, γ, K Most C	No	Iterative
Poisson hyperplanes	$n > 1$	Jump	Most γ, K	No	Iterative
Continuous spectral	All n	Gaussian	All C, γ, K	No	Kriging
Discrete spectral	$n = 1$ $n > 1$	Gaussian	Most C, γ, K Finite-range C	Yes	Kriging
Sequential indicators	All n	Discrete	Indicator C	No	Direct
Truncated Gaussian	All n	Discrete	Special C class	No	Iterative
Voronoi	All n	Mosaic	Specific C	No	Iterative
Poisson polyhedra	All n	Mosaic	Exponential C	No	Iterative
Substitution	All n	Discrete	Particular C classes	No	Iterative
Boolean	All n	Binary	Specific C class	No	Iterative

Note: C, γ and K indicate the possibility of simulating covariances, variograms and generalized covariances; n: space dimension. Nonergodic methods produce Gaussian simulations by averaging a large number of elementary simulations. Gaussian simulations can also be obtained with simulations of a jump model with very small jumps. Many variants exist, notably as regards the RF type.

the form $\lambda_1 T_k(x_1) + \cdots + \lambda_M T_k(x_M)$, where M, $\lambda_1, \ldots, \lambda_M$, and $x_1, \ldots, x_M$ are arbitrary, tends to a Gaussian distribution as $k \to \infty$, which means that any finite-dimensional distribution of $T_k(\cdot)$ tends to a multivariate Gaussian distribution.

In practice we select a finite value for k. An upper bound of the difference between the marginal distribution of T_k and the Gaussian distribution is given by the Berry–Esséen theorem (Feller, 1971) for SRFs with a zero mean, a finite variance σ^2, and a finite third-order absolute moment $m_3 = E[|S_i(x)|^3]$: Under these assumptions, there exists a constant c such that for all x and k

$$\sup_{y \in \mathbb{R}} |\Pr\{T_k(x) < \sigma y\} - G(y)| < \frac{c}{\sqrt{k}} \frac{m_3}{\sigma^3}$$

where G is the standard normal c.d.f. The constant c is greater than 0.4097. In 2011 the greatest known lower bound for c is 0.4794. Similar upper bounds can be obtained for any linear combination of the values taken by the random function at different locations. Other criteria are given by Lantuéjoul (2002, Chapter 15).

7.2   DIRECT CONDITIONAL SIMULATION OF A CONTINUOUS VARIABLE

In this section we consider two methods that directly generate conditional simulations at a finite number of points: the sequential method and its variants, and the covariance matrix decomposition method.

7.2.1   Sequential Simulation

Sequential Simulation in the General Case

Consider a vector-valued random variable $\mathbf{Z} = (Z_1, Z_2, \ldots, Z_N)'$ for which a realization of the subvector $(Z_1, Z_2, \ldots, Z_M)'$ is known and equal to $(z_1, z_2, \ldots, z_M)'$ $(0 \le M < N)$. The distribution of the vector $\mathbf{Z}$ conditional on $Z_i = z_i$, $i = 1, 2, \ldots, M$, can be factorized in the form

$$\Pr\{z_{M+1} \le Z_{M+1} < z_{M+1} + dz_{M+1}, \ldots, z_N \le Z_N < z_N + dz_N \mid z_1, \ldots, z_M\}$$
$$= \Pr\{z_{M+1} \le Z_{M+1} < z_{M+1} + dz_{M+1} \mid z_1, \ldots, z_M\}$$
$$\times \Pr\{z_{M+2} \le Z_{M+2} < z_{M+2} + dz_{M+2} \mid z_1, \ldots, z_M, z_{M+1}\}$$
$$\vdots$$
$$\times \Pr\{z_N \le Z_N < z_N + dz_N \mid z_1, \ldots, z_M, z_{M+1}, \ldots, z_{N-1}\}$$

$$(7.1)$$

We can therefore simulate the vector $\mathbf{Z}$ sequentially by randomly selecting Z_i from the conditional distribution $\Pr\{Z_i < z_i \mid z_1, \ldots, z_{i-1}\}$ for $i = M+1, \ldots, N$ and including the outcome z_i in the conditioning data set for the next step.

This procedure is absolutely general and can be used in particular for Z_i of the form $Z_i = Z(x_i)$, where $Z(x)$ is an RF and where the x_i are the sample points $(i = 1, 2, \ldots, M)$ and the points where we wish to simulate the RF $(i = M+1, \ldots, N)$. It makes possible the construction of both a nonconditional simulation ($M = 0$) and a conditional simulation ($M > 0$). The procedure can be applied to the cosimulation of several nonindependent RFs. It produces simulations that match not only the covariance but also the spatial distribution.

The practical difficulty is that in general, we do not know how to calculate the conditional probabilities involved in (7.1), except in the ideal case of a Gaussian random vector, where the method is classically employed [e.g., see Ripley (1987)]. It has been introduced in geostatistical applications by Alabert and Massonat (1990) to simulate Gaussian RFs (small-scale log-permeability variations).

Sequential Gaussian Simulation

The application of the above method to the simulation of a Gaussian RF, known as sequential Gaussian simulation (SGS), is straightforward. Indeed for a Gaussian RF *with known mean*, the conditional distribution of Z_i is Gaussian, with mean Z_i^* and variance $\sigma_{\mathrm{K}i}^2$, where Z_i^* is the simple kriging estimator of Z_i from $\{Z_j : j < i\}$, and $\sigma_{\mathrm{K}i}^2$ the associated kriging variance. In principle the statistical properties of the simulation are independent of the order in which the points x_i, $i > M$ are scanned, but in practice this is not the case because the algorithm cannot be applied rigorously when N is large. Indeed, even if the original data are few, the size of the kriging system grows with the number of simulated points since these are progressively added to the data set. Kriging at the point x_i is then carried out with the nearest points among $\{x_j : j < i\}$. If the points x_i are scanned in a random or quasi-random order, the size of the neighborhood shrinks as the simulation algorithm proceeds. Consequently, the resulting simulation is only approximate. Emery (2004a) details the properties of such simulations for a typical 2D example, a spherical variogram with a range much shorter than the simulated domain and a 10% nugget effect: The behavior at the origin as well as the sill are correctly reproduced but the junction to the sill is smoother than it should be, resulting in a slight bias at intermediate distances. In order to limit such artifacts, Deutsch and Journel (1992) recommend to scan the points x_i in a semi-random order: First simulate on a coarse grid, then on a finer grid until the final grid is reached. Emery shows that this strategy produces no improvement because the main artifacts arise at the fine-grid scale. The sole solution for limiting the artifacts is to use sufficiently large neighborhoods (the bias does not exceed 10% with neighborhoods of 20 conditioning values and is negligible with neighborhoods of 100 conditioning values).

When the mean cannot be considered known—for example, under an assumption of local rather than global stationarity—ordinary kriging can be substituted for simple kriging. The price to pay is some bias in the reproduction of the covariance model. The variogram, however, is correctly reproduced when kriging Z_i from all $Z_j, j < i$. In practice a subset of these Z_j is used, which may cause a significant bias in the variogram, larger than with SK. See Emery (2004a, 2010a) for further details.

Ordinary kriging is therefore suited to genuine intrinsic random functions whose mean and covariance are not defined (typically for an h^α variogram). The variogram is correctly reproduced by the simulations, up to the effects of the restriction of the kriging neighborhood. That approach can be generalized to Gaussian IRF$-k$ by working with intrinsic kriging.

In general the variable to be simulated is not Gaussian. In the stationary case, SGS is therefore applied after a preliminary Gaussian transformation of the data (Section 6.2). The transformed data have by design a Gaussian marginal distribution, but their compatibility with the multi-Gaussian assumption must be checked (Section 7.3.2).

In the next sections we will see more accurate—and also efficient—methods for generating Gaussian conditional simulations. We now present two special cases, both in 1D, where the sequential Gaussian simulation method is a perfect and efficient solution due to a screening effect. The autoregressive processes presented in Section 7.5.1 are another special case, of broader validity from the covariance point of view, but limited to simulations on a regular 1D or 2D grid.

Special Case 1: Gaussian Stochastic Process with an Exponential Covariance (1D)

Among stationary processes, this is the only case where the method can be applied easily without any approximation, for any set of data points and of simulated points, provided that the mean is known. Indeed, (1) as we consider the Gaussian case, the conditional distribution of Z_i given $\{Z_j : j < i\}$ is normal (Z_i^*, σ_{Ki}^2), and (2) due to the screening effect property of the exponential covariance in 1D (Section 2.5.1), the simple kriging estimator Z_i^* only involves the two adjacent data points. Constructing a nonconditional simulation is even simpler: Proceeding by increasing values of x_i, Z_i^* simply involves the previously simulated value Z_{i-1}. For example, to simulate a covariance with scale parameter a and sill σ^2 at a regular spacing Δx, we select a first value Z_1 from the marginal distribution of Z, namely a Gaussian with zero mean and variance σ^2, and then apply the recurrence relation

$$Z_i = \lambda Z_{i-1} + U_i \qquad i = 2, 3, \dots \tag{7.2}$$

where $\lambda = e^{-\Delta x/a}$ and where the U_i are independent Gaussian random variables with mean zero and variance $\sigma_K^2 = \sigma^2(1 - \lambda^2)$. Autoregressive processes generalize this construction to a slightly wider class of covariances (Section 7.5.1).

Special Case 2: Brownian Motion (1D)

The Brownian motion, also named the *Wiener process*, is usually defined as a process $Z(x)$ with independent stationary increments, such that the increment $Z(x+h) - Z(x)$ has a Gaussian distribution with mean zero and a variance proportional to $|h|$. Hence, it is a Gaussian IRF-0 with a linear variogram $\gamma(h) = b|h|$ $(b>0)$. $Z(x)$ can also be regarded as the integral over $[0, x]$ of a Gaussian white noise. Among the random processes satisfying the definition, only separable versions are considered as Brownian motions. Almost every realization is continuous but nowhere differentiable [see, e.g., Blanc-Lapierre and Fortet (1953, Section IV.5) and Doob (1953, Section VIII.2)]. This reflects the fractal behavior of the Brownian motion: As we have seen in Section 2.5.1, its Hausdorff dimension is 3/2. The simulation of a Brownian motion for a set of points $x_0 < x_1 < \cdots < x_N$ starting from an arbitrary value $z_0 = Z(x_0)$ at the origin x_0 follows immediately from its definition. It can be done iteratively using the relation

$$Z(x_i) = Z(x_{i-1}) + U_i$$

where U_i is a centered normal random variable with variance $2b\,(x_i - x_{i-1})$, independent of the past. Figure 7.4 shows a simulation at a regular interval Δx, known as a *random walk*.

It is also possible to directly generate a conditional simulation of a Brownian motion. Indeed, because it is a process with independent Gaussian increments, the following two properties hold:

1. If we consider data points $x_0 < \cdots < x_i < x_{i+1} < \cdots < x_N$ and an unknown point $x \in\,]x_i, x_{i+1}[$, we have

$$\Pr\{Z(x) < z \mid Z(x_0) = z_0, Z(x_1) = z_1, \ldots, Z(x_N) = z_N\}$$
$$= \Pr\{Z(x) < z \mid Z(x_i) = z_i, Z(x_{i+1}) = z_{i+1}\}$$

that is, the conditioning is only dependent on the two adjacent observations which are screening the influence of the other observations.

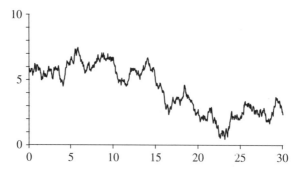

FIGURE 7.4 1D simulation of the Brownian motion or Wiener process (random walk) with variogram $\frac{1}{2}|h|$.

2. This conditional distribution is Gaussian; its mean is the ordinary kriging Z^* of $Z(x)$, which amounts in this case to a linear interpolation between x_i and x_{i+1}; its variance is the ordinary kriging variance σ_K^2, which is given in Example 4 of Section 3.4.1.

Thanks to these properties, it is easy to implement a sequential algorithm for simulating values between data points and/or refining a simulation between previously simulated points. In the case of regularly spaced points, this is a method popularized by applications of fractals under the name of *random midpoint displacement* (Voss, 1985): If $N+1$ points such that $N = 2^p$ are to be simulated at a regular interval Δx, the two extreme points are simulated first, by selecting the increment from a Gaussian distribution with mean zero and variance $\sigma^2 = 2\gamma(N\Delta x) = 2bN\Delta x$. The above properties are then used for simulating the midpoint so that the simulated points now define two intervals. This method is used again for simulating the midpoints of each new interval, and so on. After p such iterations, all points are simulated.

Sequential Simulation of Non-Gaussian Random Functions

When the random function is not Gaussian and is not the transform of a Gaussian RF, the distribution of Z_i conditional on $\{Z_j : j < i\}$, is usually not known. The simulation of Z_i is then chosen from an approximation to the conditional distribution. For example, if Z is a stationary random function whose bivariate distributions can be modeled with an isofactorial model, the pseudo-conditional distribution obtained by disjunctive kriging (Section 6.4.1) is an approximation to the conditional distribution (Emery, 2002).

The sequential indicator simulation algorithm, which transposes SGS to the simulation of indicators, is presented in Section 7.6.1. Alabert (1987) and Journel (1989) use it to simulate variables with a continuous marginal distribution. Nested indicators associated with increasing thresholds of the variable under study (e.g., the deciles) are defined, and these indicators are simulated—in fact this is a cosimulation. At any given point, what is obtained is not a specific value but instead the interval between two successive thresholds to which this value belongs. This is transformed into a specific value by random selection in the interval. This method is general but cumulates the approximations of the sequential indicator simulation algorithm and of the discretization of the marginal distribution. According to Gómez Hernández and Cassiraga (1994), this approach does not yield a satisfactory representation of long distribution tails.

7.2.2 Covariance Matrix Decomposition

In the case of a zero-mean Gaussian vector the sequential simulation in fact provides

$$Z_1 = \sigma U_1 \qquad Z_i = \sum_{j=1}^{i-1} \lambda_{ij} Z_j + \sigma_{Ki} U_i \qquad (i > 1) \qquad (7.3)$$

The λ_{ij} are the simple kriging weights of Z_i from $\{Z_j : j < i\}$, σ_{Ki}^2 is the associated kriging variance, and the successive U_i are independent standard normal random variables. Reasoning by recurrence, Z_i can also be expressed as

$$Z_i = \sum_{j=1}^{i} a_{ij} U_j \tag{7.4}$$

The matrix $\mathbf{A} = [a_{ij}]$ is a lower triangular matrix such that $\mathbf{A} \mathbf{A}' = \mathbf{C}$, where $\mathbf{C} = [C_{ij}]$ is the matrix of the $N \times N$ covariances.

Instead of performing a sequential simulation, we can first decompose the final matrix $\mathbf{C}$ into the product $\mathbf{AA}'$ [Cholesky decomposition; e.g., see Press et al. (2007, Section 2.9)], then compute $\mathbf{Z} = \mathbf{A U}$, where $\mathbf{U}$ is a vector of N independent standard normal random variables. Since $\mathbf{A}$ is lower triangular and $\mathbf{A}'$ upper triangular, this method is known as that of the *LU (lower−upper) decomposition* of the covariance matrix. Generating conditional simulations is easy, since the conditional distribution remains multivariate Gaussian (see Appendix, Section A.9).

This classic method [see Ripley (1987)] has been introduced in geostatistical applications by Davis (1987). It can be used with ordinary covariances and with covariances of locally stationary representations of IRF−k's. It is applicable as long as the Cholesky decomposition is feasible—that is, in general for at most several thousand simulated points. For larger numbers of points, matrix storage problems turn up and computing time can become excessive, since it increases as N^3. More efficient algorithms of Cholesky decomposition are available when $\mathbf{C}$ is a band matrix, as shown, for instance, by Rue and Held (2005, Section 2.4) ($\mathbf{A}$ has a lower bandwidth similar to that of $\mathbf{C}$). This is the case, for example, for a moving-average process of low order, thus with a short range (but in this specific case it is easier to simulate it by just applying the definition).

If $\mathbf{C}$ is not a band matrix but its inverse, the precision matrix $\mathbf{B}$, is a band matrix, the above method can be used to simulate a Gaussian vector $\mathbf{Y}$ with covariance matrix $\mathbf{B}$ and then take $\mathbf{Z} = \mathbf{C Y}$. Denoting by $\mathbf{L L}'$ the Cholesky decomposition of $\mathbf{B}$, the vector $\mathbf{Z}$ with covariance matrix $\mathbf{C}$ is obtained equivalently by solving $\mathbf{L}' \mathbf{Z} = \mathbf{U}$, where $\mathbf{U}$ is a vector of N independent standard normal random variables. This can be done easily because $\mathbf{L}$ is lower-triangular. This approach is advantageous when the precision matrix, rather than the covariance matrix, is known, as is the case for the Gaussian Markov random fields presented in Section 3.6.4 and, among them, for autoregressive processes.

When the vector $\mathbf{Z}$ has a high dimension, it may be more convenient to use an iterative algorithm based on the Gibbs sampler, presented in Section 7.6.3, which requires no inversion or decomposition of the covariance or the precision matrix. We now turn to methods that are more complex to implement but enable the simulation of Gaussian (and some non-Gaussian) random functions, at any point of a domain of $\mathbb{R}^n$ or at the nodes of a large regular 2D or 3D grid.

7.3 CONDITIONING BY KRIGING

In the rest of this chapter, several simulation algorithms are presented. Most of them do not provide a direct construction of the conditional simulations. However, there are general methods that can be used to transform nonconditional simulations into conditional ones.

7.3.1 Conditioning on the Data

Let us first assume that we know how to construct a nonconditional simulation— that is, a realization of a random function that has the same (generalized) covariance as the studied phenomenon but is otherwise totally unrelated to the sample data. The object of the present section is to show how to pass from a nonconditional simulation to a conditional simulation which, while retaining the structural features of the former, is calibrated on the sample data.

Consider an RF $Z(\cdot)$ known at N sample points $\{x_\alpha : \alpha = 1, 2, \ldots, N\}$. Let us assume that we have a nonconditional simulation $S(\cdot)$ independent of $Z(\cdot)$, *with the same covariance as* $Z(\cdot)$. "Conditioning" is the operation by which we can pass from $S(\cdot)$ to a simulation $T(\cdot)$ that matches the sample points.

The principle, due to G. Matheron, is quite simple. Let $Z^*(x)$ denote the kriging estimator of $Z(x)$ at the point x based on the data $Z(x_\alpha)$, and let us start from the trivial decomposition

$$Z(x) \quad = \quad Z^*(x) \quad\quad + [Z(x) - Z^*(x)]$$
$$\text{true value} = \text{kriging estimator} + \text{kriging error}$$

The kriging error is of course unknown since $Z(x)$ is not known. Now consider the same equality for $S(x)$, where $S^*(x)$ is the kriging estimator obtained as if the simulation were known only at the sample points x_α

$$S(x) = S^*(x) + [S(x) - S^*(x)]$$

This time the true value $S(x)$ is known and so is the error $S(x) - S^*(x)$. Hence the idea of substituting, in the decomposition of $Z(x)$, the unknown error by the simulation of this error, as shown in Figure 7.5; this gives $T(x)$ defined by

$$T(x) \quad = \quad Z^*(x) \quad + \quad [S(x) - Z^*(x)] \tag{7.5}$$

$$\begin{array}{ccc} \text{conditional} & \text{kriging} & \text{simulation of} \\ \text{simulation} & = \text{estimator} + & \text{kriging error} \end{array}$$

Since kriging is an exact interpolator, at a sample point we have $Z^*(x_\alpha) = Z(x_\alpha)$ and $S^*(x_\alpha) = S(x_\alpha)$, so $T(x_\alpha) = Z(x_\alpha)$. Conversely, the effect of conditioning decreases and finally vanishes when moving away from the data (at least when conditioning by SK).

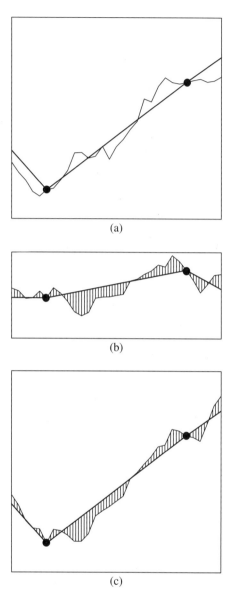

(a)

(b)

(c)

FIGURE 7.5 Conditioning a simulation: (a) real curve (unknown), sample points and kriging; (b) nonconditional simulation (known), sample points, and simulation of the kriging error; (c) kriging errors are picked from the simulation and added to the kriged curve.

There remains to verify that $T(\cdot)$ preserves the (generalized) covariance of $Z(\cdot)$ and $S(\cdot)$. It is obvious in the case of simple kriging: Since $Z(x) - Z^*(x)$ is uncorrelated with $Z^*(x')$ for all x' (Section 3.3.3), one has

$$\mathrm{Cov}(Z(x), Z(x')) = \mathrm{Cov}(Z^*(x), Z^*(x')) + \mathrm{Cov}(Z(x) - Z^*(x), Z(x') - Z^*(x'))$$

Since S and Z are independent, the covariance of T is also the sum of the covariance of Z^* and that of $S - S^*$, which equals that of $Z - Z^*$. Hence

$$\mathrm{Cov}(T(x), T(x')) = \mathrm{Cov}(Z(x), Z(x'))$$

A general proof in the framework of IRF$-k$ is presented by Delfiner (1976): Any allowable linear combination has the same variance when applied to Z, S, or T.

In the important case of Gaussian random functions, preservation of the covariance implies preservation of the spatial distribution (SRF with known mean) or of any multivariate distribution of generalized increments (IRF$-k$).

Note (and this will be important) that in (7.5), conditional simulations are only involved through kriging errors and therefore may be selected with mean zero.

When considered conditional on the $Z(x_\alpha)$, $T(\cdot)$ is no longer stationary. The method of construction (7.5) entails that

$$E[T(x) \mid Z(x_\alpha) : \alpha = 1, \dots, N] = Z^*(x),$$
$$\mathrm{Var}[T(x) \mid Z(x_\alpha) : \alpha = 1, \dots, N] = \mathrm{Var}[S(x) - S^*(x)] = \sigma_K^2(x)$$

The mean of a large number of independent conditional simulations at a given point tends to the kriging estimate, and their variance tends to the kriging variance. In figurative terms, a conditional simulation "vibrates" in between the data points within an envelope defined by the kriging standard deviation.

A conditional simulation is meant to behave like the real variable but not to estimate it: As an estimator of $Z(x)$ a simulation $T(x)$ would perform very poorly, with a variance twice the kriging variance; indeed

$$T(x) - Z(x) = [Z^*(x) - Z(x)] + [S(x) - S^*(x)]$$

so that

$$E[T(x) - Z(x)]^2 = 2\,\sigma_K^2(x)$$

A conditional simulation is indeed, in the words of the poet Jean Cocteau, "a lie which always tells the truth": It lies about the values assumed by the actual realization in between the data points, but it tells the truth about what a realization should look like.

Let us now examine two important implementation questions:

1. *Simulation of the Nugget Effect.* All properties of conditional simulations hold when the data are subject to nonsystematic measurement errors, provided that kriging is done in the same manner for $Z(x)$ and for the nonconditional simulation: (i) An error term with the same variance (and the same covariance if the errors are not independent) is added to the $S(x_\alpha)$

when calculating $S^*(x)$; (ii) the cokriging system (3.68) including error variances and covariances is used (note that the $S_{\alpha\beta}$ of this cokriging system represent the error variances and covariances). The situation is similar in the presence of a microstructure that we do not want to simulate, because it is a purely random component of the phenomenon. On the contrary, if we want to simulate it, the nugget-effect covariance associated with the microstructure must be present on the right-hand side of (3.38) and the nugget-effect variance must be added to the expression of the cokriging variance.

2. *Conditioning in Moving Neighborhood.* The proof that the substitution of errors preserves the covariance is valid for kriging with a global neighborhood. If the number of data is too large, it is necessary to use local neighborhoods. A careful design of the neighborhood search algorithm is needed to avoid the introduction of spurious discontinuities due to neighborhood changes, since these discontinuities cannot be easily distinguished from the normal jitter of most simulations. It is, for example, advisable to use large overlapping neighborhoods. An alternative solution is to use one of the approaches presented in Sections 3.6.3 to 3.6.6 for ensuring the continuity of the interpolant.

Following a terminology introduced by Journel (1986), data that are specified by a single value are often called *hard data* by contrast with *soft data*, which are defined by inequalities or by a probability distribution. Conditioning on soft data is more difficult but possible. This problem is commonly met when simulating indicator functions, and solutions are therefore presented in Sections 7.6.1 and 7.6.3.

7.3.2 Matching the Histogram

Many nonconditional simulation methods amount to a weighted moving average of a large number of independent random numbers and produce (approximately) Gaussian simulations. But the data are not obligated to be Gaussian! If the variable has a drift, it constitutes the dominant feature of the phenomenon, and the simulation reproduces it fairly well thanks to conditioning; residuals may not have the correct distribution, but this correct distribution is not known and is quite difficult to infer from the data. By contrast, in the stationary case the marginal distribution of the variable is an important feature, which can be determined empirically. This distribution is not necessarily reproduced by conditioning. For example, there may be a significant amount of negative values, even though the studied variable is necessarily positive.

The usual means to restore the observed histogram, first used by Journel (1974a,b), is to work with the random function $Y(x)$ obtained by Gaussian transformation of the variable of interest $Z(x)$, defined in Section 6.2. Let φ be the transformation expressing $Z(x)$ as a function of $Y(x)$: $Z(x) = \varphi(Y(x))$. The

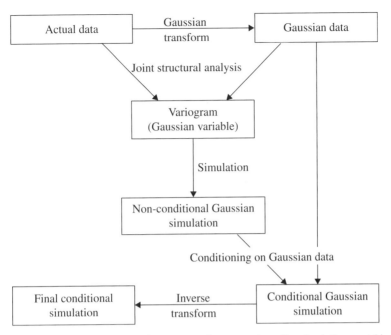

FIGURE 7.6 Flowchart of the various stages of construction of a simulation matching the histogram, the variogram, and the data.

successive steps of the generation of a conditional simulation are the following (Figure 7.6):

1. Transformation of the $Z(x_\alpha)$ data into $Y(x_\alpha)$ by the inverse transformation $Y(x_\alpha) = \varphi^{-1}(Z(x_\alpha))$.
2. Structural analysis of the $Y(x_\alpha)$ data, or, better, joint structural analysis of the $Y(x_\alpha)$ and $Z(x_\alpha)$ data, to obtain the variogram of $Y(x)$.
3. Nonconditional simulation of $Y(x)$, leading to $S_Y(x)$, using a Gaussian simulation method.
4. Conditioning of $S_Y(x)$ on the Gaussian data $Y(x_\alpha)$, leading to $T_Y(x)$.
5. Application of the transformation $T_Z(x) = \varphi(T_Y(x))$.

This procedure is perfectly suited to any RF that can be considered as the image of the transform of a zero mean Gaussian RF. Indeed, conditioning by simple kriging amounts to sampling the conditional distribution of the Gaussian RF $Y(x)$ and, after the transformation φ, the conditional distribution of the RF $Z(x)$. Conditioning can be done by ordinary kriging if the mean of $Y(x)$ may deviate from zero. In that case, ordinary kriging only provides an approximation to the conditional distribution of the Gaussian RF.

One notices that the nonconditional simulation and the conditioning concern the Gaussian RF $Y(x)$ and not $Z(x)$. Therefore structural analysis must be done on $Y(x)$. Another possibility is to interchange steps 4 and 5, namely transform the nonconditional simulation $S_Y(x)$ into a nonconditional simulation $S_Z(x) = \varphi(S_Y(x))$ and condition it on the $Z(x_\alpha)$, using of course the variogram of $Z(x)$. But this is not recommended, since (1) the optimization criterion of kriging (minimum mean square error) is well adapted to Gaussian RFs and (2) conditioning generally does not preserve the histogram when the studied random function and its nonconditional simulation are not Gaussian random functions.

The transformation of the $Z(x_\alpha)$ data into $Y(x_\alpha)$ and the modeling of the transform φ have been addressed in Section 6.2. The transformation to Gaussian data deserves special care to ensure that their variogram has unit sill, in order to avoid unrealistically large values in the Gaussian simulations in areas weakly controlled by the data. The anamorphosis or transformation function itself, and especially its tails, must be modeled carefully, because the Gaussian simulations usually include larger values than the Gaussian data; an expansion with Hermite polynomials is not recommended here.

We will conclude on this subject with three practical remarks:

1. Since we are using techniques that imply the framework of Gaussian SRFs, it is advisable to at least check, using the methods of Section 6.4.4, that the bivariate distributions of the transformed data $Y(x_\alpha)$ can be considered Gaussian:
   - The scatterplot of $(Y(x_\alpha), Y(x_\alpha + h))$, for fixed h, ought to be elliptical.
   - The madogram $\gamma_1(h)$ and the variogram $\gamma(h)$ of $Y(x)$ ought to satisfy $\gamma_1(h)/\gamma_1(\infty) = \sqrt{\gamma(h)/\gamma(\infty)}$.
   - The variograms of $Z(x)$ and $Y(x)$ ought to be in a relationship that depends on φ.

2. If the Gaussian data include a nugget effect component σ_ε^2, we may prefer to simulate the continuous part of the variogram only, as explained above. The transform φ, however, has been defined with data including that random component. $T_Z(x)$ should therefore be calculated as the mathematical expectation of $\varphi(T_Y(x) + \sigma_\varepsilon U)$, where U is an independent standard normal random variable—that is, as

$$T_Z(x) = \int \varphi\left(T_Y(x) + \sigma_\varepsilon u\right) g(u)\, du$$

   where g is the standard normal p.d.f.

3. Conditioning should be made by simple kriging. However, if the size of the data domain is modest relative to the covariance range, then the sample mean, usually close to zero by design, can be considered as unknown and ordinary kriging substituted for simple kriging. This ensures the reproduction of the variogram rather than the covariance and induces the generation of conditional simulations with a regional

mean different from the sample mean of the Gaussian data. This was the objective, but then the final simulation depends largely on the modeling of the anamorphosis, and the procedure must be checked. An intermediate solution between conditioning by SK and OK is to consider the mean of each simulation as a random variable in a Bayesian framework, the random drift model presented in Section 3.4.10. Ordinary kriging can also be adopted if strict stationarity is not assured—for example, in border areas. See Emery (2007c) for further details and an example showing the usefulness of a local conditioning by OK.

7.3.3 Probability-Field Simulation

We make a digression to present a simulation technique whose statistical properties are not established but which is used for its speed and practicality. The method is based on the well known fact that to draw a random number from the distribution with c.d.f. F it suffices to draw a uniform random number p between 0 and 1 and compute $z = F^{-1}(p)$. Letting F and z depend on location x, we have the relation $z(x) = F_x^{-1}[p(x)]$. When the local distributions F_x are known, it is therefore equivalent to simulate $Z(x)$ or the probability field (p-field) $P(x)$.

Ideally, these local F_x are the conditional distributions given hard data and possibly also soft data. In any case, $P(x)$ is not stationary. The trick is to simulate a surrogate *stationary* p-field. If Z is a stationary RF, the natural candidate is the uniform transform $U(x) = F[Z(x)]$ which hopefully has the same important spatial features of continuity and anisotropy as $P(x)$. The simulation algorithm comprises two steps: (1) a nonconditional simulation of the surrogate p-field $U(x)$, (2) the transformation $T(x) = F_x^{-1}[U(x)]$. The second step achieves automatic conditioning on hard data, since at a data point F is concentrated on a single value and F^{-1} always returns that value. By design the algorithm also reproduces the univariate distributions F_x. The spatial correlation built in $P(x)$ induces spatial correlation in the simulated $Z(x)$, which is the effect sought, but does not reproduce the nonconditional nor the conditional covariance of Z.

The covariance of the uniform transform $U(x)$ can be calculated theoretically by formula (6.24) in the case of an isofactorial model. In particular, if the bivariate distributions of Z are Gaussian, the covariance of U is related to the correlogram $\rho(h)$ of Z by the simple formula

$$\text{Cov}[U(x), U(x+h)] = \frac{1}{2\pi} \arcsin\left[\tfrac{1}{2}\rho(h)\right]$$

In the interval of variation $[-0.5, +0.5]$ the arcsine function differs little from a straight line so that, up to a scaling factor, the covariance of U is nearly the same as $\rho(h)$, and *not* smoother as might have been expected. In fact, if the distributions F_x are Gaussian with means $\mu(x)$ and standard deviations $\sigma(x)$, there is no need to go through $U(x)$ and we can simulate directly

$$T(x) = \mu(x) + \sigma(x) Y(x)$$

where $Y(x)$ is a standard Gaussian field with correlogram $\rho(h)$. Here p-field simulation amounts to a local modulation of a stationary error field. This reveals two limitations of the method. First, the covariance of T

$$\text{Cov}[T(x), T(x+h)] = \sigma(x)\, \sigma(x+h)\, \rho(h)$$

may not reflect the complexity of the nonstationarity pattern of the conditional covariance of Z. Second, when there are only hard data, the simple kriging estimate and standard deviation define $\mu(x) = z^*(x)$ and $\sigma(x) = \sigma_K(x)$; but considering the typical behavior of standard deviation maps, with cusps at data points, passing on their shape to simulations is not really desirable.

The p-field simulation technique has been proposed by Srivastava (1992) and Froidevaux (1993). A retrospective is provided by Srivastava and Froidevaux (2005). We see its value as a way to introduce spatial correlation in Monte Carlo studies in which the intervals of variation are defined from external knowledge.

7.4   TURNING BANDS

Some covariance models may be simulated directly in $\mathbb{R}^n$. But it is often simpler to use the turning bands method, which enables the construction of simulations in space from simulations on lines. Thus methods of simulation on a line are of prime interest even if the final goal is to simulate in 2D or 3D. The turning bands method was first used by Chentsov (1957) in the special case of Brownian random functions. The general principle of the method appears as a remark in Matérn (1960, p. 16), but its development for simulations is due to Matheron (1973a).

7.4.1   Presentation of the Method in the Plane

The method is usually employed for SRFs but can be used for IRF$-k$'s as well and is presented here in this context. The turning bands method consists of adding up a large number of independent simulations defined on lines scanning the plane. Up to a scale factor, the value of the simulation at a point x of the plane is the sum of the values assumed at the projections of x on the different lines by the corresponding one-dimensional simulations. More specifically, consider a system of n_D lines emanating from the origin of space and scanning

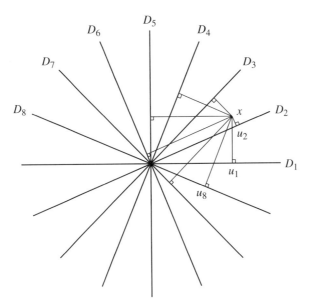

FIGURE 7.7 The principle of turning bands in 2D.

the plane regularly (Figure 7.7); the angle between two adjacent lines is π/n_D. We define the following:

- $\theta_t \in [0, \pi[$ is the angle of the line D_t with the x axis.
- u_t is the unit vector of D_t, with components $\cos \theta_t$ and $\sin \theta_t$.
- s_t represents the abscissa on D_t, centered at the origin.

Independent zero-mean nonconditional simulations $S_t(s_t)$ with (generalized) covariance $K_1(h)$ are associated to the lines D_t.

Let us consider a point $x = (x, y)$ in the plane. Its projection on D_t is a point with abscissa

$$s_t = \; <x, u_t> \; = x \cos \theta_t + y \sin \theta_t$$

The simulation at x is then defined by

$$S(x) = \frac{1}{\sqrt{n_D}} \sum_{t=1}^{n_D} S_t(s_t) \tag{7.6}$$

The elementary simulations being independent and with the same (generalized) covariance $K_1(h)$, the generalized covariance of $S(x)$ is

$$K_2(h) = \frac{1}{n_D} \sum_{t=1}^{n_D} K_1(<h, u_t>)$$

If the number of lines is large enough, the discrete sum is an approximation of the integral

$$K_2(h) = \frac{1}{\pi} \int_0^\pi K_1(|h|\cos\theta)\, d\theta \tag{7.7}$$

In practice, K_2 is given, and we have to invert relation (7.7) to obtain K_1. Brooker (1985) has shown that for $h \ge 0$, K_1 is given by

$$K_1(h) = K_2(0) + \int_0^h h(h^2 - u^2)^{-1/2} \frac{d}{du} K_2(u)\, du \tag{7.8}$$

In particular, the usual spherical covariance (of $\mathbb{R}^3$) with sill C and range a is obtained by using

$$K_1(h) = \begin{cases} C\left[1 - \dfrac{3\pi}{4}\left(\dfrac{|h|}{a} - \dfrac{1}{2}\dfrac{|h|^3}{a^3}\right)\right] & \text{if } |h| \le a, \\[4ex] C\left[1 - \dfrac{3}{2}\left(\dfrac{|h|}{a} - \dfrac{1}{2}\dfrac{|h|^3}{a^3}\right)\right]\arcsin\left(\dfrac{a}{|h|}\right) - \dfrac{3\,h^2}{4\,a^2}\sqrt{1 - \dfrac{a^2}{h^2}} & \text{if } |h| \ge a \end{cases}$$

Note that, unlike K_2, the covariance K_1 reaches zero only asymptotically when $|h| \to \infty$. For other models, see Gneiting (1998).

In principle, random directions could be used instead of regular directions. Theoretically, it even suffices to take a single line with an angle drawn at random from a uniform distribution over $[0, \pi[$ to obtain the covariance (7.7) exactly. However, such procedure will only ensure the correct covariance on average over the ensemble of simulations, while none in particular would have the desired properties. For example, simulations based on a single line exhibit a zonal anisotropy even though the parent covariance model $K_2(h)$ is isotropic (see Figure 7.3b). For a fixed value of the number of lines, it is advisable to use a regular system of lines rather than random directions, for it ensures better second-order ergodicity.

In the early implementation of the method, the simulations along each line were discretized, so that the same value $S_t(s_t)$ was assigned to a whole "band" perpendicular to D_t and containing s_t. Hence the name turning "bands" given to the method.

Note that as can be seen in Figure 7.7, $S(x)$ at point x integrates the contribution of the lines at their intersections with the circle of diameter $|x|$ going through the origin and the point x. Thus the turning bands algorithm can be seen as a back-projection. Its formalism is similar to the formalism of Radon transforms used in tomography.

7.4.2 Generalization to *n*-Dimensional Space

The easiest way for generalizing the method is to take a simulation $S_1(s)$ with (generalized) covariance $K_1(h)$ on a line and to define the direction of this line in $\mathbb{R}^n$ by a random unit vector u, namely by a point on the unit sphere. The simulation at a point x of $\mathbb{R}^n$ is then defined by

$$S(x) = S_1(<x, u>)$$

where the scalar product $<x, u>$ stands for the projection of x onto the line. In practice, just as in the plane, several lines are used rather than a single one; and since opposite directions play the same role, the directions are selected over the unit half-sphere (see next section).

Taking the expectation with respect to the random direction u, the (generalized) covariance of $S(x)$ is

$$K_n(h) = \frac{1}{S_n} \int K_1(<h, u>) \, du$$

where the integration extends over the unit sphere of $\mathbb{R}^n$ and where S_n is the surface area of this sphere. With this procedure, the covariance $K_n(h)$ is clearly isotropic though all realizations of $S(x)$ are anisotropic. Using the same notation K_n to denote this covariance as a function of $r = |h|$, we have explicitly

$$K_n(r) = \frac{2}{\sqrt{\pi}} \frac{\Gamma(\frac{n}{2})}{\Gamma(\frac{n-1}{2})} \int_0^1 K_1(vr)(1 - v^2)^{(n-3)/2} \, dv \qquad (7.9)$$

where Γ is the gamma function (A.1). If $K_1(h)$ and $K_n(h)$ have spectral densities $f_1(u)$ in $\mathbb{R}^1$ and $f_n(u)$ in $\mathbb{R}^n$, respectively, f_1 and f_n are isotropic and can be considered as functions of $\rho = |u|$. Using the same notation f_n to represent the spectral density as a function of ρ, we can translate (7.9) in spectral terms:

$$f_n(\rho) = \pi^{-n/2} \, \Gamma(n/2) \, \rho^{1-n} f_1(\rho)$$

Similar relations are obtained for the spectral measure of isotropic generalized covariances of IRF$-k$. Relation (7.9) has a very simple form when $n = 3$:

$$K_3(r) = \int_0^1 K_1(vr) \, dv = \frac{1}{r} \int_0^r K_1(u) \, du \qquad (7.10)$$

Formula (7.9) associates an isotropic covariance K_n in $\mathbb{R}^n$ to every covariance K_1 in $\mathbb{R}^1$. In fact this formula can be identified with the general form of an isotropic covariance in $\mathbb{R}^n$ (Matérn, 1960; Matheron, 1973a). Thus the correspondence between K_1 and K_n is one to one, and (7.9) can be inverted. As seen in Section 7.4.1, the inversion of (7.9) is not easy for $n = 2$ (and similarly for

n even). When $n = 3$, the inverse formula of (7.10) is particularly simple (and generally so when n is odd):

$$K_1(r) = \frac{d}{dr}[r\,K_3(r)] \tag{7.11}$$

If K_3 has a finite range a (i.e., $K_3(r) = 0$ when $r > a$), K_1 has the same property. This is generally not the case for the covariance K_1 associated with a covariance K_2 with finite range, as can be seen in (7.8) (and also for the covariance K_1 associated with K_n when n is even). Since (7.11) is simpler than (7.8), it may be better to generate a 2D simulation as the planar section of a 3D simulation than to simulate directly in 2D (this is of course limited to the case where the covariance K_2 is also a valid model in $\mathbb{R}^3$).

The correspondence between K_1 and K_n is also particularly simple, for all n, in the case of a generalized covariance of type r^α, α not even: The $K_1(v\,r)$ term in the integral (7.9) is factorized as $v^\alpha\,r^\alpha$ and K_n is of the form $B_{n\alpha}\,r^\alpha$ with

$$B_{n\alpha} = \frac{\Gamma\left(\frac{n}{2}\right)\Gamma\left(\frac{\alpha+1}{2}\right)}{\Gamma\left(\frac{1}{2}\right)\Gamma\left(\frac{\alpha+n}{2}\right)} \tag{7.12}$$

Likewise, if K_1 is a sum of $A_\alpha\,r^\alpha$ terms, K_n is a sum of $A_\alpha\,B_{n\alpha}\,r^\alpha$ terms. In particular, a polynomial generalized covariance in $\mathbb{R}^1$ generates by turning bands a polynomial generalized covariance in $\mathbb{R}^n$.

For $k = 2$ the explicit correspondence between the generalized covariance K_1 in 1D and the generalized covariances K_2 and K_3 in 2D and 3D, respectively, is given by

$$K_1(r) = -b_0 r + b_1 r^3 - b_2 r^5,$$

$$K_2(r) = -\frac{2}{\pi}b_0 r + \frac{4}{3\pi}b_1 r^3 - \frac{16}{15\pi}b_2 r^5,$$

$$K_3(r) = -\frac{1}{2}b_0 r + \frac{1}{4}b_1 r^3 - \frac{1}{6}b_2 r^5$$

Generalized covariances of type $(-1)^{k+1}\,r^{2k}\log r$ remain of that type after application of the turning bands operator and formula (7.12) applies also to that case with $\alpha = 2k$. Specifically, one has $B_{n2} = 1/n$ for the spline GC $r^2\log r$, and $B_{n4} = 3/[n\,(n+2)]$ for the GC $-r^4\log r$.

7.4.3 Efficient Selection of the Line Directions

As we have just seen, a single line of random direction produces, on the average, the desired covariance. However, such a simulation is constant in all hyperplanes orthogonal to the line so that its sample covariance is not that of the model: The simulation is not ergodic in the covariance. In practice, as in (7.6) we take the normalized sum of n_D elementary simulations corresponding to n_D lines of regularly distributed or random directions. Regarding ergodicity, for fixed n_D it is preferable to use lines with directions distributed as regularly as possible rather than randomly.

In 2D, the choice of lines is simple: discretize the directions according to $\theta_j = (j-1)\pi/n_D$, $j = 1, \ldots, n_D$; one may prefer a stratified random grid and select θ_j randomly in the interval $[(j-1)\pi/n_D, j\pi/n_D]$. One should take n_D large enough (from one hundred to a few hundred).

In 3D, the maximum number of regularly spaced directions is 15. These directions are defined by the lines joining opposite edges of an icosahedron. The first 3D simulations by turning bands were constructed from 1D simulations on these 15 lines. But these 15 special directions show up on the realizations obtained. To go beyond 15 lines, one can take random directions, or several groups of 15 lines. But drawing independent random directions from a uniform distribution on the half-sphere does not provide a very regular discretization (Figure 7.8a). Quasi-random sequences give better results in this respect (Figure 7.8b). Freulon and de Fouquet (1991) recommend the following method:

1. Simulate points with a uniform distribution on $[0, 1] \times [0, 1]$ using a quasi-random sequence (they use the two-dimensional Van der Corput, or Halton, sequence; cf. Section 7.9.4).

2. To each point with components U_1 and U_2, associate the direction of unit vector $(\cos(2\pi U_1)\sqrt{1 - U_2^2}, \sin(2\pi U_1)\sqrt{1 - U_2^2}, U_2)$, which will then be uniformly distributed over the unit half-sphere.

It is recommended to use several hundred lines.

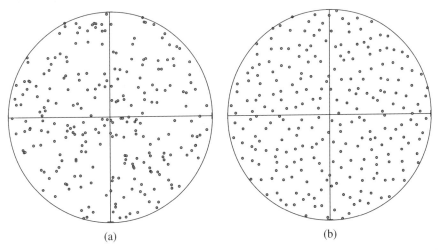

(a)   (b)

FIGURE 7.8 Schmidt equiareal projection of 256 points of the unit half-sphere, representing 256 directions in space: (a) random points; (b) quasi-random points derived from an equidistributed Halton sequence.

Finally, when using the turning bands method, the problem of simulating $S(x)$ in $\mathbb{R}^n$ boils down to the following two problems:

1. Find the covariance $K_1(h)$ defined in $\mathbb{R}^1$ associated with the isotropic covariance $K_n(r)$ to be obtained in $\mathbb{R}^n$; Table A.2 gives the covariances $K_1(h)$ associated with the most common models $K_3(r)$.
2. Simulate random functions on lines with the given covariance $K_1(h)$.

For example, to simulate the exponential covariance $K_3(r) = \sigma^2 \exp(-r/a)$ in $\mathbb{R}^3$, we have to simulate

$$K_1(h) = \sigma^2 \left(1 - \frac{|h|}{a}\right) \exp\left(-\frac{|h|}{a}\right)$$

on the lines. We recognize, up to a factor 2 for the scale parameter, the covariance (2.74) of the migration process presented in Section 2.5.3, which can be simulated very easily. This method of construction is specific to this particular model. Emery and Lantuéjoul (2006) propose specific solutions for a broad range of covariance models, which all belong to the general methods presented below.

7.5 NONCONDITIONAL SIMULATION OF A CONTINUOUS VARIABLE

We now review the main simulation methods. Some may be used directly in $\mathbb{R}^n$, others only in $\mathbb{R}$, but they may be extended to $\mathbb{R}^n$ by turning bands.

7.5.1 Autoregressive and Moving-Average Models

The two methods seen in Section 7.2 (sequential simulation and decomposition of the covariance matrix), when applied to the simulation of Gaussian RFs, differ essentially in presentation. The first one amounts to the application of (7.3) and is of autoregressive type, whereas the second, corresponding to (7.4), is based on moving averages. These methods are not applicable when the number of points is large. But in the case of stationary discrete processes, they can be adapted to summations over a finite number of terms. Box and Jenkins (1976) have developed ARMA models with a view to fitting their parameters directly from time series data, the covariance function being derived. We present these models from our slightly different perspective: The covariance of the phenomenon is known, from a structural analysis, and the parameters of the ARMA model are derived.

1D Autoregressive Process

Consider points at a regular interval Δx in 1D space, along with a "discrete-time" process $Z_t = Z(t \, \Delta x)$, t negative or positive integer. This definition recalls that we consider Z_t as a discrete version of a random process $Z(x)$ defined over the whole line. We will only consider processes that are stationary and have mean zero. Due to the orthogonality property of simple kriging, Z_t may be considered as the sum of the simple kriging estimator from all past values $Z_{t'}$, $t' < t$, and an uncorrelated random variable a_t with mean zero and variance σ_K^2 (the kriging variance):

$$Z_t = \sum_{i=1}^{\infty} \alpha_i Z_{t-i} + a_t \tag{7.13}$$

The process is said to be an autoregressive process of order p, or an AR(p) process, when only the first p weights α_j of (7.13) are nonzero:

$$Z_t = \alpha_1 Z_{t-1} + \cdots + \alpha_p Z_{t-p} + a_t \tag{7.14}$$

This process has a memory of finite length p and evolves due to random innovations a_t (the a_t are uncorrelated). Z_t has the form of a regression on its own previous values, hence the name autoregressive. This model generalizes the algorithm (7.2) presented for simulating a stochastic process with an exponential covariance.

The coefficients $\alpha_1, \ldots, \alpha_p$ of the AR(p) process are estimated very easily using the covariance $C(h)$ of the random function $Z(x)$ for $h = i \, \Delta x$ and $|i| = 0, 1, \ldots, p$. Indeed, when we multiply both sides of (7.14) by Z_{t-i} and denote $C_i = C(i \, \Delta x)$ the covariance of Z_{t-i} and Z_t, we get

$$C_i = \alpha_1 C_{i-1} + \alpha_2 C_{i-2} + \cdots + \alpha_p C_{i-p} \tag{7.15}$$

Doing the same for all $i = 1, 2, \ldots, p$ yields the p linear equations of the simple kriging system:

$$\sum_{j=1}^{p} \alpha_j C_{i-j} = C_i \qquad (i = 1, 2, \ldots, p) \tag{7.16}$$

These equations are also known as the *Yule–Walker equations*. The matrix of the system has a special pattern: The elements on symmetric diagonals are identical. Such a matrix is called a symmetric Toeplitz matrix, and an especially efficient algorithm, the Levinson algorithm, may be used to solve system (7.16) [e.g., Press et al. (2007, Section 2.8.2)].

The problem, when the covariance function is given, is whether it can be considered as that of an AR(p) process. In practice, the choice of p is made by considering the partial correlation ϕ_k, defined as the kth value α_k obtained from the Yule–Walker equations with k terms. ϕ_k can be interpreted as the kriging weight assigned to Z_{t-k} when kriging Z_t from $Z_{t-1}, \ldots, Z_{t-k}$. The covariance terms C_i are those of an AR(p) model if $\phi_k = 0$ for $k > p$. In practice, if ϕ_k does not reach zero, p is chosen such that $|\phi_k| < \varepsilon$ for $k > p$, where ε is a given threshold.

The knowledge of $C_0, C_1, \ldots, C_p$ determines the coefficients $\alpha_1, \ldots, \alpha_p$. These in turn determine C_i for $i > p$ by iterative application of relation (7.15). Box and Jenkins (1976) show that the general form of the covariance is then

$$C_i = \sum_{j=1}^{p} A_j \, (G_j)^i \qquad (i \geq 0)$$

where $G_1^{-1}, G_2^{-1}, \ldots, G_p^{-1}$ are the complex roots of the characteristic equation

$$1 - \sum_{i=1}^{p} \alpha_i z^i = 0 \qquad (z \in \mathbb{C})$$

and $A_1, A_2, \ldots, A_p$ are complex coefficients (which depend on the covariance).

In the case of an SRF, the G_i satisfy $|G_i| < 1$. Therefore the covariance is a mixture of damped exponentials (G_i real) and damped sine waves (pair of conjugate complex roots). For example, if $p = 1$, one has

$$\frac{C_i}{C_0} = \left(\frac{C_1}{C_0}\right)^i = \alpha_1^i$$

that is, an exponential model if $\alpha_1 > 0$ and a damped sine wave if $\alpha_1 < 0$ (in both cases $|\alpha_1| < 1$). Thus the method is especially suited for exponentially decreasing covariances. Other covariance models can be simulated only approximately.

An AR(p) model can be simulated with a sequential algorithm:

1. Start with a random value Z_0, drawn from the marginal distribution of $Z(\cdot)$.
2. Simulate iteratively $Z_1, Z_2, \ldots, Z_{p-1}$ by applying, respectively, the AR(1), AR(2), $\ldots$, AR($p-1$) recursions associated with the covariance of Z.
3. Simulate iteratively Z_p, Z_{p+1}, and so on, by applying the AR(p) recursion.

The random variables a_t are usually chosen Gaussian so that the simulation is Gaussian (see discussion at the end of this section).

2D Autoregressive Model

The generalization of definitions (7.13) and (7.14) to 2D is not unique because the notions of "past" and "future" have no general meaning in this space. Starting from the upper left corner of the simulated grid, Boulanger (1990) simulates in squares of increasing size from the neighboring previously simulated nodes. He proposes a methodology to find the parameters ensuring the reproduction of the covariance with a prespecified accuracy. Mignolet and Spanos (1992) propose a fairly similar method where the grid is simulated column after column.

The extension to 3D seems intractable except in special cases (e.g., separable exponential covariance).

1D Moving-Average Process

As previously, we consider points at a regular interval Δx in 1D space, along with a process $Z_t = Z(t\Delta x)$, t negative or positive integer. But Z_t is now defined as a weighted moving average of a sequence of independent identically distributed (i.i.d.) random variables a_t with mean $E[a_t] = 0$ and variance $Var[a_t] = \sigma^2$,

$$Z_t = \sum_{j=-\infty}^{+\infty} \beta_j a_{t-j} \qquad (7.17)$$

where the weights β_j are such that $\sum \beta_j^2 < \infty$. Z_t is then a stationary process with covariance

$$C_k = \sigma^2 g_k \qquad (7.18)$$

where g_k is the discrete covariogram of the weights

$$g_k = \sum_{j=-\infty}^{+\infty} \beta_j \beta_{j+k} \qquad (7.19)$$

Applications deal with the case where the number of nonzero weights is finite. Since the resulting covariance is not altered by a translation of the β_j, we may consider that the first nonzero weight is β_0 and that only the next q weights $\beta_1, \ldots, \beta_q$ are nonzero. Definition (7.17) then becomes

$$Z_t = \beta_0 \alpha_t + \beta_1 \alpha_{t-1} + \cdots + \beta_q \alpha_{t-q} \qquad (7.20)$$

and the discrete covariogram (7.19) of the weights has a range equal to $q+1$, in the sense that $g_k = 0$ for $|k| \ge q+1$.

Since the covariance is not modified if the β_j are multiplied by the same nonzero coefficient c, provided that the variance σ^2 is divided by c^2, we may also fix $\beta_0 = 1$. We then obtain the moving average process of order q, or MA(q) process, as defined by Box and Jenkins (1976). For a time process the physical interpretation is that at time t "nature" produces a "shock" or "innovation" a_t unrelated to shocks at all other times. These shocks are combined through a filter characterizing the genetic mechanism of the phenomenon, and the output is the observed signal Z_t. In the sequel we will not impose $\beta_0 = 1$.

In our approach, which differs from that of Box and Jenkins, we use such a model for a covariance $C(h)$ that has been already identified. Thus we have to derive q and the coefficients β_j such that the discrete covariance C_k given by (7.18) and (7.19) is equal to $C(k\Delta x)$. Unlike for AR models the discrete covariance C_k is not linear in $\beta_0, \beta_1, \ldots, \beta_q$ but involves products of these parameters, which makes estimation more difficult. Boulanger (1990) proves two results:

1. If $C(h)$ has finite range a, there exists a set of $q + 1$ coefficients β_j such that $C_k = C(k\Delta x)$, where $q = \lfloor a/\Delta x \rfloor$ (integer part of $a/\Delta x$).
2. If the covariance only tends to zero when h tends to infinity but has a finite integral range, it can be approximated to any prespecified precision by the covariance of an MA(q) model. The order q is taken as $q = \lfloor a'/\Delta x \rfloor$, where a' is the smallest distance such that $|C(h)| < \varepsilon$ whenever $|h| > a'$, ε being a given threshold.

There remains to find the coefficients β_j. An approximate method is often used when the covariance function is of the form $C(h) = \sigma^2 g(h)$, where $g(h)$ is known to be the covariogram of some function $w(x)$ whose support has length a (see Section 7.5.2). Since a translation of w does not change its autoconvolution function, we may suppose without loss of generality that the support of w is $[0, a]$. As an approximation we can take $\beta_j = |\Delta x|^{1/2} w(j \Delta x)$ and rescale these coefficients so that $\sigma^2 \sum_j \beta_j^2$ reproduces the desired variance $C(0)$.

That approximation is usually satisfactory provided that the discretization step Δx is small in comparison with the range. A rigorous method where the coefficients β_j are limits of series is also available (Boulanger, 1990).

In practice, MA(q) models are efficient when q has a low value, namely for covariances with a short range. In this case each innovation a_t only has a local effect on the simulation. Boulanger uses this property in a method for direct conditioning.

2D Moving-Average Model

The generalization of definition (7.17) to 2D or 3D is obvious. But again there is no unique generalization of definition (7.20). Boulanger (1990) shows that the 2D generalization of most of the results he obtained for the 1D case need a causal model, namely a lexicographic moving average, where the nonzero $\beta_{j_1 j_2}$ weights are those such that

$$j_1 = 0 \qquad \text{and} \qquad 1 \leq j_2 \leq q,$$

$$1 \leq j_1 \leq q \qquad \text{and} \qquad -q \leq j_2 \leq q$$

With this definition of a 2D MA(q) process, all the results of the 1D case can be carried over to 2D, except that now in 2D a covariance with a finite range cannot necessarily be represented *exactly* by an MA(q) process with q finite [see also Guyon (1993, Section 1.3.3)]; but as any covariance with a finite integral range, it can at least be *approximated* by such a process.

Spanos and Mignolet (1992) propose a slightly different algorithm where the moving-average domain is the rectangle defined by

$$-q_1 \le j_1 \le q_1 \qquad \text{and} \qquad -q_2 \le j_2 \le q_2$$

The coefficients are determined so as to minimize an error in the frequency domain.

One important feature of moving-average methods is their local character. Since innovations only have a local influence (defined by the parameter q), it is possible to modify a realization in a local window without changing the other regions. This feature is used by Le Ravalec et al. (2000) to match local non-linear constraints (see Section 7.8.5).

Mixed Autoregressive Moving-Average Model

There does not necessarily exist an AR(p) or MA(q) model, with p or q finite and not too high, that leads to the desired covariance. Box and Jenkins (1976) introduced a mixture of the AR(p) and MA(q) models that provides more flex-ibility. A mixed autoregressive moving-average process, or ARMA(p, q) process, includes a finite number of both autoregressive and moving-average terms

$$Z_t = \alpha_1 Z_{t-1} + \cdots \alpha_p Z_{t-p} + a_t + \beta_1 a_{t-1} + \cdots + \beta_q a_{t-q} \qquad (7.21)$$

The importance of this mixed model is that it permits a "parsimonious" representation of a linear process; that is, it requires fewer parameters than a model based on a purely autoregressive or purely moving-average process. The point is obvious if one attempts to represent an AR(p) model as an MA, or an MA(q) model as an AR: this is possible, but an infinite number of terms is required. Mignolet and Spanos (1992) and Spanos and Mignolet (1992) propose a procedure to closely approximate an autoregressive model or a moving-average model, respectively, by an ARMA model with fewer parameters (in 2D). Similarly, Samaras et al. (1985) develop the use of ARMA models with $p = q$, including in the multivariate case. Further results can be found in Guyon (1993, Chapter I).

Spatial Distribution and Conclusion

The random variables a_t involved in the definition of ARMA(p, q) models are usually chosen Gaussian, so that the simulation is Gaussian. If not, the finite-dimensional distributions of MA(q) models tend to be Gaussian by virtue of the central limit theorem, under appropriate conditions. The same applies for AR(p) models since they are equivalent to moving-average models with an infinite number of terms.

The method has been extended to the simulation of random functions that have an infinite variance, or even an infinite mean, notably stable distributions (Boulanger, 1990).

In conclusion, autoregressive moving-average models offer the following advantages:

- The multivariate distribution of the simulations can be exactly Gaussian.
- The autoregressive method can be directly used in 2D, and the moving-average method in 2D or 3D, without the need for turning bands.
- It can be applied to large grids (for fixed p and q, and once the parameters are known, the computation time is simply proportional to the number of grid nodes).
- In the case of MA(q) models, it is possible to directly produce conditional simulations.
- Moving average models allow for local updating.

Their limitations are the following:

- ARMA models cannot fit all covariance models: AR(p) models are for exponentially decreasing covariances, and MA(q) models are for covariances whose range is small compared with the simulated domain.
- Autoregressive models may present a spatial dissymmetry in 2D and cannot be used efficiently in 3D.
- The simulation is limited to grid nodes.

7.5.2   Dilution of Poisson Germs

The Poisson point process has long been used to define random function models [e.g., in $\mathbb{R}$, Blanc-Lapierre and Fortet (1953, Chapter V) and in $\mathbb{R}^n$, Matérn (1960, Chapter 3)]. The method outlined here is a dilution of Poisson points considered as germs and corresponds to processes called "moving average models with constant or stochastic weight function" by Matérn and Poisson shot noise by several authors. It is a good candidate for generating the stationary 1D simulations required by the turning bands method when the 1D covariance has a finite range, as well as for directly generating simulations in $\mathbb{R}^n$ with spatially variable parameters (anisotropy directions, ranges, and obviously sill). The approach can also be extended to IRFs and IRF$-k$'s. Let us first recall the definition and main properties of the Poisson point process.

Poisson Point Process

The Poisson point process in $\mathbb{R}^n$ corresponds exactly to the intuitive idea of points distributed in space "at random." The Poisson point process with *intensity*, or density, λ ($\lambda > 0$) is characterized by the following properties:

1. The number $N(V)$ of points inside a domain V is a Poisson random variable with parameter $\lambda \, |V|$, where $|V|$ represents the measure of V (length, surface, or volume)

$$\Pr\{N(V) = k\} = e^{-\lambda \, |V|} \frac{(\lambda \, |V|)^k}{k!},$$

$$E[N(V)] = \mathrm{Var}[N(V)] = \lambda \, |V|$$

2. If V_i, $i = 1, 2, \ldots, p$, are pairwise disjoint domains, the random variables $N(V_i)$ are mutually independent.

The Poisson point process has an important conditional property that corresponds to the notion of random points: Given $N(V) = n_V$, these points are independently and uniformly distributed over V.

Thus a Poisson point process with intensity λ can be simulated within a bounded domain V as follows:

1. Draw the number of points n_V from a Poisson distribution with mean $\lambda|V|$.
2. Draw the n_V points independently from a uniform distribution within V.

This method can easily be applied to intervals of $\mathbb{R}^n$ (segments in $\mathbb{R}^1$, rectangles in $\mathbb{R}^2$, parallelepipeds in $\mathbb{R}^3$). Combined with an acceptance-rejection technique, it allows the simulation of a Poisson point process in a complex domain assumed to be enclosed in a union of pairwise disjoint sets. In situations where the acceptance ratio is too low, which may be the case in spaces of large dimension (e.g., in design and analysis of computer experiments), iterative methods are available [e.g., Lantuéjoul (2002, Section 8.3)].

In 1D, Poisson points delimit intervals whose lengths are independent random variables with an exponential distribution with mean $\theta = 1/\lambda$. This distribution has the following conditional property: If it is known that the interval length T is larger than a given value t, the residual length $T - t$ has the same distribution as the a priori length T. In particular, the residual length from a fixed origin to the next Poisson point has the same exponential distribution as T. Hence another method for simulating a 1D Poisson point process from an arbitrary origin, which without loss of generality we will place at $x = 0$:

1. Draw intervals T_i, $i = 1, 2, \ldots$, as independent random variables from an exponential distribution with mean $\theta = 1/\lambda$.
2. Define the successive Poisson points X_i, $i = 1, 2, \ldots$, associated with these intervals by $X_1 = T_1$ and the recurrence relation $X_i = X_{i-1} + T_i$, $i = 2, 3, \ldots$.

The Poisson point process can be defined also on $]-\infty, 0]$ by inverting the method of construction. In all cases the origin $x = 0$ is not a Poisson point.

Dilution by a Fixed Function

Let $w(x)$ be a square integrable function on $\mathbb{R}^n$ and $g(h)$ its covariogram (2.30)

$$g(h) = (w*\breve{w})(h) = \int w(x)\, w(x+h)\, dx \qquad (7.22)$$

For the sake of simplicity, we will assume that $w(x)$ is integrable. We assign a constant "dose" μ to each of the points X_i of a Poisson point process with intensity λ and dilute it by the influence function w (Figure 7.9). At a point x the sum of the contributions of all X_i defines a random function

$$Z(x) = \mu \sum_i w(x - X_i) \qquad (7.23)$$

A variant is to assign independent random doses a_i with mean μ and variance σ^2, which corresponds to the definition

$$Z(x) = \sum_i a_i\, w(x - X_i) \qquad (7.24)$$

Note that (7.23) can be considered as the special case $\sigma = 0$. With this convention, $Z(x)$ is an SRF with mean m and covariance $C(h)$ given in both cases by

$$\begin{aligned} m &= \lambda\mu \int w(x)\,dx \\ C(h) &= \lambda(\mu^2 + \sigma^2)\,g(h) \end{aligned} \qquad (7.25)$$

Heuristic proofs of these results are easy. A rigorous proof can be found in Blanc-Lapierre and Fortet (1953).

Some results persist with point models other than Poisson points. Consider, for example, the transposition of definition (7.24) to a regular grid of points X_i with a random origin

$$Z(x) = \sum_{i \in \mathbb{Z}^n} a_i\, w(x - X_0 - i\,\Delta x) \qquad (7.26)$$

where the vector Δx represents the grid spacing and X_0 a random origin within a grid cell centered at the origin of the axes. If $\mu = 0$, it is easy to prove that

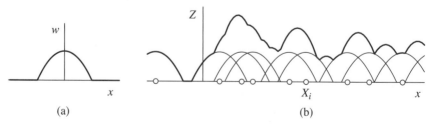

(a)   (b)

FIGURE 7.9   Construction of a simulation by dilution of Poisson germs: (a) dilution function; (b) Poisson points and the construction of the simulation.

formula (7.25) remains valid with $\lambda = 1/|\Delta x|$. This model constitutes the generalization of moving-average models to the continuous case. For fixed $X_0 = x_0$, the RF $Z(x)$ is not stationary, and therefore many independent simulations must be added to achieve stationarity. This model is very convenient to build the numerous 1D simulations involved in the use of the turning bands method when the support of w is bounded: By selecting Δx equal to the length of this support, the value of $Z(x)$ at any given x involves only one dilution point.

Now the problem is the following: Given a covariance function $C(h)$, does there exist an integrable function $w(x)$ such that $w * \breve{w} = C$? Since $w(x)$ has a Fourier transform, which we denote by $\varphi(u)$, $w * \breve{w}$ has the Fourier transform $|\varphi(u)|^2$, and therefore the answer is limited to covariances that possess a spectral density $f(u)$ (in the sense of ordinary functions). Any function $w(x)$ whose Fourier transform satisfies $|\varphi(u)|^2 = f(u)$ has $C(h)$ as autoconvolution function, as the following diagram shows, where $\mathcal{F}$ represents the Fourier transform and $\mathcal{F}^{-1}$ its inverse:

$$
\begin{array}{ccc}
w(x) & \xrightarrow{\; w * \breve{w} \;} & C(h) \\[4pt]
\mathcal{F}\downarrow\uparrow\mathcal{F}^{-1} & & \mathcal{F}\downarrow\uparrow\mathcal{F}^{-1} \\[4pt]
\varphi(u) & \xrightarrow[\; |\varphi|^2 \;]{} & f(u)
\end{array}
$$

This is the case, for example, for $w = \mathcal{F}^{-1}|f|^{1/2}$ if $|f|^{1/2}$ is integrable. This function is real and symmetric, but does not necessarily have a bounded support when the support of $C(h)$ is bounded, which would be a desirable property. Other functions may therefore be more appropriate than $\mathcal{F}^{-1}|f|^{1/2}$. In a very thorough paper, Ehm et al. (2004) seek the conditions under which a covariance with range a (i.e., such that $C(h) = 0$ whenever $|h| \geq a$) can be expressed in the form (7.22) with a real function w satisfying $w(x) = 0$ for $|x| \geq a/2$. In the 1D case, this is always possible. In n-dimensional space, $n > 1$, this is possible for isotropic covariances if and only if a series of conditions involving the zeros of the analytic continuation of the radial part of the spectral density f are met (the function w is then isotropic, up to a translation). For example, the spherical covariance, defined as the autoconvolution of the indicator function of the sphere of $\mathbb{R}^3$ with diameter a, cannot be expressed as an autoconvolution in $\mathbb{R}^2$.

Example 1: Simulation of Classic Models on the Line and in Space

Most stationary covariance models can be expressed as the autoconvolution $w * \breve{w}$ of a simple function, which makes them very easy to simulate on a line. Table A.1 gives the function w associated with some covariance models considered in $\mathbb{R}$ (i.e., no turning bands method is applied). Table A.2 gives the function w for the covariance $C_1(h)$ in $\mathbb{R}$ associated with some covariance models $C_3(h)$ in $\mathbb{R}^3$ by the turning bands relation (7.11). Isotropic covariances with a finite range a in $\mathbb{R}^3$ are associated with covariances with the same finite range in $\mathbb{R}$. The latter can be efficiently simulated in the form (7.26) with $\Delta x = a$, as proposed by Lantuéjoul (1994; 2002, Chapter 15) for the spherical covariance.

Note that in $\mathbb{R}$ the exponential covariance is obtained with a dilution function which is itself exponentially decreasing (but equal to zero for negative values of x). The consequence is that the process $Z(x)$ decreases exponentially at the same rate as the covariance, except at the Poisson points where it has a unit jump (Figure 7.2b). It is an Ambarzumian process (Matheron, 1969b).

Note also that some models can be simulated directly in $\mathbb{R}^n$, $n > 1$, as shown by Matérn (1960, p. 30): the Matérn model (2.56) can be expressed as an autoconvolution function. A special case of this model, the exponential covariance (2.51), can be obtained in $\mathbb{R}^n$ with

$$w(x) \propto (|x|/a)^{-\frac{n-1}{4}} K_{\frac{n-1}{4}}(|x|/a) \tag{7.27}$$

but $w(x)$ is infinite at the origin [K_ν is the modified Bessel function of the second kind (A.4)]. Figures 7.2c and 7.3c are 1D and 2D sections of a 3D simulation built with this dilution function, which takes the form

$$w(x) = \frac{1}{\sqrt{2\pi a^3}} \frac{\exp(-|x|/a)}{|x|/a}$$

Another procedure, based on a mixture of spherical models, will be considered later. It is more convenient than a dilution by the deterministic function (7.27), since the spherical models are the easiest ones to simulate in $\mathbb{R}^n$, as shown below.

The Gaussian model (2.54) can also be expressed as an autoconvolution by the function

$$w(x) \propto \exp(-2x^2/a^2) \tag{7.28}$$

for all n.

Example 2: Simulation of Spherical Models

The indicators of simple sets are among the most convenient dilution functions. In particular, the spherical model of $\mathbb{R}^n$ can be obtained from the indicator of a sphere with diameter a:

$$w_n(x) = \begin{cases} 1 & \text{if } |x| \leq a/2, \\ 0 & \text{if } |x| > a/2 \end{cases}$$

If a simulation has to be built over a domain V, the Poisson point process must be simulated over the dilated[2] set $V \oplus S_{\breve{w}}$, where $S_{\breve{w}}$ is the support of the function $\breve{w}(x) = w(-x)$. In $\mathbb{R}^3$ the steps are then the following:

1. Define a domain D that includes $V \oplus S_{\breve{w}}$. If the final objective is a conditional simulation, do not forget that V must include all the conditioning data points; in practice, take the smallest parallelepiped containing all the points to be simulated and all the data points, and extend it by $a/2$ on all sides (or take the smallest sphere that contains all these points, and increase its radius by $a/2$).
2. Generate N_D from a Poisson distribution with mean $\lambda |D|$.
3. Select N_D independent random points X_i from a uniform distribution within D.
4. At each of these points, set up a sphere with diameter a weighted by an independent random value with mean μ and variance σ^2.
5. Compute the value of the simulation at any given point x as the sum of the weights of the spheres containing x.

The resulting simulation has a spherical variogram with range a and sill $\frac{\pi}{6} \lambda(\mu^2 + \sigma^2)a^3$.

As seen in Section 2.5.1, the functions of $\mathbb{R}^{n-2q}$ deduced from w_n by Radon transform of order $2q$ allow the simulation of the integrated spherical models (2.48) associated with random functions which are differentiable q times in $\mathbb{R}^{n-2q}$. For example, integrating w_5 along two coordinate axes yields

$$
w_{5-2}(x) = \begin{cases} \pi\left(\dfrac{a^2}{4} - x^2\right) & \text{if } |x| \leq a/2, \\ 0 & \text{if } |x| > a/2 \end{cases}
$$

which generates the "cubic" model (2.49).

The method can be adapted to reproduce local variations in the variogram parameters. For example, a geometric anisotropy of variable amplitude can be simulated by forming ellipsoids whose shape, size, and orientation depend on location. Likewise, the second moment $\mu^2 + \sigma^2$ can be adjusted if the sill varies.

Dilution by a Stochastic Function

Figure 7.10a illustrates the construction of the spherical model of $\mathbb{R}^2$, namely the circular variogram (2.46). This construction is based on random disks of equal size. The circular variogram is cumbersome, and the spherical model of $\mathbb{R}^3$ is often used even when working in $\mathbb{R}^2$. The simulation method described above amounts to (a) picking random points in the slice delimited by the two planes

---

[2] The dilation $V \oplus S$ is the union of all translates of V by a vector of S [see Serra (1982)].

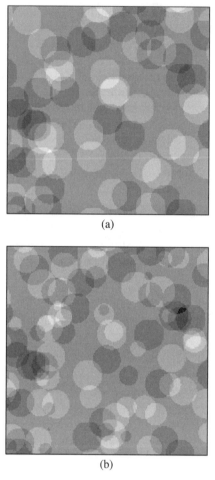

(a)

(b)

FIGURE 7.10 2D simulation of circular and spherical covariances by dilution: (a) circular covariance (identical coins); (b) spherical covariance (random coins).

parallel to the simulation plane and at distance $a/2$ from it and (b) forming spheres with constant diameter a. This is equivalent to picking random points in the plane (with a 2D intensity $\lambda_2 = a\, \lambda_3$ where λ_3 represents the 3D intensity of sphere centers) and forming disks with random diameters drawn from the c.d.f. $F_a(d)$ of the diameters of random planar sections of a sphere with diameter a

$$
F_a(d) = \begin{cases} 1 - \sqrt{1 - \dfrac{d^2}{a^2}} & \text{if } d \le a \\[2ex] 1 & \text{if } d \ge a \end{cases}
$$

This method has been popularized by Alfaro (1980) under the name of the *random coins method* (Figure 7.10b). However, it had been already shown by Matérn (1960, p. 36) that the spherical model of $\mathbb{R}^{m+n}$ can be simulated in $\mathbb{R}^n$ with

$$
F_a(d) = \begin{cases} 1 - \left(1 - \dfrac{d^2}{a^2}\right)^{m/2} & \text{if } d \le a, \\ 1 & \text{if } d \ge a \end{cases}
$$

By using other distributions F for d, the method can be extended to other covariance models with a linear behavior at the origin, and in particular to the exponential model in $\mathbb{R}^n$. For example, in $\mathbb{R}$ the exponential covariance $\exp(-|h|/a)$ can be simulated by random segments with lengths drawn from the c.d.f.

$$
F(d) = 1 - e^{-d/a}
$$

and a choice of λ, μ, and σ^2 such that $\lambda\,(\mu^2 + \sigma^2) = 1/a$. A practical problem arises here, as for other covariances that tend only asymptotically to zero so that the distribution of d extends to infinity: Even if the simulated domain is bounded, it is theoretically necessary to implement germs in the whole space $\mathbb{R}^n$, but most of the spheres associated with these germs will not intersect the simulated domain. Hammersley and Nelder (1955) have given a thorough general solution in $\mathbb{R}^n$ and detailed its use for the exponential model for $n = 1, 2, 3$. The procedure is defined so that a finite number of spheres intersect the simulated domain and is built so as to only generate these spheres. Figure 7.2d shows a simulation on the line (with random segments) and Figure 7.3d a simulation in the plane (random coins). This method allows the simulation of any completely monotone covariance in $\mathbb{R}^n$ for all n, since such a covariance can be expressed in the form (2.28) as a mixture of exponential models.

Stochastic dilution functions other than indicators of spheres can of course be used. Since any isotropic covariance model valid in $\mathbb{R}^n$ for all n can be considered as a mixture of Gaussian covariances according to relation (2.27), such a model can in principle be simulated with dilution functions (7.28) with a random scale parameter a [e.g., see Matérn (1960, p. 33) for the Matérn and Cauchy models].

1D Extension to IRFs

The simulation by dilution of Poisson germs requires, in principle, square integrable functions and only permits the simulation of ordinary covariances. This method can, however, be extended to functions $w(x)$ such that

$$
G(h) = \frac{1}{2}\int [w(x + h) - w(x)]^2 dx \tag{7.29}
$$

exists. Reasoning along the same lines as before, one obtains an IRF-0 with the variogram $\gamma(h) = \lambda\, G(h)$.

Consider now the case of an $|h|^\alpha$ model in $\mathbb{R}$. As shown by Mandelbrot and Van Ness (1968), the function $G(h) = b\, |h|^\alpha$ can be written in the form (7.29) with

$$
w(x) = \begin{cases}
0 & \text{if } x \le 0, \\
\sqrt{b/R_\alpha}\, x^{(\alpha-1)/2} & \text{if } x > 0
\end{cases}
$$

and

$$
R_\alpha = \frac{1}{2\alpha} + \frac{1}{2} \int_0^\infty \left[(1+u)^{(\alpha-1)/2} - u^{(\alpha-1)/2} \right]^2 du
$$

Mandelbrot (1975a) uses a weighting function related to the above one by antisymmetry (i.e., $w(x) = -w(-x)$ for $x < 0$; R_α is then modified). In both cases, R_α is finite provided that $0 < \alpha < 2$, which is precisely the condition for $|h|^\alpha$ to be a variogram. Since the support of w is infinite, even Poisson points at a large distance from the simulated domain have an influence on the simulated values within that domain. In practice, only Poisson points up to a certain distance from the simulated domain are considered, which results in a minor approximation. As shown by Chilès (1995), a large number of Poisson points is necessary to achieve ergodicity (typically 10,000 Poisson points).

Special Case of a Poisson Random Walk

An important special case is $\alpha = 1$. Then $w(x)$ is simply a step function, so Poisson germs falling outside the simulated domain only contribute a constant value and do not modify the increments: There is no need to simulate these germs. The RF model, known as the Poisson random walk, is usually defined on $[0, +\infty[$ without reference to dilution:

1. Consider the Poisson points $\{X_i : i = 1, 2, \ldots\}$ of a 1D Poisson point process with intensity λ defined on $[0, +\infty[$.
2. Consider a series of i.i.d. random variables $\{a_i : i = 1, 2, \ldots\}$ with mean 0 and variance σ^2.
3. Define $Z(x)$ as being equal to 0 in the interval $[0, X_1[$, jumping by a_i at the point X_i, and remaining constant in the interval $[X_i, X_{i+1}[$, $i = 1, 2, \ldots$.

$Z(x)$ can be defined for negative x values by inverting the construction method. $Z(x)$ is a step function: It increases by i.i.d. random jumps at the Poisson points and remains constant between successive discontinuity points. $Z(x+h) - Z(x)$, $h > 0$, is then equal to the sum of the $N([x, x+h[)$ random jumps that occur between x and $x+h$. By the properties of the Poisson point process and because the random jumps are independent, $Z(x)$ is a process with independent and

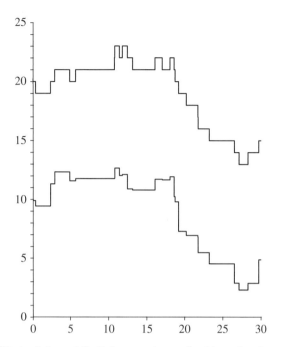

FIGURE 7.11 1D simulations of the Poisson random walk with random jumps $+1$ or -1 (top) and with Gaussian jumps (bottom). They have the same linear variogram $\frac{1}{2}|h|$ as the Brownian motion of Figure 7.4.

strictly stationary increments [e.g., Doob (1953)]. In particular, since $N([x, x+h[)$ is a Poisson random variable with parameter $\theta = \lambda |h|$ and the random jumps have mean 0 and variance σ^2, the Poisson random walk has no drift and a linear variogram $\gamma(h) = \frac{1}{2}\lambda\sigma^2|h|$.

Figure 7.11 displays such simulations for a Poisson random walk with intensity $\lambda = 1$ and random jumps equal to -1 or $+1$ (top), or to a standard normal variable (bottom); they have the same variogram as the Brownian motion of Figure 7.4. Note that $Z(x)$ is not a Gaussian IRF-0, even though the jumps are Gaussian: For example, since $Z(x)$ is piecewise constant, an increment $Z(x+h) - Z(x)$ has a nonzero probability to be equal to 0 (this probability is at least equal to $e^{-\lambda|h|}$, the probability that no Poisson point falls between x and $x+h$). Finally, note that since $Z(x)$ is a step function without accumulation points, its Hausdorff dimension is 1: $Z(x)$ has a linear variogram and yet is not fractal.

1D Extension to IRF-k with a Polynomial Generalized Covariance

The requested 1D IRF$-k$ can be obtained by linear combination of successive integrations of a process $W_0(x)$ with a linear variogram $\gamma(h) = |h|$—for example, a Poisson random walk or a Brownian motion, as we have seen in Section 4.5.8.

A special choice for the process $W_0(x)$ in an interval $[-R, R]$ is to use a single discontinuity, selected from a uniform distribution in that interval, and a step function whose value is zero in the half line containing the origin and a random value with mean zero and variance $4R$ in the other half line. Denoting by X_0 the random point and by a the random step, the pth integral of W_0 has the simple form

$$
W_p(x) = \begin{cases} \dfrac{a(x - X_0)^p}{p!} 1_{x < X_0} & \text{if } X_0 < 0, \\[3mm] \dfrac{a(x - X_0)^p}{p!} 1_{x \geq X_0} & \text{if } X_0 \geq 0 \end{cases}
$$

To be more explicit, if we limit ourselves to $k = 2$, any IRF$-k$ with generalized covariance

$$
K(h) = -b_0|h| + b_1|h|^3 - b_2|h|^5
$$

can be obtained with

$$
Z(x) = c_0 W_0(x) + c_1 W_1(x) + c_2 W_2(x)
$$

where the coefficients c_0, c_1, c_2 are deduced from b_0, b_1, b_2 by

$$
c_0 = \sqrt{b_0}, \qquad c_2 = \sqrt{120b_2}, \qquad c_1 = \sqrt{6b_1 + 2c_0 c_2}
$$

Note that the terms under the square roots are positive or zero; this results from the conditions on the coefficients b_0, b_1, b_2 to ensure that $K(h)$ is a valid generalized covariance in $\mathbb{R}$.

Obviously, this simulation (with a single discontinuity) is not ergodic. Thus a large number of such simulations, properly scaled, must be added.

Extension to n-Dimensional IRFs with Turning Bands or Poisson Hyperplanes

Simulations of a linear, power, or polynomial generalized covariance in $\mathbb{R}^n$ can be easily obtained by turning bands from independent 1D simulations generated as shown above, since the corresponding 1D generalized covariances remain of the same type.

If the directions normal to the turning bands are random and each 1D simulation contains a single germ, as seen above for a linear variogram, the method is equivalent to a simulation based on Poisson hyperplanes (see Section 7.6.5 for their definition and main properties). We will use the 2D terminology (with Poisson lines) for simplicity, but the generalization to $\mathbb{R}^n$, $n > 2$, is straightforward. Consider a disk with radius R enclosing the simulation domain and the lines of a Poisson line process that intersect this disk. Consider each of these lines as a fault subdividing the plane into two half-planes. More precisely, associate with line number i a random function $W_i(x)$ equal to zero in the half-plane containing the origin (the center of the disk) and to an independently selected random variable a_i with mean 0 and variance σ^2 in the other half-plane. Define $Z(x)$ as the sum of all the $W_i(x)$ (the sum always contains a finite number of nonzero terms). Since the intersections of a Poisson line process with a given line is a Poisson point process whose intensity λ characterizes the Poisson line process, the values taken by $Z(x)$ along any line are those of a Poisson random walk, and hence $Z(x)$ is an IRF-0 with linear variogram $\gamma(h) = \frac{1}{2}\lambda\sigma^2|h|$.

The cross-sectional profile of the faults is a step function. If we replace it by a profile proportional to $w(s)$ (s is the abscissa along the perpendicular to the fault, the origin being on the fault), we get a method that has been proposed by Mandelbrot (1975a) for simulating an $|h|^\alpha$ variogram in $\mathbb{R}^n$, but now we must also consider the Poisson lines that do not intersect the disk. Figure 7.12 shows a 2D simulation of a linear variogram.

Spatial Distribution and Conclusion

The dilution method allows the simulation of specific types of random functions. If the simulations are generated according to formula (7.23) with a constant dose $\mu = 1$ and a fixed indicator function w with support S_w, they have a Poisson marginal distribution with parameter $\theta = \lambda |S_w|$. The function w is determined by the target covariance and λ is chosen so as to lead to the desired value of θ. Lantuéjoul (2002, Section 14.1) provides an iterative algorithm to condition such simulations. Stable random functions can also be simulated using stable random numbers (Boulanger, 1990).

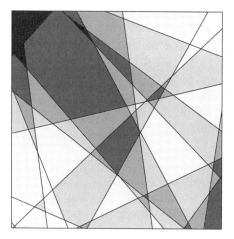

FIGURE 7.12 2D simulation with a linear variogram, constructed from Poisson lines.

Otherwise a Gaussian simulation is often desirable. One would think it can be obtained by using definition (7.24) with i.i.d. normal a_i. This is a double mistake:

1. The dilution of a Poisson point process cannot provide an RF whose marginal distribution is exactly Gaussian. Indeed, limiting the discussion to SRFs, when the support of w is finite—which it always is in practice—the expression (7.24) reduces to a sum of N i.i.d. random variables, where N is distributed as a Poisson variable with parameter $\theta = \lambda \, |S_w|$, with S_w being the support of w. Denoting by $\chi(u)$ the characteristic function of the random variable $a_i \, w(x - X_i)$, the characteristic function of $Z(x)$ is $E[\exp(i \, uZ)] = \exp[-\theta \, (1 - \chi(u))]$. It cannot be of the form $\exp(-\sigma_Z^2 u^2/2)$, which is the characteristic function of a Gaussian distribution with variance σ_Z^2 (necessarily $|\chi(u)| \leq 1$). The conclusion is the same with stochastic dilution functions.

2. On the other hand, it can be shown that the distribution of a_i that provides the best approximation of the marginal distribution of $Z(x)$ by a Gaussian is not the Gaussian but the discrete distribution $a_i = \pm\sigma$ with equal probability. The larger the λ, the better the approximation.

In fact, when $w(x)$ is an indicator function, the space is partitioned by the functions $w(x - X_i)$ into zones where the simulation is constant. To get simulations that look continuous rather than piecewise constant at medium scale and that are ergodic in the covariance, it is necessary to use a sufficiently large value for λ. Gaussian simulations are obtained as the limit when $\lambda \to \infty$ and $(\mu^2 + \sigma^2) \to 0$ so that $\lambda(\mu^2 + \sigma^2)$ remains constant (thus producing an infinite number of infinitely small "discontinuities" within any bounded domain). This is also true for simulations based on Poisson hyperplanes.

To summarize, the advantages of the dilution method are the following:

- The method can be directly used for simulating in 2D or 3D (or higher dimension) without the need for turning bands (at least for SRFs).
- The simulation can be calculated exactly at any point and not only at grid nodes.
- The method is very flexible and can be adapted to a regionalization of the covariance parameters—for example, to variations of the ratio and/or directions of an anisotropy, in the framework of a local or global model (see end of Section 2.6.1).
- It can generate a large class of non-Gaussian random functions.

Its limitations are the following:

- The simulations can be Gaussian only asymptotically.
- Finding an appropriate dilution function may be difficult if an unusual covariance model is considered.

- Approximations are necessary when the support of the dilution function is unbounded.

7.5.3 Continuous Spectral Method

Continuous Spectral Simulation of Stationary Random Functions

Since any SRF has a spectral representation (Sections 1.1.5 and 2.3.3), simulating an SRF with a given covariance can be achieved by applying (2.16)—that is, by simulating an orthogonal random spectral measure satisfying (2.17). This can be obtained by simulating two uncorrelated *real* random measures satisfying (1.9) (e.g., in $\mathbb{R}$, two uncorrelated processes with independent increments over $[0, +\infty[$). But this sets up discretization problems, and the basic simulation algorithm, first proposed by Khinchin (1934), is as follows:

Let U be a random vector of $\mathbb{R}^n$ with probability distribution $F(du)/\sigma^2$, where $\sigma^2 = \int F(du)$ (random frequency), and let Φ be a random variable with uniform distribution over $[0, 2\pi[$ (random phase), independent of U. Then the random function in $\mathbb{R}^n$ defined as

$$Z(x) = \sigma\sqrt{2}\cos(2\pi <U, x> + \Phi) \tag{7.30}$$

is stationary (wide sense) with mean 0 and covariance $C(h)$ given by (2.14):

$$C(h) = \int e^{2\pi i <u, h>} F(du) = \int \cos(2\pi <u, h>) F(du)$$

Indeed, for fixed $U = u$, the covariance of $Z(\cdot)$ is

$$C(h)|_{U=u} = \sigma^2 \cos(2\pi <u, h>)$$

Randomizing U with the distribution $F(du)/\sigma^2$ yields the covariance $C(h)$. Obviously, such simulations are not ergodic: The covariance of a particular realization is a cosine function and the desired covariance is reproduced (up to the usual fluctuations) only on the average over many realizations. Therefore, in practice, one uses a sum, appropriately scaled, of many basic simulations. This point will be examined at the end of this section.

In the isotropic case in $\mathbb{R}^n$, $n > 1$, choosing the random frequency vector U amounts to independently choosing its modulus and its orientation. It is then apparent that the continuous spectral method is equivalent to turning bands with a single random line and a 1D spectral simulation on this line. In practice, to achieve ergodicity, it is preferable to sum basic simulations based on lines with directions distributed as regularly as possible rather than randomly, which amounts to use the turning bands method with a large number of lines, and 1D spectral simulations on these lines.

Table A.1 gives the spectral density (2.18) associated with the most common covariance models in $\mathbb{R}$, which will be used for 1D simulation of these models. Table A.2 gives the spectral density of the covariances in $\mathbb{R}$ associated by turning bands with the most common models in $\mathbb{R}^3$. Mantoglou and Wilson (1982) present simulations combining turning bands and 1D spectral simulations. Early presentations of the spectral method directly in $\mathbb{R}^n$ with applications to turbulent velocity fields and hydrology are due to Shinozuka (1971) and to Mejía and Rodríguez-Iturbe (1974), respectively (with an extension to space-time phenomena).

Extension to Intrinsic Random Functions

As we have seen in Section 2.3.3, the spectral representation of an IRF-0 is very similar to that of an SRF. The main difference is that the spectral measure $F(du) \equiv \chi(du)/(4\pi^2|u|^2)$ associated with the variogram is no longer necessarily finite: The low frequencies can have infinite variance. To circumvent this problem, Emery and Lantuéjoul (2008) propose an extension of the usual algorithm (7.30) which limits the probability density of low-frequency contributions and in counterpart increases their amplitude. The basic random function is defined as

$$Z(x) = \theta(U)\cos(2\pi <U, x> + \Phi) \tag{7.31}$$

where U is a random vector of $\mathbb{R}^n$ with a symmetric probability density function $f(u)$ (random frequency), Φ is a random variable with uniform distribution over $[0, 2\pi[$ (random phase), independent of U, and $\theta(u)$ is an amplitude depending on u. Since we are interested in the simulation of isotropic random functions, $\theta(u)$ can be considered as an isotropic function. Moreover, since the kinship of definition (7.30) with turning bands remains true for definition (7.31), we only need to examine the one-dimensional case. Definition (7.31) then amounts to

$$Z(x) = \theta(U)\cos(2\pi Ux + \Phi)$$

where U is a positive random variable with probability density function $f(u)$, Φ is a random variable with uniform distribution over $[0, 2\pi[$, independent of U, and $\theta(u)$ is a function of u. Since U and Φ are independent, the increment $Z(x+h) - Z(x)$ has zero mean and a semivariance equal to

$$\frac{1}{2}\mathrm{E}[Z(x+h) - Z(x)]^2 = \mathrm{E}[\theta^2(U)\sin^2(\pi Uh)] = \int_0^\infty \theta^2(u)\frac{1 - \cos(2\pi uh)}{2}f(u)\,du$$

This variance only depends on h; hence, $Z(\cdot)$ is an IRF-0. According to the spectral representation (2.21) of a variogram, for $Z(\cdot)$ to have the variogram $\gamma(h)$ and the associated measure $\chi(du)$ it suffices to choose f and θ such that

$$\frac{1}{2}\theta^2(u)f(u)\,du = \frac{\chi(du)}{2\pi^2u^2} \tag{7.32}$$

In the sequel we will only consider power variograms or generalized covariances, as well as log-polynomial generalized covariances, which remain of the same type after application of the turning bands operator.

In $\mathbb{R}$ and for the power variogram $b\,|h|^\alpha$, $b>0$, requirement (7.32) amounts to

$$\tfrac{1}{2}\theta^2(u)f(u) = -\frac{2\Gamma(\frac{\alpha+1}{2})}{\pi^{\alpha+1/2}\Gamma(-\frac{\alpha}{2})}\frac{b}{u^{\alpha+1}} \qquad (u>0)$$

The choice of f and θ is not unique. Emery and Lantuéjoul consider the pair

$$f(u) = \frac{\sin(\pi\alpha/2)}{\pi}\frac{1}{(1+u)\,u^{\alpha/2}}, \qquad \theta(u) = \sqrt{\frac{4b\,\Gamma(\alpha+1)}{(2\pi)^\alpha}\frac{1+u}{u^{\alpha/2+1}}} \qquad (u>0)$$

This function f is the probability density function of a beta random variable U of the second kind with parameters $(1-\alpha/2, \alpha/2)$. Equivalently, U is the ratio of two independent standard gamma random variables with shape parameters $1-\alpha/2$ and $\alpha/2$, respectively, the simulation of which can be achieved by acceptance–rejection algorithms.

Extension to IRF−k

The above method can easily be extended to IRF−k with a power generalized covariance like $(-1)^{k+1}\,|h|^\alpha$, $2k<\alpha<2k+2$, because this is the generalized covariance of the kth integral of an IRF−0 with a variogram in $|h|^{\alpha-2k}$ (whose exponent is between 0 and 2) and the kth integral of a sine or cosine function remains of that type. The above method can be applied with the density f associated to $\alpha-2k$ and the function θ associated to α.

Other choices are possible for f and θ. For instance, f is the probability density function of a beta random variable U of the second kind with parameters $(1/2, 1/2)$ and therefore does not depend on α:

$$f(u) = \frac{1}{\pi(1+u)\sqrt{u}}, \qquad \theta(u) = \sqrt{-\frac{4\pi b\,\Gamma(\frac{\alpha+1}{2})}{\Gamma(-\frac{\alpha}{2})}\frac{1+u}{(\pi u)^{\alpha+1/2}}} \qquad (u>0)$$

By successive integrations, IRF−k's with a power generalized covariance in $(-1)^{k+1}\,|h|^\alpha$, $2k<\alpha<2k+2$ are obtained with

$$f(u) = \frac{1}{\pi(1+u)\sqrt{u}}, \qquad \theta(u) = \sqrt{-b\frac{\pi^{1/2-\alpha}}{2^{2k-2}}\frac{\Gamma(\alpha+1)}{\Gamma(\alpha-2k+1)}\frac{\Gamma(\frac{\alpha+1}{2}-k)}{\Gamma(k-\frac{\alpha}{2})}\frac{1+u}{u^{\alpha+1/2}}} \qquad (u>0)$$

This choice is well suited to the simulation of IRF−k's with a spline generalized covariance. Indeed, as seen in Section 4.5.6, the generalized covariance

$(-1)^{k+1} |h|^{2k} \log(|h|)$ can be obtained as a limit of power generalized covariances

$$(-1)^{k+1}|h|^{2k} \log(|h|) = \lim_{\alpha \to 0} (-1)^{k+1} \left[\frac{|h|^{2k+\alpha} - |h|^{2k}}{\alpha} \right]$$

Since a generalized covariance of order k is defined up to an even polynomial of order $2k$, an IRF$-k$ with that covariance is obtained with

$$f(u) = \frac{1}{\pi(1+u)\sqrt{u}}, \qquad \theta(u) = \sqrt{-\frac{b\,\Gamma(2k+1)}{(2\pi)^{2k-1}} \frac{1+u}{u^{2k+1/2}}} \qquad (u > 0)$$

Spatial Distribution and Conclusion

The random function defined by (7.30) or (7.31) is not second-order ergodic: its realizations are sinusoidal functions in $\mathbb{R}^1$, cylinders with a sinusoidal base in $\mathbb{R}^2$, and so on (Figures 7.2a and 7.3a); the covariances of these realizations are cosine functions so that none of them has the desired covariance $C(h)$ or generalized covariance $K(h)$ (except of course if $C(h)$ itself is a cosine covariance). Simulations of this random function will only have the correct (generalized) covariance on the average. Therefore, in practice, one uses the sum, appropriately scaled, of a large number of independent basic simulations to approximate joint normality and ergodicity. Lantuéjoul (1994) shows that in the case of an exponential covariance, several thousand basic simulations must be added to correctly reproduce the bivariate Gaussian distributions (a relative error criterion, which is very stringent, was used).

The advantages of this method are the following:

- A simulation can be produced for any covariance function (provided that its Fourier transform can be computed).
- The exact value of the simulation can be computed at any location.

Its limitations are the following:

- Ergodicity is very slow to reach.
- The method may be computationally tedious if many random frequencies are superimposed in order to approximate joint normality and ergodicity (the cosine terms are all different).

In practice, the method is suitable when the spectrum is not widely spread, namely for differentiable SRFs, whose covariance have a parabolic behavior at the origin. In the discrete case with spacing Δx, it may be more convenient to consider the corresponding infinite series $Z_n = Z(n\Delta x)$, $n \in \mathbb{Z}$, with discrete covariance $C_m = C(m\Delta x)$, $m \in \mathbb{Z}$, and its spectral representation (which is periodic with period $1/\Delta x$). In 1D and for a continuous spectrum, one has, for example, the correspondence

$$f_{\Delta x}(u) = \sum_{m=-\infty}^{+\infty} C_m \, \exp(-2\pi i \, mu \, \Delta x),$$

$$C_m = \Delta x \int_{-1/(2\Delta x)}^{+1/(2\Delta x)} \exp(2\pi i \, mu\Delta x) \, f_{\Delta x}(u) \, du$$

where i denotes the unit pure imaginary number. But, in practice, the simulated sequence is finite, and discrete Fourier transforms provide a more efficient algorithm that can be used for most covariance models and produces Gaussian simulations.

7.5.4 Discrete Spectral Method

The discrete spectral method is designed to simulate SRFs on a finite regular grid in n-dimensional space. It is often presented by considering the specific form of the covariance matrix of the vector of the values taken by the SRF at the grid nodes (Wood and Chan, 1994; Dietrich and Newsam, 1993, 1997). In 1D this covariance matrix is a Toeplitz matrix (it is constant along any line parallel to the upper-left to lower-right diagonal), and even a circulant matrix in the periodic case (each line is identical to the preceding one up to a unit circular shift). In the latter situation simplifications occur so that the method amounts to simulation algorithms based on the discrete Fourier transform. The n-D case, $n > 1$, leads to similar simplifications. To benefit from these simplifications, the covariance to be simulated is embedded in a periodic one, hence the denomination of *circulant embedding* given to the method. We will adopt another viewpoint, developed independently, which directly considers the discrete spectrum of the discrete version of the SRF. The algorithm was presented by Pardo-Iguzquiza and Chica-Olmo (1993)—earlier approximate algorithms existed—whereas Chilès and Delfiner (1996) have shown how to cope with the limitations of the method.

Discrete Spectral Representation of a Stationary Random Function in 1D

Let us first examine the 1D case. If the simulation is computed only at equally spaced points $x = x_0 + n\Delta x$, $n = 0, 1, \ldots, N-1$, over an interval of finite length $L = N\Delta x$, it is convenient to apply the formalism of discrete Fourier transforms (DFT). Since this formalism pertains to periodic series, the series $Z_n = Z(n\Delta x)$ is considered as a stationary process on $\mathbb{Z}$ with period N, for which it suffices to consider the values over a single period $\{0, 1, \ldots, N-1\}$.

The smallest observable frequency is $\Delta u = 1/(N\Delta x)$, and all other frequencies are a multiple of Δu, with a maximum of $1/(2\Delta x) = (N/2) \Delta u$. There is a dual correspondence between the series $Z_n = Z(n\Delta x)$ and its discrete Fourier transform $Y_k = Y(k\Delta u)$,

$$Y_k = \sum_{n=0}^{N-1} Z_n e^{-2\pi i\, kn/N} \tag{7.33}$$

$$Z_n = \frac{1}{N} \sum_{k=0}^{N-1} Y_k e^{2\pi i\, kn/N} \tag{7.34}$$

and between the covariance series $C_m = C(m\Delta x)$ and its discrete Fourier transform $F_k = F(k\Delta u)$,

$$F_k = \sum_{m=0}^{N-1} C_m e^{-2\pi i\, km/N} \tag{7.35}$$

$$C_m = \frac{1}{N} \sum_{k=0}^{N-1} F_k e^{2\pi i\, km/N} \tag{7.36}$$

In this section i denotes the unit pure imaginary number. Assuming $E(Z_n) = 0$, the Y_k satisfy $E(Y_k \overline{Y}_{k'}) = 0$ if $k \neq k'$, and $E|Y_k|^2 = NF_k$.

These two relationships can be obtained as a discrete approximation to the Fourier relationships (2.18) of the continuous case. But they also constitute an exact result that can be proved without any reference to the continuous case.

Since the covariance is real and symmetric, the F_k are also real and symmetric. Note that the periodicity and the symmetry about 0 imply a symmetry about $N/2$,

$$C_{N-m} = C_{-m} = C_m$$
$$F_{N-k} = F_{-k} = F_k$$

A key property of the F_k is to be nonnegative, which is equivalent to the assumption that the C_m constitute a valid discrete covariance model.

Since the Z_n are real random variables, the Y_k are complex random variables that satisfy the Hermitian symmetry $Y_{N-k} = \overline{Y}_k$. This means that their real and imaginary parts U_k and V_k satisfy $U_{N-k} = U_k$ and $V_{N-k} = -V_k$. In particular, $V_0 = 0$, and $V_{N/2} = 0$ if N is even, which will usually be assumed in the sequel. The orthogonality property is equivalent to the following two properties:

1. U_k and V_k are uncorrelated and have the same variance $\sigma_k^2 = NF_k/2$, except for $k = 0$ and for $k = N/2$ where U_k has variance $\sigma_k^2 = NF_k$ and $V_k = 0$.
2. The U_k $(k = 0, \ldots, N/2)$ and V_k $(k = 1, \ldots, N/2-1)$ are mutually uncorrelated.

Hence there are in fact N independent real variables $(U_0, \ldots, U_{N/2}, V_1, \ldots, V_{N/2-1}$ when N is even) corresponding to N data points. Thus, by simulating independent zero-mean random variables U_k and V_k with the appropriate variances σ_k^2 and combining them through (7.34), it is possible to generate a process $Z(x)$ at discrete points $x = n \, \Delta x$ so that its covariance at lags $m\Delta x$ exactly reproduces a series of covariance terms C_m. Formula (7.34) should not be generalized to calculate $Z(x)$ at points other than the nodes.

Discrete Spectral Simulation in 1D

In practice, the calculation flow chart is as follows:

$$C_m \rightarrow F_k \rightarrow Y_k \rightarrow Z_n$$

1. Compute the covariance terms $C_m = C(m\Delta x)$, $m = 0, \ldots, N/2$, and complete by symmetry.
2. Compute the Fourier transform of the C_m's using (7.35) to obtain the Fourier coefficients F_k and the corresponding variances σ_k^2.
3. Simulate U_0, U_k and V_k for $0 < k < N/2$, and $U_{N/2}$ with the variances σ_k^2; then complete by symmetry for the U_k's and by antisymmetry for the V_k's.
4. Invert the sequence $Y_k = U_k + i \, V_k$ using the inverse Fourier transform (7.34) to get the Z_n's.
5. Discard the last (or the first) $N/2 - 1$ points Z_n. There are variations on this strategy, as explained below.

The reason for chopping off part of the simulated sequence lies in the circular character of the DFT. The simulated covariance is periodic with period $L = N\Delta x$, and since the interval considered for m is $[0, N - 1]$ rather than $[-N/2, N/2 - 1]$, the graph looks like the one shown in Figure 7.13a. Here we have assumed that the covariance has a finite range a ($C(h) = 0$ for $|h| \geq a$) and that $L \geq 2a$. One sees that correlations vanish and then pick up again for lags $h > L-a$. To avoid this spurious effect due to the DFT formalism, it is necessary to limit the simulated sequence Z_n (also periodic with period L) to a length no greater than $L - a$.

The other case $L < 2a$ is depicted in Figure 7.13b. A section of length L is cut out from the covariance, symmetrically about the origin, and repeated periodically. Within the interval $[0, L]$ the periodic covariance coincides with the original in the interval $[0, L/2]$, and therefore one must restrict the simulated sequence to a length of $L/2$, or $N/2 + 1$ points.

To summarize, if we want to generate a simulation of Z_n over an effective interval length l (after discard), we must first consider a simulation over an interval of length $L = \min(l + a, 2l)$.

Limitations on Covariance Models and Solutions

The method was sometimes used with a variant of steps 1 and 2 where the values F_k are obtained by discretizing the spectral measure $F(du)$ of

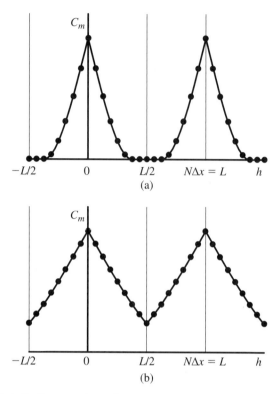

FIGURE 7.13 Periodically repeated covariance: (a) $L \geq 2a$; (b) $L < 2a$. [From Chilès and Delfiner (1997), with kind permission from Kluwer Academic Publishers.]

the continuous covariance function $C(h)$ [e.g., Shinozuka and Jan (1972), whose method is in fact intermediate between the continuous and the discrete cases as they are presented here]. Such a technique, however, differs in a subtle but important manner from the correct procedure in that the discrete spectrum F_k is not simply the value of $F(du)$ at frequency $k\Delta u$. Rather, it represents an aliased version of the spectral density in which the contributions of frequencies $(k + pN)\,\Delta u$ for all integers $p \in \mathbb{Z}$ are folded in the interval $[0, (N-1)\,\Delta u]$ and added. Through this mechanism a correct representation of N covariance terms can be achieved with just N spectral terms, whereas a far greater number of terms may be required for an adequate representation of $F(du)$. It is only for covariances with high regularity at the origin, and thus short spectral spread, that the two approaches may coincide.

There is another important difference with the continuous case: The Fourier transform of $C(h)$ automatically produces a nonnegative spectral measure $F(du)$; in the discrete case, this is not always true because the periodic repetition of C_m does not necessarily lead to a valid covariance model. There are,

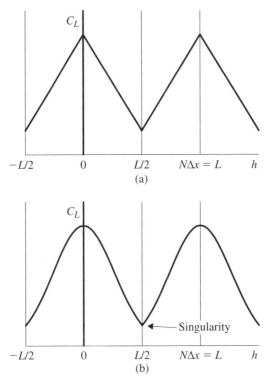

FIGURE 7.14 Periodic repetition of a truncated covariance: (a) periodic repetition of a truncated convex covariance gives a valid covariance; (b) periodic repetition of a truncated Gaussian covariance does not produce a valid covariance. [From Chilès and Delfiner (1996).]

however, two cases where the F_k are necessarily nonnegative because the periodic repetition of $C(h)$ with period L is itself a valid covariance:

1. If $L \geq 2a$ (whatever the shape of $C(h)$).
2. If $C(h)$ is a convex function for $h \in [0, L/2]$.

We have already seen that there is no problem in the first case, shown in Figure 7.13a. In the second case (Figure 7.14a), let us consider the symmetric function which is equal to $C(h) - C(L/2)$ when $|h| \leq L/2$ and to zero beyond this interval. This function decreases to zero and is convex for $h \in [0, \infty[$. It is thus a covariance function according to Pólya's theorem (Section 2.3.3), and it has a range equal to $L/2$. We are thus back to the first case. Since $C(h)$ differs from it only by the addition of the constant $C(L/2)$ when $|h| \leq L/2$, the conclusion also holds for $C(h)$ and the truncated covariance can be extended to the whole line by periodic repetition.

In other cases, such as covariances with a parabolic behavior at the origin or a hole effect, negative F_k's may be found, especially in the absence of a nugget effect (otherwise the nugget variance is included in all F_k). The problem can be best illustrated by a counterexample in the continuous case. Consider a Gaussian covariance $C(h) = \exp(-h^2/a'^2)$, which reaches zero only asymptotically when h tends to infinity. Consider now the restriction of $C(h)$ over the interval $[-L/2, L/2]$, and extend it to the whole line by periodic repetition, producing $C_L(h)$ as shown in Figure 7.14b. This "covariance" is twice differentiable at the origin and hence should be twice differentiable everywhere; in fact it is not differentiable for $|h| = L/2$ and thus cannot be a valid model for a covariance.

The problem is not so severe in the discrete case, for the number of terms F_k is finite. But negative values for the spectrum may be found. There are then three solutions:

1. Extend the length L of the simulated domain. Indeed, if the discrete covariance $C(m\Delta x)$, $m \in \mathbb{Z}$, has a finite integral range and a strictly positive spectral density, then the terms F_k are nonnegative if L is sufficiently large (Wood and Chan, 1994). We have seen that this is the case if the covariance has a finite range a and $L \geq 2a$. If the covariance reaches the value zero only asymptotically, increasing L will either solve the problem or limit its extent to some low negative values. The other two solutions can then be used.

2. Simply set the negative terms to zero. This will of course produce a bias, but this bias will be small if the above solution is also used, and it can be evaluated by inverting the corrected spectrum using formula (7.36).

3. Simulate the "time-aliased" covariance $\tilde{C}(h)$ defined as the sum of shifted versions of the original covariance [e.g., on time aliasing, see Oppenheim and Shafer (1989)]

$$\tilde{C}(h) = \sum_{p=-\infty}^{+\infty} C(h + pL)$$

$\tilde{C}(h)$ and its discretized version are periodic with period $L = N\Delta x$. Using formula (7.35), the discrete spectrum $\tilde{F}_k$ of $\tilde{C}_m$ is found to be

$$\tilde{F}_k = \sum_{p=-\infty}^{+\infty} \sum_{m=0}^{N-1} C_{m+pN} \, e^{-2\pi i \, km/N} = \sum_{m=-\infty}^{+\infty} C_m \, e^{-2\pi i \, km/N}$$

$\tilde{F}_k$ coincides with the spectrum of the complete covariance C_m (calculated at the frequencies $k \, \Delta u$) and is therefore nonnegative. Thus $\tilde{C}_m$ is always a valid covariance. When $L \geq 2a$, $\tilde{C}_m$ and C_m coincide over the interval

$[-L/2, L/2]$, which explains why there is no problem. Otherwise, $\tilde{C}(h)$ and $C(h)$ are different, but it is easy to see if the difference is acceptable. This correction ought to be used only when $C(h)$ is low for $h > L/2$. If this is not the case, it is necessary to first increase L. Experimentation shows that aliasing performs very well for differentiable covariances, whereas setting negative spectrum terms to zero works better for covariances with a linear behavior at the origin.

Extension to Intrinsic Random Functions

By design, the discrete spectral method is concerned with SRFs with periodic covariances. We will show the following extension: In $\mathbb{R}$ any IRF-0 whose variogram $\gamma(h)$ is concave over $[0, L/2]$ possesses a locally stationary representation with a periodic covariance of period L, and can therefore be simulated by the discrete spectral method at grid nodes discretizing the segment of length $L/2$.

Indeed, given a variogram $\gamma(h)$ that is concave over $[0, L/2]$, consider the function $C(h)$ defined on $[-L/2, L/2]$ by $C(h) = \gamma(L/2) - \gamma(h)$ and extended to the whole line by periodic repetition with period L. $C(h)$ is convex in the interval $[0, L/2]$, decreases from $\gamma(L/2) > 0$ to zero in this interval, and therefore is a valid periodic covariance. An SRF with covariance $C(h)$ is a locally stationary representation of the IRF-0 on $[0, L/2]$.

In practice the discrete spectral simulation will be made over a segment of length L, but only the first half is used in view of the symmetry introduced by periodic repetition. Also note that the definition of $C(h)$ on $[-L/2, L/2]$ can be replaced by $C(h) = A - \gamma(h)$, provided that $A \geq A_0$ where A_0 is the smallest value ensuring that $C(h)$ remains a covariance (A_0 is usually smaller than $\gamma(L/2)$). Since the choice of A simply affects the value F_0 of the discrete spectrum, it suffices in the application of the discrete spectral method to replace $C(h)$ by $-\gamma(h)$ and set to zero the negative value that will be found for F_0.

In view of this, it is always possible to simulate the variogram $\gamma(h) = b|h|^\alpha$ by the discrete spectral method:

- If $\alpha \leq 1$, $b|h|^\alpha$ is a concave variogram, and the discrete spectral method can be applied directly.
- If $\alpha > 1$, the *increments* of the IRF-0 have a convex covariance, and the discrete spectral method can be used to simulate these increments.

At first glance, one might wonder why we couldn't just use the fact that $\gamma(h)$ has the locally stationary representation (4.33) and simulate that by the spectral method. The reason is that the periodic repetition of this local covariance beyond the interval $[-L/2, L/2]$ yields a function that is not positive definite when $\alpha > 1$ (the Fourier series has negative terms). There remains to show that the increments $R_n = Z_n - Z_{n-1}$, $n = 1, 2, \ldots$, have a convex covariance. $Z(x)$

being an IRF-0, these increments define a stationary random sequence, whose covariance can be expressed as a function of the variogram $\gamma(h)$ of $Z(x)$ by

$$\text{Cov}(R_n, R_{n+m}) = \gamma((m-1)\Delta x) - 2\gamma(m\Delta x) + \gamma((m+1)\Delta x)$$

When $\gamma(h) = b \, |h|^\alpha$, this discrete covariance is convex for $m \in \mathbb{N}$ if and only if $\alpha > 1$, the case we are considering. The proof is purely technical and will not be reproduced. It can be obtained easily if we start from the following result:
If $a_1, \ldots, a_p$ are positive numbers, $\left[\frac{1}{p}\sum_{i=1}^{p} a_i^\alpha\right]^{1/\alpha}$ is an increasing function of α.

We can therefore construct a simulation of the increments R_n by the discrete spectral method and deduce a simulation of the Z_n by taking $Z_0 = 0$ and $Z_n = Z_{n-1} + R_n$, $n = 1, 2, \ldots$ Figure 7.15 shows 1D simulations obtained for $\alpha = 1/3, 2/3, 1, 4/3, 5/3$.

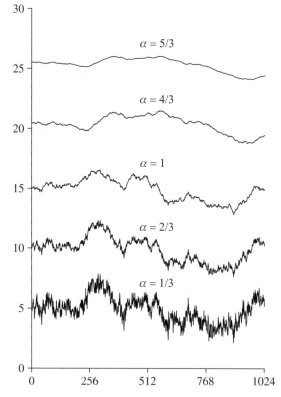

FIGURE 7.15 1D simulations of fractional Brownian motions with variograms of type $|h|^\alpha$, for $\alpha = 1/3, 2/3, 1, 4/3, 5/3$, by the discrete spectral method. The simulations are built at 1025 discrete points. All variograms are equal to 1 for $h = 512$. The fractal dimension of these simulations is $D = 2 - \alpha/2$.

Extension to IRF–k

This method can be partly extended to IRF–k with a power generalized covariance, $k > 0$, by iterating the procedure. Indeed the discrete covariance of increments of order 2 of a GC of the form $b\,|h|^{\alpha}$ is convex if $3 \leq \alpha \leq 4$. The discrete covariance of increments of order 3 of a GC of the form $-b\,|h|^{\alpha}$ is convex if $5 \leq \alpha \leq 5.672$.

The approach based on increments can also be used for an IRF–1 with the spline generalized covariance $K(h) = b\,h^2 \log |h|$. Indeed in that case the increments R_n form an intrinsic random sequence whose variogram is

$$\tfrac{1}{2}\mathrm{Var}(R_{n+m} - R_m) =$$
$$b\big[(m+1)^2 \log(m+1) - 2\,m^2 \log m + (m-1)^2 \log(m-1)\big](\Delta x)^2 \qquad (m \geq 1)$$

This is a concave function of $m \geq 1$. The same procedure can be used for simulating an IRF–2 with the generalized covariance $K(h) = -b\,h^4 \log |h|$ by starting from second-order increments (their variogram is concave). Figure 7.16 shows simulations obtained in this way. Conversely, the variogram of the third-order increments of an IRF–3 with generalized covariance $K(h) = b\,h^6 \log |h|$ is not concave and cannot be simulated in this way.

The locally stationary representations proposed by Stein (2001) for $0 < \alpha < 2\,k + \nu_0$ for some some $\nu_0 \approx 1.6915$ have the nice property that their covariance can be extended to a covariance of $\mathbb{R}$ by symmetrization and periodic repetition. They can therefore be simulated by the discrete spectral method. The case $2\,k + \nu_0 < \alpha < 2\,k + 2$ can be simulated by working in the framework of an IRF–$(k+1)$. Stein proposes an alternative when remaining at the order k. That solution, however, includes a random drift of degree $k + 1$, so that the simulations are not ergodic and honor the covariance on average only.

Extension to 2D and 3D

The spectral method can also be extended to $\mathbb{R}^2$ and $\mathbb{R}^3$ (or higher dimensions): x, Δx, N and their associated indexes simply have to be replaced by vectors with the corresponding dimension, the product kn in $e^{2\pi i k n / N}$ by the scalar product $<k, n>$, and the simple summations by multiple summations. The only problem is to correctly define the independent variables U_k and V_k due to the relations $U_{N-k} = U_k$ and $V_{N-k} = -V_k$ and the variances σ_k^2. Borgman et al. (1984) present the correct algorithm in 2D (but unfortunately apply it by replacing the discrete spectrum by a discretization of the spectral density function). Pardo-Iguzquiza and Chica-Olmo (1993) detail the algorithm in 2D

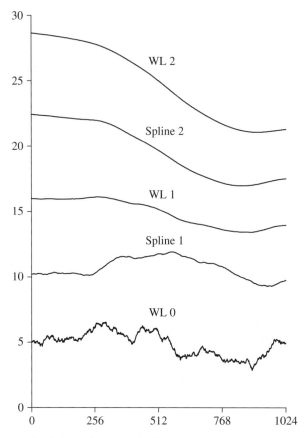

FIGURE 7.16 1D simulations of an IRF$-k$ with a generalized polynomial or spline covariance by the discrete spectral method: direct simulation of a locally stationary representation (covariance of type $(-1)^{k+1} |h|^{2k+1}$, labeled WL k) or simulation of finite differences and discrete integration (generalized covariance of type $(-1)^{k+1} |h|^{2k} \log |h|$, labeled Spline k).

and in 3D. For example, in 2D, for $N = (N_1, N_2)$, N_1 and N_2 even, and $k = (k_1, k_2)$, the independent variables are

- $U_{0\,0}, U_{0\,N_2/2}, U_{N_1/2\,0}$, and $U_{N_1/2\,N_2/2}$, with $\sigma^2_{k_1 k_2} = N_1 N_2 F_{k_1 k_2}$
- $U_{k_1 k_2}$ and $V_{k_1 k_2}$ for $\{k_1 = 0;\ k_2 = 1, \ldots, N_2/2 - 1\}, \{k_1 = 1, \ldots, N_1/2 - 1;\ k_2 = 0, \ldots, N_2 - 1\}$, $\{k_1 = N_1/2;\ k_2 = 1, \ldots, N_2/2 - 1\}$, with $\sigma^2_{k_1 k_2} = N_1 N_2 F_{k_1 k_2}/2$.

Dietrich and Newsam (1993, 1997) and Wood and Chan (1994) present the n-D case.

The method allows the simulation of several covariance components in a single step. Each component may have its own geometric anisotropy, or its own zonal anisotropy along or perpendicular to a coordinate axis. One of these components can be a nugget effect. Like in 1D the F_k are all nonnegative if the size of the simulated domain is such that the covariance reaches zero in all directions before symmetrization and periodic repetition. When this is not the case, the limitations due to negative terms for the spectrum are more serious than in the 1D case: Even convex covariances may lead to negative terms in the discrete spectrum.

Gneiting et al. (2006) explore an approach initiated by Stein (2002) for extending the possibility of an exact simulation of some covariance models $C(h)$ when standard embedding does not yield a valid covariance. Suppose that we want to simulate a 2D isotropic covariance over a square domain of size $l \times l$ in a situation where some terms of the spectrum are negative. The principle of the method is to replace the covariance $C(h)$ by another isotropic covariance $C_1(h)$ equal to $C(h)$ when $|h| < l$ and to 0 when $|h|$ exceeds some range $a_1 \geq l$. The simulation is then performed on a square of size $2a_1 \times 2a_1$, from which a square of size $l \times l$ is extracted. The method is all the more efficient as a_1 is small. Such a covariance C_1 does not always exist. Gneiting et al. (2006) give some general sufficient conditions and apply them to the Matérn and stable covariance models. Stein (2002) considers the 2D and 3D simulation of a power variogram.

When that approach is not possible, the solutions presented for the 1D case can provide quite satisfactory approximations, even for unbounded variograms. Figure 7.17 shows, for example, that the behavior of an isotropic Gaussian covariance at the origin is perfectly reproduced if $L/2$ is equal to (or greater than) the distance where the covariance is 5% of the variance and the aliased covariance is used.

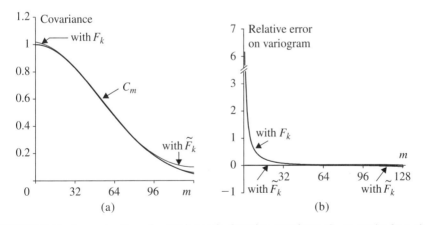

FIGURE 7.17 Approximation of a 2D isotropic Gaussian covariance C_m as results from the truncated spectrum F_k of the original covariance and from the spectrum $\tilde{F}_k$ of the aliased covariance: (a) in absolute scale; (b) in terms of relative error on the variogram. The grid is $N \times N$ with $N = 256$ and is used up to $N/2$. Note that the aliased covariance associated with $\tilde{F}_k$ perfectly reproduces the covariance C_m at the origin, which is not the case for the covariance deduced from the truncated spectrum F_k. The graphs were computed for the main directions of the grid (the approximation is better in the other directions).

Conditional Simulation

Conditioning can be achieved as usual by means of a separate kriging. Dietrich and Newsam (1996) present a conditioning implementation specific to the discrete spectral method, where the data points need not be located at grid nodes but must remain within the simulated domain. Yao (1998) has a different approach: In fact, conditioning on the value at a grid node amounts to introduce a linear constraint on the random variables $\{U_k : k = 0, \ldots, N/2\}$ and $\{V_k : k = 1, \ldots, N/2 - 1\}$. Each pair $\{U_k, V_k\}$ can be equivalently represented by a random amplitude and a random phase. Because the covariance depends only on the amplitudes, Yao proposes to condition a nonconditional simulation by iteratively modifying the phases of the pairs $\{U_k, V_k\}$.

FFT Algorithms

The direct and inverse Fourier transforms can be calculated with the fast Fourier transform algorithm (FFT). It is advantageous to select N as a power of 2. The discrete Fourier transform can then be computed in $O(N \log N)$ operations. In 2D and 3D, very large grids often have to be simulated. From a practical point of view, it may become impossible to store the whole complex array in core memory, as required by the standard FFT algorithm. Special algorithms achieve parallelism and allow versatile memory management. Algorithms have been developed that do not require N to be a power of 2, but a product of small prime numbers. This may be very useful; for example, if we need a 3D simulation with $N_1 = N_2 = N_3 = 72$, the standard algorithm requires that we extend the size to $N_1 = N_2 = N_3 = 128$, thus multiplying the global size of the array by a factor of about 6, which can be avoided by these algorithms. The interested reader is referred to the review of Press et al. (2007, Chapter 12).

Spatial Distribution and Conclusion

The method presented here can also be regarded as a moving average method [see Ripley (1987)]. In its standard application, the U_k and the V_k are selected from Gaussian distributions so that the method, by definition, produces a simulation whose multivariate distribution is Gaussian. Other choices are possible. Selecting U_k and V_k independently from a Gaussian distribution with variance $\sigma_k^2 = F_k/2$ amounts to selecting a random amplitude with mean F_k and a random phase with a uniform distribution over $[0, 2\pi[$. If we fix the amplitude to $\sqrt{F_k}$ and only randomize the phase, we produce simulations whose sample covariance coincides with the covariance C_m (for each realization and not only on the average over a large number of realizations). But this is true for the periodic covariance of the infinite sequence Z_n and not for the sample covariance of the finite sequence which is retained. Nevertheless, this makes it possible to build simulations whose sample variograms display much lower fluctuations around the simulated model than Gaussian simulations (Chilès and Delfiner, 1997).

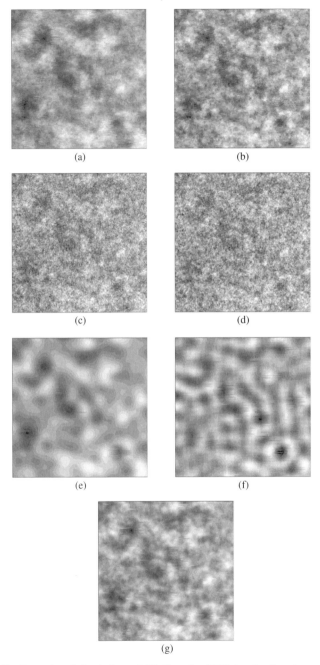

FIGURE 7.18 Examples of simulations of 2D Gaussian SRFs by the discrete spectral method: (a) spherical covariance; (b) exponential covariance; (c) stable covariance with shape parameter $\alpha = 0.5$; (d) hyperbolic covariance; (e) Gaussian covariance; (f) cardinal-sine covariance; (g) covariance $\sin\left(\frac{\pi}{2}\exp(-|h|/a)\right)$. Practical range: 1/8th the grid size. Note that (c) and (d) are almost identical.

In conclusion, the discrete spectral method is a very powerful method of simulation of Gaussian random fields on a regular grid. It offers the following advantages:

- The multivariate distribution of the simulations is exactly Gaussian (in its standard use).
- The method is completely general and allows an exact simulation of any covariance with a finite range, and either an exact simulation or an approximate one to a desired accuracy if the range is infinite.
- It can be extended to variograms and generalized covariances.
- The method can be directly used for simulating in 2D or 3D (or higher dimension), without the need for turning bands.
- It is possible to simulate nested structures, as well as geometric anisotropies, zonal anisotropies parallel or orthogonal to the axes, and a nugget effect, in a single step.
- The FFT algorithms are computationally very efficient; variants saving memory storage make it possible to process large 2D or 3D grids without losing much efficiency.

Its limitations are the following:

- Some covariance models cannot be simulated exactly: those for which the periodic repetition from the simulated domain to the whole space does not lead to a valid covariance.
- It is not possible to extend a previous simulation to a larger domain: the whole simulation must be recomputed and will not be consistent with the previous one.
- The simulation is limited to grid nodes.

These limitations are usually mild compared to the advantages. Figure 7.18 displays 2D examples of simulations of Gaussian SRFs built with the discrete spectral method.

Simulation with Wavelets

For completeness we will briefly mention simulation methods based on wavelets [e.g., Flandrin (1992) and Zeldin and Spanos (1996)]. The motivation for wavelets is to introduce spatial localization in the classic Fourier analysis. Indeed the sine waves used for Fourier analysis are perfectly localized in frequency but not at all in space, so the Fourier coefficients give no clue on the position of local features such as the contours in an image. By contrast, a wavelet decomposition uses a basis of carefully chosen functions that typically vanish outside a compact support and are obtained by scaling and shifting a basic function $\psi(x)$, the *mother wavelet*. The simplest example is the Haar function $\psi(x)$ equal to $+1$ for $0 \leq x < 1/2$, to -1 for $1/2 \leq x < 1$, and to 0 outside the interval $[0, 1[$. By dilation and translation, it yields the family of step functions

$$\psi_{jn}(x) = 2^{-j/2}\, \psi(2^{-j}x - n), \qquad j, n \in \mathbb{Z}$$

which is a discrete and orthonormal basis of $L^2(\mathbb{R})$ (square integrable functions). The support of $\psi_{jn}(x)$ is the interval of length 2^j starting at $n\, 2^j$. Thus j is a scale index, and n is a translation index (2^j is analogous to a wavelength and 2^{-j} to a frequency). All wavelets sum to zero. The Haar function is well-localized in space but not in frequency, since its Fourier transform decays only like $1/|u|$. Wavelets more regular than the Haar function are often used, notably the compactly supported Daubechies wavelets (Daubechies, 1992; Cohen, 1992; Meyer, 1993). They have higher moments that vanish as well.

Associated with the mother wavelet is the "father wavelet," or "scaling function," ϕ and the family

$$\phi_{jn}(x) = 2^{-j/2}\, \phi(2^{-j}x - n), \qquad j, n \in \mathbb{Z}$$

Here ϕ is used to calculate the *approximation* of $Z(x)$ at a given resolution 2^J, while ψ is used to represent the *details*, namely the difference between two successive approximations. Equivalent terms are *trend* and *fluctuations* [see Meyer (1993, p. 40)]. A scaling function sums to one. For the Haar function, for example, we have $\phi(x) = 1_{0 \le x < 1}$. Specifically, for any given resolution 2^J, the following decomposition holds:

$$Z(x) = 2^{-J/2} \sum_{n=-\infty}^{+\infty} a_{Jn}\, \phi_{Jn}(x) + \sum_{j=-\infty}^{J} 2^{-j/2} \sum_{n=-\infty}^{+\infty} d_{jn}\, \psi_{jn}(x)$$

The coefficients a_{Jn} and d_{jn} are obtained as scalar products of $Z(x)$ with $\phi_{Jn}(x)$ and $\psi_{jn}(x)$ but can also be computed recursively (recurrence in J or j). Recurrence relations also make it possible to exactly reconstruct $Z(x)$ if an initial (coarser) approximation a_{Jn} and the different sequences of details d_{jn} at finer scales $j \le J$ are given. This is the basis of simulations with wavelets. When $Z(x)$ is an SRF, the sequences a_j and d_j are stationary random sequences. The wavelet functions form an orthonormal basis, but contrary to spectral representations, the random sequences d_j are not uncorrelated. To simulate an SRF with a given covariance function, it is therefore necessary to derive the direct and cross-covariances of the various random sequences and to simulate them accordingly. The problem seems to be more complex than it was at the beginning. Fortunately, many covariance terms are negligible, and iterative algorithms can be developed if the mother wavelet is chosen adequately. Zeldin and Spanos (1996) present some 1D and 2D examples where the covariance function of the wavelet representation at a given scale approximates the desired covariance quite closely.

When $\psi(x)$ is the Haar function, the wavelet method is equivalent to the simulation via *local average subdivision* (Fenton and Vanmarcke, 1990): The average value over the whole domain (a segment in 1D) is simulated; then the region is subdivided in two subdomains, whose average values are simulated conditionally on the first average value, and so on.

The wavelet method has been designed from the beginning to simulate fractional Brownian random functions. The success of the algorithm, namely the possibility to efficiently truncate the infinite sum to a limited number of terms, depends critically on the adequate choice of the wavelet family; for example, see Flandrin (1992), Elliott and Majda (1994), and Sellan (1995). This method achieves rigorously what is only approximated by the random displacement method. It includes as a particular case the simulation via local average subdivision, applied to the fractional Brownian random function by Fenton and Vanmarcke (1990).

Finally, let us mention the Fourier-wavelet method (Elliot and Majda, 1994; Elliot et al., 1997). Instead of using expansions in the physical space, it deals with expansions in the spectral domain, using the Fourier transforms of the $\{\phi_{Jn}(\cdot),\ n \in \mathbb{Z}\}$ and $\{\psi_{jn}(\cdot),\ j \in \mathbb{Z},\ j \leq J,\ n \in \mathbb{Z}\}$ as complete orthonormal basis of $L^2(\mathbb{R})$. Kurbanmuradov and Sabelfeld (2006) compare the two approaches for multivariate random fields.

To summarize, wavelet algorithms are fairly complex but offer the possibility of zooming over many decades of scaling behavior in a simulation done at a low resolution (10 to 15 decades are reported for fractional Brownian random functions).

7.6 SIMULATION OF A CATEGORICAL VARIABLE

The simplest categorical variable is one that only assumes the values 0 or 1—that is, the indicator of a set, which we consider here as random. But the covariance function is an extremely poor tool for describing the geometric properties of these very special random functions. For example, the covariance does not give any information on the connectivity of the medium. In fact the covariance is the same for the random set considered and its complement (e.g., grains and pores), while their connectivities are generally very different. Richer tools have been developed in mathematical morphology, but these can be determined only if we have a continuous image of a realization of the random set. Since this book deals essentially with phenomena that are not measured continuously, nor even necessarily on a regular grid, we will not dwell long on the theory of random sets and its applications in mathematical morphology and stochastic geometry. We refer the reader to the literature on this subject—in particular, Kendall and Moran (1963), Matheron (1967, 1975a), Kendall (1974), Serra (1982), Stoyan et al. (1987), and Molchanov (2005). This presentation is limited to the main random-set models that can be of use for geostatistical simulations and to simple generalizations allowing the representation of m-valued indicators ($m > 2$; e.g., facies number) or the construction of mosaic RFs: random tessellations, Boolean models, and marked point processes.

Conversely, some indicator simulation methods derive from standard geostatistical methods, since they involve an underlying continuous variable. This continuous variable can have a physical meaning, for example when we study the indicator associated with a certain cutoff grade or contamination threshold. It can also be a conventional feature of the model; for example, nested indicators (e.g., lithological facies) can be obtained by slicing a Gaussian variable at successive levels. These methods will also be presented: truncated Gaussian method, and substitution random functions.

In this section we first consider the sequential method, as adapted to indicators, which allows a direct construction of conditional simulations, at the cost of some approximations. Next, we outline the iterative algorithms based on Markov chains, which are often used for conditioning indicator simulations or generating constrained simulations. Finally, we review the main models, except object-based methods and constrained simulations which will be studied in the next sections.

7.6.1　Sequential Indicator Simulation

Sequential indicator simulation (SIS) is the application of the general sequential simulation method to the case of an indicator function, or more generally of several nested indicators. The method has been developed by Alabert (1987) and Journel (1989). Let us consider for the moment the case of a single indicator, the generalization to the case of several nested indicators being straightforward. We recall that the principle of the sequential method is to sequentially draw the value at each new simulated point from the conditional distribution given the data and the values simulated previously. In the case of an indicator, the values are 0 or 1, and the conditional distribution is therefore defined by its conditional expectation, but this in general is not known. Alabert and Journel propose to replace it by the simple kriging estimate of the indicator. This approximation preserves the mean and the covariance structure of the RF when the data are randomized (provided, however, that kriging produces estimates that lie in the interval $[0, 1]$).

Basic Algorithm

More formally we denote by $I(x)$ the indicator, which is assumed to be an SRF with mean p, variance $\sigma^2 = p(1 - p)$, and covariance $C(h)$. Assume that $I(x)$ is known at the sample points $\{x_\alpha : \alpha = 1, \ldots, M\}$. Let x_{M+1} be the first point where we set out to simulate $I(x)$. The simple kriging estimator of $I(x_{M+1})$ from $I(x_1), \ldots, I(x_M)$ is of the form

$$I^* = p + \sum_{\alpha=1}^{M} \lambda_\alpha (I(x_\alpha) - p) \tag{7.37}$$

where the λ_α are the solutions of the system

$$\sum_{\beta=1}^{M} \lambda_\beta C(x_\beta - x_\alpha) = C(x_{M+1} - x_\alpha), \qquad \alpha = 1, \ldots, M$$

Let us draw the simulated value T_{M+1} of $I(x_{M+1})$ from the Bernoulli distribution with mean I^*. For simplicity, we refer to the set $\{I(x_1), \ldots, I(x_M)\}$ of conditioning data simply as "data." Conditionally on these data, we have

$$\mathrm{E}[T_{M+1} \mid \text{data}] = I^*,$$

$$\mathrm{E}[T_{M+1}^2 \mid \text{data}] = I^*,$$

$$\mathrm{E}[I(x_\alpha) T_{M+1} \mid \text{data}] = I(x_\alpha) I^*$$

Thus, given expression (7.37) of I^*, we obtain the following upon randomizing the data:

$$E[T_{M+1}] = E[I^*] = p,$$

$$E[T^2_{M+1}] = E[I^*] = p,$$

$$\mathrm{Var}[T_{M+1}] = p(1-p),$$

$$E[I(x_\alpha)T_{M+1}] = p^2 + \sum_{\beta=1}^{M} \lambda_\beta \, E\big[(I(x_\alpha) - p)(I(x_\beta) - p)\big]$$

The (centered) covariance of $I(x_\alpha)$ and T_{M+1} is therefore

$$\mathrm{Cov}[I(x_\alpha), T_{M+1}] = \sum_{\beta=1}^{M} \lambda_\beta \, C(x_\beta - x_\alpha) = C(x_{M+1} - x_\alpha)$$

This simulation method therefore does match the mean and the covariance function of $I(x)$. This property, proved for all the points $x_1, \ldots, x_M,\ x_{M+1}$, remains true by iteration when we add x_{M+2}, and so on, provided that at each step the point previously simulated is included in the data set.

Note that the conditional variance of T_{M+1} is

$$\mathrm{Var}[T_{M+1} \mid \mathrm{data}] = I^*(1 - I^*)$$

It is not equal to σ_K^2 as for standard conditioning but takes a value that fluctuates around σ_K^2, its expectation being σ_K^2. Indeed

$$E[\mathrm{Var}(T_{M+1}|\mathrm{data})] = E[I^*(1 - I^*)]$$

$$= p - p^2 - \sum_{\alpha=1}^{M}\sum_{\beta=1}^{M} \lambda_\alpha \lambda_\beta \, C(x_\beta - x_\alpha)$$

$$= C(0) - \sum_{\alpha=1}^{M} \lambda_\alpha \, C(x_{M+1} - x_\alpha)$$

$$= \sigma_K^2$$

The only theoretical problem with this method is that I^* can be less than 0 or greater than 1. This can occur in one dimension with very common models (e.g., the spherical covariance), as well as in a space of more dimensions for practically any covariance other than a pure nugget effect. When this occurs, we obviously set I^* to 0 or 1 accordingly, but then the covariance is no longer reproduced exactly. Note that nothing prevents us from using the algorithm with a covariance model that is not a genuine indicator covariance, but then the covariance of the simulated indicator cannot match the (inappropriate) input covariance.

Conditioning no longer matches the covariance nor even the variogram if we condition by ordinary kriging. Indeed, going through the above steps, we obtain, for example,

$$\text{Cov}[I(x_\alpha), T_{M+1}] = C(x_{M+1} - x_\alpha) - \mu$$

or

$$\tfrac{1}{2}\text{E}[I(x_\alpha) - T_{M+1}]^2 = \gamma(x_{M+1} - x_\alpha) + \mu'$$

where μ and μ' are the Lagrange parameters of the kriging systems (3.17) and (3.19).

Generalization to Nested Indicators

The method can easily be extended to nested indicators $I_1(x), \ldots, I_m(x)$ satisfying the characteristic property

$$I_i(x) = 1 \quad \Rightarrow \quad I_j(x) = 1 \ \forall j < i$$

Such indicators are obtained, for example, by applying increasing thresholds to an RF $Z(x)$:

$$I_i(x) = 1_{Z(x) < z_i}, \qquad \text{where} \quad z_1 < \cdots < z_{i-1} < z_i < \cdots < z_m \qquad (7.38)$$

These indicators can be estimated at the point x_{M+1} by cokriging, and the estimates define a cumulative distribution function provided that the usual order relationships between indicators are also satisfied by the cokriging estimates. We then draw T from this distribution. The implementation of this method raises some practical difficulties, which are discussed in Section 6.3.3 (modeling of the direct and cross-covariances, possible nonmonotonicity of the estimated c.d.f.).

When the indicators derive from an RF Z, the simulation method can be applied, through an appropriate coding shown in Figure 7.19, to account for data given in the form of an interval, an inequality, or an a priori distribution function, namely "membership" data (between 0 and 1)

$$Y(x_\alpha; z_i) = \text{Pr}\{Z(x_\alpha) < z_i \mid \text{local information at } x_\alpha\}$$

(Journel, 1989; Journel and Alabert, 1989). The local information can be the value taken on by an auxiliary variable at this point x_α such as a geological facies type, or an expert's opinion on the value $z(x_\alpha)$.

Journel and Alabert (1989) apply this methodology to an exhaustive sampling of a 2-ft $\times$ 2-ft sandstone section by 40×40 permeability data, represented in eight gray levels. SIS is used to simulate the nested indicators

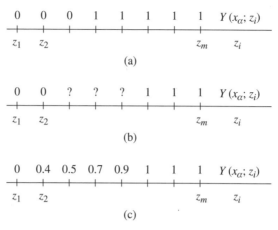

FIGURE 7.19 Coding of the information in view of sequential simulation of nested indicators associated with a continuous variable: (a) exact value (all indicators are filled in); (b) value defined by an interval (indicators in the interval are left blank); (c) value defined by a distribution (c.d.f. values are used). [From Journel (1989).]

corresponding to the gray levels, either from 16 hard data only or from these 16 hard data plus a regular grid of 12×12 soft data (the soft information indicates that the permeability value at the data point belongs to one of three large intervals that divide the whole range of permeability values). The comparison of these simulations with the exhaustive image shows that the soft data provide a clear improvement.

The spatial distribution of the simulations generated by the sequential indicator algorithm has been studied by Emery (2004b). Even if each node were simulated conditional on all previously simulated nodes, the multivariate distribution of the simulation would depend on the visiting sequence. The covariance and therefore the bivariate distributions are not reproduced correctly in the case of a random set if kriged values I^* below 0 or above 1 are found, which is almost always the case, with the special exceptions of a pure nugget effect in n-dimensional space and of the exponential covariance in 1D. Figure 7.3 i shows a 2D simulation of an exponential covariance; it does not reproduce the behavior of the variogram at the origin. The bivariate distributions of categorical variables associated with a mosaic random function or nested indicators are likewise not reproduced correctly, with similar special exceptions (Markov processes, pure nugget effect). The method treats a random set and its complement in the same way and therefore is not suitable for simulating a medium made of pores and grains, which usually play very different roles.

In conclusion, the sequential indicator method is very flexible but requires approximations (conditioning from neighboring data only, monotonicity corrections) which produce simulations without a clear status and which usually do not meet their objective. According to Daly and Caers (2010), they display a characteristic "blob" geometry and do not capture the actual shape and connectivity of the permeable formations of petroleum reservoirs. An additional shortcoming, mainly when nested indicators are simulated, is that a result is obtained even if the model defined by the indicator direct and cross-covariances is inconsistent.

7.6.2 Iterative Methods Based on Markov Chains

Simulation methods other than the sequential method do not directly provide conditional simulations of indicators. Unlike for SRFs whose realizations are continuous, there is no general conditioning algorithm as simple as adding a simulated kriging error. This algorithm can of course be used with a non-conditional indicator simulation and will indeed preserve the covariance. However, the conditional simulation will not be an indicator function: It could take any value between 0 and 1 (and even values outside this interval). Iterative methods make it possible, at least for some SRF models, to condition indicator simulations while preserving the binary character of the simulations. They are also used for the construction of simulations by simulated annealing. The basic methods rely on Markov chains and have been developed after the work of Metropolis et al. (1953) to simulate complex physical systems and solve optimization problems. The reader will find complementary results in Ripley (1987).

General Framework

Iterative methods are mainly used to produce or condition simulations on a grid. The general principle is to start with a simulated grid that does not fulfill all the requirements in terms of spatial variability or conditioning and then to update it step by step until these requirements are met. More precisely, let us consider the simulation of a random function $Z(x)$ in the following situation:

1. The points where the simulation is to be built constitute a set of N sites $x_1, \ldots, x_N$ (usually the nodes of a regular grid).
2. Z can take p discrete values $1, 2, \ldots, p$ (e.g., it is a binary variable, a categorical variable, or a class of a distribution of a continuous variable).

The realizations of $Z(x)$ satisfy $Z(x_1) = z_1, \ldots, Z(x_N) = z_N$, where $z_1, \ldots, z_N$ assume one of the p discrete values. Any particular grid of values—that is, any particular value of the *vector* $\mathbf{z} = (z_1, \ldots, z_N)'$—will be called a state and denoted by s (we do not distinguish two realizations of the random function $Z(x)$ if they coincide at the N sites). The vector $\mathbf{Z} = (Z(x_1), \ldots, Z(x_N))'$ can take p^N states s_i ($i = 1, \ldots, p^N$). Its N-dimensional probability distribution is a particular finite-dimensional distribution of the random function, and it defines the monodimensional probability distribution π that assigns the probability $\pi_i > 0$ to state s_i (we consider for the sake of simplicity that any of the p^N possible states has a nonzero probability of occurrence; otherwise, the number of possible states must be decreased accordingly). Simulating $Z(x)$ over the

N sites amounts to drawing a vector $\mathbf{Z}$ from its N-dimensional distribution, or equivalently drawing a state s from the probability distribution π. The number of possible states can be very large, and the calculation of each probability π_i impossible, so particular methods are necessary. The present section describes iterative methods to sample from π: Starting with an initial simulated state $S^{(0)}$ from an arbitrarily chosen distribution, we construct a sequence of random states $S^{(k)}$, $k = 1, 2, \ldots$, whose probability distribution converges to π when k increases. This method derives from the properties of Markov chains [e.g., see Feller (1968, Chapter XV)]: The successive $S^{(k)}$ are obtained as Markov transitions from one state to the next.

The transition from $S^{(k)}$ to $S^{(k+1)}$ is defined by the transition matrix $\mathbf{P} = [p_{ij}]$, which is independent of k:

$$p_{ij} = \Pr\{S^{(k+1)} = s_j \mid S^{(k)} = s_i\}$$

The chain defined by the matrix $\mathbf{P}$ is said irreducible if every state can be reached from every other state in a finite number of transitions. A Markov chain with transition matrix $\mathbf{P}$ has the invariant distribution π if and only if

$$\sum_i \pi_i p_{ij} = \pi_j \qquad \forall j \tag{7.39}$$

and this chain is reversible if and only if the detailed balance conditions

$$\pi_i p_{ij} = \pi_j p_{ji} \qquad \forall i \neq j \tag{7.40}$$

are satisfied. Note that (7.40) implies (7.39). The invariant distribution π of an irreducible Markov chain is unique, and the distribution of $S^{(k)}$ will converge to π (we assume that all states are ergodic, that is with a recurrence time which has a finite mean and is not systematically a multiple of some period $t > 1$).

Drawing a state s randomly from the distribution π can be achieved by selecting a Markov chain with transition probabilities satisfying (7.40), building a Markov chain $S^{(k)}$ up to a sufficiently large k value for the chain to have reached the equilibrium, and retaining this final state. Details about the stability and convergence of Markov chains can be found in Meyn and Tweedie (1993). Now the problem reduces to finding a matrix $\mathbf{P}$ satisfying (7.40). Two choices for $\mathbf{P}$ are commonly used:

1. The *Metropolis algorithm* (a generalization of the algorithm used by Metropolis et al., 1953). We start by choosing a *symmetric* transition matrix $\mathbf{Q}$. At each step k, denoting the current state by s_i, the matrix $\mathbf{Q}$ is used to generate a candidate state s_j from the conditional probability $\mathbf{q}_i$ (vector of the q_{ij} for this value of i). This candidate is accepted always if

$\pi_j \geq \pi_i$, and with probability π_j / π_i if $\pi_j < \pi_i$. Otherwise, we remain in state s_i. This defines the transition probabilities

$$p_{ij} = \min\left(1, \frac{\pi_j}{\pi_i}\right) q_{ij} \qquad \text{if } i \neq j,$$

$$p_{ii} = q_{ii} + \sum_{j \neq i} \max\left(0, 1 - \frac{\pi_j}{\pi_i}\right) q_{ij}$$

which are seen to satisfy (7.40). The invariant distribution is unique and the Markov chain converges to it if π is not constant and **Q** is irreducible.

2. *Barker's algorithm* (Barker, 1965). It is similar to the Metropolis algorithm, except that $\min(1, \pi_j / \pi_i)$ is replaced by $\pi_j / (\pi_i + \pi_j)$. Equation (7.40) is still satisfied. The invariant distribution is unique, and the Markov chain converges to it if **Q** is irreducible.

Hastings (1970) removes the symmetry condition on **Q** by letting

$$p_{ij} = \min\left(1, \frac{\pi_j\, q_{ji}}{\pi_i\, q_{ij}}\right) q_{ij} \qquad \text{if } i \neq j,$$

$$p_{ii} = q_{ii} + \sum_{j \neq i} \max\left(0, 1 - \frac{\pi_j\, q_{ji}}{\pi_i\, q_{ij}}\right) q_{ij}$$

These algorithms can be used even if π_i and π_j are not explicitly known, provided that their ratio can be calculated. An introductory exposition of the Metropolis-Hastings algorithm can be found in Chib and Greenberg (1995).

Special Case of Conditional Simulations

In the case of conditional simulations, not all states are permitted. Let us denote the space of all states by Ω and the subset of permitted states by Ω_c. The conditional probability of the permitted states is

$$\tilde{\pi}_i = \frac{\pi_i}{\pi(\Omega_c)} \quad (i \in \Omega_c) \qquad \text{with} \quad \pi(\Omega_c) = \sum_{j:\, s_j \in \Omega_c} \pi_j$$

A way to generate a conditional simulation is to restrict the transition kernel to Ω_c: starting with a state s_j belonging to Ω_c, we generate a new state s_j as above but accept it only if it belongs to Ω_c and otherwise remain in state s_i, and so on. This method is not always valid, because the transition kernel between states of Ω_c induced by this algorithm may not satisfy the conditions (7.39) and (7.40). This is the case, for example, if we can find two states of Ω_c which cannot communicate without a transition by a state not belonging to Ω_c. Conditions ensuring the correctness of the algorithm are given, for example, by Lantuéjoul (2002, Chapter 8). Simulated annealing provides a solution in the other cases. Since this is primarily an optimization algorithm, it is presented in Section 7.8.4.

Spin Exchange

A particular case of Hasting's method, introduced by Flinn (1974) to simulate phase separation on a grid with two types of atoms, is usually known as the "spin exchange" method: Each transition involves only two sites that exchange their values. The algorithm is the following:

1. Select two sites at random, independently (or the second one in a neighborhood of the first one). If the values at these sites are equal, redraw the second site.
2. Pass from the present state s_i to the state s_j defined by the exchange of the values at these sites, with the probability $\min(1, \pi_j/\pi_i)$ (Metropolis algorithm) or $\pi_j/(\pi_i + \pi_j)$ (Barker's algorithm).

The Markov chain produced is only irreducible over the set of states with the same marginal distribution: The permutations do not change the histogram of the values over the N sites. So the initial state must be selected so as to match the marginal distribution.

Local Replacement: Gibbs Sampler

Another particular case is when each transition involves a single site: $\mathbf{Q}$ is chosen such that $q_{ij} > 0$ only if s_i and s_j are states that differ in value at a single site. If s_i and s_j differ only by their values i_r and j_r at site x_r,

$$\frac{\pi_j}{\pi_i} = \frac{\Pr\{Z(x_r) = j_r, \text{rest}\}}{\Pr\{Z(x_r) = i_r, \text{rest}\}} = \frac{\Pr\{Z(x_r) = j_r \mid \text{rest}\}}{\Pr\{Z(x_r) = i_r \mid \text{rest}\}}$$

The simplest case is when $Z(x)$ can take only two different values ($p = 2$). The values i_r and j_r are then the only possible ones for $Z(x_r)$ so that

$$\Pr\{Z(x_r) = j_r \mid \text{rest}\} = \frac{\pi_j}{\pi_i + \pi_j}$$

If the site is chosen at random, Barker's algorithm reduces to drawing its value from the conditional distribution given the values at the other sites.

This method can be extended to the case where $Z(x)$ can take $p > 2$ different values, or even has a continuous distribution. The following iteration is performed:

1. Select a site, say x_r, at random or by a systematic scan of all the sites.
2. Choose the new value for this site from the conditional distribution of $Z(x_r)$ given the other values $Z(x_s)$, $s = 1, \ldots, r-1, r+1, \ldots, N$.

Usually this conditional distribution is assumed to depend on the neighboring values only. Thus we are able to simulate the joint distribution at all sites by successive simulations of conditional univariate distributions. This method was developed by Geman and Geman (1984), who call it the "Gibbs sampler" because, when the sites constitute a square lattice, $\mathbf{Z}$ can be considered as a Markov random field with a Gibbs distribution. Casella and George (1992) give an intuitive presentation of the Gibbs sampler, showing that it "can be thought of as a practical implementation of the fact that the knowledge of the conditional distributions is sufficient to determine a joint distribution (if it exists!)."

Multiple Metropolis–Hastings Algorithm

Sénégas (2002a) presents special cases and generalizations of the Metropolis and Gibbs algorithms. Let us focus on the generation of simulations conditioned on noisy data. If all states remain permitted, conditional simulations can be generated with the Metropolis algorithm by selecting the candidate state from the state distribution, whatever the current state. But a large proportion of candidate states may be rejected. Sénégas therefore extends the algorithm to a selection among a set of candidate states. Let $\pi(\mathbf{z})$ be the marginal distribution of $\mathbf{Z}$ and $\tilde{\pi}(\mathbf{z}) = \pi(\mathbf{z} \mid \text{data})$ be its conditional distribution given the noisy data. Using Bayes relation, the latter can be expressed with the conditional distribution of the data given $\mathbf{Z} = \mathbf{z}$ and with the marginal distributions of $\mathbf{Z}$ and of the data:

$$\tilde{\pi}(\mathbf{z}) = \pi(\mathbf{z} \mid \text{data}) = \frac{\pi(\text{data} \mid \mathbf{z}) \, \pi(\mathbf{z})}{\pi(\text{data})}$$

The application, developed when $\mathbf{Z}$ is a Gaussian random vector, is based on the following property: If $\mathbf{Z}_1$ and $\mathbf{Z}_2$ are two i.i.d. Gaussian random vectors with mean zero, $\mathbf{Z}_1 \cos\theta + \mathbf{Z}_2 \sin\theta$ is a random vector with the same distribution.[3] It is then easy to build a set of Gaussian random vectors by combination of two vectors only.

Denoting the current state by $\mathbf{z}^{\text{current}}$, the multiple Metropolis–Hastings algorithm of each iteration is the following:

1. Select a state vector $\mathbf{z}$ from the state distribution, independently of the current state vector.

2. Build the K candidate states

$$\mathbf{z}_k^{\text{cand}} = \mathbf{z}^{\text{current}} \cos\left(2\pi \frac{k}{K}\right) + \mathbf{z} \sin\left(2\pi \frac{k}{K}\right), \qquad k = 0, 1, \ldots, K-1.$$

[3] Let us mention a more general theorem due to Matheron (1982b): If $Z_1(x)$ and $Z_2(x)$ are two independent zero-mean unit-variance Gaussian SRFs with the same covariance $\rho(h)$ and if $\Theta(x)$ is an IRF with strictly stationary increments, independent of Z_1 and Z_2, then the random function $Z_1(x) \cos \Theta(x) + Z_2(x) \sin \Theta(x)$ is second-order stationary with Hermitian bivariate distributions with $T_n(h) = \rho^n(h) \, \mathrm{E}[(\cos(\Theta(x+h) - \Theta(x))^n]$.

3. Compute

$$p_k = \pi(\text{data} \mid \mathbf{z}_k^{\text{cand}}), \qquad k = 0, 1, \ldots, K - 1,$$

at least up to a multiplicative factor, and norm them so that they define a probability distribution.

4. Select k from that distribution and set $\mathbf{z}^{\text{new}} = \mathbf{z}_k^{\text{cand}}$.

Note that the candidate states include the current state ($k = 0$), as well as state $\mathbf{z}$ if K is a multiple of 4. Sénégas (2002a,b) applies this algorithm to stereovision. The vector $\mathbf{Z}$ represents a topographic grid and the conditioning stereographic pair amounts to a grid of the form $\mathbf{f}(\mathbf{Z}) + \boldsymbol{\varepsilon}$, where $\boldsymbol{\varepsilon}$ is an error vector with i.i.d. components.

7.6.3 Application of the Gibbs Sampler to the Simulation of a Gaussian Vector

Let us come back to the simulation of a Gaussian vector. The Gibbs sampler is the basis for several iterative methods to simulate Gaussian RFs or categorical variables linked to Gaussian RFs. Suppose that we want to simulate a Gaussian N-vector $\mathbf{Z}$ for a given distribution. In this section $\{Z_i : i = 1, \ldots, N\}$ denote the components of $\mathbf{Z}$ (Z_i can be of the form $Z(x_i)$, but this is not necessary), and a realization $\mathbf{z}$ of the random vector represents a state. Without lack of generality we assume that the Z_i's are standard normal, so $\mathbf{Z}$ has mean vector $\mathbf{0}$ and given covariance matrix $\boldsymbol{\rho} = [\rho_{ij}]$ with $\rho_{ii} = 1$.

Simulation of a Gaussian Vector with the Gibbs Sampler

Let us start with the direct application of the Gibbs sampler, which is straightforward. Indeed, after the initialization step (e.g., by the vector $\mathbf{0}$), the iterative procedure of the Gibbs sampler can be rewritten as:

1. Select a component, say i, at random or by a systematic scan of all the components.
2. Choose the new value for this component from the conditional distribution of Z_i given the other values $\{Z_j : j \neq i\}$.

The conditional distribution of Y_i is Gaussian with mean the SK estimate z_{-i}^* of Z_i from the current values of $\{Z_j : j \neq i\}$ and variance the SK variance σ_{Ki}^2. As we have seen in Section 3.6.4, the kriging estimate and variance are

$$z_{-i}^* = -\frac{1}{B_{ii}} \sum_{j \neq i} B_{ij} z_j \qquad \text{and} \qquad \sigma_{Ki}^2 = B_{ii}$$

where $\mathbf{B} = \rho^{-1}$ is the precision matrix. Step 2 then amounts to changing the current value z_i^{current} of Z_i to

$$z_i^{\text{new}} = z_{-i}^* + \sigma_{\mathrm{K}i}\, U$$

where U is an independent standard normal random variable.

This method can obviously generate conditional simulations by fixing the values assigned to the known components and scanning only the other components.

However, it is not used to simply generate a nonconditional or conditional simulation because that would require inverting the matrix ρ. If the inverse of ρ can be calculated, it is more efficient to use the Cholesky decomposition method presented in Section 7.2.2. If not, it is tempting to krige Z_i from a subset of the $\{Z_j : j \neq i\}$—for example, from the neighboring nodes when the Z_i's represent values on a grid—but the algorithm may diverge, and severely. The noticeable exception is that of Gaussian Markov random fields (Besag, 1974, 1975; Rue and Held, 2005; see Section 3.6.4), which are actually defined by their precision matrix $\mathbf{B}$. That matrix is supposed to be sparse, so an iteration of the Gibbs sampler is not computationally demanding.

In the case of the simulation of a Gaussian vector the real value of the Gibbs sampler in fact lies in the following two generalizations.

From the Gibbs Sampler to a Gibbs Propagation Algorithm

The following implementation requires neither the inversion of the covariance matrix nor the knowledge of the precision matrix and can therefore be applied to vectors of large dimension (at least several tens of thousands components). This algorithm is due to Lantuéjoul (2011), who had the brilliant idea of reversing the viewpoint. The precision matrix $\mathbf{B} = \rho^{-1}$ is also a covariance matrix. It is very easy to simulate a Gaussian random vector $\mathbf{Y}$ with mean $\mathbf{0}$ and covariance matrix $\mathbf{B}$, because the Gibbs sampler requires $\mathbf{B}^{-1}$, which is nothing but ρ. Since $\rho_{ii} = 1$, the kriging estimate of Y_i given all values y_j^{current} of Y_j for $j \neq i$ is

$$y_{-i}^* = -\sum_{j \neq i} \rho_{ij}\, y_j^{\text{current}}$$

and the kriging variance is equal to 1. In the iterative algorithm the new value of Y_i takes the simple form

$$y_i^{\text{new}} = y_{-i}^* + U$$

where U is an independent standard normal random variable. Once the final vector $\mathbf{Y}$ is obtained, it suffices to set $\mathbf{Z} = \rho\, \mathbf{Y}$ to obtain a Gaussian vector with the covariance matrix ρ, which was the target. Indeed,

$$\mathrm{E}[\mathbf{Z}\, \mathbf{Z}'] = \mathrm{E}[\rho\, \mathbf{Y}\, \mathbf{Y}'\rho] = \rho\, \mathbf{B}\, \rho = \rho$$

Note that $\mathbf{Y}$ and $\mathbf{Z}$ satisfy the relation $E[\mathbf{Y}\,\mathbf{Z}'] = \mathbf{I}$, where $\mathbf{I}$ is the identity matrix, or equivalently $E[Y_i\,Z_j] = \delta_{ij}$.

Lantuéjoul goes one step further and rewrites the algorithm by initializing both $\mathbf{Z}$ and $\mathbf{Y}$ (e.g., to $\mathbf{0}$) and, during the iterative process, by immediately propagating the change in the component Y_i to the whole vector $\mathbf{Z}$. Indeed, since $\sum_j \rho_{ij}\, y_j^{\text{current}}$ is nothing but z_i^{current} the expression of y_{-i}^* can be written

$$y_{-i}^* = y_i^{\text{current}} - z_i^{\text{current}}$$

so the change in Y_i can be written

$$y_i^{\text{new}} - y_i^{\text{current}} = U - z_i^{\text{current}}$$

This induces changes

$$z_j^{\text{new}} - z_j^{\text{current}} = \rho_{ji}(U - z_i^{\text{current}}), \qquad j = 1, \ldots, N$$

and in particular $z_i^{\text{new}} = U$. Finally the iterative algorithm can be written as follows:

1. Select a component, say i, at random or by a systematic scan of all the components.
2. Select a new value z_i^{new}.
3. Update the other components so that

$$z_j^{\text{new}} - z_j^{\text{current}} = \rho_{ji}(z_i^{\text{new}} - z_i^{\text{current}}), \qquad j \neq i$$

The vector $\mathbf{Y}$ no longer needs to be considered. Moreover, Lantuéjoul shows that the distribution of Z_i^{new} need not be independent of Z_i^{current}. It must only be standard normal and conditionally independent of $\{Z_j^{\text{current}}: j \neq i\}$ given Z_i^{current}. If we want to avoid the abrupt changes resulting from selecting Z_i^{new} independently of any other value, we can take, for example, $Z_i^{\text{new}} = rZ_i^{\text{current}} + \sqrt{1 - r^2}\,U$ with a positive correlation r.

This simulation algorithm achieves what seemed mission impossible at first: Update $\mathbf{Z}$ without inverting the covariance matrix ρ. Lantuéjoul applies it to the simulation of a Cox process conditionally on count data on different supports.

Conditioning on Indicator Values or on Inequalities

Section 7.6.1 showed how the sequential indicator simulation algorithm can be adapted to condition a simulation of an SRF $Z(x)$ on inequalities when this SRF is discretized and replaced by a series of nested indicator functions. This approach is only approximate. We describe here another method where no discretization of the interval of variation of Z is necessary. It is limited to Gaussian SRFs, but this is an important special case because some random set models are based on Gaussian SRFs. We remain in the framework of the simulation of a Gaussian random vector $\mathbf{Z}$ whose components Z_i are of the form $Z(x_i)$ for some random function $Z(x)$, but this is not mandatory. We suppose that we have:

- hard data $i = 1, \ldots, M$, where Z_i is exactly known: $Z_i = z_i$;
- soft data $i = M + 1, \ldots, N$, where Z_i is only known to belong to an interval of $\mathbb{R}$: $Z_i \in A_i$, where A_i is of the form $[a_i, b_i]$, $]-\infty, b_i]$, $[a_i, +\infty[$, or the union of a finite number of such intervals.

We have only to address the simulation of the soft data conditional on the hard and soft data. Indeed, once these components have been simulated, simulating a vector of size $N' > N$ whose additional components are unknown amounts to generating a conditional simulation given hard data for $i = 1, \ldots, N$. This can be done with the Gibbs sampler but if the Z_i's represent the values $Z(x_i)$ of an SRF $Z(x)$, some of the usual conditional simulation methods may be more efficient.

Freulon and de Fouquet (1993) propose to simulate $\{Z_i : i = M + 1, \ldots, N\}$ conditional on the hard and soft data with the Gibbs sampler. Their algorithm can be presented now as a simple adaptation of the first algorithm presented in this section. The initial state is obtained by assigning the hard data values z_i to components $i = 1, \ldots, M$, and selecting a random value within A_i for each component $i = M + 1, \ldots, N$, from the Gaussian p.d.f. truncated by A_i. The initial state is consistent with all the data (hard and soft) but not with the spatial structure. The Gibbs sampler permits the introduction of the spatial structure while still honoring the data. Specifically, the following sequence is iterated:

1. Select a component, say i, among the soft data (randomly or by a systematic scan).
2. Ignore the value of Z_i and determine its kriging estimate z_{-i}^* from the values z_j of all the other components; also compute the corresponding kriging variance.
3. Replace the current value of Z_i by the kriged value, plus a Gaussian residual with variance equal to the kriging variance, randomly selected so as to match the inequality A_i. An acceptance–rejection technique is preferred to the inverse Gaussian c.d.f. which has numerical problems for the large arguments associated with small kriging variances.

When the number of iterations becomes very large, the distribution of the simulated state tends to the conditional distribution of $\{Z_i : i = 1, \ldots, N\}$ given the hard and soft data. This result is obtained under the assumption that Z_i is kriged from all other values, otherwise the algorithm can diverge. A counterpart is that the algorithm seems to be robust when simple kriging is replaced by universal kriging [Chilès et al. (2005), in applications to 3D geological modeling by the potential field method].

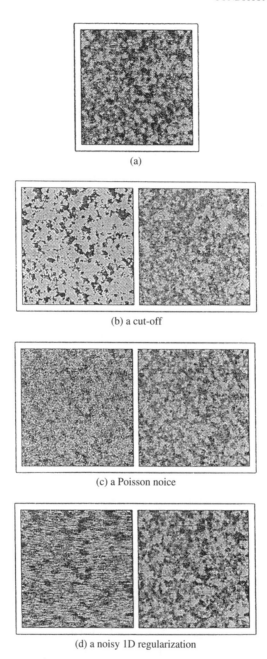

(a)

(b) a cut-off

(c) a Poisson noice

(d) a noisy 1D regularization

FIGURE 7.20 Simulation of a Gaussian SRF conditional on a truncated or noisy image: (a) initial image; (b) truncated image in black and white (threshold $y_1 = 0$) (left) and simulation of the (unknown) initial image conditional on the truncated image (right); (c) image corrupted by Poisson noise (left) and simulation of the (unknown) initial image conditional on the corrupted image (right); (d) image corrupted by noise and deformed by linear averaging along lines (left) and simulation of the (unknown) initial image conditional on the corrupted and deformed image (right). [From Freulon (1994), with kind permission from Kluwer Academic Publishers.]

An alternative would be to use the Gibbs propagation algorithm instead of the Gibbs sampler to avoid the inversion of the covariance matrix. We should then only allow small perturbations that ensure that all inequality conditions remain satisfied (this excludes situations when hard data are available). The convergence of the algorithm has not yet been assessed.

Freulon and de Fouquet (1993) demonstrate the use of the method for simulating a 2D gray-tone image conditioned on a black-and-white image. The problem is this: Produce a continuous gray-tone image with the correct covariance such that after thresholding it coincides with the given black-and-white image. Freulon (1994) extends the method to other types of conditioning: (1) on an image corrupted by Poisson noise, where the value at grid node x_i is not Z_i but an integer random value selected from a Poisson distribution with mean Z_i, and (2) on an image corrupted by noise and deformed by linear averaging along lines. The results, shown in Figure 7.20, are impressive.

As is shown next, this method can be extended to the simulation of a series of nested indicator functions that can be considered as the result of applying thresholds to a common Gaussian SRF (a single indicator function corresponding to the case of a single threshold).

7.6.4 Truncated Gaussian Simulation

Consider an indicator or a series of indicators that originate from applying one or more thresholds to a standard Gaussian RF $Y(x)$,

$$I_i(x) = 1_{y_{i-1} \le Y(x) < y_i} \quad \text{with} \quad -\infty = y_0 < y_1 < \cdots < y_{i-1} < y_i < \cdots < y_m = +\infty \tag{7.41}$$

This is known as a truncated (or thresholded) Gaussian simulation (TGS) and constitutes a straightforward generalization of the notion of y-level excursion set (the random set where $Y(x) \ge y$). The thresholds y_i are chosen so as to match the proportions p_i of the various indicators

$$y_i = G^{-1}\left(\sum_{j=1}^{i} p_j\right), \quad i = 1, \ldots, m-1$$

Once the correlogram $\rho(h)$ of the underlying Gaussian $Y(x)$ is known, the direct and cross-covariances of the various indicators are known (see Sections 2.5.3 and 6.4.3). The simulation of the indicators I_i reduces to that of Y. In applications we choose directly $\rho(h)$ such that the theoretical variograms of the indicators, deduced from $\rho(h)$ by the relation (2.76) or (6.29), fit the sample variograms well. Indeed these relations are not easily inverted, except if the cutoff is at $y = 0$. Furthermore, while any correlogram $\rho(h)$ gives a valid model of indicator covariance by application of (2.76), it is not certain that conversely, the inversion of this relation is a valid correlogram model, even if we start from an indicator covariance: indeed an indicator covariance is not necessarily the covariance of a truncated Gaussian SRF, as we have seen in Section 2.5.3. Figure 7.21 shows indicator simulations obtained by thresholding Gaussian simulations at $y = 0$. Figure 7.22a shows an example with three facies.

Let us now examine the construction of conditional simulations. If we are dealing with Y data directly, we are in the standard case of conditioning by kriging. If, on the other hand, we are dealing with indicator data, we proceed in four steps:

1. Global structural analysis of the direct and cross-variograms of the different indicators, with a fit to a model of the form (6.29) derived from a correlogram $\rho(h)$.
2. Simulation of $Y(x)$ at the data points conditional on the indicator values, using the method for conditioning on inequalities.
3. Simulation of the whole grid conditional on these simulated hard data.
4. Transformation of the simulation of Y into a simulation of the indicators.

This method can easily be adapted to a regionalization of the proportions of the different indicators and of the corresponding thresholds. There are also adaptations for including connectivity constraints (Allard, 1994). TGS was initially designed for simulating lithofacies of petroleum reservoirs (Matheron et al., 1987, 1988). Felletti (1999) applies it in the scope of a thorough geological and geostatistical study of a turbiditic system.

Truncated Pluri-Gaussian Simulation

Truncated Gaussian simulations are of diffusive type, in the sense that given (7.41), the facies i can be surrounded only by facies $i-1$ and $i+1$. The method, however, has been generalized to facies that do not follow one another in a fixed order by using two Gaussian SRFs Y_1 and Y_2. The arrangement of the facies is represented by a facies substitution diagram, as shown in Figure 7.22b. The horizontal axis represents $G(y_1)$, whereas the vertical axis represents $G(y_2)$, both varying from 0 to 1, and the diagram shows the thresholds transforming the pair (y_1, y_2) into a facies. It also shows which facies can be in contact. The areas associated with the facies correspond to their proportions if the random functions Y_1 and Y_2 are independent. In the example of Figure 7.22b all facies can be in contact. The method is known as truncated pluri-Gaussian simulation, or simply pluri-Gaussian simulation (PGS): Galli et al. (1994) and Le Loc'h and Galli (1997).

Figure 7.22c shows an example obtained with correlated Gaussian SRFs. This approach has been used to represent the complex internal architecture of carbonate depositional systems—for example, the geometric relation between algal mound formations with an irregular rounded shape and beds of sediments onlapping the mound relief (Van Buchem et al., 2000). There are other extensions of PGS to model a transition zone [oxydo-reduction front; Langlais et al. (2008)], a joint simulation of a categorical variable and a continuous variable (e.g., lithology and grade), or a joint simulation of two correlated

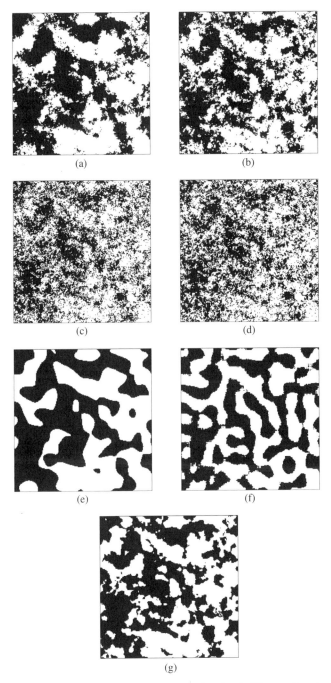

FIGURE 7.21 Examples of simulation of indicators by thresholding a Gaussian simulation. These examples correspond to the threshold $y_1 = 0$ of the simulations of Figure 7.18. Note that (g) has an exponential covariance.

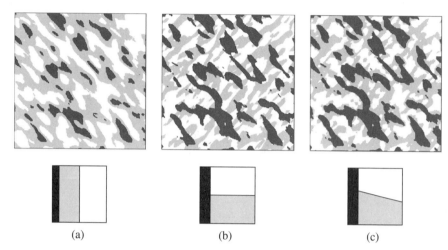

FIGURE 7.22 Truncated Gaussian and pluri-Gaussian simulations, and corresponding facies substitution diagrams. (a) TGS with three facies obtained by applying two thresholds to a single SRF: the white and black facies cannot be in contact. (b) PGS with facies defined by thresholding two independent Gaussian SRFs: each facies is in contact with the others. (c) PGS with facies obtained by thresholding two correlated SRFs: the black facies is more often in contact with the gray facies than with the white facies (the threshold separating the gray and white facies is a fixed value; the facies diagram represents it with an oblique shape to show the effect of the correlation of the two SRFs; this effect can also be obtained with two independent Gaussian SRFs Y_1 and Y_2 and a variable threshold on Y_2 depending on the value of Y_1).

categorical variables (e.g., lithology and diagenesis). The applications, originally in reservoir characterization, extend now to mining. A complete presentation of the method, its variants, and its applications is given by Armstrong et al. (2011).

When the proportions vary spatially, which is always the case in geological applications, the current practice is to model these variations empirically, as shown in the case study of Section 7.10.2. Haas and Formery (2002) propose a Bayesian approach for modeling the proportions. In the case of two facies with given proportions and a given number N of independent samples, the distribution of the numbers N_1 and N_2 of data in facies 1 and 2, respectively, follow a binomial distribution. Conversely, given the number of data in each facies and assuming complete ignorance on the proportion of facies 1 (uniform prior on [0, 1]), the posterior distribution of this proportion is a beta distribution. This can be generalized to more than two facies, with multinomial and Dirichlet distributions, respectively. It is also possible to take account of prior geological information through a prior distribution of the proportion vector. If a prior Dirichlet distribution is assumed, the posterior distribution has the remarkable property of remaining of that type. The authors extend this model to spatial variations of the proportions and propose an algorithm for simulating the proportions. The simulation of a PGS could thus include a preliminary step for sampling the proportions from the posterior distribution.

7.6.5 Tessellation

A tessellation is a partition of space into nonoverlapping cells (usually convex polygons or polyhedra). By extension, it denotes a mosaic model based on that partition.

General Principle

Let us consider a stationary partition of $\mathbb{R}^n$ in cells P_i (usually convex polyhedra). A characteristics of this partition is the function

$$P(h) = \Pr\{x \text{ and } x + h \text{ belong to the same cell}\} \qquad (7.42)$$

Now assign i.i.d. random values a_i to these cells (independently of the cells), and define $Z(x) = a_i$ for $x \in P_i$. Z is by definition a stationary mosaic random function: Two values $Z(x)$ and $Z(x + h)$ are equal with probability $P(h)$, or i.i.d. with probability $1 - P(h)$. If we denote by m and σ^2 the mean and variance of the random variables a_i and suppose that they are finite, the random function $Z(x)$ has mean m and stationary covariance

$$C(h) = \sigma^2 P(h) \qquad (7.43)$$

If the a_i are chosen Gaussian, $Z(x)$ has a Gaussian marginal distribution (but its joint distributions are not Gaussian). Categorical variables are obtained by using discrete distributions. Random sets correspond of course to random variables equal to 1 with probability p or 0 with probability $1 - p$, thus to $\sigma^2 = p(1 - p)$.

Ergodic arguments show that $P(h) = g(h)/g(0)$, where $g(h)$ is the geometric covariogram of the typical cell (i.e., the limit of the average of the geometric covariograms of the cells in a domain V that tends to infinity). Conversely, a covariance may be proportional to some geometric covariogram and yet not be associated with a random tessellation. For example, the spherical covariance cannot be the covariance of a random tessellation of $\mathbb{R}^3$ (Emery, 2010b); however, it can be the covariance of a random tessellation of $\mathbb{R}$, as the covariance of a renewal process. Let us now review some stationary tessellation models.

Renewal Processes

In 1D a renewal process is a point process for which the lengths U_i of the intervals between successive events are independently and identically distributed with a common probability distribution $F(du)$ [e.g., see Feller (1971, Chapter XI)]. There is no restriction on F other than to be concentrated on $[0, \infty[$ and to have a finite mean μ. Thus

$$\mu = \int_0^\infty u\, F(du) = \int_0^\infty [1 - F(u)]\, du$$

Since, in general, the origin is not a point of the process, the distribution of the interval separating the origin from the first event to the right plays a special role. The choice of initial distribution that ensures stationarity of the process is the density

$$f_0(u) = \frac{1}{\mu}(1 - F(u))$$

This is also the distribution of the *residual waiting time* between an arbitrary point x and the next point of the process. The probability that x and $x+h$ belong to the same interval is simply the probability that the residual waiting time exceeds h and is therefore $1 - F_0(h)$. Thus

$$P(h) = \frac{1}{\mu}\int_h^{\infty} [1 - F(u)]\, du \qquad (7.44)$$

The simplest case is the Poisson point process with intensity λ. Here F and F_0 coincide, $F(u) = F_0(u) = \exp(-\lambda u)$, $u > 0$, $\mu = 1/\lambda$, and from (7.43) $Z(x)$ has the exponential covariance

$$C(h) = \sigma^2 \exp(-\lambda|h|) \qquad (7.45)$$

with scale parameter $1/\lambda$. If the random variables a_i of the mosaic model take their values within a finite space of states (e.g., integers from 1 to N), the resulting process $Z(x)$ is a continuous-time homogeneous Markov chain [a jump process in the sense of Feller (1971, Section X.3)]. An example is shown in Figure 7.23a,b for $N = 2$.

Coming back to the general case, the covariance (7.44) is convex, positive, and linear at the origin. Conversely, any covariance function that satisfies these properties can be simulated with a renewal process by taking

$$F(u) = 1 - \frac{C'(u)}{C'(0)}, \quad \sigma^2 = C(0), \quad F_0(u) = 1 - \frac{C(u)}{C(0)}$$

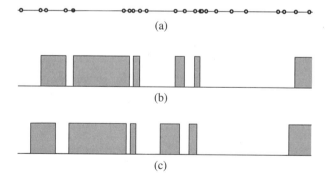

(a)

(b)

(c)

FIGURE 7.23 1D random sets associated with a Poisson point process: (a) Poisson points; (b) 2-state Markov process (base partition defined by segments between Poisson points); (c) Voronoi process (base partition defined by segments joining interval midpoints).

The following algorithm may be used to simulate the renewal process over the segment $[0, L]$:

1. Select the first point of the process to the right of the origin from the distribution F_0.
2. Simulate the next segments independently from the distribution F until the end point L is reached.

Generalization to a Concave Unbounded Variogram

We note that if $C(L)$ is not equal to 0, there is a nonzero probability that the first point on the right side of the origin falls after L. In this case the simulation is constant over $[0, L]$. To avoid this, a slight modification is introduced so that the first point always belongs to the interval $[0, L]$. To obtain this without introducing a bias on the covariance, Matheron (1988) showed that it suffices to use the previous algorithm with

$$F(u) = 1 - \frac{\gamma'(u)}{\gamma'(0)}, \qquad \sigma^2 = \gamma(L), \qquad F_0(u) = \frac{\gamma(u)}{\gamma(L)}$$

where $\gamma(h) = C(0) - C(h)$ is the variogram of $Z(x)$.

The algorithm has a broader application domain than convex covariances with a linear behavior at the origin: It can be applied to simulate any concave variogram with a linear behavior at the origin, even if this variogram is not bounded. It has been generalized by Matheron (1988) to simulate any concave variogram, even if not linear at the origin. In this case it is necessary to consider a generalization of renewal processes to point processes with accumulation points, and the simulation is built at discrete points. The simulations obtained with this algorithm are fractal, but with Hausdorff dimension $D = 2 - \alpha$ instead of $2 - \alpha/2$ for a Gaussian IRF-0 with the same variogram (fractional Brownian random function).

The generalization of renewal processes to $\mathbb{R}^n$, $n > 1$, is simple only in the case of Poisson processes. We will examine now the two types of generalization based, respectively, on Poisson point processes in $\mathbb{R}^n$ and Poisson hyperplanes.

Voronoi Tessellation

Consider points $\{x_i : i = 1, 2, \dots\}$ in $\mathbb{R}^n$, and associate to each x_i the set P_i of all points x in $\mathbb{R}^n$ that are closer to x_i than to any other x_j, $j \neq i$. In 1D the P_i's are the segments joining the midpoints of the intervals $[x_i, x_{i+1}]$ if the x_i's are ordered by increasing values. In 2D they are the convex polygons delimited by the perpendicular bisectors of the segments joining neighboring points (Figure 7.24a). These polygons are usually named Voronoi polygons (metallogeny), or in other contexts polygons of influence (mining), Thiessen polygons (hydrology), cellular networks, Dirichlet polygons [see Stoyan et al. (1987)]. More generally, in $\mathbb{R}^n$ the P_i's are convex polyhedra. If we randomize the x_i's into a stationary random point process of points X_i, the P_i's define a partition of space into a stationary random set of polyhedra, from which we can build a mosaic random function whose covariance is given by (7.43). Figure 7.24b exhibits an RF with a Gaussian marginal distribution, and Figure 7.24c exhibits a random set.

The problem is to obtain the function $P(h)$ defined by (7.42). Matérn (1960, p. 40) derives it when the X_i's form a Poisson point process in $\mathbb{R}^n$. The result has a simple expression in the 1D case (Figure 7.23c). Each Voronoi segment joins the midpoints of two consecutive intervals of the Poisson point process; hence the lengths of the Voronoi segments are gamma random variables with parameter $\alpha = 2$ and scale $b = 2\lambda$, where λ is the Poisson point intensity. Two successive Voronoi segments contain one-half of the same Poisson interval, and therefore they are not independent and do not define a renewal process.

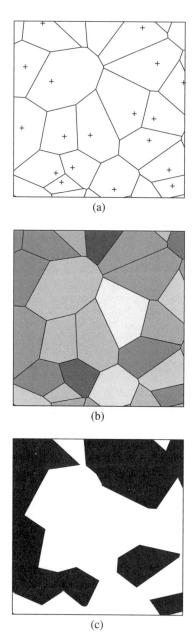

(a)

(b)

(c)

FIGURE 7.24 2D mosaic simulations associated with a Voronoi partition: (a) Voronoi polygons; (b) mosaic RF with Gaussian marginal; (c) mosaic random set ($p = 0.5$). The initial point process is made of Poisson points.

However, formula (7.44) remains applicable—it can be derived directly by a geometric argument involving only stationarity—and combined with (7.43) yields the covariance

$$C(h) = \sigma^2 (1 + \lambda|h|) \exp(-2\lambda|h|)$$

In $\mathbb{R}^2$, Poisson–Voronoi polygons, namely Voronoi polygons associated with a planar Poisson point process, pave the plane with quasi-hexagonal polygons and are therefore of great interest in many applications. Their main properties were derived by Miles (1970). Other results have been obtained by Crain (1978) by Monte Carlo techniques. Miles (1972) presents the properties of the Poisson–Voronoi polyhedra in $\mathbb{R}^3$.

The nonconditional simulation of a Voronoi tessellation is straightforward provided that border effects are taken into account. Generating simulations conditioned on point data requires different algorithms if the marginal distribution is continuous (two data points with the same value belong to the same cell) or discrete (data points with the same value may belong to different cells). Lantuéjoul (2002, Chapter 12) proposes iterative algorithms for both situations.

Poisson Hyperplanes

Instead of starting from random points in the plane or in space, one can start from "random lines" or "random planes." This permits the simulation of random functions with an exponential covariance (or a linear variogram; see Section 7.5.2).

The principle for generating Poisson lines in $\mathbb{R}^2$, Poisson planes in $\mathbb{R}^3$, and more generally Poisson hyperplanes in $\mathbb{R}^n$ is to start from a Poisson point process with intensity $\lambda_n \, d_\omega$ on a line D_ω going through the origin of space and with direction the central direction of the solid angle $d\omega$. Through each of these points we erect a line, plane, or hyperplane perpendicular to D_ω, and we repeat this for all directions ω of the unit half-sphere. Thus we obtain a network of lines or planes that appear random because no orientation or region of space is favored (generalizations are of course possible). Mathematical developments are beyond our present scope and can be found in Miles (1969) and Matheron (1971a, 1975a).

The intersection of a subspace $\mathbb{R}^m$ with a Poisson hyperplane process of $\mathbb{R}^n$ ($m < n$) produces a Poisson hyperplane process of $\mathbb{R}^m$. In particular, the intersection of a line with a Poisson hyperplane process forms a Poisson point process whose intensity is independent of the line considered. The intensity λ of the induced Poisson point processes on lines may be taken as the parameter of the Poisson hyperplane processes in $\mathbb{R}^n$ and its subspaces.[4]

Consider now the case where we want to simulate Poisson hyperplanes within a hypersphere with diameter D centered at the origin. Each hyperplane is characterized by two parameters (Figure 7.25):

1. Its distance d to the origin.
2. The direction ω of the unit vector orthogonal to the hyperplane, directed from the origin to the hyperplane.

Equivalently the hyperplane is characterized by the vector joining the origin to the projection of the origin onto the hyperplane. Now rotate the vectors associated with the Poisson hyperplanes so as to align them with the positive side of the first coordinate axis. The end points of the rotated vectors define a point process on the positive part of this axis. This process is also a Poisson point process. A geometric argument shows that its intensity is $\lambda S_n / V_{n-1}$, where S_n is the surface area of the unit-radius sphere of $\mathbb{R}^n$ and V_n its volume [as results from formulas (A.5), the ratio S_n / V_{n-1} is equal to 2, π, or 4 when n equals 1, 2, or 3, respectively]. Hence we

---

[4] Starting from different definitions, Matheron and Serra use one-half of this λ as the main parameter, and Miles uses a third parameter.

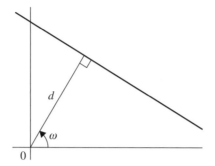

FIGURE 7.25   Parameters defining a line (2D): direction $\omega \in [0, 2\pi]$ and distance d.

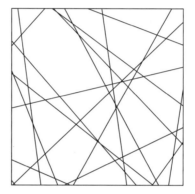

FIGURE 7.26   Simulation of Poisson lines and Poisson polygons tessellation.

have the following procedure for simulating the Poisson hyperplanes that intersect a hypersphere with diameter D:

1. Generate a 1D Poisson point process with intensity $\lambda S_n/V_{n-1}$ over the segment $[0, D/2]$.
2. To each point X_i of this Poisson process, associate a random orientation ω_i drawn independently from a uniform distribution over the unit hypersphere, and erect the hyperplane characterized by the distance $d_i = |X_i|$ and the orientation ω_i.

This model can be easily extended to a nonuniform distribution of the orientations ω_i.

Poisson Polyhedra Tessellation

Poisson lines form a partition of the plane into random convex polygons (Figure 7.26). Similarly, Poisson hyperplanes divide the space $\mathbb{R}^n$ into random convex polyhedra. Let us assign a constant value to all the points of a polyhedron and draw this constant independently from a distribution with variance σ^2. Since the intersection of the Poisson hyperplanes with any line defines a 1D Poisson point process with intensity λ, this mosaic random function is an SRF with covariance

$$C(h) = \sigma^2 \exp(-\lambda|h|) \tag{7.46}$$

This is covariance (7.45), but this time h belongs to $\mathbb{R}^n$. Figure 7.3f shows an RF with a Gaussian marginal distribution, and Figure 7.3h shows a random set. The hyperplanes stand out because the boundaries of the random sets are aligned on them. In figurative terms, Voronoi models consider crystals that grew from germs, whereas Poisson polyhedra are based on a network of faults. The typical polygons obtained in 2D have very different features than those of Voronoi polygons; in particular, the mean number of vertices of these polygons is equal to 4 instead of 6 for Voronoi polygons. Results can be found in Miles (1964, 1973, in $\mathbb{R}^2$; 1972 in $\mathbb{R}^3$; 1971 in $\mathbb{R}^n$), Matheron (1971a, 1972a, 1975a) in $\mathbb{R}^2$, $\mathbb{R}^3$, and $\mathbb{R}^n$, and also Stoyan et al. (1987).

Let us mention a variant, the *alternating random set*, which is simulated as follows: Draw at random, with equal probability, the value $m - \sigma$ or $m + \sigma$ (0 or 1 for a random set, with $m = \sigma = 1/2$) for the polygon or polyhedron that contains the origin; assign the alternate value to the polyhedra that have a face in common with this polyhedron; and iterate. $Z(x)$ and $Z(x + h)$ have therefore identical or alternating values depending on whether the number of Poisson hyperplanes intersecting the segment joining x to $x + h$ is even or odd. Simple probabilistic calculations show (Blanc-Lapierre and Fortet, 1953, pp. 187–191) that the centered covariance is

$$C(h) = \sigma^2 \exp(-2\lambda|h|)$$

By dividing the intensity by a factor of 2, we obtain the same covariance as in (7.46) for simulations that look very different (Figure 7.3j). But we have lost one degree of freedom, since the proportion associated with the value 1 is always $p = 1/2$.

Like for Voronoi tessellations, Lantuéjoul (2002, Chapter 12) proposes two iterative algorithms for generating simulations conditioned on point data for, respectively, a continuous marginal distribution and a discrete one.

Voronoi tessellation and Poisson polygon tessellation represent two extreme cases, the former without aligned edges and the latter with systematically aligned edges. The STIT tessellation (Nagel and Weiss, 2005; Nagel et al., 2008), also defined in $\mathbb{R}^n$, proposes an intermediate behavior and is of interest to model crack patterns.

7.6.6 Substitution Random Functions

Definition and Properties

Substitution random functions (Matheron, 1989; Lantuéjoul, 1991, 1993, 2002, Chapter 17) are the multidimensional generalization of the subordinated processes defined by Feller (1971, Section X.7). The shift in terminology emphasizes the stereological content of the construction, which transfers certain properties of a coding process to a random function.

An RF $Z(x)$ is said to be a substitution random function if it is of the form $Z(x) = Y(T(x))$, where $T(x)$ is a random function of $x \in \mathbb{R}^n$ and $Y(t)$ is a stochastic process of $t \in \mathbb{R}$. $T(x)$ is called the directing function, and $Y(t)$ is the coding process. Since all kinds of models for $T(x)$ and $Y(t)$ can be used, the substitution method gives rise to a broad variety of random functions. If $T(x)$ and $Y(t)$ are continuous RFs, the level sets of $T(x)$ remain level sets of $Z(x)$: The substitution simply changes the level set values. If $T(x)$ is constant over domains that partition $\mathbb{R}^n$, so is also $Z(x)$: The substitution simply changes the values assigned to the various domains. So the most interesting case is when $T(x)$ is a continuous RF and $Y(t)$ a discrete-state process. We will restrict this presentation to the case studied in detail by Lantuéjoul (1993), where $T(x)$ has strictly stationary increments, $Y(t)$ is a stationary continuous-time Markov chain with covariance $C(h)$, and $T(x)$ and $Y(t)$ are independent. This model is a generalization of the discrete diffusion random functions of Section 6.4.3, whose nice properties carry over in a straightforward manner:

- $Z(x)$ is ergodic, provided that $T(x)$ is not stationary (if it were stationary, this would restrict the range of the variations of T around the mean, and lead to the use of a small portion of the definition domain of $Y(t)$).
- $Z(x)$ has the same univariate distribution as $Y(t)$.
- $Z(x)$ is second-order stationary:

$$\mathrm{Cov}[Z(x), Z(x+h)] = \mathrm{E}[C(|T(x+h) - T(x)|)]$$

an expression that does not depend on x since T has stationary increments.

- If $Y(t)$ has an isofactorial expansion, $Z(x)$ has an isofactorial expansion with the same factors as $Y(t)$.

Examples

Lantuéjoul (1991, 1993, 2002) shows several 2D examples of such substitution random functions that demonstrate the capability of this model to represent very different morphologies. A simple model is the following:

- The directing function $T(x)$ is a 2D IRF-0 associated with random jumps on Poisson lines, defining a linear variogram, as explained in Section 7.5.2.
- The coding process $Y(t)$ is a two-state continuous-time Markov chain, whose sojourn time distributions in the states 0 and 1 have densities $b_0 \exp(-b_0 t)$ and $b_1 \exp(-b_1 t)$, respectively (more than two states can also be considered).

The intensity of Poisson lines is a scale factor, whereas the distribution of the random jumps has a large influence on the character of the simulation:

- Figure 7.27 shows a simulation of $Z(x)$ when the discontinuities have a stable distribution, namely a characteristic function $\Phi(u) = \exp(-|u|^\alpha)$ with $\alpha = 0.47$. It gives a strong impression of homogeneity, like a texture. This is due to the fact that the intensity of random lines is high and the jump distribution is heavy-tailed, so very different values of T may be assigned to two neighboring points.
- Figure 7.28a displays a similar simulation when the discontinuities take the values -1 and $+1$ only, with equal probability, as in the standard simulation of a linear variogram. Here the coding process is an eight-state continuous Markov chain. Because a very large number of Poisson lines has been used (about 50,000) and the coding process produces exponentially distributed sojourn times in the various states, structures emerge at

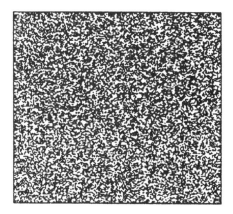

FIGURE 7.27 Simulation of a substitution RF by Markov coding of an RF whose discontinuities have a stable distribution with index $\alpha = 0.47$. [From Lantuéjoul (1993), with kind permission from Kluwer Academic Publishers.]

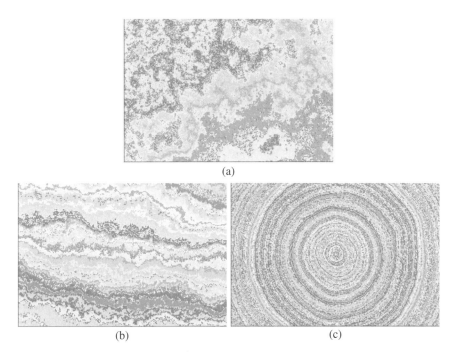

(a)

(b)   (c)

FIGURE 7.28 Simulation of a substitution RF by Markov coding of an RF with discontinuities equal to ± 1: (a) stationary isotropic RF; (b) stationary anisotropic RF; (c) nonstationary RF. [From C. Lantuéjoul, personal communication.] (See color insert)

all scales although, strictly speaking, this random function is not fractal. Figures 7.28b and 7.28c are obtained by changing the rule attributing the discontinuity values in the directing simulation. Similar color images were produced by Lantuéjoul (1991).

We note that two variants of the same algorithm can lead to very different results. The first one (Figure 7.27) gives a covariance with a finite integral range (this is true when $\alpha < 0.5$), whereas the second one (Figure 7.28a) leads to an infinite integral range.

Conditioning

The nonconditional simulation of a substitution random function follows directly from its definition, provided of course that we know how to simulate the independent underlying random functions $T(x)$ and $Y(t)$. But conditioning such a simulation is not so simple. If the conditioning data at each sample point x_α were the values $t_\alpha = T(x_\alpha)$ of the directing function and $z_\alpha = Y(t_\alpha)$ of the coding process, we could meet our objective in two conditioning steps:

1. Conditional simulation of $T(x)$ given the data $T(x_\alpha) = t_\alpha$.
2. Conditional simulation of $Y(t)$ given the data $Y(t_\alpha) = z_\alpha$.

Note that from a practical point of view, $T(x)$ must be simulated at the nodes of the simulated grid, whereas $Y(t)$ must be simulated at the points defined by the values taken by $T(x)$. In practice, however, the conditioning has to be based on the data $z_\alpha = Z(x_\alpha)$ only. Before applying the two conditioning steps, a preliminary step is to simulate the $T(x_\alpha)$. The principle is to first build a non-conditional simulation of $Y(t)$, and then simulate the $T(x_\alpha)$ under the constraints that $Y(T(x_\alpha)) = z_\alpha$. Such a constraint means that $T(x_\alpha)$ belongs to one of the intervals where $Y(t)$ is in the state z_α. This is a case of conditioning on inequalities except that we are considering unions of intervals rather than single intervals. For Gaussian RFs the solution proposed in Section 7.6.3 can be extended to the present case. Lantuéjoul (1993, 2002) presents an example of such conditioning.

7.7   OBJECT-BASED SIMULATIONS: BOOLEAN MODELS

We are considering here a class of models, the Boolean models and their variations, obtained by combining *objects* placed at random points. These models can also be considered as *marked point processes*, in the sense that they are based on a point process and marks (here the objects) attached to the points of the process (Stoyan et al., 1987, Section 4.2). They constitute a family of very flexible models and are sometimes used on the basis of a physical or genetic interpretation which defines the objects of the particular model used. They are also used without reference to any plausible physical interpretation when they produce an acceptable fit of the observations.

7.7.1   Boolean Models

Boolean Random Set

A Boolean random set corresponds to the intuitive idea of the union of randomly located objects. The main theoretical results are due to Matheron (1967, 1975a), though special forms have been used earlier [see Stoyan et al. (1987, p. 68)].

Consider a process of Poisson points X_i in $\mathbb{R}^n$, $i = 1, 2, \ldots,$ and let A_i, $i = 1, 2, \ldots,$ be i.i.d. random objects (elements of volume, or more precisely compact—i.e., closed and bounded—subsets of $\mathbb{R}^n$). The union of the A_i shifted to points X_i constitutes by definition a Boolean random set X:

$$X = \bigcup_i \tau_{X_i} A_i \qquad (7.47)$$

where τ_h denotes the operator of translation by a vector h. Definition (7.47) is the set formulation of the dilution approach (7.23).

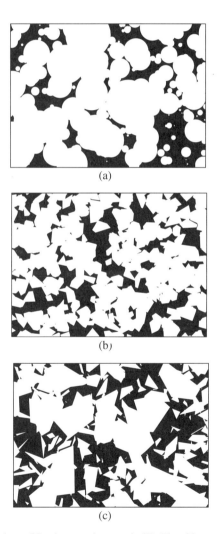

(a)

(b)

(c)

FIGURE 7.29 Realizations of Boolean random sets in 2D. The objects are: (a) disks; (b) Voronoi polygons; (c) Poisson polygons (Delfiner, 1970). [From C. Lantuéjoul, personal communication.]

The points X_i of the Poisson process are referred to as germs and the compacts A_i as primary objects. In this model the germs grow independently until they form primary objects that can overlap. The space $\mathbb{R}^n$ is thus divided into two phases, the objects X which is the union of the primary objects, and its complement, the background.

A very simple example consists of taking spheres with a given diameter for the objects. Variants presented in the framework of the dilution methods apply here also: for example, spheres with random and independent diameters (Figure 7.29a). Figures 7.29b and 7.29c present two 2D simulations where the objects are Voronoi polygons and Poisson polygons, respectively.

A Boolean model X depends on two parameters, the intensity λ of the Poisson point process and the probability distribution of the primary objects. In general, a random closed set X is characterized by its *Choquet capacity*, or *hitting functional*, namely the mapping T that associates to each compact subset B of $\mathbb{R}^n$ the probability that B hits X:

$$T(B) = \Pr\{X \cap B \neq \varnothing\}$$

In the case of Boolean random sets, instead of evaluating $T(B)$ directly, we evaluate the probability $Q(B) = 1 - T(B)$ that B is contained in the background. It can be shown that this is (Matheron, 1967; Serra 1982)

$$Q(B) = e^{-\lambda E|A \oplus \check{B}|} \qquad (A \oplus \check{B} = \bigcup_{y \in B} A_{-y})$$

where $\check{B}$ denotes the symmetric of B with respect to the origin. If B is reduced to a single point, we obtain the background probability of the Boolean model

$$q = e^{-\lambda E|A|} \tag{7.48}$$

If $B = \{x, x + h\}$ is a pair of points, then $|A \oplus \check{B}| = |A \cup A_{-h}| = 2|A| - |A \cap A_{-h}|$. Denoting by $K(h) = E|A \cap A_{-h}|$ the mean geometric covariogram of the primary objects, the bivariate distribution of the background of the Boolean model is

$$\Pr\{x \in \overline{X}, x + h \in \overline{X}\} = q^2 e^{\lambda K(h)} = q e^{-\lambda[K(0) - K(h)]} \tag{7.49}$$

Therefore the covariance of the background, which is also that of the objects, is

$$C(h) = q\left[e^{-\lambda[K(0) - K(h)]} - e^{-\lambda K(0)}\right] \tag{7.50}$$

If B is a segment or a ball, the explicit calculation of $|A \oplus \check{B}|$ is possible only if A is almost certainly convex. If so, it can be shown, for example (take for B a segment), that the distribution of background intercepts is exponential. We note that, contrary to dilution SRFs, the mean and the covariance of a Boolean random set do not depend linearly on λ. The intensity λ is therefore an essential parameter of a Boolean random set.

These results are useful for testing the validity of a Boolean model with convex primary objects and for attempting the statistical inference of its parameters. When we study a random set in $\mathbb{R}^n$, we frequently examine the intersection of the random set with a hyperplane. The overall study is made easier by the fact that the intersection of a Boolean random set with a hyperplane is still a Boolean random set. On the other hand, in the subspace of this hyperplane, this intersection can be a Boolean random set of convex objects without this being the case in $\mathbb{R}^n$. In practice, even if we have a complete image, statistical inference is tricky due to edge effects, and various methods have been developed to deal with this [see Serra (1982, Section XIII.B.8), Stoyan et al. (1987, Section 3.3), and Lantuéjoul and Schmitt (1991) for a comparison of more recent methods]. Complementary information is available in the book of Molchanov (1997) devoted to the Boolean model.

Boolean random sets can be extended easily to a regionalization of the intensity λ as well as to a regionalization of the distribution of the primary objects. A typical problem is that different intensity functions and different distributions for the primary objects can lead to the same model. Schmitt and Beucher (1997) and Beucher et al. (2005) address the inference of the intensity of a model with a regionalized intensity from a 2D image and in the 3D case of well data, respectively.

Boolean Random Functions

A generalization of Boolean random sets to RFs is to set up at each point X_i of the Poisson point process a realization of a nonnegative random function f_i, called primary function, drawn independently from one point to another. Instead of cumulating the effects of these functions as for dilution RFs, we now take the maximum and let

$$Z(x) = \sup_i f_i(x - X_i) \tag{7.51}$$

The graph of the Boolean random function $Z(x)$ is the envelope of the subgraphs of the primary functions (Figure 7.30). Each primary function delimits a geometric object in $\mathbb{R}^{n+1}$ defined as the space below the graph of the function. If these objects are convex, it is possible to calculate the covariance of $Z(x)$. In particular, if they are cylinders, the covariance includes

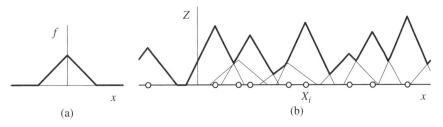

FIGURE 7.30 Construction of a simulation of a Boolean RF (1D): (a) primary function (a cone); (b) resulting simulation with cones with random height.

a linear term at the origin. If they are cones, $Z(x)$ is much more regular, and its covariance is parabolic at the origin. More generally, consider the random primary object defined by the cutoff z on the primary function f,

$$A_z = \{x \in \mathbb{R}^n : f(x) \geq z\}$$

The union of all such primary objects set up at Poisson points X_i forms a Boolean random set whose background probability is by (7.48),

$$F(z) = \Pr\{Z(x) < z\} = \exp(-\lambda \mathrm{E}|A_z|)$$

Furthermore, it can be shown that the bivariate distribution function is given by

$$F_h(z_1, z_2) = \Pr\{Z(x) < z_1, Z(x+h) < z_2\} = F(z_1)F(z_2)\exp(\lambda \mathrm{E}|A_{z_1} \cap \tau_h A_{z_2}|)$$

which generalizes (7.49).

The case of cylindrical objects corresponds to functions f_i of the form

$$f_i(x) = a_i 1_{A_i}(x)$$

and can already be used to describe a variety of situations. The A_i can belong to several classes of objects, that we number 1, 2, and so on. If a_i is the class number of the object A_i, we obtain a simulation where the objects of a class override the objects of the lower classes. However, if the a_i are i.i.d. random variables, there is no hierarchy between the objects. In both cases, if the a_i take a finite number of possible values, corresponding, for example, to a facies number, two adjacent facies do not necessarily have consecutive facies numbers (Figure 7.31). The bivariate distributions of such simulations follow an isofactorial model with orthogonal indicator residuals (see Section 6.4.3).

One of the attractions of Boolean random functions is that they are well-adapted to the change of support by the maximum (dilation in the case of random sets). This type of change of support occurs, for example, if the a_i represent a pollutant level rather than a facies number (although this level may depend on the facies) and if we characterize the pollution of a zone by the highest local concentration in this zone.

Simulating such RFs is very simple [relation (7.51)]. The difficulty is elsewhere, namely the design of criteria to establish that this model is compatible with the available observations and the identification of its parameters (intensity of the Poisson point process, choice of the family of primary functions, and distribution of its parameters). Boolean RFs have been introduced by Jeulin and Jeulin (1981) to model the roughness of metallic fractures. The theory is developed by Serra (1988). The characterization and quantification of Boolean RFs from images is studied by Préteux and Schmitt (1988). A presentation can be found in Chautru (1989). Further references and variations about this model are given by Jeulin (1991).

Dead-Leaves Models and Sequential Random Functions

If the objects of a Boolean random set are of nonzero size and if the intensity λ tends to infinity, the objects end up occupying all the space. This is the basis for the dead-leaves model (Matheron, 1968b; Serra, 1982, Section XIII.C.2; Lantuéjoul, 2002, Chapter 14). In this terminology the objects are leaves, which progressively cover the ground, hiding parts of leaves fallen previously. Looked at from above at a fixed time, this defines a partition of the space whose classes are the visible portions of the leaves not entirely hidden by the subsequent ones (Figure 7.32a). A mosaic model may then be defined from this tessellation, whose covariance is proportional to $K(h)/[2K(0) - K(h)]$, where $K(h)$ is the

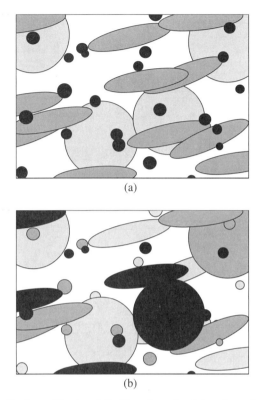

(a)

(b)

FIGURE 7.31 Realizations of Boolean RFs. The subgraphs of the primary functions are cylinders with three types of base (ellipse, large disk, small disk): (a) with hierarchy—the cylinder height (i.e., the value attached to the base) is a function of the base type (here highest for small disks, and lowest for large disks); (b) without hierarchy—the cylinder height is independent of the base type.

mean geometric covariogram of the primary leaves. A variation is the multi-dead-leaves model (Jeulin, 1979) where the leaves belong to different populations (based on type of tree, or color), which defines a multiphase tessellation (Figure 7.32b). The simplest way to simulate a dead-leaves model is to reverse time, as proposed by Kendall and Thönnes (1999): Simulate the last fallen leaf, then the previous one, and so on. The simulation is ended when the ground is no longer visible. This is a typical case of exact simulation, in the sense of Propp and Wilson (1996), because the simulation does not require an infinite number of iterations to perfectly sample the desired model. [Note that the simulation is biased if we do not reverse time and stop the simulation once the ground is completely covered by leaves, because the ultimate leaf that covers the last hole has more chances to be large; see Lantuéjoul (2002, Section 14.3).]

A similar generalization of Boolean random functions to dead-leaves random functions is proposed by Jeulin (1989). All these models can be considered as sequential random functions if we look at their evolution as the leaves are falling. Jeulin (1991, 1997) presents additional models of sequential random functions as well as methods for their statistical inference from an image.

(a)

(b)

FIGURE 7.32 Realizations of mosaic random functions derived from dead-leaves models: (a) single-dead-leaves model (black poplar) with independent assignment of a value to each leaf; (b) multi-dead-leaves model with value assignment depending on leaf species (alder, elm, oak, poplar). (See color insert)

Let us also mention the *transparent dead leaves*, which form a family of models ranging from the dead-leaves model to Gaussian random functions (Galerne, 2010): A random value is associated with each leaf. When a leaf is added, the new value of the simulation in the domain covered by the leaf is a fraction $\alpha \in \,]0, 1]$ of the value attached to the leaf, plus the fraction $1 - \alpha$ of the current value of the simulation. α is a transparency coefficient. The dead-leaves model corresponds to $\alpha = 1$, whereas a Gaussian random function is obtained when α tends to 0.

7.7.2 Stationary Point Processes

Procedures to construct random sets or random functions from Poisson point processes can be generalized to other stationary point processes. For example, we can define dilution random functions and Voronoi tessellations based on an arbitrary distribution of points. But this generalization is mainly applied to Boolean models and is therefore presented here. There are so many models of point processes that we will not try to be exhaustive.

The interested reader is referred to the point process literature, such as Bartlett (1963, 1975), Krickeberg (1974), and Ripley (1977, 1981) and to the random set literature, such as Serra (1982) and Stoyan et al. (1987). We will only present some simple models shown in Figure 7.33 and a structural tool for identifying them: the K-function.

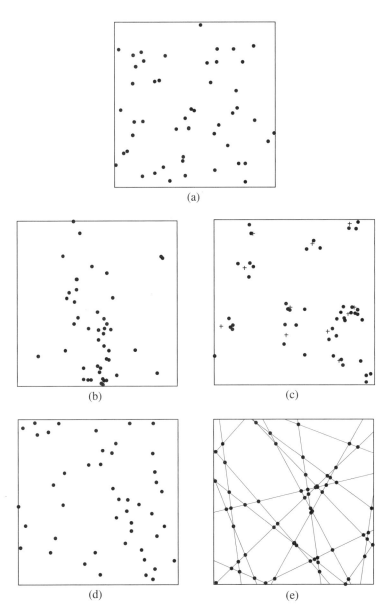

FIGURE 7.33   Simulation of point processes: (a) Poisson points; (b) Cox process (the intensity is a lognormal RF); (c) Poisson cluster process (the crosses are the cluster centers); (d) hard-core process; (e) Poisson alignments.

Cox Process

The Poisson point process can be generalized immediately to the case where the intensity λ is not constant but is a measurable function $\lambda(x)$ with positive values. The number of points falling in V is a Poisson random variable with mean

$$\lambda(V) = \int_V \lambda(x)\, dx$$

These points are distributed independently according to the probability density function $\lambda(x)/\lambda(V)$, $x \in V$. The covariance between $N(V)$ and $N(V')$ is

$$\text{Cov}[N(V), N(V')] = \lambda(V \cap V')$$

This process is not stationary. To make it stationary, it suffices to replace the regionalized intensity $\lambda(x)$ by a nonnegative stationary random function $\wedge(x)$ called the *potential*. We then obtain the Cox process or *doubly stochastic Poisson process* (Cox, 1955; Figure 7.33b). $N(V)$ is distributed as a Poisson with random parameter

$$\wedge(V) = \int_V \wedge(x)\, dx$$

The probabilities and the noncentered moments are obtained by randomizing the expressions obtained for the Poisson process with regionalized intensity. In the end, and denoting respectively by m and $C(h)$ the mean and the covariance function of the potential $\wedge$, we get

$$\text{E}\big(N(V)\big) = m\,|V|,$$

$$\text{Cov}\big(N(V), N(V')\big) = m|V \cap V'| + \int_V \int_{V'} C(y - x)\, dx\, dy$$

Then, even if V and V' do not overlap, $N(V)$ and $N(V')$ are generally not independent. Their covariance is the sum of two terms: The first one accounts for the Poisson distribution of the points once the intensity is fixed, while the second captures the intensity fluctuations. This hierarchic character is not rare in applications. A typical example, studied by Kleingeld and Lantuéjoul (1993), is that of an alluvial deposit of precious stones where the gems are trapped in anfractuosities of the terrain. Where the terrain is regular, one expects to find few precious stones, whereas in rugged terrain one may find a cluster of stones.

The Cox process has a number of remarkable properties. For example, if $N(V)$ follows a negative binomial distribution, then the potential follows a gamma distribution [Feller (1971, pp. 55–57); the negative binomial distribution is termed "the limiting form of the Pólya distribution"]. Similarly, if $N(V)$ follows a Sichel distribution, then the potential follows an inverse Gaussian distribution [see Sichel (1973) for the Sichel distribution; Jørgensen (1982) for the inverse Gaussian distribution; and also Matheron (1981d)].[5] These properties allow Kleingeld et al. (1997) to carry out the statistical inference of a Cox process. Simulations of point processes often have to be generated conditional on count data. While this is straightforward for the Poisson point process, at least when the count data are relative to disjoint supports [otherwise, see Lantuéjoul (2002, Section 11.2)], this is not a trivial task for the Cox process. The authors propose an iterative algorithm based on the discrete Gaussian model and the Gibbs sampler when the potential is the transform of a Gaussian random function. Lantuéjoul (2011) extends it to count data relative to different support sizes.

---

[5]The Sichel distribution and the generalizations proposed by Matheron are very useful for modeling discrete distributions with a very long tail, typically the distribution of the number of precious stones in a given volume.

Shooting Process or Cluster Process

The shooting process or cluster process was defined by Neyman and Scott (1958) for representing clusters of galaxies (they also considered clusters of clusters and their evolution with time, which we will not do). The properties of this model were studied by Neyman and Scott (1972). Each point X_i of a primary Poisson process with intensity λ_0 is considered as a target at which ν shots are fired. The number of shots is a random variable, and the impact of each shot is a secondary point, randomly and independently located around the target. The shooting process is formed of all the secondary points (Figure 7.33c). Another terminology, often used when modeling natural phenomena, is *parent–daughter model*: The primary points represent the location of parents (e.g., plants) and the secondary points are their daughters (e.g., seed fall locations). Note that this process can also be seen as a particular case of the Boolean model where each random object associated with a Poisson point is in fact a cluster of objects.

This model has been generalized to cluster processes with regionalized or random intensity (Deverly, 1984b). But this only makes sense if we specify the scale of the phenomena taken into account by the two aspects (clusters and regionalization), for example, by giving the clusters a local extension and considering that the intensity changes slowly in space. This problem is already there when we choose to use the Cox process or the cluster process, since we can consider the presence of clusters as zones of high intensity, and vice versa. It can be shown that a cluster process where the number of points per cluster has a Poisson distribution is a Cox process.

Hard-Core Models

Hard-core models represent some kind of inhibition between the points. There are many models. A simple one is introduced by Matérn (1960, p. 47). Consider a realization of a Poisson point process with intensity λ_0 as a primary process. As a realization of the hard-core process, keep every point of the primary process whose nearest neighbor (in the primary process) is at a distance at least equal to a given minimum distance R (Figure 7.33d).

Poisson Alignments

Serra (1982) proposes this model to represent aligned points. A realization of this process in 2D is composed of all the intersections of the lines of a realization of Poisson lines (Figure 7.33e). This model can be extended to $\mathbb{R}^n$ by considering the intersections of hyperplanes of dimension $n - 1$.

A Structural Tool: The K-Function

For all the above point processes, we know how to calculate the mean and the variance of $N(V)$ and also the covariance of $N(V)$ and $N(V_h)$. This covariance can be used as a structural tool to differentiate these models and fit their parameters, but it is not very powerful. Fortunately, there are other tools which derive from the fact that the spatial distribution of a point process can also be determined by its *Palm distribution*, where the point process is observed from a moving point of the process [for an exact definition of the Palm distribution, see Stoyan et al. (1987, Section 4.4) or Cressie (1991, Section 8.3.4)]. An example of such tool is the distribution of the distance from a point of the process to its nearest or kth nearest neighbor, particularly useful when the point process is isotropic. We will focus on the main tool used in this case, which is the K-function.

The point processes presented above are isotropic provided that the random intensity function (Cox process) or the clusters (shooting process) are isotropic. In this case Krickeberg (1974) and Ripley (1976) propose to replace the covariance by the K-function, which has a more geometric interpretation and permits an easier discrimination of the various models. The knowledge of this function, which is also called the "second reduced moment function,"

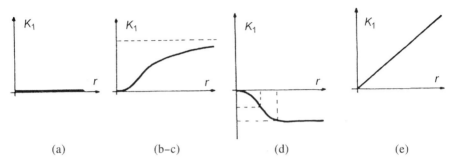

FIGURE 7.34 Shape of the function $K_1(r)$ for different models of point processes: (a) Poisson points; (b–c) Cox process and cluster process; (d) hard-core process; (e) Poisson alignments. [From Serra (1982), with permission of Academic Press Limited, London.]

is equivalent to that of the covariance. First, consider the points of the point process that belong to the sphere of radius r centered at an arbitrary point x (this point almost surely does not coincide with a point of the process); the count of these points has for mathematical expectation $\lambda K_0(r)$ with $K_0(r) = V_n \, r^n$, where V_n is the volume of the unit-radius sphere of $\mathbb{R}^n$ (as usual, λ is the intensity of the point process). Consider now a typical point x_i of the point process, and let $\lambda K(r)$ be the expected number of points of the process within the sphere of radius r centered at the point X_i, which is itself not counted. $\lambda K(r)$ usually differs from $\lambda K_0(r)$, a noticeable exception being the Poisson point process. Hence $\lambda [K(r) - K_0(r)]$ is the average excess of points belonging to the sphere of radius r centered at a point of the process (this point excluded) over and above the average number.

Figure 7.34 shows the behavior of $K_1(r) = K(r) - K_0(r)$ for the various models presented above. They are very different, except again for the Cox process and the cluster process which cannot be easily differentiated: For these models, K_1 is positive, whereas it is equal to zero for a Poisson point process and is negative for a hard-core process. Since the function $K_1(r)$ can be estimated directly [see Ripley (1976) and Cressie (1991)] and its theoretical expression for the various models is known, it is a good structural tool. Note that like the covariance, the K—or K_1—function summarizes the variances and covariances of point counts, namely second-order moments, and thus it does not represent all the information about a stationary isotropic point process: Two different processes can have the same K-function. Baddeley and Silverman (1984) show a 2D example of a point process that is very different from a Poisson point process and yet has the same K-function.

When studying a random set from discrete sampling, the boundaries of the individual objects are usually unknown, which makes the inference of the underlying point process difficult.

7.7.3 Conditional Simulation of a Boolean Model

Conditioning a Boolean Random Set: The Example of Fracture Networks

The simulation of a Boolean random set is straightforward once its parameters are known: intensity of the Poisson point process and probability distribution of the primary objects (type, size, orientation). But conditioning a simulation by observations is less straightforward, except in simple cases, as we now show for fracture networks.

When modeling 3D fracture networks, fractures are usually considered as planar objects. Fracture networks are often represented by Poisson models:

Poisson planes for fractures that are infinite at the scale of the study, random disks for finite fractures [i.e., a 3D Boolean model of 2D disks (Baecher et al., 1977)]. The latter model has been generalized to a parent–daughter model with a regionalized intensity [i.e., a 3D Boolean model of disk clusters built on a Cox process (Chilès, 1989a,b)]. The fractures are usually divided into several sets corresponding to different average fracture orientations. Each set is characterized by a distribution of the fracture orientation (the unit vector normal to the fracture) and of the fracture size (disk diameter). The data are fracture intersections with boreholes or fracture traces on drift walls or outcrops. The orientation of each fracture is measured, but its extension is not always known. These data are affected by several types of bias (censoring, truncation, size). Specific methods have been developed to handle fracture data. 3D fracture network simulations are often used for fluid flow studies. Since actual fractures are not pairs of parallel plates, flow is often seen to be channelized. 2D Boolean models of random segments are therefore simulated within each fracture, and flow is studied in the resulting 3D channel network, as shown in Figure 7.35 (Long and Billaux, 1987; Billaux, 1989). The reader is referred to Chilès and de Marsily (1993) for an overview of these methods with references to other authors. More general marked-point processes are also used [e.g., Wen and Sinding-Larsen (1997)]. These models have been developed for networks of joints (discontinuities created without lateral displacement) in fairly isotropic media, typically granitic formations. Other models are used for subvertical

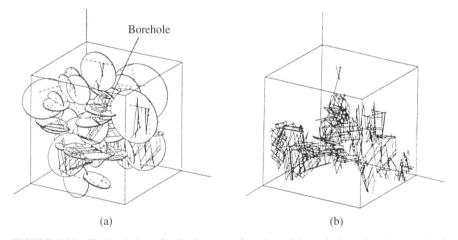

(a)   (b)

FIGURE 7.35 3D simulation of a Boolean set of random disks and channels: (a) network of random disks (for reasons of legibility, not all the disks encountering the simulated domain are shown), and random segments simulated on each disk; (b) network of random channels (only the channels connected to a small borehole located at the center of the block are shown). [Reprinted from Billaux (1989), in V. Maury and D. Fourmaintraux, eds., Rock at great depth—Rock mechanics and rock physics at great depth/Mécanique des roches et physique des roches en condition de grande profondeur/Felsmechanik und Felsphysik in großer Tiefe/Proceedings of an international symposium, Pau, 28-31.08.89. 1989, 1620 pp., 3 vols.]

joints controlled by the bedding of stratiform formations (Chilès et al., 2000). Fault networks often have a different organization (Castaing et al., 1996). Other approaches try to incorporate mechanical rules and to model the fracturing process. Our purpose is not to cover the large field of fractured media modeling but simply to illustrate the conditional simulation of Poisson fracture networks.

Conditioning simulations of Poisson planes or disks on observed fractures is easy because all points, lines, or planes of a Poisson process are mutually independent. The procedure consists in independently simulating the fractures that intersect the surveyed lines or surfaces (step 3) and those that do not (steps 1 and 2):

1. Construct a nonconditional simulation.
2. Reject all the fractures that intersect the surveyed lines or surfaces, and retain all those that do not intersect.
3. Add the actual fractures that have been observed.

The simplest case is the Poisson plane model because fractures are infinite. The conditioning procedure is represented in Figure 7.36, in 2D for simplicity,

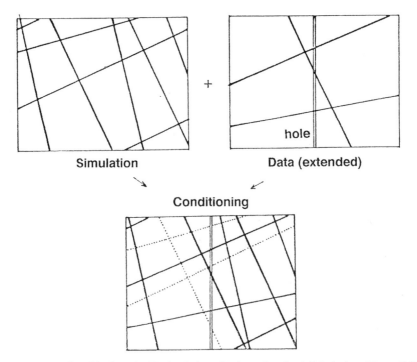

FIGURE 7.36 Conditioning of a 3D simulation of Poisson lines by drill-hole data. [From Chilès (1989a,b), with kind permission from Battelle Press and Kluwer Academic Publishers.]

the data being the intersections with a borehole (location of the intersections and fracture orientation). This procedure has been suggested by Andersson et al. (1984). In the case of the random disk model, step 3 introduces a slight complication because the extension of a fracture must be determined from an observed intersection or trace. In other words, we have to simulate the whole fracture from the joint distribution of disk diameter and disk center location conditional on the observation (point or trace). This conditional distribution results from geometric probability calculations. Rejection techniques make it possible to condition a fracture on additional observations, for example, the intersection by another borehole or a trace on another drift wall, but the user must be able to "correlate" the two intersections of the fracture. An example, based on actual data surveyed on drift walls, is presented by Chilès et al. (1992).

The same approach can be used with a Boolean model of disk clusters, but since the random objects associated with the Poisson points are the clusters, the rejection defined in step 2 and the completion of the observed intersections or traces in step 3 must be applied at this level (see Figure 7.37). This is also feasible (Chilès, 1989a,b). The main constraint is practical: The user must be able to define from the data which fractures belong to the same cluster.

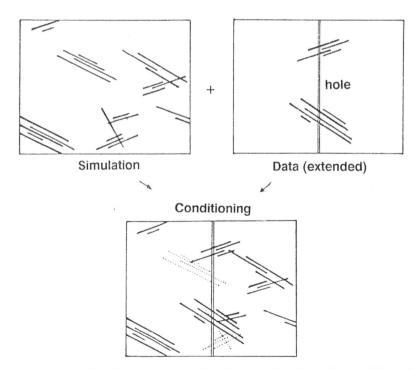

FIGURE 7.37 Conditioning of a 3D simulation of clusters of random disks by drill-hole data. [From Chilès (1989a, b), with kind permission from Battelle Press and Kluwer Academic Publishers.]

Conditional Simulation a Boolean Random Set in the General Case

Conditioning a simulation of fracture network in $\mathbb{R}^n$ is possible for two reasons: (1) The primary objects are elements of a subspace of $\mathbb{R}^n$ so that any point belonging to the fracture network belongs to a single primary object (we neglect fracture intersections); (2) we can determine that two observations are from the same object (fracture or cluster, according to the model). In the general case, one point of the random set may belong to several objects. Moreover, several data points may belong to a common object, but we have no simple means of knowing this. Therefore the method used for fracture networks cannot be transposed to the general case [see, however, Haldorsen (1983) and Chessa (1995) for a similar approach for simulating the geometry of a petroleum reservoir conditional on the observations in wells].

An iterative Markov procedure, suggested by G. Matheron in 1990 and presented by Lantuéjoul (1997, 2002, Chapter 13), gives a general solution, provided that the objects have a strictly positive measure and the number of conditioning points, known to belong to the background or the objects, is finite. This algorithm is valid for Boolean sets based on Poisson point processes with a regionalized intensity $\lambda(x)$ that is integrable: $\int \lambda(x)dx = \lambda_1 < \infty$. The standard Boolean random set ($\lambda(x)$ constant) does not fulfill this condition, but as we will simulate it in a finite domain V, we can limit the support of $\lambda(\cdot)$ to $V \oplus \check{A}$ if all the objects are identical to A, or to $V \oplus \check{S}$ where S is a sphere large enough to include any elementary object.

For a simple presentation of the algorithm, it is convenient to define a λ-object as a primary object with a random location according to the density $\lambda(x)/\lambda_1$. Clearly, a Boolean random set is the union of a Poisson number (with mean value λ_1) of independent λ-objects. The principle of the algorithm is the following: At time $t = 0$, start with an initial simulation, composed of primary objects, that is simply compatible with the data (but does not claim to represent a Boolean random set). Then, denoting by $\varphi(t)$ the pattern of objects at time t and by $|\varphi(t)|$ the number of objects of this pattern, in each elementary time interval $]t, t + dt]$, add a random object with probability $\lambda_1 dt$, remove an object with probability $|\varphi(t)|dt$, and do nothing with probability $1 - (\lambda_1 + |\varphi(t)|)\, dt$. The objects are removed and added randomly but under the constraint that the simulation always remains consistent with the data. The implementation of the algorithm is the following (Lantuéjoul, 1997, 2002, Chapter 13):

1. *Initialization.* Set $t = 0$, and simulate a pattern φ of λ-objects that respects the conditions on the objects and on the background.

2. *Moving Forward in Time.* Simulate an exponential holding time T with parameter $\lambda_1 + |\varphi|$, and add T to t (where $|\varphi|$ is the number of points of the pattern φ).

3. *Transition to the Next State.* Simulate a random variable U equal to $+1$ with probability $\lambda_1/(\lambda_1 + |\varphi|)$, or -1 with the complementary probability $|\varphi|/(\lambda_1 + |\varphi|)$, and

- if $U = +1$, simulate a λ-object and insert it in φ provided this does not violate the conditioning on the background;
- if $U = -1$, randomly select one of the $|\varphi|$ present λ-objects and remove it, provided that this does not violate the conditioning on the objects.

4. Go to (2).

Note that this algorithm tends to equalize $|\varphi|$ with λ_1: when $|\varphi| < \lambda_1$, objects are added more often than removed, and the other way around when $|\varphi| > \lambda_1$. Time plays no role here but will be useful for generalizations.

The following procedure can be used to initialize the process:

1. Set $\varphi = \varnothing$
2. Simulate a λ-object A.
3. If A does not hit the conditioning points that belong to the background and hits at least one of the conditioning points belonging to the objects not yet covered by φ, insert A in φ; else go to (2).
4. If φ does not cover all the conditioning points belonging to the objects, go to (2).

The initialization step requires a finite though possibly large number of operations, unless the data are not compatible with the model (the number of iterations needed gives an indication on the adequacy of the model to the data). The main algorithm never ends but, in practice, must be stopped after a finite time, which raises problems of convergence. First note that the general algorithm can be used even in the nonconditional case (no conditioning point), starting from an empty set. Then $|\varphi|$ evolves according to a birth-and-death process that is known to converge to the desired Poisson variable (Feller, 1968, Section XVII.7). The proof of the convergence of the conditional algorithm is established in Lantuéjoul (1997). In both cases the convergence rate is an exponentially decreasing function of time. Figure 7.38 shows four conditional simulations of a Boolean set of channels.[6]

If the objects are not identical, some random objects centered in the dilated part of $V \oplus \check{S}$ will not intersect V. This may involve the simulation of many useless objects if the size distribution of the objects is very broad. The algorithm proposed by Lantuéjoul (2002) simulates only the necessary objects. To this end, when a Poisson point is implemented at location x, the random object is taken in the distribution of the random object under the condition that it intersects V. The density of the Poisson process at x is decreased accordingly.

---

[6] If the objects (e.g., the channels) are regarded as a permeable phase, there is a critical value of the proportion of area occupied by the objects for which the system *percolates*, in the sense that a connected component much larger than the others suddenly connects two opposite boundaries. This proportion is called the *percolation threshold* and has been studied extensively by Allard (1993) and Allard et al. (1993).

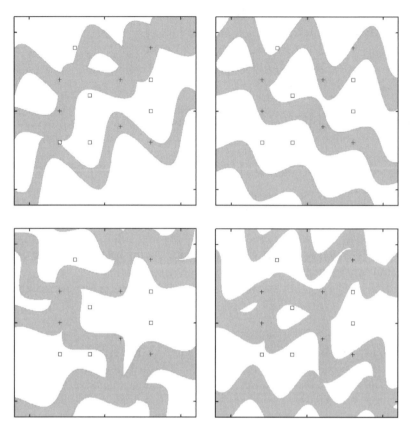

FIGURE 7.38 Conditional simulations of Boolean channels modeled as sinusoidal strips, constrained by six points in the objects and six points in the background.

Conditional Simulation of Other Boolean Models

The Markov procedure of the above algorithm is a very simple one: We pass from one state to the next by adding or removing one object, and the objects are independent. A similar algorithms allows the simulation of Boolean random functions (Lantuéjoul, 2002, Chapter 14). The algorithm of Boolean random sets lends itself to several generalizations while remaining in the convenient setting of Markov models. *Spatial birth-and-death processes*, in particular, model more complex patterns of points or objects than Boolean models (Preston, 1975; Stoyan et al., 1987, Section 5.5.5; Møller, 1989; Guyon, 1993, Section 6.5; Lantuéjoul, 1997).

7.8 BEYOND STANDARD CONDITIONING

Reproducing the histogram and the covariance of a phenomenon and conditioning on sample points is not always sufficient to provide realistic simulations. Indeed, we have seen that different random function models sharing the same histogram and covariance can display very different behaviors. When we select a random function model, we implicitly specify all finite-dimensional distributions. If the data permit, for example if we have a complete image, it is sensible to check that some of these distributions are compatible with the data. If this is not the case we have to look for a suitable alternative random function model. Unfortunately the geostatistician's toolbox may not contain one. To overcome that problem some approaches avoid the need of an explicit random function model, either by simulating the physical processes generating the field of interest, or by exploiting a training image. Adaptive methods may also provide a solution honoring given constraints: a standard simulation matching only the histogram and the covariance, for example, is adapted iteratively until it matches the imposed constraints.

An important contribution of geostatistics, and especially of conditional simulations, has been the development of tools for converting observations such as, for example, seismic cubes or production history into information about the physical parameters of the system, a process known as inversion. Stochastic inversion has been considered for a long time in hydrogeology, geophysics, and petroleum engineering. In hydrogeology, see the synthesis paper of Carrera et al. (2005) and, from a geostatistical perspective, the review of Chilès (2001). Stochastic inverse modeling is now a vast domain beyond the scope of this book, with Tarantola (2005) as a major reference. We will simply focus on three contributions of stochastic methods to inverse modeling: simulated annealing, which is a popular optimization method, the gradual deformation method, which offers much flexibility, and the Bayesian framework used to incorporate general geological knowledge when the probabilistic model cannot be completely inferred from the data. We will not revisit data assimilation, presented in Section 5.8.4, which updates simulations ("ensemble members") of a physical system by assimilating data provided by repeated surveys.

7.8.1 Stochastic Process-Based Simulation

Given the difficulty to define random function models providing realistic realizations of complex geological environments, an alternative is to model the geological, physical, or chemical processes that generated the geological architectures we observe now. Early on, Matheron (1969b) and Jacod and Joathon (1971, 1972) developed *stochastic process-based models* where the sedimentation process is governed by a differential equation relating sediment thickness to influx of sedimentary material and sea depth. A variety of situations can be simulated reflecting the choice of the parameters (rate of influx of sedimentary material, compaction, subsidence, meandering): for example, continuous layers, lenses, and migration of channels. Very realistic images are now generated with simulators mimicking the erosion, transport and deposition of clastic sediments (Tetzlaff and Harbaugh,1989; Watney et al., 1999) or the meandering of channels (Cojan et al., 2005; Pyrcz et al., 2009). Stochastic process-based models can be conditioned, at least approximately, on global and local constraints (facies proportions, drill hole data). The genetic hypotheses give an explanatory value to the model and not just a descriptive one.

7.8.2 Multipoint Simulation

Like process-based models, multipoint simulation (MPS) avoids the need of choosing a random function model and identifying its parameters. To this end, it adopts a nonparametric approach based on *training images*. It was first designed for simulating geological facies in a petroleum reservoir when a training image of an analogue of the reservoir is available. The objective is to generate a numeric model inheriting the geometry of facies associations seen in the training image while honoring the data on the reservoir. When only two facies are considered, the studied variable $Z(x)$ is the indicator of one of the facies, otherwise it is a categorical variable representing the facies number. Let $T(x)$ represent the similar variable in the training image. The domains where Z and T are defined are different, but we only assume that the two variables display similar spatial variations. The simulation algorithm uses the sequential method presented in Section 7.2.1. To simulate $Z_i = Z(x_i)$ knowing the simulation at points $x_j, j < i$ (hard data and previously simulated points), the algorithm searches the training image for all configurations identical to $\{x_1, x_2, \ldots, x_i\}$ up to a translation vector τ and satisfying, $T(x_1 + \tau) = z_1, T(x_2 + \tau) = z_2, \ldots, T(x_{i-1} + \tau) = z_{i-1}$. The sample conditional histogram of $T(x_i + \tau)$ is then computed for these configurations, $Z(x_i)$ is assigned a random value z_i drawn from that histogram, and the outcome z_i is included in the conditioning data set for the next iteration.

In practice, like for usual sequential simulations, conditioning is limited to a subset of the current data. Indeed, the training image, even if large, is finite, so that it will provide no configuration fitting all the conditioning data as soon as i gets large. The geometrical configuration defined by this subset is called a *template*. The resulting simulation of course depends on the choice of the templates. From the computing time point of view, it depends also on the implementation. The method was first proposed by Guardiano and Srivastava (1993) for simulating a sandstone formed by alternating fine, relative impervious sediments and coarser, more permeable sediments. The simulation of this medium by MPS, with neighborhoods of at most 16 points, led to images which were closer to the morphology of the training image than realizations obtained by sequential Gaussian simulation but not yet fully convincing. Since this early application, MPS has benefited from many improvements.

Scanning the image each time a pixel is to be simulated being computationally slow, Strebelle (2002) proposed an algorithm where training data events are classified and stored in a search tree prior to simulation. Data events considered correspond to given templates, and among them to only those combinations of indicator or categorical values present in the training image. If some templates lead to too poor statistics, simpler templates are used. Harding et al. (2005) introduced a sub-grid approach to save computing time and better simulate large structures: A coarse grid is simulated first, then a finer grid, and so on, until the final grid is simulated. Templates differ from one grid to the next.

It is not always easy to adequately reproduce the spatial continuity of the training image with a pixel by pixel procedure. Zhang et al. (2006), as well as Arpat and Caers (2007), patch a pattern of several pixels rather than only the central pixel i. To reduce the dimensionality of the training data events, they are classified on the basis on a limited number of criteria (Zhang et al.) or indexed in a list (Arpat and Caers). A distance function is introduced to compare a data event in the training image with the conditioning data, and the training data event with the smallest distance is selected. That approach also allows the simulation of continuous variables.

Mariethoz and Renard (2010) revert to a pixel-by-pixel simulation but use a different algorithm: To simulate a pixel, they define a random path through the training image and select the first similar training event whose distance to the conditioning data event is smaller than some threshold. This is equivalent to scanning the whole training image for all data events similar to the conditioning data event and then selecting one of them, while not requiring the preliminary storage of the training image events in a tree structure.

A number of other improvements have been tailored to specific needs. Because of the sequential procedure, multipoint simulations can easily be constrained by auxiliary variables—for instance, local facies proportions, or local density and anisotropy of objects (Chugunova and Hu, 2008). They can also meet spatial constraints (global facies proportions, connectivity between wells). Finally, MPS can be coupled with the gradual deformation method or the probability perturbation method to integrate hydrodynamic data (production data of a petroleum reservoir). The interested reader is referred to the comprehensive reviews of Hu and Chugunova (2008) and Daly and Caers (2010). Figure 7.39 shows a 2D gray-tone training image of an arrangement of

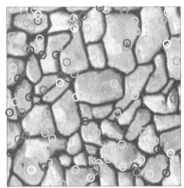

FIGURE 7.39 Training image (left) and multipoint simulation with 100 randomly located conditioning data (right). Positions of conditioning data are marked by circles whose gray levels indicate the values of the data. [From Mariethoz and Renard (2010).]

stones and an MPS simulation based on 100 conditioning points (Mariethoz and Renard, 2010). Although the result shows some differences with the training image—in particular, the presence of smaller stones—the resemblance is pretty impressive!

Ideally, the method requires a number of training images of a much larger size than the domain to be simulated. However, it is sometimes used with a single, possibly incomplete, training image. Contrary to random function models, which implicitly include an infinite number of training images in an infinite domain, the empirical conditional distributions of the MPS approach do not generally constitute a consistent set of conditional distributions. Indeed an MPS can be interpreted as a Markov random field, as noted by Daly (2005), but unlike classic Markov RF modeling in which the conditional distributions are obtained by an inference procedure that ensures a consistent mathematical model, the assignment of the conditional distribution is purely empirical for MPS, which usually leads to theoretical inconsistencies. The exact status of MPS is therefore unclear. Each simulation reproduces the training images up to a point depending on the templates selected. There are all intermediates between simulations obtained by patching large patterns of the training image and simulations based on small templates and taking more freedom with the training image. There is of course no guarantee that a set of such simulations spans the domain of the possible realizations of the phenomenon. From that point of view, MPS belongs more to stochastic imaging than to stochastic simulation.

Despite these shortcomings, MPS offers much flexibility for generating simulations that display complex features such as those encountered in geology and constrain them by auxiliary information. This requires of course that the training images are relevant and representative of the spatial heterogeneity and that their essential features can be characterized by statistics defined on a limited point configuration. Outcrops provide natural training images, but their extension to 3D is not straightforward. For this reason, 3D training images are often standard nonconditional simulations—for instance, Boolean simulations based on realistic objects and association rules—so that MPS is used as a proxy for a true conditioning. Very realistic training images are also generated with stochastic process-based simulators.

7.8.3 Stochastic Seismic Inversion

Bortoli et al. (1993), Haas (1993), and Haas and Dubrule (1994) propose a method for constructing 3D simulations of acoustic impedance (the product ρV of rock density ρ and seismic velocity V) that are consistent with both seismic and well data. Well logs provide acoustic impedance measurements with a high vertical resolution while seismic provides spatial coverage but with a much lower resolution due to frequency limitations.

The principle of the method is to simulate by geostatistical means a vertical impedance profile at each seismic grid node (x, y) in 2D space, and from this profile derive a synthetic seismic trace that is compared to the real seismic trace. If the agreement is good, the impedance profile is accepted and incorporated in the data; else another impedance profile is tried. Specifically, the following sequential algorithm is used:

1. Select a node (x, y) at random among those where the impedance profile is unknown.
2. Construct n_S simulations of acoustic impedance along the profile to the vertical of the node (x, y) conditionally on the traces already known.
3. Transform the n_S impedance profiles into synthetic seismic traces obtained by convolving the reflection coefficients derived from acoustic impedance with the source wavelet.
4. Compare each of the n_S simulated seismic traces with the measured trace (the criterion used is the correlation coefficient or the variance of discrepancies); the impedance profile that provides the simulated trace closest to the measured trace is selected.
5. Add the selected profile to the data, and if all the profiles have not yet been simulated, return to step 1 for another profile.

The validity of the above rests on three main assumptions: (i) that the convolution model applies, (ii) that the source wavelet is constant, and (iii) that geostatistical stationarity is achieved. In the end this method of simulation performs an inversion of the seismic data. For quality results the various steps must be carefully calibrated and validated. In particular, considering uncertainties on seismic traces, the choice of the required degree of similarity between the simulated synthetic trace and the real trace is critical. The degree of similarity obtained is a function of n_S. Using too high a value for n_S would lead to overconstraining the simulation. In the present case a value between 10 and 100 seems sufficient, as we can see in Figure 7.40 showing a simulated seismic section for three different values of n_S (1, 10, 100) and also the real seismic section. The real seismic is poorly reproduced by the simulated seismic when $n_S = 1$—in other words, for an ordinary simulation (the correlation coefficient is less than 0.4)—whereas it is well reproduced, except for a few local anomalies, with $n_S = 100$ (the correlation coefficient reaches 0.8).

Stochastic inversion has evolved with the introduction of linear approximations to the Zoeppritz equations for seismic waves by Buland and Omre (2003a). It is no longer an optimization scheme that requires several trials to generate a single realization but is instead a process generating at once several realizations that honor the seismic measurements. It also enables pre-stack inversion, and has been extended from models based on a regular sampling of impedance in time to layered models operating directly in a stratigraphic grid (Escobar et al., 2006; Williamson et al., 2007).

Sancevero et al. (2005) compare stochastic seismic inversion with standard deterministic inversion on a synthetic turbiditic reservoir. Deterministic inversion

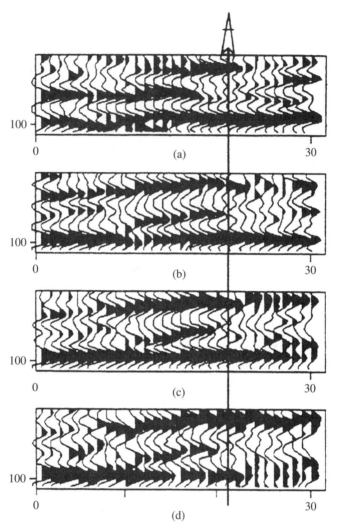

FIGURE 7.40 Geostatistical inversion of seismic data and comparison with reality on a seismic section: (a) simulation with $n_S = 1$; (b) simulation with $n_S = 10$; (c) simulation with $n_S = 100$; (d) real section. [From Haas (1993).]

is much cheaper to run but only reproduces thick sand bodies and tends to exaggerate reservoir connectivity. Ashley (2005) reaches the same conclusions with other examples and likens deterministic inversion to kriging, with the same limitations when cutoffs are applied, and he likens stochastic inversion to conditional simulations. Phrased differently, deterministic inversion is an inversion at the seismic scale constrained by well logs, whereas stochastic inversion is an inversion at the log scale constrained by seismic. Stochastic inversion incorporates the variability below seismic resolution and is a valuable tool for fields with critical flow heterogeneities at that scale. Realizations constrained by seismic provide realistic views of the spatial distribution of heterogeneities and give access to information about possible connected volumes. The word "views" is important here because a particular realization is not a good local predictor, no matter how carefully selected, and should not be used as a basis for a deterministic reservoir model.

Now acoustic impedance is not the final target. It is useful through its correlation with reservoir properties such as porosity, lithology, and fluid content. Regression equations may be established from borehole data, or more complex methods based on rock physics models may be used (Caldwell and Hamman, 2004; Bachrach, 2006). Direct petrophysical inversion methods in terms of reservoir properties have been proposed as an alternative to elastic inversion of impedances (Bornard et al., 2005; Buland et al., 2008). This has several advantages, notably that it is easier to introduce constraints in the petrophysical world of porosity and the likes than in the petro-elastic world of density and velocities. The argument for elastic inversion is that it remains close to the seismic data and permits intermediate quality checks. Which approach is better is an ongoing debate among geophysicists.

7.8.4   Simulated Annealing

Simulated annealing is a powerful, but heuristic, means of building conditional simulations subject to complex constraints (Hegstad et al., 1994). Simulated annealing was introduced by Kirkpatrick et al. (1983) as a device to obtain improved solutions of a variety of combinatorial optimization problems. Annealing is a manufacturing process by which a molten metal is cooled very slowly to produce a stress-free solid. At high temperature the molecules of the molten metal move relatively freely and reorder themselves into a very low energy crystal structure. Nature is able to find this minimum energy state for slowly cooled systems. If the liquid metal is cooled quickly, it does not reach this state but ends up in a polycrystalline or amorphous state with a higher energy. So the essence of the process is slow cooling, allowing ample time for redistribution of the atoms as they lose mobility. In statistical physics the p.d.f. of a state s with potential energy $U(s)$ at absolute temperature $T > 0$ is of the form

$$f_T(s) \propto \exp\left(-\frac{U(s)}{kT}\right) \tag{7.52}$$

where Boltzmann's constant k relates temperature to energy. Such a distribution is known as a Gibbs, or Bolzmann, distribution. A system in thermal equilibrium at temperature T has its energy probabilistically distributed among all different states s. Even at a low temperature, there is a positive probability for the system to be in a high energy state. It is this probability that gives the system a chance to get out of a local energy minimum and to find a more global one later. However, the probability for the system to be in a high-energy state decreases when T tends to 0, since the p.d.f. f_T concentrates progressively on the set $S_{\min}$ of the states s that have the minimum energy $U_{\min}$. The annealing process achieves a configuration of atoms with energy near or at the minimum of U when T returns to a low value.

Simulated annealing mimics the metallurgical process to optimize a grid of values so that it honors given constraints. A first image of the grid, which does not honor the constraints, is iteratively relaxed: A new image is proposed by swapping two pixels or using another algorithm exploring the possible images independently of the constraints, and it is selected with an acceptance–rejection technique; as in the metallurgical process, the change for the new image will more often than not reduce the energy. In order to develop the analogy with annealing, the energy function $U(s)$ associated with any configuration or state s of the image is defined as some measure of the difference between the desired features and those of the candidate realization s. The analogy with (7.52) suggests defining a p.d.f. on all images by

$$f_\lambda(s) \propto \pi(s)\, e^{-\lambda U(s)}$$

where $\pi(s)$ is the prior distribution of states independently of the constraints and λ plays the role of $1/T$ and so increases progressively to infinity.

Sampling from f_λ is made in two steps:

1. Using the notation of Section 7.6.2 and denoting the current state by s_i, we choose a new state s_j according to the transition matrix **P** associated with the distribution π; to this end a Metropolis-type algorithm is used (which implies a first acceptance test).
2. The change from state s_i to state s_j is confirmed if $U(s_j) < U(s_i)$, and it is randomized with probability $\exp(-\lambda[U(s_j) - U(s_i)])$ in the opposite case.

When λ remains fixed, an iterative use of this algorithm samples the states s from f_λ without being attracted by a local minimum. By "cooling" progressively, namely by increasing λ sufficiently slowly, one approaches a sampling of the set $S_{\min}$. In practice, λ is changed at each step. Convergence to a uniform sampling of $S_{\min}$ is ensured if $\lambda_k = (1/c) \log (k+1)$ is taken for the value of λ at step k. In this formula, c represents the minimum increase in energy required to get out of any local minimum of $U(s)$ that is not a global minimum and enter the neighborhood of a state with a lower energy (Hajek, 1988). Such convergence to infinity for λ_k is very slow. In practice, $\lambda_k = \lambda_0 /\alpha^k$ is commonly used (e.g., $\alpha = 0.90$ or 0.95), but convergence is no longer assured. Millions of iterations are often required to approach a low-energy state. Therefore simulated annealing will be efficient only if $U(s_j)$ can be obtained easily by updating $U(s_i)$.

Simulated annealing was designed to solve optimization problems such as the famous traveling salesman problem of finding the shortest itinerary for visiting his clients [e.g., Press et al. (2007, Section 10.12)]. Along this line, it is used for the optimization of sampling design [see, e.g., Romary et al. (2011) for the design of an air survey network when explanatory variables such as road traffic and population density are available, the objective being the minimization of the average kriging variance]. In a domain closer to the simulation of random functions, Geman and Geman (1984) applied simulated annealing combined with the Gibbs sampler to the restoration of images that have been degraded by an additive or multiplicative noise, a nonlinear transformation, and/or blurring.

In the pioneering applications of simulated annealing to aquifers and oil reservoirs, the distribution π was simply the marginal distribution [Farmer (1992) for rock facies; Deutsch (1993) and Langlais and Doyle (1993) for permeability; Srivastava (1994) for the indicator of a fracture network]. The initial state was obtained by independently selecting the value of each site from that marginal distribution, and the candidate states were obtained by swapping two pixels or changing the value at a single pixel. Spatial features were therefore enforced through a function $U(s)$ made of terms representing square differences between the sample variogram value of state s for a given lag and the desired variogram value taken from a variogram model or from the sample variogram of a training image. This a "brute force" use of simulated annealing because, in principle, the statistical properties of the random function should be captured by the prior model π and not by the energy function. Such use gives poor results at a very high computational cost. For example, the variogram lags entered in the energy function are overfitted, whereas the variogram lags not considered in $U(s)$ may become unrealistic. Simulated annealing can be advantageously replaced by sound simulation methods when the random function model is specified, or by multi-point simulation if one intends to simply reproduce a training image.

Some of the early applications also sought to constrain simulations on complex data. For example, to condition on the effective permeability derived from a well test, Deutsch (1993) supplemented the variogram terms of $U(s)$ by an additional term representing the square difference between the effective permeability computed numerically for state s and the true effective permeability. Additional information on these early applications can be found in Deutsch and Cockerham (1994) and Ouenes et al. (1994). Since the spatial features are only considered through the energy function, the exact status of the simulations remains unclear. When several simulations are generated, it is hard to know if a correct level of fluctuation is reached and if all possible situations have been correctly sampled.

To correctly sample states satisfying additional constraints, π should be the distribution of states prior to considering the constraints; this means that the initial state and the candidate states are nonconditional simulations, or conditional simulations in the usual sense if conditioning point-support data are available. The energy function is of the form

$$U(s) = [\boldsymbol{\mu}(s) - \boldsymbol{\mu}^{\text{obs}}]' \, \mathbf{V}^{-1} \, [\boldsymbol{\mu}(s) - \boldsymbol{\mu}^{\text{obs}}] \tag{7.53}$$

where the vector $\mu(s) - \mu^{\text{obs}}$ measures the discrepancy between some geological or engineering properties $\mu_p(s)$ calculated on the simulation and their desired values μ_p^{obs} observed or measured by physical experiments and $\mathbf{V}$ is the covariance matrix of measurement errors. In applications to aquifers and oil reservoirs, $\mu(s)$ is obtained by running a flow simulator; this requires a significant computational effort that cannot be repeated for millions of iterations. Simulated annealing can therefore be applied only with a limited number of unknown variables and an optimized sequence of states. Romary (2009, 2010) develops such an approach for constraining simulations of permeability or rock facies of synthetic oil reservoirs on production data (water cut curves). This is made in a Gaussian context, modeling either log-permeability as a Gaussian SRF or rock facies as a truncated Gaussian RF. Several improvements to the usual algorithms are introduced:

- *Dimension Reduction.* The Karhunen–Loève theorem (Loève, 1955) states that a stochastic process $Z(x)$ can be represented in a compact domain $\mathcal{D}$ by an infinite linear combination of orthogonal functions:

$$Z(x) = \sum_{p=1}^{\infty} X_p \, \phi_p(x), \qquad x \in \mathcal{D}$$

  where the functions ϕ_p depend on the domain $\mathcal{D}$ and on the covariance of Z and where the X_p are random variables. If Z is Gaussian, the X_p are independent. By truncating the infinite sum to some maximum value, $Z(x)$ can be approximated with a random vector $\mathbf{X}$ of moderate dimension.
- *Adaptation to the Temperature.* The transition matrix is adapted to the temperature to allow global changes at high temperature and local changes at low temperature.
- *Parallel Interacting Markov Chains* [Iba (2001) in statistical physics; Andrieu et al. (2007) in statistical mathematics]. The principle is to run several chains in parallel, each at a different temperature, and to allow them to exchange information, swapping their current states. This is not only a means to speed up the calculations by parallel computing, but also to improve the sampling of states.

Finally, let us mention that simulated annealing may succeed in conditioning simulations when simpler conditioning methods do not apply. A case in point is the conditioning of random functions with a negative binomial marginal distribution and Meixner isofactorial bivariate distributions, with an illustration in forestry (Emery, 2005c).

7.8.5 Gradual Deformation

Simulated annealing is very computationally demanding, especially when the energy function cannot be updated easily, because it requires a large number of iterations. This is unfortunately often the case when we want to constrain on global characteristics such as dynamic production data of a petroleum reservoir or an aquifer.

Hu (2000a) proposes an alternative method that is approximate but very efficient. The idea is, at each iteration, to propose a continuous series of candidate states instead of a single one, as well as to select the candidate state minimizing the objective function. In comparison with simulated annealing and with the multiple Metropolis–Hastings algorithm of Section 7.6.2, the method is approximate because (i) it selects the candidate state with the lowest objective function rather than randomly from a likelihood function, and (ii) it does not include a cooling process.

The method was first developed for Gaussian random functions. Suppose that we want to model a grid of values represented by a Gaussian random vector $\mathbf{Z} = (Z_1, \ldots, Z_N)'$ with mean vector $\mathbf{m} = \mathbf{0}$ for simplicity and covariance matrix $\mathbf{C}$. Like the multiple Metropolis–Hastings algorithm, the method is based on the fact that if $\mathbf{Z}_1$ and $\mathbf{Z}_2$ are two i.i.d. Gaussian random vectors with mean zero, $\mathbf{Z}_1 \cos t + \mathbf{Z}_2 \sin t$ follows the same distribution.

A state is a realization $\mathbf{z}$ of the random vector $\mathbf{Z}$. In order to account for indirect data such as production data, we define an objective function $U(\mathbf{z})$ to be minimized, similar to the energy function $U(s)$ defined by (7.53) for simulated annealing. Denoting the current state by $\mathbf{z}^{\text{current}}$, the algorithm of each iteration is the following:

1. Select a state vector $\mathbf{z}$ from the distribution of states, independently of the current state vector.

2. Consider the continuous series of candidate states:

$$\mathbf{z}^{\text{cand}}(t) = \mathbf{z}^{\text{current}} \cos t + \mathbf{z} \sin t, \qquad t \in [0, 2\pi[$$

3. Determine the value t_{opt} minimizing $U(\mathbf{z}^{\text{cand}}(t))$ and set $\mathbf{z}^{\text{current}} = \mathbf{z}^{\text{cand}}(t_{\text{opt}})$.

The value t_{opt} can always be found by discretizing t and evaluating the function U for each value of t. It is in fact usually obtained by numerical techniques that require fewer calculations of $U(\mathbf{z})$ than a systematic or random scan—for example, gradient-based methods if U is differentiable with respect to t [see Hu and Le Ravalec-Dupin (2004)]. The process is stopped when U is low enough for the fit to be considered good.

A variant is to independently select several states $\mathbf{z}_1, \ldots, \mathbf{z}_p$, and to select the linear combination $\mathbf{z}^{\text{cand}}(t_1, \ldots, t_p) = \lambda_0 \mathbf{z}^{\text{current}} + \lambda_1 \mathbf{z}_1 + \cdots + \lambda_p \mathbf{z}_p$ with

$$\lambda_0 = \prod_{j=1}^{p} \cos t_j, \qquad \lambda_i = \sin t_i \prod_{j=i+1}^{p} \cos t_j, \quad i = 1, \ldots, p-1, \qquad \lambda_p = \sin t_p$$

leading to the lowest U. This is more satisfactory than the standard method, but the determination of the optimal linear combination is more difficult.

Since the method is approximate, it may lead to a significant structural drift from the initial stochastic model when the number of iterations increases. Hu (2002) proposes improvements to the standard method. He also proposes a modified algorithm to deform conditional simulations.

The state $\mathbf{z}$ selected at step 1 can be generated by any nonconditional simulation method. However, a moving-average simulation method has the advantage that the gradual deformation can be applied locally and not necessarily globally. Indeed, using for simplicity the 1D notation, suppose that Z_i is simulated as

$$Z_i = \sum_{j=-p}^{p} \beta_j U_{i+j}$$

where the U_j are independent standard normal random variables. Simulating the vector $\mathbf{Z}$ of the Z_i's then amounts to simulating the vector $\mathbf{U}$ of the U_j's. Since U_j only has a local influence on the Z_i's and since the U_j's are independent, we can fix the local state in an area where we have a correct fit to indirect data and continue the deformation in other areas. Le Ravalec et al. (2000) propose a variant of the discrete Fourier method enabling an efficient calculation of moving averages.

This approach is obviously applicable to any random function expressed as the transform of a Gaussian random function: The gradual deformation is applied to the Gaussian random function, and the objective function is expressed as a function of the Gaussian. Typical applications concern the calibration of simulations to well-pressure data in the study of a permeability field with a lognormal spatial distribution (Hu, 2000a; Hu and Le Ravalec-Dupin, 2004) or geological facies in the framework of the truncated Gaussian model (Le Ravalec et al., 2000).

The gradual deformation method has applications other than Gaussian simulations. Let us consider the elementary case where we want to deform the position of a random point over the unit square: Its coordinates are defined by two independent random variables U and V with uniform distribution on $[0, 1]$. We can transform them into independent random variables X and Y with standard normal distribution by letting $X = G^{-1}(U)$ and $Y = G^{-1}(V)$, where G is the standard normal c.d.f. With two independent random points (U_1, V_1) and (U_2, V_2), we can define a gradual deformation $X(t) = X_1 \cos t + X_2 \sin t$, $Y(t) = Y_1 \cos t + Y_2 \sin t$ and transform it into a gradual migration $U(t) = G(X(t))$, $V(t) = G(Y(t))$. The point $(U(t), V(t))$ describes a trajectory that is symmetric with respect to the center of the square, as shown in Figure 7.41, which achieves a strong mixing of possible states. More generally, any random variable can be subjected to gradual deformation through a preliminary Gaussian transformation. This enables the gradual deformation of a Boolean model by gradual migration of the Poisson points or gradual change of object sizes (Hu, 2000b; Hu and Jenni, 2005). Similarly, Verscheure et al. (2012) condition fractal simulations of subseismic faults on production history. The main difficulty is to

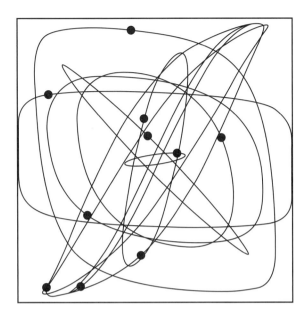

FIGURE 7.41 Trajectories of gradual deformation of a set of i.i.d. points (black dots) uniformly located in the unit square. [From Hu (2000b).]

develop a model with smooth deformation: A slight migration or extension of a simulated fault may have a large impact on the local connectivity of the fault network, not to speak about the suppression or the addition of a fault.

Gradual deformation may also be used for inference of parameters such as variogram range or object size. These parameters are treated as random variables, transformed to Gaussian, and calibrated by the gradual deformation method (Hu, 2000a), which constitutes a simplified alternative to a Bayesian approach to parameter uncertainty.

In summary, the gradual deformation method can provide a series of simulations honoring complex constraints. The status of these simulations is not clear but, due to the strong mixing properties of the method, they should explore distinct areas of the state space, without guaranteeing a thorough sampling of the conditional distribution. We leave aside the debate of deciding whether it is better to apply a fully consistent method such as simulated annealing, requiring great simplification of the system size and the physical model, or to use a more heuristic method such as gradual deformation requiring less simplification.

7.8.6 Bayesian Approach

We sometimes have so few data that we cannot infer the covariance or variogram with any confidence, which is already a problem if we intend to perform kriging, and even more so if we want to construct a simulation. This situation is common in the study of petroleum reservoirs. Bayesian methods provide a framework for dealing with such problem, using knowledge input by the geologist. It is assumed that (1) from geological considerations the geologist can specify the type of reservoir, namely, in geostatistical terms, the type of RF model describing the reservoir properties of interest (e.g., lithology, porosity, permeability, oil saturation), (2) this RF model is of a type that we know how to simulate, and (3) an expert, reasoning by analogy or otherwise, can specify a prior distribution for the RF parameters.

The reservoir is usually represented by a vector-valued random variable of high dimensionality which will be denoted by R. The set of parameters of the distribution of R is a vector-valued random variable of low dimensionality which will be denoted by Θ. Our model is then a set of RFs indexed by the possible values θ of Θ. Let us give some examples:

1. R is a Boolean model of channels, namely a set of random points with their marks (length, width, direction, curvature of the channel). Θ comprises the Poisson intensity (fixed or regionalized) and the parameters of the distribution of the channel marks.

2. R comprises an indicator or categorical RF (facies number) and a Gaussian RF (logarithm of permeability), discretized at the nodes of a grid. Θ comprises the parameters of the categorical variable (proportion of each facies, range) and those of the Gaussian RF (logarithmic mean and variance of permeability in each facies, nugget effect, ranges).

3. *Model with Constraints.* A binary function $I(r)$ specifies whether or not the reservoir $R = r$ is possible. The situation is similar to the previous ones except that the p.d.f. of R is multiplied by $I(r)$, up to a normalizing factor.

Denoting by $f(\theta)$ the prior p.d.f. of Θ and by $f(r \mid \theta)$ the prior p.d.f. of R when $\Theta = \theta$ is fixed, the prior p.d.f. of R is

$$f(r) = \int f(r \mid \theta) f(\theta) \, d\theta$$

A nonconditional simulation of R is simply a vector r selected from this p.d.f. In practice it is usually difficult to express $f(r)$ analytically and to simulate R directly, and the simulation is generated in two steps:

1. Select θ from the p.d.f. $f(\theta)$, which is a standard problem of sampling from a vector-valued random variable of low dimensionality.

2. Select r from the conditonal p.d.f. $f(r \mid \theta)$; that is, generate a nonconditional simulation with known parameters θ, which is a standard problem.

This amounts to sampling both R and Θ. In applications we have a set of observations, represented by a vector-valued random variable referenced to as *data*, and we are interested in simulations conditional on these data. For the same reason as above, we will sample R from the posterior p.d.f. $f(r \mid \text{data})$ by sampling both R and Θ from $f(r, \theta \mid \text{data})$. From the Bayes's rule, this posterior p.d.f. is

$$f(r, \theta \mid \text{data}) = \frac{f(\text{data} \mid r, \theta) f(r, \theta)}{f(\text{data})} \qquad (7.54)$$

Given the reservoir properties $R = r$, the distribution of the data does not depend on θ so $f(\text{data} \mid r, \theta)$ is a function of r only. Equation (7.54) is thus of the form

$$f(r, \theta \mid \text{data}) = \text{const} \times f(\text{data} \mid r) f(r \mid \theta) f(\theta) \qquad (7.55)$$

where const denotes $1/f(\text{data})$ which usually cannot be computed (it would require the double integration of the right-hand side in θ and r). This method requires, however, that we know $f(\text{data} \mid r)$ for any r. Since the observations usually comprise direct data of the reservoir properties in wells and indirect observations such as seismic or production history, this amounts to postulating that we are able to solve the forward problem—that is, compute the indirect observations knowing the reservoir properties everywhere. The results depend on unknown data acquisition parameters (e.g., the depth of investigation of a logging tool, or the exact signature of a seismic source), so they are represented by a probability density function, namely $f(\text{data} \mid r)$. Actually, the determination of $f(\text{data} \mid r)$ is not a simple matter, especially when scale effects are present.

Now the problem is to sample from (7.55). A standard technique for drawing from a probability distribution known up to a normalizing constant is the acceptance–rejection method. If we know a majorization $A(\text{data})$ of $f(\text{data} \mid r)$ over all possible values of r, we can construct a conditional simulation of R by adding the following third step to steps 1 and 2:

3. Let $p = f(\text{data} \mid r)/A(\text{data})$; accept r with the probability p, or reject it with the complementary probability $1 - p$ and go back to step 1.

But this method is very inefficient. When p is low, which it is when $f(\text{data} \mid r)$ is not flat, a large number of simulations obtained with great effort end up being rejected. For example, with standard conditional simulations where observations are exact values of the reservoir properties, data is a subset of R. There is a very low, if not zero, probability that a nonconditional simulation (r, θ) fits the observations, so step 3 will almost surely lead to rejection. The conclusion is the same if only some of the observations are exact values of reservoir properties. Other algorithms must be used.

In simple cases the simulation can be carried out with standard techniques. For example, Doyen et al. (1994), assuming θ known, define a sequential algorithm that combines a local prior distribution of lithology (sand/shale indicator) obtained by indicator kriging with a function representing the p.d.f. of a local seismic attribute (indirect variable) given the lithological facies. Such simplification is not possible in more complex situations, and another method is used, the Metropolis–Hastings algorithm. This requires to cast the problem in terms similar to those of Section 7.6.2. For example, the reservoir (a state in the terminology of Markov fields) is represented by a multivariate grid of values, or by an array of coordinates, sizes, and orientations of Boolean objects. Constraints like those in Example 3 are accounted for by simulated annealing techniques. It may be a good idea to initialize the iterative process with a conventional simulation conditioned on the observations of reservoir properties. The method must be tailored to each application and is clearly computationally intensive despite numerous approximations. Details can be found in Omre and Tjelmeland (1997).

This approach has been initiated by Haldorsen and Lake (1984) in a simplified version, and it has been developed further by Georgsen and Omre (1993) (combination of marked point processes and Gaussian RFs, including spatial interaction), Eide et al. (1997) (Gaussian RFs, uncertainty in the data acquisition parameters), Lia et al. (1997) (a variety of data types and RF models), to cite only a few. An achievement of this method is to produce, in principle, images that are consistent with the observations and constitute a reliable basis for quantitative evaluations. However, we basically get what we put in. If there are plenty of data, the results are reliable regardless of the prior distribution; if there are few data—which is the working assumption here—the results mainly reflect the prior model. This may be dangerous if the prior model is not consistent with the available data.

Recent work has focused the attention on seismic processing. Indeed in petroleum reservoir evaluation there are usually large amounts of seismic data with good spatial coverage. Seismic data only capture contrasts in the media and appear with considerable convolution; moreover, the signal-to-noise ratio may be poor. Bayesian approaches are well-suited for reservoir characterization based on these types of unprecise data. In a series of papers, Buland and Omre (2003a–c) and Buland et al. (2003), Bayesian Gaussian inversion, corresponding to Bayesian kriging, is used for seismic inversion into continuous reservoir properties like elastic material properties. The concept is further generalized into Bayesian skew-Gaussian inversion in Karimi et al. (2010) where marginal p.d.f.'s with skewness can be modeled.

Recent developments in seismic inversion into categorical lithology-fluid classes can be found in the review paper of Bosch et al. (2010). Further work along this line including spatial coupling of the lithology-fluid classes is presented in Ulvmoen and Omre (2010), Ulvmoen et al. (2010), and Rimstad and Omre (2010).

7.9 ADDITIONAL TOPICS

7.9.1 Simulation of Anisotropies and Space-Dependent Variograms

So far we have considered only simulations with a stationary and, for some methods such as turning bands, isotropic covariance. Let us now examine some generalizations.

Simulation of Anisotropies

A covariance (ordinary or generalized) may have several components, and each may have its own anisotropy. Some methods allow a direct construction of simulations with the desired covariance; this is the case of the sequential method and the discrete spectral method (provided any zonal anisotropy is oriented parallel to the axes). In other cases the RFs studied can be considered as the sum of independent RFs associated with the various components of the covariance, so we simulate each component separately. The simulation of the anisotropies is not a problem:

- Simulating a zonal anisotropy is the same as making a simulation in a subspace of the observation space.
- A geometric anisotropy can always be reduced to the isotropic case by means of a simple geometric transformation; however, some methods, such as the dilution of Poisson germs, can be used to directly simulate anisotropies using anisotropic dilution functions (i.e., ellipsoids will give a spherical covariance with a geometric anisotropy).

These solutions cannot be used with object-based models displaying a more complex anisotropy than a global geometric anisotropy: For such models the anisotropy must be modeled at the level of the objects.

Simulation of a Locally Stationary Random Function

When we study a domain of large extent, it is rarely homogeneous and with the same covariance everywhere. But generally, we can consider the phenomenon as locally stationary (or locally intrinsic), with a covariance that changes slowly in space. In the neighborhood of a point x_0, we have, for example, a variogram of the form

$$\gamma_{x_0}(h) = c(x_0)\, \gamma_0\left(\frac{h}{a(x_0)}\right)$$

where the sill $c(x_0)$ and the scale parameter $a(x_0)$ vary slowly with x_0 at the scale of the distances considered for h. It is important to match these structural parameters.

In the case of Gaussian variables, the variations of the sill (or more generally of a multiplicative factor) can be handled easily: To obtain a nonconditional simulation displaying these variations, it suffices to perform a simulation of the basic model with variogram $\gamma_0(h)$, then multiply it by $\sqrt{c(x)}$ at each point simulated. The variations of the scale parameter are more difficult to deal with, unless we can directly use a dilution method. To simulate, for example, a spherical model with variable local range $a(x_0)$, we dilute the Poisson germs by spheres of suitable diameter: The germ i located at the point $X_i = x$ is diluted by a sphere with diameter $a(x)$.

In the case of object-based models, we can in the same manner locally adapt the Poisson point density and/or the size of the objects.

7.9.2 Change of Support

In applications we often need simulations of block values rather than point simulations. For example, the simulation of mining methods is based on simulations of average grades over blocks whose size corresponds to the selection unit.

Discretization by a Point Simulation

Some simulation methods define the simulation continuously, but it is usually not possible to integrate it analytically over a block. In practice, the simplest method for simulating a block is usually to discretize it into a fine enough grid and simulate the nodes by point simulation. The simulated block value is the mean of its simulated discretization points. The resulting approximation is acceptable if the variability is small at the scale of the discretization grid. This approach offers two advantages:

- We get both the point and the block simulations. For a simulation of mining methods, we can therefore simulate at the same time an additional reconnaissance survey (e.g., pre-mining drill-holes) and mining blocks (selection units).
- It is very fast to generate a simulation for a new block size, or simulate blocks of variable dimension (e.g., patchwork grid for hydrogeologic calculations).

Direct Simulation of Blocks

If we have to simulate a large number of small blocks, the discretization method may become prohibitive. In the case of a Gaussian SRF we can easily perform a direct simulation of the blocks, assuming that they are all of the same size v. From the point covariance $C(h)$ of $Z(x)$, we can deduce the regularized covariance $C_v(h)$ of the block values $Z_v(x)$. Simulating blocks with the point model $C(h)$ simply amounts to simulating points with the regularized model $C_v(h)$. We can also derive the covariances between point data $Z(x_\alpha)$ and block values $Z_v(x)$ and therefore condition by a separate kriging.

In the case of a non-Gaussian SRF we can also propagate the change of support to the probability distribution if the discrete Gaussian model of change of support is applicable (cf. Section 6.8). In this model the SRFs Z and Z_v are considered as deriving from two SRFs Y and Y_v with Gaussian marginal distributions by two transformations φ and φ_v. The characteristics of the SRF Z_v (transformation function φ_v, covariance functions of Z_v and of Y_v) are deduced from those of the point-support SRF Z (transformation function φ, covariance functions of Z and of Y, themselves related). We therefore have all the elements needed to simulate Y_v and then Z_v.

7.9.3 Upscaling

Variables such as permeability (or hydraulic conductivity) are not additive: The permeability of a block is not the mean value of the permeabilities of the samples taken in the block. There is a problem of scale. As a consequence, it is usually not possible to simulate block hydraulic conductivity conditional on data at another scale. One notable exception is when block permeability can be considered as the geometric average of "point-support" permeability in the block. If we take as main variable the logarithm of permeability, the log-permeability of the block is the average of point-support log-permeability in the block: Upscaling of permeability K amounts to change of support for log K, so block log-permeability can be simulated directly with the methods dedicated to a change of support. It makes sense to work with log-permeability rather than permeability because flow equations can be rewritten with log K rather than K.

The assumption that the change of scale on permeability follows a geometric average is never true for a finite-size block but can be an acceptable approximation for large blocks in specific circumstances. Indeed, regarding "point-support" permeability as a realization of a stationary and ergodic random function $K(x)$ (more generally, it is a tensor) and considering the case of uniform flow, Matheron (1967, ch. VI) shows the emergence of an effective permeability K_{eff} when the domain becomes large enough with respect to the range of the permeability variogram. Such effective permeability is an intrinsic quantity independent of the conditions at the domain boundaries. It is always comprised between the harmonic and arithmetic means of the point-support permeabilities (Wiener bounds):

$$[\mathrm{E}(K^{-1})]^{-1} \leq K_{\mathrm{eff}} \leq \mathrm{E}(K)$$

Exact expressions are available in special cases only. One of them is when K is a lognormal RF with isotropic covariance; then K_{eff} is given by

$$\log K_{\mathrm{eff}} = \mathrm{E}(\log K) + \left(\frac{1}{2} - \frac{1}{n}\right)\sigma^2 \qquad (7.56)$$

where n is the space dimension and σ^2 the variance of log K. This formula shows that K_{eff} increases with n: As the number of dimensions increases, a greater number of possible flow paths are offered to the fluid to go around low permeability zones.[7] In 2D, this equation reduces to an averaging of log K.

[7] Formula (7.56) is true in 1D independently of the lognormal assumption. It has been established in 2D by Matheron (1967). In 3D the result was conjectured by Landau and Lifshitz (1960) in electrodynamics and by Matheron (1967) in hydrology; Nœtinger (2000) shows that it is a mean-field approximation.

The 2D case is worth of interest because in sedimentary environments hydraulic conductivity is usually integrated on the height of the aquifer, so modeling hydraulic conductivity in 3D amounts to modeling transmissivity in 2D.

Formula (7.56) can be rewritten in the equivalent form

$$K_{\text{eff}} = [\mathrm{E}(K^\alpha)]^{1/\alpha} \qquad (7.57)$$

where $\alpha = -1$ in 1D (harmonic mean), $\alpha = 0$ in 2D (geometric mean), and $\alpha = 1/3$ in 3D. In the anisotropic case, the averaging exponent α may be related to a global anisotropy ratio involving the vertical to horizontal permeability ratio K_v/K_h and the ratio of correlation lengths L_v/L_h (Nœtinger and Haas, 1996). Given the stratified nature of sedimentary geology, the exponent α is generally closer to 0.7 than to 0.33.

In engineering practice one has to work at the intermediate scale of blocks, where the conditions required for the emergence of an effective permeability are not satisfied. A *block permeability* can be defined as the equivalent permeability of a homogeneous block giving the same flow through the actual block (Rubin and Gómez-Hernández, 1990). This permeability depends on the boundary conditions and thus is not an intrinsic property of the block. It can be obtained from point-support simulations, for specified boundary conditions, either by solving the flow equations numerically for all blocks of interest, or by using an approximation such as the popular *power averaging* formula (Journel et al., 1986)

$$K_{\mathrm{b}}(V) = \left[\frac{1}{V} \int_V K(x)^\alpha \, dx \right]^{1/\alpha} \qquad -1 \le \alpha \le 1$$

where the exponent α is fitted to match numerically computed blocks permeabilities. In the case of a scalar isotropic lognormal permeability field, consistency with the effective permeability formula is achieved with $\alpha = 1 - 2/n$. Except when $\alpha = 0$ (geometric mean) or $\alpha = 1$ in 2D (arithmetic mean), $K_{\mathrm{b}}(V)$ cannot be simulated directly: it is necessary to simulate $K(x)$ on a fine grid and then perform the upscaling.

Finally let us mention that block permeabilities are sought because they constitute the basic input to flow models at the scale of the aquifer. However, finite difference or finite element methods rely in fact on *inter-block permeabilities*, involved in the expression of the flow traversing the face common to two adjacent blocks. In that case a better approach is to calculate directly the inter-block permeabilities without going through the step of block permeabilities: see Kruel-Romeu and Nœtinger (1995), Roth (1995), and Roth et al. (1996).

We refer the reader to the 1996 special issue of the *Journal of Hydrology* dedicated to the determination of effective parameters in subsurface flow and transport modeling. It notably includes review papers of Wen and Gómez-Hernández on hydraulic conductivity and of Sánchez-Vila et al. on transmissivity. Other references are the review of Renard and de Marsily (1997) and the more recent article of Nœtinger et al. (2005).

7.9.4 Cosimulation

So far we have considered the simulation of a single variable, although on occasions we have mentioned possible ways of cosimulating several variables and have given examples. Of course, the simulation of several variables requires the knowledge of the multivariate RF model. While for most classic models nonconditional simulation is easy, conditioning on the data is case-dependent. Let us give two examples.

Parallel Simulations

In a multi-Gaussian context, possibly after a preliminary transformation, the sequential method can be applied directly to simulate several variables. However, a two-step approach is often easier. Consider, for example, a p-variate model in which all covariances are proportional. It suffices to start from p independent simulations, which can be constructed sequentially, or better, in parallel, if the algorithm lends itself to that (turning bands, discrete spectral method), and then to combine them linearly so as to restore correlations between variables. Conditioning is done afterward by cokriging, and it can be parallelized as well. This method can be readily generalized to the case of the linear model of coregionalization presented in Section 5.6.5, since the studied random functions can be expressed as linear combinations of independent random functions.

Cascaded Simulations

A problem often encountered in geological applications is to simulate a categorical variable representing a facies type, along with one or several material properties that depend on the facies type. A favorable case is when there is a strong relationship between the physical property and the facies type and when we can determine this facies type fairly reliably. A typical example is a formation comprising two lithologies with very different permeabilities (e.g., sand and shale). The simulation is then done in two steps: (1) univariate conditional simulation of the facies type (indicator simulation, truncated Gaussian simulation, or Boolean model), and (2) within each facies type, conditional simulation of the material property using only the measurements in that facies type. The implementation of the first step assumes that for each measurement of the material property we know to which facies it belongs. If not so, we must first assign a facies type to each measurement, conditionally on the observed value (e.g., using a Bayesian method). If several simulations are constructed, this assignment must be redone each time.

The other case is when the material property is independent of the facies type; then both variables are simulated separately. In practice, of course, the situation is rarely so clear-cut: there is some degree of dependence of the material property on the facies type, and we must find an adapted method.

We may also have to deal with two or three categorical variables displaying some relationship—for example, a lithological code, a diagenesis code, and a seismic attribute. The truncated pluri-Gaussian simulation method has been extended to consistently simulate these codes (Armstrong et al., 2011, Chapter 2).

7.9.5   Generation of Pseudorandom or Quasi-random Numbers

Simulations of random functions are in fact based on simulations of random variables. The construction of a simulation therefore implies that we have a generator of pseudorandom numbers, or numbers that can be considered as realizations of i.i.d. random variables. The problem is handled in practice in two steps (although there are of course variations):

- Generation of pseudorandom numbers with a uniform distribution over [0, 1].
- Generation of numbers with any given distribution from the previous numbers.

We usually choose a generator enabling the reproduction of a sequence of random numbers generated earlier. It is then possible to reproduce a simulation already constructed (but not saved), or to appreciate the repercussions of a modification of the structural parameters (scale parameter, sill) by constructing several simulations corresponding to different parameters while reusing the same pseudorandom numbers.

Pseudorandom Numbers with a Uniform Distribution

We expect from a pseudorandom-number generator that it delivers a sequence of numbers $x_i \in [0, 1]$ such that the p-tuples $(x_i, x_{i+1}, \ldots, x_{i+p-1})$ are uniformly distributed over $[0, 1]^p$ for $p \leq p_{max}$, where p_{max} depends on the application. And this should be the case whatever the length of the simulated series. In the first edition of this book we presented the widely used linear congruential method. It produces pseudorandom numbers that are in fact rationals of the form $x_i = y_i/m$, where m is a fixed integer, and where the y_i form a sequence of integers less than m defined by the recurrence relation

$$y_{i+1} \equiv [a\, y_i + b] \qquad (\mathrm{mod}\ m)$$

The sequence is initialized by y_0. The multiplier a, shift b, and initial value y_0 are integers less than m. Clearly the sequence y_i can take at most m distinct values. After a certain time we encounter a number already generated, and the same sequence is then reproduced over and over. The sequence therefore has a period $p \leq m$. Besides it can have a pseudoperiod q, a divisor of p, for which it is reproduced up to a constant (modulo m). This algorithm is very simple, and it is very easy to access any pseudorandom number in the series. But it has weaknesses with respect to the above requirement. We gave some criteria for choosing values a and b enabling us to "fix" these weaknesses, without totally eliminating them.

Much better algorithms are available now and we refer the interested reader to Press et al. (2007, Section 7.1). In view of the rapidly expanding digital storage capacity, these pseudo-random number generators may well be replaced by databases of pseudorandom numbers produced by physical white noise generators such as dongles.

Pseudorandom Numbers with a Given Distribution

We generally start with pseudorandom numbers distributed uniformly over [0, 1]. These values x_i can be transformed into pseudorandom numbers z_i from a c.d.f. F by inversion of F—that is, by taking $z_i = F^{-1}(x_i)$. This method has the practical advantage that a single uniform pseudorandom number suffices to generate a number from the desired distribution. The ith pseudorandom number from distribution F is associated with the ith number of the x_i sequence.

It is not always easy to invert the c.d.f. F, especially at the extremes of the distribution. Other methods are available: acceptance–rejection, composition of several distributions, quotient of uniform variables, and the like, plus an abundance of algorithms specific to some distributions. A fairly complete presentation can be found in Devroye (1986), Knuth (1997), and Press et al. (2007, Section 7.3). Lantuéjoul (2002, Section 7.2) summarizes the main methods useful for geostatistical applications.

Quasi-random Numbers

Uniform random numbers permit the simulation of random points in a given domain D by drawing the various coordinates independently. Random points produce sequences of points $x_1, x_2, \ldots,$ that are *equidistributed* in D, in the sense that the proportion of points among $x_1, x_2, \ldots, x_i$ that belong to $D' \subset D$ tends to $|D'|/|D|$ when i tends to infinity, this for any D'. But random points are not homogeneously distributed: Clusters can be observed as well as zones without any point. When selecting the directions of the lines in the turning-bands method, for example, we do not really need random points but simply equidistributed points on the unit half-sphere; these can be generated from equidistributed points over a square (Section 7.4.3).

There exist sequences of equidistributed points that are more homogeneously distributed than random points. The construction usually relies on the intuitive principle that the ith point of the sequence must be located within D as far as possible from the $(i-1)$ previous ones. Press et al. (2007, Section 7.8) present several sequences that are equidistributed over a square, and more generally over an n-dimensional hypercube.

A simple example is the *Halton sequence*, the n-dimensional generalization of the *Van der Corput sequence*, which is defined as follows: The space dimensionality being n consider the first n prime numbers p_j ($p_1 = 2$, $p_2 = 3$, $p_3 = 5$, etc.). The number $i \in \mathbb{N}$ is written in base p_j:

$$i = a_{j0} + a_{j1}\, p_j + \cdots + a_{jk}\, p_j^k + \cdots$$

We set

$$u_j(i) = \frac{a_{j0}}{p_j} + \frac{a_{j1}}{p_j^2} + \cdots + \frac{a_{jk}}{p_j^{k+1}} + \cdots$$

The sequence of n-tuples $x_i = (u_1(i), \ldots, u_n(i))$, $i > 0$, is equidistributed over $]0, 1[^n$.

7.9.6 Check of the Simulations

Just like the real phenomenon, a simulation is not a random function but a regionalized variable that we regard as a realization of a random function. Its actual structural characteristics (histogram, variogram, etc.), evaluated from its values at the nodes of the discretization grid, differ more or less from the characteristics of the theoretical model or those of the sample data. Therefore it is useful to check that the characteristics of the simulation are not too far from those sought.

Checking the simulations is all the more useful as their construction relies in practice on pseudorandom numbers whose behavior cannot be totally guaranteed. Furthermore, conditioning can introduce anomalies if the spatial distribution of the simulated RF is not really compatible with the data. We present below some general-purpose tests designed to verify that the simulations are technically correct. A validation of the physical correctness is naturally required as well, but it is application-specific.

Check of a Single Simulation

Statistical tests designed for independent random variables are not applicable here because the values taken by a simulation are obviously correlated. But we can at least perform simple checks at various levels: for example, Gaussian nonconditional simulation, Gaussian conditional simulation, and final simulation. The checks concern the histogram of simulated values, their variogram, and even the scatterplots between the values assumed at x and $x+h$ for a few values of h. These graphs have a regional character at the scale of the simulated domain, since they are calculated from a fine discretization of this domain. We are therefore able to calculate the fluctuation variance of the spatial mean of the simulation [according to (2.33), it depends only on the covariance] and, in some cases (at least the Gaussian one), of the spatial variance of the simulation and of its variogram (cf. Section 2.9.2). This enables us to see if the deviations of these characteristics from their theoretical values are acceptable.

Check of a Set of Simulations

Several simulations of the same phenomenon are often generated in order to exhibit multiple images of what reality might be or determine the conditional distribution of some complex characteristics. In addition to the above checks on individual simulations, two methods can be applied to make sure that the different simulations are really (conditionally) uncorrelated:

1. *Check of the Normed Sum.* Consider N independent simulations (conditional or not) $S_i(x)$ with the same variogram $\gamma(h)$. Their normed sum

$$S(x) = \frac{1}{\sqrt{N}} \sum_{i=1}^{N} S_i(x)$$

still has the variogram $\gamma(h)$ (but the histogram becomes more Gaussian—if it was not already Gaussian). We can check this using the methods mentioned above.

2. *Check of Simulated Errors.* This test is concerned with the conditional simulations obtained by adding to the kriging estimate a simulation of the kriging error. Using the notations of Section 7.3.1 (Z, real field; S_i, nonconditional simulation number i; T_i, conditional simulation number i; Z^*, kriging of Z; S_i^*, kriging of S_i; σ_K^2, kriging variance), the field

$$\varepsilon_i(x) = \frac{T_i(x) - Z^*(x)}{\sigma_K(x)} = \frac{S_i(x) - S_i^*(x)}{\sigma_K(x)}$$

is a standard error. If the nonconditional simulations $S_i(x)$ are Gaussian, so is $\varepsilon_i(x)$. This often remains true to a first approximation in the non-Gaussian case, since the $\varepsilon_i(x)$ are mixtures of random variables with the same distribution. For a fixed i, the errors $\varepsilon_i(x)$ are obviously spatially correlated. On the other hand, for a fixed x, the errors $\varepsilon_i(x)$, $i = 1, \ldots, N$, must be independent, centered, standardized, and possibly Gaussian. We can easily verify this for a few points of the simulation using classic tests (e.g., the χ^2 test). An example is given by Chilès (1977).

7.10 CASE STUDIES

7.10.1 Simulation of a Nickel Deposit

Conditional simulations were initially motivated by the study of mineral deposits (Journel, 1973, 1974a,b). Kriging enables estimation of the resources in place but not the evaluation of the recoverable resources, which is of prime importance in feasibility studies. The nonlinear methods of Chapter 6 ignore aspects that are rather simple from the mining engineer's point of view; for example, that in an open pit a block cannot be exploited without extracting the blocks situated above it. The difficulty of evaluating recoverable resources is due to the fact that this concept involves the mining method, its degree of selectivity, and its flexibility, none of which can be reduced to linear operations; it is also due to the fact that the result depends heavily on the local fluctuations of the mineralization. To determine the optimal mining method and the corresponding recoverable resources, we would need to have knowledge of the deposit at every point, which of course is never the case. A conditional simulation is then very useful. It provides a numerical model that can be known at every point. This model is used to simulate various mining methods or ore processes, perform sensitivity studies, and optimize choices. To the extent that the numerical model provided by the conditional simulation reproduces not only the global aspect of the actual deposit but also its local variability, we can reasonably consider that the method that proves optimal on the numerical model remains optimal, by and large, for the real mineral deposit.

The simulation of mining methods based on a numerical model produced by a conditional simulation constitutes a subject that was developed around 1980 and is beyond the scope of this book. The interested reader is referred to Deraisme and Dumay (1979), Da Rold et al. (1980), Deraisme et al. (1983, 1984), Kim et al. (1982), Luster (1986), and Vieira et al. (1993). Deraisme and van Deursen (1995) apply this approach to a standard civil engineering problem, the volumetric computation of dredging projects, and quantify the bias incurred by cutting on a kriged estimate rather than on simulations. The area estimate for blasting work above the cutting level is 273,500 m^2 based on the kriged model, whereas conditional simulations of the seabed and the geological layers give an interval of 268,000 to 391,000 m^2, with an average of about

310,000 m^2. Dimitrakopoulos et al. (2005), Whittle and Bozorgebrahimi (2005), and Albor Consuegra and Dimitrakopoulos (2009) present original applications of multiple geostatistical simulations to the design of an open pit. Dimitrakopoulos (2011) reviews the developments of the last decade in mine planning. The following example, while focused on the description of the methodology for generating conditional simulations, gives a glimpse of the contribution of the approach to the choice of a mining method.

Orebody and Objective of the Simulation

The Tiébaghi nickel orebody (New Caledonia) is sampled by several hundred drill-holes with an approximate grid spacing of 70 m × 70 m. The ore can be subdivided into six categories, from top to bottom:

- The cover: C.
- Three categories of laterites: L1, L2, L3.
- Two categories of garnierites: G1, G2.

Below we find the bedrock. Figure 7.42 shows a typical drill-hole log.

The studied zone forms a plateau. The global resources in place have been estimated by kriging and are known with good precision, namely within 8%, using a conventional confidence interval of ±2 standard deviations. Local resources, however, are poorly estimated; a 70-m × 70-m panel can hardly be evaluated better than up to 50%. Yet the choice of the mining method depends on even more local characteristics, such as fluctuations of the top of the garnierites or of the bedrock. It is not possible, even by halving the grid spacing (a fourfold increase in the number of drill-holes), to estimate these surfaces with reasonable precision. Hence the idea of constructing a conditional simulation. We can then test various mining methods (bulk or selective mining, homogenization stock piles, single or multiple working areas, etc.) and search for the one that best matches the ore supply requirements of the processing plant (homogeneity of the ore, regularity of supply) and also the economic criteria (cutoff grade, undesirable side product).

The simulation is constructed over a fine grid (6.25 m × 6.25 m), which enables us to visualize local fluctuations and, by grouping of 4 × 4 points, to simulate with good precision the mean characteristics of the 25-m × 25-m panels. For each category of ore, we simulate its thickness and also the mean grades of the various constituents (Ni, Co, Mg, Al). The grades are obtained in fact by simulating the associated accumulations (i.e., average grade × thickness). Naturally, thicknesses and accumulations are highly correlated and call for the use of multivariate methods described in Chapter 5. Appropriate adaptations are required to account for very special bivariate distributions of thickness and metal accumulation (not reducible to the bi-Gaussian case). We focus here on the simulation of thicknesses alone, which prove to be

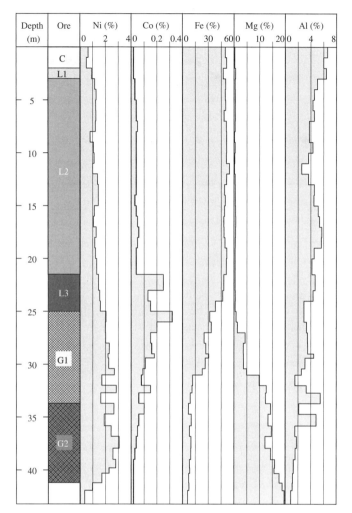

FIGURE 7.42 Typical drill-hole log in the Tiébaghi nickel orebody. [From Chilès (1984), with kind permission from Kluwer Academic Publishers.]

uncorrelated between ore categories. The task then involves the construction of six different simulations. Chilès (1984) gives a more complete presentation of the simulation procedure, including for grades.

Gaussian Transformation

To reproduce the sample distribution of the thickness of a given ore, we simulate the Gaussian variable associated with this thickness. Thus we start by transforming the initial data into Gaussian data, and in the end we do the reverse transformation on the Gaussian simulation.

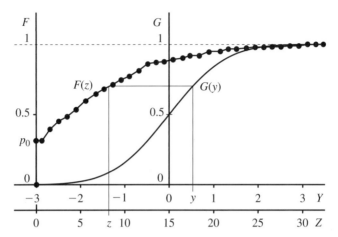

FIGURE 7.43 Graphic transformation of thickness of the L2 lateritic ore. [From Chilès (1984), with kind permission from Kluwer Academic Publishers.]

The transformation function is deduced from the cumulated histogram. Figure 7.43 shows as an example the histogram of the thickness of the L2 ore, along with its graphic fit. We observe a discontinuity at the origin, due to the fact that 32% of the time the L2 ore is absent. This creates a difficulty for the transformation of the "waste" data into a Gaussian: To this zero thickness we must associate a Gaussian value less than the limit $y_0 = G^{-1}(0.32)$, but which one? (G is the c.d.f. of the standard Gaussian.) We will come back to this later.

Variography

Simulations are much more sensitive than kriging to the behavior of the variogram near the origin. There is no hope of obtaining a realistic simulation at a 6.25-m grid spacing if we do not know the variogram at less than 70 m. Fortunately, some complementary data are available besides the 70-m spacing data:

- A cross of drill-holes at 25-m spacing. This is located at the edge of the plateau and must be corrected for a proportional effect.
- About 30 old drill-holes, located at about 5 m from drill-holes of the 70-m grid. Since they are regularly spaced over the whole plateau, they provide for a good estimation of the variogram at 5-m lag.

The variography is simplified by the fact that the variograms of laterites turn out to be similar, as do those of garnierites. In order to compute the variograms of the Gaussians, the zero thicknesses are transformed into the conditional expectation of a Gaussian $E[Y \mid Y < y_0]$. This entails a bias in the sample variogram. This bias can be calculated theoretically in the scope of a bivariate Gaussian model for Y and is taken into account in the fit of a variogram model. The main variability occurs in the first 200 meters for the laterites and in the first 100 meters for the garnierites. The variograms of original thicknesses provide a reference for checking the quality of the final simulation (after conditioning and reverse transformation).

Construction of the Simulation

This involves three main steps:

1. *Construction of Nonconditional Simulations of the Gaussians.* This step is straightforward; it is carried out independently for each ore category.
2. *Conditioning on the Data.* Ideally the conditioning kriging should be carried out with a global neighborhood. This prevents the parasitic discontinuities caused by neighborhood changes, which make it impossible to know whether the fluctuations of the conditional simulation represent something real or are a mere artifact. Here the number of data precluded conditioning with a global neighborhood. Therefore the domain was divided into subzones, each being kriged with a global neighborhood. Some overlap of the subzones was allowed to avoid discontinuities at the boundaries. Conditioning on zero thicknesses— that is, on Gaussian data for which all we know is that they are less than the threshold u_0—can be performed rigorously using the algorithm presented in Section 7.6.3. At the time of the study, this method was not yet known and zero thicknesses were just replaced, after Gaussian transformation, by the conditional expectation of the Gaussian.
3. *Reverse Transformation to Original Thicknesses.* We now have conditional simulations of the Gaussians. By reverse transformation we obtain the conditional simulations of the thicknesses. In this direction the transformation is straightforward.

Check of the Simulation

Checks are made at each step. In the final stage we check that histograms of sample thicknesses are well-reproduced by the simulation, and we also check that the same is true for the variograms. Figure 7.44 displays the sample variogram of the total thickness of laterites along with that of the conditional simulation. They seem fully compatible (the slight nugget effect seen on the sample variogram, due to the uncertainty in the exact position of the interfaces, is not simulated).

A more general check is provided by a comparison between simulation and kriging. Indeed the characteristics of the real orebody lie within a range of ± 2 standard deviations about the kriged values (referring to the conventional 95% interval). The simulation must fall in the same interval. To check this, the mean thicknesses of the different ores are estimated directly by kriging. As Table 7.2

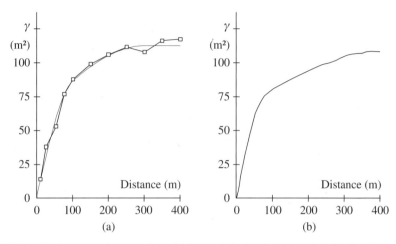

FIGURE 7.44 Sample variogram of the thickness of the laterites (a) and regional variogram of the conditional simulation (b). [From Chilès (1984), with kind permission from Kluwer Academic Publishers.]

TABLE 7.2 Comparison of Average Thicknesses of the Simulation and of Kriging (in Meters)

Ore	Simulation	Kriging	Standard Error
C	11.1	11.0	0.5
L1	4.5	5.1	0.5
L2	6.1	6.4	0.6
L3	3.7	4.3	0.6
G1	10.2	9.9	0.6
G2	2.9	3.1	0.4

shows, the simulation is fully compatible with kriging. We note that it corresponds to an outcome relatively weak in laterites.

Results

Figure 7.45 displays a vertical section of the orebody obtained by kriging, and the same vertical section derived from the conditional simulation. As in kriging, the simulation matches the data (the two sections coincide at drill-holes locations). However, the striking difference is the character of these two sections. If we based the mining method on the kriged section, we would be making a serious mistake: The regular beds would suggest the use of an unselective bulk mining technique (e.g., a dragline or a bucket wheel excavator), while if reality exhibits variations similar to those shown by the simulation, more mobile and more selective mining techniques are to be considered, at least for the garnierites (shovels or backhoes).

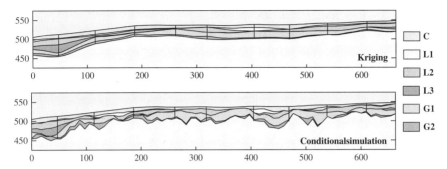

FIGURE 7.45 Example of a vertical section through the Tiébaghi nickel orebody. Top: kriged section; bottom: simulated section. [From Chilès (1984), with kind permission from Kluwer Academic Publishers.] (See color insert)

7.10.2 Simulation of an Oil Reservoir

After mining, the first applications of conditional simulations were in hydrology (Delfiner and Delhomme, 1975) and petroleum (Delfiner and Chilès, 1977). Applications to oil reservoirs grew with methods for simulating lithological variations (Alabert, 1987; Matheron et al., 1987). A typical simulation of an oil reservoir is a three-step procedure (Dubrule, 1993):

1. Simulation of the genetic units (fluvial channels, mouth bars, turbiditic lobes).
2. Simulation of facies types within these units (sand, shale, sandstone).
3. Simulation of petrophysical variables within each facies type (porosity, permeability, saturation, etc.).

These simulations reflect the heterogeneities and the complexity of the reservoir. Their main application is realistic reservoir fluid flow and oil recovery simulation from which a range of likely economic scenarios can be derived. Examples of the complete sequence of steps for such simulation exercice are given by Hewett and Behrens (1988) and Morelon et al. (1991), and a variety of case studies are described in the book edited by Yarus and Chambers (1994).

Step 1 is modeled with object-based methods: genetic units are represented as objects of random shape, size, and orientation randomly distributed in space. Early applications by Delhomme and Giannesini (1979) modeled sandstone reservoir units, drains, and shale breaks in the Hassi Messaoud field (Algeria). More recent applications abound [e.g., references cited by Dubrule (1993)]. These methods offer the advantage of flexibility: The distribution of the parameters can easily be adapted to local conditions. An alternative is to use stochastic process-based models.

Step 3 involves the techniques developed for continuous variables. The first application was the volumetric evaluation of the Lacq reservoir (Delfiner and

TABLE 7.3 Bulk Rock Volume Estimates Based on 27 Wells

OWC	Bulk Rock Volume (10^6 m^3)			
	Kriging	Simulations	σ Simulations	"True" Value
620 m	302.8	324.6	44.7	331.4
630 m	362.7	394.9	54.5	397.6
640 m	430.0	473.2	65.3	470.6

Note: Volume from kriged depth; mean and standard deviation of 30 simulations; mean of 30 simulations based on 84 wells, regarded as the true value.

Chilès, 1977), which generalized a method based on a shortcut to simulations proposed by Haas and Jousselin (1976). Since the reservoir is bounded by the top of a dome and truncated below by an aquifer, its estimation is a nonlinear problem, namely the estimation of the integral of $[z_{OWC} - Z(x)]1_{Z(x) < z_{OWC}}$ where $Z(x)$ is the depth to the reservoir top, counted positively downward, and z_{OWC} is the depth of the oil–water contact. Volume calculations were therefore based on conditional simulations. Table 7.3 shows the results obtained using the first 25 wells drilled plus two other wells selected to constrain simulations realistically ("pseudo" wells are often necessary to ensure closure of the reservoir). Because there were in fact 84 wells drilled, these could be used to establish "true" values.

One can see that simulations give better estimates than kriging, which systematically underestimates volumes. It is also possible to compute volume histograms and determine the traditional P90, P50, P10 quantiles (Pr{Volume > P90} = 90%). The method can be extended to the calculation of oil in place by introducing porosity and fluid saturations. This type of probabilistic reserves evaluation has become common in the oil industry. Small-scale permeability variations can also be generated within the reservoir (Alabert and Massonnat, 1990). A commonly used technique is the so-called *cloud transform* which adds a stochastic component to the log-permeability estimates obtained by regression on porosity, thereby maintaining the heterogeneity of the permeability field (Deghani et al., 1997; Aly et al., 1999; Alqassab, 1999; Kolbjørnsen and Abrahamsen, 2005). This amounts to simulating permeability K conditional on porosity, but in a pure regression model, without any spatial continuity on K.

The following example is focused on step 2, the simulation of facies types. A common approach for such applications is sequential indicator simulation: Journel and Gómez-Hernández (1989) use it to simulate the sand/shale distribution in a clastic sequence, and Alabert and Massonnat (1990) use it to model the spatial arrangement of sedimentary bodies (channels, turbiditic lobes, slumps, laminated facies, etc.); Doyen et al. (1994) extend the method in a Bayesian setting to incorporate seismic data in addition to well data. The example presented below is based on the truncated pluri-Gaussian simulation method (Section 7.6.4).

Objective of the Simulation

The example considered by Matheron et al. (1987, 1988) is the Brent cliff (Great Britain). It has been reworked for this presentation. This cliff comprises four lithofacies types: shale, sandy shale, shaly sandstone, and sandstone exhibiting local permeabilities that differ by orders of magnitude. They have been clearly identified on a photograph of the cliff and in 35 neighboring wells. The objective is to construct a 3D model based on these wells, by conditional simulation, for the purpose of studying flows. (To this effect, a permeability depending on the facies outcome is assigned to every node of the simulation.) The data are preprocessed to make the beds horizontal.

Let us denote by F_i the facies and by $I_i(x)$ their indicator functions, the index i corresponding to shale, sandy shale, shaly sandstone, and sandstone, respectively, when i varies from 1 to 4.

Relations Between Facies

The first hurdle when we wish to implement the pluri-Gaussian method is to identify the relations between facies and summarize them in a facies substitution diagram. A useful tool is the comparison of cross-variograms of indicators with simple variograms. Let γ_i denote the variogram of I_i and let γ_{ij} be the cross-variogram of I_i and I_j. Their ratio can be interpreted as a conditional probability. Indeed, since we are considering indicators, for $j \neq i$ we have

$$\frac{\gamma_{ij}(h)}{\gamma_i(h)} = \frac{E[I_i(x+h) - I_i(x)][I_j(x+h) - I_j(x)]}{E[I_i(x+h) - I_i(x)]^2}$$

$$= -\Pr\{x \text{ or } x + h \in F_i \text{ and the other } \in F_j \mid x \text{ or}$$

$$x + h \in F_i \text{ and the other } \notin F_i\}$$

If the facies F_i and F_j are neighbors, $-\gamma_{ij}(h) / \gamma_i(h)$ as well as $-\gamma_{ij}(h)/\gamma_j(h)$ should decrease with h, at least for short h; this is what we observe for sandy shale and shaly sandstone, as well as for shaly sandstone and sandstone (Figure 7.46a). On the contrary, if facies F_i and F_j are not in contact, these expressions should increase with h; this is the case for sandy shale and sandstone (Figure 7.46b). Finally, if facies F_i is superimposed on the others, the ratios $\gamma_{ij}(h)/\gamma_i(h)$, $j \neq i$, should show no significant variation with distance h; this is what we observe for shale (Figure 7.46c). We will therefore model the four facies with two Gaussian SRFs: A threshold on Y_1 will separate shale (facies F_1) from the other facies, whereas two thresholds on Y_2 will delimit the other three facies conditional on the fact that we are not in F_1. This is synthetized in the substitution diagram of Figure 7.47, where the areas attached to each facies correspond to their average proportions. The ordering F_2-F_3-F_4 (or equivalently F_4-F_3-F_2) is natural here because moving from F_2 to F_4 usually requires a transition by F_3.

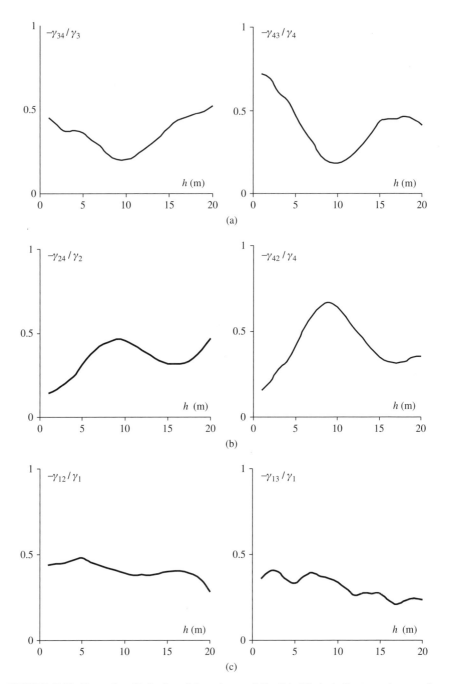

FIGURE 7.46 Examples of behavior of the ratio $-\gamma_{ij}(h)/\gamma_i(h)$ of facies indicator variograms for some pairs (i, j): (a) pairs (3, 4) and (4, 3), showing a diffusive transition between shaly sandstone and sandstone; (b) pairs (2, 4) and (4, 2), showing the absence of contact of sandy shale and sandstone; (c) pairs (1, 2) and (1, 3), showing that shale is superimposed on the other facies. [From H. Beucher, personal communication.]

Local Proportions

Petroleum reservoirs cannot be represented by a stationary model. There are at least vertical variations of facies proportions. In the present case, sandstones, for example, are located essentially in the lower half, as shown in Figure 7.47, which displays the proportions $p_i(x)$ of the four facies as a function of vertical position of point x. Lateral variations are not considered here but are usually present when modeling large domains. The thresholds, also functions of location, or at least of vertical position, can then be determined. In this case we have

- threshold on Y_1 separating F_1 from the other facies: $y_1(x) = G^{-1}(p_1(x))$
- threshold on Y_2 separating F_2 from F_3: $y_{21}(x) = G^{-1}(p_2(x)/(1 - p_1(x)))$
- threshold on Y_2 separating F_3 from F_4: $y_{22}(x) = G^{-1}((p_2(x) + p_3(x))/$
  $(1 - p_1(x)))$

Variography

Since Y_1 separates F_1 from the other facies, its variogram is linked to the variogram of I_1. If the proportion p_1 were constant, we would choose the correlogram $\rho_1(h)$ of Y_1 so that the variogram of the indicator I_1, obtained by application of relation (2.76), fits the sample variogram of the indicator. We must take the variations of the proportion into account, and hence of the threshold. Let us drop the facies index for the moment. We are considering a Gaussian SRF $Y(x)$ and the indicator $I(x) = 1_{Y(x) < y(x)}$, where $y(x) = G^{-1}(p(x))$ is the threshold associated with the proportion $p(x)$ at point x (here depending on the vertical coordinate only). That indicator can be expanded into normalized Hermite polynomials by applying (A.10):

$$I(x) = G(y(x)) + g(y(x)) \sum_{n=1}^{\infty} \frac{\chi_{n-1}(y(x))}{\sqrt{n}} \chi_n(Y(x))$$

The nonstationary covariance of the indicator at two data points x_α and x_β is therefore

$$C(x_\alpha, x_\beta) = g(y_\alpha) g(y_\beta) \sum_{n=1}^{\infty} \frac{\chi_{n-1}(y_\alpha) \chi_{n-1}(y_\beta)}{n} [\rho(x_\beta - x_\alpha)]^n \qquad (7.58)$$

In practice we calculate the sample variogram or covariance of the indicator for a given h using all pairs of sample points at a distance h apart, as we do in the stationary case. The corresponding theoretical value is obtained by averaging

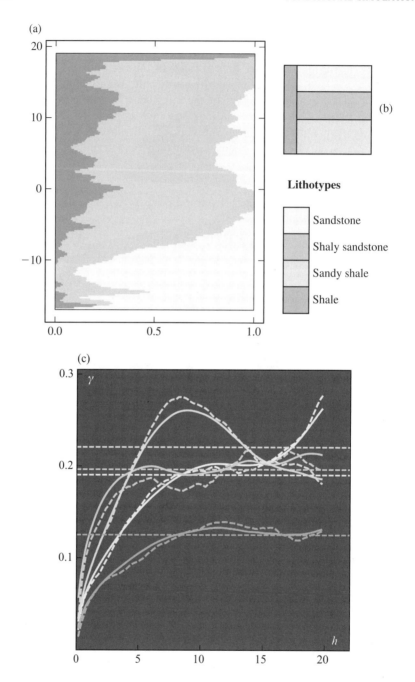

FIGURE 7.47 Identification of the parameters of the pluri-Gaussian simulation model: (a) vertical proportion curves; (b) facies substitution diagram; (c) sample variograms of facies indicators and fits derived from the variograms of the two Gaussian SRFs. [Output from Isatis®. From H. Beucher, personal communication.] (See color insert)

(7.58) for the same pairs. It is a function of $\rho(h)$. The correlogram ρ is then selected so as to obtain a correct fit of the indicator sample variogram.

Indicators I_2, I_3, and I_4 are the product of an indicator on Y_1 (the complementary to I_1) and an indicator on Y_2. The nonstationary covariance of these indicators is the product of two expressions similar to (7.58). ρ_2 is selected so as to obtain a satisfactory fit of the three indicator variograms of I_2, I_3, and I_4.

Figure 7.47 shows the sample vertical indicator variograms obtained from the borehole data and their fit associated to spherical models for the Gaussians (two components, with vertical ranges of 4 and 15 m for Y_1, 5 and 20 m for Y_2, and horizontal ranges 20 times larger). Note the nonstationary behavior induced by proportion variations. The anisotropy ratios have been obtained from the cliff data (they have a lower resolution than borehole data but give the variogram behavior at short horizontal distance).

Simulation

The initial study (1987) was carried out with three facies only. A single Gaussian SRF was necessary. The study used a separable exponential covariance because that model has advantageous Markov properties which simplify the calculation of the simulations. It is no longer necessary to stick to this model. The simulations have thus been generated with the usual method. This implies the following steps:

1. Transform the indicator data to interval data about Y_1 and Y_2 (note that no information is provided about Y_2 when $I_1 = 1$).
2. Simulate $Y_1(x_\alpha)$ at all data points (use the Gibbs sampler).
3. Simulate $Y_2(x_\beta)$ at the data points where interval data are available (facies F2, F3, F4).
4. Simulate $Y_1(x)$ conditional on the $Y_1(x_\alpha)$ (standard conditional simulation).
5. Simulate $Y_2(x)$ conditional on the $Y_2(x_\beta)$.
6. Transform $(Y_1(x), Y_2(x))$ into a facies according to the thresholds $y_i(x)$.

Step 1 is performed only once, while the others are repeated for each simulation.

Results

Figure 7.48 displays vertical cross-sections of two conditional simulations. The simulation reproduces the main features seen on the cliff: strong lateral continuity of the sandstone and shaly sandstone bed at the base with local thinning of the sandstone, more shaly intermediate streaks, scattered sandstone channels near the top, local occurrences of shale in the middle and upper parts. It is important to check if the simulations are representative of the reality as regards flow properties. A calculation of the equivalent permeability on the image of the real cliff and on simulations of 1987 showed deviations of only

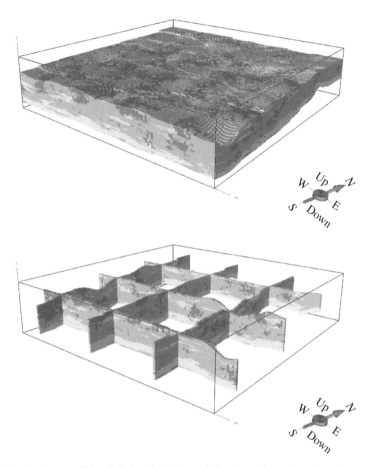

FIGURE 7.48 3D conditional facies simulation of the Brent formation: 3D view and fence diagram. Size of simulated domain: 1 km × 1 km × 30 m. [Output from Isatis®. From H. Beucher, personal communication.] (See color insert)

about 15%, a reasonably good result (Matheron et al., 1987). This equivalent permeability has not been calculated on the new simulations, but the deviation should be reduced since we now have more freedom in the choice of the covariance model than in the 1987 study.

Appendix

This appendix gathers some classic definitions and results used in this book, as well as simulation formulas for a few covariance models.

A.1 MEASURE THEORY DEFINITIONS

Measures are encountered in many places of this book, especially in the context of spectral theory. For readers not familiar with measure theory, we recall a few definitions, but without going into mathematical details. Good sources are Royden (1968) and Neveu (1970).

Essentially a measure is a set function—that is, a function which assigns a number to certain sets. The classic example is the Lebesgue measure that associates to each interval of the line its length, or to a subset in 2D or 3D its area or its volume. But we need a more general definition in order, for example, to assign a positive value to an isolated point (an atom).

σ-Algebra. A collection of subsets $\mathcal{A}$ of a set Ω is a σ-algebra if it contains the empty set $\emptyset$ and is closed with respect to complementation and with respect to countable unions:

$$A \in \mathcal{A} \quad \Rightarrow \quad A^c \in \mathcal{A}; \qquad A_i \in \mathcal{A} \quad \Rightarrow \quad \overset{\infty}{\underset{i=1}{\cup}} A_i \in \mathcal{A}$$

From De Morgan's laws it follows that $\mathcal{A}$ is also closed with respect to countable intersections.

Borel Sets. The collection $\mathcal{B}$ of Borel sets of $\mathbb{R}^n$ is the smallest σ-algebra containing all of the open sets of $\mathbb{R}^n$ (or equivalently, all finite or infinite intervals of the line, or n-dimensional rectangles in $\mathbb{R}^n$).

Measurable Function—Random Variable. Given a space $(\Omega, \mathcal{A})$, a real-valued function f : $\Omega \rightarrow \mathbb{R}$ is said to be measurable if $f^{-1}(B) \in \mathcal{A}$ for every Borel set B of $\mathbb{R}$. By definition a random variable is a measurable function on $(\Omega, \mathcal{A})$. When the domain space Ω is $\mathbb{R}$ or $\mathbb{R}^n$ (and $\mathcal{A}$ is $\mathcal{B}$), measurable functions include all ordinary functions of practical interest.

Measure. A measure μ on a σ-algebra $\mathcal{A}$ of subsets of Ω is a set function mapping $\mathcal{A}$ onto $]-\infty, +\infty[$ such that $\mu(\emptyset) = 0$ and

Geostatistics: Modeling Spatial Uncertainty, Second Edition. J.P. Chilès and P. Delfiner.
© 2012 John Wiley & Sons, Inc. Published 2012 by John Wiley & Sons, Inc.

$$\mu\left(\overset{n}{\underset{i=1}{\cup}} A_i\right) = \sum_{i=1}^{n} \mu(A_i) \qquad (\sigma\text{-additivity})$$

for any sequence A_i of pairwise disjoint sets in $\mathcal{A}$. The measure μ is *positive* if $\mu(A) \geq 0$ $(A \in \mathcal{A})$ and *bounded* if

$$\sup\{|\mu(A)| : A \in \mathcal{A}\} < \infty$$

The measure μ is bounded if and only if $\mu(\Omega) < \infty$. A positive measure μ is called *finite* if $\mu(\Omega) < \infty$.

Some authors name "measure" a positive measure and the others "signed measures." It is also possible to define a complex measure as a set function assigning a complex number to each $A \in \mathcal{A}$.

Examples: The Lebesgue measure $\mu(A) = |A|$ is a positive measure, but it is not bounded. A probability measure is a bounded positive measure such that $\mu(\Omega) = 1$.

Dirac Measure. The Dirac measure δ assigns the value 1 to any set B that contains the origin of coordinates and 0 to any set that doesn't. It corresponds to a unit mass placed at the point 0. An equivalent definition is to consider δ as a pseudofunction $\delta(\cdot)$ summing to one and equal to zero everywhere, except at 0 where it is infinite (Dirac delta function). For any continuous function φ we have

$$\delta(\varphi) = \int \delta(t)\,\varphi(t)\,dt = \varphi(0)$$

Likewise, the measure δ_x corresponds to a unit mass at the point x, and we have

$$\delta_x(\varphi) = \int \delta(x - t)\,\varphi(t)\,dt = \varphi(x)$$

The above shows that δ is the identity operator for convolution, in the sense that $\delta * \varphi = \varphi$.

Support. The support of a measure μ on $\mathcal{B}$ is the smallest closed set F of $\mathcal{B}$ such that μ vanishes in the complement of F. More precisely $\mu(O) = 0$ for any open set $O \subset \mathcal{B} - F$. The support of the Dirac measure is a single point, and that of the Lebesgue measure the whole space.

Absolute Continuity (Radon–Nikodym Theorem). A measure ν is absolutely continuous with respect to the positive measure μ if $\mu(A) = 0 \Rightarrow \nu(A) = 0$. Then the measure ν has a *density function* f such that $\nu(A) = \int_A f(x)\,\mu(dx)$ for any $A \in \mathcal{A}$.

When μ is the Lebesgue measure, absolute continuity of ν means, in particular, that any point $\{x\}$ has measure zero $\nu(\{x\}) = 0$ (no atom).

Almost Surely, Almost Everywhere. We say that an event E occurs "almost surely" if $\Pr\{E\} = 1$, and that a property holds "almost everywhere" if the set of points where it fails to hold is of measure zero.

Orthogonal Complex Random Measure. The set function $\xi(B)$ associating a finite variance complex random variable to each $B \in \mathcal{B}$ is an orthogonal random measure if it satisfies the following properties:

$$\begin{cases} \mathrm{E}\left[\xi(B)\,\overline{\xi(B')}\right] = 0 & \text{if} \quad B, B' \in \mathcal{B}, \quad B \cap B' = \varnothing \\ \xi(B \cup B') = \xi(B) + \xi(B') & \text{if} \quad B, B' \in \mathcal{B}, \quad B \cap B' = \varnothing \\ B_n \downarrow \varnothing \text{ in } \mathcal{B} & \Rightarrow \quad \mathrm{E}|\xi(B_n)|^2 \to 0 \end{cases}$$

Examples

1. Consider a Poisson point process with intensity λ; the number of points $N(B)$ that fall in B minus its expected value defines the random orthogonal measure $\xi(B) = N(B) - \lambda|B|$.

2. According to Chentsov (1957), Gaussian white noise is a random measure $\xi(\cdot)$ such that (a) $\xi(B)$ is normally distributed with mean 0 and variance $\lambda|B|$, and (b) when B and B' do not intersect $\xi(B)$ and $\xi(B')$ are independent and $\xi(B \cup B') = \xi(B) + \xi(B')$.

3. In 1D, let U be a random variable with a symmetric probability distribution $F(du)$, and let Φ be uniform on $[0, 2\pi[$ and independent of U. The complex measure $\xi(\cdot)$ defined by

$$\sqrt{2}\,\xi(B) = \exp(i\,\Phi)\,\delta_U(B) + \exp(-i\,\Phi)\,\delta_{-U}(B)$$

is random and orthogonal and satisfies $E|\xi(du)|^2 = F(du)$. It is the random spectral measure of an SRF whose covariance is the Fourier transform of $F(du)$.

A.2   GAMMA FUNCTION (EULER'S INTEGRAL)

The gamma function is defined for $x > 0$ by

$$\Gamma(x) = \int_0^\infty e^{-u} u^{x-1}\, du \tag{A.1}$$

It satisfies the recurrence relation

$$\Gamma(x+1) = x\,\Gamma(x)$$

Hence for a positive integer value $x = n$, one has

$$\Gamma(n+1) = n!$$

For $x < 0$ the integral (A.1) diverges but the gamma function can be calculated, for x noninteger, by the reflection formula

$$\Gamma(x)\,\Gamma(1-x) = \frac{\pi}{\sin \pi x}$$

For the particular value $x = 1/2$, this formula gives

$$\Gamma\left(\tfrac{1}{2}\right) = \sqrt{\pi}$$

Another useful formula is the duplication formula

$$\Gamma(2x) = \Gamma(x)\,\Gamma\left(x + \tfrac{1}{2}\right)\frac{2^{2x-1}}{\sqrt{\pi}}$$

Finally let us also mention the link with the beta function

$$B(x, y) = \int_0^1 u^{x-1}(1-u)^{y-1}\, du = \frac{\Gamma(x)\,\Gamma(y)}{\Gamma(x+y)} \qquad x > 0, \quad y > 0.$$

A.3 BESSEL FUNCTIONS

Since there are some variations about the definition of Bessel functions, we give the definition of the functions used in this book. It corresponds to the definitions used by Olver (1972).

The *Bessel function of the first kind* with index ν is defined by the development

$$J_\nu(x) = \left(\frac{x}{2}\right)^\nu \sum_{k=0}^\infty \frac{(-1)^k}{k!\,\Gamma(\nu+k+1)} \left(\frac{x}{2}\right)^{2k} \tag{A.2}$$

Two particular cases are

$$J_{-1/2}(x) = \sqrt{\frac{2}{\pi x}}\cos x, \qquad J_{1/2}(x) = \sqrt{\frac{2}{\pi x}}\sin x$$

The *modified Bessel function of the first kind* is defined by the development

$$I_\nu(x) = \left(\frac{x}{2}\right)^\nu \sum_{k=0}^\infty \frac{1}{k!\,\Gamma(\nu+k+1)} \left(\frac{x}{2}\right)^{2k} \tag{A.3}$$

The *modified Bessel function of the second kind* is defined by

$$K_\nu(x) = \frac{\pi}{2}\frac{I_{-\nu}(x)-I_\nu(x)}{\sin \pi\nu} = K_{-\nu}(x) \tag{A.4}$$

A.4 UNIT SPHERE OF $\mathbb{R}^n$

The surface area S_n and the volume V_n of the unit-radius sphere of $\mathbb{R}^n$ are given by

$$S_n = \frac{2\pi^{n/2}}{\Gamma(n/2)} \qquad V_n = \frac{S_n}{n} = \frac{\pi^{n/2}}{\Gamma\left(\frac{n}{2}+1\right)} \tag{A.5}$$

For example, $S_1 = 2$, $S_2 = 2\pi$, $S_3 = 4\pi$, $V_1 = 2$, $V_2 = \pi$, and $V_3 = \frac{4}{3}\pi$.

A.5 GAUSSIAN DISTRIBUTION AND HERMITE POLYNOMIALS

A.5.1 Gaussian Distribution

The standard Gaussian (synonym: standard normal) probability density function is defined by

$$g(y) = \frac{1}{\sqrt{2\pi}}\exp\left(-\frac{y^2}{2}\right)$$

It has mean 0, variance 1, and dispersion indicator $S = 1/\sqrt{\pi}$.

A.5.2 Hermite Polynomials

- The orthogonal polynomials associated with the standard Gaussian are the Hermite polynomials $H_n(y)$. They are defined by the Rodrigues formula

$$H_n(y)\, g(y) = \frac{d^n}{dy^n} g(y) \qquad (A.6)$$

These polynomials satisfy

$$\int H_m(y)\, H_n(y)\, g(y)\, dy = n!\, \delta_{mn}$$

An orthonormal basis for the Gaussian distribution is defined by the normalized polynomials

$$\chi_n(y) = \frac{1}{\sqrt{n!}} H_n(y)$$

- For $n \geq 1$ relation (A.6) can be written as

$$H_n(y)\, g(y) = \frac{d}{dy}\left(H_{n-1}(y)\, g(y) \right) \qquad (n \geq 1) \qquad (A.7)$$

Since $\frac{d}{dy} g(y) = -y\, g(y)$, we deduce the explicit form of Hermite polynomials:

$$H_n(y) = \sum_{i=0}^{\lfloor n/2 \rfloor} (-1)^{n-i} \frac{n!}{2^i\, i!\, (n-2i)!}\, y^{n-2i}$$

where $\lfloor n/2 \rfloor$ stands for the integer part of $n/2$. From this expression it follows that the derivative of $H_n(y)$ is proportional to $H_{n-1}(y)$:

$$\frac{d}{dy} H_n(y) = -n\, H_{n-1}(y) \qquad (A.8)$$

- Applying definition (A.7) to H_{n+1}, we get

$$H_{n+1}(y)\, g(y) = \frac{d}{dy}\left(H_n(y)\, g(y) \right) = \left(-y\, H_n(y) + \frac{d}{dy} H_n(y) \right) g(y)$$

whence

$$H_{n+1}(y) = -y\, H_n(y) + \frac{d}{dy} H_n(y)$$

and given (A.8),

$$H_{n+1}(y) = -y\, H_n(y) - n\, H_{n-1}(y)$$

This recurrence relation allows an easy calculation of the $H_n(y)$ from the first two polynomials $H_0(y) = 1$ and $H_1(y) = -y$. Thus, for example, $H_2(y) = y^2 - 1$, $H_3(y) = -y^3 + 3y$, and so on. Notice that the definition of H_n given here differs from the most common one by a factor $(-1)^n$.

- Relation (A.7) is equivalent to

$$s_n(y) = \int_{-\infty}^{y} H_n(u)\, g(u)\, du = \begin{cases} G(y) & \text{if } n = 0 \\ H_{n-1}(y)\, g(y) & \text{if } n > 0 \end{cases} \qquad (A.9)$$

This formula permits an easy calculation of the coefficients of the expansion of $1_{y < y_0}$:

$$1_{y < y_0} = G(y_0) + g(y_0) \sum_{n=1}^{\infty} \frac{H_{n-1}(y_0)}{n!} H_n(y) \tag{A.10}$$

- In applications we also need to know the incomplete integral:

$$S_{mn}(y) = \int_{-\infty}^{y} H_m(u) \, H_n(u) \, g(u) \, du$$

Using relation (A.7) and integration by parts, and then relation (A.8), it can be shown that

$$S_{mn}(y) = m \, S_{m-1,n-1}(y) + H_m(y) \, H_{n-1}(y) \, g(y) \tag{A.11}$$

To initialize the recurrence calculation, we use the symmetry $S_{mn}(y) = S_{nm}(y)$ so as to reduce the problem to the case $n \geq m$, and we start from $S_{0,\,n-m}(y) = s_{n-m}(y)$, which is given by (A.9).

- Let us also mention a useful expansion for studying lognormal distributions:

$$\exp(\lambda y) = \exp\left(\frac{1}{2}\lambda^2\right) \sum_{n=0}^{\infty} (-1)^n \frac{\lambda^n}{n!} H_n(y) \tag{A.12}$$

- Other useful formulas derived from Hochstrasser (1972, p. 786) and Matheron (1974c) are the integrals

$$\int_{-\infty}^{+\infty} H_n(a+y) \, g(y) \, dy = (-1)^n a^n$$

$$\int_{-\infty}^{+\infty} H_n(by) \, g(y) \, dy = \begin{cases} \dfrac{(2k)!}{2^k k!} (b^2 - 1)^k & \text{if } n = 2k \\ 0 & \text{if } n = 2k+1 \end{cases}$$

$$\int_{-\infty}^{+\infty} H_n(a+by) \, g(y) \, dy = (1 - b^2)^n H_n\left(\frac{a}{\sqrt{1-b^2}}\right), \qquad |b| < 1$$

- Studying multivariate Hermite polynomials, Withers (2000) derives a simple and powerful formula, which in our univariate case amounts to

$$H_n(y) = (-1)^n \, E[y + i\,U]^n$$

where U is a standard normal random variable and $i^2 = -1$, a formula already mentioned by Hochstrasser.

Other properties of Hermite polynomials for geostatistical applications can be found in Matheron (1974c, pp. 64–68) and in the literature on orthogonal polynomials (Szegö, 1939; Hochstrasser, 1972).

A.6 GAMMA DISTRIBUTION AND LAGUERRE POLYNOMIALS

A.6.1 Gamma Distribution

The standard gamma probability density functions are defined on $]0, \infty[$ by

$$g_\alpha(y) = \frac{1}{\Gamma(\alpha)} e^{-y} y^{\alpha-1} \qquad (y > 0)$$

where α is a positive parameter and $\Gamma(\cdot)$ is the gamma function (A.1).

The shape of the density g_α depends on α: If $\alpha < 1$, g_α is a decreasing function, unbounded at the origin; if $\alpha > 1$, g_α is a bell-shaped curve, which tends to a Gaussian distribution for large values of α; in the intermediate case $\alpha = 1$, g_1 is the exponential density.

The gamma distribution has moments of all orders:

$$E[Y^n] = \frac{\Gamma(\alpha + n)}{\Gamma(\alpha)}$$

In particular,

$$E[Y] = \alpha, \quad \text{Var}[Y] = \alpha$$

Its dispersion indicator is

$$S = \frac{1}{\sqrt{\pi}} \frac{\Gamma(\alpha + \frac{1}{2})}{\Gamma(\alpha)} \tag{A.13}$$

Its Laplace transform is

$$E[e^{-\lambda Y}] = \frac{1}{(1 + \lambda)^\alpha}$$

The standard gamma distribution corresponds to the case $b = 1$ of the general gamma distribution with parameter $\alpha > 0$ and scale $b > 0$, whose density is

$$g_{\alpha, b}(y) = \frac{b^\alpha}{\Gamma(\alpha)} e^{-by} y^{\alpha - 1} \qquad (y > 0)$$

If Y has the distribution $g_{\alpha, b}$, then bY has the distribution g_α.

The sum of two independent gamma variables with the same scale parameter is also a gamma varibale. Specifically

$$g_{\alpha, b} * g_{\beta, b} = g_{\alpha + \beta, b}$$

A.6.2 Laguerre Polynomials

- The orthogonal polynomials associated with the standard gamma distribution are the Laguerre polynomials $L_n^\alpha(y)$, defined on $]0, \infty[$ like g_α. They are defined by the Rodrigues formula

$$L_n^\alpha(y) g_\alpha(y) = \frac{\Gamma(\alpha)}{\Gamma(\alpha + n)} \frac{d^n}{dy^n} [y^n g_\alpha(y)] = \frac{d^n}{dy^n} g_{\alpha + n}(y) \tag{A.14}$$

These polynomials are orthogonal over the distribution g_α but are not normalized:

$$\int_0^\infty L_m^\alpha(y) L_n^\alpha(y) g_\alpha(y) \, dy = \frac{\Gamma(\alpha) n!}{\Gamma(\alpha + n)} \delta_{mn}$$

To get an orthonormal basis, we thus take

$$\chi_n^\alpha(y) = \sqrt{\frac{\Gamma(\alpha + n)}{\Gamma(\alpha) n!}} L_n^\alpha(y)$$

- From definition (A.14) we get for $n \geq 1$,

$$L_n^\alpha(y) \, g_\alpha(y) = \frac{d}{dy} \left(L_{n-1}^{\alpha+1}(y) \, g_{\alpha+1}(y) \right) \tag{A.15}$$

Using the relations

$$g_{\alpha+1}(y) = \frac{y}{\alpha} g_\alpha(y) \quad \text{and} \quad \frac{d}{dy} g_\alpha(y) = g_{\alpha-1}(y) - g_\alpha(y)$$

we deduce by recurrence that the explicit form of the nth Laguerre polynomial is

$$L_n^\alpha(y) = \sum_{i=0}^{n} (-1)^i \binom{n}{i} \frac{\Gamma(\alpha)}{\Gamma(\alpha+i)} y^i$$

and that its derivative is, up to a multiplicative factor, the Laguerre polynomial of degree $n-1$ associated with the distribution with parameter $\alpha + 1$:

$$\frac{d}{dy} L_n^\alpha(y) = -\frac{n}{\alpha} L_{n-1}^{\alpha+1}(y) \tag{A.16}$$

- Laguerre polynomials with parameter α can easily be calculated from $L_0^\alpha(y) = 1$ and $L_1^\alpha(y) = 1 - \frac{y}{\alpha}$ via the recurrence relation:

$$(\alpha + n) L_{n+1}^\alpha(y) = (2n + \alpha - y) L_n^\alpha(y) - n L_{n-1}^\alpha(y)$$

- From (A.15) we get

$$s_n^\alpha(y) = \int_0^y L_n^\alpha(u) \, g_\alpha(u) \, du = \begin{cases} G_\alpha(y) & \text{if} \quad n = 0 \\ L_{n-1}^{\alpha+1}(y) \, g_{\alpha+1}(y) & \text{if} \quad n > 0 \end{cases} \tag{A.17}$$

- The incomplete integral

$$s_{mn}^\alpha(y) = \int_0^y L_m^\alpha(u) \, L_n^\alpha(u) \, g_\alpha(u) \, du$$

can be computed by recurrence, but in a somewhat more complex way than in the Gaussian case because the recurrence also involves the parameters of the gamma distribution. Setting (A.15) into the definition of $S_{mn}^\alpha(y)$, integrating by parts, and then using (A.16), we obtain

$$S_{mn}^\alpha(y) = \frac{m}{\alpha} S_{m-1, n-1}^{\alpha+1}(y) + L_m^\alpha(y) \, L_{n-1}^{\alpha+1}(y) \, g_{\alpha+1}(y) \tag{A.18}$$

To initialize the recurrence calculation, we use the symmetry $S_{mn}^\alpha(y) = S_{nm}^\alpha(y)$ so as to reduce the problem to the case $n \geq m$, and we start from $S_{0, n-m}^{\alpha+m}(y) = s_{n-m}^{\alpha+m}(y)$.

A.7   NEGATIVE BINOMIAL DISTRIBUTION AND MEIXNER POLYNOMIALS

A.7.1   Negative Binomial Distribution

The integer random variable Y follows a negative binomial distribution if the probability that Y is equal to i is

$$p_i = (1-p)^\alpha \frac{\Gamma(\alpha+i)}{\Gamma(\alpha)} \frac{p^i}{i!} \qquad i \in \mathbb{N}$$

where $\alpha > 0$ and $0 < p < 1$ are parameters and $\Gamma(\cdot)$ is the gamma function (A.1).

The negative binomial distribution may be considered as the discrete equivalent of the gamma distribution. The shape of the distribution defined by the p_i depends on α. In particular, if $\alpha < 1$, the p_i are decreasing. For $\alpha = 1$ we get the geometric, or Pascal, distribution.

The negative binomial distribution has for generating function

$$E[s^Y] = \left(\frac{1-p}{1-ps}\right)^\alpha$$

The negative binomial distribution has moments of all orders and, in particular,

$$E[Y(Y-1)\cdots(Y-n+1)] = \frac{\Gamma(\alpha+n)}{\Gamma(\alpha)}\left(\frac{p}{1-p}\right)^n,$$

$$E[Y] = \alpha p/(1-p), \qquad \mathrm{Var}[Y] = \alpha p/(1-p)^2$$

Its dispersion indicator is

$$S = \alpha \frac{p}{(1-p)^2} F\left(\alpha+1, \frac{1}{2}; 2; -4\frac{p}{(1-p)^2}\right) \tag{A.19}$$

where F represents the hypergeometric function (Lantuéjoul, personal communication).

A.7.2   Meixner Polynomials

The orthogonal polynomials associated with the negative binomial distribution with parameters α and p are the Meixner polynomials $M_n^{\alpha p}(i)$. These can easily be calculated from $M_0^{\alpha p}(i) = 1$ and $M_1^{\alpha p}(i) = 1 - \frac{1-p}{p}\frac{i}{\alpha}$ via the recurrence

$$p(\alpha+n)M_{n+1}^{\alpha p}(i) = [n(1+p) + \alpha p - i(1-p)]\,M_n^{\alpha p}(i) - n\,M_{n-1}^{\alpha p}(i) \tag{A.20}$$

Meixner polynomials satisfy the symmetry relation $M_n^{\alpha p}(i) = M_i^{\alpha p}(n)$. An orthonormal basis is obtained by taking

$$\chi_n^{\alpha p}(i) = \sqrt{\frac{\Gamma(\alpha+n)}{\Gamma(\alpha)}\frac{p^n}{n!}}\, M_n^{\alpha p}(i)$$

A.8   MULTIVARIATE NORMAL DISTRIBUTION

Let $\mathbf{Y} = (Y_1, \ldots, Y_p)'$ be a p-dimensional random vector with mean vector $\boldsymbol{\mu}$ and covariance matrix $\boldsymbol{\Sigma}$, that is,

$$\boldsymbol{\mu} = (\mu_1, \ldots, \mu_p)', \qquad \boldsymbol{\Sigma} = E(\mathbf{Y} - \boldsymbol{\mu})(\mathbf{Y} - \boldsymbol{\mu})'$$

If $\mathbf{Y}$ is multivariate normal (= Gaussian), its probability density is

$$f(\mathbf{Y}) = \frac{1}{|\boldsymbol{\Sigma}|^{1/2}(2\pi)^{p/2}} \exp[-\tfrac{1}{2}(\mathbf{Y} - \boldsymbol{\mu})' \, \boldsymbol{\Sigma}^{-1} \, (\mathbf{Y} - \boldsymbol{\mu})]$$

where $|\boldsymbol{\Sigma}|$ is the determinant of the $p \times p$ matrix $\boldsymbol{\Sigma}$. Every linear combination $Y = \lambda_1 Y_1 + \ldots + \lambda_p Y_p$ is normally disributed with mean $\mu = \sum_{i=1}^{p} \lambda_i \mu_i$ and variance $\sigma^2 = \sum_{i=1}^{p} \sum_{j=1}^{p} \lambda_i \lambda_j \sigma_{ij}$. This property characterizes the multivariate normal distribution.

Suppose now that $\mathbf{Y}$, $\boldsymbol{\mu}$ and $\boldsymbol{\Sigma}$ are partitioned as follows:

$$\mathbf{Y} = \begin{bmatrix} \mathbf{Y}_1 \\ \mathbf{Y}_2 \end{bmatrix}, \qquad \boldsymbol{\mu} = \begin{bmatrix} \boldsymbol{\mu}_1 \\ \boldsymbol{\mu}_2 \end{bmatrix}, \qquad \boldsymbol{\Sigma} = \begin{bmatrix} \boldsymbol{\Sigma}_{11} & \boldsymbol{\Sigma}_{12} \\ \boldsymbol{\Sigma}_{21} & \boldsymbol{\Sigma}_{22} \end{bmatrix}$$

where $\mathbf{Y}_1$ is a $q \times 1$ vector and all other vectors and matrix are dimensioned accordingly. Then the conditional distribution of $\mathbf{Y}_1$ given $\mathbf{Y}_2$ is a q-variate normal with mean vector and covariance matrix:

$$E(\mathbf{Y}_1 \mid \mathbf{Y}_2) = \boldsymbol{\mu}_1 + \boldsymbol{\Sigma}_{12} \boldsymbol{\Sigma}_{22}^{-1} (\mathbf{Y}_2 - \boldsymbol{\mu}_2), \qquad \mathrm{Cov}(\mathbf{Y}_1 \mid \mathbf{Y}_2) = \boldsymbol{\Sigma}_{11} - \boldsymbol{\Sigma}_{12} \boldsymbol{\Sigma}_{22}^{-1} \boldsymbol{\Sigma}_{21}$$

Two remarkable properties:

1. The regression of $\mathbf{Y}_1$ on $\mathbf{Y}_2$ is a linear function of $\mathbf{Y}_2$
2. The covariance matrix of $\mathbf{Y}_1$ given $\mathbf{Y}_2$ does not depend on $\mathbf{Y}_2$ (homoscedasticity)

As a consequence, if $Y_0, Y_1, \ldots, Y_N$ are jointly normal and with means 0, the regression of Y_0 on $Y_1, \ldots, Y_N$ is of the form

$$E(Y_0 \mid Y_1, \ldots, Y_N) = \lambda_1 Y_1 + \cdots + \lambda_N Y_N$$

and the conditional variance about this value does not depend on $Y_1, \ldots, Y_N$.

For bivariate normal random variables X and Y with correlation coefficient ρ, we have

$$E(Y \mid X) = \mu_Y + \rho \frac{\sigma_Y}{\sigma_X} (X - \mu_X), \qquad \mathrm{Var}(Y \mid X) = (1 - \rho^2) \sigma_Y^2$$

For a trivariate normal X, Y, Z conditioned on Z, a useful relation is

$$\mathrm{Cov}(X, Y \mid Z) = (\rho_{XY} - \rho_{XZ} \rho_{YZ}) \sigma_X \sigma_Y$$

This covariance is also the covariance of the residuals from the regressions of X on Z and of Y on Z. Dividing this by the residuals standard deviations gives the partial correlation

$$\rho_{XY.Z} = (\rho_{XY} - \rho_{XZ} \rho_{YZ}) / \sqrt{(1 - \rho_{XZ}^2)(1 - \rho_{YZ}^2)}$$

A.9 LOGNORMAL DISTRIBUTION

By definition, the random variable Z is lognormal if $Y = \log Z$ is normal. Denoting by m and σ^2 the mean and variance of Y, the mean of Z is

$$M = \mathrm{E}[\exp(Y)] = \exp\left(m + \tfrac{1}{2}\sigma^2\right) \tag{A.21}$$

Likewise, the vector-valued random variable $\mathbf{Z} = (Z_1, \ldots, Z_p)'$ is lognormal if the vector-valued random variable $\mathbf{Y} = (Y_1, \ldots, Y_p)'$, where $Y_i = \log Z_i$, is normal (i.e., has a *multivariate Gaussian distribution*).

The vector-valued random variable $\mathbf{Z}$ being assumed lognormal, let us consider the random variable $Z = Z_1^{\lambda_1} \cdots Z_p^{\lambda_p}$. It is of the form $Z = \exp(Y)$ with $Y = \lambda_1 Y_1 + \cdots + \lambda_p Y_p$. The vector $\mathbf{Y}$ having a multivariate normal distribution, the random variable Y is normal, and Z is lognormal.

This property, together with formula (A.21), allows, for example, the calculation of the noncentered moments of a lognormal random variable and the noncentered covariance of two lognormal variables. In particular, the variance of $Z = \exp(Y)$ is

$$\Sigma^2 = \mathrm{Var}(Z) = M^2[\exp(\sigma^2) - 1] \tag{A.22}$$

and if Y_1 and Y_2 are bivariate Gaussian with means m_1 and m_2, variances σ_1^2 and σ_2^2, and correlation coefficient ρ, the centered covariance of $Z_1 = \exp(Y_1)$ and $Z_2 = \exp(Y_2)$ is

$$\mathrm{Cov}(Z_1, Z_2) = M_1 M_2[\exp(\rho\, \sigma_1 \sigma_2) - 1] \tag{A.23}$$

where M_1 and M_2 are the means of Z_1 and Z_2.

Still in the bivariate lognormal case, it is interesting to look at the regression of Z_2 on Z_1. The conditional distribution of Y_2 given Y_1 is normal with mean and variance

$$\mathrm{E}(Y_2 \mid Y_1) = m_2 + \rho\frac{\sigma_2}{\sigma_1}(Y_1 - m_1) \qquad \mathrm{Var}(Y_2 \mid Y_1) = (1 - \rho^2)\sigma_2^2$$

Therefore the conditional distribution of Z_2 given Z_1 is lognormal with mean and variance given by

$$\begin{aligned}
\mathrm{E}(Z_2 \mid Z_1) &= \gamma_2(Z_1/\gamma_1)^{\rho\, \sigma_2/\sigma_1} \exp[(1 - \rho^2)\sigma_2^2/2] \\
\mathrm{Var}(Z_2 \mid Z_1) &= [\mathrm{E}(Z_2 \mid Z_1)]^2 \, [\exp((1 - \rho^2)\sigma_2^2) - 1]
\end{aligned} \tag{A.24}$$

where $\gamma_1 = \exp(m_1)$ and $\gamma_2 = \exp(m_2)$. The regression curve is a power function and not a straight line, except when $\rho\, \sigma_2/\sigma_1 = 1$.

A.10 SIMULATION FORMULAS

Useful formulas for the simulation of common covariance models in $\mathbb{R}^1$ and $\mathbb{R}^3$ are presented in Tables A.1 and A.2, respectively.

TABLE A.1   Formulas for the Simulation of Common Covariance Models in $\mathbb{R}$[1]

Covariance in $\mathbb{R}$	Dilution Function	Spectral Density

Triangle model

$$C_1(h) = \begin{cases} C\left(1 - \dfrac{h}{a}\right) & \text{if } h \leq a \\[2mm] 0 & \text{if } h \geq a \end{cases}$$

$$w_1(s) = \begin{cases} \sqrt{\dfrac{C}{a}} & \text{if } |s| \leq \dfrac{a}{2} \\[2mm] 0 & \text{if } |s| > \dfrac{a}{2} \end{cases}$$

$$f_1(u) = 2Ca\,\frac{1 - \cos v}{v^2}$$

Spherical model

$$C_1(h) = \begin{cases} C\left(1 - \dfrac{3}{2}\dfrac{h}{a} + \dfrac{1}{2}\dfrac{h^3}{a^3}\right) & \text{if } h \leq a \\[2mm] 0 & \text{if } h \geq a \end{cases}$$

$$w_1(s) = \begin{cases} \sqrt{\dfrac{3}{4}\dfrac{C}{a}\left(1 - 2\dfrac{s}{a}\right)} & \text{if } |s| \leq \dfrac{a}{2} \\[2mm] 0 & \text{if } |s| > \dfrac{a}{2} \end{cases}$$

$$f_1(u) = 3Ca\,\frac{(1 - \cos v)^2 + (v - \sin v)^2}{v^4}$$

Cubic model

$$C_1(h) = \begin{cases} C\left(1 - 7\dfrac{h^2}{a^2} + \dfrac{35}{4}\dfrac{h^3}{a^3} - \dfrac{7}{2}\dfrac{h^5}{a^5} + \dfrac{3}{4}\dfrac{h^7}{a^7}\right) & \text{if } h \leq a \\[2mm] 0 & \text{if } h \geq a \end{cases}$$

$$w_1(s) = \begin{cases} \sqrt{\dfrac{105}{64}\dfrac{C}{a}\left(1 + 2\dfrac{s}{a} - 4\dfrac{s^2}{a^2} - 8\dfrac{s^3}{a^3}\right)} & \text{if } |s| \leq \dfrac{a}{2} \\[2mm] 0 & \text{if } |s| > \dfrac{a}{2} \end{cases}$$

$$f_1(u) = 105Ca\,\frac{\left[4v\sin v + (6 - v^2)\cos v - 6\right]^2 + \left[(6 - v^2)\sin v - 4v\cos v - 2v\right]^2}{v^8}$$

Exponential model

$$C_1(h) = C\exp\left(-\frac{h}{a}\right)$$

$$w_1(s) = \begin{cases} 0 & \text{if } s < 0 \\[2mm] \sqrt{2\dfrac{C}{a}}\,\exp\left(-\dfrac{s}{a}\right) & \text{if } s \geq 0 \end{cases}$$

$$f_1(u) = \frac{2Ca}{1 + v^2}$$

Gaussian model

$$C_1(h) = C\exp\left(-\frac{h^2}{a^2}\right)$$

$$w_1(s) = \sqrt{\sqrt{\frac{4}{\pi}\frac{C}{a}}\,\exp\left(-2\frac{s^2}{a^2}\right)}$$

$$f_1(u) = \sqrt{\pi}\,Ca\,\exp\left(-\frac{v^2}{4}\right)$$

For each covariance $C_1(h)$ in $\mathbb{R}$ the table gives a dilution function $w_1(s)$ satisfying $w_1 * \overset{*}{w}_1 = C_1$, and the spectral density $f_1(u)$ of $C_1(h)$. Here h stands for $|h|$, and $f_1(u)$ is expressed as a function of $v = 2\pi a u$.

TABLE A.2 Formulas for the Simulation of Common Covariance Models in $\mathbb{R}^3$ by Turning Bands

Covariance in $\mathbb{R}$	Covariance in $\mathbb{R}^3$	Dilution Function	Spectral Density

Spherical model

$$C_3(r) = \begin{cases} C\left(1 - \dfrac{3r}{2a} + \dfrac{1}{2}\dfrac{r^3}{a^3}\right) & \text{if } r \le a \\[2mm] 0 & \text{if } r \ge a \end{cases}$$

$$C_1(h) = \begin{cases} C\left(1 - 3\dfrac{h}{a} + 2\dfrac{h^3}{a^3}\right) & \text{if } h \le a \\[2mm] 0 & \text{if } h \ge a \end{cases}$$

$$w_1(s) = \begin{cases} \sqrt{12\dfrac{C}{a}\dfrac{s}{a}} & \text{if } |s| \le \dfrac{a}{2} \\[3mm] 0 & \text{if } |s| > \dfrac{a}{2} \end{cases}$$

$$f_1(u) = \frac{12Ca}{v^4}\left[2\sin\frac{v}{2} - v\cos\frac{v}{2}\right]^2$$

Cubic model

$$C_3(r) = \begin{cases} C\left(1 - 7\dfrac{r^2}{a^2} + \dfrac{35}{4}\dfrac{r^3}{a^3} - \dfrac{7}{2}\dfrac{r^5}{a^5} + \dfrac{3}{4}\dfrac{r^7}{a^7}\right) & \text{if } r \le a \\[2mm] 0 & \text{if } r \ge a \end{cases}$$

$$C_1(h) = \begin{cases} C\left(1 - 21\dfrac{h^2}{a^2} + 35\dfrac{h^3}{a^3} - 21\dfrac{h^5}{a^5} + 6\dfrac{h^7}{a^7}\right) & \text{if } h \le a \\[2mm] 0 & \text{if } h \ge a \end{cases}$$

$$w_1(s) = \begin{cases} \sqrt{210\dfrac{C}{a}\left(\dfrac{1}{2}\dfrac{s}{a} - 2\dfrac{s^3}{a^3}\right)} & \text{if } |s| \le \dfrac{a}{2} \\[3mm] 0 & \text{if } |s| > \dfrac{a}{2} \end{cases}$$

$$f_1(u) = \frac{840Ca}{v^8}\left[(12 - v^2)\sin\frac{v}{2} - 6v\cos\frac{v}{2}\right]^2$$

Exponential model

$$C_3(r) = C\exp\left(-\frac{r}{a}\right)$$

$$C_1(h) = C\left(1 - \frac{h}{a}\right)\exp\left(-\frac{h}{a}\right)$$

$$w_1(s) = \begin{cases} 0 & \text{if } s < 0 \\[2mm] 2\sqrt{\dfrac{C}{a}\left(1 - \dfrac{s}{a}\right)\exp\left(-\dfrac{h}{a}\right)} & \text{if } s \ge 0 \end{cases}$$

$$f_1(u) = \frac{4Cav^2}{(1+v^2)^2}$$

Gaussian model

$$C_3(r) = C\exp\left(-\frac{r^2}{a^2}\right)$$

$$C_1(h) = C\left(1 - 2\frac{h^2}{a^2}\right)\exp\left(-\frac{h^2}{a^2}\right)$$

$$w_1(s) = \sqrt{\sqrt{\frac{256}{\pi}\frac{C}{a}\frac{s}{a}}\exp\left(-2\frac{s^2}{a^2}\right)}$$

$$f_1(u) = \sqrt{\pi}Cav^2\exp\left(-\frac{v^2}{2}\right)$$

For each covariance $C_3(r)$ in $\mathbb{R}^3$ the table gives: the associated covariance $C_1(h)$ to be simulated in $\mathbb{R}^1$, a dilution function $w_1(s)$ satisfying $w_1 * \breve{w}_1 = C_1$, and the spectral density $f_1(u)$ of $C_1(h)$. Here h stands for $|h|$, and $f_1(u)$ is expressed as a function of $v = 2\pi a u$.

References

The following references include technical reports of the "Centre de Géostatistique, Fontainebleau", mostly from Georges Matheron (in French). These reports, and also published material when permission was granted by the copyright owner, can be freely downloaded from the website of MINES ParisTech, Center of Geosciences and Geoengineering, Geostatistics:

http://www.geosciences.mines-paristech.fr/en > Geostatistics > Documents and Software > On-line library

Aanonsen, S. I., G. Nævdal, D. S. Oliver, A. C. Reynolds, and B. Vallès (2009). The ensemble Kalman filter in reservoir engineering—A review. *SPE Journal*, **14(3)**, 93–412.

Abrahamsen, P. (1993). Bayesian kriging for seismic depth conversion of a multi-layer reservoir. In *Geostatistics Tróia '92*, A. Soares, ed. Kluwer, Dordrecht, Netherlands, Vol. 1, pp. 385–398.

Adler, R. J., and J. E. Taylor (2007). *Random Fields and Geometry*. Springer, New York.

Agterberg, F. P. (2007). Mixtures of multiplicative cascade models in geochemitry. *Nonlinear Processes in Geophysics*, **14(3)**, 201–209.

Ahrens, L. H. (1954). The lognormal distribution of the elements (A fundamental law of geochemistry and its subsidiary). *Geochimica et Cosmochimica Acta*, **5(2)**, 49–73.

Aitchison, D. (1986). *The Statistical Analysis of Compositional Data*. Chapman & Hall, London.

Alabert, F. (1987). *Stochastic Imaging of Spatial Distributions Using Hard and Soft Information*. M.Sc. thesis, Stanford University, California.

Alabert, F., and G. J. Massonnat (1990). Heterogeneity in a complex turbiditic reservoir: Stochastic modeling of facies and petrophysical variability. *SPE paper* 20604, 65th Annual Technical Conference and Exhibition of the Society of Petroleum Engineers, New Orleans, Louisiana, pp. 775–790.

Albor Consuegra, F. R., and R. Dimitrakopoulos (2009). Stochastic mine design optimisation based on simulated annealing: pit limits, production schedules, multiple orebody scenarios and sensitivity analysis. *Mining Technology: Transactions of the Institute of Mining and Metallurgy, Section A*, **118(2)**, 79–90.

Alfaro, M. (1979). *Etude de la robustesse des simulations de fonctions aléatoires.* Doctoral thesis, E.N.S. des Mines de Paris.

Alfaro, M. (1980). The random coin method: Solution to the problem of the simulation of a random function in the plane. *Journal of the International Association for Mathematical Geology*, **12(1)**, 25–32.

Alfaro, M. (1984). Statistical inference of the semivariogram and the quadratic model. In *Geostatistics for Natural Resources Characterization*, G. Verly, M. David, A. G. Journel, and A. Maréchal, eds. Reidel, Dordrecht, Part 1, pp. 45–53.

Alfaro, M. (1993). Naturalistic simulations of deposits. Application to the problem of the kriging variance. Presented at the Conference *Geostatistics for the Next Century*, Montreal, June 1993.

Alfsen, E. M. (1971). *Compact Convex Sets and Boundary Integrals.* Springer, Berlin.

Allard, D. (1993). *Connexité des ensembles aléatoires: Application à la simulation des réservoirs pétroliers hétérogènes.* Doctoral thesis, E.N.S. des Mines de Paris.

Allard, D. (1994). Simulating a geological lithofacies with respect to connectivity information using the truncated Gaussian model. In *Geostatistical Simulations*, M. Armstrong and P.A. Dowd, eds. Kluwer, Dordrecht, pp. 197–211.

Allard, D., and Heresim Group (1993). On the connectivity of two random set models: The truncated Gaussian and the Boolean. In *Geostatistics Tróia '92*, Vol. 1, A. Soares, ed. Kluwer, Dordrecht, pp. 467–478.

Allègre, C. J., and E. Lewin (1995). Scaling laws and geochemical distributions. *Earth and Planetary Science Letters*, **132**, 1–13.

Almeida, A. S., and A. G. Journel (1994). Joint simulation of multiple variables with a Markov-type coregionalization model. *Mathematical Geology*, **26(5)**, 565–588.

Alqassab, H. M. (1999). Constraining permeability field to engineering data: An innovative approach in reservoir characterization. *Saudi Aramco Journal of Technology*, Fall 1999, 43–53.

Aly, A., W. J. Lee, A. Datta-Gupta, K. Mowafi, M. Prida, and M. Latif (1999). Application of geostatistical modeling in an integrated reservoir simulation study of the Lower Bahariya reservoir, Egypt. *SPE paper 53118-MS*, Middle East Oil Show and Conference, Bahrain.

Andersson, J., A. M. Shapiro, and J. Bear (1984). A stochastic model of a fractured rock conditioned by measured information. *Water Resources Research*, **20(1)**, 79–88.

Andrieu, C., A. Jasra, A. Doucet, and P. Del Moral (2011). On nonlinear Markov chain Monte Carlo. *Bernoulli*, **17(3)**, 987–1014.

Angerer, E., P. Lanfranchi, and S. Rogers (2003). Integrated fracture characterization from seismic processing to reservoir modeling. In *Proceedings of the 73rd SEG International Exposition & Annual Meeting*, Dallas, Texas, pp. 43–46.

Antoine, J. N., and J. P. Delhomme (1993). A method to derive dips from bedding boundaries in boreholes images. *SPE Formation Evaluation*, June, pp. 96–102.

Arfken, G. (1985). *Mathematical Methods for Physicists*, 3rd ed. Academic Press, Orlando, FL.

Armstrong, M. (1994). Is research in mining geostats as dead as a dodo? In *Geostatistics for the Next Century*, R. Dimitrakopoulos, ed. Kluwer, Dordrecht, pp. 303–312.

Armstrong, M., and P. Delfiner (1980). Towards a more robust variogram: A case study on coal. Technical Report N-671, Centre de Géostatistique, Fontainebleau, France.

Armstrong, M., and R. Jabin (1981). Variogram models must be positive-definite. *Journal of the International Association for Mathematical Geology*, **13(5)**, 455–459.

Armstrong, M., and G. Matheron (1986a). Disjunctive kriging revisited: Part I. *Mathematical Geology*, **18(8)**, 711–728.

Armstrong, M., and G. Matheron (1986b). Disjunctive kriging revisited: Part II. *Mathematical Geology*, **18(8)**, 729–742.

Armstrong, M., and G. Matheron (1986c). Isofactorial models for granulodensimetric data. *Mathematical Geology*, **18(8)**, 743–757.

Armstrong, M., and H. Wackernagel (1988). The influence of the covariance function on the kriged estimator. *Sciences de la Terre, Série Informatique Géologique*, **27(II)**, 245–262.

Armstrong, M., A. G. Galli, H. Beucher, G. Le Loc'h, D. Renard, B. Doligez, R. Eschard, and F. Geffroy (2011). *Plurigaussian Simulations in Geosciences*, 2nd ed. Springer, Berlin. First edition (2003).

Arnold, L. (1973). *Stochastische Differentialgleichungen*. Oldenburg, München, Germany. Translation (1974): *Stochastic Differential Equations: Theory and Applications*. John Wiley & Sons, New York.

Arpat, B., and J. Caers (2007). Conditional simulations with patterns. *Mathematical Geology*, **39(2)**, 177–203.

Ashley, F. (2005). Limitations of deterministic and advantages of stochastic seismic inversion. *CSEG Recorder*, **30(2)**, 5–11.

Atkinson, A. B. (1970). On the measurement of inequality. *Journal of Economic Theory*, **2(3)**, 244–263.

Aug, C. (2004). *Modélisation géologique 3D et caractérisation des incertitudes par la méthode du champ de potentiel*. Doctoral Thesis, E.N.S. des Mines de Paris.

Averkov, G., and G. Bianchi (2009). Confirmation of Matheron's conjecture on the covariogram of a planar convex body. *Journal of the European Mathematical Society*, **11(6)**, 1187–1202.

Ayyub, B. M. (2001). *Elicitation of Expert Opinions for Uncertainty and Risks*. CRC Press, Boca Raton, FL.

Bachrach, R. (2006). Joint estimation of porosity and saturation using stochastic rock-physics modeling. *Geophysics*, **71(5)**, O53–O63.

Baddeley, A. J., and B. W. Silverman (1984). A cautionary example on the use of second-order methods for analyzing point patterns. *Biometrics*, **40**, 1089–1093.

Baecher, G. B., N. A. Lanney, and H. H. Einstein (1977). Statistical description of rock properties and sampling. In *Proceedings of the 18th U.S. Symposium on Rock Mechanics*, AIME, New York, 5C1, pp. 1–8.

Baghdadi, N., S. Cavelier, J. P. Chilès, B. Bourgine, T. Toutin, C. King, P. Daniels, J. Perrin, and C. Truffert (2005). Merging of airborne elevation data and Radarsat data to develop a digital elevation model. *International Journal of Remote Sensing*, **26(1)**, 141–166.

Bailly, J. S., P. Monestiez, and P. Lagacherie (2006). Modelling spatial variability along drainage networks with geostatistics. *Mathematical Geosciences*, **38(5)**, 515–539.

Banerjee, S., A. E. Gelfand, A. O. Finley, and H. Sang (2008). Gaussian predictive process models for large spatial data sets. *Journal of the Royal Statistical Society, Series B*, **70(4)**, 825–848.

Bardossy, A., I. Bogardi, and W. E. Kelly (1990). Kriging with imprecise (fuzzy) variograms. I: Theory. II: Application. *Mathematical Geology*, **22(1)**, 63–79 (I), 81–94 (II).

Barker, A. A. (1965). Monte Carlo calculation of the radial distribution function for a proton-electron plasma. *Australian Journal of Physics*, **18(2)**, 119–133.

Barnes, R. J., and T. B. Johnson (1984). Positive kriging. In *Geostatistics for Natural Resources Characterization*, Part 1, G. Verly, M. David, A. G. Journel, and A. Maréchal, eds. Reidel, Dordrecht, pp. 231–244.

Bartlett, M.S. (1963). The spectral analysis of point processes. *Journal of the Royal Statistical Society, Series B*, **25(2)**, 264–296.

Bartlett, M.S. (1975). *The Statistical Analysis of Spatial Pattern*. Chapman & Hall, London.

Bastin, G., B. Lorent, C. Duque, and M. Gevers (1984). Optimal estimation of the average areal rainfall and optimal selection of rain gauge locations. *Water Resources Research*, **20(4)**, 463–470.

Baume, O., B. Gauvreau, M. Bérangier, F. Junker, H. Wackernagel, and J. P. Chilès (2009). Geostatistical modeling of sound propagation: Principles and a field application experiment. *Journal of the Acoustical Society of America*, **126(6)**, 2894–2904.

Beckman, P. (1973). *Orthogonal Polynomials for Engineers and Physicists*. Golem Press, Boulder, CO.

Beers, W. C. M. van, and J. P. C. Kleijnen (2004). Kriging interpolation in simulation: A survey. In *Proceedings of the 2004 Winter Simulation Conference*, R. G. Ingalls, M. D. Rossetti, J. S. Smith, and B. A. Peters, eds. IEEE, Piscataway, NJ, pp. 113–121.

Beirlant, J., Y. Goegebeur, J. Segers, and J. Teugels (2004). *Statistics of Extremes. Theory and Applications*. Wiley, Chichester, England.

Benzécri, J. P., et al. (1973). *L'analyse des données*. Tome 1, *La taxinomie*; Tome 2, *L'analyse des correspondances*. Dunod, Paris.

Bertino, L., G. Evensen, and H. Wackernagel (2002). Combining geostatistics and Kalman filtering for data assimilation in an estuarine system. *Inverse Problems*, **18(1)**, 1–23.

Bertino, L., G. Evensen, and H. Wackernagel (2003). Sequential data assimilation techniques in oceanography. *International Statistical Review*, **71(2)**, 223–241.

Besag, J. (1974). Spatial interaction and the statistical analysis of lattice systems (with discussion). *Journal of the Royal Statistical Society, Series B*, **36(2)**, 192–236.

Besag, J. (1975). Statistical analysis of non-lattice data. *Journal of the Royal Statistical Society, Series D*, **24(3)**, 179–195.

Besag, J., and D. Mondal (2005). First-order intrinsic autoregressions and the de Wijs process. *Biometrika*, **92(4)**, 909–920.

Beucher, H., M. Benito García-Morales, and F. Geffroy (2005). Fitting the Boolean parameters in a non-stationary case. In *Geostatistics Banff 2004*, Vol. 2, O. Leuangthong and C. V. Deutsch, eds. Springer, Dordrecht, pp. 569–578.

Bez, N. (2002). Global fish abundance estimation from regular sampling: The geostatistical transitive method. *Canadian Journal of Fisheries and Aquatic Sciences*, **59(12)**, 1921–1931.

Billaux, D. (1989). Influence de la connectivité d'un réseau de fractures sur sa réponse à un essai de pompage. In *Proceedings of ISRM/SPE International Symposium Rock Mechanics and Rock Physics at Great Depth*, V. Maury and D. Fourmaintraux, eds. Balkema, Rotterdam, Netherlands, pp. 473–480.

Blanchin, R., and J. P. Chilès (1992). Geostatistical modelling of geological layers and optimization survey design for the Channel tunnel. In *Computer Graphics in Geology: Three-Dimensional Computer Graphics in Modeling Geologic Structures and Simulating Geologic Processes*, R. Pflug and J. W. Harbaugh, eds. Springer-Verlag, Berlin, pp. 251–256.

Blanchin, R., and J. P. Chilès (1993a). The Channel Tunnel: Geostatistical prediction of the geological conditions and its validation by the reality. *Mathematical Geology*, **25(7)**, 963–974.

Blanchin, R., and J. P. Chilès (1993b). Channel Tunnel: Geostatistical prediction facing the ordeal of reality. In *Geostatistics Tróia '92*, Vol. 2, A. Soares, ed. Kluwer, Dordrecht, pp. 757–766.

Blanchin, R., J. P. Chilès, and F. Deverly (1989). Some applications of geostatistics to civil engineering. In *Geostatistics*, Vol. 2, M. Armstrong, ed. Kluwer, Dordrecht, pp. 785–795.

Blanc-Lapierre, A., and R. Fortet (1953). *Théorie des fonctions aléatoires*. Masson, Paris.

Boisvert, J. B., J. G. Manchuk, and C. V. Deutsch (2009). Kriging in the presence of locally varying anisotropy using non-euclidean distances. *Mathematical Geosciences*, **41(5)**, 585–601.

Bordessoule, J. L., C. Demange, and J. Rivoirard (1988). Estimation des réserves géologiques en uranium par variables utiles à résidu autokrigeable. *Sciences de la Terre, Série Informatique Géologique*, **28**, 27–51.

Bordessoule, J. L., C. Demange, and J. Rivoirard (1989). Using an orthogonal residual between ore and metal to estimate in-situ resources. In *Geostatistics*, Vol. 2, M. Armstrong, ed. Kluwer, Dordrecht, pp. 923–934.

Borgman, L., M. Taheri, and R. Hagan (1984). Three-dimensional frequency-domain simulation of geological variables. In *Geostatistics for Natural Resources Characterization*, Part 1, G. Verly, M. David, A. G. Journel, and A. Maréchal, eds. Reidel, Dordrecht, pp. 517–541.

Bornard, R., F. Allo, T. Coléou, Y.Freudenreich, D. H. Caldwell, and J. G. Hamman (2005). Petrophysical seismic inversion to determine more accurate and precise reservoir properties. *SPE Paper* 94144-MS, SPE Europec/EAGE Annual Conference, Madrid, Spain, pp. 2242–2246.

Bortoli, L. J., F. Alabert, A. Haas, and A. G. Journel (1993). Constraining stochastic images to seismic data. In *Geostatistics Tróia '92*, Vol. 1, A. Soares, ed. Kluwer, Dordrecht, pp. 325–337.

Bosch, M., T. Mukerji, and E. F. Gonzalez (2010). Seismic inversion for reservoir properties combining statistical rock physics and geostatistics: A review. *Geophysics*, **75(5)**, 75A165–75A176.

Boulanger, F. (1990). *Modélisation et simulation de variables régionalisées, par des fonctions aléatoires stables*. Doctoral thesis, E.N.S. des Mines de Paris.

Bourgault, G., and D. Marcotte (1993). Spatial filtering under the linear coregionalization model. In *Geostatistics Tróia '92*, Vol. 1, A. Soares, ed. Kluwer, Dordrecht, pp. 237–248.

Box, G. E. P., and G. M. Jenkins (1976). *Time Series Analysis: Forecasting and Control*, revised edition, Holden-Day, Oakland, CA. First edition (1970).

Bradley, R., and J. Haslett (1992). High-interaction diagnostics for geostatistical models of spatially referenced data. *The Statistician*, **41(3)**, 371–380.

Bras, R. L., and I. Rodríguez-Iturbe (1985). *Random Functions and Hydrology*. Addison-Wesley, Reading, MA.

Brochu, Y., and D. Marcotte (2003). A simple approach to account for radial flow and boundary conditions when kriging hydraulic head fields for confined aquifers. *Mathematical Geology*, **35(2)**, 111–139.

Brooker, P. I. (1985). Two-dimensional simulation by turning bands. *Journal of the International Association for Mathematical Geology*, **17(1)**, 81–90.

Buhmann, M. D. (2003). *Radial Basis Functions: Theory and Implementations*. Cambridge University Press, New York.

Buland, A., and H. Omre (2003a). Bayesian linearized AVO inversion. *Geophysics*, **68(1)**, 185–198.

Buland, A., and H. Omre (2003b). Bayesian wavelet estimation from seismic and well data. *Geophysics*, **68(6)**, 2000–2009.

Buland, A., and H. Omre (2003c). Joint AVO inversion, wavelet estimation and noise-level estimation using a spatially coupled hierarchical Bayesian model. *Geophysical Prospecting*, **51(6)**, 531–550.

Buland, A., O. Kolbjørnsen, and H. Omre (2003). Rapid spatially coupled AVO inversion in the Fourier domain. *Geophysics*, **68(3)**, 824–836.

Buland, A., O. Kolbjørnsen, R. Hauge, O. Skjæveland, and K. Duffaut (2008). Bayesian lithology and fluid prediction from seismic prestack data. *Geophysics*, **73(3)**, C13–C21.

Burgers, G., P. J. van Leeuwen, and G. Evensen (1998). Analysis scheme in the ensemble Kalman filter. *Monthly Weather Review*, **126(6)**, 1719–1724.

Buse, A. (1973). Goodness of fit in generalized least squares estimation. *The American Statistician*, **27(3)**, 106–108.

Calcagno, P., J. P. Chilès, G. Courrioux, and A. Guillen (2008). Geological modelling from field data and geological knowledge. Part I. Modelling method coupling 3D potential-field interpolation and geological rules. *Physics of the Earth and Planetary Interiors*, **171(1–4)**, 147–157.

Caldwell, D. H., and J. G. Hamman (2004). IOI—A method for fine-scale, quantitative description of reservoir properties from seismic. Paper B027, EAGE 66th Conference and Exhibition, Paris.

Carlier, A. (1964). *Contribution aux méthodes d'estimation des gisements d'uranium*. Doctoral thesis, Université de Paris, Série A, No. 4226, No. d'ordre 5077.

Carr, J. R., and E. D. Deng (1987). Comparison of two techniques for applying disjunctive kriging: the Gaussian anamorphosis model versus the direct statistical inference of the bivariate distributions. *Mathematical Geology*, **19(1)**, 57–68.

Carr, J. R., D. E. Myers, and C. E. Glass (1985). Co-kriging—A computer program. *Computers & Geosciences*, **11(2)**, 111–127.

Carrera, J., A. Alcolea, A. Medina, J. Hidalgo, and L. J. Slooten (2005). Inverse problem in hydrogeology. *Hydrogeology Journal*, **13(1)**, 206–222.

Casella, G., and E. I. George (1992). Explaining the Gibbs sampler. *The American Statistician*, **46(3)**, 167–174.

Castaing, C., M. A. Halawani, F. Gervais, J. P. Chilès, A. Genter, B. Bourgine, G. Ouillon, J. M. Brosse, P. Martin, A. Genna, and D. Janjou (1996). Scaling relationships in intraplate fracture systems related to Red Sea rifting. *Tectonophysics*, **261(4)**, 291–314.

Champigny, N., and M. Armstrong (1989). Estimation of fault-controlled deposits. In *Geostatistics*, Vol. 2, M. Armstrong, ed. Kluwer, Dordrecht, pp. 971–983.

Chautru, J. M. (1989). The use of Boolean random functions in geostatistics. In *Geostatistics*, Vol. 1, M. Armstrong, ed. Kluwer, Dordrecht, pp. 201–212.

Chauvet, P. (1982). The variogram cloud. In *Proceedings of the 17th APCOM International Symposium, Golden, Colorado*, pp. 757–764.

Chauvet, P. (1987). *Eléments d'analyse structurale des FAI−k à 1 dimension.* Doctoral thesis, E.N.S. des Mines de Paris.

Chauvet, P., J. Pailleux, and J. P. Chilès (1976). Analyse objective des champs météorologiques par cokrigeage. *La Météorologie, Sciences et Techniques, 6ème Série*, **4**, 37–54.

Chentsov, N. N. (1957). Lévy Brownian motion for several parameters and generalized white noise. *Theory of Probability and Its Applications*, **2(2)**, 265–266.

Chessa, A. (1995). *Conditional Simulation of Spatial Stochastic Models for Reservoir Heterogeneity.* Doctoral thesis, Technische Universiteit Delft, Holland.

Chib, S., and E. Greenberg (1995). Understanding the Metropolis-Hastings algorithm. *The American Statistician*, **49(4)**, 327–335.

Chihara, T. S. (1978). *An Introduction to Orthogonal Polynomials.* Gordon and Breach, New York.

Chilès, J. P. (1974). Analyse classique et analyse en FAI−k de données topographiques de la région de Noirétable. Technical Report N-394, Centre de Géostatistique, Fontainebleau, France.

Chilès, J. P. (1976). How to adapt kriging to non-classical problems: Three case studies. In *Advanced Geostatistics in the Mining Industry*, M. Guarascio, M. David, and C. Huijbregts, eds. Reidel, Dordrecht, pp. 69–89.

Chilès, J. P. (1977). *Géostatistique des phénomènes non stationnaires (dans le plan).* Doctoral thesis, Université de Nancy-I, France.

Chilès, J. P. (1978). L'inférence statistique automatique des FAI−k. Technical Report N-584, Centre de Géostatistique, Fontainebleau, France.

Chilès, J. P. (1979a). La dérive à la dérive. Technical Report N-591, Centre de Géostatistique, Fontainebleau, France.

Chilès, J. P. (1979b). Le variogramme généralisé. Technical Report N-612, Centre de Géostatistique, Fontainebleau, France.

Chilès, J. P. (1984). Simulation of a nickel deposit: Problems encountered and practical solutions. In *Geostatistics for Natural Resources Characterization*, Part 2, G. Verly, M. David, A. G. Journel, and A. Maréchal, eds. Reidel, Dordrecht, pp. 1015–1030.

Chilès, J. P. (1989a). Three-dimensional geometric modeling of a fracture network. In *Geostatistical, Sensitivity, and Uncertainty Methods for Ground-Water Flow and Radionuclide Transport Modeling*, B. E. Buxton, ed. Battelle Press, Columbus, OH, pp. 361–385.

Chilès, J. P. (1989b). Modélisation géostatistique de réseaux de fractures. In *Geostatistics*, Vol. 1, M. Armstrong, ed. Kluwer, Dordrecht, pp. 57–76.

Chilès, J. P. (1995). Quelques méthodes de simulation de fonctions aléatoires intrinsèques. In *Compte Rendu des Journées de Géostatistique*, C. de Fouquet, ed. Cahiers de Géostatistique, Fasc. 5, Ecole des Mines de Paris, pp. 97–112.

Chilès, J. P. (2001). On the contribution of hydrogeology to advances in geostatistics. In *geoENV III – Geostatistics for Environmental Applications*, P. Monestiez, D. Allard, and R. Froidevaux, eds. Kluwer, Dordrecht, pp. 1–16.

Chilès, J. P. (2012). Validity range of the discrete Gaussian change-of-support model. Technical Report, Center of Geosciences and Geoengineering, MINES ParisTech, Fontainebleau, France.

Chilès, J. P., and P. Delfiner (1975). Reconstitution par krigeage de la surface topographique à partir de différents schémas d'échantillonnage photogrammétrique. *Bulletin de la Société Française de Photogrammétrie*, **57**, 42–50.

Chilès, J. P., and P. Delfiner (1996). Using FFT techniques for simulating Gaussian random fields. In *Proceedings of the Conference on Mining Geostatistics, Berg-en-dal, Kruger National Park, South Africa, 19–22 September 1994*, Geostatistical Association of South Africa, pp. 131–140.

Chilès, J. P., and P. Delfiner (1997). Discrete exact simulation by the Fourier method. In *Geostatistics Wollongong '96*, Vol. 1, E. Y. Baafi and N. A. Schofield, eds. Kluwer, Dordrecht, pp. 258–269.

Chilès, J. P., and R. Gable (1984). Three-dimensional modelling of a geothermal field. In *Geostatistics for Natural Resources Characterization*, Part 2, G. Verly, M. David, A. G. Journel, and A. Maréchal, eds. Reidel, Dordrecht, pp. 587–598.

Chilès, J. P., and S. Gentier (1993). Geostatistical modeling of a single fracture. In *Geostatistics Tróia '92*, Vol. 1, A. Soares, ed. Kluwer, Dordrecht, pp. 5–108.

Chilès, J. P., and A. Guillen (1984). Variogrammes et krigeages pour la gravimétrie et le magnétisme. *Sciences de la Terre, Série Informatique Géologique*, **20**, 455–468.

Chilès, J. P., and C. Lantuéjoul (2005). Prediction by conditional simulation. In *Space, Structure, and Randomness—Contributions in Honor of Georges Matheron in the Fields of Geostatistics, Random Sets, and Mathematical Morphology*, M. Bilodeau, F. Meyer, and M. Schmitt, eds. Springer, New York, pp. 39–68.

Chilès, J. P., and H. T. Liao (1993). Estimating the recoverable reserves of gold deposits: Comparison between disjunctive kriging and indicator kriging. In *Geostatistics Tróia '92*, Vol. 2, A. Soares, ed. Kluwer, Dordrecht, pp. 1053–1064.

Chilès, J. P., and G. de Marsily (1993). Stochastic models of fracture systems and their use in flow and transport modeling. In *Flow and Contaminant Transport in Fractured Rock*, J. Bear, C. F. Tsang, and G. de Marsily, eds. Academic Press, San Diego, CA, Chapter 4, pp. 169–236.

Chilès, J. P., R. Gable, and R. H. Morin (1991). Contribution de la géostatistique à l'étude de la thermicité des rides océaniques du Pacifique. In *Compte Rendu des Journées de Géostatistique*, C. de Fouquet, ed. Cahiers de Géostatistique, Fasc. 1, Ecole des Mines de Paris, pp. 51–61.

Chilès, J. P., F. Guérin, and D. Billaux (1992). 3D stochastic simulation of fracture network and flow at Stripa conditioned on observed fractures and calibrated on measured flow rates. In *Rock Mechanics*, J. R. Tillerson and W. R. Wawersik, eds. Balkema, Rotterdam, pp. 533–542.

Chilès, J. P., B. Bourgine, and I. Niandou (1996). Designing an additional sampling pattern to determine the variogram at short distances. In *Proceedings of the Conference on Mining Geostatistics, Berg-en-dal, Kruger National Park, South Africa, 19–22 September 1994*, Geostatistical Association of South Africa, pp. 200–212.

Chilès, J. P., B. Bourgine, C. Castaing, and A. Genter (2000). Stochastic modelling and simulation of fracture networks in petroleum and geothermal reservoirs. In *Geostatistics 2000 Cape Town*, Vol. 1, W. J. Kleingeld and D. G. Krige, eds. Geostatistical Association of Southern Africa, pp. 413–423.

Chilès, J. P., C. Aug, A. Guillen, and T. Lees (2005). Modelling the geometry of geological units and its uncertainty in 3D from structural data: The potential-field method. In *Orebody Modelling and Strategic Mine Planning–Uncertainty Management Models*, R. Dimitrakopoulos, ed. Spectrum Series 14, The Australasian Institute of Mining and Metallurgy, Carlton, Victoria, pp. 329–336.

Christakos, G. (1992). *Random Field Models in Earth Sciences*. Academic Press, San Diego, CA.

Christakos, G. (2000). *Modern Spatiotemporal Geostatistics*. Oxford University Press, New York.

Chugunova, T. L., and L. Y. Hu (2008). Multiple-point simulations constrained by continuous auxiliary data. *Mathematical Geology*, **40(2)**, 133–146.

Chung, H., and J. J. Alonso (2002). Design of a low-boom supersonic business jet using cokriging approximation models. *AIAA paper* 2002–5598, 9th AIAA/ISSMO Symposium on Multidisciplinary Analysis and Optimization Proceedings, American Institute of Aeronautics and Astronautics.

Clark, I. (1979). Does geostatistics work? In *Proceedings of the 16th APCOM International Symposium*, T. J. O'Neil, ed. Society of Mining Engineers of the AIME, New York, pp. 213–225.

Cohen, A. (1992). *Ondelettes et traitement numérique du signal*. Masson, Paris.

Cojan, I., O. Fouché, S. Lopez, and J. Rivoirard (2005). Process-based reservoir modelling in the example of meandering channel. In *Geostatistics Banff 2004*, Vol. 2, O. Leuangthong and C.V. Deutsch, eds. Springer, Dordrecht, pp. 611–620.

Coléou, T. (2002). Time-lapse filtering and improved repeatability with automatic factorial cokriging (AFACK). In *Extended Abstracts of the 64th EAGE Conference & Exhibition, Florence, Italy*, Paper A–18.

Coles, S., J. Heffernan, and J. Tawn (1999). Dependence measures for extreme value analyses. *Extremes*, **2(4)**, 339–365.

Cooley, D., P. Naveau, and P. Poncet (2006). Variograms for spatial max-stable random fields. In *Dependence in Probability and Statistics*, P. Bertail, P. Doukhan, and P. Soulier, eds. Springer, New York, pp. 373–390.

Courrioux, G., C. Lajaunie, J. P. Chilès, and J. Lazarre (1998). Foliation fields and 3D geological modeling. In *Proceedings of 3D Modeling of Natural Objects, A Challenge for 2000's*, E.N.S. de Géologie, Nancy, France.

Cox, D. R. (1955). Some statistical models connected with series of events. *Journal of the Royal Statistical Society, Series B*, **17(2)**, 129–164.

Cox, D. R., and D.V. Hinkley (1974). *Theoretical Statistics*. Chapman & Hall, London.

Crain, I. K. (1978). The Monte-Carlo simulation of random polygons. *Computers & Geosciences*, **4(2)**, 131–141.

Cramér, H. (1940). On the theory of stationary random processes. *Annals of Mathematics*, **41(1)**, 215–230.

Cramér, H. (1942). On harmonic analysis in certain functional spaces. *Arkiv för Matematik, Astronomi och Fysik (Uppsala)*, **28B(12)**, 1–7.

Cramér, H. (1945). *Mathematical Methods of Statistics*. Almqvist & Wiksells, Uppsala, Sweden. Also (1974): Princeton University Press, Princeton, NJ.

Craven, P., and G. Wahba (1978). Smoothing noisy data with spline functions: Estimating the correct degree of smoothing by the method of generalized cross-validation. *Numerische Mathematik*, **31(4)**, 377–403.

Cressie, N. (1985). Fitting variogram models by weighted least squares. *Journal of the International Association for Mathematical Geology*, **17(5)**, 563–586.

Cressie, N. (1987). A nonparametric view of generalized covariances for kriging. *Mathematical Geology*, **19(5)**, 425–449.

Cressie, N. (1991). *Statistics for Spatial Data*. John Wiley & Sons, New York. Reprinted (1993).

Cressie, N., and D. M. Hawkins (1980). Robust estimation of the variogram: I. *Journal of the Mathematical Association for Mathematical Geology*, **12(2)**, 115–125.

Cressie, N., and H. C. Huang (1999). Classes on nonseparable, spatio-temporal stationary covariance functions. *Journal of the American Statistical Association*, **94(448)**, 1330–1340.

Cressie, N., and G. Johannesson (2008). Fixed rank kriging for very large spatial data sets. *Journal of the Royal Statistical Society, Series B*, **70(1)**, 209–226.

Cressie, N., and J. Kornak (2003). Spatial statistics in the presence of location error with application to remote sensing of the environment. *Statistical Science*, **18(4)**, 436–456.

Cressie, N., and C. K. Wikle (2011). *Statistics for Spatio-Temporal Data*. John Wiley & Sons, New York.

Cressie, N., J. Frey, B. Harch, and M. Smith (2006). Spatial prediction on a river network. *Journal of Agricultural, Biological and Environmental Statistics*, **11(2)**, 127–150.

Cressman, G. P. (1959). An operational objective analysis system. *Monthly Weather Review*, **87(10)**, 367–374.

Crum, M. M. (1956). On positive-definite functions. *Proceedings of the London Mathematical Society, Third Series*, **S3–6(24)**, 548–560.

Dagbert, M., M. David, D. Crozel, and A. Desbarats (1984). Computing variograms in folded strata-controlled deposits. In *Geostatistics for Natural Resources Characterization*, Part 1, G. Verly, M. David, A. G. Journel, and A. Maréchal, eds. Reidel, Dordrecht, pp. 71–89.

Daly, C. (2001). Stochastic vector and tensor fields applied to strain modeling. *Petroleum Geoscience*, **7(S)**, S97–S104.

Daly, C. (2005). Higher order models using entropy, Markov random fields and sequential simulation. In *Geostatistics Banff 2004*, Vol. 1, O. Leuangthong and C. V. Deutsch, eds. Springer, Dordrecht, pp. 215–224.

Daly, C., and J. Caers (2010). Multi-point geostatistics—An introductory overview. *First Break*, **28(9)**, 39–47.

Daly, C., D. Jeulin, and C. Lajaunie (1989). Application of multivariate kriging to the processing of noisy images. In *Geostatistics*, Vol. 2, M. Armstrong, ed. Kluwer, Dordrecht, pp. 749–760.

Dantzig, G.B. (1963). *Linear Programming and Extensions*. Princeton University Press, Princeton, NJ.

Da Rold, C., J. Deraisme, and R. Dumay (1980). Etude méthodologique de la simulation d'une exploitation de nickel de Nouvelle-Calédonie sur un modèle géostatistique de gisement. *Revue de l'Industrie Minérale*, **62(12)**, 651–663.

Daubechies, I. (1992). *Ten Lectures on Wavelets*. Society for Industrial and Applied Mathematics, Philadelphia.

David, M. (1977). *Geostatistical Ore Reserve Estimation*. Elsevier, Amsterdam.

David, M. (1988). *Handbook of Applied Advanced Geostatistical Ore Reserve Estimation*. Elsevier, Amsterdam.

David, M., D. Marcotte, and M. Soulié (1984). Conditional bias in kriging and a suggested correction. In *Geostatistics for Natural Resources Characterization*, Part 1, G. Verly, M. David, A.G. Journel, and A. Maréchal, eds. Reidel, Dordrecht, pp. 217–230.

Davis, M. W. (1987). Production of conditional simulations via the LU triangular decomposition of the covariance matrix. *Mathematical Geology*, **19(2)**, 91–98.

Dehghani, K., K. A. Edwards, and P. M. Harris (1997). Modeling of waterflood in a vuggy carbonate reservoir. *SPE paper* 38910, SPE Annual Technical Conference and Exhibition, San Antonio, Texas.

Delfiner, P. (1970). Le schéma booléen-poissonien. Technical Report N-212, Centre de Morphologie Mathématique, Fontainebleau, France.

Delfiner, P. (1973). Analyse objective du géopotentiel et du vent géostrophique par krigeage universel. Technical Report No. 321, Etablissement d'Etudes et de Recherches Météorologiques, Direction de la Météorologie Nationale, Paris. New print (1975): *La Météorologie, 5e Série*, **25**, 1–58.

Delfiner, P. (1976). Linear estimation of nonstationary spatial phenomena. In *Advanced Geostatistics in the Mining Industry*, M. Guarascio, M. David, and C. Huijbregts, eds. Reidel, Dordrecht, pp. 49–68.

Delfiner, P. (1977). *Shift Invariance under Linear Models*. Ph.D. dissertation, Princeton University, Princeton, NJ.

Delfiner, P., and J. P. Chilès (1977). Conditional simulation: a new Monte-Carlo approach to probabilistic evaluation of hydrocarbon in place. *SPE paper* 6985. Also: Technical Report N-526, Centre de Géostatistique, Fontainebleau, France.

Delfiner, P., and J. P. Delhomme (1975). Optimum interpolation by kriging. In *Display and Analysis of Spatial Data*, J. C. Davis and M. J. McCullagh, eds. John Wiley & Sons, London, pp. 96–114.

Delfiner, P., and R. O. Gilbert (1978). Combining two types of survey data for estimating geographical distribution of plutonium in Area 13. In *Selected*

Environmental Plutonium Research Reports of the NAEG, Vol. 2, NVO-192. Department of Energy, NTIS, Springfield, Virginia, pp. 361–404.

Delfiner, P., J. P. Delhomme, and J. Pélissier-Combescure (1983). Application of geostatistical analysis to the evaluation of petroleum reservoirs with well logs. In *Proceedings of the SPWLA 24th Annual Logging Symposium, Calgary.*

Delhomme, A. E. K., and J. F. Giannesini (1979). New reservoir description technics improve simulation results in Hassi-Messaoud field—Algeria. *SPE paper* 8435, 54th Annual Fall Technical Conference and Exhibition of the Society of Petroleum Engineers, Las Vegas, Nevada.

Delhomme, J. P. (1976). *Applications de la théorie des variables régionalisées dans les sciences de l'eau.* Doctoral thesis, Université Pierre & Marie Curie—Paris VI. Simplified text and abstract (1978, same title): *Bulletin du B.R.G.M. (deuxième série)*, Section III, No. 4, 341–375.

Delhomme, J. P. (1978). Kriging in the hydrosciences. *Advances in Water Resources*, **1(5)**, 251–266.

Delhomme, J. P. (1979). Kriging under boundary conditions. Presented at the American Geophysical Union Fall Meeting, San Francisco.

Delhomme, J. P., M. Boucher, G. Meunier, and F. Jensen (1981). Apport de la géostatistique à la description des stockages de gaz en aquifère. *Revue de l'Institut Français du Pétrole*, **36(3)**, 309–327.

Demange, C., C. Lajaunie, C. Lantuéjoul, and J. Rivoirard (1987). Global recoverable reserves: Testing various changes of support models on uranium data. In *Geostatistical Case Studies*, G. Matheron and M. Armstrong, eds. Reidel, Dordrecht, pp. 135–147.

Deraisme, J., and P. van Deursen (1995). Geostatistical modeling of multiple layers and its application to volumetric computation of dredging projects. In *Applications of Statistics and Probability—Civil Engineering Reliability and Risk Analysis*, Vol. 2, M. Lemaire, J. L. Favre, and A. Mébarki, eds. Balkema, Rotterdam, pp. 1221–1227.

Deraisme, J., and R. Dumay (1979). Geostatistics and mining processes. In *Proceedings of the 16th International APCOM Symposium*, T. J. O'Neil, ed. Society of Mining Engineers of the AIME, New York, pp. 149–162.

Deraisme, J., and J. Rivoirard (2009). Histogram modelling and simulations in the case of skewed distributions with a 0-effect: Issues and new developments. In *Proceedings of the IAMG 2009 Conference, Stanford, California*, IAMG.

Deraisme, J., C. de Fouquet, and H. Fraisse (1983). Méthodologie de simulation d'exploitation souterraine dans les mines métalliques—Application à un gisement d'uranium en Australie. *Industrie Minérale—Les Techniques*, **83(10)**, 513–523.

Deraisme, J., C. de Fouquet, and H. Fraisse (1984). Geostatistical orebody model for computer optimization of profits from different underground mining methods. In *Proceedings of the 18th International APCOM Symposium*, Institution of Mining and Metallurgy, London, pp. 583–590.

Deraisme, J., J. Rivoirard, and P. Carrasco-Castelli (2008). Multivariate uniform conditioning and block simulations with discrete Gaussian model: Application to Chuquicamata deposit. In *Geostats 2008*, Vol. 1, J. M. Ortiz and X. Emery, eds. Gecamin, Santiago, Chile, pp. 69–78.

Desassis, N., and D. Renard (2011). Automatic variogram modeling by iterative least squares—Univariate and multivariate cases. Technical Report, Center of Geosciences and Geoengineering, MINES ParisTech, Fontainebleau, France.

Desbarats, A. J., and R. Dimitrakopoulos (2000). Geostatistical simulation of regionalized pore-size distributions using min/max autocorrelation factors. *Mathematical Geology*, **32(8)**, 919–942.

Deutsch, C.V. (1993). Conditioning reservoir models to well test information. In *Geostatistics Tróia '92*, Vol. 1, A. Soares, ed. Kluwer, Dordrecht, pp. 505–518.

Deutsch, C.V. (2002). *Geostatistical Reservoir Modeling*. Oxford University Press, New York.

Deutsch, C.V., and P. W. Cockerham (1994). Practical considerations in the application of simulated annealing to stochastic simulation. *Mathematical Geology*, **26(1)**, 67–82.

Deutsch, C. V., and A. G. Journel (1992). *GSLIB: Geostatistical Software Library and User's Guide*. Oxford University Press, New York. Second edition (1998).

Deverly, F. (1984a). Geostatistical approach to mining sampling. In *Proceedings of the 18th APCOM International Symposium*, Institution of Mining and Metallurgy, London, pp. 379–388.

Deverly, F. (1984b). *Echantillonnage et géostatistique*. Doctoral thesis, E.N.S. des Mines de Paris.

Devroye, L. (1986). *Non-Uniform Random Variate Generation*. Springer, New York.

Diamond, P., and M. Armstrong (1984). Robustness of variograms and conditioning of kriging matrices. *Journal of the Mathematical Association for Mathematical Geology*, **16(8)**, 809–822.

Dietrich, C. R., and G.N. Newsam (1993). A fast and exact method for multidimensional Gaussian stochastic simulations. *Water Resources Research*, **29(8)**, 2861–2869.

Dietrich, C. R., and G. N. Newsam (1996). Fast and exact method for multidimensional Gaussian stochastic simulations: extension to conditional simulations. *Water Resources Research*, **32(6)**, 1643–1652.

Dietrich, C. R., and G. N. Newsam (1997). Fast and exact simulation of stationary Gaussian processes through circulant embedding of the covariance matrix. *SIAM Journal on Scientific Computing*, **18(4)**, 1088–1107.

Diggle, P. J., and P. J. Ribeiro, Jr. (2007). *Model-based Geostatistics*. Springer.

Dimitrakopoulos, R. (2011). Strategic mine planning under uncertainty—Stochastic optimization for strategic mine planning: A decade of developments. *Journal of Mining Science*, **47(2)**, 138–150.

Dimitrakopoulos, R., and X. Luo (1997). Joint space-time modeling in the presence of trends. In *Geostatistics Wollongong '96*, Vol. 1, E.Y. Baafi and N.A. Schofield, eds. Kluwer, Dordrecht, pp. 138–149.

Dimitrakopoulos, R., L. Martinez, and S. Ramazan (2005). Optimising open pit design with simulated orebodies and Whittle Four-X—A maximum upside/minimum downside approach. In *Orebody Modelling and Strategic Mine Planning—Uncertainty and Risk Management Models*, R. Dimitrakopoulos, ed. Spectrum Series 14, The Australasian Institute of Mining and Metallurgy, Carlton, Victoria, pp. 181–186.

Dolloff, J., B. Lofy, A. Sussman, and C. Taylor (2006). Strictly positive definite correlation functions. In *Signal Processing, Sensor Fusion, and Target Recognition XV*, I. Kadar, ed. Proceedings of SPIE, Vol. 6235, pp. 1–18.

Dong, A. (1990). *Estimation géostatistique des phénomènes régis par des équations aux dérivées partielles*. Doctoral thesis, E.N.S. des Mines de Paris.

Doob, J. L. (1953). *Stochastic Processes*. John Wiley & Sons, New York. Reprinted (1990).

Dousset, P. E., and L. Sandjivy (1987). Analyse krigeante des données géochimiques multivariables prélevées sur un site stannifère en Malaisie. *Sciences de la Terre, Série Informatique Géologique*, **26**, 1–22.

Dowd, P. A. (1971). Applications of geostatistics. Internal Report, Zinc Corporation, N. B.H.C., Broken Hill, Australia.

Dowd, P. A. (1982). Lognormal kriging—The general case. *Journal of the International Association for Mathematical Geology*, **14(5)**, 475–499.

Dowd, P. (1984). The variogram and kriging: robust and resistant estimators. In *Geostatistics for natural resources characterization*, G. Verly, M. David, A. G. Journel, and A. Maréchal, eds. Reidel, Dordrecht, Part 1, pp. 91–106.

Dowd, P. A. (1989). Generalized cross-covariances. In *Geostatistics*, Vol. 1, M. Armstrong, ed. Kluwer, Dordrecht, pp. 151–162.

Dowd, P. A., and A. G. Royle (1977). Geostatistical applications in the Athabasca tar sands. In *Proceedings of the 15th APCOM International Symposium*, Australasian Institute of Mining and Metallurgy, Parkville, Australia, pp. 235–242.

Doyen, P. M., D. E. Psaila, and S. Strandenes (1994). Bayesian sequential indicator simulation of channel sands from 3-D seismic data in the Oseberg Field, Norwegian North Sea. *SPE paper* 28382, 69th Annual Technical Conference and Exhibition of the Society of Petroleum Engineers, New Orleans, Louisiana, pp. 197–211.

Doyen, P. M., L. D. den Boer, and W. R. Pillet (1996). Seismic porosity mapping in the Ekofisk field using a new form of collocated cokriging. *SPE paper* 36498, Annual Technical Conference and Exhibition, Denver, CO.

Dubois, D. (2006). Possibility theory and statistical reasoning. *Computational Statistics & Data Analysis*, **51(1)**, 47–69.

Dubrule, O. (1981). *Krigeage et splines en cartographie automatique. Application à des exemples pétroliers*. Doctoral thesis, E.N.S. des Mines de Paris.

Dubrule, O. (1983a). Two methods with different objectives: splines and kriging. *Journal of the International Association for Mathematical Geology*, **15(2)**, 245–257.

Dubrule, O. (1983b). Cross validation of kriging in a unique neighborhood. *Journal of the International Association for Mathematical Geology*, **15(6)**, 687–699.

Dubrule, O. (1993). Introducing more geology in stochastic reservoir modelling. In *Geostatistics Tróia '92*, Vol. 1, A. Soares, ed. Kluwer, Dordrecht, pp. 351–369.

Dubrule, O., and C. Kostov (1986). An interpolation method taking into account inequality constraints: I. Methodology. *Mathematical Geology*, **18(1)**, 33–51.

Duchon, J. (1975). Fonctions splines de type plaque mince en dimension 2. *Séminaire d'Analyse Numérique*, No. 231, Université Scientifique et Médicale de Grenoble, France.

Duchon, J. (1976). Fonctions splines et espérances conditionelles de champs gaussiens. *Annales Scientifiques de l'Université de Clermont-Ferrand 2 (France), Série Mathématiques*, **61(14)**, 19–27.

Efron, B. (1979). Bootstrap methods: Another look at the jackknife. *The Annals of Statistics*, **7(1)**, 1–26.

Egozcue, J. J., V. Pawlowsky-Glahn, G. Mateu-Figueras, and C. Barceló-Vidal (2003). Isometric logratio transformations for compositional data analysis. *Mathematical Geology*, **35(3)**, 279–300.

Ehm, W., T. Gneiting, and D. Richards (2004). Convolution roots of radial positive definite functions with compact support. *Transactions of the American Mathematical Society*, **356(11)**, 4655–4685.

Eide, A. L., H. Omre, and B. Ursin (1997). Stochastic reservoir characterization conditioned on seismic data. In *Geostatistics Wollongong '96*, Vol. 1, E. Y. Baafi and N. A. Schofield, eds. Kluwer, Dordrecht, pp. 442–453.

Elliott, F. W., Jr., and A. J. Majda (1994). A wavelet Monte Carlo method for turbulent diffusion with many spatial scales. *Journal of Computational Physics*, **113(1)**, 82–111.

Elliott, F. W., Jr., D. J. Horntrop, and A. J. Majda (1997). A Fourier-wavelet Monte Carlo method for fractal random fields. *Journal of Computational Physics*, **132(2)**, 384–408.

Embrechts, P., C. Klüppelberg, and T. Mikosch (1997). *Modelling Extremal Events for Insurance and Finance*. Springer, Berlin.

Emery, X. (2002). Conditional simulation on nongaussian random functions. *Mathematical Geology*, **34(1)**, 79–100.

Emery, X. (2004a). Testing the correctness of the sequential algorithm for simulating Gaussian random fields. *Stochastic Environment Research and Risk Assessment*, **18(6)**, 401–413.

Emery, X. (2004b). Properties and limitations of sequential indicator simulation. *Stochastic Environment Research and Risk Assessment*, **18(6)**, 414–424.

Emery, X. (2005a). Simple and ordinary multigaussian kriging for estimating recoverable reserves. *Mathematical Geology*, **37(3)**, 295–319.

Emery, X. (2005b). Conditional simulation of random fields with bivariate gamma isofactorial distributions. *Mathematical Geology*, **37(4)**, 419–445.

Emery, X. (2005c). Geostatistical simulation of random fields with bivariate isofactorial distributions by adding mosaic models. *Stochastic Environmental Research and Risk Assessment*, **19(5)**, 348–360.

Emery, X. (2006a). A disjunctive kriging program for assessing point-support conditional distributions. *Computers & Geosciences*, **32(7)**, 965–983.

Emery, X. (2006b). Two ordinary kriging approaches to predicting block grade distributions. *Mathematical Geology*, **38(7)**, 801–819.

Emery, X. (2007a). On some consistency conditions for geostatistical change-of-support models. *Mathematical Geology*, **39(2)**, 205–223.

Emery, X. (2007b). Reducing fluctuations in the sample variogram. *Stochastic Environmental Research and Risk Assessment*, **21(4)**, 391–403.

Emery, X. (2007c). Conditioning simulations of Gaussian random fields by ordinary kriging. *Mathematical Geology*, **39(6)**, 607–623.

Emery, X. (2008). Change of support for estimating local block grade distributions. *Mathematical Geology*, **40(6)**, 671–688.

Emery, X. (2010a). Multi-Gaussian kriging and simulation in the presence of an uncertain mean value. *Stochastic Environment Research and Risk Assessment*, **24(2)**, 211–219.

Emery, X. (2010b). On the existence of mosaic and indicator random fields with spherical, circular, and triangular variograms. *Mathematical Geosciences*, **42(8)**, 969–984.

Emery, X., and C. Lantuéjoul (2006). TBSIM: A computer program for conditional simulation of three-dimensional Gaussian random fields via the turning bands method. *Computers & Geosciences*, **32(10)**, 1615–1628.

Emery, X., and C. Lantuéjoul (2008). A spectral approach to simulating intrinsic random fields with power and spline generalized covariances. *Computational Geosciences*, **12(1)**, 121–132.

Emery, X., and J. M. Ortiz (2005a). Internal consistency and inference of change-of-support isofactorial models. In *Geostatistics Banff 2004*, Vol. 2, O. Leuangthong and C.V. Deutsch, eds. Springer, Dordrecht, pp. 1057–1066.

Emery, X., and J. M. Ortiz (2005b). Histogram and variogram inference in the multigaussian model. *Stochastic Environmental Research and Risk Assessment*, **19(1)**, 48–58.

Emery, X., and J. M. Ortiz (2007). Weighted sample variograms as a tool to better assess the spatial variability of soil properties, *Geoderma*, **140(1–2)**, 81–89.

Emery, X., and J. M. Ortiz (2011). A comparison of random field models beyond bivariate distributions. *Mathematical Geosciences*, **43(2)**, 183–202.

Escobar, I., P. Williamson, A. Cherrett, P. M. Doyen, R. Bornard, R. Moyen, and T. Crozat (2006). Fast geostatistical stochastic inversion in a stratigraphic grid. In *Expanded Abstracts, 76th SEG Annual International Meeting*, pp. 2067–2071.

Evensen, G. (1994). Sequential data assimilation with a nonlinear quasi-geostrophic model using Monte Carlo methods to forecast error statistics. *Journal of Geophysical Research*, **99(C5)**, 10143–10162.

Evensen, G. (2009). *Data Assimilation—The Ensemble Kalman Filter*, 2nd ed. Springer, Berlin.

Evertsz, C. J. G., and B. B. Mandelbrot (1989). Multifractal measures. In *Chaos and Fractals*, H. O. Peitgen, H. Jürgens, and D. Saupe, eds. Springer, New York, pp. 921–953.

Falk, M., and R. Michel (2006). Testing for tail independence in extreme value models. *Annals of the Institute of Statistical Mathematics*, **58(2)**, 261–290.

Farmer, C. L. (1992). Numerical rocks. In *Mathematics of Oil Recovery*, P. R. King, ed. Oxford University Press, Oxford, England, pp. 437–447.

Feder, J. (1988). *Fractals*. Plenum Press, New York.

Feller, W. (1968). *An Introduction to Probability Theory and Its Applications*, Vol. 1, 3rd ed. John Wiley & Sons, New York. First edition (1950).

Feller, W. (1971). *An Introduction to Probability Theory and Its Applications*, Vol. 2, 2nd ed. John Wiley & Sons, New York. First edition (1966).

Felletti, F. (1999). *Quantificazione e previsione delle variazioni di facies in una successione torbiditica. Integrazione tra analisi geologica e geostatistica (Fm. di Castagnola, Bacino terziario piemontese)*. Doctoral thesis, Università degli Studi di Milano.

Fenton, G. A., and E. H. Vanmarcke (1990). Simulation of random fields via local average subdivision. *Journal of Engineering Mechanics*, **116(8)**, 1733–1749.

Fisher, R. A., and L. H. C. Tippett (1928). Limiting forms of the frequency distribution of the largest or smallest member of a sample. *Mathematical Proceedings of the Cambridge Philosophical Society*, **24(2)**, 180–190.

Flandrin, P. (1992). Wavelet analysis and synthesis of fractional Brownian motion. *IEEE Transactions on Information Theory*, **35(1)**, 197–199.

Flinn, P. A. (1974). Monte Carlo calculation of phase separation in a two-dimensional Ising system. *Journal of Statistical Physics*, **10(1)**, 89–97.

Formery, P. (1964). *Cours de géostatistique*. Ecole des Mines de Paris.

Fouquet, C. de, and C. Bernard-Michel (2006). Modèles géostatistiques de concentrations ou de débits le long des cours d'eau. *Comptes Rendus de l'Académie des Sciences, Géoscience*, **338(5)**, 307–318.

Francois-Bongarçon, D. (1998). Extensions to the demonstration of Gy's formula. *Exploration and Mining Geology*; **7(1–2)**, 149–154.

François-Bongarçon, D. (2004). Theory of sampling and geostatistics: An intimate link. *Chemometrics and Intelligent Laboratory Systems*, **74(1)**, 143–148.

Freulon, X. (1994). Conditional simulation of a Gaussian random vector with non linear and/or noisy observations. In *Geostatistical Simulations*, M. Armstrong and P. A. Dowd, eds. Kluwer, Dordrecht, pp. 57–71.

Freulon, X., and C. de Fouquet (1991). Pratique des bandes tournantes à 3D. In *Compte Rendu des Journées de Géostatistique*, C. de Fouquet, ed. Cahiers de Géostatistique, Fasc. 1, Ecole des Mines de Paris, pp. 101–117.

Freulon, X., and C. de Fouquet (1993). Conditioning a Gaussian model with inequalities. In *Geostatistics Tróia '92*, Vol. 1, A. Soares, ed. Kluwer, Dordrecht, pp. 201–212.

Friedman, A. (1975–1976). *Stochastic Differential Equations and Applications*, Vols. 1 and 2. Academic Press, New York.

Froidevaux, R. (1993). Probability field simulation. In *Geostatistics Tróia '92*, Vol. 1, A. Soares, ed. Kluwer, Dordrecht, pp. 73–83.

Fry, N. (1979). Random point distributions and strain measurements in rocks. *Tectonophysics*, **60(1–2)**, 89–105.

Fujiwara, M. (2008). Identifying interactions among salmon populations from observed dynamics. *Ecology*, **89(1)**, 4–11.

Furrer, R., M. G. Genton, and D. W. Nychka (2006). Covariance tapering for interpolation of large spatial datasets. *Journal of Computational and Graphical Statistics*, **15(3)**, 502–523.

Galerne, B. (2010). *Stochastic Image Models and Texture Synthesis*. Doctoral thesis, Ecole normale supérieure de Cachan, France.

Galli, A., F. Gerdil-Neuillet, and C. Dadou (1984). Factorial kriging analysis: a substitute to spectral analysis of magnetic data. In *Geostatistics for Natural Resources Characterization*, Part 1, G. Verly, M. David, A. G. Journel, and A. Maréchal, eds. Reidel, Dordrecht, pp. 543–557.

Galli, A., H. Beucher, G. Le Loc'h, B. Doligez, and Heresim Group (1994). The pros and cons of the truncated Gaussian method. In *Geostatistical Simulations*, M. Armstrong and P. A. Dowd, eds. Kluwer, Dordrecht, pp. 217–233.

Gandin, L. S. (1963). *Ob"ektivnyi analiz meteorologicheskikh polei*. Gidrometeologicheskoe Izdatel'stvo, Leningrad. Translation (1965): *Objective Analysis of Meteorological Fields*. Israel Program for Scientific Translations, Jerusalem.

Garreta, V., P. Monestiez, and J. M. Ver Hoef (2010). Spatial modelling and prediction on river networks: Up model, down model or hybrid? *Environmetrics*, **21(5)**, 439–456.

Gautschi, W. (1972). Error function and Fresnel integrals. In *Handbook of Mathematical Functions with Formulas, Graphs, and Mathematical Tables*, M. Abramowitz and I. A. Stegun, eds. John Wiley & Sons, New York, pp. 295–329.

Gel, Y., A. E. Raftery, and T. Gneiting (2004). Calibrated probabilistic mesoscale weather field forecasting: The geostatistical output perturbation method. *Journal of the American Statistical Association*, **99(467)**, 575–583.

Gel'fand, I. M., and G. E. Shilov (1958). *Obobchennie funktsii i deistviia nad nimi*. Editions Technico-Littéraire, Moscow. Translation (1964): *Generalized Functions*. Vol. 1: *Properties and Operations*. Academic Press, New York.

Gel'fand, I. M., and N. Y. Vilenkin (1961). *Nekotorye primenenia garmonitsheskovo analisa*. Editions de Littérature de Physique et de Mathématique, Moscow. Translation (1964): *Generalized Functions*, Vol. 4: *Applications of Harmonic Analysis*. Academic Press, New York.

Gel'fand, I. M., M. I. Graev, and N. Y. Vilenkin (1962). *Obobchennie funktsii, Vypusk 5: Integral'naya geometriya svyazannye s nei voprosy teorii predstavlenii*. Moscow. Translation (1966): *Generalized Functions*, Vol. 5: *Integral Geometry and Representation Theory*. Academic Press, New York.

Geman, S., and D. Geman (1984). Stochastic relaxation, Gibbs distributions and the Bayesian restoration of images. *IEEE Transactions on Pattern Analysis and Machine Intelligence*, **6(6)**, 721–741.

Genton, M. G. (1998a). Highly robust variogram estimation. *Mathematical Geology*, **30(2)**, 213–221.

Genton, M. G. (1998b). Variogram fitting by generalized least squares using an explicit formula for the covariance structure. *Mathematical Geology*, **30(4)**, 323–345.

Genton, M. G., and P. J. Rousseeuw (1995). The change-of-variance function of M-estimators of scale under general contamination. *Journal of Computational and Applied Mathematics*, **64(1–2)**, 69–80.

Georgsen, F., and H. Omre (1993). Combining fibre processes and Gaussian random functions for modelling fluvial reservoirs. In *Geostatistics Tróia '92*, Vol. 1, A. Soares, ed. Kluwer, Dordrecht, pp. 425–439.

Gikhman, I. I., and A.V. Skorokhod (1972). *Stochastic Differential Equations*. Springer, Berlin.

Gilbert, R. O. (1987). *Statistical Methods for Environmental Pollution Monitoring*. Van Nostrand Reinhold, New York.

Gini, C. (1921). Measurement of inequality of incomes. *The Economic Journal*, **31(121)**, 124–126.

Gnedenko, B. (1943). Sur la dimension limite du terme maximum d'une série aléatoire. *Annals of Mathematics*, **44(3)**, 423–453.

Gneiting, T. (1997). Normal scale mixtures and dual probability densities. *Journal of Statistical Computation and Simulation*, **59(4)**, 375–384.

Gneiting, T. (1998). Closed form solutions of the two-dimensional turning bands equation. *Mathematical Geology*, **30(4)**, 379–390.

Gneiting, T. (1999a). Isotropic correlation functions on d-dimensional balls. *Advances in Applied Probability*, **31(3)**, 625–631.

Gneiting, T. (1999b). Correlation functions for atmospheric data analysis. *Quaterly Journal of the Royal Meteorological Society, Part A*, **125(559)**, 2449–2464.

Gneiting, T. (2002). Nonseparable, stationary covariance functions for space–time data. *Journal of the American Statistical Association*, **97(458)**, 590–600.

Gneiting, T., and P. Guttorp (2010). Continuous-parameter spatio-temporal processes. In *Handbook of Spatial Statistics*, A. E. Gelfand, P. Diggle, M. Fuentes, and P. Guttorp, eds. Chapman & Hall, London, pp. 427–436.

Gneiting, T., and M. Schlather (2004). Stochastic models that separate fractal dimension and the Hurst effect. *SIAM Review*, **46(2)**, 269–282.

Gneiting, T., Z. Sasvári, and M. Schlather (2001). Analogies and correspondences between variograms and covariance functions. *Advances in Applied Probability*, **33(3)**, 617–630.

Gneiting, T., H. Ševčiková, D. B. Percival, M. Schlather, and Y. Jiang (2006). Fast and exact simulation of large Gaussian lattice systems in $\mathbb{R}^2$: Exploring the limits. *Journal of Computational and Graphical Statistics*, **15(3)**, 483–501.

Gneiting, T., M. G. Genton, and P. Guttorp (2007a). Geostatistical space-time models, stationarity, separability, and full symmetry. In *Statistical Methods for Spatio-Temporal Systems*, B. Finkenstädt, L. Held, and V. Isham, eds. Chapman & Hall/CRC, Boca Raton, FL, pp. 151–175.

Gneiting, T., F. Belabdaoui, and A. E. Raftery (2007b). Probabilistic forecasts, calibration and sharpness. *Journal of the Royal Statistical Society, Series B*, **69(2)**, 243–268.

Gneiting, T., W. Kleiber, and M. Schlather (2010). Matérn cross-covariance functions for multivariate random fields. *Journal of the American Statistical Association*, **105(491)**, 1167–1177.

Goldberger, A. S. (1962). Best linear unbiased prediction in the generalized regression model. *Journal of the American Statistical Association*, **57**, 369–375.

Golubov, B. I. (1981). On Abel–Poisson type and Riesz means. *Analysis Mathematica*, **7(3)**, 161–184.

Gómez-Hernández, J. J., and E. F. Cassiraga (1994). Theory and practice of sequential simulation. In *Geostatistical Simulations*, M. Armstrong and P. A. Dowd, eds. Kluwer, Dordrecht, pp. 111–124.

Gonçalves, M.A. (2001). Characterization of geochemical distributions using multi-fractal models. *Mathematical Geology*, **33(1)**, 41–61.

Gonzalez-Casanova, P., and R. Alvarez (1985). Splines in geophysics. *Geophysics*, **50(12)**, 2831–2848.

Goovaerts, P. (1994). Comparative performance of indicator algorithms for modeling conditional probability distribution functions. *Mathematical Geology*, **26(3)**, 389–411.

Goovaerts, P. (1997). *Geostatistics for Natural Resources Characterization*. Oxford University Press, New York.

Goovaerts, P., and P. Sonnet (1993). Study of spatial and temporal variations of hydrochemical variables using Factorial Kriging Analysis. In *Geostatistics Tróia '92*, Vol. 1, A. Soares, ed. Kluwer, Dordrecht, pp. 745–756.

Goovaerts, P., P. Sonnet, and A. Navarre (1993). Factorial kriging analysis of sprin-water contents in the Dyle River basin, Belgium. *Water Resources Research*, **29(7)**, 2115–2125.

Gorbachev, D. V. (2001). Extremum problem for periodic functions supported in a ball (Russian). *Matematicheskie Zametki*, **69(3)**, 346–352; translation in *Mathematical Notes*, **69(3–4)**, 313–319.

Goulard, M. (1989). Inference in a coregionalization model. In *Geostatistics*, Vol. 1, M. Armstrong, ed. Kluwer, Dordrecht, pp. 397–408.

Goulard, M., and M. Voltz (1992). Linear coregionalization model: Tools for estimation and choice of cross-variogram matrix. *Mathematical Geology*, **24(3)**, 269–286.

Gribov, A. and K. Krivoruchko (2004). Geostatistical mapping with continuous moving neighbourhood. *Mathematical Geology*, **36(2)**, 267–281.

Grzebyk, M. (1993). *Ajustement d'une corégionalisation stationnaire*. Doctoral thesis, E.N.S. des Mines de Paris.

Grzebyk, M., and H. Wackernagel (1994). Multivariate analysis and spatial/temporal scales: Real and complex models. In *Proceedings of XVIIth International Biometric Conference, Hamilton, Ontario*, Vol. 1, pp. 19–33.

Guarascio, M. (1976). Improving the uranium deposits estimations (the Novazza case). In *Advanced Geostatistics in the Mining Industry*, M. Guarascio, M. David, and C. Huijbregts, eds. Reidel, Dordrecht, pp. 351–367.

Guardiano, F. B., and R. M. Srivastava (1993). Multivariate geostatistics: Beyond bivariate moments. In *Geostatistics Tróia '92*, Vol. 1, A. Soares, ed. Kluwer, Dordrecht, pp. 133–144.

Guibal, D., and A. Remacre (1984). Local estimation of the recoverable reserves: Comparing various methods with the reality on a porphyry copper deposit. In *Geostatistics for Natural Resources Characterization*, Part 1, G. Verly, M. David, A. G. Journel, and A. Maréchal, eds. Reidel, Dordrecht, pp. 435–448.

Guttorp, P., and T. Gneiting (2006). Studies in the history of probability and statistics XLIX: On the Matérn correlation family. *Biometrika*, **93(4)**, 989–995.

Guyon, X. (1993). *Champs aléatoires sur un réseau—Modélisations, statistique et applications*. Masson, Paris. Translation (1995): *Random Fields on a Network—Modeling, Statistics and Applications*. Springer, New York.

Gy, P. (1975). *Théorie et pratique de l'échantillonnage des matières morcelées*. Editions Pierre Gy, Cannes, France.

Gy, P. (1979). *Sampling of Particulate Materials—Theory and Practice*. Elsevier, Amsterdam, Netherlands.

Haan, L. de, and A. Ferreira (2006). *Extreme Value Theory. An Introduction*. Springer, New York.

Haas, A. (1993). Simulation de réservoirs pétroliers par inversion géostatistique. In *Compte Rendu des Journées de Géostatistique*, C. de Fouquet, ed. Cahiers de Géostatistique, Fasc. 3, Ecole des Mines de Paris, pp. 87–99.

Haas, A., and O. Dubrule (1994). Geostatistical inversion—A sequential method of stochastic reservoir modeling constrained by seismic data. *First Break*, **12(11)**, 561–569.

Haas, A., and P. Formery (2002). Uncertainties in facies proportion estimation I. Theoretical framework: The Dirichlet distribution. *Mathematical Geology*, **34(6)**, 679–702.

Haas, A., and C. Jousselin (1976). Geostatistics in the petroleum industry. In *Advanced Geostatistics in the Mining Industry*, M. Guarascio, M. David, and C. Huijbregts, eds. Reidel, Dordrecht, pp. 333–347.

Haas, A., G. Matheron, and J. Serra (1967). Morphologie mathématique et granulométries en place, I; II. *Annales des Mines*, I: No. 11, 735–753; II: No. 12, 767–782.

Haas, A., P. Biver, and K. Altisen (1998). Constrained kriging and simulation of a well layering towards a geophysical thickness map. In *ECMOR VI Proceedings*, 6th European Conference on Mathematics and Oil Recovery, Peebles, Scottland, Paper C-31.

Hajek, B. (1988). Cooling schedules for optimal annealing. *Mathematics of Operations Research*, **13(2)**, 311–329.

Haldorsen, H. H. (1983). *Reservoir Characterization Procedures for Numerical Simulation*. Ph.D. dissertation, University of Texas, Austin.

Haldorsen, H. H., and L. W. Lake (1984). A new approach to shale management in field-scale models. *SPE Journal*, August 1984, 447–457.

Halmos, P. R. (1951). *Introduction to Hilbert Space and the Theory of Spectral Multiplicity*. Chelsea, New York.

Hammersley, J. M., and J. A. Nelder (1955). Sampling from an isotropic Gaussian process. *Mathematical Proceedings of the Cambridge Philosophical Society*, **51(4)**, 652–662.

Handcock, M. S. (1994). Measuring the uncertainty in kriging. In *Geostatistics for the Next Century*, R. Dimitrakopoulos, ed. Kluwer, Dordrecht, pp. 436–447.

Handcock, M. S., and J. R. Wallis (1994). An approach to statistical spatial-temporal modeling of meteorological fields (with discussion). *Journal of the American Statistical Association*, **89(426)**, 368–390.

Harding, A., S. Strebelle, M. Levy, J. Thorne, D. Xie, S. Leigh, and R. Preece (2005). Reservoir facies modelling: New advances in MPS. In *Geostatistics Banff 2004*, Vol. 2, O. Leuangthong and C. V. Deutsch, eds. Springer, Dordrecht, pp. 559–568.

Hartman, L., and O. Hössjer (2008). Fast kriging of large data sets with Gaussian Markov fields. *Computational Statistics & Data Analysis*, **52(5)**, 2331–2349.

Haslett, J. (1989). Geostatistical neighbourhoods and subset selection. In *Geostatistics*, Vol. 2, M. Armstrong, ed. Kluwer, Dordrecht, pp. 569–577.

Hastings, W. K. (1970). Monte Carlo sampling methods using Markov chains and their applications. *Biometrika*, **57(1)**, 97–109.

Hawkins, D. M., and N. Cressie (1984). Robust kriging—A proposal. *Journal of the International Association for Mathematical Geology*, **16(1)**, 3–18.

Hegstad, B. K., H. Omre, H. Tjelmeland, and K. Tyler (1994). Stochastic simulation and conditioning by annealing in reservoir description. In *Geostatistical Simulations*, M. Armstrong and P. A. Dowd, eds. Kluwer, Dordrecht, pp. 43–55.

Hepp, V., and A. C. Dumestre (1975). CLUSTER, a method for selecting the most probable dip results from dipmeter surveys. *SPE paper* 5543, 50th Annual Technical Conference and Exhibition of the Society of Petroleum Engineers, Dallas, Texas.

Hertog, D. den, J. P. C. Kleijnen, and A. Y. D. Siem (2006). The correct kriging variance estimated by bootstrapping. *Journal of the Operational Research Society*, **57(4)**, 400–409.

Herzfeld, U. C. (1989). Variography of submarine morphology: Problems of deregularization, and cartographical implications. *Mathematical Geology*, **21(7)**, 693–713.

Hewett, T. A., and R. A. Behrens (1988). Conditional simulation of reservoir heterogeneity with fractals. *SPE paper* 18326, 63rd Annual Technical Conference and Exhibition of the Society of Petroleum Engineers, Houston, Texas, pp. 645–660.

Higdon D. (2002). Space and space-time modeling using process convolutions. In *Quantitative Methods for Current Environmental Issues*, C. W. Anderson, V. Barnett, P. C. Chatwin, and A. H. El-Shaarawi, eds. Springer, London, pp. 37–54.

Higdon, D., J. Swall, and J. Kern (1999). Non-stationary spatial modeling. In *Bayesian Statistics 6: Proceedings of the Sixth Valencia International Meeting*, J. M. Bernardo, J. O. Berger, A. P. Dawid, and A. F.M. Smith, eds. Oxford University Press, New York, pp. 761–768.

Hochstrasser, U. W. (1972). Orthogonal polynomials. In *Handbook of Mathematical Functions with Formulas, Graphs, and Mathematical Tables*, M. Abramowitz and I. A. Stegun, eds. John Wiley & Sons, New York, pp. 771–802.

Høst, G., H. Omre, and P. Switzer (1995). Spatial interpolation errors for monitoring data. *Journal of the American Statistical Association*, **90(431)**, 853–861.

Hu, L. Y. (1988). *Mise en œuvre du modèle gamma pour l'estimation des distributions spatiales*. Doctoral thesis, E.N.S. des Mines de Paris.

Hu, L. Y. (2000a). Gradual deformation and iterative calibration of Gaussian-related stochastic models. *Mathematical Geology*, **32(1)**, 87–108.

Hu, L. Y. (2000b): Gradual deformation of non-Gaussian stochastic simulations. In *Geostatistics 2000 Cap Town*, Vol. 1, W. J. Kleingeld and D. G. Krige, eds. Geostatistical Association of Southern Africa, pp. 94–103.

Hu, L. Y. (2002). Combination of dependent realizations within the gradual deformation method. *Mathematical Geology*, **34(8)**, 953–963.

Hu, L. Y., and T. Chugunova (2008). Multiple-point geostatistics for modeling subsurface heterogeneity: A comprehensive review. *Water Resources Research*, **44**, W11413.

Hu, L. Y., and S. Jenni (2005). History matching of object-based stochastic reservoir models. *SPE paper* 81503-PA, *SPE Journal*, **10(3)**, 312–323.

Hu, L. Y., and C. Lantuéjoul (1988). Recherche d'une fonction d'anamorphose pour la mise en oeuvre du krigeage disjonctif isofactoriel gamma. *Sciences de la Terre, Série Informatique Géologique*, **28**, 145–173.

Hu, L.Y., and M. Le Ravalec-Dupin (2004). An improved gradual deformation method for reconciling random and gradient searches in stochastic optimizations. *Mathematical Geology*, **36(6)**, 703–719.

Huang, C., H. Zhang, and S. M. Robeson (2011). On the validity of covariance and variogram functions on the sphere. *Mathematical Geosciences*, **43(6)**, 721–733.

Huber, P. J. (1964). Robust estimation of a location parameter. *Annals of Mathematical Statistics*, **35(1)**, 73–101.

Huber, P. J. (1981). *Robust Statistics*. John Wiley & Sons, New York. (Republished in paperback, 2004.)

Huijbregts, C., and G. Matheron (1971). Universal kriging (an optimal method for estimating and contouring in trend surface analysis). In *9th International Symposium on Techniques for Decision-Making in the Mineral Industry*, J. I. McGerrigle, ed. Canadian Institute of Mining and Metallurgy, pp. 159–169.

Iba, Y. (2001). Extended ensemble Monte Carlo. *International Journal of Modern Physics C*, **12(5)**, 623–656.

Isaaks, E. H. (1984). Indicator simulation: Application to the simulation of a high grade uranium mineralization. In *Geostatistics for Natural Resources Characterization*,

Part 2, G. Verly, M. David, A. G. Journel, and A. Maréchal, eds. Reidel, Dordrecht, pp. 1057–1069.

Isaaks, E. H., and R. M. Srivastava (1989). *An Introduction to Applied Geostatistics.* Oxford University Press, New York.

Jackson, M., A. Maréchal, et al. (1979). Recoverable reserves estimated by disjunctive kriging: A case study. In *Proceedings of the 16th APCOM International Symposium*, T. J. O'Neil, ed. Society of Mining Engineers of the AIME, New York, pp. 240–249.

Jacod, J., and P. Joathon (1971). Use of random-genetic models in the study of sedimentary processes. *Journal of the International Association for Mathematical Geology*, **3(3)**, 219–233.

Jacod, J., and P. Joathon (1972). Conditional simulation of sedimentary cycles in three dimensions. In *Proceedings of the International Sedimentary Congress, Heidelberg, August 1971*, D. F. Merriam, ed. Plenum Press, New York, pp. 139–165.

Jaquet, O. (1989). Factorial kriging analysis applied to geological data from petroleum exploration. *Mathematical Geology*, **21(7)**, 683–691.

Jarrow, R., and A. Rudd (1983). *Options Pricing.* Irwin, Homewood, IL.

Jenkins, G. W., and D. G. Watts (1968). *Spectral Analysis and Its Applications.* Holden-Day, San Francisco, CA.

Jeulin, D. (1979). *Morphologie mathématique et propriétés physiques des agglomérés de minerai de fer et de coke métallurgique.* Doctoral thesis, E.N.S. des Mines de Paris.

Jeulin, D. (1989). Morphological modeling of images by sequential random functions. *Signal Processing*, **16(4)**, 403–431.

Jeulin, D. (1991). *Modèles morphologiques de structures aléatoires et de changement d'échelle.* Doctoral thesis, Université de Caen, France.

Jeulin, D. (1997). Dead leaves models: From space tessellation to random functions. In *Advances in Theory and Applications of Random Sets*, D. Jeulin, ed. World Scientific, Singapore, pp. 137–156.

Jeulin, D., and P. Jeulin (1981). Synthesis of rough surfaces by random morphological models. *Stereologica Iugoslavia*, **3(Suppl. 1)**, 239–246.

Jordan, M. I., ed. (1998). *Learning in Graphical Models.* Kluwer, Dordrecht.

Jordan, M. I. (2004). Graphical models. *Statistical Science*, **19(1)**, 140–155.

Jørgensen, B. (1982). *Statistical Properties of the Generalized Inverse Gaussian Distribution.* Springer, New York.

Journel, A. G. (1973). Simulation de gisements miniers. *Industrie Minérale—Mine*, **73(4)**, 221–226.

Journel, A. G. (1974a). Geostatistics for conditional simulation of orebodies. *Economic Geology*, **69(5)**, 673–687.

Journel, A. G. (1974b). *Simulations conditionnelles: Théorie et pratique.* Doctoral thesis, Université de Nancy-I, France.

Journel, A. G. (1980). The lognormal approach to predicting local distributions of selective mining unit grades. *Journal of the International Association for Mathematical Geology*, **12(4)**, 285–303.

Journel, A. G. (1982). The indicator approach to estimation of spatial distributions. In *Proceedings of the 17th APCOM International Symposium*, T. B. Johnson and R. J. Barnes, eds. Society of Mining Engineers of the AIME, New York, pp. 793–806.

Journel, A. G. (1983). Nonparametric estimation of spatial distributions. *Journal of the International Association for Mathematical Geology*, **15(3)**, 445–468.

Journel, A. G. (1984). The place of non-parametric geostatistics. In *Geostatistics for Natural Resources Characterization*, Part 1, G. Verly, M. David, A. G. Journel, and A. Maréchal, eds. Reidel, Dordrecht, pp. 307–335.

Journel, A. G. (1986). Constrained interpolation and qualitative information—The soft kriging approach. *Journal of the International Association for Mathematical Geology*, **18(3)**, 269–286.

Journel, A. G. (1989). Imaging of spatial uncertainty: A non-Gaussian approach. In *Geostatistical, Sensitivity, and Uncertainty Methods for Ground-Water Flow and Radionuclide Transport Modeling*, B. E. Buxton, ed. Battelle Press, Columbus, OH pp. 585–599.

Journel, A. G. (1999). Markov models for cross-covariances. *Mathematical Geology*, **31(8)**, 955–964.

Journel, A. G., and F. Alabert (1989). Non-Gaussian data expansion in the Earth Sciences. *Terra Nova*, **1(2)**, 123–134.

Journel, A. G., and C. V. Deutsch (1993). Entropy and spatial disorder. *Mathematical Geology*, **25(3)**, 329–355.

Journel, A. G., and J. J. Gómez-Hernández (1989). Stochastic imaging of the Wilmington clastic sequence. *SPE paper* 19857, 64th Annual Technical Conference and Exhibition of the Society of Petroleum Engineers, San Antonio, Texas, pp. 591–606.

Journel, A. G., and C. J. Huijbregts (1978). *Mining Geostatistics*. Academic Press, London.

Journel, A. G., and D. Posa (1990). Characteristic behavior and order relations for indicator variograms. *Mathematical Geology*, **22(8)**, 1011–1025.

Journel, A. G., C. Deutsch, and A.J. Desbarats (1986). Power averaging for block effective permeability. *SPE paper* 15128, 56th California Regional Meeting of the Society of Petroleum Engineers, Oakland, California, pp. 329–334.

Jowett, G. H. (1955a). The comparison of means of industrial time series. *Applied Statistics*, **4(1)**, 32–46.

Jowett, G. H. (1955b). Sampling properties of local statistics in stationary stochastic series. *Biometrika*, **42(1–2)**, 160–169.

Jowett, G. H. (1955c). The comparison of means of sets of observations from sections of independent stochastic series. *Journal of the Royal Statistical Society, Series B*, **17(2)**, 208–227.

Jun, M., and M. L. Stein (2008). Nonstationary covariance models for global data. *The Annals of Applied Statistics*, **2(4)**, 1271–1289.

Karimi, O., H. Omre, and M. Mohammadzadeh (2010). Bayesian closed-skew Gaussian inversion of seismic AVO data for elastic material properties. *Geophysics*, **75(1)**, R1–R11.

Karlin, S., and J. L. McGregor (1957). The differential equations of birth-and-death processes, and the Stieltjes moment problem. *Transactions of the American Mathematical Society*, **85(2)**, 489–546.

Karlin, S., and J. McGregor (1960). Classical diffusion processes and total positivity. *Journal of Mathematical Analysis and Applications*, **1(2)**, 163–183.

Kendall, D. G. (1974). Foundations of a theory of random sets. In *Stochastic Geometry*, E.F. Harding and D. G. Kendall, eds. John Wiley & Sons, New York, pp. 322–376.

Kendall, M. G. (1970). *Rank Correlation Methods*, 4th ed. Griffin, London.

Kendall, M. G., and P. A. P. Moran (1963). *Geometrical Probability*. Griffin, London.

Kendall, W. S., and E. Thönnes (1999). Perfect simulation in stochastic geometry. *Pattern Recognition*, **32(9)**, 1569–1586.

Kent, J. T., M. Mohammadzadeh, and A. M. Mosamam (2011). The dimple in Gneiting's spatial-temporal covariance model. *Biometrika*, **98(2)**, 489–494.

Khinchin, A. Y. (1934). Korrelationstheorie der stationären stochastischen Prozesse. *Mathematische Annalen*, **109(4)**, 604–615.

Kim, Y. C., P. K. Medhi, and A. Arik (1982). Investigation of in-pit ore-waste selection procedures using conditionally simulated orebodies. In *Proceedings of the 17th APCOM International Symposium*, T.B. Johnson and R.J. Barnes, eds. Society of Mining Engineers of the AIME, New York, pp. 121–142.

Kimeldorf, G., and G. Wahba (1970). A correspondence between Bayesian estimation of stochastic processes and smoothing by splines. *Annals of Mathematical Statistics*, **41(2)**, 495–502.

Kirkpatrick, S., C. D. Gelatt, , Jr., and M. P. Vecchi (1983). Optimization by simulated annealing. *Science*, **220(4598)**, 671–680.

Kitanidis, P. K. (1983). Statistical estimation of polynomial generalized covariance functions and hydrologic applications. *Water Resources Research*, **19(4)**, 909–921.

Kitanidis, P. K. (1985). Minimum-variance unbiased quadratic estimation of covariances of regionalized variables. *Journal of the International Association for Mathematical Geology*, **17(2)**, 195–208.

Kitanidis, P. K., and R. W. Lane (1985). Maximum likelihood parameter estimation of hydrologic spatial processes by the Gauss–Newton method. *Journal of Hydrology*, **79(1–2)**, 53–71.

Kitanidis, P. K., and E. G. Vomvoris (1983). A geostatistical approach to the inverse problem in groundwater modeling (steady state) and one-dimensional simulations. *Water Resources Research*, **19(3)**, 677–690.

Kleijnen, J. P. C. (2008). *Design and Analysis of Simulation Experiments*. Springer, Berlin.

Kleijnen, J. P. C. (2009). Kriging metamodeling in simulation: A review. *European Journal of Operational Research*, **192(3)**, 707–716.

Kleingeld, W. J. (1987). *La géostatistique pour des variables discrètes*. Doctoral thesis, E.N.S. des Mines de Paris.

Kleingeld, W. J., and C. Lantuéjoul (1993). Sampling of orebodies with a highly dispersed mineralization. In *Geostatistics Tróia '92*, Vol. 2, A. Soares, ed. Kluwer, Dordrecht, pp. 953–964.

Kleingeld, W. J., C. Lantuéjoul, C. F. Prins, and M. L. Thurston (1997). The conditional simulation of a Cox process with application to deposits with discrete particles. In *Geostatistics Wollongong '96*, Vol. 2, E. Y. Baafi and N. A. Schofield, eds. Kluwer, Dordrecht, pp. 683–694.

Knuth, D. E. (1997). *The Art of Computer Programming*, Vol. 2: *Seminumerical Algorithms*, 3rd ed. Addison-Wesley, Reading, MA.

Kolbjørnsen, O., and P. Abrahamsen (2005). Theory of the cloud transform for applications. In *Geostatistics Banff 2004*, Vol. 1, O. Leuangthong and C. V. Deutsch, eds. Springer, Dordrecht, pp. 45–54.

Kolmogorov, A. N. (1940a). Kurven im Hilbertschen Raum die gegenüber eine einparametrigen Gruppe von Bewegungen invariant sind. *Comptes Rendus (Doklady) de l'Académie des Sciences de l'URSS*, **26(1)**, 6–9.

Kolmogorov, A. N. (1940b). Wienersche Spiralen und einige andere interessante Kurven im Hilbertschen Raum. *Comptes Rendus (Doklady) de l'Académie des Sciences de l'URSS*, **26(2)**, 115–118.

Kolmogorov, A. N. (1941a). The local structure of turbulence in incompressible viscous fluid at very large Reynolds' numbers. *Comptes Rendus (Doklady) de l'Académie des Sciences de l'URSS*, **30(4)**, 301–305. Reprinted (1961) in *Turbulence: Classic Papers on Statistical Theory*, S. K. Friedlander and L. Topping, eds. Interscience Publishers, New York, pp. 151–155.

Kolmogorov, A. N. (1941b). Dissipation of energy in the locally isotropic turbulence. *Comptes Rendus (Doklady) de l'Académie des Sciences de l'URSS*, **32(1)**, 16–18. Reprinted (1961) in *Turbulence: Classic Papers on Statistical Theory*, S.K. Friedlander and L. Topping, eds. Interscience Publishers, New York, pp. 159–161.

Kolmogorov, A. N. (1941c). Interpolation und Extrapolation von stationären zufälligen Folgen. *Izvestiia Akademii Nauk SSSR, Seriia Matematicheskaia*, **5(1)**, 3–14.

Kolountzakis, M. N., and S. G. Révész (2003). On a problem of Turán about positive definite functions. *Proceedings of the American Mathematical Society*, **131(11)**, 3423–3430.

Kolovos, A., G. Christakos, D. T. Hristopulos, and M. L. Serre (2004). Methods for generating nonseparable spatio-temporal covariance models with potential environmental applications. *Advances in Water Resources*, **27(8)**, 815–830.

Koopmans, L. H. (1974). *The Spectral Analysis of Time Series*. Academic Press, New York.

Kostov, C., and O. Dubrule (1986). An interpolation method taking into account inequality constraints: II. Practical approach. *Mathematical Geology*, **18(1)**, 53–73.

Krickeberg, K. (1974). Moments of point processes. In *Stochastic Geometry*, E. F. Harding and D. G. Kendall, eds. John Wiley & Sons, New York, pp. 89–113.

Krige, D. G. (1951). A statistical approach to some basic mine valuation problems on the Witwatersrand. *Journal of the Chemical, Metallurgical and Mining Society of South Africa*, December, 119–139.

Krige, D. G. (1952). A statistical analysis of some of the borehole values in the Orange Free State goldfield. *Journal of the Chemical, Metallurgical and Mining Society of South Africa*, September, 47–64.

Krige, D. G. (1978). *Lognormal-de Wijsian Geostatistics for Ore Evaluation*. South African Institute of Mining and Metallurgy, Johannesburg.

Krige, D. G. (1997). A practical analysis of the effects of spatial structure and of data available and accessed, on conditional biases in ordinary kriging. In *Geostatistics Wollongong '96*, Vol. 1, E. Y. Baafi and N. A. Schofield, eds. Kluwer, Dordrecht, pp. 799–810.

Kruel-Romeu, R., and B. Nœtinger (1995). Calculation of internodal transmissivities in finite difference models of flow in heterogeneous porous media. *Water Resources Research*, **31(4)**, 943–959.

Kurbanmuradov, O., and K. Sabelfeld (2006). Stochastic spectral and Fourier-wavelet methods for vector Gaussian random fields. *Monte Carlo Methods and Applications*, **12(5–6)**, 395–445.

Kyriakidis, P. C., and A. G. Journel (1999). Geostatistical space-time models: A review. *Mathematical Geology*, **31(6)**, 651–684.

Lajaunie, C. (1990). Comparing some approximate methods for building local confidence intervals for predicting regionalized variables. *Mathematical Geology*, **22(1)**, 123–144.

Lajaunie, C., and R. Béjaoui (1991). Sur le krigeage des fonctions complexes. Technical ReportN-23/91/G, Centre de Géostatistique, Fontainebleau, France.

Lajaunie, C., and C. Lantuéjoul (1989). Setting up the general methodology for discrete isofactorial models. In *Geostatistics*, Vol. 1, M. Armstrong, ed. Kluwer, Dordrecht, pp. 323–334.

Lajaunie, C., G. Courrioux, and L. Manuel (1997). Foliation fields and 3D cartography in geology: Principles of a method based on potential interpolation. *Mathematical Geology*, **29(4)**, 571–584.

Landau, L. D., and E. M. Lifshitz (1960). *Electrodynamics of Continuous Media*. Pergamon Press, Oxford, England.

Langlais, V. (1989). On the neighborhood search procedure to be used when kriging with constraints. In *Geostatistics*, Vol. 2, M. Armstrong, ed. Kluwer, Dordrecht, pp. 603–614.

Langlais, V. (1990). *Estimation sous contraintes d'inégalités*. Doctoral thesis, E.N.S. des Mines de Paris.

Langlais, V., and J. Doyle (1993). Comparison of several methods of lithofacies simulation on the fluvial gypsy sandstone of Oklahoma. In *Geostatistics Tróia '92*, Vol. 1, A. Soares, ed. Kluwer, Dordrecht, pp. 299–310.

Langlais, V., H. Beucher, and D. Renard (2008). In the shade of the truncated Gaussian simulation. In *Geostats 2008*, Vol. 2, J. M. Ortiz and X. Emery, eds. Gecamin, Santiago, Chile, pp. 799–808.

Langsæter, A. (1926). Om beregning av middelfeilen ved regelmessige linjetakseringer. *Meddelelser fra Det norske Skogforsøksvesen*, **2(7)**, 5–47.

Lantuéjoul, C. (1988). On the importance of choosing a change of support model for global reserves estimation. *Mathematical Geology*, **20(8)**, 1001–1019.

Lantuéjoul, C. (1990). *Cours de sélectivité*. Lecture Notes C-140, Centre de Géostatistique, Fontainebleau, France.

Lantuéjoul, C. (1991). Ergodicity and integral range. *Journal of Microscopy*, **161(3)**, 387–403.

Lantuéjoul, C. (1993). Substitution random functions. In *Geostatistics Tróia '92*, Vol. 1, A. Soares, ed. Kluwer, Dordrecht, pp. 37–48.

Lantuéjoul, C. (1994). Non conditional simulation of stationary isotropic multigaussian random functions. In *Geostatistical Simulations*, M. Armstrong and P. A. Dowd, eds. Kluwer, Dordrecht, pp. 147–177.

Lantuéjoul, C. (1997). Conditional simulation of object-based models. In *Advances in Theory and Applications of Random Sets*, D. Jeulin, ed. World Scientific, Singapore, pp. 271–288.

Lantuéjoul, C. (2002). *Geostatistical Simulation: Models and Algorithms*. Springer, Berlin.

Lantuéjoul (2011). Three-dimensional conditional simulation of a Cox process using multisupport samples – Methodological aspects. Technical Report, Centre of Geosciences and Geoengineering, Mines ParisTech, Fontainebleau, France.

Lantuéjoul, C., and M. Schmitt (1991). Use of two new formulae to estimate the Poisson intensity of a Boolean model. In *Proceedings of 13rd GRETSI Symposium, Juan-les-Pins, France*, Vol. 2, pp. 1045–1048.

Lantuéjoul, C., J. N. Bacro, and L. Bel (2011). Storm processes and stochastic geometry. *Extremes*, **14(4)**, 413–428.

Lasky, S. G. (1950). How tonnage and grade relations help predict ore reserves. *Engineering and Mining Journal*, **151(4)**, 81–85.

Laslett, G. M. (1994). Kriging and splines: An empirical comparison of their predictive performance in some applications. *Journal of the American Statistical Association*, **89(426)**, 391–409.

Lefranc, M., B. Beaudoin, J. P. Chilès, D. Guillemot, C. Ravenne, and A. Trouiller (2008). Geostatistical characterization of Callovo-Oxfordian clay variability from high-resolution log data. *Physics and Chemistry of the Earth*, **33(Suppl. 1)**, S2–S13.

Le Loc'h, G., and A. Galli (1997). Truncated plurigaussian method: Theoretical and practical points of view. In *Geostatistics Wollongong '96*, Vol. 1, E. Y. Baafi and N. A. Schofield, eds. Kluwer, Dordrecht, pp. 211–222.

Le Ravalec, M., B. Noetinger, and L. Y. Hu (2000). The FFT moving average (FFT-MA) generator: An efficient numerical method for generating and conditioning Gaussian simulations. *Mathematical Geology*, **32(6)**, 701–723.

Lia, O., H. Omre, H. Tjelmeland, L. Holden, and T. Egeland (1997). Uncertainties in reservoir production forecasts. *AAPG Bulletin*, **81(5)**, 775–802.

Liao, H. T. (1990). *Estimation des réserves récupérables des gisements d'or—Comparaison entre krigeage disjonctif et krigeage des indicatrices*. Doctoral thesis, Université d'Orléans; Document du BRGM, No. 202 (1991), Bureau de Recherches Géologiques et Minières, Orléans, France.

Lindgren, F., H. Rue, and J. Lindström (2011). An explicit link between Gaussian fields and Gaussian Markov random fields: the stochastic partial differential equation approach. *Journal of the Royal Statistical Society, Series B*, **73(4)**, 423–498.

Loader, C., and P. Switzer (1992). Spatial covariance estimation for monitoring data. In *Statistics in Environmental and Earth Sciences*, A. Walden and P. Guttorp, eds. Griffin, Santa Ana, CA, pp. 52–69.

Loaiciga, H. A., and M. A. Marino (1990). Error analysis and stochastic differentiability in subsurface flow modeling. *Water Resources Research*, **25(12)**, 2897–2902.

Loève, M. (1955). *Probability Theory*. Princeton University Press, Princeton, NJ.

Long, J. C. S., and D. M. Billaux (1987). From field data to fracture network modeling: An example incorporating spatial structure. *Water Resources Research*, **23(7)**, 1201–1216.

Loquin, J., and D. Dubois (2010). Kriging and epistemic uncertainty: A critical discussion. In *Methods for Handling Imperfect Spatial Information*, R. Jeansoulin, O. Papini, H. Prade, and R. Schockaert, eds. Springer, Berlin, pp. 269–305.

Luster, G. R. (1986). Homogeneization and proportioning of mined materials: Applications of conditional simulation of coregionalization. In *Proceedings of the 19th APCOM International Symposium*, R.V. Ramani, ed. Society of Mining Engineers, Littleton, CO, pp. 163–172.

Ma, Y. Z., and J. J. Royer (1988). Local geostatistical filtering: Application to remote sensing. *Sciences de la Terre, Série Informatique Géologique*, **27**, 17–36.

Machuca-Mory, D.F. (2010). *Geostatistics with Location-Dependent Statistics*. Ph.D. dissertation, University of Alberta, Edmonton.

Madsen, K., H. B. Nielsen, and O. Tingleff (1999). Methods for non-linear least squares problems. Technical Report H38, Department of Mathematical Modelling, Technical University of Denmark, Lyngby, Denmark.

Mallet, J. L. (2004). Space-time mathematical framework for sedimentary geology. *Mathematical Geology*, **36(1)**, 1–32.

Mallet, J. L. (2008). *Numerical Earth Models*. EAGE Publications BV.

Mandelbrot, B. B. (1967). Sporadic random functions and conditional spectral analysis: Self similar examples and limits. In *Proceedings of the Fifth Berkeley Symposium on Mathematical Statistics and Probability*, Vol. 3, L. Le Cam and J. Neyman, eds. University of California Press, Berkeley, Ca., pp. 155–179.

Mandelbrot, B. B. (1975a). Fonctions aléatoires pluri-temporelles: approximation poissonienne du cas brownien et généralisations. *Comptes Rendus Hebdomadaires des Séances de l'Académie des Sciences de Paris, Série A*, **280**, 1075–1078.

Mandelbrot, B. B. (1975b). *Les objets fractals: Forme, hasard, et dimension*. Flammarion, Paris. Third edition (1989).

Mandelbrot, B. B. (1977). *Fractals: Form, Chance, and Dimension*. Freeman, San Francisco.

Mandelbrot, B. B. (1982). *The Fractal Geometry of Nature*. Freeman, New York.

Mandelbrot, B. B. (1985). Self-affine fractals and fractal dimension. *Physica Scripta*, **32(4)**, 257–260.

Mandelbrot, B. B., and J. W. Van Ness (1968). Fractional Brownian motions, fractional noises and applications. *SIAM Review*, **10(4)**, 422–437.

Mantoglou, A., and J. Wilson (1982). The turning bands method for simulation of random fields using line generation by a spectral method. *Water Resources Research*, **18(5)**, 1379–1394.

Marbeau, J. H., and J. P. Marbeau (1989). Formal computation of integrals for 3-D kriging. In *Geostatistics*, Vol. 2, M. Armstrong, ed. Kluwer, Dordrecht, pp. 773–784.

Marbeau, J. P. (1976). *Géostatistique forestière*. Doctoral thesis, E.N.S. des Mines de Paris.

Marchant, B. P., and R. M. Lark (2007). Optimized sample schemes for geostatistical surveys. *Mathematical Geology*, **39(1)**, 113–134.

Marcotte, D. (1996). Fast variogram computation with FFT. *Computers & Geosciences*, **22(10)**, 1175–1186.

Marcotte, D., and M. Chouteau (1993). Gravity data transformation by kriging. In *Geostatistics Tróia '92*, Vol. 1, A. Soares, ed. Kluwer, Dordrecht, pp. 249–260.

Maréchal, A. (1976). The practice of transfer functions: Numerical methods and their application. In *Advanced Geostatistics in the Mining Industry*, M. Guarascio, M. David, and C. Huijbregts, eds. Reidel, Dordrecht, pp. 253–276.

Maréchal, A. (1982). Local recovery estimation for co-products by disjunctive kriging. In *Proceedings of the 17th APCOM International Symposium*, T. B. Johnson and R. J. Barnes, eds. Society of Mining Engineers of the AIME, New York, pp. 562–571.

Maréchal, A. (1984). Kriging seismic data in presence of faults. In *Geostatistics for Natural Resources Characterization*, Part 1, G. Verly, M. David, A. G. Journel, and A. Maréchal, eds. Reidel, Dordrecht, pp. 385–420.

Mariethoz, G., and P. Renard (2010). Reconstruction of incomplete data sets or images using direct sampling. *Mathematical Geosciences*, **42(3)**, 245–268.

Mariethoz, G., P. Renard, and J. Straubhaar (2010). The Direct Sampling method to perform multiple-point geostatistical simulations. *Water Resources Research*, **46**, W11536.

Maronna, R. A., R. D. Martin, and V. J. Yohai (2006). *Robust Statistics: Theory and Methods*. John Wiley & Sons, Chichester, England.

Marshall, R. J., and K. V. Mardia (1985). Minimum norm quadratic estimation of components of spatial covariance. *Journal of the International Association for Mathematical Geology*, **17(5)**, 517–525.

Marsily, G. de (1986). *Quantitative Hydrogeology: Groundwater Hydrology for Engineers*. Academic Press, San Diego, CA.

Marsily, G. de, F. Delay, J. Gonçalvès, P. Renard, V. Teles, S. Violette (2005). Dealing with spatial heterogeneity. *Hydrogeology Journal*, **13(1)**, 161–283.

Matérn, B. (1960). *Spatial Variation—Stochastic Models and Their Application to Some Problems in Forest Surveys and Other Sampling Investigations*. Meddelanden från Statens Skogsforskningsinstitut, Vol. 49, No 5, Almaenna Foerlaget, Stockholm. Second edition (1986), Springer, Berlin.

Matheron, G. (1955). Applications des méthodes statistiques à l'évaluation des gisements. *Annales des Mines*, No. 12, 50–75.

Matheron, G. (1962–1963). *Traité de géostatistique appliquée, Tome I; Tome II: Le krigeage*. I: Mémoires du Bureau de Recherches Géologiques ct Minières, No. 14 (1962), Editions Technip, Paris; II: Mémoires du Bureau de Recherches Géologiques et Minières, No. 24 (1963), Editions B.R.G.M., Paris.

Matheron, G. (1965). *Les variables régionalisées et leur estimation. Une application de la théorie des fonctions aléatoires aux Sciences de la Nature*. Masson, Paris.

Matheron, G. (1967). *Eléments pour une théorie des milieux poreux*. Masson, Paris.

Matheron, G. (1968a). *Osnovy Prikladnoï Geostatistiki* (Treatise of Geostatistics). Mir, Moscow.

Matheron, G. (1968b). Schéma booléen séquentiel de partition aléatoire. Technical Report N-83, Centre de Géostatistique, Fontainebleau, France.

Matheron, G. (1969a). *Le krigeage universel*. Cahiers du Centre de Morphologie Mathématique de Fontainebleau, Fasc. 1, Ecole des Mines de Paris.

Matheron, G. (1969b). Les processus d'Ambarzoumian et leur application en géologie. Technical Report N-131, Centre de Géostatistique, Fontainebleau, France.

Matheron, G. (1970). *La théorie des variables régionalisées et ses applications.* Cahiers du Centre de Morphologie Mathématique de Fontainebleau, Fasc. 5, Ecole des Mines de Paris. Translation (1971): *The Theory of Regionalized Variables and Its Applications.*

Matheron, G. (1971a). Les polyèdres poissoniens isotropes. *Actes du 3e Colloque Européen sur la Fragmentation, Cannes, France,* II-9, 509–534. *Zerkleinern 3. Vortrage Symposium, Cannes,* Dechema Monographien, Vol. 69, No. 1292–1326, Part 2, Verlag Chemie GmbH, pp. 575–600.

Matheron, G. (1971b). La théorie des fonctions aléatoires intrinsèques généralisées. Note Géostatistique N° 117. Technical Report N-252, Centre de Géostatistique, Fontainebleau, France.

Matheron, G. (1972a). Ensembles fermés aléatoires, ensembles semi-markoviens et polyèdres poissoniens. *Advances in Applied Probability,* **4(3)**, 508–541.

Matheron, G. (1972b). Inférence statistique pour les covariances généralisées (dans $\mathbb{R}^1$). Technical Report N-281, Centre de Géostatistique, Fontainebleau, France.

Matheron, G. (1973a). The intrinsic random functions and their applications. *Advances in Applied Probability,* **5(3)**, 439–468.

Matheron, G. (1973b). Le krigeage disjonctif. Technical Report N-360, Centre de Géostatistique, Fontainebleau, France.

Matheron, G. (1974a). Effet proportionnel et lognormalité ou: le retour du serpent de mer. Technical Report N-374, Centre de Géostatistique, Fontainebleau, France.

Matheron, G. (1974b). Représentations stationnaires et représentations minimales pour les FAI–k. Technical Report N-377, Centre de Géostatistique, Fontainebleau, France.

Matheron, G. (1974c). Les fonctions de transfert des petits panneaux. Technical Report N-395, Centre de Géostatistique, Fontainebleau, France.

Matheron, G. (1975a). *Random Sets and Integral Geometry.* John Wiley & Sons, New York.

Matheron, G. (1975b). Compléments sur les modèles isofactoriels. Technical Report N-432, Centre de Géostatistique, Fontainebleau, France.

Matheron, G. (1975c). Modèles isofactoriels discrets et modèle de Walsh. Technical Report N-449, Centre de Géostatistique, Fontainebleau, France.

Matheron, G. (1976a). A simple substitute for conditional expectation: The disjunctive kriging. In *Advanced Geostatistics in the Mining Industry,* M. Guarascio, M. David, and C. Huijbregts, eds. Reidel, Dordrecht, pp. 221–236.

Matheron, G. (1976b). Forecasting block grade distributions: the transfer functions. In *Advanced Geostatistics in the Mining Industry,* M. Guarascio, M. David, and C. Huijbregts, eds. Reidel, Dordrecht, pp. 237–251.

Matheron, G. (1978). *Estimer et choisir.* Cahiers du Centre de Morphologie Mathématique de Fontainebleau, Fasc. 7, Ecole des Mines de Paris. Translation (1989): *Estimating and Choosing—An Essay on Probability in Practice.* Springer, Berlin.

Matheron, G. (1979a). Comment translater les catastrophes. La structure des F.A.I. générales. Technical Report N-617, Centre de Géostatistique, Fontainebleau, France.

Matheron, G. (1979b). Recherche de simplification dans un problème de cokrigeage. Technical Report N-628, Centre de Géostatistique, Fontainebleau, France.

Matheron, G. (1980). Modèles isofactoriels pour l'effet zéro. Technical Report N-659, Centre de Géostatistique, Fontainebleau, France.

Matheron, G. (1981a). Splines and kriging: Their formal equivalence. In *Down-to-Earth-Statistics: Solutions Looking for Geological Problems*, D. F. Merriam, ed. Syracuse University Geological Contributions, Syracuse, New York, pp. 77–95.

Matheron, G. (1981b). Remarques sur le changement de support. Technical Report N-690, Centre de Géostatistique, Fontainebleau.

Matheron, G. (1981c). Remarques sur le krigeage et son dual. Technical Report N-695, Centre de Géostatistique, Fontainebleau, France.

Matheron, G. (1981d). Quatre familles discrètes. Technical Report N-703, Centre de Géostatistique, Fontainebleau, France.

Matheron, G. (1982a). Pour une analyse krigeante des données régionalisées. Technical Report N-732, Centre de Géostatistique, Fontainebleau, France.

Matheron, G. (1982b). La destructuration des hautes teneurs et le krigeage des indicatrices. Technical Report N-761, Centre de Géostatistique, Fontainebleau, France.

Matheron, G. (1984a). Modèle isofactoriel et changement de support. *Sciences de la Terre, Série Informatique Géologique*, **18**, 71–123.

Matheron, G. (1984b). The selectivity of the distributions and the "second principle of geostatistics." In *Geostatistics for Natural Resources Characterization*, Part 1, G. Verly, M. David, A. G. Journel, and A. Maréchal, eds. Reidel, Dordrecht, pp. 421–433.

Matheron, G. (1984c). Isofactorial models and change of support. In *Geostatistics for Natural Resources Characterization*, Part 1, G. Verly, M. David, A. G. Journel, and A. Maréchal, eds. Reidel, Dordrecht, pp. 449–467.

Matheron, G. (1984d). Changement de support en modèle mosaïque. *Sciences de la Terre, Série Informatique Géologique*, **20**, 435–454.

Matheron, G. (1984e). Une méthodologie générale pour les modèles isofactoriels discrets. *Sciences de la Terre, Série Informatique Géologique*, **21**, 1–64.

Matheron, G. (1985). Comparaison de quelques distributions du point de vue de la sélectivité. *Sciences de la Terre, Série Informatique Géologique*, **24**, 1–21.

Matheron, G. (1986). Sur la positivité des poids de krigeage. Technical Report N-30/86/G, Centre de Géostatistique, Fontainebleau, France.

Matheron, G. (1987a). Suffit-il, pour une covariance, d'être de type positif? *Sciences de la Terre, Série Informatique Géologique*, **26**, 51–66.

Matheron, G. (1987b). A simple answer to an elementary question. *Mathematical Geology*, **19(5)**, 455–457.

Matheron, G. (1988). Simulation de fonctions aléatoires admettant un variogramme concave donné. *Sciences de la Terre, Série Informatique Géologique*, **28**, 195–212.

Matheron, G. (1989). Two classes of isofactorial models. In *Geostatistics*, Vol. 1, M. Armstrong, ed. Kluwer, Dordrecht, pp. 309–322.

Matheron, G. (1993). Une conjecture sur la covariance d'un ensemble aléatoire (Notes prises par E. Castelier). In *Compte Rendu des Journées de Géostatistique*, C. de Fouquet, ed. Cahiers de Géostatistique, Fasc. 3, Ecole des Mines de Paris, pp. 107–113.

Matheron, G., H. Beucher, C. de Fouquet, A. Galli, D. Guérillot, and C. Ravenne (1987). Conditional simulation of the geometry of fluvio-deltaic reservoirs. *SPE paper* 16753, 62nd Annual Technical Conference and Exhibition of the Society of Petroleum Engineers, Dallas, Texas, pp. 123–131.

Matheron, G., H. Beucher, C. de Fouquet, A. Galli, and C. Ravenne (1988). Simulation conditionnelle à trois faciès dans une falaise de la formation du Brent. *Sciences de la Terre, Série Informatique Géologique*, **28**, 213–249.

McLain, D. H. (1980). Interpolation methods for erroneous data. In *Mathematical Methods in Computer Graphics and Design*, K.W. Brodlie, ed. Academic Press, London, pp. 87–104.

Mejía, J. M., and I. Rodríguez-Iturbe (1974). On the synthesis of random field sampling from the spectrum: An application to the generation of hydrologic spatial processes. *Water Resources Research*, **10(4)**, 705–711.

Metropolis, N., A. W. Rosenbluth, M. N. Rosenbluth, A. H. Teller, and E. Teller (1953). Equations of state calculations by fast computing machines. *Journal of Chemical Physics*, **21(6)**, 1087–1092.

Meyer, Y. (1993). *Wavelets. Algorithms and Applications*. Society for Industrial and Applied Mathematics, Philadelphia.

Meyn, S. P., and R. L. Tweedie (1993). *Markov Chains and Stochastic Stability*. Springer, London.

Micchelli, C. A. (1986). Interpolation of scattered data: Distance matrices and conditionally positive definite functions. *Constructive Approximation*, **2**, 11–22.

Mignolet, M. P., and P. D. Spanos (1992). Simulation of homogeneous two-dimensional random fields: Part I—AR and ARMA models. *Journal of Applied Mechanics*, **59(2)**, 260–269.

Miles, R. E. (1964). Random polygons determined by random lines in a plane, I; II. *Proceedings of the National Academy of Sciences of USA*, I: **52(4)**, 901–907; II: **52(5)**, 1157–1160.

Miles, R. E. (1969). Poisson flats in Euclidean spaces. Part I: A finite number of random uniform flats. *Advances in Applied Probability*, **1(2)**, 211–237.

Miles, R. E. (1970). On the homogeneous planar Poisson point process. *Mathematical Biosciences*, **6(1)**, 85–127.

Miles, R. E. (1971). Poisson flats in Euclidean spaces. Part II: Homogeneous Poisson flats and the complementary theorem. *Advances in Applied Probability*, **3(1)**, 1–43.

Miles, R. E. (1972). The random division of space. *Advances in Applied Probability*, **4(Suppl.)**, 243–266.

Miles, R. E. (1973). The various aggregates of random polygons determined by random lines in a plane. *Advances in Mathematics*, **10(2)**, 256–290.

Molchanov, I. (1997). *Statistics of the Boolean Model for Practitioners and Mathematicians*. John Wiley & Sons, Chichester.

Molchanov, I. (2005). *Theory of Random Sets*. Springer, London.

Møller, J. (1989). On the rate of convergence of spatial birth-and-death processes. *Annals of the Institute of Statistical Mathematics*, **41(3)**, 565–581.

Monestiez, P., and P. Switzer (1991) Semiparametric estimation of nonstationary spatial covariance models by metric multidimensional scaling. Technical Report SIMS 165,

Department of Statistics and Department of Applied Earth Sciences, Stanford University, California.

Monestiez, P., L. Dubroca, E. Bonnin, J. P. Durbec, and C. Guinet (2005a). Comparison of model based geostatistical methods in ecology: Application to fin whale spatial distribution in northwestern Mediterranean Sea. In *Geostatistics Banff 2004*, Vol. 2, O. Leuangthong and C. V. Deutsch, eds. Springer, Berlin, pp. 777–786.

Monestiez, P., J. S. Bailly, P. Lagacherie, and M. Voltz (2005b). Geostatistical modelling of spatial processes on directed trees: Application to fluvisol extent. *Geoderma*, **128(3–4)**, 179–191.

Monestiez, P., L. Dubroca, E. Bonnin, J. P. Durbec, and C. Guinet (2006). Geostatistical modelling of spatial distribution of Balaenoptera physalus in the Northwestern Mediterranean Sea from sparse count data and heterogeneous observation efforts. *Ecological Modelling*, **193(3–4)**, 615–628.

Monin, A. S., and A. M. Yaglom (1965). *Statisticheskaya gidromekhanika—Mekhanika Turbulenosti*. Nauka Press, Moscow. Translation, edition updated (1971, 1975): *Statistical Fluid Mechanics: Mechanics of Turbulence*. MIT Press, Cambridge, MA, 2 Vols.

Morelon, I. F., B. Doligez, D. R. Guerillot, D. Rahon, and Y. Touffait (1991). An application of a 3D geostatistical imaging to reservoir fluid flow simulations. *SPE paper* 22312, Sixth Petroleum Computer Conference of the Society of Petroleum Engineers, Dallas, Texas, pp. 223–231.

Mostad, P. F., T. Egeland, N. L. Hjort, A. G. Kraggerud, and P. Y. Biver (1997). Variogram estimation in a Bayesian framework. In *Geostatistics Wollongong '96*, Vol. 1, E. Y. Baafi and N.A. Schofield, eds. Kluwer, Dordrecht, pp. 223–233.

Mosteller, F., and J. W. Tukey (1977). *Data Analysis and Regression*. Addison-Wesley, Reading, MA.

Muge, F. H., and G. Cabeçadas (1989). A geostatistical approach to eutrophication modelling. In *Geostatistics*, Vol. 1, M. Armstrong, ed. Kluwer, Dordrecht, pp. 445–457.

Müller, W. G. (2007). *Collecting Spatial Data—Optimum Design of Experiments for Random Fields*, 3rd ed. Springer, Berlin.

Myers, D. E. (1982). Matrix formulation of co-kriging. *Journal of the International Association for Mathematical Geology*, **14(3)**, 249–257.

Myers, D. E. (1983). Estimation of linear combinations and co-kriging. *Journal of the International Association for Mathematical Geology*, **15(5)**, 633–637.

Nagel, W., and V. Weiss (2005). Crack STIT tessellations: characterization of the stationary random tessellations stable with respect to iteration. *Advances in Applied Probability*, **37(4)**, 859–883.

Nagel, W., J. Mecke, J. Ohser, and V. Weiss (2008). A tessellation model for crack patterns on surfaces. *Image Analysis and Stereology*, **27(2)**, 73–78.

Narboni, P. (1979). Application de la méthode des variables régionalisées à deux forêts du Gabon. Technical Report, Note Statistique No. 18, Centre Technique Forestier Tropical, Nogent-sur-Marne, France. Simplified text and abstract: *Bois et Forêts des Tropiques*, No. 188, 47–68.

Naveau, P., A. Guillou, D. Cooley, and J. Diebolt (2009). Modelling pairwise dependence of maxima in space. *Biometrika*, **96(1)**, 1–17.

Neumann, J. von, and I.J. Schoenberg (1941). Fourier integrals and metric geometry. *Transactions of the American Mathematical Society*, **50(2)**, 226–251.

Neveu, J. (1970). *Bases mathématiques du calcul des probabilités*. Masson, Paris.

Neyman, J., and E. L. Scott. (1958). Statistical approach to problems of cosmology. *Journal of the Royal Statistical Society, Series B*, **20(1)**, 1–43.

Neyman, J., and E. L. Scott (1972). Processes of clustering and applications. In *Stochastic Point Processes, Statistical Analysis, Theory and Applications*, P. A. W. Lewis, ed. John Wiley & Sons, New York, pp. 646–681.

Nielsen, A. A., K. Conradsen, J. L. Pedersen, and A. Steenfelt (2000). Maximum autocorrelation factorial kriging. In *Geostats 2000 Cape Town*, Vol. 2, W. J. Kleingeld and D. G. Krige, eds. Geostatistical Association of Southern Africa, pp. 538–547.

Nœtinger, B. (2000). Computing the effective permeability of log-normal permeability fields using renormalization methods. *Comptes Rendus de l'Académie des Sciences de Paris, Series IIA, Earth and Planetary Science*, **331(5)**, 353–357.

Nœtinger, B., and A. Haas (1996). Permeability averaging for well tests in 3D stochastic reservoir models. *SPE paper* 36653-MS, 1996 SPE Annual Technical Conference and Exhibition, Denver, Colorado.

Nœtinger, B., V. Artus, and G. Zargar (2005). The future of stochastic and upscaling methods in hydrogeology. *Hydrogeology Journal*, **13(1)**, 184–201.

Obukhov, A.M. (1941). O raspredelenii energii v spektre turbulentnogo potoka (and expanded abstract in German: Über die Energieverteilung im Spektrum des Turbulenzstromes). *Izvestiya Akademii Nauk SSSR, Seriya Geografitcheskaya i Geofizicheskaya*, No. 4–5, 454–456.

Obukhov, A. M. (1949a). Struktura temperaturnogo polya v turbulentnom potoke (Structure of the temperature field in a turbulent flow). *Izvestiya Akademii Nauk SSSR, Seriya Geografitcheskaya i Geofizicheskaya*, **13(1)**, 58–69.

Obukhov, A. M. (1949b). Lokalnaya struktura atmosphernoy turbulentnosty (Local structure of atmospheric turbulence). *Doklady Akademii Nauk SSSR*, **67(4)**, 643–646.

Olver, F. W. J. (1972). Bessel functions of integer order. In *Handbook of Mathematical Functions with Formulas, Graphs, and Mathematical Tables*, M. Abramowitz and I. A. Stegun, eds. John Wiley & Sons, New York, pp. 355–433.

Omre, H. (1987). Bayesian kriging: Merging observations and qualified guesses in kriging. *Mathematical Geology*, **19(1)**, 25–39.

Omre, H., and K. B. Halvorsen (1989). The Bayesian bridge between simple and universal kriging. *Mathematical Geology*, **21(7)**, 767–786.

Omre, H., and H. Tjelmeland (1997). Petroleum geostatistics. In *Geostatistics Wollongong '96*, Vol. 1, E. Y. Baafi and N. A. Schofield, eds. Kluwer, Dordrecht, pp. 41–52.

Oppenheim, A. V., and R. W. Schafer (1989). *Discrete-Time Signal Processing*. Prentice-Hall, Englewood Cliffs, NJ.

Orfeuil, J. P. (1972). Simulation du Wiener-Lévy et de ses intégrales. Technical Report N-290, Centre de Géostatistique, Fontainebleau, France.

Ouenes, A., S. Bhagavan, P. H. Bunge, and B. J. Travis (1994). Application of simulated annealing and other global optimization methods to reservoir description: Myths and realities. *SPE paper* 28415, 69th Annual Technical Conference and Exhibition of the Society of Petroleum Engineers, New Orleans, Louisiana, pp. 547–561.

Pan, G., D. Gaard, K. Moss, and T. Heiner (1993). A comparison between cokriging and ordinary kriging: Case study with a polymetallic deposit. *Mathematical Geology*, **25(3)**, 377–398.

Pardo-Iguzquiza, E., and M. Chica-Olmo (1993). The Fourier integral method: An efficient spectral method for simulation of random fields. *Mathematical Geology*, **25(2)**, 177–217.

Parker, H. M. (1984). Trends in geostatistics in the mining industry. In *Geostatistics for Natural Resources Characterization*, Part 2, G. Verly, M. David, A. G. Journel, and A. Maréchal, eds. Reidel, Dordrecht, pp. 915–934.

Parker, H. M. (1991). Statistical treatment of outlier data in epithermal gold deposit reserve estimation. *Mathematical Geology*, **23(2)**, 175–199.

Parker, H. M., A. G. Journel, and W. C. Dixon (1979). The use of conditional lognormal probability distribution for the estimation of open-pit ore reserves in stratabound uranium deposits—A case study. In *Proceedings of the 16th APCOM International Symposium*, T. J. O'Neil, ed. Society of Mining Engineers of the AIME, New York, pp. 133–148.

Pawlowsky-Glahn, V., ed. (2005). Special issue: Advances in compositional data, *Mathematical Geology*, **37(7)**, 671–850.

Pawlowsky-Glahn, V., and R. A. Olea (2004). *Geostatistical Analysis of Compositional Data*. Oxford University Press, New York.

Pawlowsky, V., R. A. Olea, and J. C. Davis (1994). Additive logratio estimation of regionalized compositional data: An application to the calculation of oil reserves. In *Geostatistics for the Next Century*, R. Dimitrakopoulos, ed. Kluwer, Dordrecht, pp. 371–382.

Pelletier, B., P. Dutilleul, G. Larocque, and J. W. Fayles (2004). Fitting the linear model of coregionalization by generalized least squares. *Mathematical Geology*, **36(3)**, 323–343.

Petitgas, P. (1993). Geostatistics for fish stock assessments: A review and an acoustic application. *ICES Journal of Marine Science*, **50(3)**, 285–298.

Pilz, J. (1994). Robust Bayes linear prediction of regionalized variables. In *Geostatistics for the Next Century*, R. Dimitrakopoulos, ed. Kluwer, Dordrecht, pp. 464–475.

Pilz, J., G. Spoeck, and M. G. Schimek (1997). Taking account of uncertainty in spatial covariance estimation. In *Geostatistics Wollongong '96*, Vol. 1, E. Y. Baafi and N. A. Schofield, eds. Kluwer, Dordrecht, pp. 302–313.

Polus-Lefebvre, E., C. de Fouquet, C. Bernard-Michel, N. Flipo, and M. Poulin (2008). Geostatistical model for concentrations or flow rates in streams: Some results. In *Geostats 2008*, Vol. 2, J. M. Ortiz and X. Emery, eds. Gecamin, Santiago, Chile, pp. 871–880.

Pólya, G. (1949). Remarks on characteristic functions. In *Proceedings of the (First) Berkeley Symposium on Mathematical Statistics and Probability*, J. Neyman, ed. University of California Press, Berkeley, CA, pp. 115–123.

Press, W. H., S. A. Teukolsky, W. T. Vetterling, and B. P. Flannery (2007). *Numerical Recipes: The Art of Scientific Computing*, 3rd ed. Cambridge University Press.

Preston, C. (1975). Spatial birth-and-death processes. *Bulletin of the International Statistical Institute*, **46(2)**, 371–391.

Préteux, F., and M. Schmitt (1988). Boolean texture analysis and synthesis. In *Image Analysis and Mathematical Morphology*, Vol. 2: *Theoretical Advances*, J. Serra, ed. Academic Press, London, Chapter 18, pp. 377–400.

Propp, J. G., and D. B. Wilson (1996). Exact sampling with coupled Markov chains and applications to statistical mechanics. *Random Structures and Algorithms*, **9(1–2)**, 223–252.

Puente, C. E., and R. L. Bras (1986). Disjunctive kriging, universal kriging, or no kriging: Small sample results with simulated fields. *Mathematical Geology*, **18(3)**, 287–305.

Pukelsheim, F. (1994). The three sigma rule. *The American Statistician*, **48(2)**, 88–91.

Pyrcz, M. J., J. B. Boisvert, and C. V. Deutsch (2009). ALLUVSIM: A program for event-based stochastic modeling of fluvial depositional systems. *Computers & Geosciences*, **35(8)**, 1671–1685.

Rabbani, M., and P. W. Jones (1991). *Digital Image Compression Techniques*. SPIE— The International Society for Optical Engineering, Bellingham, WA.

Rao, C. R. (1970). Estimation of heteroscedastic variances in a linear model. *Journal of the American Statistical Association*, **65(329)**, 161–172.

Rao, C. R. (1971a). Estimation of variance and covariance components—MINQUE theory. *Journal of Multivariate Analysis*, **1(3)**, 257–275.

Rao, C. R. (1971b). Minimum variance quadratic unbiased estimation of variance components. *Journal of Multivariate Analysis*, **1(4)**, 445–456.

Rao, C. R. (1973). *Linear Statistical Inference and Its Applications*. John Wiley & Sons, New York.

Raspa, G., R. Bruno, and P. Kokkiniotis (1989). An application of disjunctive kriging: Using the negative binomial model with different change of support models. In *Geostatistics*, Vol. 2, M. Armstrong, ed. Kluwer, Dordrecht, pp. 935–945.

Reilly, C., and A. Gelman (2007). Weighted classical variogram estimation for data with clustering. *Technometrics*, **49(2)**, 184–194.

Remacre, A. Z. (1984). Conditionnement uniforme. *Sciences de la Terre, Série Informatique Géologique*, **18**, 125–139.

Remacre, A. Z. (1987). Conditioning by the panel grade for recovery estimation of non-homogeneous orebodies. In *Geostatistical Case Studies*, G. Matheron and M. Armstrong, eds. Reidel, Dordrecht, pp. 135–147.

Renard, D., and P. Ruffo (1993). Depth, dip and gradient. In *Geostatistics Tróia '92*, Vol. 1, A. Soares, ed. Kluwer, Dordrecht, pp. 167–178.

Renard, P., and G. de Marsily (1997). Calculating equivalent permeability: A review. *Advances in Water Resources*, **20(5–6)**, 253–278.

Rendu, J. M. (1979). Kriging, logarithmic kriging, and conditional expectation: Comparison of theory with actual results. In *Proceedings of the 16th APCOM International Symposium*, T. J. O'Neil, ed. Society of Mining Engineers of the AIME, New York, pp. 199–212.

Rendu, J. M. (1984). Geostatistical modelling and geological controls. In *Proceedings of the 18th APCOM International Symposium*, Institution of Mining and Metallurgy, London, pp. 467–476.

Rendu, J. M., and L. Readdy (1980). Geology and the semivariogram—A critical relationship. In *Proceedings of the 17th APCOM International Symposium*, T. B.

Johnson and R. J. Barnes, eds. Society of Mining Engineers of the AIME, New York, pp. 771–783.

Resnick, S. I. (1987). *Extreme Values, Regular Variation, and Point Processes*. Springer, New York.

Resnick, S. I. (2007). *Heavy-Tail Phenomena. Probabilistic and Statistical Modeling*. Springer, New York.

Rimstad, K., and H. Omre (2010). Impact of rock-physics depth trends and Markov random fields on hierarchical Bayesian lithology/fluid prediction. *Geophysics*, **75(4)**, R93–R108.

Ripley, B. D. (1976). The second-order analysis of stationary point processes. *Journal of Applied Probability*, **13(2)**, 255–266.

Ripley, B. D. (1977). Modelling spatial patterns. *Journal of the Royal Statistical Society, Series B*, **39(2)**, 172–212.

Ripley, B. D. (1981). *Spatial Statistics*. John Wiley & Sons, New York.

Ripley, B.D. (1987). *Stochastic Simulation*. John Wiley & Sons, New York.

Rivoirard, J. (1984). *Le comportement des poids de krigeage*. Doctoral thesis, E.N.S. des Mines de Paris.

Rivoirard, J. (1987). Two key parameters when choosing the kriging neighborhood. *Mathematical Geology*, **19(8)**, 851–856.

Rivoirard, J. (1988). Modèles à résidus d'indicatrices autokrigeables. *Sciences de la Terre, Série Informatique Géologique*, **28**, 303–326.

Rivoirard, J. (1989). Models with orthogonal indicator residuals. In *Geostatistics*, Vol. 1, M. Armstrong, ed. Kluwer, Dordrecht, pp. 91–107.

Rivoirard, J. (1990). A review of lognormal estimators for in situ reserves. *Mathematical Geology*, **22(2)**, 213–221.

Rivoirard, J. (1994). *Introduction to Disjunctive Kriging and Non-Linear Geostatistics*. Oxford University Press, New York.

Rivoirard, J. (2001). Which models for collocated cokriging? *Mathematical Geology*, **33(2)**, 117–131.

Rivoirard, J. (2002). On the structural link between variables in kriging with external drift. *Mathematical Geology*, **34(7)**, 797–808.

Rivoirard, J. (2004). On some simplifications of cokriging neighborhood. *Mathematical Geology*, **36(8)**, 899–915.

Rivoirard, J., and T. Romary (2011). Continuity for kriging with moving neighborhood. *Mathematical Geosciences*, **43(4)**, 469–481.

Rivoirard, J., J. Simmonds, K. G. Foote, P. Fernandes, and N. Bez (2000). *Geostatistics for Estimating Fish Abundance*. Blackwell Science, Oxford, England.

Rivoirard, J., C. Demange, X. Freulon, A. Lécureuil, and N. Bellot (2012). A top-cut model for deposits with heavy-tailed grade distribution. *Mathematical Geosciences*, **44**.

Rockafellar, R. T. (1970). *Convex Analysis*. Princeton University Press, Princeton, NJ.

Rodríguez-Iturbe, I., and J.M. Mejía (1974). The design of rainfall networks in time and space. *Water Resources Research*, **10(4)**, 713–728.

Romary, T. (2009). Integrating production data under uncertainty by parallel interacting Markov chains on a reduced dimensional space. *Computational Geosciences*, **13(1)**, 103–122.

Romary, T. (2010). History matching of approximated lithofacies models under uncertainty. *Computational Geosciences*, **14(2)**, 343−355.

Romary, T., C. de Fouquet, and L. Malherbe (2011). Sampling design for air quality measurement surveys: An optimization approach. *Atmospheric Environment*, **45(21)**, 3613−3620.

Rossi, M. E., and H. M. Parker (1994). Estimating recoverable reserves: Is it hopeless? In *Geostatistics for the Next Century*, R. Dimitrakopoulos, ed. Kluwer, Dordrecht, pp. 259−276.

Rossi, R. E., D. J. Mulla, A. G. Journel, and E. H. Franz (1992). Geostatistical tools for modeling and interpreting ecological spatial dependence. *Ecological Monographs*, **62(2)**, 277−314.

Roth, C. (1995). *Contribution de la géostatistique à la résolution du problème inverse en hydrogéologie*. Doctoral thesis, E.N.S. des Mines de Paris; Document du BRGM, No. 241, Bureau de Recherches Géologiques et Minières, Orléans, France.

Roth, C., J. P. Chilès, and C. de Fouquet (1996). Adapting geostatistical transmissivity simulations to finite differences flow simulators. *Water Resources Research*, **32(10)**, 3237−3242.

Rothschild, M., and J. E. Stiglitz (1973). Some further results on the measurement of inequality. *Journal of Economic Theory*, **6(2)**, 188−204.

Rouhani, S., and T. J. Hall (1989). Space−time kriging of groundwater data. In *Geostatistics*, Vol. 2, M. Armstrong, ed. Kluwer, Dordrecht, pp. 639−651.

Rouhani, S., and H. Wackernagel (1990). Multivariate geostatistical approach to space−time data analysis. *Water Resources Research*, **26(4)**, 585−591.

Rousseeuw, P. J., and C. Croux (1993). Alternatives to the median absolute deviation. *Journal of the American Statistical Association*, **88(424)**, 1273−1283.

Royden, H. L. (1968). *Real Analysis*. Collier-Macmillan, London.

Royer, J. F. (1975). Comparaison des méthodes d'analyse objective par interpolation optimale et par approximations successives. Technical Report No. 365, Etablissement d'Etudes et de Recherches Météorologiques, Direction de la Météorologie, Paris.

Royer, J. J. (1988). Comparaison de quelques méthodes de déconvolution. Apport de la géostatistique. *Sciences de la Terre, Série Informatique Géologique*, **28**, 327−354.

Rozanov, Yu. A. (1982). *Markov Random Fields*. Springer, New York.

Rubin, Y., and J. J. Gómez-Hernández (1990). A stochastic approach to the problem of upscaling of conductivity in disordered media: Theory and unconditional numerical simulations. *Water Resources Research*, **26(4)**, 691−701.

Rudin, W. (1970). An extension theorem for positive-definite functions. *Duke Mathematical Journal*, **37(1)**, 49−53.

Rue, H., and L. Held (2005). *Gaussian Markov Random Fields*. Chapman & Hall/CRC, Boca Raton, FL.

Rutherford, I. D. (1972). Data assimilation by statistical interpolation of forecast error fields. *Journal of the Atmospheric Sciences*, **29(5)**, 809−815.

Sabourin, R. (1976). Application of two methods for an interpretation of the underlying variogram. In *Advanced Geostatistics in the Mining Industry*, M. Guarascio, M. David, and C. Huijbregts, eds. Reidel, Dordrecht, pp. 101−109.

Sacks, J., W. J. Welch, T. J. Mitchell, and H. P. Wynn (1989). Design and analysis of computer experiments. *Statistical Science*, **4(4)**, 409–453.

Samaras, E., M. Shinozuka, and A. Tsurui (1985). ARMA representation of random processes. *Journal of Engineering Mechanics*, **111(3)**, 449–461.

Sampson, P., and P. Guttorp (1992). Nonparametric estimation of nonstationary spatial covariance structure. *Journal of the American Statistical Association*, **87(417)**, 108–119.

Sancevero, S. S., A. Z. Remacre, R. de Souza, and E. Cesário Mundim (2005). Comparing deterministic and stochastic seismic inversion for thin-bed reservoir characterization in a turbidite synthetic reference model of Campos Basin, Brazil. *The Leading Edge*, **24(11)**, 1168–1172.

Sánchez-Vila, X., J. Carrera, and J. P. Girardi (1996). Scale effects in transmissivity. *Journal of Hydrology*, **183(1–2)**, 1–22.

Sandjivy, L. (1984). The factorial kriging analysis of regionalized data. Its applications to geochemical prospecting. In *Geostatistics for Natural Resources Characterization*, Part 1, G. Verly, M. David, A.G. Journel, and A. Maréchal, eds. Reidel, Dordrecht, pp. 559–571.

Sans, H., and J. R. Blaise (1987). Comparing estimated uranium grades with production figures. In *Geostatistical Case Studies*, G. Matheron and M. Armstrong, eds. Reidel, Dordrecht, pp. 169–185.

Santaló, L. A. (1976). *Integral Geometry and Geometric Probability*. Addison-Wesley, Reading, MA.

Sarmanov, I. O. (1968). A generalized symmetric gamma correlation. *Soviet Mathematics Doklady*, **9(2)**, 547–550.

Schelin, L., and S. Sjöstedt-de Luna (2010). Kriging predicting intervals based on semiparametric bootstrap. *Mathematical Geosciences*, **42(8)**, 985–1000.

Schlather, M. (2002). Models for stationary max-stable random fields. *Extremes*, **5(1)**, 33–44.

Schlather, M. (2010). Some covariance models based on normal scale mixtures. *Bernoulli*, **16(3)**, 780–797.

Schlather, M., and J. A. Tawn (2003). A dependence measure for multivariate and spatial extreme values: Properties and inference. *Biometrika*, **90(1)**, 139–156.

Schlatter, T. W. (1975). Some experiments with a multivariate statistical objective analysis scheme. *Monthly Weather Review*, **103(3)**, 246–257.

Schmitt, M., and H. Beucher (1997). On the inference of the Boolean model. In *Geostatistics Wollongong '96*, Vol. 1, E. Y. Baafi and N. A. Schofield, ed. Kluwer, Dordrecht, pp. 200–210.

Schoenberg, I. J. (1938a). Metric spaces and completely monotone functions. *Annals of Mathematics*, **39(4)**, 811–831.

Schoenberg, I. J. (1938b). Metric spaces and positive definite functions. *Transactions of the American Mathematical Society*, **44(3)**, 522–536.

Schoenberg, I. J. (1942). Positive definite functions on spheres. *Duke Mathematics Journal*, **9(1)**, 96–108.

Schwartz, L. (1950–1951). *Théorie des distributions*. Publications de l'Institut de Mathématique de l'Université de Strasbourg, 2 Vols. New edition (1966), Hermann, Paris.

Séguret, S. (1989). Filtering periodic noise by using trigonometric kriging. In *Geostatistics*, Vol. 1, M. Armstrong, ed. Kluwer, Dordrecht, pp. 481–491.

Séguret, S., and P. Huchon (1990). Trigonometric kriging: A new method for removing the diurnal variation from geomagnetic data. *Journal of Geophysical Research*, **95(B13)**, 21,383–21,397.

Sellan, F. (1995). Synthèse de mouvements browniens fractionnaires à l'aide de la transformation par ondelettes. *Comptes Rendus de l'Académie des Sciences de Paris, Série I*, **321(3)**, 351–358.

Sénégas, J. (2002a). *Méthodes de Monte Carlo en vision stéréoscopique. Application à l'étude de modèles numériques de terrain*. Doctoral thesis, E.N.S. des Mines de Paris.

Sénégas, J. (2002b). A Markov chain Monte Carlo approach to stereovision. In *Computer Vision—ECCV 2002*, A. Heyden, G. Sparr, M. Nielsen, and P. Johansen, eds. Springer, Berlin, pp. 97–111.

Serra, J. (1967). *Echantillonnage et estimation locale des phénomènes de transition miniers*, 2 Vols. Doctoral thesis, Université de Nancy, France.

Serra, J. (1968). Les structures gigognes: Morphologie mathématique et interprétation métallogénique. *Mineralium Deposita*, **3(2)**, 135–154.

Serra, J. (1982). *Image Analysis and Mathematical Morphology*. Academic Press, London.

Serra, J. (1988). Boolean random functions. In *Image Analysis and Mathematical Morphology*, Vol. 2: Theoretical Advances, J. Serra, ed. Academic Press, London, Chapter 15, pp. 317–342.

Serra, O. (1985). *Sedimentary Environments from Wireline Logs*. Schlumberger, Paris, New York.

Shapiro, D. E., and P. Switzer (1989). Extracting time trends from multiple monitoring sites. Technical Report SIMS 132, Department of Statistics, Stanford University, California.

Shinozuka, M. (1971). Simulation of multivariate and multidimensional random processes. *The Journal of the Acoustical Society of America*, **49(1B)**, 357–367.

Shinozuka, M, and C. M. Jan (1972). Digital simulation of random processes and its applications. *Journal of Sound and Vibration*, **25(1)**, 111–128.

Shorrocks, A. F. (1983). Ranking income distributions. *Economica*, **50(197)**, 3–17.

Sichel, H. S. (1973). Statistical valuation of diamondiferous deposits. In *Proceedings of the 11th APCOM International Symposium*, M. D. G. Salamon and F. H. Lancaster, eds. The South African Institute of Mining and Metallurgy, Johannesburg, South Africa, pp. 17–25.

Simard, R. (1980). *Etude de la représentativité des levés gravimétriques et de son influence sur l'interprétation*. Doctoral thesis, Université de Lausanne, Switzerland.

Simon, E., and L. Bertino (2009). Application of the Gaussian anamorphosis to assimilation in a 3-D coupled physical-ecosystem model of the North Atlantic with the EnKF: A twin experiment. *Ocean Science*, **5(4)**, 495–510.

Sinclair, A. J., and M. Vallée (1994). Improved sampling control and data gathering for improved mineral inventories and production control. In *Geostatistics for the Next Century*, R. Dimitrakopoulos, ed. Kluwer, Dordrecht, pp. 323–329.

Smith, R. L. (1990). Max-stable processes and spatial extremes. Technical Report, Department of Mathematics, University of Surrey, Guildford, England.

Snowden, V. (1994). Improving predictions by studying reality. In *Geostatistics for the Next Century*, R. Dimitrakopoulos, ed. Kluwer, Dordrecht, pp. 330–337.

Sobczyk, K. (1991). *Stochastic Differential Equations. With Applications to Physics and Engineering*. Kluwer, Dordrecht.

Sølna, K., and P. Switzer (1996). Time-trend estimation for a geographic region. *Journal of the American Statistical Association*, **91(434)**, 577–589.

Solow, A. R. (1985). Bootstrapping correlated data. *Journal of the International Association for Mathematical Geology*, **17(7)**, 769–775.

Spanos, P. D., and M. P. Mignolet (1992). Simulation of homogeneous two-dimensional random fields: Part II—MA and ARMA models. *Journal of Applied Mechanics*, **59(2)**, 270–277.

Spector, A., and B. K. Bhattacharyya (1966). Energy density spectrum and autocorrelation function of anomalies due to simple magnetic models. *Geophysical Prospecting*, **14(3)**, 242–272.

Spector, A., and F. S. Grant (1970). Statistical models for interpreting aeromagnetic data. *Geophysics*, **35(2)**, 293–302.

Srivastava, R. M. (1992). Reservoir characterization with probability field simulation. *SPE paper* 24753, 67th Annual Technical Conference and Exhibition, Washington, D.C., pp. 927–938.

Srivastava, R. M. (1994). An annealing procedure for honouring change of support statistics in conditional simulation. In *Geostatistics for the Next Century*, R. Dimitrakopoulos, ed. Kluwer, Dordrecht, pp. 277–290.

Srivastava, R. M., and R. Froidevaux (2005). Probability field simulation: A retrospective. In *Geostatistics Banff 2004*, Vol. 1, O. Leuangthong and C. V. Deutsch, eds. Springer, Dordrecht, pp. 55–64.

Starks, T. H., and J. M. Fang (1982). On the estimation of the generalized covariance function. *Journal of the International Association for Mathematical Geology*, **14(1)**, 57–64.

Steffens, F. E. (1993). Geostatistical estimation of animal abundance in the Kruger National Park, South Africa. In *Geostatistics Tróia '92*, Vol. 2, A. Soares, ed. Kluwer, Dordrecht, pp. 887–897.

Stein, M. L. (1986a). A modification of minimum norm quadratic estimation of a generalized covariance function for use with large data sets. *Mathematical Geology*, **18(7)**, 625–633.

Stein, M. L. (1986b). A simple model for spatial–temporal processes. *Water Resources Research*, **22(13)**, 2107–2110.

Stein, M. L. (1999). *Interpolation of Spatial Data: Some Theory for Kriging*. Springer, New York.

Stein, M. L. (2001). Local stationarity and simulation of self-affine intrinsic random functions. *IEEE Transactions on Information Theory*, **47(4)**, 1385–1390.

Stein, M. L. (2002). Fast and exact simulation of fractional Brownian surfaces. *Journal of Computational and Graphical Statistics*, **11(3)**, 587–599.

Stein, M. L. (2005). Space–time covariance functions. *Journal of the American Statistical Association*, **100(469)**, 310–321.

Stoyan, D., W. S. Kendall, and J. Mecke (1987). *Stochastic Geometry and Its Applications*. John Wiley & Sons, New York.

Strassen, V. (1965). The existence of probability measures with given marginals. *Annals of Mathematical Statistics*, **36(2)**, 423–439.

Strebelle, S. (2002). Conditional simulation of complex geological structures using multiple point statistics. *Mathematical Geology*, **34(1)**, 1–22.

Subramanyam, A., and H. S. Pandalai (2001). A characterization of symmetric isofactorial models. *Mathematical Geology*, **33(1)**, 103–114.

Sullivan, J. (1984). Conditional recovery estimation through probability kriging— Theory and practice. In *Geostatistics for Natural Resources Characterization*, Part 1, G. Verly, M. David, A. G. Journel, and A. Maréchal, eds. Reidel, Dordrecht, Holland, pp. 365–384.

Suppe, J. (1985). *Principles of Structural Geology*. Prentice-Hall, Englewood Cliffs, NJ.

Switzer, P. (1977). Estimation of spatial distributions from point sources with application to air pollution measurement. *Bulletin of the International Statistical Institute*, **47(2)**, 123–137.

Switzer, P. (1984). Inference for spatial autocorrelation functions. In *Geostatistics for Natural Resources Characterization*, Part 1, G. Verly, M. David, A. G. Journel, and A. Maréchal, eds. Reidel, Dordrecht, pp. 127–140.

Switzer, P. (1989). Non-stationary spatial covariances estimated from monitoring data. In *Geostatistics*, Vol. 1, M. Armstrong, ed. Kluwer, Dordrecht, pp. 127–138.

Switzer, P. (1993). The spatial variability of prediction errors. In *Geostatistics Tróia '92*, Vol. 1, A. Soares, ed. Kluwer, Dordrecht, pp. 261–272.

Switzer, P., and A. A. Green (1984). Min/max autocorrelation factors for multivariate spatial imagery. Technical Report SWI NSF 06, Department of Statistics, Stanford University, California.

Switzer, P., and H. Parker (1976). The problem of ore versus waste discrimination. In *Advanced Geostatistics in the Mining Industry*, M. Guarascio, M. David, and C. Huijbregts, eds. Reidel, Dordrecht, pp. 203–218.

Switzer, P., and H. Xiao (1988). Approximate prediction intervals for spatial interpolation using bivariate information: An experiment. Technical Report SIMS 118, Department of Statistics and Department of Applied Earth Sciences, Stanford University, California.

Szegö, G. (1939). *Orthogonal Polynomials*. Colloquium Publications, American Mathematical Society, Providence, RI. Fourth edition (1975).

Tarantola, A. (2005). *Inverse Problem Theory and Methods for Model Parameter Estimation*. Society for Industrial and Applied Mathematics, Philadelphia.

Taylor, G. I. (1938). The spectrum of turbulence. *Proceedings of the Royal Society, Series A*, **164(919)**, 476–490.

Tetzlaff, D. M., and J. W. Harbaugh (1989). *Simulating Clastic Sedimentation*. Van Nostrand Reinhold, New York.

Tolosana-Delgado, R., V. Pawlowsky-Glahn, and J. J. Egozcue (2008). Indicator kriging without order relation violations. *Mathematical Geosciences*, **40(3)**, 327–347.

Tukey, J. W. (1960). A survey of sampling from contaminated distributions. In *Contributions to Probability and Statistics*, I. Olkin, ed. Stanford University Press, CA, pp. 448–485.

Tukey, J. W. (1977). *Exploratory Data Analysis*. Addison-Wesley, Reading, MA.

Turcotte, D. L. (1986). A fractal approach to the relationship between ore grade and tonnage. *Economic Geology*, **81(6)**, 1528–1532.

Turcotte, D. L. (1997). *Fractals and Chaos in Geology and Geophysics*, 2nd ed. Cambridge University Press, New York.

Ulvmoen, M., and H. Omre (2010). Improved resolution in Bayesian lithology/fluid inversion from prestack seismic data and well observations: Part 1—Methodology. *Geophysics*, **75(2)**, R21–R35.

Ulvmoen, M., H. Omre, and A. Buland (2010). Improved resolution in Bayesian lithology/fluid inversion from prestack seismic data and well observations: Part 2—Real case study. *Geophysics*, **75(2)**, B73–B82.

Van Buchem F. S. P., B. Doligez, R. Eschard, O. Lerat, G. M. Grammer, and C. Ravenne (2000). Stratigraphic architecture and stochastic reservoir simulation of a mixed carbonate/siliciclastic platform (Upper Cretaceous, Paradox basin, USA). In *Genetic Stratigraphy on the Exploration and the Production Scales*, P. W. Homewood and G. P. Eberli, eds. Bulletin du Centre de Recherche Elf Exploration-Production, Mémoire 24, pp. 109–128.

Ver Hoef, J. M., and N. Cressie (1993). Multivariable spatial prediction. *Mathematical Geology*, **25(2)**, 219–240.

Ver Hoef, J. M., and E. E. Peterson (2010). A moving average approach for spatial statistical models of stream networks. *Journal of the American Statistical Association*, **105(489)**, 6–18.

Ver Hoef, J. M., E. Peterson, and D. Theobald (2006). Spatial statistical models that use flow and stream distance. *Environmental and Ecological Statistics*, **13(4)**, 449–464.

Verly, G. (1983). The multigaussian approach and its applications to the estimation of local reserves. *Journal of the International Association for Mathematical Geology*, **15(2)**, 259–286.

Verly, G. (1984). The block distribution given a point multivariate normal distribution. In *Geostatistics for Natural Resources Characterization*, Part 1, G. Verly, M. David, A. G. Journel, and A. Maréchal, eds. Reidel, Dordrecht, pp. 495–515.

Verly, G. (1986). Multigaussian kriging—A complete case study. In *Proceedings of the 19th APCOM International Symposium*, R. V. Ramani, ed. Society of Mining Engineers, Littleton, CO, pp. 283–298.

Verscheure, M., A. Fourno, and J. P. Chilès (2012). Joint inversion of fracture model properties for CO_2 storage monitoring or oil recovery history matching. *Oil & Gas Science and Technology*, **66**.

Vieira, J. L., J. F. Travassos, A. Gonçalves, N. Santos, and F. H. Muge (1993). Simulation of medium and short term mine production scheduling using geostatistical modelling. In *Geostatistics Tróia '92*, Vol. 2, A. Soares, ed. Kluwer, Dordrecht, pp. 1013–1027.

Vistelius, A. B. (1960). The skew frequency distributions and the fundamental law of the geochemical processes. *The Journal of Geology*, **68(1)**, 1–22.

Voss, R. F. (1985). Random fractal forgeries. In *Fundamental Algorithms for Computer Graphics*, R. A. Earnshaw, ed. Springer, Berlin, pp. 13–16 and 805–835.

Vysochanskiĭ, D. F., and Yu. I. Petunin (1980). Justification of the 3σ rule for unimodal distributions (Russian). *Theory of Probability and Mathematical Statistics*, **21**, 25–36.

Wackernagel, H. (1985). *L'inférence d'un modèle linéaire en géostatistique multivariable*. Doctoral thesis, E.N.S. des Mines de Paris.

Wackernagel, H. (1988). Geostatistical techniques for interpreting multivariate spatial information. In *Quantitative Analysis of Mineral and Energy Resources*, C. F. Chung et al., eds. Reidel, Dordrecht, pp. 393–409.

Wackernagel, H. (2003). *Multivariate Geostatistics—An Introduction with Applications*, 3rd ed. Springer, Berlin. First edition (1995).

Wackernagel, H., R. Webster, and M. A. Oliver (1988). A geostatistical method for segmenting multivariate sequences of soil data. In *Classification and Related Methods of Data Analysis*, H. H. Bock, ed. Elsevier, New York, pp. 641–650.

Wahba, G. (1990). *Spline Models for Observational Data*. Society for Industrial and Applied Mathematics, Philadelphia.

Walvoort, D. J. J., and J. J. de Gruijter (2001). Compositional kriging: A spatial interpolation method for compositional data. *Mathematical Geology*, **33(8)**, 951–966.

Watney, W. L., E. C. Rankey, and J. Harbaugh (1999). Perspectives on stratigraphic simulation models: Current approaches and future opportunities. In *Numerical Experiments in Stratigraphy*, Vol. 62, J.W. Harbaugh, W.L. Watney, E.C. Rankey, R. Slingerland, and R.H. Goldstein, eds. SEPM Special Publication, Society for Sedimentary Geology, Tulsa, OK, pp. 3–21.

Watson, G. S. (1984). Smoothing and interpolation by kriging and with splines. *Journal of the International Association for Mathematical Geology*, **16(6)**, 601–615.

Webster, R., and M. A. Oliver (1989). Optimal interpolation and isarithmic mapping of soil properties. VI. Disjunctive kriging and mapping the conditional probability. *Journal of Soil Science*, **40(3)**, 497–512.

Webster, R., and M. A. Oliver (1990). *Statistical Methods in Soil and Land Resource Survey*. Oxford University Press, Oxford, England.

Wen, R., and R. Sinding-Larsen (1997). Stochastic modeling and simulation of small faults by marked point processes and kriging. In *Geostatistics Wollongong '96*, Vol. 1, E. Y. Baafi and N. A. Schofield, eds. Kluwer, Dordrecht, pp. 398–414.

Wen, X. H., and J. J. Gómez-Hernández (1996). Upscaling hydraulic conductivities in heterogeneous media: An overview. *Journal of Hydrology*, **183(1–2)**, ix–xxxii.

Whittle, D., and E. Bozorgebrahimi (2005). Hybrid pits—Linking conditional simulation and Lerchs–Grossmann through set theory. In *Orebody Modelling and Strategic Mine Planning—Uncertainty and Risk Management Models*, R. Dimitrakopoulos, ed. Spectrum Series 14, The Australasian Institute of Mining and Metallurgy, Carlton, Victoria, pp. 37–42.

Whittle, P. (1954). On stationary processes in the plane. *Biometrika*, **41(3–4)**, 434–449.

Whittle, P. (1963). Stochastic processes in several dimensions. *Bulletin of the International Statistical Institute*, **40**, 974–994.

Wiener, N. (1942). *The Extrapolation, Interpolation, and Smoothing of Stationary Time Series. With Engineering Applications*. Report of the Services 19, Research Project DIC-6037, MIT, Cambridge, Massachusetts. Printed in book form (1949): John Wiley & Sons, New York.

Wijs, H. J. de (1951). Statistics of ore distribution. Part I: Frequency distribution of assay values. *Geologie en Mijnbouw (Journal of the Royal Netherlands Geological and Mining Society)*, New Series, **13(11)**, 365–375.

Wijs, H. J. de (1953). Statistics of ore distribution. Part II: Theory of binomial distribution applied to sampling and engineering problems. *Geologie en Mijnbouw (Journal of the Royal Netherlands Geological and Mining Society), New Series*, **15(1)**, 12–24.

Wijs, H. J. de (1976). Models for the estimation of world ore reserves. *Geologie en Mijnbouw*, **55(1–2)**, 46–50.

Wilkinson, J. H. (1965). *The Algebraic Eigenvalue Problem*. Oxford University Press, Oxford, England.

Williamson, P. R., A. J. Cherrett, and R. Bornard (2007). Geostatistical stochastic elastic inversion—An efficient method for integrating seismic and well data constraints. Paper D025, EAGE 69th Conference & Exhibition, London.

Withers, C. S. (2000). A simple expression for the multivariate Hermite polynomials. *Statistics & Probability Letters*, **47(2)**, 165–169.

Wood, A. T. A., and G. Chan (1994). Simulation of stationary Gaussian processes in $[0, 1]^d$. *Journal of Computational and Graphical Statistics*, **3(4)**, 409–432.

Wood, G., M. A. Oliver, and R. Webster (1990). Estimating soil salinity by disjunctive kriging. *Soil Use and Management*, **6(3)**, 97–104.

Xu, W., T. T. Tran, R. M. Srivastava, and A. G. Journel (1992). Integrating seismic data in reservoir modeling: The collocated cokriging alternative. *SPE paper* 24742, 67th Annual Technical Conference and Exhibition, Washington, D.C.

Yaglom, A. M. (1957). Nekotorye klassy slucajnych polej v *n*-mernom prostranstve, rodstvennye stacionarnym slucajnym processam. *Teorija Verojatnostej i ee Primenenija*, **2(3)**, 292–338. Translation (1957): Some classes of random fields in *n*-dimensional space, related to stationary random processes. *Theory of Probability and Its Applications*, **2(3)**, 273–320.

Yaglom, A. M. (1962). *An Introduction to the Theory of Stationary Random Functions*, rev. ed. Prentice-Hall, Englewood Cliffs, NJ. Reprint (1973), Dover, New York.

Yaglom, A. M. (1987). *Correlation Theory of Stationary and Related Random Functions*, Vol. I: *Basic Results*; Vol. II: Supplementary Notes and References. Springer, New York.

Yaglom, A. M., and M. S. Pinsker (1953). Random processes with stationary increments of order *n*. *Doklady Akademii Nauk SSSR*, **90(5)**, 731–734.

Yao, T. (1998). Conditional spectral simulation with phase identification. *Mathematical Geology*, **30(3)**, 285–308.

Yao, T., and T. Mukerji (1997). Application of factorial kriging to improve seismic data integration. In *Geostatistics Wollongong '96*, Vol. 1, E. Y. Baafi and N. A. Schofield, eds. Kluwer, Dordrecht, pp. 350–361.

Yarus, J. M., and R. L. Chambers, eds. (1994). *Stochastic Modeling and Geostatistics. Principles, Methods, and Case Studies*. American Association of Petroleum Geologists, Tulsa, OK.

Yates, S. R., and M. V. Yates (1988). Disjunctive kriging as an approach to management decision making. *Soil Science Society of America Journal*, **52(6)**, 1554–1558.

Yates, S. R., A. W. Warrick, and D. E. Myers (1986). Disjunctive kriging. 1. Overview of estimation and conditional probability. 2. Examples. *Water Resources Research*, **22(5)**, 1: 615–621; 2: 623–630.

Yfantis, E. A., M. Au, and F. S. Makri (1994). Image compression and kriging (with discussion). In *Geostatistics for the Next Century*, R. Dimitrakopoulos, ed. Kluwer, Dordrecht, pp. 156–170.

Young, D. S. (1982). Development and application of disjunctive kriging model: Discrete Gaussian model. In *Proceedings of the 17th APCOM International Symposium*, T. B. Johnson and R. J. Barnes, eds. Society of Mining Engineers of the AIME, New York, pp. 544–561.

Zadeh, L. A. (1965). Fuzzy sets. *Information and Control*, **8(3)**, 338–353.

Zadeh, L. A., and J. R. Ragazzini (1950). An extension of Wiener's theory of prediction. *Journal of Applied Physics*, **21(7)**, 645–655.

Zeldin, B. A., and P. D. Spanos (1996). Random field representation and synthesis using wavelet bases. *Journal of Applied Mechanics*, **63(4)**, 946–952.

Zhang, T., P. Switzer, and A. Journel (2006). Filter-based classification of training image patterns for spatial simulation. *Mathematical Geology*, **38(1)**, 63–80.

Zimmerman, D. L. (1989). Computationally efficient restricted maximum likelihood estimation of generalized covariance functions. *Mathematical Geology*, **21(7)**, 655–672.

Zimmermann, H. J. (2001). *Fuzzy Set Theory and Its Applications*, 4th ed. Kluwer, Boston.

Zinn, B., and C. F. Harvey (2003). When good statistical models of aquifer heterogeneity go bad: A comparison of flow, dispersion and mass transfer in multigaussian and connected conductivity fields. *Water Resources Research*, **39(3)**, 1051–1068.

Index

Geostatistics: Modeling Spatial Uncertainty, Second Edition. J.P. Chilès and P. Delfiner.
© 2012 John Wiley & Sons, Inc. Published 2012 by John Wiley & Sons, Inc.

WILEY SERIES IN PROBABILITY AND STATISTICS
Established by Walter A. Shewhart and Samuel S. Wilks

Editors: *David J. Balding, Noel A. C. Cressie, Garrett M. Fitzmaurice,*
Harvey Goldstein, Iain M. Johnstone, Geert Molenberghs, David W. Scott,
Adrian F. M. Smith, Ruey S. Tsay, Sanford Weisberg
Editors Emeriti: *Vic Barnett, J. Stuart Hunter, Joseph B. Kadane, Jozef L. Teugels*

The *Wiley Series in Probability and Statistics* is well established and authoritative. It covers many topics of current research interest in both pure and applied statistics and probability theory. Written by leading statisticians and institutions, the titles span both state-of-the-art developments in the field and classical methods.

Reflecting the wide range of current research in statistics, the series encompasses applied, methodological and theoretical statistics, ranging from applications and new techniques made possible by advances in computerized practice to rigorous treatment of theoretical approaches.

This series provides essential and invaluable reading for all statisticians, whether in academic, industry, government, or research.

*Now available in a lower priced paperback edition in the Wiley Classics Library.
†Now available in a lower priced paperback edition in the Wiley–Interscience Paperback Series.

CHILÈS and DELFINER · Geostatistics: Modeling Spatial Uncertainty, *Second Edition*

CHOW and LIU · Design and Analysis of Clinical Trials: Concepts and Methodologies, *Second Edition*

CLARKE · Linear Models: The Theory and Application of Analysis of Variance

CLARKE and DISNEY · Probability and Random Processes: A First Course with Applications, *Second Edition*

* COCHRAN and COX · Experimental Designs, *Second Edition*

COLLINS and LANZA · Latent Class and Latent Transition Analysis: With Applications in the Social, Behavioral, and Health Sciences

CONGDON · Applied Bayesian Modelling

CONGDON · Bayesian Models for Categorical Data

CONGDON · Bayesian Statistical Modelling, *Second Edition*

CONOVER · Practical Nonparametric Statistics, *Third Edition*

COOK · Regression Graphics

COOK and WEISBERG · An Introduction to Regression Graphics

COOK and WEISBERG · Applied Regression Including Computing and Graphics

CORNELL · A Primer on Experiments with Mixtures

CORNELL · Experiments with Mixtures, Designs, Models, and the Analysis of Mixture Data, *Third Edition*

COVER and THOMAS · Elements of Information Theory

COX · A Handbook of Introductory Statistical Methods

* COX · Planning of Experiments

CRESSIE · Statistics for Spatial Data, *Revised Edition*

CRESSIE and WIKLE · Statistics for Spatio-Temporal Data

CSÖRGO and HORVÁTH · Limit Theorems in Change Point Analysis

DANIEL · Applications of Statistics to Industrial Experimentation

DANIEL · Biostatistics: A Foundation for Analysis in the Health Sciences, *Eighth Edition*

* DANIEL · Fitting Equations to Data: Computer Analysis of Multifactor Data, *Second Edition*

DASU and JOHNSON · Exploratory Data Mining and Data Cleaning

DAVID and NAGARAJA · Order Statistics, *Third Edition*

* DEGROOT, FIENBERG, and KADANE · Statistics and the Law

DEL CASTILLO · Statistical Process Adjustment for Quality Control

DᴇMARIS · Regression with Social Data: Modeling Continuous and Limited Response Variables

DEMIDENKO · Mixed Models: Theory and Applications

DENISON, HOLMES, MALLICK and SMITH · Bayesian Methods for Nonlinear Classification and Regression

DETTE and STUDDEN · The Theory of Canonical Moments with Applications in Statistics, Probability, and Analysis

DEY and MUKERJEE · Fractional Factorial Plans

DILLON and GOLDSTEIN · Multivariate Analysis: Methods and Applications

DODGE · Alternative Methods of Regression

* DODGE and ROMIG · Sampling Inspection Tables, *Second Edition*

* DOOB · Stochastic Processes

DOWDY, WEARDEN, and CHILKO · Statistics for Research, *Third Edition*

DRAPER and SMITH · Applied Regression Analysis, *Third Edition*

DRYDEN and MARDIA · Statistical Shape Analysis

DUDEWICZ and MISHRA · Modern Mathematical Statistics

DUNN and CLARK · Basic Statistics: A Primer for the Biomedical Sciences, *Third Edition*

DUPUIS and ELLIS · A Weak Convergence Approach to the Theory of Large Deviations

---

*Now available in a lower priced paperback edition in the Wiley Classics Library.

†Now available in a lower priced paperback edition in the Wiley–Interscience Paperback Series.

---

*Now available in a lower priced paperback edition in the Wiley Classics Library.

†Now available in a lower priced paperback edition in the Wiley–Interscience Paperback Series.

*Now available in a lower priced paperback edition in the Wiley Classics Library.
†Now available in a lower priced paperback edition in the Wiley–Interscience Paperback Series.

*Now available in a lower priced paperback edition in the Wiley Classics Library.

†Now available in a lower priced paperback edition in the Wiley–Interscience Paperback Series.

MURTHY, XIE, and JIANG · Weibull Models

MYERS, MONTGOMERY, and ANDERSON-COOK · Response Surface Methodology: Process and Product Optimization Using Designed Experiments, *Third Edition*

MYERS, MONTGOMERY, VINING, and ROBINSON · Generalized Linear Models. With Applications in Engineering and the Sciences, *Second Edition*

† NELSON · Accelerated Testing, Statistical Models, Test Plans, and Data Analyses

† NELSON · Applied Life Data Analysis

NEWMAN · Biostatistical Methods in Epidemiology

OCHI · Applied Probability and Stochastic Processes in Engineering and Physical Sciences

OKABE, BOOTS, SUGIHARA, and CHIU · Spatial Tesselations: Concepts and Applications of Voronoi Diagrams, *Second Edition*

OLIVER and SMITH · Influence Diagrams, Belief Nets and Decision Analysis

PALTA · Quantitative Methods in Population Health: Extensions of Ordinary Regressions

PANJER · Operational Risk: Modeling and Analytics

PANKRATZ · Forecasting with Dynamic Regression Models

PANKRATZ · Forecasting with Univariate Box-Jenkins Models: Concepts and Cases

PARDOUX · Markov Processes and Applications: Algorithms, Networks, Genome and Finance

* PARZEN · Modern Probability Theory and Its Applications

PEÑA, TIAO, and TSAY · A Course in Time Series Analysis

PIANTADOSI · Clinical Trials: A Methodologic Perspective

PORT · Theoretical Probability for Applications

POURAHMADI · Foundations of Time Series Analysis and Prediction Theory

POWELL · Approximate Dynamic Programming: Solving the Curses of Dimensionality, *Second Edition*

POWELL and RYZHOV · Optimal Learning

PRESS · Bayesian Statistics: Principles, Models, and Applications

PRESS · Subjective and Objective Bayesian Statistics, *Second Edition*

PRESS and TANUR · The Subjectivity of Scientists and the Bayesian Approach

PUKELSHEIM · Optimal Experimental Design

PURI, VILAPLANA, and WERTZ · New Perspectives in Theoretical and Applied Statistics

† PUTERMAN · Markov Decision Processes: Discrete Stochastic Dynamic Programming

QIU · Image Processing and Jump Regression Analysis

* RAO · Linear Statistical Inference and Its Applications, *Second Edition*

RAO · Statistical Inference for Fractional Diffusion Processes

RAUSAND and HØYLAND · System Reliability Theory: Models, Statistical Methods, and Applications, *Second Edition*

RAYNER · Smooth Tests of Goodnes of Fit: Using R, *Second Edition*

RENCHER · Linear Models in Statistics

RENCHER · Methods of Multivariate Analysis, *Second Edition*

RENCHER · Multivariate Statistical Inference with Applications

* RIPLEY · Spatial Statistics

* RIPLEY · Stochastic Simulation

ROBINSON · Practical Strategies for Experimenting

ROHATGI and SALEH · An Introduction to Probability and Statistics, *Second Edition*

ROLSKI, SCHMIDLI, SCHMIDT, and TEUGELS · Stochastic Processes for Insurance and Finance

ROSENBERGER and LACHIN · Randomization in Clinical Trials: Theory and Practice

ROSS · Introduction to Probability and Statistics for Engineers and Scientists

ROSSI, ALLENBY, and MCCULLOCH · Bayesian Statistics and Marketing

† ROUSSEEUW and LEROY · Robust Regression and Outlier Detection

*Now available in a lower priced paperback edition in the Wiley Classics Library.

†Now available in a lower priced paperback edition in the Wiley–Interscience Paperback Series.